DATE DUE

VITAMINS AND HORMONES

VOLUME 72

Editorial Board

TADHG P. BEGLEY

ANTHONY R. MEANS

BERT W. O'MALLEY

LYNN RIDDIFORD

ARMEN H. TASHJIAN, JR.

PLANT HORMONES

VITAMINS AND HORMONES
ADVANCES IN RESEARCH AND APPLICATIONS

Editor-in-Chief

GERALD LITWACK

Professor and Chair Emeritus
Department of Biochemistry and Molecular Pharmacology
Thomas Jefferson University Medical College
Philadelphia, Pennsylvania
Visiting Scholar
Department of Biological Chemistry
David Geffen School of Medicine at UCLA
Toluca Lake, California

VOLUME 72

AMSTERDAM • BOSTON • HEIDELBERG • LONDON
NEW YORK • OXFORD • PARIS • SAN DIEGO
SAN FRANCISCO • SINGAPORE • SYDNEY • TOKYO
Academic Press is an imprint of Elsevier

ELSEVIER

Elsevier Academic Press
525 B Street, Suite 1900, San Diego, California 92101-4495, USA
84 Theobald's Road, London WC1X 8RR, UK

This book is printed on acid-free paper. ∞

Copyright © 2005, Elsevier Inc. All Rights Reserved.

No part of this publication may be reproduced or transmitted in any form or by any means, electronic or mechanical, including photocopy, recording, or any information storage and retrieval system, without permission in writing from the Publisher.

The appearance of the code at the bottom of the first page of a chapter in this book indicates the Publisher's consent that copies of the chapter may be made for personal or internal use of specific clients. This consent is given on the condition, however, that the copier pay the stated per copy fee through the Copyright Clearance Center, Inc. (www.copyright.com), for copying beyond that permitted by Sections 107 or 108 of the U.S. Copyright Law. This consent does not extend to other kinds of copying, such as copying for general distribution, for advertising or promotional purposes, for creating new collective works, or for resale.
Copy fees for pre-2005 chapters are as shown on the title pages. If no fee code appears on the title page, the copy fee is the same as for current chapters.
0083-6729/2005 $35.00

Permissions may be sought directly from Elsevier's Science & Technology Rights Department in Oxford, UK: phone: (+44) 1865 843830, fax: (+44) 1865 853333, E-mail: permissions@elsevier.com. You may also complete your request on-line via the Elsevier homepage (http://elsevier.com), by selecting "Customer Support" and then "Obtaining Permissions."

For all information on all Elsevier Academic Press publications visit our Web site at www.books.elsevier.com

ISBN-13: 978-0-12-709872-2
ISBN-10: 0-12-709872-0

PRINTED IN THE UNITED STATES OF AMERICA
05 06 07 08 09 9 8 7 6 5 4 3 2 1

Working together to grow libraries in developing countries

www.elsevier.com | www.bookaid.org | www.sabre.org

ELSEVIER   BOOK AID International   Sabre Foundation

CONTENTS

CONTRIBUTORS XIII
PREFACE XVII

1

THE GENOMIC VIEW OF GENES RESPONSIVE TO THE ANTAGONISTIC PHYTOHORMONES, ABSCISIC ACID, AND GIBBERELLIN

JUNSHI YAZAKI AND SHOSHI KIKUCHI

I. Introduction 2
II. Comprehensive Analysis of ABA-Responsive Genes by Genomics Approach 5
III. Comprehensive Analysis for GA-Responsive Genes by the Genomics Approach 12
IV. Specificity of Gene Responses to Both Hormones 16
V. cis-Element Analysis 16
VI. Conclusion and Future Prospects 21
References 23

2

GRAVITROPIC BENDING AND PLANT HORMONES

SONIA PHILOSOPH-HADAS, HAYA FRIEDMAN, AND SHIMON MEIR

I. Introduction   32
II. The Gravitropic Bending of Plants   34
III. Role of Auxin   36
IV. Role of Ethylene   50
V. Role of Gibberellins   56
VI. Role of Abscisic Acid   58
VII. Role of Cytokinins   60
VIII. Role of Brassinosteroids, Jasmonates, and Salicylic Acid   63
IX. Concluding Remarks and Future Prospects   65
   References   66

3

HORMONAL REGULATION OF SEX EXPRESSION IN PLANTS

SEIJI YAMASAKI, NOBUHARU FUJII, AND HIDEYUKI TAKAHASHI

I. Introduction   80
II. Differentiation of Flower Buds and Sex Expression in Maize   82
III. Genetic and Hormonal Regulation of Sex Expression in Maize   83
IV. Differentiation of Flower Bud and Sex Expression in Cucumber   87
V. Hormonal Regulation of Sex Expression in Cucumber   88
VI. Genetic Regulation of Sex Expression in Cucumber   94
VII. Environmental Regulation of Sex Expression   96
VIII. Identity of Floral Organs and Regulation of Sex Expression   97
IX. Programmed Cell Death and Sex Determination   102
X. Conclusions   103
   References   104

4

Plant Peroxisomes

Shoji Mano and Mikio Nishimura

I. Introduction 112
II. Discovery, Structure, and Localization in Cells 113
III. Peroxisome Functions in Higher Plants 114
IV. Transition of Peroxisome Function 122
V. Peroxisome Biogenesis 123
VI. Challenge of Comprehensive Approach in the Postgenomic Era 140
VII. Few Final Considerations and Perspectives 143
References 143

5

Regulatory and Functional Interactions of Plant Growth Regulators and Plant Glutathione S-Transferases (GSTs)

Ann Moons

I. Classification, Structure, and Subcellular Localization of Plant GSTs 157
II. Functions of Plant GSTs 161
III. The Transcriptional Regulation of GST Gene Expression 174
IV. Potential Involvement of Plant Growth Regulators in Post-transcriptional Regulations of GST Gene Expression 195
V. Conclusions 197
References 198

6

Auxins

Catherine Perrot-Rechenmann and Richard M. Napier

I. Introduction 204
II. Generation of IAA 205

III. Cell Division and Tissue Culture   209
IV. Transport of IAA: Introduction   210
V. Embryonic Patterning   215
VI. Vascular Patterning   215
VII. Auxin Perception, Receptors, and Signaling   216
VIII. Organ Patterning   221
IX. Apical Dominance and Branching   222
X. Tropisms and Epinasty   222
XI. Adventitious Rooting and Wound Responses   223
XII. Fruit Growth   224
XIII. Herbicides   225
References   226

7

Regulatory Networks of the Phytohormone Abscisic Acid

Zhen Xie, Paul Ruas, and Qingxi J. Shen

I. Introduction   236
II. Signaling Pathways   238
III. Cross-talk of ABA and GA   257
IV. Conclusions   258
References   259

8

Cytokinin Biosynthesis and Regulation

Hitoshi Sakakibara

I. Introduction   272
II. Structural Variation of Cytokinin   273
III. Biosynthesis of Cytokinin   275
IV. Regulation of Cytokinin Biosynthesis in Higher Plants   280
References   283

9

GIBBERELLIN METABOLISM AND SIGNALING

STEPHEN G. THOMAS, IVO RIEU, AND CAMILLE M. STEBER

I. Introduction   290
II. Gibberellin Biosynthesis   295
III. Gibberellin Signal Transduction   312
IV. Cross-talk with Other Hormone-Signaling Pathways   323
V. Perspectives   326
References   326

10

NITRIC OXIDE SIGNALING IN PLANTS

ALLAN D. SHAPIRO

I. Introduction   340
II. Nitric Oxide Biosynthesis   341
III. NO Metabolism and Transport   350
IV. NO Function in Plants   358
V. Conclusions   384
References   386

11

ETHYLENE BIOSYNTHESIS AND SIGNALING: AN OVERVIEW

ANNELIES DE PAEPE AND DOMINIQUE VAN DER STRAETEN

I. Introduction   400
II. Biosynthesis of Ethylene: Mechanism and Regulation   401
III. Ethylene Signaling   405

IV. Transcriptional Regulation of Ethylene Response   417
V. Cross-talk in Plant Hormone Signaling   419
VI. Ethylene in Plant Disease Resistance and Abiotic Stresses   421
VII. Conclusions   422
   References   423

12

JASMONATE: AN OXYLIPIN SIGNAL WITH MANY ROLES IN PLANTS

JOHN BROWSE

I. Introduction   432
II. The Biochemistry of Jasmonate Synthesis   433
III. Jasmonate Signaling in Insect Defense   435
IV. New Defense Roles and Signal Integration   441
V. Jasmonate Regulates Reproductive Development   446
VI. Ubiquitination by SCF^{COI1} is an Early and Essential Step in Jasmonate Signaling   449
   References   451

13

PLANT SEX PHEROMONES

HIROYUKI SEKIMOTO

I. Introduction   458
II. Brown Algal Pheromones   459
III. The *Volvox* Sex-Inducing Pheromone   460
IV. The Sex Pheromones of *Closterium*   463
V. Other Green Algal Pheromones   467
VI. Spermatozoid Attractant(s) in Mosses   468
VII. Pheromones in Ferns   469
VIII. Attractants in the Pollen Tube   470
IX. Conclusions   472
   References   472

14

PLANT BRASSINOSTEROID HORMONES

TADAO ASAMI, TAKESHI NAKANO, AND SHOZO FUJIOKA

I. Biosynthesis and Metabolism of Brassinosteroids   480
II. Signal Transduction of Brassinosteroids   485
III. Conclusions   498
    References   498

15

TERPENOIDS AS PLANT ANTIOXIDANTS

J. GRAßMANN

I. Introduction   506
II. Plant Antioxidants   507
III. Terpenoids   507
    References   529

INDEX   537

CONTRIBUTORS

Numbers in parenthesis indicate the pages on which the authors' contributions begin.

Tadao Asami (479) Discovery Research Institute, RIKEN, 2-1 Hirosawa, Wako, Saitama 351-0198, Japan.

John Browse (431) Institute of Biological Chemistry, Washington State University, Pullman, Washington 99164.

Annelies De Paepe (399) Unit Plant Hormone Signaling and Bio-imaging, Department of Molecular Genetics, Ghent University, K.L. Ledeganckstraat 35, B-9000 Gent, Belgium.

Haya Friedman (31) Department of Postharvest Science of Fresh Produce, Agricultural Research Organization, The Volcani Center, Bet-Dagan 50250, Israel.

Nobuharu Fujii (79) Graduate School of Life Sciences, Tohoku University, 2-1-1 Katahira, Aoba-ku, Sendai 980-8577, Japan.

Shozo Fujioka (479) Discovery Research Institute, RIKEN, 2-1 Hirosawa, Wako, Saitama 351-0198, Japan.

J. Graßmann (505) Institute of Vegetable Science—Quality of Vegetal Foodstuff, Life Science Center Weihenstephan, Dürnast 2, 85350 Freising, Germany.

Shoshi Kikuchi (1) Department of Molecular Genetics, National Institute of Agrobiological Sciences (NIAS), 2-1-2 Kannon-dai, Tsukuba, Ibaraki 305–8602, Japan.

Richard M. Napier (203) Warwick HRI, University of Warwick, Wellesbourne, Warwick, CV35 9EF, United Kingdom.

Shoji Mano (111) Department of Cell Biology, National Institute for Basic Biology, Okazaki 444-8585, Japan; Department of Molecular Biomechanics, School of Life Science, The Graduate University for Advanced Studies, Okazaki 444-8585, Japan.

Shimon Meir (31) Department of Postharvest Science of Fresh Produce, Agricultural Research Organization, The Volcani Center, Bet-Dagan 50250, Israel.

Ann Moons (155) National Research Council Canada, Biotechnology Research Institute, Montreal Canada, H4P 2R2.

Takeshi Nakano (479) Discovery Research Institute, RIKEN, 2-1 Hirosawa, Wako, Saitama 351-0198, Japan.

Mikio Nishimura (111) Department of Cell Biology, National Institute for Basic Biology, Okazaki 444-8585, Japan; Department of Molecular Biomechanics, School of Life Sciences, The Graduate University for Advanced Studies, Okazaki 444-8585, Japan.

Catherine Perrot-Rechenmann (203) ISV-CNRS, 91198 Gif-sur-Yvette, Cedex, France.

Sonia Philosoph-Hadas (31) Department of Postharvest Science of Fresh Produce, Agricultural Research Organization, The Volcani Center, Bet-Dagan 50250, Israel.

Ivo Rieu (289) IACR Rothamsted Research, CPI Division, Harpenden, Hertfordshire, AL5 2JQ, United Kingdom.

Paul Ruas (235) Department of Biological Sciences, University of Nevada, Las Vegas, Nevada 89154.

Hitoshi Sakakibara (271) Plant Science Center, RIKEN, 1-7-22 Suehiro, Tsurumi, Yokohama 230-0045, Japan.

Hiroyuki Sekimoto (457) Institute of Life Sciences, Graduate School of Arts and Sciences, University of Tokyo, 3-8-1 Komaba, Meguro, Tokyo 153-8902, Japan; Department of Chemical and Biological Sciences, Faculty of Science, Japan Women's University, 2-8-1 Mejirodai, Bunkyo, Tokyo 112-8681, Japan.

Allan D. Shapiro (339) Biotechnology Program, Florida Gulf Coast University, Fort Myers, Florida 33965-6565.

Qingxi J. Shen (235) Department of Biological Sciences, University of Nevada, Las Vegas, Nevada 89154.

Camille M. Steber (289) USDA-ARS and Department of Crop and Soil Science, Washington State University, Pullman, Washington 99164.

Hideyuki Takahashi (79) Graduate School of Life Sciences, Tohoku University, 2-1-1 Katahira, Aoba-ku, Sendai 980-8577, Japan.

Stephen G. Thomas (289) IACR Rothamsted Research, CPI Division, Harpenden, Hertfordshire, AL5 2JQ, United Kingdom.

Dominique Van Der Straeten (399) Unit Plant Hormone Signaling and Bio-imaging, Department of Molecular Genetics, Ghent University, K.L. Ledeganckstraat 35, B-9000 Gent, Belgium.

Zhen Xie (235) Department of Biological Sciences, University of Nevada, Las Vegas, Nevada 89154.

Seiji Yamasaki (79) Faculty of Education, Fukuoka University of Education, 1-1 Akamabunkyomachi, Munakata, Fukuoka 811-4192, Japan.

Junshi Yazaki (1) Department of Molecular Genetics, National Institute of Agrobiological Sciences (NIAS), 2-1-2 Kannon-dai, Tsukuba, Ibaraki 305–8602, Japan.

PREFACE

In recent years there has not been a thematic volume of this Serial devoted to plant hormones although an occasional review has appeared in some of the eclectic volumes dating back to 1991. In the last 20 years, research on plant hormones has advanced rapidly so that this is an appropriate time to review many of those advances. There are 15 reviews comprising this volume which, it is hoped, covers most, if not all, the important topics on this subject.

To begin with, J. Yazaki and S. Kikuchi report on: "The Genomic View of Genes Responsive to the Antagonistic Phytohormones, Abscisic Acid, and Gibberellin." S. Philosoph-Hadas, H. Friedman, and S. Meir review: "Gravitropic Bending and Plant Hormones." "Hormonal Regulation of Sex Expression in Plants" is the topic contributed by S. Yamasaki, N. Fijii and H. Takahashi. S. Mano and M. Nishimura discuss: "Plant Peroxisomes". Next, A. Moons writes on: "Regulatory and Functional Interactions of Plant Growth Regulators and Glutathione S-Transferases (GSTs)."

Moving on to more specific topics, C. Perrot-Rechenmann and R. M. Napier submit a review on: "Auxins." Z. Xie, P. Ruas, and Q. J. Shen write on: "Regulatory Networks of the Phytohormone Abscisic Acid." Then, H. Sakakibara reports on: "Cytokinin Biosynthesis and Regulation." S. G. Thomas, I. Rieu, and C. M. Steber review: "Gibberellin Metabolism and Signaling." A. D. Shapiro writes on: "Nitric Oxide Signaling in Plants." This is followed by: "Ethylene Biosynthesis and Signaling: An Overview" written by D. P. Annelies and V. D. Straeten Dominique. J. Browse reviews: "Jasmonate: An Oxylipin Signal with Many Roles in Plants." H. Sekimoto follows with a paper on: "Plant Sex Pheromones." Penultimately, T. Asami,

T. Nakano, and S. Fujioka contribute: "Plant Brassinosteroid Hormones." The final manuscript is from J. Graßmann on: "Terpenoids as Plant Antioxidants."

Vitamins and Hormones aggressively continues to attempt to keep pace in the fields of vitamin research and hormone-related topics with a broad view of the definitions of these subjects.

Although there have been a few contributions on the subject of plant hormones included in eclectic volumes of this Serial during the last 13 years, this is the first time a volume has been dedicated to this topic in its entirety. One reason for this timely review is that the field of plant hormones has made enormous progress in this period rivalling the complexity of hormone action in the mammalian literature. Of course, most of the hormones reviewed are unique to plants as well as their actions through receptors. The genetic aspects and the receptorology are reminiscent of the mammalian systems. The well known hormones are reviewed including cytokinins, abscisic acid, gibberellin, and auxin. In addition, there are reviews on nitric oxide, brassinosteroids, jasmonate, ethylene, and pheromones. Other topics include: Genes that are regulated by abscisic acid and gibberellin, functional differentiation and transition of peroxisomes, plant antioxidants, gravitropic bending, and the actions of plant hormones on glutathione transferase. This up-to-date volume will be welcomed by researchers in plant biology and molecular biology as well as graduate students and undergraduates who have interest in horticulture and other aspects of agriculture as well as students of modern biology who wish to access current information on plant hormone research in one volume.

Gerald Litwack
Toluca Lake, California
March, 2005

1

THE GENOMIC VIEW OF GENES RESPONSIVE TO THE ANTAGONISTIC PHYTOHORMONES, ABSCISIC ACID, AND GIBBERELLIN

JUNSHI YAZAKI AND SHOSHI KIKUCHI

Department of Molecular Genetics, National Institute of Agrobiological Sciences (NIAS), 2-1-2 Kannon-dai, Tsukuba, Ibaraki 305–8602, Japan

I. Introduction
   A. Interaction Between Gibberellin and Abscisic Acid
   B. Genome Sequencing and Collection of Full-Length cDNAs from Plants
   C. High-Throughput Expression Analysis
II. Comprehensive Analysis of ABA-Responsive Genes by Genomics Approach
   A. Specificity of Genes with Altered Levels of Expression in Response to ABA
III. Comprehensive Analysis for GA-Responsive Genes by the Genomics Approach
   A. Specificity of Genes with Altered Expression Levels in Response to GA
IV. Specificity of Gene Responses to Both Hormones
V. cis-Element Analysis
   A. Specificity of cis-Elements for Phytohormone-Responsive Genes
VI. Conclusion and Future Prospects
References

We now have the various genomics tools for monocot (*Oryza sativa*) and a dicot (*Arabidopsis thaliana*) plant. Plant is not only a very important agricultural resource but also a model organism for biological research. It is important that the interaction between ABA and GA is investigated for controlling the transition from embryogenesis to germination in seeds using genomics tools. These studies have investigated the relationship between dormancy and germination using genomics tools. Genomics tools identified genes that had never before been annotated as ABA- or GA-responsive genes in plant, detected new interactions between genes responsive to the two hormones, comprehensively characterized *cis*-elements of hormone-responsive genes, and characterized *cis*-elements of rice and *Arabidopsis*. In these research, ABA- and GA-regulated genes have been classified as functional proteins (proteins that probably function in stress or PR tolerance) and regulatory proteins (protein factors involved in further regulation of signal transduction). Comparison between ABA and/or GA-responsive genes in rice and those in *Arabidopsis* has shown that the *cis*-element has specificity in each species. *cis*-Elements for the dehydration–stress response have been specified in *Arabidopsis* but not in rice. *cis*-Elements for protein storage are remarkably richer in the upstream regions of the rice gene than in those of *Arabidopsis*. © 2005 Elsevier Inc.

I. INTRODUCTION

A. INTERACTION BETWEEN GIBBERELLIN AND ABSCISIC ACID

The interaction between gibberellin (GA) and abscisic acid (ABA) is an important factor controlling the stages of seed development from embryogenesis to germination. The effects of these hormones are antagonistic, in that ABA promotes seed dormancy whereas GA promotes seed germination. Seed germination and seed dormancy are thus antithetical phenomena. The genes expressed mainly in seeds and flowers are commonly regulated by ABA and GA (Heck *et al.*, 1993; Jacobsen *et al.*, 1996). During germination of cereal grains, the embryo secretes GA to the aleurone layer, where the GA promotes the expression of several genes encoding hydrolytic enzymes (Lovegrove and Hooley, 2000). Expression of these genes is repressed by ABA during seed development, in dormant seeds, and in seedlings under unfavorable conditions. To shed light on the control of seed germination and seed dormancy by ABA and GA, we need to determine the member genes and gene products acting as components in the signal transduction of the two hormones; and at which step in the two-signal transduction process does cross-talk occur. High-throughput technology is now available and adaptable to the resolution of these phenomena. Some researchers have used this technology to perform comprehensive analyses of genes responsive to ABA and GA in *Arabidopsis* and rice (Ogawa *et al.*, 2003; Seki *et al.*, 2002; Yazaki *et al.*, 2004b). They have tried taking a genomic approach to the physiological response to ABA and GA by using genome sequences, full-length cDNA collections, and useful databases for computer modeling analysis.

B. GENOME SEQUENCING AND COLLECTION OF FULL-LENGTH cDNAs FROM PLANTS

The completion of sequencing of the primary genome of *Arabidopsis thaliana* in December 2000 was a major milestone. The genome was the first plant genome and only the third high-quality finished eukaryotic genome to be sequenced (The Arabidopsis Genome Initiative, 2000), after those of the fruit fly *Drosophila melanogaster* (Adams *et al.*, 2000; Myers *et al.*, 2000) and the nematode *Caenorhabditis elegans* (The *C. elegans* Sequencing Consortium, 1998). Most of the analyses reported in this landmark genome sequence paper (The Arabidopsis Genome Initiative, 2000) were based on computational gene predictions. Empirical knowledge of what regions in the genome truly encode genes was limited. However, when over 10,000 full-length cDNA sequences of *Arabidopsis* became available, only 32% of the predicted gene models were found to be incorrect (Yamada *et al.*, 2003). The "correct" gene model, as determined by expression data, is only one of many models that are possible from a given transcription unit (Allona *et al.*, 1998), but at least it is known to exist, whereas predicted gene models may not. When multiple copies of full-length cDNAs are available, internal alternative splicing is also found (Zavolan *et al.*, 2003). Thus, an important complement to the raw genomic sequence is a large collection of expressed sequences. Large collections of full-length cDNA have been sequenced in *Arabidopsis* (Yamada *et al.*, 2003), rice (Kikuchi *et al.*, 2003), *C. elegans* (Reboul *et al.*, 2003), *Drosophila* (Stapleton *et al.*, 2002), mice (Carninci *et al.*, 2003), and humans (Imanishi *et al.*, 2004; Mammalian Gene Collection Program Team, 2002).

A second plant genome sequence, that of rice, was released in 2002 by two groups. A 4× shotgun coverage of *Oryza sativa* ssp. *indica* (Yu *et al.*, 2002) covered 361 Mb of the estimated 466 Mb. *Oryza sativa* ssp. *japonica* was sequenced to 5× shotgun coverage, resulting in 390 Mb of assembled sequence (Goff *et al.*, 2002). Monsanto had earlier generated 259 Mb of *japonica* sequence from 3391 bacterial artificial chromosome (BAC) clones (Barry, 2001), but these data were initially available to, and searchable by, only registered users. This draft sequence was provided to the International

Rice Genome Sequencing Project (IRGSP) for completion of high-quality genome sequencing (Sasaki *et al.*, 2002). When the predicted rice genes are compared with those of *Arabidopsis*, almost half of the genes seem to be unique to rice, and rice is now clearly the second model plant.

The use of transformation techniques is routine (Hiei *et al.*, 1994), and there are now large collections of mapping populations (Harushima *et al.*, 1998) and sequenced indexed collections of mutant lines (An *et al.*, 2003). Full-length cDNA collections have been made and sequenced (Kikuchi *et al.*, 2003), and genomics tools such as high-density microarrays are being developed (Yazaki *et al.*, 2004b). Because rice and *Arabidopsis* are developmentally separated by about 200 million years, together they provide a system for studying the evolution of plant signaling and developmental pathways. Already, comparison of flowering time pathways between the facultative long-day plant *Arabidopsis* and the short-day plant rice has revealed common molecular components (Cremer and Coupland, 2003). It is important for the rice genome to be completely finished to the quality of that of *Arabidopsis* (Palmer and McCombie, 2002); this goal is expected to be reached by the end of 2004 (http://demeter.bio.bnl.gov/Shanghai_summary.html).

C. HIGH-THROUGHPUT EXPRESSION ANALYSIS

In all organisms, new approaches are required to identify clones and confirm the expression of the remaining genes that are expressed at low levels and/or that are tissue specific. One approach is simply the confirmation of expression of hypothetical genes via RT-PCR of various cDNA populations and then to extend cDNAs via $5'$ and $3'$ RACE (Xiao *et al.*, 2002). This one-by-one approach is time consuming and is biased toward sequences initially detected by gene prediction algorithms. Another approach is the use of high-density oligonucleotide microarrays (Shoemaker *et al.*, 2001; Yamada *et al.*, 2003) and massively parallel signature sequencing (MPSS; Hoth *et al.*, 2002; Meyers *et al.*, 2004), a technology that has developed rapidly. A huge amount of information has also been gathered from the use of high-throughput technologies in the area of bioinformatics research, such as methods of data filtering (Kadota *et al.*, 2001), normalizing (Li and Hung Wong, 2001), and clustering (Hastie *et al.*, 2000). This technology will become increasingly important, because technologies such as the use of microarrays and genome tiling arrays currently utilize not only gene expression profiling but also other genome profiling techniques such as epigenetics analysis (Hatada *et al.*, 2002) and DNA-protein–binding profiling (Katou *et al.*, 2003). The use of plant microarrays has recently become common in rice (Akimoto-Tomiyama *et al.*, 2003; Yazaki *et al.*, 2003), barley (Negishi *et al.*, 2002), *Arabidopsis* (Yamada *et al.*, 2003), maize (Lee *et al.*, 2002), and zinnia (Demura *et al.*, 2002); it is now a standard investigative approach in plants and other organisms.

II. COMPREHENSIVE ANALYSIS OF ABA-RESPONSIVE GENES BY GENOMICS APPROACH

ABA is an essential phytohormone that controls many aspects of plant growth and developmental processes, including seed dormancy (controlling germination), stomatal movement, and plant tolerance to environmental stress (drought, salinity, low temperature), in response to changes in water status. At the molecular level, ABA-dependent changes in gene expression and posttranslational modification of gene product underpin these processes (Leung and Giraudat, 1998; Schroeder *et al.*, 2001). Analysis of the expression of ABA-inducible genes in plants has indicated that cross-talk and specificity exist among ABA-dependent signal pathways, environmental stresses, and plant developmental processes (Ingram and Bartels, 1996; Shinozaki and Yamaguchi-Shinozaki, 2000; Zhu, 2002). The presence of cross-talk and specificity indicates the existence of complex regulatory mechanisms involved in these physiological conditions such as controlling germination, environmental stress, and stomatal movement (Shinozaki and Yamaguchi-Shinozaki, 2000; Zhu, 2002). Microarray technology allows the effective determination of transcriptional abundance of many, or all, transcriptional units on a genome, in contrast with only one by expression analysis. This technology is capable of revealing unexpected cross-talk in the ABA response network. Data from high-throughput expression analyses of ABA responses in a number of plants have already been analyzed (Table I). These data have been reported in *Arabidopsis* and rice. Genes for ABA response obtained by using high-throughput technology are categorized in Table II by functional annotation.

A. SPECIFICITY OF GENES WITH ALTERED LEVELS OF EXPRESSION IN RESPONSE TO ABA

1. Functional Proteins (Proteins That Probably Function in Stress Tolerance)

This group includes proteins that directly protect against environmental stresses. These proteins probably function by protecting seed cells from drought and include the enzymes required for biosynthesis of various osmoprotectants, late-embryogenesis–abundant (LEA) proteins, antifreeze proteins, chaperones, and detoxification enzymes (Bray, 1997; Ingram and Bartels, 1996; Shinozaki and Yamaguchi-Shinozaki, 1997).

a. Protein of Proteins Storage Genes

Genes for storage proteins such as lectin (Wilkins and Raikhel, 1989), globulin (Heck *et al.*, 1993), and the LEA family (Moons *et al.*, 1997), which play roles in development after germination (lectin, globulin) and drought tolerance (LEA), have been reported as ABA responsive. All the expression

TABLE I. Reports on ABA-Responsive or GA-Responsive Genes Identified by High-Throughput Methods of Gene Expression Profiling

Author	Description	Target source	Platform	Organism	Number of responsive genes	QRT-PCR, Northern blot	All data published at GEO
Hoth et al. (2002)	ABA-responsive gene	Seedlings treated with 50 μM ABA for 8 or 9 h	Massively parallel signature sequencing (MPSS)	Arabidopsis thaliana	ABA 2319	Yes	No
Ogawa et al. (2003)	GA-responsive gene	Seed (GA-deficient mutant ga1–3) treated with 2 μM GA for 1, 3, 6, or 12 h	Oligonucleotide-based Arabidopsis 8200 DNA microarray	Arabidopsis thaliana	GA 357	Yes	No
Rabbani et al. (2003)	ABA-responsive gene	Root or leaf treated with 100 μM ABA for 5, 10, or 24 h	Rice 1700 EST array	Oryza sativa	ABA 45	Yes	No
Seki et al. (2002)	ABA/salt/dehydration/ low temperature-responsive gene	Seedlings treated with 100 μM ABA	Arabidopsis 7000 full-length cDNA array	Arabidopsis thaliana	ABA 245	Yes (data not shown)	No
Yang et al. (2004)	GA/BR-responsive gene	Shoots treated with 5 μM GA for 24 h	Rice 4000 EST array	Oryza sativa	GA 96	Yes	No
Yazaki et al. (2003)	ABA/GA-responsive gene	Callus treated with 50 μM ABA for 6 h or 3 days	Rice 9000 EST (full insert) array	Oryza sativa	ABA 190, GA 319	Yes	No
Yazaki et al. (2004b)	ABA/GA-responsive gene	Callus treated with 50 μM ABA for 3 days	Rice 22000 long oligo (60-mer) array derived from rice full-length cDNA	Oryza sativa	ABA 200, GA 301	Yes	Yes

TABLE II. Characterization of ABA-Responsive Genes Identified by High-Throughput Expression Profiling

Category	Hoth et al. (2002)	Rabbani et al. (2003)	Seki et al. (2002)	Yazaki et al. (2003)	Yazaki et al. (2004)
Functional proteins					
Genes for protein storage	ERD10, RAB18-like protein, LEA proteins	Lectin, LEA protein	ERD10, RD17, LEA proteins	Lectin, globulin 2, group 3 LEA, hydrophobic LEA-like protein, Osem	Hydrophobic LEA-like protein, group 3 late embryogenesis abundant protein, lectin, globulin-like protein
Stress-responsive genes	Low-temperature–induced protein 78, dehydrin	Dehydrin, metallothionein, osmoprotectant-related protein, cold-acclimation protein	Dehydrin, heat shock protein, senescence-related protein, osmoprotectant-related protein	Water stress-regulated gene, osmotin protein gene	Cold-regulated gene, low-temperature gene
Pathogen-related–protein genes	Disease-resistance–response protein		Beta-1,3-glucanase, pathogen-inducible alpha-dioxygenase	Thionin, endchitinase, PBZ1	Chitinase, elicitor-responsive gene-3, glutathione S-transferase

(*Continues*)

TABLE II. (Continued)

Category	Hoth et al. (2002)	Rabbani et al. (2003)	Seki et al. (2002)	Yazaki et al. (2003)	Yazaki et al. (2004)
Regulatory proteins					
Protein kinase/protein phosphatase	Protein phosphatase 2C, protein phosphatase 2A, receptor-like protein kinase, MAP kinase, serine/threonine kinase	Protein phosphatase 2C, phosphoglycerate kinase	Serine/threonine protein kinase, receptor-like protein kinase	MAPK4 protein, protein phosphatase 2A	Putative tyrosine phosphatase, ankyrin kinase, protein phosphatase 2C
Transcriptional factor	Homeodomain transcriptional factor, bHLH, NAM	NAC-type DNA-binding protein, bZIP DNA-binding protein, C2H2-type zinc finger DNA-binding protein	DREB, WRKY, MYB	NBS-LRR-like protein	R2R3MYB-domain protein, homeodomain leucine zipper protein, WRKY transcription factor 35, OsNAC3 protein

Category					
Protein-degradation gene	Ubiquitin-conjugating enzyme E2, RING zinc-finger protein		Ubiquitin-conjugating enzyme		
Other proteins					
Other hormone-responsive gene	Ethylene-responsive transcriptional coactivator, ethylene-responsive element binding factor, putative auxin-induced protein		Myrosinase-binding protein (JA-regulated gene)		Auxin transport protein, IAA (auxin indole-3-acetic acid)-glucose synthetase
Photosynthesis	COP8	Chlorophyll *a/b*-binding protein, thioredoxin	Chlorophyll *a/b*-binding protein, thioredoxin	Light-induced-protein gene	Chlorophyll *a/b*-binding protein
Protease/proteinase inhibitor	Trypsin inhibitor, protease inhibitor2, cysteine proteinase-like protein	Papain cysteine protease, protease inhibitor	Cysteine proteinase, trypsin inhibitor	Bowman Birk trypsin inhibitor, proteinase inhibitor, cysteine endopeptidase	C13 cysteine proteinase precursor, oryzacystatin
Lipid transfer protein	Nonspecific lipid-transfer protein precursor		Lipid-transfer protein	Lipid transfer protein, phospholipid transfer protein	Lipid transfer protein

profiles listed in Table I show these genes to be ABA responsive. The LEA protein family genes, including those for Osem (Hattori et al., 1995), group 3 LEA protein, RAB24 protein, water-stress–regulated gene, aldose-reductase–related protein (Bartels et al., 1991), and glucose and ribitol dehydrogenase (Alexander et al., 1994), are not only ABA responsive but also antagonistically GA responsive (Yazaki et al., 2003). The levels of expression of these genes in each reports verified the accuracy of the high-throughput expression analysis.

b. Proteins of Stress-Responsive Genes

One-by-one analyses of gene expression have revealed cross-talk between ABA-responsive genes and genes for tolerance to stresses such as low temperature, high salinity, and dehydration (Shinozaki and Yamaguchi-Shinozaki, 2000). This cross-talk was also reported in all the expression profiles for ABA-responsive genes listed in Table I. Genome-wide analysis using cDNA microarrays (Table I, Rabbani et al., 2003; Seki et al., 2002) has revealed cross-talk between ABA-responsive and low-temperature–responsive or high-salinity–responsive or drought-responsive genes, indicating the existence of a substantial common regulatory system; there is greater cross-talk between genes for drought and high-salinity stress and between those for drought and the ABA-signaling process than between cold and high-salinity stress genes or between cold and ABA-signaling–process genes. In addition, response genes for other stresses associated with senescence, osmosis, and heat shock have been reported in genome-wide expression analyses (Table I, Rabbani et al., 2003; Seki et al., 2002; Yazaki et al., 2003); these results indicate cross-talk between ABA signaling and each stress.

c. Pathogen-Related Proteins

High-throughput technology has revealed unexpected cross-talk between the ABA-signaling pathway and genes for pathogen-related protein (*PR* gene). The reports (Table I) of genome-wide expression analysis of *PR* genes, such as those for beta-1,3-glucanase, disease-resistance–response protein, chitinase, thionin, probenazole-inducible protein, and peroxidase, show cross-talk between ABA response and pathogen response. The downregulation of genes for PR proteins by ABA treatment suggests that seeds do not require high levels of protection from threats such as pathogens in the external environment (Table I, Hoth et al., 2002; Yazaki et al., 2004b).

2. Regulatory Proteins (Protein Factors Involved in Further Regulation of Signal Transduction)

This group includes proteins that regulate gene expression and signal transduction in the stress response—probably protein kinases or protein phosphatase, transcriptional factors, and protein-degradation proteins.

a. Protein Kinases/Protein Phosphatases

Protein phosphorylation-dephosphorylation events represent key steps in signal transduction. Protein kinase genes such as those for serine/threonine kinase, MAP kinase, and receptor-like protein kinase, have been induced or repressed by ABA treatment in all studies that have used high-throughput technology. Also, protein phosphatase genes such as those for protein phosphatase 2C and 2A are responsive to ABA treatment. The product of the protein phosphatase 2C gene in *Arabidopsis* has been suggested to act in the ABA-signaling network, thus correlating ABA-induced expression of a gene with its function in the signaling pathway (Leung *et al.*, 1994; Meyer *et al.*, 1994). Also, identification of a gene homologue in *Arabidopsis* for the response of protein phosphatase 2C to ABA treatment in rice suggests that the action of protein phosphatase 2C is a common mechanism in the ABA-signaling networks of the two plant species. In the presence of ABA, dephosphorylation events may activate or inhibit transcription factors. However, target proteins for dephosphorylation have not yet been identified.

b. Transcriptional Factors and DNA-Binding Proteins

Protein biosynthesis is required for ABA induction of some genes (Shinozaki and Yamaguchi-Shinozaki, 1996). The proteins that need to be synthesized include transcriptional factors (TFs), which, in turn, amplify gene expression as a secondary response. High-throughput technology has identified various genes that encode TFs or DNA-binding proteins, such as *DREB*, *WRKY*, *MYB*, and *NAC* (Table I, Hoth *et al.*, 2002; Rabbani *et al.*, 2003; Seki *et al.*, 2002; Yazaki *et al.*, 2004b), whose expression is regulated by ABA. This result suggests that a cascade of TFs may mediate stimulus-dependent gene expression in the ABA response.

c. Proteins of Protein-Degrading Genes

Regulated proteolysis mediated by the ubiquitin-proteasome system is a key regulatory component of many cellular processes, including cell cycle control, transcription, and receptor desensitization (Ciechanover, 1998; Kirschner, 1999). Lopez-Molina *et al.* (2001) demonstrated that the bZIP factor ABI5 is rapidly degraded in the absence of ABA, but not in the presence of ABA. This degradation can be blocked by inhibitors of the 26S proteasome, suggesting that the ubiquitin-proteasome system plays a role in ABA signaling. Table II summarizes the regulation of genes coding for proteins putatively involved in regulated proteolysis as "protein-degradation gene." Proteolysis-related genes were not identified in rice. In *Arabidopsis*, ABA might trigger the controlled degradation of a variety of cellular regulatory proteins via the ubiquitin pathway.

d. Other Proteins

Genes for proteases, including cysteine protease, are thought to be required for protein turnover and recycling of amino acids. Protease inhibitors, including trypsin inhibitors, may have a defensive role against the plants' own proteases (Thomas and Stoddart, 1980). Lipid transfer proteins may have a function in repairing stress-induced damage in membranes or changes in the lipid composition of membranes, perhaps to regulate permeability to toxic ions and the fluidity of the membrane (Holmberg and Bulowb, 1998; Torres-Schumann *et al.*, 1992).

III. COMPREHENSIVE ANALYSIS FOR GA-RESPONSIVE GENES BY THE GENOMICS APPROACH

Data from high-throughput expression analyses have already been analyzed for a number of plant GA responses listed in Table I. These data have been reported in *Arabidopsis* and rice. Genes for the GA response obtained using high-throughput technology are categorized in Table III by functional annotation.

A. SPECIFICITY OF GENES WITH ALTERED EXPRESSION LEVELS IN RESPONSE TO GA

1. Functional Proteins (Proteins That Probably Function in Development And Stress Tolerance)

a. Cell Elongation and Cell Division

Genes for cell elongation and cell division such as expansin, xyloglucan endotransglycosylase, and cell wall invertase have been reported to be GA responsive (Campbell and Braam, 1999; Cosgrove, 2000; Lee *et al.*, 2001; Rose *et al.*, 2002). All the expression profiles listed in Table I (Ogawa *et al.*, 2003; Yang *et al.*, 2004; Yazaki *et al.*, 2004b) for GA-responsive genes also show these genes to be GA responsive.

b. Stress-Responsive Genes

Although little is known about the relationship between GA and stress-related genes, ABA has been demonstrated to be involved in gene activation in response to various abiotic and biotic stresses. Exogenous ABA can induce the transcription of genes for salt (Garcia *et al.*, 1998), chitinase (Hong and Hwang, 2002), HSP (Campbell *et al.*, 2001), and MT-like proteins (Dong and Dunstan, 1996). Considering the antagonistic effects of GA and ABA, it is reasonable to assume that GA downregulates some stress-related genes, as identified in microarray results listed in Table I. However, the fact that GA upregulates some stress-related genes in reports of Table I suggests that it has an antagonistic function by alternative splicing variants likely in humans can generate both an activator and a repressor (Delmas *et al.*, 1992).

c. Pathogen-Related–Protein Genes

The detection of many kinds of genes for pathogen-related proteins suggests that GA response pathways have cross-talk with pathogen-related pathways. Further, salicylic acid (SA), a hormone known to mediate disease response, was shown to positively or negatively regulate cell enlargement and division (Vanacker *et al.*, 2001), two physiological processes known to be controlled by GA. The expression profiles of pathogen-related genes in GA treatment in rice and *Arabidopsis* reveal cross-talk among SA, GA, and ABA. Detection of such changes in the expression of these defense-related genes by GA treatment suggests that at germination the plant is more vulnerable and the defense-related genes are therefore upregulated.

2. Regulatory Proteins

a. Protein Kinases/Protein Phosphatases

Protein kinase genes that function as key steps in signal transduction, such as MAPK-like protein, MAPKK, and receptor-like protein kinase, were induced and repressed by GA as well as by ABA in all studies using high-throughput technology. The action of a protein kinase gene in the GA signaling network suggests that GA-induced expression of such genes is correlated with their functions in the protein kinase related signaling pathway. However, none of the expression profiles listed in Table I (Ogawa *et al.*, 2003; Yang *et al.*, 2004; Yazaki *et al.*, 2004b) provided genes for protein phosphatase.

b. Signal Transduction (Including Transcriptional Factors)

Genes for MYB proteins, which constitute a diverse class of DNA-binding proteins and function in the regulation of the cell cycle, in cellular morphogenesis, and in controlling secondary metabolism (Jin and Martin, 1999); the gene for MADS box-like protein, which functions in the construction of the reproductive organs (Riechmann and Meyerowitz, 1997); and a gene for proteins of the NAC family, which are involved in the development of plant tissue (Kikuchi *et al.*, 2000), are all regulated by GA. The reports listed in Table I suggest that the genes have functions related to seed germination. In addition, high-throughput technology has provided other genes that encode TFs or DNA-binding proteins, such as DOF, leucine zipper protein, and bHLH (Table I, Ogawa *et al.*, 2003; Yang *et al.*, 2004; Yazaki *et al.*, 2004b), and whose expression is regulated by GA. This suggests that a cascade of TFs may mediate stimulus-dependent gene expression in the GA response.

TABLE III. Characterization of GA-Responsive Genes Identified by High-Throughput Expression Profiling

Category	Ogawa et al. (2003)	Yang et al. (2004)	Yazaki et al. (2003)	Yazaki et al. (2004)
Functional proteins				
Cell elongation and cell division	Xyloglucan endotransglycosylase, beta-galactosidase, alpha-expansin	Xyloglucan endotransglycosylase-like protein	Sucrose-UDP glucosyltransferase, elongation factor 2, beta-expansin, arabinoxylan arabinofuranohydrolase	Xyloglucan endotransglycosylase-like protein, endo-xyloglucan transferase, cell wall invertase, expansin
Stress-responsive genes (detoxification)	Heat shock protein 60, thaumatin, metallothionein-like protein, glutathione S-transferase	Glutathione peroxidase-like protein, heat shock protein, salT, metallothionein-like protein	Heat shock protein, anaerobically inducible early gene 2, metallothionein-like protein, phenylalanine ammonia-lyase	Glutathione transferase, metallothionein-like protein, glucose-6-phosphate dehydrogenase
Pathogen-related-protein genes	beta-1,3-glucanase, peroxidase, endochitinase	Pathogenesis-related protein, PBZ1, chitinase	Pathogenesis-related thaumatin-like protein, thionin	Thionin, beta-1,3-glucanase, chitinase, elicitor-responsive gene-3, peroxidase

Regulatory proteins				
Protein kinase/protein phosphatase	Receptor-like protein kinase, protein kinase, cyclin-dependent kinase inhibitor protein, protein phosphatase 2C	Pyruvate dehydrogenase kinase 1	MAPK-like protein, calmodulin	Ankyrin kinase, mitogen-activated protein kinase kinase
Signal transduction (including transcriptional factors)	bHLH transcription factor, myb transcription factor, DOF zinc finger protein, NAC domain protein, Homeobox-leucine zipper protein	MYB124	Leucine zipper protein, ES43 for steroid hormone receptors, susceptibility homeodomain transcription factor	MADS box protein, OsNAC6, homeodomain leucine zipper protein
Other proteins				
Other hormone-responsive genes	Auxin-induced protein, small-auxin-up-RNA (SAUR)-like protein, auxin transport protein, ethylene response sensor, IAA-inducible gene	Salicylic acid carboxyl methyltransferase		Auxin transport protein, putative IAA1 protein

IV. SPECIFICITY OF GENE RESPONSES TO BOTH HORMONES

The modeling of seed dormancy and germination with microarray data shows that many response genes have functions related to both hormones. By using microarray platforms, hybridization, and data normalization, Yazaki *et al.* (2004a,b) showed that such genes in rice were responsive to both hormones under the same experimental conditions (Table IV). Almost all genes responsive to both hormones were regulated antagonistically. Rice genes upregulated by ABA and downregulated by GA included those of the LEA family, cysteine protease inhibitor, low-temperature–response gene, and homeodomain leucine zipper protein; this suggests that their functions are related to the stress response in seed dormancy. Defense-related genes, such as those for PR proteins, were upregulated by GA and downregulated by ABA. This suggests that the seed does not require high levels of protection from threats such as pathogens in the external environment, but that at germination the plant is more vulnerable and the defense-related genes are therefore upregulated.

In *Arabidopsis*, there have been no previous comprehensive microarray analyses, under the same experimental conditions, for genes with antagonistic responses to both hormones. Comparison of genes responsive to the two hormones among the results of three comprehensive expression analyses listed in Table I (Hoth *et al.*, 2002; Ogawa *et al.*, 2003; Seki *et al.*, 2002) generated genes for cross-talk response between the hormones (Table IV). *Arabidopsis* genes upregulated by ABA and downregulated by GA include LEA protein, trypsin inhibitor, dehydrin, low-temperature–response gene, NAC protein, and protein phosphatase 2C, suggesting that the functions of these genes are related to the stress response in seed dormancy, as has been found in rice. The expression pattern of genes for PR proteins and for cell growth-related proteins in *Arabidopsis* (upregulation by GA and downregulation by ABA) are similar to that in rice. Therefore, highly effective expression analysis is possible by computer modeling using various array platforms or different high-throughput methods that employ different gene codes, such as the GenBank accession or the MIPS code.

V. CIS-ELEMENT ANALYSIS

A. SPECIFICITY OF *cis*-ELEMENTS FOR PHYTOHORMONE-RESPONSIVE GENES

Plant science entered a new era after the completion of the entire genome sequencing of *Arabidopsis* and rice that represent model systems for dicot and monocot plants, respectively. Each genome project has been producing not only genome sequences but also a huge number of ESTs and full-length cDNAs for these plants. These two kinds of sequence data have provided an opportunity to perform genome-wide *cis*-element analyses with the aim of gaining a comprehensive understanding of genetic mechanisms. Some researchers have performed *cis*-element analyses using sequences of ABA- and GA-responsive genes.

1. *cis*-Elements of ABA-Responsive Genes

Almost all the genes induced in plants by ABA have *cis*-elements for the ABA response, including ABRE (Marcotte *et al.*, 1989) and DRE (Seki *et al.*, 2001). However, the results of microarray analysis have demonstrated the existence of genes that do not contain recognized *cis*-elements for the ABA response, such as ABRE, DRE, DRE-like, MYB, MYC (Iwasaki *et al.*, 1995; Abe *et al.*, 1997), or coupling elements (Shen and Ho, 1995), in their promoter regions. These results point to the existence of novel *cis*-acting elements involved in ABA-inducible gene expression. It may be possible to find new *cis*-elements or multiple functions of already known *cis*-elements by additional investigation of these genes. Members of the recently identified family of WRKY transcription factors have been implicated in some stress responses related to the ABA signal pathway (Eulgem *et al.*, 2000). In rice and *Arabidopsis*, the WRKY transcription factor, a protein known to mediate the pathogen-induced defense program, is induced by ABA. Not only do these data reveal cross-talk between the ABA response pathway and the pathogen-induced defense program, they also suggest that the W box–binding motif sequences related to binding the WRKY protein are novel *cis*-acting forms involved in ABA-induced gene expression. Comparison between ABA-responsive genes in rice and those in *Arabidopsis* has shown that the *cis*-element has specificity in each species. *cis*-Elements for the dehydration-stress response, including ACGTATERD1 and ABRELATERD1 (Simpson *et al.*, 2003), have been specified in *Arabidopsis* but not in rice. The specificities of *Arabidopsis cis*-elements for dehydration stress might be derived from differences in the growth environments of each species. On the other hand, *cis*-elements for protein storage, including RYREPEATGMGY2 (Lelievre *et al.*, 1992) and RYREPEATBNNAPA (Ezcurra *et al.*, 2000), are remarkably richer in the upstream regions of the rice gene than in those of *Arabidopsis*. These specificities of protein storage in rice might be related to the fact that the structure of the rice seed is different from that of the *Arabidopsis* seed.

These differences between rice and *Arabidopsis* in *cis*-elements for protein storage and dehydration stress response may have accumulated through differences in the organization of each plant or through evolutionary responses to the growth environment. Comparative analysis of *cis*-elements among various species to detect features characteristic of plants may become

TABLE IV. Categorization of Genes Whose Expression Levels Are Altered by ABA and GA Treatment

Category	Rice ABA/GA[a]	Arabidopsis ABA/GA[b]
ABA up, GA down		
Storage protein	Group 3 LEA protein, hydrophobic LEA-like protein, globulin-like protein, RAB24 protein, embryo globulin 1, Osem gene	NAP-similar protein, late-embryogenesis–abundant protein (LEA14)
Stress	Water-stress–regulated gene, cold-regulated gene, heat shock protein, low-temperature–response gene	Dehydrin RAB18-like protein, low-temperature–induced protein 78, low-temperature–induced 65-kD protein
Cell growth	Aldose reductase-related protein, glucose dehydrogenase, lipid transfer protein mRNA, myo-inositol phosphate synthase	Xyloglucan endotransglycosylase, myo-inositol 1-phosphate synthase, sucrose synthetase, root cap protein 2
Development	Glutamate decarboxylase, OsGAD1, homocysteine S-methyltransferase-4	
Proteinase/inhibitor	Oryzacystatin, cysteine protease inhibitor	Trypsin inhibitor
Transcriptional factors	Homeodomain leucine zipper protein	Salt-tolerance zinc finger protein, NAC domain protein, putative OsNAC6 protein
Other hormones		Auxin-induced protein, putative auxin-induced GH3, UDP-glucose (indole-3-acetate)
Phosphatase/kinase		Protein phosphatase 2C (AtP2C-HA)

	Other proteins	RNA-binding protein, malate oxidoreductase, histone H1
ABA down, GA up		
	Defense	Cationic peroxidase, elicitor-responsive gene-3, peroxidase, chitinase, senescence-associated protein 6	Thaumatin-like protein, peroxidase, putative glucan synthase
	Cell growth	Root cap protein 2, clone cepe7.pk0027.e11	Putative endoxyloglucan glycosyltransferase, nodulin-like protein, putative expansin
	Proteinase	Seedling for cysteine proteinase	
	Stress	Ankyrin-kinase	
	Transcriptional factors	MADS box-like protein	Putative bHLH transcription factor, putative zinc finger protein
	Other proteins		Histone H3, cytochrome P450, ribosomal protein S11, class 1 nonsymbiotic hemoglobin
ABA up, GA up		
	Stress		Dehydrin(AtCOR47), putative seed imbibition protein
	Transcriptional factors		Putative myb-related transcription factor (MYB3), translation factor EF-1 alpha
	Cell growth		ATCER2, fatty acid elongation factor, glutamate synthase, FAS1
	Other proteins		Male sterility 2-like protein, ribonucleoprotein

[a] Microarray data for a set of transcripts previously reported as ABA- and GA-responsive of rice, was compared (Yazaki *et al.*, 2003, 2004b).
[b] Expression profile of *Arabidopsis* ABA- and GA-responsive gene obtained from high-throughput technology previously reported and modified here, was compared (Hoth et al., 2002; Ogawa et al., 2003; Seki et al., 2002).

useful for the more efficient creation of transgenic plants. These comparisons may also support predictions of the physiological functions of the products of unknown genes.

2. *cis*-Elements of GA-Responsive Genes

Analysis of alpha-amylase gene promoters in barley aleurone cells has identified a *cis*-element (GA-responsive element; GARE) that is necessary and sufficient for GA responsiveness (Skriver *et al.*, 1991). A GA-inducible transcription factor, HvGAMYB, binds specifically to GARE to activate transcription (Gubler *et al.*, 1995). Ogawa *et al.* (2003) investigated whether the GA-regulated genes selected in their study of *Arabidopsis* contained known GA-responsive *cis*-elements (Ogawa *et al.*, 2003). The occurrence of GARE in GA-responsive genes was about 20%. This result suggests that GARE does not serve as a major *cis*-element for GA-inducible gene expression in *Arabidopsis*. The *cis*-element motif responsible for the GA induction of *LEAFY* in *Arabidopsis* was found in only one gene among 230 GA-responsive genes identified in a study by Blazquez and Weigel (2000), suggesting that this sequence does not serve as a common *cis*-element for GA responsiveness in *Arabidopsis*. Those researchers also found that the ABA-responsive element (ABRE) for ABA-induced gene expression (Busk, 1998) is present in approximately 25% of GA downregulated genes. In germinating lettuce seeds, exogenous GA_3 treatment has been shown to reduce endogenous ABA levels when the seeds are incubated in the dark after far-red light treatment, which inhibits germination (Toyomasu *et al.*, 1994). To determine whether the downregulation of ABRE-containing genes by GA_4 results from reduction in ABA content, Ogawa *et al.* (2003) performed ABA measurements. Their results suggested that the downregulation of ABRE-containing genes by exogenous GA_4 was not caused by a decrease in endogenous ABA levels.

A basic domain/Leu zipper transcription factor, ABI5, binds to ABRE (Kim *et al.*, 2002), and another transcription factor, ABI3, binds to a sequence called a Sph/RY element (Ezcurra *et al.*, 2000; Suzuki *et al.*, 1997), which often is found adjacent to ABREs that are positive regulators of the ABA response (Finkelstein *et al.*, 2002). FIERY1 (FRY1), an inositol polyphosphate 1-phosphatase, functions as a negative regulator of ABRE-containing ABA-inducible genes (Xiong *et al.*, 2001). QRT-PCR of the report (Ogawa *et al.*, 2003) showed that ABI3 and ABI5 mRNA levels decreased and FRY1 transcript abundance increased with GA treatment. These results suggest that genes for three transcriptional factors (ABI3, ABI5, and FRY1) may partly mediate the downregulation of ABRE-containing genes by GA. Comparison between GA-responsive genes in rice and those in *Arabidopsis* gives a result similar to that of ABA-responsive genes (Yazaki *et al.*, 2004b).

3. Function Assignment of Genes Responsive to ABA, GA, or Both by cis-Element Analysis in Rice

Comparison of *cis*-elements among genes for phytohormone response supports the assignment of functions to unknown genes and new functions to known genes in the hormone signal network. Yazaki *et al.* (2004b) characterized *cis*-elements of genes responsive to ABA, GA, or both, and speculated on the role of each of these gene groups in both hormone signal networks (Table V).

The upstream regions of ABA up/GA down genes were remarkably rich in *cis*-elements for storage proteins, including ACGTABREMOTIFA2O-SEM (Hattori *et al.*, 2002) and six other kinds. Also, *cis*-elements for storage protein were specified as elements of genes upregulated only by ABA. Genes downregulated only by GA were rich in *cis*-elements for defense, ethylene response, and development. GA up/ABA down genes were remarkably rich in *cis*-elements for amylase. Genes that were ABA-downregulated, GA-upregulated, or both had *cis*-elements for the stress (defense) response. These results suggest that antagonistic-response genes have *cis*-elements for protein storage and amylase, which are related to seed dormancy and germination, respectively. In contrast, genes responsive to only one of the two hormones are rich in elements for stress and defense responses, which promote seed dormancy and germination, respectively. Yazaki *et al.* (2004b) speculated that plant cells are particularly susceptible to stress from the external environment during germination and that germination is in fact a self-stress phenomenon. In addition, their results suggest that plant cells are resistant to stress from the external environment during seed dormancy. *cis*-Elements for storage proteins were included in the ABA up/GA down genes, demonstrating that these genes are related to seed dormancy. In contrast, *cis*-elements for amylase were included in the GA up/ABA down genes, which are involved in germination.

VI. CONCLUSION AND FUTURE PROSPECTS

High-throughput technology has enabled genome-wide functional expression analysis of plants. In the arena of ABA and GA research, advanced researchers have used high-throughput technology to elucidate the signaling networks of these hormones. These researchers have shown that the technology is powerful for identification of large numbers of genes that have never before been annotated as ABA or GA responsive. It can be used to detect new interactions between genes responsive to the two hormones, to comprehensively characterize *cis*-elements of hormone-responsive genes, and to characterize the *cis*-elements of rice and *Arabidopsis*. Also, reports of expression analysis using high-throughput tools (References list in Table I) have revealed that the technology of functional genomics can be used to

TABLE V. Specification of *cis*-Elements of ABA- and GA-Responsive Genes in Rice

Gene group	Category	*cis*-Element specificities
Comparison of *cis*-element numbers among genes upregulated by ABA, downregulated by GA, and both upregulated by ABA and downregulated by GA		
Upregulated by ABA, downregulated by GA	Development	MYBPZM
	Light	IBOX, BOXIIPCCHS
	Other hormone	CATATGGMSAUR, MARTBOX
	Protein storage	RYREPEATVFLEB4, ACGTABREMOTIFA2OSEM, DPBFCOREDCDC3, CACGTGMOTIF, RYREPEATLEGUMINBOX, TATABOX3, RYREPEATGMGY2
Downregulated by GA	Defense	SEBFCONSSTPR10A
	Development	ACGTCBOX
	Light	TBOXATGAPB
	Other hormone	ERELEE4
	Stress (dehydration)	MYCATRD22
Upregulated by ABA	Light	TBOXATGAPB
	Storage protein	RAV1BAT, CANBNNAPA
	Stress (dehydration)	MYCATRD22
Comparison of *cis*-element numbers among genes upregulated by GA, downregulated by ABA, and both upregulated by GA and downregulated by ABA		
Upregulated by GA, downregulated by ABA	Amylase	AMYBOX1, TATCCAOSAMY
	Light	IBOX, -10PEHVPSBD
	Stress (various)	MYBST1
Upregulated by GA	Defense	SEBFCONSSTPR10A
	Light	BOXIINTPATPB
	Stress (various)	MYBPLANT, TATABOXOSPAL
Downregulated by ABA	Defense	SEBFCONSSTPR10A
	Development	ACGTABOX
	Stress (various)	LTRECOREATCOR15, PALBOXAPC

cis-Elements specificities: The column indicates *cis*-elements that are rich in the upstream regions of genes in each group. Details of each *cis*-element can be found in PLACE (http://www.dna.affrc.go.jp/htdocs/PLACE/).

identify many new genes and transcriptional regulators faster and more accurately than ever before. Comparison of genes for responses to both hormones in *Arabidopsis* and rice indicates that the responses have similar molecular mechanisms in both dicots and monocots. However, comparison of the cis-elements of the genes responsive to both hormones in both species suggests that they play species-specific roles in the hormone-signal network at a transcriptional level, and also suggests that there are interspecific differences in organization and in evolutionary response to the growth environment. For the most effective analysis of gene expression, all of the expression profiles obtained with high-throughput tools need to be registered on the public database of the Gene Expression Omnibus (GEO) at NCBI (Edgar et al., 2002). For the most effective functional analysis at the genome level, all available genomic information needs to be integrated in databases such as the Rice PIPELINE (Yazaki et al., 2004a), RARGE. The systematic connection of powerful tools and information on functional genomics will allow researchers in the life sciences—not just the plant sciences but also scientists working with other organisms—to expand their research in new directions.

REFERENCES

Abe, H., Yamaguchi-Shinozaki, K., Urao, T., Iwasaki, T., Hosokawa, D., and Shinozaki, K. (1997). Role of *Arabidopsis* MYC and MYB homologs in drought- and abscisic acid-regulated gene expression. *Plant Cell* **9**, 1859–1868.

Adams, M.D., Celniker, S. E., Holt, R. A., Evans, C. A., Gocayne, J. D., Amanatides, P. G., Scherer, S. E., Li, P. W., Hoskins, R. A., Galle, R. F., George, R. A., Lewis, S. E., Richards, S., Ashburner, M., Henderson, S. N., Sutton, G. G., Wortman, J. R., Yandell, M. D., Zhang, Q., Chen, L. X., Brandon, R. C., Rogers, Y. H., Blazej, R. G., Champe, M., Pfeiffer, B. D., Wan, K. H., Doyle, C., Baxter, E. G., Helt, G., Nelson, C. R., Gabor Miklos, G. L., Abril, J. F., Agbayani, A., An, H. J., Andrews-Pfannkoch, C., Baldwin, D., Ballew, R. M., Basu, A., Baxendale, J., Bayraktaroglu, L., Beasley, E. M., Beeson, K. Y., Benos, P. V., Berman, B. P., Bhandari, D., Bolshakov, S., Borkova, D., Botchan, M. R., Bouck, J., Brokstein, P., Brottier, P., Burtis, K. C., Busam, D. A., Butler, H., Cadieu, E., Center, A., Chandra, I., Cherry, J. M., Cawley, S., Dahlke, C., Davenport, L. B., Davies, P., Pablos, Bd, Delcher, A., Deng, Z., Mays, A. D., Dew, I., Dietz, S. M., Dodson, K., Doup, L. E., Downes, M., Dugan-Rocha, S., Dunkov, B. C., Dunn, P., Durbin, K. J., Evangelista, C. C., Ferraz, C., Ferriera, S., Fleischmann, W., Fosler, C., Gabrielian, A. E., Garg, N. S., Gelbart, W. M., Glasser, K., Glodek, A., Gong, F., Gorrell, J. H., Gu, Z., Guan, P., Harris, M., Harris, N. L., Harvey, D., Heiman, T. J., Hernandez, J. R., Houck, J., Hostin, D., Houston, K. A., Howland, T. J., Wei, M. H., Ibegwam, C., Jalali, M., Kalush, F., Karpen, G. H., Ke, Z., Kennison, J. A., Ketchum, K. A., Kimmel, B. E., Kodira, C. D., Kraft, C., Kravitz, S., Kulp, D., Lai, Z., Lasko, P., Lei, Y., Levitsky, A. A., Li, J., Li, Z., Liang, Y., Lin, X., Liu, X., Mattei, B., McIntosh, T. C., McLeod, M. P., McPherson, D., Merkulov, G., Milshina, N. V., Mobarry, C., Morris, J., Moshrefi, A., Mount, S. M., Moy, M., Murphy, B., Murphy, L., Muzny, D. M., Nelson, D. L., Nelson, D. R., Nelson, K. A., Nixon, K., Nusskern, D. R., Pacleb, J. M., Palazzolo, M., Pittman, G. S., Pan, S., Pollard, J., Puri, V., Reese, M. G., Reinert, K., Remington, K., Saunders, R. D., Scheeler, F., Shen, H., Shue, B. C., Kiamos, I., Simpson, M., Skupski, M. P., Smith, T., Spier, E., Spradling, A. C., Stapleton, M., Strong, R., Sun, E., Svirskas, R., Tector, C., Turner, R., Venter, E., Wang, A. H., Wang, X., Wang, Z. Y., Wassarman, D. A., Weinstock, G. M., Weissenbach, J., Williams, S. M., Woodage, T., Worley, K. C., Wu, D., Yang, S., Yao, Q. A., Ye, J., Yeh, R. F., Zaveri, J. S., Zhan, M., Zhang, G., Zhao, Q., Zheng, L., Zheng, X. H., Zhong, F. N., Zhong, W., Zhou, X., Zhu, S., Zhu, X., Smith, H. O., Gibbs, R. A., Myers, E. W., Rubin,

G. M., and Venter, J. C. (2000). The Genome Sequence of *Drosophila melanogaster. Science* **287**, 2185–2195.

Akimoto-Tomiyama, C., Sakata, K., Yazaki, J., Nakamura, K., Fujii, F., Shimbo, K., Yamamoto, K., Sasaki, T., Kishimoto, N., Kikuchi, S., Shibuya, N., and Minami, E. (2003). Rice gene expression in response to *N*-acetylchitooligosaccharide elicitor: Comprehensive analysis by DNA microarray with randomly selected ESTs. *Plant Mol. Biol.* **52**, 537–551.

Alexander, R., Alamillo, J. M., Salamin, F., and Bartel, D. (1994). A novel embryo-specific barley cDNA clone encodes a protein with homologies to bacterial glucose and ribitol dehydrogenase. *Planta* **192**, 519–525.

Allona, I., Quinn, M., Shoop, E., Swope, K., Cyr, S. S., Carlis, J., Riedl, J., Retzel, E., Campbell, M. M., Sederoff, R., and Whetten, R. W. (1998). Analysis of xylem formation in pine by cDNA sequencing. *Proc. Natl. Acad. Sci. USA* **95**, 9693–9698.

An, S., Park, S., Jeong, D. H., Lee, D. Y., Kang, H. G., Yu, J. H., Hur, J., Kim, S. R., Kim, Y. H., Lee, M., Han, S., Kim, S. J., Yang, J., Kim, E., Wi, S. J., Chung, H. S., Hong, J. P., Choe, V., Lee, H. K., Choi, J. H., Nam, J., Kim, S. R., Park, P. B., Park, K. Y., Kim, W. T., Choe, S., Lee, C. B., and An, G. (2003). Generation and analysis of end sequence database for T-DNA tagging lines in rice. *Plant Physiol.* **133**, 2040–2047.

Barry, G. F. (2001). The use of the monsanto draft rice genome sequence in research. *Plant Physiol.* **125**, 1164–1165.

Bartels, D., Engelhardt, K., Roncarati, R., Schneider, K., Rotter, M., and Salamini, F. (1991). An ABA and GA modulated gene expressed in the barley embryo encodes an aldose reductase related protein. *EMBO J.* **10**, 1037–1043.

Blazquez, M. A., and Weigel, D. (2000). Integration of floral inductive signals in *Arabidopsis*. *Nature* **404**, 889–892.

Bray, E. A. (1997). Plant responses to water deficit. *Trends in Plant Sci.* **2**, 48–54.

Busk, P. K. (1998). Regulation of abscisic acid-induced transcription. *Plant Mol. Biol.* **37**, 425–435.

Campbell, J., Klueva, N., Nieto-Sotelo, J., Ho, T., and Nguyen, H. T. (2001). Cloning of new members of heat shock protein HSP101 gene family in wheat (*Triticum aestivum* (L.) Moench) inducible by heat, dehydration, and ABA. *Biochim. Biophys. Acta – Gene Struct. Expr.* **1517**, 270–277.

Campbell, P., and Braam, J. (1999). Xyloglucan endotransglycosylases: Diversity of genes, enzymes and potential wall-modifying functions. *Trends Plant Sci.* **4**, 361–366.

Carninci, P., Waki, K., Shiraki, T., Konno, H., Shibata, K., Itoh, M., Aizawa, K., Arakawa, T., Ishii, Y., Sasaki, D., Bono, H, Kondo, S., Sugahara, Y., Saito, R., Osato, N., Fukuda, S., Sato, K., Watahiki, A., Hirozane-Kishikawa, T., Nakamura, M., Shibata, Y., Yasunishi, A., Kikuchi, N., Yoshiki, A., Kusakabe, M., Gustincich, S., Beisel, K., Pavan, W., Aidinis, V., Nakagawara, A., Held, W. A., Iwata, H., Kono, T., Nakauchi, H., Lyons, P., Wells, C., Hume, D. A., Fagiolini, M., Hensch, T. K., Brinkmeier, M., Camper, S., Hirota, J., Mombaerts, P., Muramatsu, M., Okazaki, Y., Kawai, J., and Hayashizaki, Y. (2003). Targeting a complex transcriptome: The construction of the mouse full-length cDNA encyclopedia. *Genome Res.* **13**, 1273–1289.

Ciechanover, A. (1998). The ubiquitin-proteasome pathway: On protein death and cell life. *EMBO J.* **17**, 7151–7160.

Cosgrove, D. J. (2000). Loosening of plant cell walls by expansins. *Nature* **407**, 321–326.

Cremer, F., and Coupland, G. (2003). Distinct photoperiodic responses are conferred by the same genetic pathway in *Arabidopsis* and in rice. *Trends Plant Sci.* **8**, 405–407.

Delmas, V., Laoide, B. M., Masquilier, D., de Groot, R. P., Foulkes, N. S., and Sassone-Corsi, P. (1992). Alternative usage of initiation codons in mRNA encoding the cAMP-responsive-element modulator generates regulators with opposite functions. *Proc. Natl. Acad. Sci. USA* **89**, 4226–4230.

Demura, T., Tashiro, G., Horiguchi, G., Kishimoto, N., Kubo, M., Matsuoka, N., Minami, A., Nagata-Hiwatashi, M., Nakamura, K., Okamura, Y., Sassa, N., Suzuki, S., Yazaki, J., Kikuchi, S., and Fukuda, H. (2002). Visualization by comprehensive microarray analysis of gene expression programs during transdifferentiation of mesophyll cells into xylem cells. *Proc. Natl. Acad. Sci. USA* **99,** 15794–15799.

Dong, J., and Dunstan, D. (1996). Expression of abundant mRNAs during somatic embryogenesis of white spruce [*Picea glauca* (Moench) Voss]. *Planta* **199,** 459–466.

Edgar, R., Domrachev, M., and Lash, A. E. (2002). Gene Expression Omnibus: NCBI gene expression and hybridization array data repository. *Nucl. Acids. Res.* **30,** 207–210.

Eulgem, T., Rushton, P., Robatzek, S., and Somssich, I. (2000). The WRKY superfamily of plant transcription factors. *Trends Plant Sci.* **5,** 199–206.

Ezcurra, I., Wycliffe, P., Nehlin, L., Ellerstrom, M., and Rask, L. (2000). Transactivation of the *Brassica napus* napin promoter by ABI3 requires interaction of the conserved B2 and B3 domains of ABI3 with different *cis*-elements: B2 mediates activation through an ABRE, whereas B3 interacts with an RY/G-box. *The Plant Journal* **24,** 57–66.

Finkelstein, R. R., Gampala, S. S. L., and Rock, C. D. (2002). Abscisic acid signaling in seeds and seedlings. *Plant Cell* **14,** S15–S45.

Garcia, A., Engler, J. A., Claes, B., Villarroel, R., Van Montagu, M., Gerats, T., and Caplan, A. (1998). The expression of the salt-responsive gene *salT* from rice is regulated by hormonal and developmental cues. *Planta* **207,** 172–180.

Goff, S. A., Ricke, D., Lan, T. H., Presting, G., Wang, R., Dunn, M., Glazebrook, J., Sessions, A., Oeller, P., Varma, H., Hadley, D., Hutchison, D., Martin, C., Katagiri, F., Lange, B. M., Moughamer, T., Xia, Y., Budworth, P., Zhong, J., Miguel, T., Paszkowski, U., Zhang, S., Colbert, M., Sun, Wl, Chen, L., Cooper, B., Park, S., Wood, T. C., Mao, L., Quail, P., Wing, R., Dean, R., Yu, Y., Zharkikh, A., Shen, R., Sahasrabudhe, S., Thomas, A., Cannings, R., Gutin, A., Pruss, D., Reid, J., Tavtigian, S., Mitchell, J., Eldredge, G., Scholl, T., Miller, R. M., Bhatnagar, S., Adey, N., Rubano, T., Tusneem, N., Robinson, R., Feldhaus, J., Macalma, T., Oliphant, A., and Briggs, S. (2002). A draft sequence of the rice genome (*Oryza sativa* L. ssp. *japonica*). *Science* **296,** 92–100.

Gubler, F., Kalla, R., Roberts, J. K., and Jacobsen, J. V. (1995). Gibberellin-regulated expression of a myb gene in barley aleurone cells: Evidence for Myb transactivation of a high-pI [alpha]-amylase gene promoter. *Plant Cell* **7,** 1879–1891.

Harushima, Y., Yano, M., Shomura, A., Sato, M., Shimano, T., Kuboki, Y., Yamamoto, T., Lin, S. Y., Antonio, B. A., Parco, A., Kajiya, H., Huang, N., Yamamoto, K., Nagamura, Y., Kurata, N., Khush, G. S., and Sasaki, T. (1998). A high-density rice genetic linkage map with 2275 markers using a single F2 population. *Genetics* **148,** 479–494.

Hastie, T., Tibshirani, R., Eisen, M., Alizadeh, A., Levy, R., Staudt, L., Chan, W., Botstein, D., and Brown, P. (2000). 'Gene shaving' as a method for identifying distinct sets of genes with similar expression patterns. *Genome Biol.* **1,** research0003.1–research0003.21.

Hatada, I., Kato, A., Morita, S., Obata, Y., Nagaoka, K., Sakurada, A., Sato, M., Horii, A., Tsujimoto, A., and Matsubara, K. (2002). A microarray-based method for detecting methylated loci. *J. Hum. Genet.* **47,** 448–451.

Hattori, T., Terada, T., and Hamasuna, S. (1995). Regulation of the Osem gene by abscisic acid and the transcriptional activator VP1: Analysis of *cis*-acting promoter elements required for regulation by abscisic acid and VP1. *Plant J.* **7,** 913–925.

Hattori, T., Totsuka, M., Hobo, T., Kagaya, Y., and Yamamoto-Toyoda, A. (2002). Experimentally determined sequence requirement of ACGT-containing abscisic acid response element. *Plant Cell Physiol.* **43,** 136–140.

Heck, G., Chamberlain, A., and Ho, T. (1993). Barley embryo globulin 1 gene, Beg1: Characterization of cDNA, chromosome mapping and regulation of expression. *Mol. Gene Genet.* **239,** 209–218.

Hiei, Y., Ohta, S., Komari, T., and Kumashiro, T. (1994). Efficient transformation of rice (*Oryza sativa* L.) mediated by *Agrobacterium* and sequence analysis of the boundaries of the T-DNA. *Plant J.* **6,** 271–282.

Holmberg, N., and Bulowb, L. (1998). Improving stress tolerance in plants by gene transfer. *Trends Plant Sci.* **3,** 61–66.

Hong, J. K., and Hwang, B. K. (2002). Induction by pathogen, salt and drought of a basic class II chitinase mRNA and its in situ localization in pepper (*Capsicum annuum*). *Physiologia Plantarum* **114,** 549–558.

Hoth, S., Morgante, M., Sanchez, J. P., Hanafey, M. K., Tingey, S. V., and Chua, N. H. (2002). Genome-wide gene expression profiling in *Arabidopsis thaliana* reveals new targets of abscisic acid and largely impaired gene regulation in the abi1-1 mutant. *J. Cell. Sci.* **115,** 4891–4900.

Imanishi, T., Itoh, T., Suzuki, Y., O'Donovan, C., Fukuchi, S., Koyanagi, K. O., Barrero, R. A., Tamura, T., Yamaguchi-Kabata, Y., Tanino, M., Yura, K., Miyazaki, S., Ikeo, K., Homma, K., Kasprzyk, A., Nishikawa, T., Hirakawa, M., Thierry-Mieg, J., Thierry-Mieg, D., Ashurst, J., Jia, L., Nakao, M., Thomas, M. A., Mulder, N., Karavidopoulou, Y., Jin, L., Kim, S., Yasuda, T., Lenhard, B., Eveno, E., Suzuki, Y., Yamasaki, C., Takeda, Ji, Gough, C., Hilton, P., Fujii, Y., Sakai, H., Tanaka, S., Amid, C., Bellgard, M., Bonaldo, Md F., Bono, H., Bromberg, S. K., Brookes, A. J., Bruford, E., Carninci, P., Chelala, C., Couillault, C., Souza, S. J., Debily, M. A., Devignes, M. D., Dubchak, I., Endo, T., Estreicher, A., Eyras, E., Fukami-Kobayashi, K., Gopinath, R., Graudens, E., Hahn, Y., Han, M., Han, Z. G., Hanada, K., Hanaoka, H., Harada, E., Hashimoto, K., Hinz, U., Hirai, M., Hishiki, T., Hopkinson, I., Imbeaud, S., Inoko, H., Kanapin, A., Kaneko, Y., Kasukawa, T., Kelso, J., Kersey, P., Kikuno, R., Kimura, K., Korn, B., Kuryshev, V., Makalowska, I., Makino, T., Mano, S., Mariage-Samson, R., Mashima, J., Matsuda, H., Mewes, H. W., Minoshima, S., Nagai, K., Nagasaki, H., Nagata, N., Nigam, R., Ogasawara, O., Ohara, O., Ohtsubo, M., Okada, N., Okido, T., Oota, S., Ota, M., Ota, T., Otsuki, T., Piatier-Tonneau, D., Poustka, A., Ren, S. X., Saitou, N., Sakai, K., Sakamoto, S., Sakate, R., Schupp, I., Servant, F., Sherry, S., Shiba, R., Shimizu, N., Shimoyama, M., Simpson, A. J., Soares, B., Steward, C., Suwa, M., Suzuki, M., Takahashi, A., Tamiya, G., Tanaka, H., Taylor, T., Terwilliger, J. D., Unneberg, P., Veeramachaneni, V., Watanabe, S., Wilming, L., Yasuda, N., Yoo, H. S., Stodolsky, M., Makalowski, W., Go, M., Nakai, K., Takagi, T., Kanehisa, M., Sakaki, Y., Quackenbush, J., Okazaki, Y., Hayashizaki, Y., Hide, W., Chakraborty, R., Nishikawa, K., Sugawara, H., Tateno, Y., Chen, Z., Oishi, M., Tonellato, P., Apweiler, R., Okubo, K., Wagner, L., Wiemann, S., Strausberg, R. L., Isogai, T., Auffray, C., Nomura, N., Gojobori, T., and Sugano, S. (2004). Integrative annotation of 21,037 human genes validated by full-length cDNA clones. *PLoS Biol.* **2,** e162–0.

Ingram, J., and Bartels, D. (1996). The molecular basis of dehydration tolerance in plants. *Ann. Rev. Plant Physiol. Plant Mol. Biol.* **47,** 377–403.

Iwasaki, T., Yamaguchi-Shinozaki, K., and Shinozaki, K. (1995). Identification of a *cis*-regulatory region of a gene in *Arabidopsis thaliana* whose induction by dehydration is mediated by abscisic acid and requires protein synthesis. *Mol. Gen. Genet.* **247,** 391–398.

Jin, H., and Martin, C. (1999). Multifunctionality and diversity within the plant MYB-gene family. *Plant Mol. Biol.* **41,** 577–585.

Katou, Y., Kanoh, Y., Bando, M., Noguchi, H., Tanaka, H., Ashikari, T., Sugimoto, K., and Shirahige, K. (2003). S-phase checkpoint proteins Tof1 and Mrc1 form a stable replication-pausing complex. *Nature* **424,** 1078–1083.

Kadota, K., Miki, R., Bono, H., Shimizu, K., Okazaki, Y., and Hayashizaki, Y. (2001). Preprocessing implementation for microarray (PRIM): An efficient method for processing cDNA microarray data. *Physiol. Genomics* **4,** 183–188.

Kikuchi, K., Ueguchi-Tanaka, M., Yoshida, K. T., Nagato, Y., Matsusoka, M., and Hirano, H. Y. (2000). Molecular analysis of the NAC gene family in rice. *Mol. Genet. Genomics* **262,** 1047–1051.

Kikuchi, S., Satoh, K., Nagata, T., Kawagashira, N., Doi, K., Kishimoto, N., Yazaki, J., Ishikawa, M., Yamada, H., Ooka, H., Hotta, I., Kojima, K., Namiki, T., Ohneda, E., Yahagi, W., Suzuki, K., Li, C. J., Ohtsuki, K., Shishiki, T., Otomo, Y., Murakami, K., Iida, Y., Sugano, S., Fujimura, T., Suzuki, Y., Tsunoda, Y., Kurosaki, T., Kodama, T., Masuda, H., Kobayashi, M., Xie, Q., Lu, M., Narikawa, R., Sugiyama, A., Mizuno, K., Yokomizo, S., Niikura, J., Ikeda, R., Ishibiki, J., Kawamata, M., Yoshimura, A., Miura, J., Kusumegi, T., Oka, M., Ryu, R., Ueda, M., Matsubara, K., Kawai, J., Carninci, P., Adachi, J., Aizawa, K., Arakawa, T., Fukuda, S., Hara, A., Hashidume, W., Hayatsu, N., Imotani, K., Ishii, Y., Itoh, M., Kagawa, I., Kondo, S., Konno, H., Miyazaki, A., Osato, N., Ota, Y., Saito, R., Sasaki, D., Sato, K., Shibata, K., Shinagawa, A., Shiraki, T., Yoshino, M., and Hayashizaki, Y. (2003). Collection, mapping, and annotation of over 28,000 cDNA clones from japonica Rice. *Science* **301**, 376–379.

Kim, S. Y., Ma, J., Perret, P., Li, Z., and Thomas, T. L. (2002). *Arabidopsis* ABI5 subfamily members have distinct DNA-binding and transcriptional activities. *Plant Physiol.* **130**, 688–697.

Kirschner, M. (1999). Intracellular proteolysis. *Trends Cell Biol.* **9**, M42–M45.

Lee, J., Williams, M., Tingey, S., and Rafalski, J. (2002). DNA array profiling of gene expression changes during maize embryo development. *Funct. Integr. Genomics* **2**, 13–27.

Lee, Y., Choi, D., and Kende, H. (2001). Expansins: Ever-expanding numbers and functions. *Curr. Opin. Plant Biol.* **4**, 527–532.

Lelievre, J., Oliveira, L., and Nielsen, N. (1992). 5'-CATGCAT-3' elements modulate the expression of glycinin RT genes. *Plant Physiol.* **98**, 387–391.

Leung, J., Bouvier-Durand, M., Morris, P., Guerrier, D., Chefdor, F., and Giraudat, J. (1994). *Arabidopsis* ABA response gene ABI1: Features of a calcium-modulated protein phosphatase. *Science* **264**, 1148–1152.

Leung, J., and Giraudat, J. (1998). Abscisic acid signal transduction. *Annu. Rev. Plant Physiol. Plant Mol. Biol.* **49**, 199–222.

Li, C., and Hung Wong, W. (2001). Model-based analysis of oligonucleotide arrays: Model validation, design issues and standard error application. *Genome Biol.* **2**, research0032.1–research0032.11.

Lopez-Molina, L., Mongrand, S., and Chua, N. H. (2001). A postgermination developmental arrest checkpoint is mediated by abscisic acid and requires the ABI5 transcription factor in *Arabidopsis*. *Proc. Natl. Acad. Sci. USA* **98**, 4782–4787.

Lovegrove, A., and Hooley, R. (2000). Gibberellin and abscisic acid signalling in aleurone. *Trends Plant Sci.* **5**, 102–110.

Mammalian Gene Collection Program TeamStrausberg, R. L., Feingold, E. A., Grouse, L. H., Derge, J. G., Klausner, R. D., Collins, F. S., Wagner, L., Shenmen, C. M., Schuler, G. D., Hotta, I., Kojima, K., Namiki, T., Ohneda, E., Yahagi, W., Suzuki, K., Li, C. J., Ohtsuki, K., Shishiki, T., Otomo, Y., Murakami, K., Iida, Y., Sugano, S., Fujimura, T., Suzuki, Y., Tsunoda, Y., Kurosaki, T., Kodama, T., Masuda, H., Kobayashi, M., Xie, Q., Lu, M., Narikawa, R., Sugiyama, A., Mizuno, K., Yokomizo, S., Niikura, J., Ikeda, R., Ishibiki, J., Kawamata, M., Yoshimura, A., Miura, J., Kusumegi, T., Oka, M., Ryu, R., Ueda, M., Matsubara, K., Kawai, J., Carninci, P., Adachi, J., Aizawa, K., Arakawa, T., Fukuda, S., Hara, A., Hashidume, W., Hayatsu, N., Imotani, K., Ishii, Y., Itoh, M., Kagawa, I., Kondo, S., Konno, H., Miyazaki, A., Osato, N., Ota, Y., Saito, R., Sasaki, D., Sato, K., Shibata, K., Shinagawa, A., Shiraki, T., Yoshino, M., and Hayashizaki, Y. (2002). Generation and initial analysis of more than 15,000 full-length human and mouse cDNA sequences. *Proc. Natl. Acad. Sci. USA* **99**, 16899–16903.

Marcotte, W. R., Jr., Russell, S. H., and Quatrano, R. S. (1989). Abscisic acid-responsive sequences from the Em gene of wheat. *Plant Cell* **1**, 969–976.

Meyer, K., Leube, M., and Grill, E. (1994). A protein phosphatase 2C involved in ABA signal transduction in *Arabidopsis thaliana*. *Science* **264**, 1452–1455.

Meyers, B. C., Vu, T. H., Tej, S. S., Ghazal, H., Matvienko, M., Agrawal, V., Ning, J., and Haudenschild, C. D. (2004). Analysis of the transcriptional complexity of *Arabidopsis thaliana* by massively parallel signature sequencing. *Nat. Biotech.* **22,** 1006–1011.

Moons, A., De Keyser, A., and Van Montagu, M. (1997). A group 3 LEA cDNA of rice, responsive to abscisic acid, but not to jasmonic acid, shows variety-specific differences in salt stress response. *Gene* **191,** 197–204.

Myers, E. W., Sutton, G. G., Delcher, A. L., Dew, I. M., Fasulo, D. P., Flanigan, M. J., Kravitz, S. A., Mobarry, C. M., Reinert, K. H., Remington, K. A., Anson, E. L., Bolanos, R. A., Chou, H. H., Jordan, C. M., Halpern, A. L., Lonardi, S., Beasley, E. M., Brandon, R. C., Chen, L., Dunn, P. J., Lai, Z., Liang, Y., Nusskern, D. R., Zhan, M., Zhang, Q., Zheng, X., Rubin, G. M., Adams, M. D., and Venter, J. C. (2000). A whole-genome assembly of *Drosophila*. *Science* **287,** 2196–2204.

Negishi, T., Nakanishi, H., Yazaki, J., Kishimoto, N., Fujii, F., Shimbo, K., Yamamoto, K., Sakata, K., Sasaki, T., Kikuchi, S., Mori, S., and Nishizawa, N. K. (2002). cDNA microarray analysis of gene expression during Fe-deficiency stress in barley suggests that polar transport of vesicles is implicated in phytosiderophore secretion in Fe-deficient barley roots. *Plant J.* **30,** 83–94.

Ogawa, M., Hanada, A., Yamauchi, Y., Kuwahara, A., Kamiya, Y., and Yamaguchi, S. (2003). Gibberellin biosynthesis and response during *Arabidopsis* seed germination. *Plant Cell* **15,** 1591–1604.

Palmer, L., and McCombie, W. R. (2002). On the importance of being finished. *Genome Biol.* **3,** comment 2010.

Rabbani, M. A., Maruyama, K., Abe, H., Khan, M. A., Katsura, K., Ito, Y., Yoshiwara, K., Seki, M., Shinozaki, K., and Yamaguchi-Shinozaki, K. (2003). Monitoring expression profiles of rice genes under cold, drought, and high-salinity stresses and abscisic acid application using cDNA microarray and RNA gel-blot analyses. *Plant Physiol.* **133,** 1755–1767.

Reboul, J., Vaglio, P., Rual, J., Lamesch, P., Martinez, M., Armstrong, C., Li, S., Jacotot, L., Bertin, N., Janky, R., Moore, T., Hudson, J. J., Hartley, J., Brasch, M., Vandenhaute, J., Boulton, S., Endress, G., Jenna, S., Chevet, E., Papasotiropoulos, V., Tolias, P., Ptacek, J., Snyder, M., Huang, R., Chance, M., Lee, H., Doucette-Stamm, L., Hill, D., and Vidal, M. (2003). *C. elegans* ORFeome version 1.1: Experimental verification of the genome annotation and resource for proteome-scale protein expression. *Nat. Genet.* **34,** 35–41.

Riechmann, J., and Meyerowitz, E. (1997). MADS domain proteins in plant development. *Biol. Chem.* **378,** 1079–1101.

Rose, J. K. C., Braam, J., Fry, S. C., and Nishitani, K. (2002). The XTH family of enzymes involved in xyloglucan endotransglucosylation and endohydrolysis: Current perspectives and a new unifying nomenclature. *Plant Cell Physiol.* **43,** 1421–1435.

Sasaki, T., Matsumoto, T., Yamamoto, K., Sakata, K., Baba, T., Katayose, Y., Wu, J., Niimura, Y., Cheng, Z., Nagamura, Y., Antonio, B. A., Kanamori, H., Hosokawa, S., Masukawa, M., Arikawa, K., Chiden, Y., Yamagata, H., Yamane, H., Yoshiki, S., Yoshihara, R., Yukawa, K., Zhong, H., Iwama, H., Endo, T., Ito, H., Hahn, J. H., Kim, H. I., Eun, M. Y., Yano, M., Jiang, J., and Gojobori, T. (2002). The genome sequence and structure of rice chromosome 1. *Nature* **420,** 312–316.

Schroeder, J. I., Kwak, J. M., and Allen, G. J. (2001). Guard cell abscisic acid signalling and engineering drought hardiness in plants. *Nature* **410,** 327–330.

Seki, M., Narusaka, M., Abe, H., Kasuga, M., Yamaguchi-Shinozaki, K., Carninci, P., Hayashizaki, Y., and Shinozaki, K. (2001). Monitoring the expression pattern of 1300 *Arabidopsis* genes under drought and cold stresses by using a full-length cDNA microarray. *Plant Cell* **13,** 61–72.

Seki, M., Ishida, J., Narusaka, M., Fujita, M., Nanjo, T., Umezawa, T., Kamiya, A., Nakajima, M., Enju, A., Sakurai, T., Satou, M., Akiyama, K., Yamaguchi-Shinozaki, K., Carninci, P., Kawai, J., Hayashizaki, Y., and Shinozaki, K. (2002). Monitoring the expression pattern of around 7,000 *Arabidopsis* genes under ABA treatments using a full-length cDNA microarray. *Funct. Integr. Genomics* **2**, 282–291.

Shen, Q., and Ho, T. (1995). Functional dissection of an abscisic acid (ABA)-inducible gene reveals two independent ABA-responsive complexes each containing a G-box and a novel *cis*-acting element. *Plant Cell* **7**, 295–307.

Shinozaki, K., and Yamaguchi-Shinozaki, K. (1996). Molecular responses to drought and cold stress. *Curr. Opin. Biotechnol.* **7**, 161–167.

Shinozaki, K., and Yamaguchi-Shinozaki, K. (1997). Gene expression and signal transduction in water-stress response. *Plant Physiol.* **115**, 327–334.

Shinozaki, K., and Yamaguchi-Shinozaki, K. (2000). Molecular responses to dehydration and low temperature: Differences and cross-talk between two stress signaling pathways. *Curr. Opin. Plant Biol.* **3**, 217–223.

Shoemaker, D. D., Schadt, E. E., Armour, C. D., He, Y. D., Garrett-Engele, P., McDonagh, P. D., Loerch, P. M., and Leonardson, A. (2001). Experimental annotation of the human genome using microarray technology. *Nature* **409**, 922–927.

Simpson, S. D., Nakashima, K., Narusaka, Y., Seki, M., Shinozaki, K., and Yamaguchi-Shinozaki, K. (2003). Two different novel *cis*-acting elements of *erd1*, a *clpA* homologous *Arabidopsis* gene function in induction by dehydration stress and dark-induced senescence. *Plant J.* **33**, 259–270.

Skriver, K., Olsen, F., Rogers, J., and Mundy, J. (1991). *cis*-acting DNA elements responsive to gibberellin and its antagonist abscisic acid. *Proc. Natl. Acad. Sci. USA* **88**, 7266–7270.

Stapleton, M., Liao, G., Brokstein, P., Hong, L., Carninci, P., Shiraki, T., Hayashizaki, Y., Champe, M., Pacleb, J., Wan, K., Yu, C., Carlson, J., George, R., Celniker, S., and Rubin, G. M. (2002). The *Drosophila* gene collection: Identification of putative full-length cDNAs for 70% of *D. melanogaster* genes. *Genome Res.* **12**, 1294–1300.

Suzuki, M., Kao, C. Y., and McCarty, D. R. (1997). The conserved B3 domain of VIVIPAROUS1 has a cooperative DNA binding activity. *Plant Cell* **9**, 799–807.

The *C. elegans* Sequencing Consortium (1998). Genome sequence of the nematode *C. elegans*: A platform for investigating biology. *Science* **282**, 2012–2018.

The *Arabidopsis*, Genome Initiative (2000). Analysis of the genome sequence of the flowering plant *Arabidopsis thaliana*. *Nature* **408**, 796–815.

Thomas, H., and Stoddart, J. L. (1980). Leaf senescence. *Annu. Rev. Plant Physiol.* **31**, 83–111.

Torres-Schumann, S., Godoy, J., and Pintor-Toro, J. (1992). A probable lipid transfer protein gene is induced by NaCl in stems of tomato plants. *Plant Mol. Biol.* **18**, 749–757.

Toyomasu, T., Yamane, H., Murofushi, N., and Inoue, Y. (1994). Effects of exogenously applied gibberellin and red light on the endogenous levels of abscisic acid in photoblastic lettuce seeds. *Plant Cell Physiol.* **35**, 127–129.

Vanacker, H., Lu, H., Rate, D. N., and Greenberg, J. T. (2001). A role for salicylic acid and NPR1 in regulating cell growth in *Arabidopsis*. *Plant J.* **28**, 209–216.

Wilkins, T. A., and Raikhel, N. V. (1989). Expression of rice lectin is governed by two temporally and spatially regulated mRNAs in developing embryos. *Plant Cell* **1**, 541–549.

Xiao, Y. L., Malik, M., Whitelaw, C. A., and Town, C. D. (2002). Cloning and sequencing of cDNAs for hypothetical genes from chromosome 2 of *Arabidopsis*. *Plant Physiol.* **130**, 2118–2128.

Xiong, L., Lee, B.-h., Ishitani, M., Lee, H., Zhang, C., and Zhu, J. K. (2001). FIERY1 encoding an inositol polyphosphate 1-phosphatase is a negative regulator of abscisic acid and stress signaling in *Arabidopsis*. *Genes Dev.* **15**, 1971–1984.

Yamada, K., Lim, J., Dale, J. M., Chen, H., Shinn, P., Palm, C. J., Southwick, A. M., Wu, H. C., Kim, C., Nguyen, M., Pham, P., Cheuk, R., Karlin-Newmann, G., Liu, S. X., Lam, B., Sakano, H., Wu, T., Yu, G., Miranda, M., Quach, H. L., Tripp, M., Chang, C. H., Lee, J. M., Toriumi, M., Chan, M. M. H., Tang, C. C., Onodera, C., Deng, J. M., Akiyama, K., Ansari, Y., Arakawa, T., Banh, J., Banno, F., Bowser, L., Brooks, S., Carninci, P., Chao,

Q., Choy, N., Enju, A., Goldsmith, A. D., Gurjal, M., Hansen, N. F., Hayashizaki, Y., Johnson-Hopson, C., Hsuan, V. W., Iida, K., Karnes, M., Khan, S., Koesema, E., Ishida, J., Jiang, P. X., Jones, T., Kawai, J., Kamiya, A., Meyers, C., Nakajima, M., Narusaka, M., Seki, M., Sakurai, T., Satou, M., Tamse, R., Vaysberg, M., Wallender, E. K., Wong, C., Yamamura, Y., Yuan, S., Shinozaki, K., Davis, R. W., Theologis, A., and Ecker, J. R. (2003). Empirical analysis of transcriptional activity in the *Arabidopsis* genome. *Science* **302**, 842–846.

Yang, G., Jan, A., Shen, S., Yazaki, J., Ishikawa, M., Shimatani, Z., Kishimoto, N., Kikuchi, S., Matsumoto, H., and Komatsu, S. (2004). Microarray analysis of brassinosteroids- and gibberellin-regulated gene expression in rice seedlings. *Mol. Genet. Genomics* **271**, 468–478.

Yazaki, J., Kishimoto, N., Fujii, F., Nagata, Y., Hashimoto, A., Shimbo, K., Shimatani, Z., Kojima, K., Suzuki, K., Yamamoto, M., Yamamoto, M., Honda, S., Endo, A., Yoshida, Y., Sato, Y., Takeuchi, K., Toyoshima, K., Miyamoto, C., Wu, J., Sasaki, T., Sakata, K., Yamamoto, K., Iba, K., Oda, T., Otomo, Y., Murakami, K., Matsubara, K., Kawai, J., Carninci, P., Hayashizaki, Y., and Kikuchi, S. (2003). Genomics approach to abscisic-acid and gibberellin-responsive genes in rice. *DNA Res.* **10**, 249–261.

Yazaki, J., Kojima, K., Suzuki, K., Kishimoto, N., and Kikuchi, S. (2004a). The rice PIPELINE: A unification tool for plant functional genomics. *Nucl. Acids. Res.* **32**, D383–D387.

Yazaki, J., Shimatani, Z., Hashimoto, A., Nagata, Y., Fujii, F., Kojima, K., Suzuki, K., Taya, T., Tonouchi, M., Nelson, C., Nakagawa, A., Otomo, Y., Murakami, K., Matsubara, K., Kawai, J., Carninci, P., Hayashizaki, Y., and Kikuchi, S. (2004b). Transcriptional profiling of genes responsive to abscisic acid and gibberellin in rice: Phenotyping and comparative analysis between rice and *Arabidopsis*. *Physiol. Genomics* **17**, 87–100.

Yu, J., Hu, S., Wang, J., Wong, G. K.-S., Li, S., Liu, B., Deng, Y., Dai, L., Zhou, Y., Zhang, X., Cao, M., Liu, J., Sun, J., Tang, J., Chen, Y., Huang, X., Lin, W., Ye, C., Tong, W., Cong, L., Geng, J., Han, Y., Li, L., Li, W., Hu, G., Huang, X., Li, W., Li, J., Liu, Z., Li, L., Liu, J., Qi, Q., Liu, J., Li, L., Li, T., Wang, X., Lu, H., Wu, T., Zhu, M., Ni, P., Han, H., Dong, W., Ren, X., Feng, X., Cui, P., Li, X., Wang, H., Xu, X., Zhai, W., Xu, Z., Zhang, J., He, S., Zhang, J., Xu, J., Zhang, K., Zheng, X., Dong, J., Zeng, W., Tao, L., Ye, J., Tan, J., Ren, X., Chen, X., He, J., Liu, D., Tian, W., Tian, C., Xia, H., Bao, Q., Li, G., Gao, H., Cao, T., Wang, J., Zhao, W., Li, P., Chen, W., Wang, X., Zhang, Y., Hu, J., Wang, J., Liu, S., Yang, J., Zhang, G., Xiong, Y., Li, Z., Mao, L., Zhou, C., Zhu, Z., Chen, R., Hao, B., Zheng, W., Chen, S., Guo, W., Li, G., Liu, S., Tao, M., Wang, J., Zhu, L., Yuan, L., and Yang, H. (2002). A draft sequence of the rice genome (*Oryza sativa* L. ssp. *indica*). *Science* **296**, 79–92.

Zavolan, M., Kondo, S., Schonbach, C., Adachi, J., Hume, D. A., RIKEN GER GroupMembers, G. S. L., Hayashizaki, Y., and Gaasterland, T. (2003). Impact of alternative initiation, splicing, and termination on the diversity of the mRNA transcripts encoded by the mouse transcriptome. *Genome Res.* **13**, 1290–1300.

Zhu, J. K. (2002). Salt and drought stress signal transduction in plants. *Annu. Rev. Plant Biol.* **53**, 247–273.

FURTHER READING

Jacobsen, S. E., and Olszewski, N. E. (2003). Gibberellins regulate the abundance of RNAs with sequence similarity to proteinase inhibitors, dioxygenases and dehydrogenases. *Planta* **198**, 78–86.

2

GRAVITROPIC BENDING AND PLANT HORMONES

SONIA PHILOSOPH-HADAS, HAYA FRIEDMAN, AND SHIMON MEIR

Department of Postharvest Science of Fresh Produce, Agricultural Research Organization, The Volcani Center, Bet-Dagan 50250, Israel

    I. Introduction
   II. The Gravitropic Bending of Plants
  III. Role of Auxin
      A. Detection of Auxin Asymmetry Across Gravistimulated Organs
      B. Auxin Transport in Gravitropism
      C. Auxin Sensitivity and Response in Gravitropism
      D. Role of Auxin in Gravitropism: Summary
  IV. Role of Ethylene
      A. Involvement of Ethylene in Root Gravitropism
      B. Involvement of Ethylene in Shoot Gravitropism
      C. Possible Modes of Ethylene Action in Gravitropism of Roots and Shoots
   V. Role of Gibberellins
      A. Roots
      B. Shoots
  VI. Role of Abscisic Acid
 VII. Role of Cytokinins
      A. Shoots

B. Roots
VIII. Role of Brassinosteroids, Jasmonates, and Salicylic Acid
  A. Brassinosteroids
  B. Jasmonates
  C. Salicylic Acid
 IX. Concluding Remarks and Future Prospects
     References

Gravitropism is a complex multistep process that redirects the growth of roots and various above-ground organs in response to changes in the direction of the gravity vector. The anatomy and morphology of these graviresponding organs indicates a certain spatial separation between the sensing region and the responding one, a situation that strongly suggests the requirement of phytohormones as mediators to coordinate the process. The Cholodny–Went hypothesis suggested auxin as the main mediator of gravitropism. So far, ample evidence has been gathered with regard to auxin asymmetrical detection, polar and lateral transport involving influx and efflux carriers, response signaling pathway, and possible modes of action in differential cell elongation, supports its major role in gravitropism at least in roots. However, it is becoming clear that the participation of other hormones, acting in concert with auxin, is necessary as well. Of particular importance is the role of ethylene in shoot gravitropism, possibly associated with the modulation of auxin transport or sensitivity, and the key role implicated for cytokinin as the putative root cap inhibitor that controls early root gravitropism. Therefore, the major advances in the understanding of transport and signaling of auxin, ethylene, and cytokinin may shed light on the possibly tight and complicated interactions between them in gravitropism. Not much convincing evidence has been accumulated regarding the participation of other phytohormones, such as gibberellins, abscisic acid, brassinosteroids, jasmonates, and salicylic acid, in gravitropism. However, the emerging concept of cooperative hormone action opens new possibilities for a better understanding of the complex interactions of all phytohormones and their possible synergistic effects and involvement in the gravitropic bending process. © 2005 Elsevier Inc.

I. INTRODUCTION

Gravitropism describes the curvatures of roots and shoots that orient themselves to grow either directly along or against the prevailing gravity vector, respectively. The early experimental and pioneering work of Darwin (1880) suggested the involvement of a transported signaling molecule, which was later identified as the plant growth hormone auxin. Consequently, as proposed by the Cholodny–Went hypothesis (Went and Thimann, 1937) in shoots and roots undergoing tropic curvatures, auxin was suggested to accumulate differentially on either side of the organ such that the resulting differential auxin accumulation would reduce root growth and increase shoot growth, thereby evoking curvature in the appropriate direction. Similar differential growth patterns were also detected in some other tropic processes such as phototropism (Liscum, 2002), hydrotropism (Eapen *et al.*, 2005), epinasty (Kang, 1979), and apical hook formation (Schwark and Schierle, 1992).

The apparently simple bending curvature of plants in response to a change in the direction of the gravity vector is a complicated and multistep process, which has attracted the attention of biologists for more than a century. The anatomy and morphology of the main graviresponding organs, roots and shoots, indicates a spatial separation between the sensing region and the responding one. This situation strongly suggests that a mediator is required to coordinate the process. The Cholodny–Went hypothesis suggested auxin as the main mediator of the process (Went and Thimann, 1937), but over the years the participation of other hormones and factors was investigated. Indeed, if mediation of gravitropism is hormonal, it should involve several growth regulators because no tissue is controlled solely by a single hormone. For example, stem elongation that occurs in gravitropism is promoted by several hormones, including indole-3-acetic acid (IAA), gibberellin (GA), brassinosteroids (BR), and ethylene, but their interaction and specific role in mediating this process is still obscure. Although not much convincing evidence has been accumulated regarding the participation of these phytohormones and others in gravitropism, the emerging concept of cooperative hormone action or hormone cross-talks (Klee, 2003) opens new possibilities for better understanding their complex relations and their possible synergistic effects and involvement in the gravitropic process.

Phytohormones act as chemical messengers at target sites to regulate rates and amounts of growth of cells and tissues of various plant organs. Knowledge of the involvement of hormones in the process of gravitropism was gained through studies on the asymmetrical endogenous distribution of hormones that precedes the growth response, the effect of exogenous application of the hormones or inhibitors of hormone biosynthesis and function, an analysis of the phenotype and response of hormone-deficient or gravitropic-deficient mutants, an interaction with other hormones in regulating the gravitropic response, as well as studies of hormone transport, conjugation, and metabolism (Kaufman *et al.*, 1995; Pickard, 1985; Wilkins, 1979). In the last decade, genetic and molecular analysis of mutants has provided valuable insights into the molecular mechanisms underlying the action of phytohormones in mediating gravitropism (Lomax, 1997; Masson et al., 2002; Morita and Tasaka, 2004; Muday, 2001; Muday and DeLong, 2001).

Using the latest advances in molecular biology, structural and functional genomics, microscopy, and cellular techniques, it was possible to partially confirm older findings and predictions and enormously advance our basic understanding of gravitropism. Considerable progress has been achieved by combining spaceflight facilities and plant cell biology with the development of *Arabidopsis* as a plant model and the use of gravitropic mutants. The important breakthroughs led to a clearer picture and were summarized in excellent reviews (Blancaflor and Masson, 2003; Boonsirichai et al., 2002; Friml, 2003; Masson et al., 2002; Morita and Tasaka, 2004; Muday, 2001; Yamamoto, 2003). Nevertheless, many points need further research, in particular the elucidation of the molecular mechanisms responsible for the integration of the complex regulatory processes involved in gravitropism. This chapter will focus on the current knowledge and thinking of hormonal control of the gravitropic bending of plants in roots and in above-ground, graviresponding organs (coleoptiles, pulvini, hypocotyls, and stem inflorescence). Accumulated past knowledge will be integrated with significant new evidence regarding hormonal control of gravitropism into a more comprehensive picture.

II. THE GRAVITROPIC BENDING OF PLANTS

The gravitropic response mechanism can be divided into several sequential components, including perception of the change in the gravity vector, transduction, and asymmetrical growth response. The first step of gravity perception depends on the sedimentation of dense starch-rich plastids known as amyloplasts in the root cap columella cells and the endodermal cell layer of stems and hypocotyls (Kiss, 2000; Sack, 1991). The second component of the gravitropic response mechanism is transduction, in which the development of hormone asymmetry is obtained (Kaufman et al., 1995; Muday, 2001). In the third step, a curvature response is established that allows the organ to resume growth at a defined set angle from the gravity vector (Masson et al., 2002). This gravitropic pathway may also involve several nonhormonal regulatory messengers, such cytosolic Ca^{2+} ions (Plieth and Trewavas, 2002), inositol 1,4,5-trisphosphate (IP_3) (Perera et al., 1999), protein phosphorylation (Chang et al., 2003; Rashotte et al., 2001), phospholipase A2 (PLA2) (Lee et al., 2003), the cytoskeleton network (Blancaflor, 2002; Friedman et al., 2003b; Hou et al., 2004), pH (Fasano et al., 2001; Scott and Allen, 1999), reactive oxygen species (ROS) (Joo et al., 2001), and nitric oxide (NO) (Hu et al., 2005).

The mechanism of gravitropism is in part genetically different in each organ (roots, hypocotyls, and inflorescence stems), although some genetic components of the mechanism are shared among various organs. Since roots and shoots exhibit opposite gravitropism, the difference between these organs is not surprising. However, the partial overlap found in the genetic mechanisms that underlie hypocotyl and inflorescence stem gravitropism, although both organs can be classified as shoots that exhibit negative gravitropism, was unexpected (Masson et al., 2002). When referring to shoots, we include all above-ground graviresponding organs that exhibit negative gravitropism, including stems of herbaceous plants (Meicenheimer and Nackid, 1994) or wood trees (Hellgren et al., 2004), stemlike organs (Rorabaugh and Salisbury, 1989), inflorescence stems (Friedman et al., 1998; Fukaki et al., 1996; Philosoph-Hadas et al., 1995, 1996; Woltering, 1991), epicotyls (Migliaccio and Galston, 1987), hypocotyls (Migliaccio and Rayle, 1989), coleoptiles (Philippar et al., 1999), stem leaf sheath (Kaufman et al., 1985, 1995), or internodal (Long et al., 2002) pulvini. Other organs known to be gravitropically sensitive include stamens, flower peduncles, various fruits and leaves, but relatively little study has been devoted to these organs.

Considerable evidence exists showing that the root cap is the site of gravity perception (Boonsirichai et al., 2002), and therefore it needs a means of communication between the gravity-sensing cells and those that respond further back in the elongation zone. However, in coleoptiles, hypocotyls, epicotyls, and inflorescence stems, gravity can be perceived along the whole length of the responding region (Fukaki et al., 1998; Tasaka et al., 1999); therefore, there is no need to postulate any longitudinal transfer of the message (Firn and Digby, 1980; Firn et al., 2000; Fukaki et al., 1998; Masson et al., 2002), although a lateral transfer of the message cannot be excluded. Such a different spatial relationship between gravity sensing and responding tissues of shoots and roots suggests the existence of possible mechanistic differences between roots and shoots in gravity signal transduction and transmission.

The Cholodny–Went model (Went and Thimann, 1937), which was generated to explain plant tropisms, dominated the thinking and experimentation in this issue for more than 70 years. This theory, originally generated to explain the phototropic movement of coleoptiles, claims that lack of auxin limits growth and tropisms are brought about by the lateral migration of auxin to the extending side. As such it made successful predictions about gravity-induced auxin redistribution and growth regulation in a number of representative monocot and dicot shoots. This hypothesis also was extended to roots, in which auxin as a growth inhibitor was suggested to accumulate in the nongrowing flank. Despite many variants having evolved over the years, the basic conceptual model proposed that plant tropisms were caused by the lateral redistribution of one (or more) plant growth substances that control the differential elongation at the two sides of an organ.

However, it became apparent that the conceptual unity the model gave to roots and shoots gravitropism was inconsistent with some anatomical, morphological, and physiological differences that these organs showed. Hence, while the original concepts of the Cholodny–Went model best explain root gravitropism, the mechanistic ideas of the model on shoot gravitropism are inadequate (Firn et al., 2000). Thus, new models for hormonal control are required to describe the complex growth behavior during shoot gravitropism.

III. ROLE OF AUXIN

Auxin was the earliest hormone to be implicated in gravitropism, and for many years its role in the graviresponse as the primary hormone regulating this process was dominated by the Cholodny–Went hypothesis (Went and Thimann, 1937). Auxin is thought to act as the transducing signal between the sites of gravity perception, in the starch parenchyma cells surrounding the vascular tissue in shoots and the columella cells of root caps, and asymmetric growth of the epidermal cells of the elongation zone(s) of each organ. The approaches by which auxin has been implicated in gravitropism include identification of auxin gradients across the gravistimulated organ, use of inhibitors of auxin transport that block gravitropism, analysis of differential activation of auxin-responsive gene expression or differential occurrence of various auxin-induced responses related to growth, and isolation of mutants altered in auxin transport or response as well as in their gravitropic response. While differential auxin transport or accumulation has been difficult to demonstrate in many cases (Trewavas, 1992), it has become clear through the use of *Arabidopsis* mutants that transport of and response to auxin is prerequisite for the development of tropic curvatures (Blancaflor and Masson, 2003; Chen et al., 1999; Lomax, 1997; Masson et al., 2002; Muday, 2001, 2002; Palme and Galweiler, 1999; Rosen et al., 1999).

A. DETECTION OF AUXIN ASYMMETRY ACROSS GRAVISTIMULATED ORGANS

The Cholodny–Went theory has been challenged because the magnitude of and/or kinetics for the auxin asymmetry, as measured by bioassay or analytical techniques, is widely considered to be insufficient to promote the differential growth observed (Evans, 1991; Firn and Digby, 1980; Trewavas, 1992). Indeed, a problem has always been to determine auxin at its site of action during gravistimulation, as auxin determination in extracts is of little value as long as the most responsive cells are not definitely known (Konings, 1995). Thus, when bulk extractable auxin is determined, it is still probable that the auxin, which is actually involved locally, may be masked. In the past, analysis of auxin (and other growth substances) distribution during gravistimulation included agar-diffusion experiments out of cut surfaces of excised stimulated plant parts. Alternatively, detection of radioactivity distribution within the gravireacting organ after exogenous application of the labeled hormone was also largely employed. Most of the works on identification of auxin gradients across the gravistimulated organ employed detection of radiolabeled auxin (Kaufman *et al.*, 1995; Pickard, 1985; Wilkins, 1979; for example, Brock *et al.*, 1991; Harrison and Pickard, 1989; Lee and Evans, 1985; Muday *et al.*, 1995; Rashotte *et al.*, 2000; Young *et al.*, 1990). Most of these reports provided consistent evidence for the asymmetric distribution of IAA in coleoptiles, but the results with hypocotyls, internodes, or primary roots were sometimes contradictory (Firn and Digby, 1980; Pickard, 1985; Wilkins, 1979). Thus, it was not possible to conclude from these data whether any asymmetry detected is the cause or the result of differential growth of upper and lower halves of the gravireacting organs (Muday, 2001).

Only a few studies reported more precise auxin detection methods to follow auxin levels or redistribution in gravistimulated plant organs by using either mass spectrometry (GC–MS) (Bandurski *et al.*, 1984; Friml *et al.*, 2002a; Hellgren *et al.*, 2004; Kaufman *et al.*, 1985; Ottenschlager *et al.*, 2003; Palme and Galweiler, 1999; Philippar *et al.*, 1999; Philosoph-Hadas *et al.*, 1999, 2001) or immunoassays with specific antibodies (Mertens and Weiler, 1983). Although more reliable than the radioactivity distribution analysis, the same reservations still hold, as auxin was not determined at its specific site of action in the tissue. This problem seems to have been solved in the last decade when the asymmetrical distribution of various auxin-induced genes was employed as a means for indirect detection of auxin in gravistimulated tissues (Kamada *et al.*, 2002; Li *et al.*, 1991, 1999; McClure and Guilfoyle, 1989). In the last several years, this *in situ* auxin detection method was improved, and auxin-inducible gene reporter systems, such as *DR5::GUS* or *DR5::GFP*, were used for roots and hypocotyls (Friml *et al.*, 2002b; Hou *et al.*, 2004; Ottenschlager *et al.*, 2003; Rashotte *et al.*, 2001). In this method, auxin levels are indirectly visualized within the tissue using the synthetic *DR5::GUS* auxin-response promoter (Ulmasov *et al.*, 1997), which was derived from an early auxin-response gene, *GH3* (Hagen and Guilfoyle, 2002; Li *et al.*, 1991, 1999), whose activity was found to correlate with direct auxin measurements (Friml *et al.*, 2002a). This seems to be an accurate and powerful though indirect determination of auxin in the tissue at the site of action, with high spatial resolution. However, two disadvantages of this method were noticed: (1) the *DR5::GUS* expression was delayed in its timing of appearance as compared to the lateral transport of radiolabeled IAA (Rashotte *et al.*, 2001; Young *et al.*, 1990); (2) from the auxin contents determined in root tips with *DR5::GUS/DR5::GFP* auxin concentration can be estimated to be between 0.3 and 9 μM, and lower concentrations cannot be detected (Ottenschlager et al., 2003; Rashotte et al., 2001). Hence, for the determination of auxin gradients in shoots or during phototropism, the *DR5* promoter does not seem sensitive enough, and it was suggested that there is a need to develop a new auxin reporter that is about 10 times more sensitive than *DR5::GFP* (Yamamoto, 2003).

A large amount of experimental evidence (Kaufman et al., 1995; Konings, 1995; Muday, 2001; Pickard, 1985; Wilkins, 1979), based on the various methods of auxin determination previously detailed, suggests the occurrence of an asymmetrical distribution of auxin across the organ following gravistimulation of roots (Evans, 1991; Evans et al., 1992; Ottenschlager et al., 2003; Parker, 1991; Rashotte et al., 2001; Young and Evans, 1996; Young et al., 1990) and various above-ground organs. These included maize coleoptiles (Bandurski et al., 1984; Mertens and Weiler, 1983; Parker and Briggs, 1990; Philippar et al., 1999); dandelion peduncles (Clifford et al., 1985); oat shoot pulvini (Kaufman et al., 1987, 1995); hypocotyls of tomato (Harrison and Pickard, 1989; Rice and Lomax, 2000), soybean (McClure and Guilfoyle, 1989), or sunflower (Migliaccio and Rayle, 1989); tobacco shoots (Li et al., 1991); and snapdragon inflorescence (Philosoph-Hadas et al., 1999, 2001).

Lateral redistribution of radiolabeled IAA in favor of the lower side of the root tip occurred in roots upon gravistimulation before the gravitropic curvature response was apparent, and it was dependent upon active metabolism and Ca^{2+} ions (Evans et al., 1992; Friml et al., 2002b; Ottenschlager et al., 2003; Young and Evans, 1996; Young et al., 1990). In *Arabidopsis* roots, it was also demonstrated that an endogenous gravitropic auxin gradient was developed even in the presence of an exogenous source of auxin (Ottenschlager et al., 2003). This finding actually resolved one of the strong criticism against the Cholodny–Went hypothesis, as exogenously applied auxin was expected to dissipate the endogenous auxin gradient (Firn et al., 2000).

Lateral redistribution of radiolabeled IAA was found also in maize coleoptiles (Parker and Briggs, 1990). Additionally, gradients in endogenousfree IAA determined by GC–MS have been observed in favor of the lower flank across gravistimulated oat and maize pulvini (Kaufman et al., 1985), maize coleoptiles (Philippar et al., 1999), and snapdragon inflorescence stems (Philosoph-Hadas et al., 1999, 2001). Using the GUS reporter gene under the control of an auxin-responsive promoter, it was demonstrated that gravity stimulation produces asymmetrical patterns of auxin-induced gene expression across tobacco shoots (Li et al., 1991) and *Arabidopsis* roots (Li et al., 1999; Rashotte et al., 2001). The magnitude of the auxin gradient across the gravistimulated organ ranged between twofold and fourfold in shoots or coleoptiles (Li et al., 1991; Philosoph-Hadas et al., 1999, 2001; Philippar et al., 1999).

Despite these convincing data, differential auxin accumulation has been difficult to demonstrate in many cases (Trewavas, 1992), and the observed changes in free auxin levels in various roots and shoots systems were often rather small (Clifford et al., 1985), nonexistent despite determination by accurate and sensitive methods (Hellgren et al., 2004; Mertens and Weiler, 1983), or transient (Philosoph-Hadas et al., 1999, 2001). The results showing that no auxin gradient exists in various systems were interpreted as indicating that auxin does not play a role in the gravitropic response of pea and maize roots or sunflower hypocotyls, and displacement of IAA may be restricted to the coleoptile system only, which represents an exception (Mertens and Weiler, 1983). Failure to detect any asymmetrical redistribution of auxin in the past has led to hypotheses involving other growth regulators, such as Gibberellins (GA) in shoots and abscisic acid (ABA) in roots (Wilkins, 1979) (see Sections V and VI). Similarly, the GC–MS analysis of endogenous IAA in tree stems of Aspen and Scot pine, which demonstrated that the gravity-induced wood is formed without any obvious alterations in IAA balance, was interpreted to indicate a role for signals other than IAA in the reaction wood response (Hellgren et al., 2004). Alternatively, failure to detect auxin asymmetry could be due to the existence of a transient auxin gradient, as demonstrated for snapdragon inflorescence stems (Philosoph-Hadas et al., 1999, 2001). It should be noted that while the IAA gradient in favor of the lower flank of maize coleoptiles persisted during 60 min of gravistimulation (Philippar et al., 1999), the IAA gradient generated across the snapdragon inflorescence shoots dissipated following 30 min of gravistimulation (Philosoph-Hadas et al., 2001). This implies that the gravity-induced IAA redistribution in favor of the lower stem half is a dynamic process that occurs only at specific times during gravistimulation, probably at the early stages.

The timing of the appearance of the auxin gradient relative to gravitropic bending was not always consistent with the Cholodny–Went theory. In contrast to the detection of IAA gradient prior to bending in coleoptiles (Parker and Briggs, 1990; Philippar et al., 1999) and snapdragon spikes (Philosoph-Hadas et al., 1999, 2001), it appeared during or after the curvature in many cases, for example, in tobacco stems (Li et al., 1991) or *Arabidopsis* roots (Rashotte et al., 2001). In the latter two systems, the *DR5::GUS* expression gradient appeared 5–6 h after gravistimulation, while the lateral distribution of radiolabeled IAA in maize roots was detected much earlier (Young et al., 1990). Such a delay may reflect either a lag in activation of the machinery needed for gene expression or a higher threshold for detection of auxin levels (Muday, 2001). The possibility that the differential GUS expression is a result rather than the cause of gravitropic bending is excluded by the findings showing that inhibition of auxin transport prevented the gravitropic response (Rashotte et al., 2001). However, by using the *DR5::GFP* system in *Arabidopsis* roots, it was possible to visualize gravity-induced lateral movements of auxin from the columella to the lateral root cap cells already within 15 min after gravistimulation (Ottenschlager *et al.*, 2003).

B. AUXIN TRANSPORT IN GRAVITROPISM

The study of auxin involvement in gravitropism went beyond the search for its differential distribution and revealed the components contributing to this redistribution. Modification in auxin transport therefore has a crucial role in auxin redistribution and formation of an auxin gradient across a gravistimulated organ. IAA, is synthesized in young tissues of shoots and transported basipetally into more mature regions of the plant where it regulates cell division, expansion, and differentiation (Friml, 2003). The auxin is transported acropetally into the root cap, and there the transported auxin together with *de novo* synthesized is transported to the cortical and epidermal tissue in the root periphery where it is transported basipetally to the root base (Boonsirichai *et al.*, 2002; Lomax *et al.*, 1998; Masson *et al.*, 2002; Muday, 2001). The polar auxin transport is a cell-to-cell movement, which is an important aspect of auxin functions, and is mediated by cellular influx and efflux carriers. Hence, the basipetal or acropetal polarity of auxin transport is governed by specific distribution of the auxin efflux carrier complexes as described later. The lateral auxin transport across gravistimulated shoots and roots (Brock and Kaufman, 1988a; Brock *et al.*, 1991; Masson *et al.*, 2002) is also an important component of auxin mediation of cell growth.

Several components of the auxin transport system operating in various organs have been identified, and a detailed description of the mutants and their defects in auxin transport is summarized in several excellent reviews (Berleth *et al.*, 2004; Friml, 2003; Masson *et al.*, 2002). Most of the auxin transport components described so far are operating in roots, and mutations within these genes exhibit specific defects in root gravitropism.

1. Involvement of Auxin Influx Carriers in the Gravitropic Response

Auxin enters cells by passive diffusion in its uncharged form, or by import through a transmembrane influx carrier when the molecule is charged. *AUX1* encodes a putative transmembrane protein with sequence similarity to fungal permease, which most likely operates as an influx carrier to enable the transport of protonated auxin (Bennett *et al.*, 1996). AUX1 was found to be localized on the upper side of protophloem root cells, consistent with a phloem unloading. It also localizes on the columella cells and lateral root cap tissue and therefore may be responsible for the movement of auxin into the columella cells and its transport from the columella cells to the peripheral tissues (Swarup *et al.*, 2001). Mutation in *AUX1* causes a defect in auxin supply to the root tip due to disruption of the basipetal auxin transport from the shoot tip. This may explain the reduction in the gravitropic root response of the mutant (Marchant *et al.*, 1999; Swarup *et al.*, 2001). However, mutation in *AUX1* did not affect shoot gravitropic response despite its possible involvement in auxin loading of the phloem (Friml, 2003; Marchant *et al.*, 1999). This is probably because the shoot gravitropic response is autonomous, and there is no need for longitudinal movement of auxin to execute the shoot gravitropic response (Friml, 2003; Masson *et al.*, 2002). The mechanism of AUX1 action is still not clear, despite data suggesting that it acts as a part of a multimeric complex (Boonsirichai *et al.*, 2002; Parry *et al.*, 2001). Nevertheless, perhaps the studies showing that mutation in sterol biosynthesis (Parry *et al.*, 2001) and that treatment with chromosaponin I rescued the agravitropic phenotype of *aux*1 (Parry *et al.*, 2001; Rahman *et al.*, 2001) suggest that membrane structure has a major effect on AUX1 function.

It is possible that other auxin influx carriers may exist and function in auxin redistribution during gravistimulation. One such carrier is the gene defective in *rgr1/axr4* mutant. A mutation in this gene resembles *aux1* mutant phenotype and is rescued by application of naphthalene-acetic acid (NAA) but not by IAA or 2,4-dichlorophenoxy-acetic acid, similar to the agravitropic phenotype of *aux1* (Masson *et al.*, 2002; Yamamoto and Yamamoto, 1999). Additionally, neutral amino acid transporters may function in auxin transport (Parry *et al.*, 2001). Also, the *AUX*1 sequence shows high homology to three other *Arabidopsis* sequences, *LAX*1–3. However, the role of these three genes in gravity transduction has not been elucidated (Parry *et al.*, 2001). It is possible that some of these potential auxin influx carriers may act in shoot to enable the lateral transport of auxin.

2. Involvement of Auxin Efflux Carriers in the Gravitropic Response

The polar auxin transport is mediated by efflux carriers. Several lines of evidence strongly support a role of auxin efflux carriers for *PIN* gene family. Although eight genes of this family have been identified in *Arabidopsis*, information on their involvement in auxin transport in shoots and roots and localization of their proteins exists only for *PIN1*, *PIN2/AGR1/EIR1/WAV6*, *PIN3*, and *PIN4* (Friml and Palme, 2002). The *AtPIN* paralogs share high homology among themselves, containing two highly hydrophobic domains and six transmembrane domains linked by hydrophilic regions. They also exhibit sequence similarities with prokaryotic and eukaryotic transporters (Chen *et al.*, 1998; Luschnig *et al.*, 1998; Muller *et al.*, 1998; Palme and Galweiler, 1999; Utsuno *et al.*, 1998). Their structure and localization in polar-auxin-transport cells in a polar manner, with known direction of auxin transport, support their role as efflux carriers. In *Arabidopsis* inflorescence, AtPIN1 is localized at the lower side of elongated parenchymal and cambial cell, which is consistent with its involvement in basipetal auxin movement (Galweiler *et al.*, 1998). In roots it is localized in the pith, where acropetal movement of auxin occurs (Friml and Palme, 2002; Friml *et al.*, 2002a). In protophloem cells, it is localized in the lower side of the cells opposite to AUX1 localization (Swarup *et al.*, 2001). AtPIN4 localizes on the apical side of the quiescent center cells, and its nonpolar localization was observed on the first layer of the columella cells (Friml *et al.*, 2002a). This localization is indicative of connecting acropetal auxin transport in the pith to the root cap cell. Despite the involvement of these two carriers (AtPIN1 and AtPIN4) in auxin transport, mutations within their genes do not exhibit any alteration of the gravitropic response.

On the other hand, mutations in AtPIN2 and AtPIN3 exhibited both root and hypocotyl defects in the gravitropic response, when the shoot phenotype of *Atpin2* was restricted to very young etiolated hypocotyls (Chen *et al.*, 1998; Friml *et al.*, 2002b; Luschnig *et al.*, 1998; Muller *et al.*, 1998; Utsuno *et al.*, 1998). AtPIN2 is most likely involved in basipetal auxin transport. Its localization in roots in the upper side of cells in the peripheral tissues, and its localization in hypocotyls of etiolated seedlings during the first few days of growth, is consistent with this transport (Muller *et al.*, 1998). Abolishment of the differential distribution of *DR5::GUS*, indicative of auxin accumulation, in the lower flank of gravistimulated roots of the *Atpin2* mutant, supports the idea that basipetal auxin transport executed by AtPIN2 in roots is necessary for root gravitropism (Luschnig *et al.*, 1998).

AtPIN3 is most likely involved in lateral auxin movement in both shoots and roots. It is localized in the lateral inner side of shoot endodermal cells, and in the root pericycle cells facing the central pith and also uniformly surrounding the columella cells (Friml *et al.*, 2002b). In gravistimulated roots AtPIN3 relocalizes within 2 min to the lateral side of columella cells, facing the lower side. This pattern of AtPIN3 localization is consistent with its role in creation of differential auxin gradient across the root cap upon gravistimulation (Friml, 2003; Friml *et al.*, 2002b). AtPIN3 probably is not the only protein involved in lateral transport since the mutation in this gene caused only a mild modification of the gravitropic bending, both in hypocotyls and roots. In addition, its localization towards the center of the shoot cannot explain the accumulation of auxin at the lower flank (Friml, 2003; Morita and Tasaka, 2004). It is therefore, possible that other auxin efflux carriers participate in lateral auxin transport.

Studies on multidrug-resistance (MDR) genes that encode P-glycoprotein ABC transporters, revealed that particularly AtMDR1 participates in auxin transport in hypocotyls (Noh *et al.*, 2001). So far, evidence suggests that this protein does not function directly in auxin transport but probably affects the PIN proteins localization. Mutations within the *AtMDR1* and a related gene, *AtPGP1* disrupted AtPIN1 localization, thereby resulting in impaired basipetal auxin transport. The enhanced gravitropic phenotype in hypocotyls of these mutants probably results from lateral auxin transport due to accumulation of auxin in the hypocotyl cells (Noh *et al.*, 2003).

3. Modulation of Auxin Transport Following Gravistimulation

The role of auxin transport in gravitropism was understood by extensive studies examining the effect of auxin transport inhibitors on the gravitropic response and by showing that mutations in genes encoding components of the auxin transport machinery altered root gravitropism. Application of auxin transport inhibitors to roots completely inhibited root gravitropism even without inhibiting regular growth (Bjorkman and Leopold, 1987; Muday and Haworth, 1994; Rashotte et al., 2000). However, in shoots, inhibitors of auxin transport, which inhibited the gravitropic response, blocked the growth response only partially (Kaufman et al., 1995). Factors such as Ca^{2+} ions (Migliaccio and Galston, 1987; Plieth and Trewavas, 2002), light (Buer and Muday, 2004; Jensen et al., 1998), phosphorylation (Christensen et al., 2000), and flavonoids (Buer and Muday, 2004), were reported to modify gravitropism and auxin transport, but their modes of action are still not clear. On the other hand, membrane cycling and the actin cytoskeleton were shown to modulate auxin transport by modifying the localization of auxin transport components (Muday, 2000).

Auxin transport is dependent on membrane trafficking which is affected by the cytoskeleton and N-1-naphthylphthalamic acid (NPA)-binding protein(s). PIN1 localization was found to recycle between internal compartment and the plasma membrane, and its localization was dependent on active vesicle movements. Blocking vesicle movement either by mutation in the *GNOM* gene encoding an ADP ribosylation factor, the guanine nucleotide exchange factor, or by brefeldin A (BFA) treatment, also destroyed PIN1 localization (Geldner et al., 2001, 2003; Steinmann et al., 1999). Removing BFA resulted in an immediate proper localization of PIN1. The localization of PIN1 was also dependent on the actin cytoskeleton. Both polar auxin transport and PIN1 localization were disturbed by cytochalasin, a drug causing fragmentation of actin filaments (Butler et al., 1998; Geldner et al., 2001). The actin was also involved in the internalization of PIN1 following BFA treatment and the recovery of its localization after BFA wash out (Geldner et al., 2001). Treatment with auxin transport inhibitor, 2,3,5-triiodobenzoic acid (TIBA), also prevented the cycling of PIN1 protein (Geldner et al., 2003). Interestingly, actin was found to interact with NPA-binding protein (Butler et al., 1998). Taken together, it is possible that movement of vesicles loaded with PIN1 is dependent on the actin cytoskeleton. NPA-binding protein(s) may provide the bridge between the vesicles and actin, thereby indicating that the function of NPA-binding protein(s) is necessary for vesicle movement and recycling of at least PIN1.

Several candidate NPA-binding proteins, known to be involved in actin-dependent-vesicular cycling, have been identified (Muday, 2002), but their role in auxin transport has not been established yet. Since it was found that AtMDR1 proteins expressed in yeast can bind NPA, and mutation of this gene disrupted PIN1 localization (Noh et al., 2001), it was suggested that AtMDR1 and possibly the related protein AtPGP1, are part of the export carrier complex. Nevertheless, since in the *Atmdr1* gene NPA can still block IAA transport, it is suggested that other proteins exist (Muday, 2002). Another protein that may be involved in PIN1 localization is BIG, which has a predicted size of 560 kD (Gil et al., 2001). The *tir3/doc1* mutant defective in this gene, exhibits reduced root gravitropism (Masson et al., 2002). The mutant is also insensitive to NPA application, and may therefore function in the NPA-dependent regulation of PIN1 cycling (Gil et al., 2001).

Auxin transport is enhanced by phosphorylation and flavonoids levels. A mutation in *PINOID* (*PID*), which encodes a protein kinase, reduced auxin transport (Christensen et al., 2000), and a mutation in *RCN1*, a phosphatase 2A regulatory subunit, results in reduced activity and increased basipetal IAA transport in roots (Rashotte et al., 2001). Hence, phosphorylation seems to stimulate the polar auxin transport (Muday and DeLong, 2001). Reduced auxin transport is also caused by mutation in *transparent testa4* (*tt4*), which reduces flavonoids levels (Buer and Muday, 2004). In both *rcn1* and *tt4* mutants, increased auxin transport was correlated with a defect in gravitropism. Hence, inability of the roots to respond properly to the change in the gravity vector is probably caused by inability to block basipetal auxin transport at the upper flank (Buer and Muday, 2004; Muday, 2002; Rashotte et al., 2001). Gravistimulation enhanced flavonoid synthesis at a stage after perception, but no differential accumulation of flavonoids has been detected. It is possible that flavonoids act during gravistimulation to modulate auxin transport. A gravity-induced phosphorylation of specific proteins has been detected during gravistimulation of oat shoot pulvini (Chang et al., 2003), but it is still not clear if this phosphorylation is related to auxin transport.

Rapid changes in efflux carrier localization are needed to provide immediate changes in auxin transport upon gravistimulation. Indeed, an immediate change in localization in the root cap due to gravistimulation was reported for PIN3 (Friml et al., 2002b). However, it is still not clear if other efflux carriers also modify their location due to gravistimulation in roots and shoots. It should be emphasized that as of today, no data is available on potential lateral auxin efflux carriers in inflorescence shoots.

C. AUXIN SENSITIVITY AND RESPONSE IN GRAVITROPISM

Auxin promotes cell elongation in shoots and inhibits it in roots. This reflects a dramatic difference in sensitivity of these two organs to auxin (Kaufman et al., 1995). It has been shown in a number of studies that graviresponding plant organs are characterized by changes in apparent IAA-sensitivities (Chen et al., 1998; Evans, 1991; Ishikawa and Evans,

1993; Kim and Kaufman, 1995; Rorabaugh and Salisbury, 1989; Salisbury, 1993; Salisbury *et al.*, 1985, 1988; Stinemetz, 1996), although interaction changes of IAA with its receptors have not been demonstrated (Funke and Edelmann, 2000). The responsiveness of the tissue to auxin was usually measured by monitoring the growth of the lower and the upper flanks in response to exogenous auxin (Stinemetz, 1996), or auxin-related functions such as activation of proton pump on the plasma membranes (Kim and Kaufman, 1995). It is well known that auxin controls cell expansion by regulating the activity of plasma membrane proton pumps, resulting in cell wall acidification and increased extensibility. Based on work with soybean hypocotyls and oat shoot pulvini (Firn and Digby, 1980; Kaufman *et al.*, 1995; Rorabaugh and Salisbury, 1989), it was concluded that the lower flank was more sensitive than the upper flank to auxin, leading to differential activation of the H^+-ATPase pump on the plasma membranes (Kim and Kaufman, 1995). In addition, it was well established that upon gravistimulation there is a differential activation of auxin-responsive gene expression (del Pozo and Estelle, 1999).

1. Involvement of Components of the Auxin-Response Signaling Pathway in Gravitropism

Additional support for the involvement of auxin in the gravitropic response comes from analysis of two groups of mutants that are defective in auxin-response pathway. One group of mutants appears to be involved in the regulation of auxin-responsive genes, and the other group affects protein degradation.

a. Mutants Affecting Auxin-Responsive Genes

The current knowledge on auxin signal transduction was extensively reviewed (Hare *et al.*, 2003; Leyser, 2002). In short, auxin leads to transcription activation that most likely involves the participation of genes encoding AUX/IAAs and ARFs (auxin-response factors) proteins. The *AUX/IAA* gene family in *Arabidopsis* consists of at least 29 genes encoding short-lived nuclear proteins, the transcription of which is rapidly upregulated by auxin. AUX/IAAs form heterodimers with themselves and with the ARF family that consists of 10 genes in *Arabidopsis* (Kim *et al.*, 1997; Ulmasov *et al.*, 1999). The current proposed model is that ARFs permanently bind to auxin-upregulated genes, and under low auxin concentration they form dimers with AUX/IAA, thereby shutting down transcription. Upon auxin accumulation, AUX/IAA is rapidly degraded, and this process allows ARF–ARF dimerization and transcription (Leyser, 2002). Mutations in the *AUX/IAA* genes, *IAA*3, *IAA*17, and *IAA*14 conferred abnormal gravitropic response in the hypocotyls and roots of the mutants *shy2*, *axr3*, and *slr*, respectively (Leyser, 1996; Masson *et al.*, 2002; Tian and Reed, 1999). On the other hand, a mutation in *IAA*19 resulted in the formation of the *msg2* mutant, which exhibited an abnormal gravitropic response only in hypocotyls (Tatematsu *et al.*, 2004). So far, only the mutation in *IAA*7 resulted in reduced gravitropic response of the mutant *axr2* in roots, hypocotyls, and inflorescence (Masson *et al.*, 2002; Nagpal *et al.*, 2000; Timpte *et al.*, 1995). The fact that mutations in different *AUX/IAA* genes cause a specific phenotype either in roots, in hypocotyls, or in all organs, including inflorescence, may indicate that the auxin response during gravitropism is executed by different sets of genes, and elucidation of auxin-signaling network is necessary for understanding the precise molecular mechanism of auxin response in gravitropism.

The mutation in ARF7 caused a defect only in differential growth of hypocotyls (Harper *et al.*, 2000). Mutants of this gene were isolated independently as *msg1*, which displayed insensitivity to exogenous auxin only in hypocotyls (Watahiki and Yamamoto, 1997), and as *nph4*, which did not display a phototropic response (Liscum and Briggs, 1996). The ARF7 protein was found to interact with IAA19 in a two-hybrid yeast system, and the expression of IAA19 was reduced in *nph4/msg1* mutant, suggesting that ARF7 interacts with IAA19 in the negative regulation of differential growth (Tatematsu *et al.*, 2004). It was reported that the analysis of single *arf* T-DNA insertion mutants of 18 different ARFs did not reveal additional gravitropic mutants, indicating that the ARFs functions may overlap (Okushima *et al.*, 2005). However, the *arf7arf19* double mutant exhibited abnormal gravitropic response in both hypocotyls (like in the *arf7* single mutant) and roots (Okushima *et al.*, 2005).

b. Mutants Affecting Protein Degradation

Degradation of AUX/IAA is executed via the ubiquitin-dependent proteolysis system, described in several excellent reviews (del Pozo and Estelle, 1999; Dharmasiri and Estelle, 2004; Hare *et al.*, 2003; Leyser, 2002). Degradation via this pathway requires the ubiquitination of proteins to signal them for degradation by the proteasome system. Major components of this pathway are E1 and E2 ubiquitin activating and conjugation enzymes, and the E3 ligase which transfers the ubiquitin to target proteins. The E3 ligase is active as a SCF^{TIR1} complex, composed of several proteins including F-Box, ASK1, Cullin, and RBX1. The Cullin protein undergoes a modification by an ubiquitin-related protein of the Rub family. Like ubiquitin, the Rub proteins are also activated by an array of enzymes. Mutations within F-Box, Cullin, or Rub activating enzyme in *Arabidopsis* are responsible for the auxin-insensitive phenotype in *tir1*, *axr6*, and *axr1* mutants, respectively (Hellgren *et al.*, 2004; Leyser, 2002; Masson *et al.*, 2002). These mutants also exhibited a reduced gravitropic response in roots (Hobbie and Estelle, 1995; Hobbie *et al.*, 2000). It is surprising that so far no altered shoot phenotypes have been found in mutants of this pathway.

2. Possible Modes of Auxin Action in Inducing the Graviresponse

The activation of auxin-responsive genes described earlier, in turn encodes multiple proteins that activate various cellular responses to the hormone. Auxin controls cell elongation by regulating multiple cellular processes, including ion homeostasis, enzyme activation, cytoskeletal organization, and wall extensibility (Cosgrove, 1997; Leyser, 2002). Several hypotheses were suggested to account for the nature of auxin action. Basically, mediating elements such as K^+-channels, wall-loosening factors, enzymes such as H^+-ATPase, invertase and phospholipase A2 (PLA2), pH, actin cytoskeleton, and endogenous signaling elements including reactive oxygen species (ROS), nitric oxide (NO), and cGMP, have been suggested to be involved in the execution of auxin action, as described later. However, little is known about the molecular mechanisms controlling these processes.

a. Auxin-Induced Differential Expression of K^+-Channels

Auxin was found to directly upregulate the expression of a gene encoding for inwardly rectifying K^+-channels in *Zea mays* coleoptiles (*ZMK*1), and asymmetrical changes in auxin abundance observed during coleoptile gravitropism regulated parallel changes in *ZMK*1 expression that preceded curvature (Philippar *et al.*, 1999). Such auxin-induced changes in the K^+-channel gene expression could provide a mechanism to facilitate dynamic changes in K^+ accumulation (and therefore in cell turgor), which drive coleoptile differential elongation. Auxin was reported to upregulate transcripts of two *Arabidopsis* inward rectifiers, KAT1 and KAT2, associated with growth (Philippar *et al.*, 2004). Together, these findings point to K^+-channel genes as downstream candidate genes regulated by auxin in growth and tropism.

b. Auxin-Induced Secretion of Wall-Loosening Factors

In an attempt to question and elucidate the physiological relevance of the lateral changes of IAA distribution during gravitropic bending, Edelmann (2001) has proposed an alternative model which depends on IAA but is independent from its lateral redistribution. This proposed model implies that gravistimulation temporarily inhibits infiltration of IAA-induced, secreted wall-loosening factors into the outer cell walls of the epidermal cells in the upper flank, thereby inhibiting growth of the upper organ flank (Edelmann, 2001; Funke and Edelmann, 2000).

c. Auxin-Induced pH Changes and the Participation of the Actin Cytoskeleton

Changes in cytoplasmic and apoplastic pH in response to gravistimulation were well documented (Fasano *et al.*, 2001; Scott and Allen, 1999). A report on *Arabidopsis* roots suggests an interaction between gravity-induced auxin gradient and alkalinization of the columella cytoplasm together with the actin cytoskeleton, which downregulates gravitropism by continuously resetting the gravitropic-signaling system (Hou et al., 2004).

d. Auxin-Induced Signaling Elements Associated with Oxidation

Several reports suggest that the gravitropic response of roots is mediated by the differential formation of auxin-induced ROS (Joo et al., 2001) or the endogenous signaling molecules nitric oxide (NO) and cGMP (Hu et al., 2005), which is a second messenger generated in response to NO. The data also show that ROS scavengers reduced the gravitropic bending in roots of both Z. mays (Joo et al., 2001) and soybean (Hu et al., 2005). This confirms the previous findings showing that genes involved in oxidative bursts form the largest functional category of gravity-regulated genes, as gravistimulation induced an oxidative burst (Moseyko et al., 2002). All these findings suggest that auxin acts in root gravitropism via induction of NO and/or ROS, which seem to play key roles in the process. However, the lack of complete inhibition by inhibitors of NO and cGMP synthesis indicates that additional mechanisms may mediate auxin effects in roots (Hu et al., 2005).

e. Auxin-Induced Phospholipase A2 (PLA2)

Auxin-mediated cell elongation in gravitropism may operate via upregulation of PLA2, a phospholipase enzyme that hydrolyzes membrane glycophospholipids. This was based on findings showing that the *Arabidopsis AtPLA2* gene was strongly upregulated in auxin-treated tissues and in the curving region of gravistimulated inflorescence stems (Lee et al., 2003).

f. Auxin-Induced Invertase

The differential distribution of gene expression (Wu et al., 1993) and activity (Philosoph-Hadas et al., 1996) of the sucrose cleaving invertase enzyme was demonstrated during gravitropism of oat pulvini and snapdragon stems. This indicates that this enzyme functions during the cell wall loosening period, leading to differential growth. A report (Long et al., 2002) confirms the gravity-induced asymmetrical distribution of invertase gene expression across gravistimulated maize pulvinis, and further shows that the abundance of the invertase RNA is regulated by auxin. Also, both invertase RNA and auxin gradients occurred in parallel during gravitropism (Long et al., 2002). These results imply that auxin may act via invertase in the gravitropic response, leading to differential growth.

3. Additional Possible Auxin-Related Pathways Involved in Gravitropism

Several gravitropic mutants that exhibit auxin resistance still await characterization in terms of their significance to the mechanism of auxin-related gravitropism. The *clg1* (chirality and gravitropism) is a root-gravitropism mutation that displays a wild-type root growth inhibition in response to IAA but is resistant to the synthetic auxins, 2,4-D or NAA and to the auxin-transport inhibitors, TIBA and NPA (Ferrari *et al.*, 2000). The *rib1* mutant was also found to be insensitive to auxin transport inhibitor (Poupart and Waddell, 2000). This mutant responded normally to IAA but was found to be insensitive to indole-3-butyric acid (IBA), and had a variable root response to a change in the gravitational vector. The role of this gene in the mechanism of gravitropism is still not clear.

Gravitropism in roots may also be modulated by posttranscriptional regulation. A mutation in *HYL1* resulted in altered kinetics of root gravitropism and reduced response to auxin. However, this mutation also showed other structural phenotypes, increased resistance to cytokinins, and hypersensitivity of seed germination to ABA. The gene encodes a double-stranded RNA binding, nuclear protein (Lu and Fedoroff, 2000). It was hypothesized that this protein regulates multiple hormones in various processes, including gravitropism, by affecting the expression of genes induced by other hormone (s) (Lu and Fedoroff, 2000).

D. ROLE OF AUXIN IN GRAVITROPISM: SUMMARY

Auxins have long been implicated as the central regulators of gravitropic growth responses, as proposed by the Cholodny–Went theory. The evidence gathered at all levels of detection, transport, and response seem quite convincing that auxin is prerequisite for the development of gravitropic curvatures in roots and shoots. Criticism against the Cholodny–Went model (Firn *et al.*, 2000) should not be ignored, particularly regarding the different behavior of roots and shoots. There are several lines of evidence indicating that both roots and shoots exhibit phases in the gravitropic process that are auxin-independent. In roots, several results indicate that the classical model is inadequate to account for key features of gravitropism during the early phase of root curvature (Chen *et al.*, 2002; Fasano *et al.*, 2001; Firn *et al.*, 2000; Wolverton *et al.*, 2002). Hence, it was suggested that the early phases of graviresponse, which involve differential elongation on opposite flanks of the distal root elongation zone, might be independent of this auxin gradient, and cells may either sense gravity by themselves or receive an electric signal from the root cap (Fasano *et al.*, 2001; Wolverton *et al.*, 2002). This, however, does not diminish the major role that auxin still plays in the process. It seems, that the gravity-induced auxin gradient generated across the root appears to drive at least some of the curvature responses to gravistimulation, as mutations affecting auxin transport and/or responses interfere with gravitropism (Masson *et al.*, 2002). The fact that most of the mutations in auxin transport were impaired in root gravitropism strengthens this conclusion.

In shoots, the relatively low number of mutants found in the transport pathway supports the view that if auxin movement is part of the gravitropic response mechanism, it must be localized (Firn *et al.*, 2000). Since our knowledge about transport mechanisms is still lacking, it is possible that other transport components possibly involved in lateral auxin transport may be necessary for shoot gravitropism.

Auxin redistributions following gravistimulation have been demonstrated over the last few decades in numerous studies. In spite of the criticism raised with respect to the velocity and the amount of IAA redistribution compared with the velocity of gravitropic growth, the accuracy of the methods employed for analysis of IAA levels and metabolites and the tissue specificity, the asymmetrical distribution of auxin is considered to be the main causative factor for differential growth in gravitropism, as summarized by Yamamoto (2003) and Morita and Tasaka (2004). However, it is becoming clear that the auxin gradients across the gravistimulated organ seem to be transient, and auxin asymmetry does not persist during the whole length of the gravitropic response. Such a temporal auxin asymmetry seems to be enough for activation of auxin-responsive genes and subsequent downstream events leading to the bending response.

The nature of the functions of the downstream genes or events that auxin regulates has been slowly revealed in the last few years, and several possible modes of auxin action in inducing differential growth following gravistimulation were demonstrated. Yet, we are far from gaining a full picture of this complicated growth response.

IV. ROLE OF ETHYLENE

Ethylene has profound effects upon plant growth and development, playing important roles throughout the entire life of the plant, including germination, abscission, senescence, stress, and pathogen responses (Abeles *et al.*, 1992; Schaller and Kieber, 2002). As such it is reasonable to assume that ethylene is involved also in the gravitropic response. Indeed, after auxin, which is thought to be the central player in gravitropism, ethylene is the hormone that was mostly investigated in this process.

Ethylene is also considered as a mediator of many of the effects of other plant hormones, acting as a "second messenger" in these cases (Schaller and Kieber, 2002). Indeed, several hormones such as auxin, cytokinin, and brassinosteroids were shown to highly elevate ethylene production in etiolated *Arabidopsis* seedlings, thereby inducing the triple-response phenotype (Cary *et al.*, 1995; Woeste *et al.*, 1999). However, although ethylene is intimately involved in growth at the cellular level, its influence on graviresponses may not be a specific one. Thus, it is not yet established whether ethylene is only acting downstream of other plant hormones, mainly auxin, or it is an equal partner in signaling. In the later case, the physiological response would be the result of a complex interaction between the different signaling pathways. Ample evidence has been accumulated so far, suggesting that ethylene and auxin actions are connected and that both of them are involved in the gravitropic response.

A. INVOLVEMENT OF ETHYLENE IN ROOT GRAVITROPISM

According to the Cholodny–Went theory, auxin accumulated in the lower side of the gravistimulated root will inhibit root elongation and cause downward bending. Based on the wide range of IAA–ethylene interactions, showing stimulation of ethylene synthesis by IAA and inhibition of IAA synthesis and transport by ethylene (Burg and Burg, 1967), it seems that the IAA-induced root growth inhibition is exerted via ethylene (Pickard, 1985).

Several lines of evidence suggest that ethylene is involved in root gravitropism (Pickard, 1985). The most convincing evidence shows that exogenous ethylene strongly inhibited elongation and curvature of gravistimulated roots (Hensel and Iversen, 1980), or delayed root curvature and the development of the gravity-induced asymmetric auxin distribution across the root (Lee *et al.*, 1990a). In addition, inhibition of ethylene action retarded the gravitropic response in pea and lima bean roots (Chadwick and Burg, 1970), and inhibition of ethylene synthesis or action reduced gravicurvature in maize roots (Lee *et al.*, 1990a; Mulkey and Vaughan, 1986; Mulkey *et al.*, 1982). Also, modulation of growth by IAA-induced ethylene was shown to modify the horizontal growth of *diageotropica* (*dgt*) ethylene-requiring tomato mutant roots to vertical growth, suggesting that ethylene is required to maintain normal orientation of root growth (Zobel, 1973, 1974).

Several modes of action were suggested for ethylene in root gravitropism, most of them are related to auxin, and both hormones seem to be required for the process. Ethylene was suggested to modify curvature not by mediating the primary differential growth response causing root curvature, but rather by affecting gravity-induced lateral transport of auxin (Lee *et al.*, 1990a; Pickard, 1985). Similarly, in submerged maize and pea roots, ethylene was shown not to act at graviperception, but at the subsequent signal transmission or differential growth stages (Hoson *et al.*, 1996). In these two systems, root gravitropism was inhibited by submergence, via increased internal ethylene levels or by exogenous ethylene, and root curvature could be recovered by clinostat rotation following submergence (Hoson *et al.*, 1996).

Work with mutants has provided new evidence regarding the interaction between ethylene and auxin in root gravitropism and the effects of ethylene on auxin transport in gravitropic systems. Several agravitropic mutants that are defective in both auxin and ethylene sensitivity have been discovered. Mutations in the auxin influx protein AUX1, and in the auxin efflux carrier, EIR1/PIN2/WAV6/AGR1, caused insensitivity to both hormones. Also, treatment with the auxin transport inhibitor, NPA inhibited both root curvature and ethylene effects (Lee *et al.*, 1990a). This insensitivity is probably because auxin transport was affected (Masson et al., 2002). Thus, ethylene insensitivity in the roots of *aux1* and *agr1/eir1/pin2/wav6* mutant plants may relate to the role of ethylene in regulating auxin transport (Roman et al., 1995; Suttle, 1988), and further pointing to the importance of the interactions between these two plant hormone signaling pathways. Similarly, roots of the *clg1* mutant that were more resistant to various auxins, also displayed a notable resistance to ethylene (Ferrari et al., 2000). The authors suggested that ethylene resistance in *clg1* supports a role for this hormone in the control of root gravitropism not necessarily connected with auxin physiology and possibly in the regulation of the connected signal transduction pathway (Ferrari et al., 2000).

Additional evidence for a possible interaction between ethylene and auxin transport in root gravitropism has been provided (Buer et al., 2003; Vandenbussche et al., 2003). The first work showed that the gravity-dependent growth response forming root looping was suppressed by ethylene, and this effect was suggested to be mediated by auxin transport (Buer et al., 2003). The second work demonstrated that the *alh1* [1-aminocyclopropane-1-carboxylic acid (ACC)-related long hypocotyl 1] *Arabidopsis* mutant, which was less sensitive to ethylene and auxin, displayed a faster root response to gravity. These results indicate that *alh1* is altered in the crosstalk between ethylene and auxins, probably at the level of auxin transport (Vandenbussche et al., 2003).

Root gravitropism was only slightly affected by the ethylene-insensitive *ein2-1* mutation (Roman et al., 1995). In contrast to these findings, it was reported that the *wei2* and *wei3* (weak ethylene insensitive) *Arabidopsis* mutants, which corresponds to previously unidentified ethylene pathway genes, showed hormone insensitivity only in roots and were neither affected in their response to auxin nor in their response to gravity (Alonso et al., 2003). This may indicate that the *wei*-related branch of the ethylene-response pathway does not play a role in the gravitropic response of roots. However, the fact that these two downstream mutants are specifically affected in their response to ethylene may indicate that they function at steps connecting the general ethylene-response pathway (the cascade from ETR1 to EIN3) to the process of auxin-mediated growth (Alonso et al., 2003).

Additional evidence suggests that ethylene may be involved in the gravitropic response of roots via its effect on starch metabolism. Thus, exogenous ethylene reduced starch levels in *Arabidopsis* root columella cells and the magnitude of curvature (Guisinger and Kiss, 1999), and microgravity-grown soybean plants, which do not curve, exhibited enhanced ethylene production and decreased starch production (Klymchuk et al., 1999; Kuznetsov et al., 2001). Since amyloplasts are believed to be necessary for gravity sensing (Blancaflor and Masson, 2003; Kiss et al., 2000), these results may imply that ethylene accumulation can modify the perception stage.

B. INVOLVEMENT OF ETHYLENE IN SHOOT GRAVITROPISM

Ethylene was found to redirect shoot gravitropic response. In addition, many plant organs show a transient burst of ethylene production when transferred from a vertical to a horizontal position (Abeles et al., 1992; Harrison and Pickard, 1984; Osborne, 1976; Pickard, 1985; Wheeler et al., 1986; Wright et al., 1978). Such an ethylene burst was also observed following clinostat or space shuttle experiments (Guisinger and Kiss, 1999; Hilaire et al., 1996; Kiss et al., 1999; Salisbury and Wheeler, 1980). This complicates the understanding of the role of ethylene in shoot gravitropism because a stress-induced ethylene burst might at first overload the system, thereby inhibiting the gravitropic response (Zobel, 1974). The ethylene burst was considered to be the result of either gravistimulation-induced stress, as indicated previously, or elevated levels of free IAA that occur in the lower side of grass nodes (Wright et al., 1978). Most of the studies concluded that this increased ethylene production following changes in stem orientation has no regulatory role in the gravitropic response of shoots.

In several shoot systems, apart from the ethylene burst occurring immediately after the change in orientation, the gravitropic response has been reported to be accompanied by significantly higher amounts of ethylene produced by the lower half of the horizontally positioned stem (Ballatti and Willemoes, 1989; Clifford and Oxlade, 1989; Clifford et al., 1983; Friedman et al., 1998, 2003a, 2005; Kaufman et al., 1985; Philosoph-Hadas et al., 1995, 1996, 2001; Wheeler et al., 1986; Woltering, 1991; Woltering et al., 2004). This gravity-induced ethylene gradient was accompanied in many cases by a gradient of the ethylene precursor, ACC, its conjugated form, malonyl-ACC (MACC), and/or asymmetrical expression of ethylene biosynthesis enzymes, ACC-synthase (ACS) and/or ACC-oxidase (ACO) across the gravistimulated organ (Andersson-Gunneras et al., 2003; Friedman et al., 2005; Philosoph-Hadas et al., 1995, 1996, 2001; Saito et al., 2005; Woltering, 1991; Woltering et al., 2004). This implies that the ethylene biosynthetic pathway in many plants responds very strongly to gravistimulation. The ethylene gradient was abolished by various treatments that also inhibited gravistimulation of snapdragon spikes (Friedman et al., 1998, 2003a; Philosoph-Hadas et al., 1996, 2001) or *Ornithogalum* and other flowering shoots (Friedman et al., 2005; Philosoph-Hadas et al., 1995). These data suggest that this differential ethylene production is an integral component of shoot gravitropic response.

The physiological role of the differentially produced ethylene in asymmetrical growth is still controversial (Madlung et al., 1999) and remains unclear. Results obtained with snapdragon spikes show that the gravity-induced auxin and ethylene asymmetries coincide neither in their timing of appearance nor in their patterns (Philosoph-Hadas et al., 1999, 2001), as it was reasonable to assume. Findings by Woltering *et al.* (2004) showed a strong and tissue-specific upregulation of only one of the ACS genes (*Am-ACS3*) in snapdragon cut spikes in response to gravistimulation or IAA treatment. Thus, the *Am-ACS3* was abundantly and exclusively expressed in the bending zone cortex at the lower stem flank within 2 h of gravistimulation, but was neither expressed in vertical stems nor in other parts of gravistimulated stems, leaves, or flower. A similar gene- and cell-type-specific, auxin-inducibility of *ACS* genes in roots was demonstrated (Tsuchisaka and Theologis, 2004). The pattern of *Am-ACS3* expression and responsiveness to IAA strongly suggest that *ACS3* is responsible for the observed differential ethylene production in gravistimulated snapdragon stems, which may reflect changes in auxin signaling (Woltering *et al.*, 2004).

In some studies the ethylene gradient across the stem preceded the visual gravitropic response or was apparent in parallel to its initiation (Clifford and Oxlade, 1989; Philosoph-Hadas *et al.*, 1996; Wheeler and Salisbury, 1980, 1981; Woltering *et al.*, 2004). In other cases the increased ethylene production in the lower stem flank occurred hours after the gravitropic response has been initiated (Clifford *et al.*, 1983; Kaufman *et al.*, 1985, 1995). This led to the conclusion that ethylene may modulate but not initiate the gravitropic response in shoots. Additionally, in only a few cases (Ballatti and Willemoes, 1989; Philosoph-Hadas *et al.*, 1996, 1999; Wheeler and Salisbury, 1980, 1981; Wheeler *et al.*, 1986) could the inhibitors of ethylene synthesis or action block the gravitropic response. Several of these experiments that showed positive results (Wheeler and Salisbury, 1980, 1981) were criticized (Pickard, 1985). However, it should be noted that ethylene can both inhibit and promote stem growth (Abeles *et al.*, 1992), and different plant species may respond to the same level of ethylene in different ways. This may explain the large diversity obtained in the literature regarding the different results of various graviresponding plant systems to ethylene. Taken together, the role of ethylene gradient in shoot gravitropism is still not clear.

Another way to examine the role of ethylene in shoot gravitropism was to characterize the gravitropic response of several ethylene-insensitive mutants. If ethylene plays a primary role in shoot gravitropism, then an ethylene-insensitive mutant should be agravitropic. Seedlings of an *Arabidopsis* ethylene-insensitive mutant, *ain*1–1 (*ACC-insensitive*), which is genetically distinct from previously identified ethylene resistance loci, showed a slower gravitropic response as compared to WT seedlings (Van Der Straeten *et al.*, 1993). Similarly, the ethylene-insensitive *ein*2–1 mutant was shown to be clearly defective in the negative gravitropism of the hypocotyls (Golan *et al.*, 1996). Our results show that also the gravitropic response of inflorescence stems of the *Arabidopsis ein*2–1 mutant was significantly reduced compared to the WT plants, and the gravitropic response of inflorescence stems of other ethylene-insensitive mutants (*etr*1–1, *etr*1–2, *etr*1–3) was only slightly delayed compared to WT plants (Friedman *et al.*, unpublished results). The *nr* (never ripe) tomato mutant, which is ethylene-insensitive in both seedlings and mature stages, displayed only a slightly retarded gravitropic response (Madlung *et al.*, 1999). Since *ain*1, *ein*2, and other mutants were slightly reduced in their gravitropic response, the results further suggest that the response to ethylene modifies the gravitropic bending, and ethylene signaling is probably required for shoot bending. However, since all these mutants did not display a complete agravitropic response as expected but displayed a delayed gravitropic curvature, the data indicate that ethylene is not an absolute requirement for the gravitropic response but acts as a modulator.

Although the accumulated data suggest that ethylene does not play a primary role in the gravitropic response of tomato hypocotyls (Harrison and Pickard, 1986; Madlung *et al.*, 1999), these data cannot rule out the possibility that low ethylene levels are necessary for full gravitropic response, as first suggested by Zobel (1973) for the *dgt* tomato mutant. Thus, very low (5 nl l^{-1}) concentrations of ethylene were demonstrated to restore a normal gravitropic response in *dgt* hypocotyls (Zobel, 1973), suggesting that ethylene may act downstream of auxin in the gravitropic signal transduction. Also, moderate levels of ethylene specifically inhibited gravicurvature of tomato hypocotyls (Madlung *et al.*, 1999), and high ethylene levels (100 μl l^{-1}) were able to redirect 7-day-old etiolated pea plants to grow downward rather than upward following gravistimulation (Burg and Kang, 1993). Also, the gravity competence of rye leaves was restored by exogenous or endogenous ethylene (provided by ACC) (Edelmann, 2002; Edelmann *et al.*, 2002). In addition, ethylene-responsive element binding factors (EREBF) were found to increase at early stages of seedlings gravistimulation (Moseyko *et al.*, 2002). This further supports the notion that gravitropism involves ethylene action, although the contribution of this response to the bending response is not yet clear. These data point out that in spite of the reservations on reports favoring a role for ethylene in shoot gravitropism (Harrison and Pickard, 1986; Pickard, 1985), ethylene seems to be necessary for this process.

C. POSSIBLE MODES OF ETHYLENE ACTION IN GRAVITROPISM OF ROOTS AND SHOOTS

Several hypotheses were suggested for the possible modes of action of ethylene in the gravitropic response: (1) Ethylene may, at least partly, mediate growth through its effect on auxin transport via the PIN function, since the *pin* mutants, defective in polar auxin transport, showed a decreased sensitivity to ethylene (Chen *et al.*, 1998). (2) Ethylene may increase the tissue sensitivity (or responsiveness) to auxin. Analysis of the *Arabidopsis nph*4/*msg*1/*tir*5 locus (referred to as *nph*4) has suggested that the *NHP4*, encoding the auxin-response factor ARF7 (Harper *et al.*, 2000), functions as a modulator of auxin-dependent differential growth in stem gravitropism (Watahiki and Yamamoto, 1997). However, despite the clear requirement of NPH4/ARF7 in the regulation of differential growth, ethylene was shown to enhance the sensitivity or activity of this factor (Harper *et al.*, 2000). Thus, in the presence of ethylene, one or more ARFs may be functioning to enhance auxin sensitivity. These findings further illustrate the close connection between auxin and ethylene in the control of differential growth during gravistimulation. Similar to these findings in *Arabidopsis*, we have also found indications in snapdragon that ethylene can modulate shoot responsiveness to endogenous or exogenous IAA (Philosoph-Hadas *et al.*, 2003). The results show that in the absence of ethylene, the *Am-AUX/IAA*1 gene was differentially expressed in favor of the upper flank, rather than in favor of the lower flank of control stems. This suggests that ethylene enhanced the responsiveness of the lower stem flank to IAA, thereby probably contributing to bending maintenance for an extended period (Philosoph-Hadas *et al.*, 2003). (3) Ethylene may play a role in maintaining the reestablished vertical position by affecting, for example, stiffness of the bend stem section (Woltering *et al.*, 2004). Thus, ethylene may stimulate H_2O_2 production, increasing peroxidase activity during gravitropic bending thereby facilitating the processes involved in cell wall and stem stiffening and maintenance of the reestablished vertical orientation. This hypothesis is based on the findings showing that asymmetric application of H_2O_2 induced bending in maize primary roots (Joo *et al.*, 2001). Additional support for this idea came from data showing that gravistimulation induced an oxidative burst, with genes involved in oxidative bursts forming the largest functional category of gravity-regulated genes (Moseyko *et al.*, 2002).

In conclusion, the accumulated evidence indicates that ethylene has a role in shoot gravitropism, which is mainly associated with modulation of auxin transport or sensitivity. However, this role is still not fully elucidated. The major advances in the understanding of auxin transport and signaling may shed more light on the possibly tight and complicated interactions between auxin and ethylene. These new hypotheses may help in the future to better understand the special role of ethylene in shoot gravitropism.

V. ROLE OF GIBBERELLINS

A. ROOTS

Knowing the crucial role of GAs in cell elongation, it was reasonable to suggest that this hormone may be involved in gravistimulation. Thus, stimulation of root gravitropism by GA has been reported previously (Konings, 1995). Asymmetrical distributions of GA in favor of the upper halves of *Vicia faba* root tips or maize roots following gravistimulation were also demonstrated (el-Antably and Larsen, 1974; Webster and Wilkins, 1974; Wilkins, 1979). However, a more careful analysis using immunoassays of the kinetics of redistribution of endogenous GA_3 and GA_7 during gravistimulation did not show any significant lateral asymmetry in root tips of *Z. mays* L. or *V. faba* L. (Mertens and Weiler, 1983). Additionally, GA-deficient mutants, which showed reduced elongation of root cells (Barlow *et al.*, 1991), were found to be as graviresponsive as normal roots (Juniper, 1976). Although few measurements of GA asymmetries have been reported (Jackson and Barlow, 1981), there is no evidence for a crucial role of GAs in root gravitropism, and no such role has been further established (Konings, 1995). However, it is still possible that GA may have an indirect effect in root gravitropism, as treatment of cress (*Lepidium sativum* L.) roots with GA and kinetin resulted in loss of their gravisensitivity, which was attributed to the loss of amyloplastic starch and to the concomitant loss in the polar organization of the statocyte (Busch and Sievers, 1990).

B. SHOOTS

Evidence against the involvement of GAs in shoot differential growth, which causes gravitropic curvature, was presented, showing that increasing GA concentrations did not abolish the growth differentials (Firn and Digby, 1980). No significant lateral asymmetrical redistribution of endogenous GA_3 and GA_7 could be detected by immunoassay during gravistimulation of sunflower hypocotyls (Mertens and Weiler, 1983), and exogenously supplied GA induced no gravitropic growth in nodal regions of graviresponding *Tradescantia fluminensis* (Wandering Jew) (Funke and Edelmann, 2000). All these data indicate that the involvement of GA in the bending response of shoots is questionable. However, lateral GAs gradients as the cause of gravitropism were reported to occur in excised sunflower shoot tips that were insensitive to applied auxin, with higher concentrations of $[3,4-^3H]GA_1$ present at the bottom of gravistimulated stems (Wilkins, 1979).

GA also appears to play some role in modulating the response of cereal shoot pulvini to gravistimulus. Changes in the levels of endogenous-free GAs and GA conjugates using radioactive distribution were shown to be correlated with gravistimulated bending in intact oat leaf sheath pulvini (Pharis *et al.*, 1981). Thus, GA conjugates predominated in the upper halves, while the free GA_3-, GA_4-, and GA_7-like GAs were in greater abundance in the lower half, but these gradients were observed only after bending had occurred (at 24 h) (Pharis *et al.*, 1981). Such GA gradients could not be obtained in earlier stages (at 4 h) following gravistimulation of this organ (Kaufman *et al.*, 1985). Since GAs are not transported downward in gravistimulated pulvini, these results imply that this GA asymmetry should be aroused from local pools, either from differential synthesis of GA/GA-conjugates or from differential movement of free GAs and GA conjugates to the respective pulvinus halves (Kaufman et al., 1985). Additional evidence regarding the apparent role of GAs in the gravitropic response of cereal pulvini showed that exogenous application of GA_3 to the lower sides of horizontal segments significantly enhanced pulvinus growth and segment curvature, although exogenous GA_3 had no effect on these parameters in vertical segments (Brock and Kaufman, 1988a). A similar response to GA_3 was obtained also with dark-pretreated segments but only in the presence of sucrose (Brock and Kaufman, 1988b). These results indicate that gravistimulation induces changes in pulvinus responsiveness to GA_3, in addition to IAA, and that GA_3 may increase amyloplast movement, which is required for this response (Brock and Kaufman, 1988a,b). Also, the differential localization of the free GA_1 metabolites between upper and lower halves of WT maize pulvini was not apparent in its pleiogravitropic mutant, *lazy* (Rood et al., 1987). These results seem to suggest a role for GAs in the differential shoot growth following gravitropism. However, this further indicates that the linear gradient of hormone concentration may predominantly be the result of local changes in GA level rather than a product of hormonal movement into or across the pulvinus (Rood et al., 1987; reviewed by Kaufman et al., 1985).

Another interesting graviresponding system in which GA seems to play a certain role are woody plants with weeping branches. Growth direction and secondary xylem formation in woody stems depend on a gravity stimulus on Earth. It was shown that exogenous GA induces tension wood in weeping Japanese cherries (*Prunus jamasakura*) branches, thereby causing their upright growth (Nakamura et al., 2001). Under simulated microgravity conditions, GA promoted the secondary xylem development in Japanese flowering cherry tree seedlings, and the upper side of the inclined stem had much higher levels of GA_1 than the lower side (Sugano et al., 2003). These results suggest that stem morphogenesis in woody plants with weeping branches is mediated by differential GA-induced secondary xylem formation.

In conclusion, it seems that there is little evidence for a crucial role of GAs in root and shoot gravitropism apart from their role in directing growth in very specific graviresponding shoot-like systems, such as cereal shoot pulvini and weeping branches of woody trees.

VI. ROLE OF ABSCISIC ACID

The inhibitory effects of ABA on growth led to the suggestion that it may have a possible role in root gravitropism (Wilkins, 1979), but no ABA role was indicated for shoot gravitropism. However, the role of ABA as a major growth inhibitor was seriously questioned (Konings, 1995). Thus, while ABA was found to play a role in the graviresponse of roots that require light for their graviresponse (Feldman et al., 1985; Leopold and LaFavre,

1989), most of the accumulated data since 1985 suggest that ABA is unlikely to have a role in roots curved in darkness as an important growth inhibitor transported from the root cap to the extension zone (Konings, 1995). This conclusion is mainly based on the following observations: (1) Exogenously applied ABA promotes rather than inhibits root growth, and its inhibitory effect is gained only at concentrations significantly higher than those thought to naturally occur (Jackson and Barlow, 1981); (2) Roots of ABA-deficient plants obtained either by chemical blocking their ABA synthesis or by specific mutations (Moore, 1990) still respond to gravity. Similarly, roots of some mutants defective in response to ABA, curved like the WT; (3) Replacement of a removed half-cap by ABA had no effect on root curvature (Lee *et al.*, 1990b); (4) No transverse gradient of endogenous ABA was detected in the root cap (Young and Evans, 1996); (5) No polar basipetal transport from the cap to the root extension zone has been observed in roots that do not require light for curvature (Jackson and Barlow, 1981; Young and Evans, 1996). Another report that supports this conclusion showed that no significant lateral asymmetry of endogenous ABA could be detected by immunoassay, neither in gravitropically reacting root tips of *Z. mays* and *V. faba* nor in sunflower hypocotyls (Mertens and Weiler, 1983). Regarding this accumulated evidence, it seems therefore that ABA, once thought to be also involved in root gravitropism regulation, has clearly no role.

Although the role of ABA in gravitropism was controversial (Konings, 1995), several observations may shed more light on the possible role of ABA in gravitropic reactions. It is possible that the putative role that ABA might play in gravitropism is masked by its positive role in hydrotropism (Takahashi *et al.*, 2002). For many years the hydrotropic response of roots could not be separated from gravitropism, but this was overcome by using agravitropic mutants (Eapen *et al.*, 2005). It was found that both tropisms share the root cap cells acting as a sensor, as well as the participation of auxin and calcium but in different mechanisms (Eapen *et al.*, 2005). Based on studies with various auxin and ABA mutants, it was hypothesized that ABA may serve as a regulator of auxin transport in root hydrotropic response (Eapen *et al.*, 2005), and a similar interaction may exist also in root gravitropism (Takahashi *et al.*, 2002). This is based on new evidence showing that some ABA mutants are agravitropic (Lu and Fedoroff, 2000), that roots of ABA-deficient (*aba1–1*) and ABA-resistant (*abi2–1*) mutants showed a reduced gravitropic response (Takahashi *et al.*, 2002), and that auxin-signaling components were affected by ABA (Eapen *et al.*, 2003). Therefore, additional study of mutants affected in ABA and auxin signaling should be performed to determine if ABA plays a role in gravitropism.

A possible role for ABA in gravitropism may also be implied via its antagonistic interactions with ethylene on growth (Finkelstein and Rock, 2002). Studies show the inhibitory effects of high ABA on growth, while low endogenous ABA levels in unstressed plants were suggested to promote growth. Studies in maize and tomato indicate that the stunted growth of ABA-deficient plants is due to a failure to inhibit ethylene production, reflecting another antagonistic interaction between ABA and ethylene (Finkelstein and Rock, 2002).

Another possible line of research regarding a role for ABA in gravitropism may be via its interaction with the cytoskeleton that regulates gravitropism (Blancaflor, 2002). The signaling effect of ABA in guard cells that involves activation of K^+ efflux into the cell to regulate turgor is mediated by phospholipase D (PLD) activity (Jacob et al., 1999). The gravistimulation and auxin-induced asymmetrical expression of K^+-channels (Philippar et al., 1999) and activation of intermediates of the phospholipids signaling system (Perera et al., 1999), as well as the close association between PLD and the cytoskeleton (Gardiner et al., 2001), imply that ABA might have a role in plant gravitropism by mediating these processes.

In conclusion, while no direct evidence supports a role of ABA in root or shoot gravitropism, findings regarding its role in hydrotropism, as well as its interactions with ethylene and auxin, may imply a role for ABA in mediating gravitropism-related processes.

VII. ROLE OF CYTOKININS

Cytokinins are hormones that regulate cell division and development and play essential and crucial roles in various aspects of plant growth (Kieber, 2002). Unlike other phytohormones, the regulatory roles of cytokinins are not well understood because of the lack of biosynthetic and signaling mutants. This is probably the reason why cytokinins were not frequently implicated as hormones involved in the gravitropic response. However, several exciting reports show that this situation is now changing, with the identification of key elements of cytokinin action and metabolism, including the cytokinin oxidase/dehydrogenase (*CKX*) gene family (Werner et al., 2003), and the histidine kinase 4 cytokinin receptor (Nishimura et al., 2004). Thus, it was firmly established using the cytokinin-deficient plants that there is a positive regulatory role for cytokinins in the shoot and a negative regulatory role in the root (Werner et al., 2003), which may shed light on the possible role of cytokinin in gravitropic responses.

A. SHOOTS

Several studies have presented indirect evidence that suggests the involvement of cytokinin in the gravitropic response of shoots. These include studies showing the interaction of cytokinin with the *dgt* tomato mutant (Coenen et al., 2003; Lomax, 1997), with the *Arabidopsis msg*1 (*massugu*1; Japanese for "straight") mutant (Watahiki and Yamamoto, 1997), and with the red-light effect on the gravitropic response of *Arabidopsis* seedlings (Golan *et al.*, 1996).

The *dgt* tomato plants exhibit many pleiotropic developmental alterations in addition to the greatly reduced gravitropic response. Many of the phenotypic traits of *dgt* are more often associated with cytokinin rather than with auxin, and exogenous application of cytokinin to WT plants phenocopies the wide array of phenotypic abnormalities associated with the *dgt* mutant (Coenen and Lomax, 1998; Coenen *et al.*, 2003; Lomax, 1997). Thus, exogenous application of cytokinin to WT plants inhibited the gravitropic response in a manner similar to *dgt* and also reduced the rate of lateral auxin transport. This correlation suggests a role for cytokinin in modulating auxin movement and thus gravitropism. Such an interaction between cytokinin, auxin, and the gravitropic response mechanism was further supported by the findings showing that several gravitropic and auxin-insensitive mutants (*aux1*, *axr1*, and *axr3*) also exhibit cross-resistance to cytokinin. The possibility that cytokinin acts via ethylene cannot be excluded, and therefore these mutants display insensitivity to all three hormones.

The results obtained on the interaction of gravitropic mutants with light have provided a new insight regarding the role of cytokinin in gravitropism of shoots, as cytokinin can replace light in several aspects of the photomorphogenesis of dicot seedlings. Thus, cytokinin could restore negative gravitropism to the hypocotyls of *Arabidopsis* seedlings grown under continuous red light, and this cytokinin effect was shown to be mediated via increased ethylene production (Golan *et al.*, 1996). Similarly, it was demonstrated earlier that cytokinins inhibit hypocotyl elongation in dark-grown *Arabidopsis* seedlings (Su and Howell, 1995), an effect that was due to the cytokinin-induced ethylene production (Cary *et al.*, 1995). It seems, therefore, that cytokinin could modulate gravitropism by the same pathway as hypocotyl growth inhibition (through ethylene) or by another mechanism (related to the light effect). Further support for the possible cross-talk between cytokinin and ethylene is implicated by their mutual effects on their synthesis and signaling pathways (Nishimura *et al.*, 2004; Schaller, 1997), which may affect also gravitropism.

It should be noted that the attractive suggestion that cytokinin acts to reduce the pool of active auxin (Coenen and Lomax, 1998; Lomax, 1997) was not supported by the findings obtained with cytokinin-deficient plants (Werner *et al.*, 2003). If the model of Lomax (1997) is true, then a reduction in auxin content observed in the shoots of cytokinin-deficient plants (Werner *et al.*, 2003) is unexpected. A possible explanation may lie in the fact that the auxin reduction in these cytokinin-deficient plants may result from the reduction in the size of the shoot apical meristem and young leaves, which are the major sites of auxin biosynthesis (Werner *et al.*, 2003). This indicates that more complicated mechanisms than mutual regulation of metabolism are necessary to explain the interactions of the two hormones.

A study of *Arabidopsis* plants (Smets *et al.*, 2005) shows that cytokinins promote hypocotyl elongation in the light only when ethylene action or IAA transport is blocked. A possible cross-talk between the signaling pathways of the three hormones is proposed, which suggests that in the light, cytokinins interact with the ethylene-signaling pathway and conditionally upregulate ethylene and auxin synthesis. Thus, it is possible that similar to the cytokinin effect on gravitropism under red-light conditions which is ethylene-mediated (Golan *et al.*, 1996), cytokinin may also interact with ethylene under white light to affect hypocotyl growth and possibly their gravitropic response.

B. ROOTS

Very few studies were reported on the involvement of cytokinins in root gravitropism. Indirect evidence for the effect of kinetin applied with GA for 30 h on the graviresponsiveness of cress (*L. sativum* L.) roots was indicated, as the treatment caused complete destarching of amyloplasts, which did not cause sedimentation and destruction of the polar arrangement of cell organelles in statocytes (Busch and Sievers, 1990). The loss of structural polarity was accompanied by loss of graviresponsiveness, although root growth still occurred. These results may indicate that both GA and cytokinin are necessary at low concentrations for maintaining the statocytes competent for gravisensitivity of roots (Busch and Sievers, 1990).

A breakthrough finding suggested a key role for cytokinin in root gravitropism (Aloni *et al.*, 2004). Knowing that cytokinin has a negative regulatory role in root growth and taking advantage of cytokinin-deficient transgenic plants Werner *et al.* (2003) clearly demonstrated that cytokinin functions as an inhibitor of tropic root elongation during the gravity response. Thus, a gravity-induced asymmetrical cytokinin distribution was demonstrated, which obviously caused initiation of a downward curvature near the root apex during the early rapid phase of the gravity response. This was obtained by inhibiting elongation at the lower side and promoting growth at the upper side of the distal elongation zone closely behind the root cap. As such, cytokinin could serve as the putative root cap inhibitor of auxin-induced growth on the lower surface of horizontal roots, which for many years was thought to be ABA, although seriously criticized (Konings, 1995) (see Section VI). The work of Aloni *et al.* (2004) provided the final proof for the nature of this predicted inhibitor and suggested that both cytokinin (from the root cap) and auxin (from the young leaves) are the key hormones regulating root gravitropism.

In conclusion, the regulatory role of cytokinin on shoot gravitropism is not yet established, but it seems to be ethylene-mediated and/or operating via reduction of auxin transport. On the other hand, cytokinin seems to play a key regulatory role in root gravitropism together with auxin, as elucidated

(Aloni et al., 2004). However, in both systems, a close and complex interaction between cytokinin and other hormones seems to take place.

VIII. ROLE OF BRASSINOSTEROIDS, JASMONATES, AND SALICYLIC ACID

A. BRASSINOSTEROIDS

Brassinosteroids (BRs) are now regarded to be essential substances for growth and development in plants, and their occurrence has been demonstrated in all plant organs (Clouse and Sasse, 1998; Kim et al., 2000). The involvement of BRs in the gravitropic response was suggested in association with auxin, as both phytohormones are interactive in some aspects of plant growth and development. Exogenous application of BRs was reported to enhance the gravitropic curvature of bean (Meudt, 1987) or tomato (Park, 1998) hypocotyls, and of lamina joint of rice (Yamamuro et al., 2000), indicating that BRs might participate in regulation of shoot gravitropism. The occurrence of BR in maize primary roots was also demonstrated for the first time by showing enhancement of the gravitropic response of maize roots in an IAA-dependent manner, following exogenous application of a brassinolide or an endogenous BR isolated from maize (Kim et al., 2000). These results suggest that BRs might be involved in auxin-mediated processes of root gravitropism. It was shown that this brassinolide-stimulated gravitropic response of primary maize roots was mediated, in part, via increased ethylene production, but this stimulation is partially different from the ethylene stimulation of the gravitropic response (Chang et al., 2004). This suggests that BR effects on the gravitropic response mechanism are related to the effects of auxin and ethylene participating in the process. An interesting regulatory interaction between BRs and auxin that affects ethylene via increased expression of the *ACS* multigene family was demonstrated in mung beans hypocotyls (Yi et al., 1999). Thus, more efforts should be invested in order to define whether BRs act on gravitropism independently or via interaction with auxin and/or ethylene.

In an attempt to understand the molecular mechanism underlying the possible interactions between BRs and auxins, an array of hormone-regulated genes was analyzed in *Arabidopsis* (Goda et al., 2004). Among the 637 genes found to be induced by either one of these hormones, only 48 genes were regulated by the two hormones in common, suggesting that most of the actions of each hormone are mediated by gene expression that is unique to each (Goda et al., 2004). However, several works suggest a possible interaction of BRs with auxin signaling (Nakamura et al., 2003) and auxin transport (Nakamura et al., 2004) elements, such as IAA5, IAA19, SAUR-AC1, and PIN, which were reported to participate also in the gravitropic response of shoots and roots (Masson *et al.*, 2002). Such analyses may therefore shed light on the role of BRs in gravitropism and their interaction with other hormones.

B. JASMONATES

To the best of our knowledge, there were no specific reports regarding a possible involvement of jasmonates in gravitropism. However, evidence suggests that jasmonate and auxin use a similar signaling mechanism. Developments have helped situate jasmonate regulators in the context of general hormonal/developmental signaling. Most remarkably, a jasmonate-insensitive mutant has been found to be attributable to a new allele of the auxin-signaling gene *AUR1*. The allele, *axr1–24*, causes reduced sensitivity not only to methyl jasmonate and IAA but also to an ethylene precursor, a cytokinin analogue, a brassinolide, and ABA. This mutant is defective in the modification of E3 ubiquitin ligase, and therefore, it is possible that this pathway may be involved in many major stress and developmental pathways in plants (Tiryaki and Staswick, 2002).

C. SALICYLIC ACID

The inhibitory effect of SA on the gravitropic response was demonstrated in maize coleoptiles and roots, and circumstantial evidence suggested that this bending inhibition may be caused by inhibiting ethylene production (Medvedev and Markova, 1991). We have reported on a specific inhibitory effect of SA on the upward gravitropic bending of cut snapdragon, lupinus, and anemone flowering shoots (Friedman *et al.*, 2003a). The results show that SA inhibited bending of various cut flowering shoots in a concentration-dependent manner. SA did not inhibit amyloplast sedimentation but prevented the differential ethylene production and differential growth (Friedman *et al.*, 2003a). The inhibitory effect of SA on the gravitropic bending of snapdragon stems was explained on the basis of its possible interference with the gravity-induced auxin changes. Thus, SA may act by affecting ROS formation induced by auxin, since SA is well known to augment ROS formation both in the extracellular and intracellular environments (Chen *et al.*, 1993; Kawano *et al.*, 1998). It has been suggested that auxin-induced ROS formation may function as downstream components in the auxin-mediated signal transduction pathway of root gravitropism (Joo *et al.*, 2001). Accordingly, it is possible that SA may inhibit bending by abolishing a similar ROS gradient that might be formed across the gravi-stimulated stem, as reported for roots (Joo *et al.*, 2001).

In conclusion, BRs, jasmonates, and SA seem to act on gravitropism via interaction with auxin and/or ethylene, with SA possibly affecting oxidative responses involved in gravitropism.

IX. CONCLUDING REMARKS AND FUTURE PROSPECTS

The directional movements (tropisms) in response to changes in the gravity vector shape the plants by employing complex physiological processes and, thereby, many components of plant cells. Our understanding of root gravitropism and its hormonal regulation has been more advanced as compared with shoot gravitropism. In particular, the findings on subcellular localization and distribution of auxin influx and efflux carriers (Friml *et al.*, 2002b; Ottenschlager *et al.*, 2003), provide a strong support, at least in roots, for the Cholodny–Went hypothesis for the auxin-based tropic responses, which has been disputed for more than 70 years (Yamamoto, 2003). However, it is still not known whether this hypothesis is also applicable to shoot gravitropism (Firn *et al.*, 2000), particularly when the term "shoots" includes various above-ground organs. It is becoming clear now that shoots have a different mode of gravitropic response than roots, as their gravity perception takes place not in the tips but in the cells of the endodermis, which substantiates the lack of longitudinal signal transmission (Masson *et al.*, 2002). Therefore, a modified model should be proposed to explain shoot gravitropism and its hormonal regulation. The determination of auxin efflux and influx carriers involved in shoot gravitropism, especially those operating in inflorescence stems, seems to be an important research task in this direction.

It is becoming apparent, after so many years of research that auxin, although playing a major role in the process, is not the only hormone involved in regulation of gravitropism, and other phytohormones, particularly ethylene and cytokinin, are important participants in concert with auxin. Ample evidence has been documented so far showing that growth substances can modulate the gravitropic orientation of plants by creating localized asymmetry, which is not fully understood. Thus, elucidating the roles of the gravistimulation-induced ethylene gradient in shoots and of cytokinin gradient in roots and their combined regulation with auxin should improve our understanding of the overall control of gravitropism in these organs.

Evidence is accumulating that certain mutations can simultaneously influence the response to more than one hormone. One striking example is the *ein2* mutant that consistently shows up in screens of different hormones, such as ethylene, cytokinin, ABA, and auxin (Klee, 2003). This certainly points to the existence of regulatory interactions between two or more signal transduction pathways of these hormones in plants. Thus, an important conclusion that emerges from this chapter is that interactions among hormones and cross-talk in their signaling pathways, which are still mostly obscure, may play a role in regulating plant gravitropism. The challenge now is to determine the mechanisms of cross-talk, or how all of these pathways interact and how they are coordinately regulated.

The main research goal of gravitropism still remains to determine at the molecular level how amyloplast sedimentation modifies auxin transport leading to its asymmetrical distribution and to unravel the signal cascade in the statocytes (Morita and Tasaka, 2004). However, while hormones are mostly considered to regulate the gravitropic signal transduction and cell elongation responses, the possibility that they may be needed in some other capacities in gravitropism is pointed out by several studies and seems a research line worth pursuing. These studies demonstrated that GA plus cytokinin (Busch and Sievers, 1990) or ethylene (Guisinger and Kiss, 1999) reduced both levels of starch in amyloplasts and the gravitropic response, while IAA restored starch levels together with gravitropism to a *lazy*-1 mutant (Firn *et al.*, 2000). Future studies should clarify the possible interactions between the various hormones with amyloplasts and the cytoskeleton upon gravistimulation, promising new insights into mechanisms that may control early perception in gravitropism. Also, little is known about how the auxin signal is interpreted at a molecular level to give rise to differential growth patterns. Therefore, further analyses of several important factors in this pathway, such as the transcriptional activator NPH4/ARF7 (Harper *et al.*, 2000), or determination of the role of ROS in auxin action (Joo *et al.*, 2001) may provide important clues to clarify this basic issue. Contribution from the Agricultural Research Organization, The Volcani Center, Bet-Dagan Israel, No. 435/05.

REFERENCES

Abeles, F. B., Morgan, P. W., and Saltveit, M. E., Jr. (1992). "Ethylene in Plant Biology" Academic Press, San Diego.

Aloni, R., Langhans, M., Aloni, E., and Ullrich, C. I. (2004). Role of cytokinin in the regulation of root gravitropism. *Planta* **220**, 177–182.

Alonso, J. M., Stepanova, A. N., Solano, R., Wisman, E., Ferrari, S., Ausubel, F. M., and Ecker, J. R. (2003). Five components of the ethylene-response pathway identified in a screen for weak ethylene-insensitive mutants in *Arabidopsis. Proc. Natl. Acad. Sci. USA* **100**, 2992–2997.

Andersson-Gunneras, S., Hellgren, J. M., Bjorklund, S., Regan, S., Moritz, T., and Sundberg, B. (2003). Asymmetric expression of a poplar ACC oxidase controls ethylene production during gravitational induction of tension wood. *Plant J.* **34**, 339–349.

Ballatti, P. A., and Willemoes, J. G. (1989). Role of ethylene in the geotropic response of bermudagrass (*Cynodon dactylon* L. pers.) stolons. *Plant Physiol.* **91**, 1251–1254.

Bandurski, R. S., Schulze, A., and Momonoki, Y. (1984). Gravity-induced asymmetric distribution of a plant growth hormone. *Physiologist* **27**, S123–S126.

Barlow, P. W., Brain, P., and Parker, J. S. (1991). Cellular growth in roots of a gibberellin-deficient mutant of tomato (*Lycopersicon esculentum* Mill.) and its wild type. *J. Exp. Bot.* **42**, 339–351.

Bennett, M. J., Marchant, A., Green, H. G., May, S. T., Ward, S. P., Millner, P. A., Walker, A. R., Schulz, B., and Feldmann, K. A. (1996). Arabidopsis *AUX*1 gene: A permease-like regulator of root gravitropism. *Science* **273**, 948–950.

Berleth, T., Krogan, N. T., and Scarpella, E. (2004). Auxin signals—turning genes on and turning cells around. *Curr. Opin. Plant Biol.* **7**, 553–563.

Bjorkman, T., and Leopold, A. C. (1987). Effect of inhibitors of auxin transport and of calmodulin on a gravisensing-dependent current in maize roots. *Plant Physiol.* **84**, 847–850.

Blancaflor, E. B. (2002). The cytoskeleton and gravitropism in higher plants. *J. Plant Growth Regul.* **21**, 120–136.

Blancaflor, E. B., and Masson, P. H. (2003). Plant gravitropism. Unraveling the ups and downs of a complex process. *Plant Physiol.* **133**, 1677–1690.

Boonsirichai, K., Guan, C., Chen, R., and Masson, P. H. (2002). Root gravitropism: An experimental tool to investigate basic cellular and molecular processes underlying mechanosensing and signal transmission in plants. *Annu. Rev. Plant Biol.* **53**, 421–447.

Brock, T. G., Kapen, E. H., Ghosheh, N. S., and Kaufman, P. B. (1991). Dynamics of auxin movement in the gravistimulated leaf-sheath pulvinus of oat (*Avena sativa*). *J. Plant Physiol.* **138**, 57–62.

Brock, T. G., and Kaufman, P. B. (1988a). Altered growth response to exogenous auxin and gibberellic acid by gravistimulation in pulvini of *Avena sativa*. *Plant Physiol.* **87**, 130–133.

Brock, T. G., and Kaufman, P. B. (1988b). Effect of dark pretreatment on the kinetics of response of barley pulvini to gravistimulation and hormones. *Plant Physiol.* **88**, 10–12.

Buer, C. S., and Muday, G. K. (2004). The transparent *testa4* mutation prevents flavonoid synthesis and alters auxin transport and the response of *Arabidopsis* roots to gravity and light. *Plant Cell* **16**, 1191–1205.

Buer, C. S., Wasteneys, G. O., and Masle, J. (2003). Ethylene modulates root-wave responses in *Arabidopsis*. *Plant Physiol.* **132**, 1085–1096.

Burg, S. P., and Burg, E. A. (1967). Auxin-stimulated ethylene formation: Its relationship to auxin-inhibited growth, root geotropism and other plant processes. *In* "Biochemistry and Physiology of Plant Growth Substances" (E. F. Wightman and G. Setterfield, Eds.), pp. 1275–1294. Runge Press, Ottawa, Canada.

Burg, S. P., and Kang, B. G. (1993). Gravity-dependent ethylene action. *In* "Cellular and Molecular Aspects of Plant Hormone Ethylene" (A. Latche, A. Bleecker, and J. C. Pech, Eds.), pp. 335–340. Kluwer Academic Publishers, Dordrecht, The Netherlands.

Busch, M. B., and Sievers, A. (1990). Hormone treatment of roots causes not only a reversible loss of starch but also of structural polarity in statocytes. *Planta* **181**, 358–364.

Butler, J. H., Hu, S., Brady, S. R., Dixon, M. W., and Muday, G. K. (1998). *In vitro* and *in vivo* evidence for actin association of the naphthylphthalamic acid-binding protein from zucchini hypocotyls. *Plant J.* **13**, 291–301.

Cary, A., Liu, W., and Howell, S. (1995). Cytokinin action is coupled to ethylene in its effects on the inhibition of root and hypocotyl elongation in *Arabidopsis thaliana* seedlings. *Plant Physiol.* **107**, 1075–1108.

Chadwick, A. V., and Burg, S. P. (1970). Regulation of root growth by auxin–ethylene interaction. *Plant Physiol.* **45**, 192–200.

Chang, S. C., Cho, M. H., Kim, S. K., Lee, J. S., Kirakosyan, A., and Kaufman, P. B. (2003). Changes in phosphorylation of 50 and 53 kDa soluble proteins in graviresponding oat (*Avena sativa*) shoots. *J. Exp. Bot.* **54**, 1013–1022.

Chang, S. C., Kim, Y.-S., Lee, J. Y., Kaufman, P. B., Kirakosyan, A., Yun, H. S., Kim, T.-W., Kim, S. Y., Cho, M. H., Lee, J. S., and Kim, S. K. (2004). Brassinolide interacts with auxin and ethylene in the root gravitropic response of maize (*Zea mays*). *Physiol. Planta.* **121**, 666–673.

Chen, R., Guan, C., Boonsirichai, K., and Masson, P. H. (2002). Complex physiological and molecular processes underlying root gravitropism. *Plant Mol. Biol.* **49**, 305–317.

Chen, R., Hilson, P., Sedbrook, J., Rosen, E., Caspar, T., and Masson, P. H. (1998). The *Arabidopsis thaliana AGRAVITROPIC*1 gene encodes a component of the polar-auxin-transport efflux carrier. *Proc. Natl. Acad. Sci. USA* **95**, 15112–15117.

Chen, R., Rosen, E., and Masson, P. H. (1999). Gravitropism in higher plants. *Plant Physiol.* **120**, 343–350.

Chen, Z., Silva, H., and Klessig, D. F. (1993). Active oxygen species in the induction of plant systemic acquired resistance by salicylic acid. *Science* **262**, 1883–1886.

Christensen, S. K., Dagenais, N., Chory, J., and Weigel, D. (2000). Regulation of auxin response by the protein kinase *PINOID*. *Cell* **100**, 469–478.

Clifford, P. E., Mousdale, D. M. A., Lund, S. J., and Oxlade, E. L. (1985). Differences in auxin level detected across geostimulated dandelion peduncles: Evidence supporting a role for auxin in geotropism. *Ann. Bot.* **55**, 293–296.

Clifford, P. E., and Oxlade, E. L. (1989). Ethylene production, georesponse, and extension growth in dandelion peduncles. *Can. J. Bot.* **67**, 1927–1929.

Clifford, P. E., Reid, D. M., and Pharis, R. P. (1983). Endogenous ethylene does not initiate but may modify geobending—a role for ethylene in autotropism. *Plant Cell Environ.* **6**, 433–436.

Clouse, S. D., and Sasse, J. M. (1998). Brassinosteroids: Essential regulators of plant growth and development. *Annu. Rev. Plant Physiol. Plant Mol. Biol.* **49**, 427–451.

Coenen, C., Christian, M., Luthen, H., and Lomax, T. L. (2003). Cytokinin inhibits a subset of *diageotropica*-dependent primary auxin responses in tomato. *Plant Physiol.* **131**, 1692–1704.

Coenen, C., and Lomax, T. L. (1998). The *diageotropica* gene differentially affects auxin and cytokinin responses throughout development in tomato. *Plant Physiol.* **117**, 63–72.

Cosgrove, D. J. (1997). Cellular mechanisms underlying growth asymmetry during stem gravitropism. *Planta* **203**, S130–S135.

Darwin, C. (1880). "Power of Movement in Plants." John Murray, London.

del Pozo, J. C., and Estelle, M. (1999). Function of the ubiquitin-proteosome pathway in auxin response. *Trends Plant Sci.* **4**, 107–112.

Dharmasiri, N., and Estelle, M. (2004). Auxin signaling and regulated protein degradation. *Trends Plant Sci.* **9**, 302–307.

Eapen, D., Barroso, M. L., Campos, M. E., Ponce, G., Corkidi, G., Dubrovsky, J. G., and Cassab, G. I. (2003). A no hydrotropic response root mutant that responds positively to gravitropism in *Arabidopsis*. *Plant Physiol.* **131**, 536–546.

Eapen, D., Barroso, M. L., Ponce, G., Campos, M. E., and Cassab, G. I. (2005). Hydrotropism: Root growth responses to water. *Trends Plant Sci.* **10**, 45–50.

Edelmann, H. G. (2001). Lateral redistribution of auxin is not the means for gravitropic differential growth of coleoptiles: A new model. *Physiol. Plant.* **112**, 119–126.

Edelmann, H. G. (2002). Ethylene perception generates gravicompetence in gravi-incompetent leaves of rye seedlings. *J. Exp. Bot.* **53**, 1825–1828.

Edelmann, H. G., Gudi, G., and Kuhnemann, F. (2002). The gravitropic setpoint angle of drakgrown rye seedlings and the role of ethylene. *J. Exp. Bot.* **53**, 1627–1634.

el-Antably, H. M., and Larsen, P. (1974). Redistribution of endogenous gibberellins in geotropically stimulated roots. *Nature* **250**, 76–77.

Evans, M. L. (1991). Gravitropism: Interaction of sensitivity modulation and effector redistribution. *Plant Physiol.* **95**, 1–5.

Evans, M. L., Young, L. M., and Hasenstein, K. H. (1992). The role of calcium in the regulation of hormone transport in gravistimulated roots. *Adv. Space Res.* **12**, 211–218.

Fasano, J., Swanson, S., Blancaflor, E., Dowd, P., Kao, T., and Gilroy, S. (2001). Changes in root cap pH are required for the gravity response of the *Arabidopsis* root. *Plant Cell* **13**, 907–921.

Feldman, L. J., Arroyave, N. J., and Sun, P. S. (1985). Abscisic acid, xanthoxin and violaxanthin in the caps of gravistimulated maize roots. *Planta* **166,** 483–489.

Ferrari, S., Piconese, S., Tronelli, G., and Migliaccio, F. (2000). A new *Arabidopsis thaliana* root gravitropism and chirality mutant. *Plant Sci.* **158,** 77–85.

Finkelstein, R. R., and Rock, C. D. (2002). Abscisic acid biosynthesis and response. *In* "The *Arabidopsis* Book" (C. R. Somerville and E. M. Meyerowitz, Eds.), pp. 48, doi/10.1199/tab.0058. American Society of Plant Biologists, Rockville, MD(http://www.aspb.org/publications/arabidopsis/).

Firn, R., and Digby, J. (1980). The establishment of tropic curvatures in plants. *Annu. Rev. Plant Physiol.* **31,** 131–148.

Firn, R. D., Wagstaff, C., and Digby, J. (2000). The use of mutants to probe models of gravitropism. *J. Exp. Biol.* **51,** 1323–1340.

Friedman, H., Meir, S., Halevy, A. H., and Philosoph-Hadas, S. (2003a). Inhibition of the gravitropic bending response of flowering shoots by salicylic acid. *Plant Sci.* **165,** 905–911.

Friedman, H., Meir, S., Rosenberger, I., Halevy, A. H., Kaufman, P. B., and Philosoph-Hadas, S. (1998). Inhibition of the gravitropic response of snapdragon spikes by the calcium-channel blocker lanthanum chloride. *Plant Physiol.* **118,** 483–492.

Friedman, H., Meir, S., Rosenberger, I., Halevy, A. H., and Philosoph-Hadas, S. (2005). Calcium antagonists inhibit bending and differential ethylene production of gravistimulated *Ornithogalum* 'Nova' cut flower spikes. *Postharvest Biol. Technol.* **36,** 9–20.

Friedman, H., Vos, J. W., Hepler, P. K., Meir, S., Halevy, A. H., and Philosoph-Hadas, S. (2003b). The role of actin filaments in the gravitropic response of snapdragon flowering shoots. *Planta* **216,** 1034–1042.

Friml, J. (2003). Auxin transport—shaping the plant. *Curr. Opin. Plant Biol.* **6,** 7–12.

Friml, J., Benkova, E., Blilou, I., Wisniewska, J., Hamann, T., Ljung, K., Woody, S., Sandberg, G., Scheres, B., Jurgens, G., and Palme, K. (2002a). AtPIN4 mediates sink-driven auxin gradients and root patterning in *Arabidopsis*. *Cell* **108,** 661–673.

Friml, J., and Palme, K. (2002). Polar auxin transport–old questions and new concepts? *Plant Mol. Biol.* **49,** 273–284.

Friml, J., Wisniewska, J., Benkova, E., Mendgen, K., and Palme, K. (2002b). Lateral relocation of auxin efflux regulator PIN3 mediates tropism in *Arabidopsis*. *Nature* **415,** 806–809.

Fukaki, H., Fujisawa, H., and Tasaka, M. (1996). SGR1, SGR2, and SGR3: Novel genetic loci involved in shoot gravitropism in *Arabidopsis thaliana*. *Plant Physiol.* **110,** 945–955.

Fukaki, H., Wysocka-Dillr, J., Kato, T., Fujisawa, H., Benfey, P. N., and Tasaka, M. (1998). Genetics evidence that the endodermis is essential for shoot gravitropism in *Arabidopsis thaliana*. *Plant J.* **14,** 425–430.

Funke, M., and Edelmann, H. G. (2000). Auxin-dependent cell wall depositions in the epidermal periplasmic space of graviresponding nodes of *Tradescantia fluminensis*. *J. Exp. Bot.* **51,** 579–586.

Galweiler, L., Guan, C., Muller, A., Wilsman, E., Mendgen, K., Yephremov, A., and Palme, K. (1998). Regulation of polar auxin transport by *AtPIN1* in *Arabidopsis* vascular tissue. *Science* **282,** 2226–2230.

Gardiner, J. C., Harper, J. D. I., Weerakoon, N. D., Collings, D. A., Ritchie, S., Gilroy, S., Cyr, R. J., and Marc, J. (2001). A 90-kD phospholipase D from tobacco binds to microtubules and the plasma membrane. *Plant Cell* **13,** 2143–2158.

Geldner, N., Anders, N., Wolters, H., Keicher, J., Kornberger, W., Muller, P., Delbarre, A., Ueda, T., Nakano, A., and Jurgens, G. (2003). The *Arabidopsis GNOM* ARF-GEF mediates endosomal recycling, auxin transport, and auxin-dependent plant growth. *Cell* **112,** 219–230.

Geldner, N., Friml, J., Stierhof, Y. D., Jurgens, G. J., and Palme, K. (2001). Auxin transport inhibitors block PIN1 cycling and vesicle trafficking. *Nature* **413,** 425–428.

Gil, P., Dewey, E., Friml, J., Zhao, Y., Snowden, K. C., Putterill, J., Palme, K., Estelle, M., and Chory, J. (2001). BIG: A calossin-like protein required for polar auxin transport in *Arabidopsis. Genes Dev.* **15**, 1985–1997.

Goda, H., Sawa, S., Asami, T., Fujioka, S., Shimada, Y., and Yoshida, S. (2004). Comprehensive comparison of auxin-regulated and brassinosteroid-regulated genes in *Arabidopsis. Plant Physiol.* **134**, 1555–1573.

Golan, A., Tepper, M., Soudry, E., Horwitz, B. A., and Gepstein, S. (1996). Cytokinin, acting through ethylene, restores gravitropism to *Arabidopsis* seedlings grown under red light. *Plant Physiol.* **112**, 901–904.

Guisinger, M. M., and Kiss, J. Z. (1999). The influence of microgravity and spaceflight on columella cell ultrastructure in starch-deficient mutants of *Arabidopsis. Am. J. Bot.* **86**, 1357–1366.

Hagen, G., and Guilfoyle, T. (2002). Auxin-responsive gene expression: Genes, promoters and regulatory factors. *Plant Mol.Biol.* **49**, 373–385.

Hare, P. D., Seo, H. S., Yang, J.-Y., and Chua, N.-H. (2003). Modulation of sensitivity and selectivity in plant signaling by proteasomal destabilization. *Curr. Opin. Plant Biol.* **6**, 453–462.

Harper, R. M., Stowe-Evans, E. L., Luesse, D. R., Muto, H., Tatematsu, K., Watahiki, M. K., Yamamoto, K., and Liscum, E. (2000). The *NPH4* locus encodes the auxin response factor ARF7, a conditional regulator of differential growth in aerial *Arabidopsis* tissue. *Plant Cell* **12**, 757–770.

Harrison, M. A., and Pickard, B. G. (1984). Burst of ethylene upon horizontal placement of tomato seedlings. *Plant Physiol.* **75**, 1167–1169.

Harrison, M. A., and Pickard, B. G. (1986). Evaluation of ethylene as a mediator of gravitropism by tomato hypocotyls. *Plant Physiol.* **80**, 592–595.

Harrison, M. A., and Pickard, B. G. (1989). Auxin asymmetry during gravitropism by tomato hypocotyls. *Plant Physiol.* **89**, 652–657.

Hellgren, J. M., Olofsson, K., and Sundberg, B. (2004). Patterns of auxin distribution during gravitational induction of reaction wood in poplar and pine. *Plant Physiol.* **135**, 212–220.

Hensel, W., and Iversen, T. H. (1980). Ethylene production during clinostat rotation and effect on root geotropism. *Z. Pflanzenphysiol.* **97**, 343–352.

Hilaire, E., Peterson, B. V., Guikema, J. A., and Brown, C. S. (1996). Clinorotation affects morphology and ethylene production in soybean seedlings. *Plant Cell Physiol.* **37**, 929–934.

Hobbie, L., and Estelle, M. (1995). The axr4 auxin-resistant mutants of *Arabidopsis thaliana* define a gene important for root gravitropism and lateral root initiation. *Plant J.* **7**, 211–220.

Hobbie, L., McGovern, M., Hurwitz, L. R., Pierro, A., Liu, N. Y., Bandyopadhyay, A., and Estelle, M. (2000). The *axr6* mutants of *Arabidopsis thaliana* define a gene involved in auxin response and early development. *Development* **127**, 23–32.

Hoson, T., Kamisaka, S., and Masuda, Y. (1996). Suppression of gravitropic response of primary roots by submergence. *Planta* **199**, 100–104.

Hou, G., Kramer, V. L., Wang, Y. S., Chen, R., Perbal, G., Gilroy, S., and Blancaflor, E. B. (2004). The promotion of gravitropism in *Arabidopsis* roots upon actin disruption is coupled with the extended alkalinization of the columella cytoplasm and a persistent lateral auxin gradient. *Plant J.* **39**, 113–125.

Hu, X., Neill, S. J., Tang, Z., and Cai, W. (2005). Nitric oxide mediates gravitropic bending in soybean roots. *Plant Physiol.* **137**, 663–670.

Ishikawa, H., and Evans, M. L. (1993). The role of the distal elongation zone in the response of maize roots to auxin and gravity. *Plant Physiol.* **102**, 1203–1210.

Jackson, M. B., and Barlow, P. W. (1981). Root geotropism and the role of growth regulators from the cap: A re-examination. *Plant Cell Environ.* **4**, 107–123.

Jacob, T., Ritchie, S., Assmann, S., and Gilroy, S. (1999). Abscisic acid signal transduction in guard cells is mediated by phospholipase D activity. *Proc. Natl. Acad. Sci. USA* **96,** 12192–12197.

Jensen, P. J., Hangarter, R. P., and Estelle, M. (1998). Auxin transport is required for hypocotyl elongation in light-grown but not dark-grown *Arabidopsis*. *Plant Physiol.* **116,** 455–462.

Joo, J. H., Bae, Y. S., and Lee, J. S. (2001). Role of auxin-induced reactive oxygen species in root gravitropism. *Plant Physiol.* **126,** 1055–1060.

Juniper, B. E. (1976). Geotropism. *Ann. Rev. Plant Physiol.* **27,** 385–406.

Kamada, M., Fujii, N., Higashitani, A., and Takahashi, H. (2002). Gravity-induced asymmetry of localization of auxin-carrier proteins (CS-AUX1 and CS-PIN1) for peg formation in cucumber seedlings. *Biol. Sci. Space* **16,** 153–154.

Kang, B. G. (1979). Epinasty. *In* "Encyclopedia of Plant Physiology, New Series " (W. Haupt and M. E. Feinleib, Eds.), vol. 17, pp. 647–667. Springer-Verlag, Heildelberg.

Kaufman, P. B., Brock, T. G., Song, I., Rho, Y. B., and Ghosheh, N. S. (1987). How cereal grass shoots perceive and respond to gravity. *Am. J. Bot.* **74,** 1446–1457.

Kaufman, P. B., Pharis, R. P., Reid, D. M., and Beall, F. D. (1985). Investigation into the possible regulation of negative gravitropic curvature in intact *Avena sativa* plants and in isolated stem segments by ethylene and gibberellins. *Plant Physiol.* **65,** 237–244.

Kaufman, P. B., Wu, L. L., Brock, T. G., and Kim, D. (1995). Hormones and the orientation of growth. *In* "Plant Hormones: Physiology, Biochemistry and Molecular Biology" (P. J. Davis, Ed.), pp. 547–571. Kluwer Academic Publishers, Dordrecht, The Netherlands.

Kawano, T., Sahashi, N., Takahashi, K., Uozumi, N., and Muto, S. (1998). Salicylic acid induces extracellular superoxide generation followed by an increase in cytosolic calcium ion in tobacco suspension culture: The earliest events in salicylic acid signal transduction. *Plant Cell Physiol.* **39,** 721–730.

Kieber, J. J. (2002). Cytokinins. *In* "The *Arabidopsis* Book" (C. R. Somerville and E. M. Meyerowitz, Eds.), p. 25, doi/10.1199/tab.0063. American Society of Plant Biologists, Rockville, MD (http://www.aspb.org/publications/arabidopsis/).

Kim, D., and Kaufman, P. B. (1995). Basis for changes in the auxin-sensitivity of *Avena sativa* (oat) leaf-sheath pulvini during the gravitropic response. *J. Plant Physiol.* **145,** 113–120.

Kim, J., Harter, K., and Theologies, A. (1997). Protein–protein interactions among the Aux/IAA proteins. *Proc. Natl. Acad. Sci. USA* **94,** 11786–11791.

Kim, S. K., Chang, S. C., Lee, E. J., Chung, W. S., Kim, Y. S., Hwang, S., and Lee, J. S. (2000). Involvement of brassinosteroids in the gravitropic response of primary root of maize. *Plant Physiol.* **123,** 997–1004.

Kiss, J. Z. (2000). Mechanisms of the early phases of plant gravitropism. *Crit. Rev. Plant Sci.* **19,** 551–573.

Kiss, J. Z., Brinckmann, E., and Brillouet, C. (2000). Development and growth of several strains of *Arabidopsis* seedlings in microgravity. *Int. J. Plant Sci.* **161,** 55–62.

Kiss, J. Z., Edelmann, R. E., and Wood, P. C. (1999). Gravitropism of hypocotyls of wild-type and starch-deficient *Arabidopsis* seedlings in spaceflight studies. *Planta* **209,** 96–103.

Klee, H. (2003). Hormones are in the air. *Proc. Natl. Acad. Sci. USA* **100,** 10144–10145.

Klymchuk, D. O., Brown, C. S., and Chapman, D. K. (1999). Ultrastructural organization of cells in soybean root tips in microgravity. *J. Gravit. Physiol.* **6,** P97–P98.

Konings, H. (1995). Gravitropism of roots: An evaluation of progress during the last three decades. *Acta Bot. Neerl.* 195–223.

Kuznetsov, O. A., Brown, C. S., Levine, H. G., Piastuch, W. C., Sanwo-Lewandowski, M. M., and Hasenstein, K. H. (2001). Composition and physical properties of starch in microgravity-grown plants. *Adv. Space Res.* **28,** 651–658.

Lee, H. Y., Bahn, S. C., Kang, Y.-M., Lee, K. H., Kim, H. J., Noh, E. K., Palta, J. P., Shin, J. S., and Ryu, S. B. (2003). Secretory low molecular weight phospholipase A2 plays important roles in cell elongation and shoot gravitropism in *Arabidopsis. Plant Cell* **15,** 1990–2002.

Lee, J. S., Chang, W. K., and Evans, M. L. (1990a). Effects of ethylene on the kinetics of curvature and auxin redistribution in gravistimulated roots of *Zea mays. Plant Physiol.* **94,** 1770–1775.

Lee, J. S., and Evans, M. L. (1985). Polar transport of auxin across gravistimulated roots of maize and its enhancement by calcium. *Plant Physiol.* **77,** 824–827.

Lee, J. S., Hasenstein, K. H., Mulkey, T. J., Yang, R. L., and Evans, M. L. (1990b). Effects of abscisic acid and xanthoxin on elongation and gravitropism in primary roots of *Zea mays. Plant Sci.* **68,** 17–26.

Leopold, A. C., and LaFavre, A. K. (1989). Interactions between red light, abscisic acid, and calcium in gravitropism. *Plant Physiol.* **89,** 875–878.

Leyser, H. M. O. (1996). Mutations in the *AXR3* gene of *Arabidopsis* results in altered auxin response including ectopic expression from the *SAUR-AC*1 promoter. *Plant J.* **10,** 403–413.

Leyser, H. M. O. (2002). Molecular genetics of auxin signaling. *Ann. Rev. Plant Biol.* **53,** 377–398.

Li, Y., Hagen, G., and Guilfoyle, T. J. (1991). An auxin-responsive promoter is differentially induced by auxin gradients during tropisms. *Plant Cell* **3,** 1167–1175.

Li, Y., Wu, Y. H., Hagen, G., and Guilfoyle, T. (1999). Expression of the auxin-inducible GH3 promotor: GUS fusion gene as a useful marker for auxin physiology. *Plant Cell Physiol.* **40,** 675–682.

Liscum, E. (2002). Phototropism: Mechanisms and outcomes. *In* "The *Arabidopsis* Book" (C. R. Somerville and E. M. Meyerowitz, Eds.), p. 21, doi/10.1199/tab.0042. American Society of Plant Biologists, Rockville, MD (http://www.aspb.org/publications/arabidopsis/).

Liscum, E., and Briggs, W. R. (1996). Mutations of *Arabidopsis* in potential transduction and response components of the phototropic signaling pathway. *Plant Physiol.* **112,** 291–296.

Lomax, T. L. (1997). Molecular genetic analysis of plant gravitropism. *Gravit. Space Biol. Bull.* **10,** 75–82.

Lomax, T. L., Muday, G. K., and Rubery, P. (1998). Auxin transport. *In* "Plant Hormones—Physiology, Biochemistry, and Molecular Biology" (P. Davies, Ed.), pp. 509–530. Kluwer Academic Publishers, Dordrecht.

Long, J. C., Zhao, W., Rashotte, A. M., Muday, G. K., and Huber, S. C. (2002). Gravity-stimulated changes in auxin and invertase gene expression in maize pulvinal cells. *Plant Physiol.* **128,** 591–602.

Lu, C., and Fedoroff, N. (2000). A mutation in the *Arabidopsis HYL*1 gene encoding a dsRNA binding protein affects responses to abscissic acid, auxin, and cytokinin. *Plant Cell* **12,** 1351–2365.

Luschnig, C., Gaxiola, R. A., Grisafi, P., and Fink, G. R. (1998). EIR1, a root-specific protein involved in auxin transport, is required for gravitropism in *Arabidopsis thaliana. Genes Dev.* **12,** 2175–2187.

Madlung, A., Behringer, F. J., and Lomax, T. L. (1999). Ethylene plays multiple nonprimary roles in modulating the gravitropic response in tomato. *Plant Physiol.* **120,** 897–906.

Marchant, A., Kargul, J., May, S. J., Muller, P., Delbarre, A., Perrot-Rechenmann, C., and Bennett, M. J. (1999). AUX1 regulates root gravitropism in *Arabidopsis* by facilitating auxin uptake within root apical tissues. *EMBO J.* **18,** 2066–2073.

Masson, P. H., Tasaka, M., Morita, M. T., Guan, C., Chen, R., and Boonsirichai, K. (2002). *Arabidopsis thaliana*: A model for the study of root and shoot gravitropism. *In* "The *Arabidopsis* Book" (C. R. Somerville and E. M. Meyerowitz, Eds.), p. 23, doi/10.1199/tab.0043. American Society of Plant Biologists, Rockville, MD (http://www.aspb.org/publications/arabidopsis/).

McClure, B. A., and Guilfoyle, T. (1989). Rapid redistribution of auxin-regulated RNAs during gravitropism. *Science* **243,** 91–93.

Medvedev, S. S., and Markova, I. V. (1991). Involvement of salicylic acid in gravitropism in plants. *Doklady Bot. Sci.* **316**, 1014–1016.

Meicenheimer, R. D., and Nackid, T. A. (1994). Gravitropic response of *Kalenchoe* stems. *Int. J. Plant. Sci.* **155**, 395–404.

Mertens, R., and Weiler, E. M. (1983). Kinetic studies on the redistribution of endogenous growth regulators in gravireacting plant organs. *Planta* **158**, 339–348.

Meudt, W. T. (1987). Investigations on mechanism of brassinosteroid response.VI. Effect of brassinolide on gravitropism of bean hypocotyls. *Plant Physiol.* **83**, 195–198.

Migliaccio, F., and Galston, A. W. (1987). On the nature and origin of the calcium asymmetry arising during gravitropic response in etiolated pea epicotyls. *Plant Physiol.* **85**, 542–547.

Migliaccio, F., and Rayle, D. L. (1989). Effect of asymmetric auxin application on *Helianthus* hypocotyl curvature. *Plant Physiol.* **91**, 466–468.

Moore, R. (1990). Abscisic acid is not necessary for gravitropism in primary roots of *Zea mays*. *Ann. Bot.* **66**, 281–283.

Morita, M. T., and Tasaka, M. (2004). Gravity sensing and signaling. *Curr. Opin. Plant Biol.* **7**, 712–718.

Moseyko, N., Zhu, T., Chang, H.-W., Wang, X., and Feldman, L. J. (2002). Transcription profiling of the early gravitropic response in *Arabidopsis* using high-density oligonucleotide probe microarrays. *Plant Physiol.* **130**, 720–728.

Muday, G. K. (2000). Maintenance of asymmetric cellular localization of an auxin transport protein through interaction with the actin cytoskeleton. *J. Plant Growth Regul.* **19**, 385–396.

Muday, G. K. (2001). Auxins and tropisms. *J. Plant Growth Regul.* **20**, 226–243.

Muday, G. K. (2002). An emerging model of auxin transport regulation. *Plant Cell* **14**, 293–299.

Muday, G. K., and DeLong, A. (2001). Polar auxin transport: Controlling where and how much. *Trends Plant Sci.* **6**, 535–542.

Muday, G. K., and Haworth, P. (1994). Tomato root growth, gravitropism, and lateral development: Correlation with auxin transport. *Plant Physiol. Biochem.* **32**, 193–203.

Muday, G. K., Lomax, T. L., and Rayle, D. L. (1995). Characterization of the growth and auxin physiology of roots of the tomato mutant, *diageotropica*. *Planta* **195**, 548–453.

Mulkey, T. J., Kuzmanoff, K. M., and Evans, M. L. (1982). Promotion of growth and hydrogen ion efflux by auxin in roots of maize pretreated with ethylene biosynthesis inhibitors. *Plant Physiol.* **70**, 186–188.

Mulkey, T. J., and Vaughan, M. A. (1986). Auxin and root gravitropism: The state of our knowledge. *In* "Plant Growth Substances 1985" (M. Bopp, Ed.), pp. 241–245. Springer-Verlag, Berlin.

Muller, A., Guan, C., Galweiler, L., Tanzler, P., Huijser, P., Marchant, A., Parry, G., Bennett, M., Wisman, E., and Palme, K. (1998). AtPIN2 defines a locus of *Arabidopsis* for root gravitropism control. *EMBO J.* **17**, 6903–6911.

Nagpal, P., Walker, L. M., Young, J. C., Sonawala, A., Timpte, C., Estelle, M., and Reed, J. W. (2000). *AXR2* encodes a member of the Aux/IAA protein family. *Plant Physiol.* **123**, 563–573.

Nakamura, A., Goda, H., Shimada, Y., and Yoshida, S. (2004). Brassinosteroid selectively regulates *PIN* gene expression in *Arabidopsis*. *Biosci. Biotech. Biochem.* **68**, 952–954.

Nakamura, A., Shimada, Y., Goda, H., Fujiwara, M. T., Asami, T., and Yoshida, S. (2003). AXR1 is involved in BR-mediated elongation and *SAUR-AC1* gene expression in *Arabidopsis*. *FEBS Lett.* **553**, 28–32.

Nakamura, T., Negishi, Y., Funada, R., and Yamada, Y. (2001). Sedimentable amyloplasts in starch sheath cells of woody stems of Japanese cherry. *Adv. Space Res.* **27**, 957–960.

Nishimura, C., Ohashi, Y., Sato, S., Kato, T., Tabata, S., and Ueguchi, C. (2004). Histidine kinase homologs that act as cytokinin receptors possess overlapping functions in the regulation of shoot and root growth in *Arabidopsis*. *Plant Cell* **16**, 1365–1377.

Noh, B., Bandyopadhyay, A., Peer, W. A., Spalding, E. P., and Murphy, A. S. (2003). Enhanced gravi- and phototropism in plant *mdr* mutants mislocalizing the auxin efflux protein PIN1. *Nature* **423**, 999–1002.

Noh, B., Murphy, A. S., and Spalding, E. P. (2001). Multidrug resistance-like genes of *Arabidopsis* required for auxin transport and auxin-mediated development. *Plant Cell* **13**, 2441–2454.

Okushima, Y., Overvoorde, P. J., Arima, K., Alonso, J. M., Chan, A., Chang, C., Ecker, J. R., Hughes, B., Lui, A., Nguyen, D., Onodera, C., Quach, H., Smith, A., Yu, G., and Theologis, A. (2005). Functional genomic analysis of the *AUXIN RESPONSE FACTOR* gene family members in *Arabidopsis thaliana*: Unique and overlapping functions of ARF7 and ARF19. *Plant Cell* **17**, 444–463.

Osborne, D. J. (1976). Hormones and the growth of plants in response to gravity. *Life Sci. Space Res.* **14**, 37–46.

Ottenschlager, I., Wolff, P., Wolverton, C., Bhalerao, R. P., Sandberg, G., Ishikawa, H., Evans, M., and Palme, K. (2003). Gravity-regulated differential auxin transport from columella to lateral root cap cells. *Proc. Natl. Acad. Sci. USA* **100**, 2987–2991.

Palme, K., and Galweiler, L. (1999). PIN-pointing the molecular basis of auxin transport. *Curr. Opin. Plant Biol.* **2**, 375–381.

Park, W. J. (1998). Effect of epibrassinolide on hypocotyl growth of the tomato mutant *diageotropica*. *Planta* **207**, 120–124.

Parker, K. E. (1991). Auxin metabolism and transport during gravitropism. *Physiol. Plant.* **82**, 477–482.

Parker, K. E., and Briggs, W. R. (1990). Transport of indole-3-acetic acid during gravitropism in intact maize coleoptiles. *Plant Physiol.* **94**, 1763–1769.

Parry, G., Marchant, A., May, S., Swarup, R., Swarup, K., James, N., Graham, N., Allen, T., Martucci, T., Yemm, A., Napier, R., Manning, K., Graham King, G., and Bennett, M. (2001). Quick on the uptake: Characterization of a family of plant auxin influx carriers. *J. Plant Growth Regul.* **20**, 217–225.

Perera, I. Y., Heilmann, I., and Boss, W. F. (1999). Transient and sustained increases in inositol 1,4,5-trisphosphate precede the differential growth response in gravistimulated maize pulvini. *Proc. Natl. Acad. Sci. USA* **96**, 5838–5843.

Pharis, R. P., Legge, R. L., Noma, M., Kaufman, P. B., Ghosheh, N. S., La Croix, J. D., and Heller, K. (1981). Changes in endogenous gibberellins and the metabolism of GA_4 after geostimulation in shoots of the oat plants (*Avena sativa*). *Plant Physiol.* **67**, 892–897.

Philippar, K., Fuchs, I., Luthen, H., Hoth, S., Bauer, C. S., Haga, K., Thiel, G., Ljung, K., Sandberg, G., Bottger, M., Becker, D., and Hedrich, R. (1999). Auxin-induced K^+ channel expression represents an essential step in coleoptile growth and gravitropism. *Proc. Natl. Acad. Sci. USA* **96**, 12186–12191.

Philippar, K., Ivashikina, N., Ache, P., Christian, M., Luthen, H., Palme, K., and Hedrich, R. (2004). Auxin activates *KAT*1 and *KAT*2, two K^+-channel genes expressed in seedlings of *Arabidopsis thaliana*. *Plant J.* **37**, 815–827.

Philosoph-Hadas, S., Berkovitz-Simantov, R., Friedman, H., Meir, S., and Halevy, A. H. (2003). Role of ethylene in modulating auxin action during the gravitropic response of cut snapdragon spikes. *In* "Biology and Biotechnology of the Plant Hormone Ethylene III" (M. Vendrell, H. Klee, J. C. Pech, and F. Romojaro, Eds.), pp. 311–312. IOS Press, Amsterdam.

Philosoph-Hadas, S., Friedman, H., Berkovitz-Simantov, R., Rosenberger, I., Woltering, E. J., Halevy, A. H., and Meir, S. (1999). Involvement of ethylene biosynthesis and action in regulation of the gravitropic response of cut flowers. *In* "Biology and Biotechnology of the Plant Hormone Ethylene II" (C. C. A. K. Kanellis, H. Klee, A. B. Bleecker, J. C. Pech, and D. Grierson, Eds.), pp. 151–156. Kluwer Academic Publishers, Dordrecht, The Netherlands.

Philosoph-Hadas, S., Friedman, H., Meir, S., Berkovitz-Simantov, R., Rosenberger, I., Halevy, A. H., Kaufman, P. B., Balk, P., and Woltering, E. J. (2001). Gravitropism in cut flower stalks of snapdragon. *Adv. Space Res.* **27,** 921–932.

Philosoph-Hadas, S., Meir, S., Rosenberger, I., and Halevy, A. H. (1995). Control and regulation of the gravitropic response of cut flowering stems during storage and horizontal transport. *Acta Hort.* **405,** 343–350.

Philosoph-Hadas, S., Meir, S., Rosenberger, I., and Halevy, A. H. (1996). Regulation of the gravitropic response and ethylene biosynthesis in gravistimulated snapdragon spikes by calcium chelators and ethylene inhibitors. *Plant Physiol.* **110,** 301–310.

Pickard, B. G. (1985). Roles of hormones, protons and calcium in geotropism. *In* "Encyclopedia of Plant Physiology" (R. P. Pharis and D. M. Reid, Eds.), vol. II, pp. 193–281. Springer-Verlag, Berlin.

Plieth, C., and Trewavas, A. J. (2002). Reorientation of seedlings in the earth's gravitational field induces cytosolic calcium transients. *Plant Physiol.* **129,** 786–796.

Poupart, J., and Waddell, C. S. (2000). The *rib1* mutant is resistant to indole-3-butyric acid, an endogenous auxin in *Arabidopsis*. *Plant Physiol.* **124,** 1739–1751.

Rahman, A., Ahamed, A., Amakawa, T., Goto, N., and Tsurumi, S. (2001). Chromosaponin I specifically interacts with AUX1 protein in regulating the gravitropic response of *Arabidopsis* roots. *Plant Physiol.* **125,** 990–1000.

Rashotte, A. M., Brady, S. R., Reed, R. C., Ante, S. J., and Muday, G. K. (2000). Basipetal auxin transport is required for gravitropism in roots of *Arabidopsis*. *Plant Physiol.* **122,** 481–490.

Rashotte, A. M., DeLong, A., and Muday, G. K. (2001). Genetic and chemical reductions in protein phosphatase activity alter auxin transport, gravity response, and lateral root growth. *Plant Cell* **13,** 1683–1697.

Rice, M. S., and Lomax, T. L. (2000). The auxin-resistant *diageotropica* mutant of tomato responds to gravity via an auxin-mediated pathway. *Planta* **210,** 906–913.

Roman, G., Lubarsky, B., Kieber, J. J., Rothenberg, M., and Ecker, J. R. (1995). Genetic analysis of ethylene signal transduction in *Arabidopsis thaliana*: Five novel mutant loci integrated into a stress response pathway. *Genetics* **139,** 1393–1409.

Rood, S. B., Kaufman, P. B., Abe, H., and Pharis, R. P. (1987). Gibberellins and gravitropism in maize shoots: Endogenous gibberellin-like substances and movement and metabolism of [^{3}H]Gibberellin A$_{20}$. *Plant Physiol.* **83,** 645–651.

Rorabaugh, P. A., and Salisbury, F. B. (1989). Gravitropism in higher plant shoots. VI. Changing sensitivity to auxin in gravistimulated soybean hypocotyls. *Plant Physiol.* **91,** 1329–1338.

Rosen, E., Chen, R., and Masson, P. H. (1999). Root gravitropism: A complex response to a simple stimulus? *Trends Plant Sci.* **4,** 407–412.

Sack, F. D. (1991). Plant gravity sensing. *Int. Rev. Cytol.* **127,** 193–252.

Saito, Y., Yamasaki, S., Fujii, N., and Takahashi, H. (2005). Possible involvement of CS-ACS1 and ethylene in auxin-induced peg formation of cucumber seedlings. *Ann. Bot.* **95,** 413–422.

Salisbury, F. B. (1993). Gravitropism: Changing ideas. *Hortic. Rev.* **15,** 233–287.

Salisbury, F. B., Gillespie, L., and Rorabaugh, P. (1988). Gravitropism in higher plant shoots. V. Changing sensitivity to auxin. *Plant Physiol.* **88,** 1186–1194.

Salisbury, F. B., Rorabaugh, P. A., and White, R. (1985). Evidences for changes in sensitivity to auxin and in cell-wall properties during gravitropic bending of dicot stems. *Physiologist* **28,** S95–S96.

Salisbury, F. B., and Wheeler, R. M. (1980). Interpreting plant responses to clinostating I. Mechanical stress and ethylene. *Plant Physiol.* **67,** 677–685.

Schaller, G. E. (1997). Ethylene and cytokinin signalling in plants: The role of two-component systems. *Essays Biochem.* **32,** 101–111.

Schaller, G. E., and Kieber, J. J. A. (2002). Ethylene. *In* "The Arabidopsis Book" (C. R. Somerville and E. M. Meyerowitz, Eds.), p. 17, doi/10.1199/tab.0071. American Society of Plant Biologists, Rockville, MD (http://www.aspb.org/publications/arabidopsis/).

Schwark, A., and Schierle, J. (1992). Interaction of ethylene and auxin in the regulation of hook growth I. The role of auxin in different growing regions of the hypocotyl hook of *Phaseolus vulgaris*. *J. Plant Physiol.* **140**, 562–570.

Scott, A. C., and Allen, N. S. (1999). Changes in cytosolic pH within *Arabidopsis* root columella cells play a key role in the early signaling pathway for root gravitropism. *Plant Physiol.* **121**, 1291–1298.

Smets, R., Le, J., Prinsen, E., Verbelen, J.-P., and Van Onckelen, H. (2005). Cytokinin-induced hypocotyl elongation in light-grown *Arabidopsis* plants with inhibited ethylene action or indole-3-acetic acid transport. *Planta* **22**, 39–47.

Steinmann, T., Geldner, N., Grebe, M., Mangold, S., Jackson, C. L., Paris, S., Galweiler, L., Palme, K., and Jurgens, G. (1999). Coordinated polar localization of auxin efflux carrier PIN1 by GNOM ARF GEF. *Science* **286**, 316–318.

Stinemetz, C. L. (1996). Changes in IAA responsiveness in the elongation region of graviresponding mung bean roots. *Plant Growth Regul.* **20**, 245–251.

Su, W., and Howell, S. (1995). The effects of cytokinin and light on hypocotyl elongation in *Arabidopsis* seedlings are independent and additive. *Plant Physiol.* **108**, 1423–1430.

Sugano, M., Kaneyasu, S., and Nakamura, T. (2003). Gravitational and hormonal control in secondary xylem formation of Japanese flowering cherry. *Biol. Sci. Space* **17**, 245–246.

Suttle, J. C. (1988). Effect of ethylene treatment on polar IAA transport, net IAA uptake and specific binding of N-1-naphthylphthalamic acid in tissues and microsomes isolated from etiolated pea epicotyls. *Plant Physiol.* **88**, 795–799.

Swarup, R., Friml, J., Marchant, A., Ljung, K., Sandberg, G., Palme, K., and Bennett, M. (2001). Localization of the auxin permease AUX1 suggests two functionally distinct hormone transport pathways operate in the *Arabidopsis* root apex. *Genes Dev.* **15**, 2648–2653.

Takahashi, N., Goto, N., Okada, K., and Takahashi, H. (2002). Hydrotropism in abscisic acid, *wavy*, and gravitropic mutants of *Arabidopsis thaliana*. *Planta* **216**, 203–211.

Tasaka, M., Takehide, K., and Fukaki, H. (1999). The endodermis and shoot gravitropism. *Trends Plant Sci.* **4**, 103–107.

Tatematsu, K., Kumagai, S., Muto, H., Sato, A., Watahiki, M. K., Harper, R. M., Liscum, E., and Yamamoto, K. T. (2004). *MASSUGU*2 encodes Aux/IAA19, an auxin-regulated protein that functions together with the transcriptional activator NPH4/ARF7 to regulate differential growth responses of hypocotyl and formation of lateral roots in *Arabidopsis thaliana*. *Plant Cell* **16**, 379–393.

Tian, Q., and Reed, J. W. (1999). Control of auxin-regulated root development by the *Arabidopsis thaliana SHY2/IAA3* gene. *Development* **126**, 711–721.

Timpte, C., Lincoln, C., Pickett, F. B., Turner, J., and Estelle, M. (1995). The *AXR1* and *AUX1* genes of *Arabidopsis* function in separate auxin-response pathways. *Plant J.* **8**, 561–569.

Tiryaki, I., and Staswick, P. E. (2002). An *Arabidopsis* mutant defective in jasmonate response is allelic to the auxin-signaling mutant *axr1*. *Plant Physiol.* **130**, 887–894.

Trewavas, A. J. (1992). What remains of the Cholodny–Went theory? Introduction. *Plant Cell Environ.* **15**, 761.

Tsuchisaka, A., and Theologis, A. (2004). Unique and overlapping expression patterns among the *Arabidopsis* 1-amino-cyclopropane-1-carboxylate synthase gene family members. *Plant Physiol.* **136**, 2982–3000.

Ulmasov, T., Hagen, G., and Guilfoyle, T. J. (1999). Dimerization and DNA binding of auxin response factors. *Plant J.* **19**, 309–319.

Ulmasov, T., Murfett, J., Hagen, G., and Guilfoyle, T. J. (1997). Aux/1AA proteins repress expression of reporter genes containing natural and highly active synthetic auxin response elements. *Plant Cell* **9**, 1963–1971.

Utsuno, K., Shikanai, T., Yamada, Y., and Hashimoto, T. (1998). *Agr*, an Agravitropic locus of *Arabidopsis thaliana*, encodes a novel membrane-protein family member. *Plant Cell Physiol.* **39**, 1111–1118.

Van Der Straeten, D., Djudzman, A., Van Caeneghem, W., Smalle, J., and Van Montagu, M. (1993). Genetic and physiological analysis of a new locus in *Arabidopsis* that confers resistance to 1-aminocyclopropane-1-carboxylic acid and ethylene and specifically affects the ethylene signal transduction pathway. *Plant Physiol.* **102**, 401–408.

Vandenbussche, F., Smalle, J., Le, J., Saibo, N. J., De Paepe, A., Chaerle, L., Tietz, O., Smets, R., Laarhoven, L. J., Harren, F. J., Van Onckelen, H., Palme, K., Verbelen, J. P., and Van Der Straeten, D. (2003). The *Arabidopsis* mutant *alh1* illustrates a cross talk between ethylene and auxin. *Plant Physiol.* **131**, 1228–1238.

Watahiki, M. K., and Yamamoto, K. T. (1997). The *massugu1* mutation of *Arabidopsis* identified with failure of auxin-induced growth curvature of hypocotyl confers auxin insensitivity to hypocotyl and leaf. *Plant Physiol.* **115**, 419–426.

Webster, J. H., and Wilkins, M. B. (1974). Lateral movement of radioactivity from [^{14}C] gibberellic acid (GA$_3$) in roots and coleoptiles of *Zea mays* L. seedlings during geotropic stimulation. *Planta* **121**, 303–308.

Went, E. W., and Thimann, K. V. (1937). "Phytohormones." MacMillan, New York, NY.

Werner, T., Motyka, V., Laucou, V., Smets, R., Van Onckelen, H., and Schmulling, T. (2003). Cytokinin-deficient transgenic *Arabidopsis* plants show multiple developmental alterations indicating opposite functions of cytokinins in the regulation of shoot and root meristem activity. *Plant Cell* **15**, 2532–2550.

Wheeler, R. M., and Salisbury, F. B. (1980). Gravitropism in plant stems may require ethylene. *Science* **209**, 1126–1127.

Wheeler, R. M., and Salisbury, F. B. (1981). Gravitropism in higher plant shoots. I. A role for ethylene. *Plant Physiol.* **67**, 686–690.

Wheeler, R. M., White, R. G., and Salisbury, F. B. (1986). Gravitropism in higher plant shoots. IV. Further studies on participation of ethylene. *Plant Physiol.* **82**, 534–542.

Wilkins, M. B. (1979). Growth-control mechanisms in gravitropism. *In* "Physiology of Movements" (W. Haupt and M. E. Feinleib, Eds.), pp. 601–625. Springer-Verlag, Berlin.

Woeste, K. E., Vogel, J. P., and Kieber, J. J. (1999). Factors regulating ethylene biosynthesis in etiolated *Arabidopsis thaliana* seedlings. *Physiol. Plant.* **105**, 478–484.

Woltering, A. J., Balk, P. A., Mariska, A., Nijenhuis de Vries, A., Faivre, M., Ruys, G., Somhorst, D., Philosoph-Hadas, S., and Freidman, H. (2004). An auxin-responsive 1-aminocyclopropane-1-carboxylate synthase is responsible for differential ethylene production in gravistimulated *Antirrhinum majus* L. flower stems. *Planta* **220**, 403–413.

Woltering, E. J. (1991). Regulation of ethylene biosynthesis in gravistimulated *Kniphofia* (hybrid) flower stalks. *J. Plant Physiol.* **138**, 443–449.

Wolverton, C., Ishikawa, H., and Evans, M. L. (2002). The kinetics of root gravitropism: Dual motors and sensors. *J. Plant Growth Regul.* **21**, 102–112.

Wright, L. Z., Mousdale, D. M. A., and Osborne, D. J. (1978). Evidence for a gravity-regulated level of endogenous auxin controlling cell elongation and ethylene production during geotropic bending in grass nodes. *Biochem. Physiol. Pflanz.* **172**, 581–596.

Wu, L. L., Song, I., Kim, D., and Kaufman, P. B. (1993). Molecular basis of the increase in invertase activity elicited by gravistimulation of oat-shoot pulvini. *J. Plant Physiol.* **142**, 179–183.

Yamamoto, K. T. (2003). Happy end in sight after 70 years of controversy. *Trends Plant Sci.* **8**, 359–360.

Yamamoto, M., and Yamamoto, K. T. (1999). Effects of natural and synthetic auxins on the gravitropic growth habit of roots in two auxin-resistant mutants of *Arabidopsis*, *axr1* and *axr4*: Evidence for defects in the auxin influx mechanism of *axr4*. *J. Plant Res.* **112**, 391–396.

Yamamuro, C., Ihara, Y., Wu, X., Noguchi, T., Fujioka, S., Takatsuto, S., Ashikari, M., Kitano, H., and Matsuoka, M. (2000). Loss of function of a rice brassinosteroid *insensitive 1* homolog prevents internode elongation and bending of the lamina joint. *Plant Cell* **12**, 1591–1606.

Yi, H. C., Joo, S., Nam, K. H., Lee, J. S., Kang, B. G., and Kim, W. T. (1999). Auxin and brassinosteroid differentially regulate the expression of three members of the 1-aminocyclopropane-1-carboxylate synthase gene family in mung bean (*Vigna radiata* L.). *Plant Mol. Biol.* **41**, 443–454.

Young, L. M., and Evans, M. L. (1996). Patterns of auxin and abscisic acid movement in the tips of gravistimulated primary roots of maize. *Plant Growth Regul.* **20**, 253–258.

Young, L. M., Evans, M. L., and Hertel, R. (1990). Correlations between gravitropic curvature and auxin movement across gravistimulated roots of *Zea mays*. *Plant Physiol.* **92**, 792–796.

Zobel, R. W. (1973). Some physiological characteristics of the ethylene-requiring tomato mutant *diageotropica*. *Plant Physiol.* **52**, 385–389.

Zobel, R. W. (1974). Control of morphogenesis in the ethylene-requiring tomato mutant, *diageotropica*. *Can. J. Bot.* **52**, 735–740.

3

HORMONAL REGULATION OF SEX EXPRESSION IN PLANTS

SEIJI YAMASAKI,* NOBUHARU FUJII,[†] AND
HIDEYUKI TAKAHASHI[†]

*Faculty of Education, Fukuoka University of Education
1-1 Akamabunkyomachi, Munakata, Fukuoka 811-4192, Japan
[†]Graduate School of Life Sciences, Tohoku University
2-1-1 Katahira, Aoba-ku, Sendai 980-8577, Japan

I. Introduction
II. Differentiation of Flower Buds and Sex Expression in Maize
III. Genetic and Hormonal Regulation of Sex Expression in Maize
IV. Differentiation of Flower Bud and Sex Expression in Cucumber
V. Hormonal Regulation of Sex Expression in Cucumber
    A. Ethylene as a Primary Plant Hormone that Regulates Sex Expression
    B. Ethylene Biosynthesis and Sex Expression
    C. Ethylene Responses and Sex Expression
VI. Genetic Regulation of Sex Expression in Cucumber
VII. Environmental Regulation of Sex Expression
VIII. Identity of Floral Organs and Regulation of Sex Expression

    IX. Programmed Cell Death and Sex Determination
    X. Conclusions
       References

I. INTRODUCTION

Unlike most animal species that can travel in search for food and breeding partners, higher plants cannot move from their rooting position. As a consequence, it is likely that the present lifestyle of higher plants reflects their practical adaptation to the environment. To survive stressful environmental conditions that prevail on uplands, higher plants have evolved various organs such as roots, stems, leaves, and flowers. Flowers are the sexual organs that are used to produce progeny, and they have evolved mechanisms for promoting genetic diversity. For most species of flowering plants, cross-pollination (allogamy) is the best way to avoid inbreeding depression and to promote hybrid vigor.

Of the approximately 120,000 known species of flowering plants, 72% produce bisexual flowers while 28% produce unisexual flowers (Yampolsky and Yampolsky, 1922). Approximately one-tenth of flowering plant species are absolutely dioecious or monoecious (4 and 7%, respectively). Seven percent of species show intermediate forms of sexual dimorphism, including gynodioecy and androdioecy, whereas 10% of species contain both unisexual and bisexual flowers. Dioecious species produce male or female flowers on separate individual plants. Monoecious species produce male and female flowers on the same plant. Although the percentage of species producing unisexual flowers is lower than that of the species producing bisexual flowers, some well-known and strategically important plants are either dioecious or monoecious. For example, maize (*Zea mays*), cucumber (*Cucumis sativus*), melon (*Cucumis melo*), and fig (*Ficus carica*) are monoecious species, while asparagus (*Asparagus officinalis*), spinach (*Spinacia oleracea*), hemp (*Cannabis sativa*), mercury (*Mercurialis annua*), and hop (*Humulus lupulus*) are dioecious. For these plants, spatial separation of the sexual organs promotes allogamy.

In dioecious plants, there are three different types of sex determination. Firstly, there is the active-Y system. In dioecious melandrium (*Silene latifolia* = *Melandrium album*) and cannabis (*C. sativa*), the Y chromosome is the primary factor that defines sex (Ainsworth *et al.*, 1998; Dellaporta and Calderon-Urrea, 1993); male plants are heterogametic, displaying XY, whereas female plants are homogametic, displaying XX. The Y chromosome contains genes necessary for suppressing female expression, inducing stamen development and maturing anthers. Higher X copy number overcomes the Y chromosome masculinization effect, suggesting that the factors that suppress male expression exist on the X chromosome. Secondly, there is the X-to-autosome (X:A) balance system. The sex expression of approximately 10 dioecious species that belong to *Rumex acetosa* is classified under this system. In the X:A system, female plants are denoted 2n = 12 + XX (X:A = 1.0), whereas male plants are denoted 2n = 12 + XY$_1$Y$_2$ (X:A = 0.5). Polyploid analysis indicates that female flowers are induced when the X-to-autosome ratio is 1.0 or higher, whereas male flowers are induced when the X-to-autosome ratio is 0.5 or lower. Where ratios fall between 0.5 and 1.0, intersex (partial male/female) or hermaphrodite plants occur. The X-to-autosome ratio also applies to the sex expression of dioecious hop (*H. lupulus*). Thus, sex expression in these dioecious plants has been thoroughly investigated and elucidated with respect to sex chromosomes. Finally, dioecious plants of the third category do not have sex chromosome. For example, sex inheritance in dioecious mercury (*M. annua*) cannot be explained by heterochromosome and it has been proposed that three genes (A, B_1, and B_2) control sex expression of mercury plants (Louis, 1989). Mercury plants that harbor a dominant A gene and one of the dominant B genes, i.e., $A/- B_1/-$ and $A/- B_2/-$, exhibit male predominance. In contrast, femaleness in mercury plants is increased by the presence of either a dominant A gene or one of the dominant B genes alone (i.e., $A/A\ b_1/b_1\ b_2/b_2$, $a/a\ B_1/B_1\ b_2/b_2$, and $a/a\ b_1/b_1\ B_2/B_2$). Since there is an excellent previous review of sex determination in mercury plants (Durand and Durand, 1991), it is not discussed further in this review.

In higher plants, it is known that plant hormones such as auxin, cytokinin, gibberellins, abscisic acid, and ethylene pleiotropically regulate plant growth and development. These plant hormones also affect sex differentiation in some monoecious and dioecious plants. For example, auxin induces female flower expression in cucumber, melon, and hemp but promotes male flower expression in hop and mercury. Similarly, gibberellins induce female flowers in maize but male flowers in cucumber, melon, asparagus, and hemp. Therefore, as exemplified by these, certain plant hormones can have opposing effects on sex differentiation for different plants. In cucumber, melon, and hemp, ethylene and auxin mediate feminization, whereas gibberellins induce masculinization. In maize, asparagus, and hop, only one plant hormone is effective in inducing sexualization. In conclusion, there is no plant hormone that exhibits a generalized effect in determining the sexuality of monoecious and dioecious plants. The corollary of this observation is that each monoecious and dioecious plant must possess its own mechanism for hormonally regulating sex expression. In particular, monoecious plants that lack sex chromosomes have been the subject of several studies on the evolution and developmental mechanisms of sexual differentiation in higher plants.

This review focuses mainly on the hormonal regulation of sex expression in monoecious plants, with particular emphasis on maize (*Z. mays*) and cucumber (*C. sativus*). We also discuss the mutual and different aspects of the regulatory systems that control their sex expression and present a genetic model of sex expression in maize and cucumber plants.

II. DIFFERENTIATION OF FLOWER BUDS AND SEX EXPRESSION IN MAIZE

The monoecious plant, maize, differentiates tassels at its terminal end and ears at the lateral end, respectively. The processes of tassel development and ear inflorescence have been described extensively (Bonnett, 1940; Cheng *et al.*, 1983; Veit *et al.*, 1993). In brief, the flowering program begins with the conversion of the vegetative shoot meristem into the inflorescence meristem (Bonnett, 1948). The basic inflorescence structure of both tassels and ears is called a spikelet (Fig. 1). The maize spikelet comprises a pair of glumes (bracts) and contains two florets—one primary and one secondary (Fig. 1). Each floret differentiates into a lemma, a palea, two lodicules,

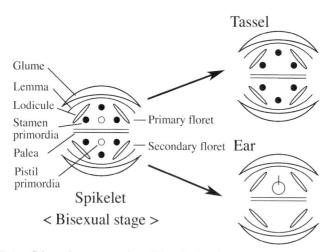

FIGURE 1. Schematic representation of the spikelet of tassels and ears in wild-type maize. The maize spikelet consists of a pair of glumes (bracts) and contains two florets (primary and secondary). Each floret differentiates a lemma, a palea, two lodicules, primordia of three stamens, and a pistil primordium. Development of the tassel and ear inflorescence are almost identical from the initiation of the flowering program until the bisexual stage. Sex differentiation in maize occurs subsequent to this common bisexual stage by selective abortion of the preformed floral organs. (Modified from DeLong *et al.* [1993]; © Cell Press.)

primordia of three stamens, and a pistil primordium composed of three fused carpels (Fig. 1; defined as the bisexual stage). Development of the tassel and ear inflorescence are very similar from the initiation of the flowering program until this bisexual stage. Sex differentiation in maize occurs subsequent to this common bisexual stage by the selective abortion of preformed floral organs. In tassels, the primordia of three stamens in each primary and secondary floret continue to develop, whereas pistil primordia in each primary and secondary floret abort (Fig. 1). Thus, both primary and secondary florets in the spikelet have functional stamens. In the ears, pistil primordia develop, whereas primordia of three stamens abort in primary florets (Fig. 1). The secondary florets abort in their entirety (Fig. 1). Thus, each spikelet contains a single functional ovule. Therefore, the process of sex differentiation in maize involves the arrest of preformed sexual organs (Calderon-Urrea and Dellaporta, 1999; Cheng et al., 1983; Dellaporta and Calderon-Urrea, 1993, 1994; Kellogg and Birchler, 1993).

III. GENETIC AND HORMONAL REGULATION OF SEX EXPRESSION IN MAIZE

In maize, several mutations that affect sex expression have been identified (Dellaporta and Calderon-Urrea, 1993, 1994; Irish, 1996; Lebel-Hardenack and Grant, 1997). The *dwarf (d)* mutants, *d1, d2, d3, d5, anther ear* (*an1*), and *D8* exhibit reduced internodal lengths and become andromonoecious. They differentiate hermaphrodites in the primary florets of the ears and male flowers in the tassels and secondary florets of the ears (Dellaporta and Calderon-Urrea, 1994; Irish, 1996; Wu and Cheung, 2000; Fig. 2). Among the *dwarf* mutants described earlier, *d1, d3,* and *d5* exhibit phenotypes that are typical of a block in gibberellins biosynthesis. This was confirmed biochemically by measuring endogenous gibberellin concentrations and by conducting metabolic studies with labeled substrates (Fujioka et al., 1988; Rood et al., 1980; Spray et al., 1984). The main pathway for gibberellin biosynthesis in higher plants, specifically maize, is described in Fig. 3. The *D1* gene is known to control each of the three steps, GA_{20} to GA_5 (Fig. 3I), GA_{20} to GA_1 (Fig. 3II), and GA_5 to GA_3 (Fig. 3III) (Spray et al., 1996). The simplest explanation for this multistep control by *D1* is that the gene product is a multifunctional enzyme that catalyzes the 2,3-dehydrogenation (GA_{20} to GA_5), 3β-hydroxylation (GA_{20} to GA_1), and 3β-hydroxylation (GA_5 to GA_3) steps by rearranging the double bond (Spray et al., 1996). The *D3* gene encodes a predicted protein with significant sequence similarity to cytochrome P450 enzymes and is thought to catalyze an early 13-hydroxylase pathway (GA_{12} to GA_{53}) (Winkler and Helentjaris, 1995). The *an1* mutant, which shows a semidwarf phenotype, also has a lesion in the gibberellin biosynthesis pathway. Furthermore, the product of the *AN1* gene is involved

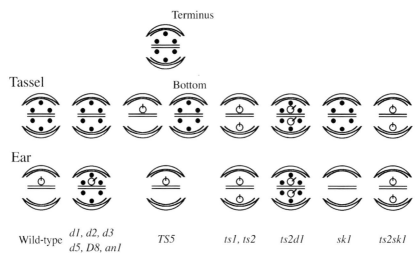

FIGURE 2. Schematic representation of florets in mutants maize plants. *d, dwarf; an1, anther ear; ts, tasselseed; sk1, silkless.* In *TS5* mutants, some primary florets have pistils at the bottom of tassels. Symbols for floral parts correspond to those of Fig. 1. Lodicules are omitted in this figure. (Modified from Dellaporta and Calderon-Urrea [1994]; © American Association for the Advancement of Science.)

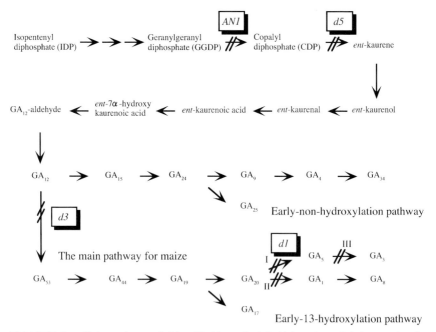

FIGURE 3. Major pathway of gibberellin biosynthesis in higher plants. It is known that the early-13-hydroxylation pathway is not unique to higher plants but is also the main pathway for maize plants. The symbol (//) indicates the location of the defect in each maize mutant (*AN1, d1, d3,* and *d5*).

in the synthesis of copalyl diphosphate (CDP) (the first tetracyclic intermediate in the pathway) from geranylgeranyl diphosphate (GGDP) (Bensen et al., 1995). The *D8* gene encodes the maize ortholog of *Arabidopsis GIBBERELLIN INSENSITIVE* and wheat *Reduced height-1* genes that act as repressors of GA-regulated processes (Peng et al., 1999). These transcription factors negatively regulate gibberellin responses in higher plants (Richards et al., 2001). Overall, the *dwarf* mutants described earlier indicate that gibberellins are required for the arrest of stamen primordia in ear florets and they therefore play an important role in the induction of femaleness in maize plants.

Plants in which the sexual conversion of the florets in the tassels is abnormal are classified as *tasselseed* (*ts*) mutants. Five *ts* loci (*ts1*, *ts2*, *ts4*, *Ts5*, and *ts6*) have been identified to date. *Ts5* is a semidominant mutation, and its mutant plants bear some female flowers that only develop pistils in the primary florets of the tassels (Emerson, 1932; Fig. 2). The *ts1* and *ts2* mutants, which are recessive mutants, form pistils in the primary and secondary florets, in both ears and tassels, and develop no stamens (Irish et al., 1994; Fig. 2). Therefore, the *TS1* and *TS2* genes are thought to mediate the arrest of pistils in the tassel florets (Dellaporta and Calderon-Urrea, 1994). Calderon-Urrea and Dellaporta (1999) showed that *TS2* mRNA does not accumulate in the *ts1* mutant and hypothesized that either the *TS1* gene positively regulates the transcription of the *TS2* gene or that the *TS1* gene is necessary for the stability of *TS2* mRNA. Cloning of the *TS2* gene revealed that it encodes an alcohol dehydrogenase-like protein that has similarity to steroid dehydrogenases (DeLong et al., 1993). In the *ts2d1* double mutants, primary and secondary florets in both ears and tassels have well-developed stamens and pistils (Dellaporta and Calderon-Urrea, 1994; Irish et al., 1994; Fig. 2). This indicates that the *TS2* gene and gibberellins act independently and additively. Also, it has been reported that maize defective in gibberellin biosynthesis bears well-developed stamens in the florets of the ears and that maize with feminized tassels has increased gibberellin-like activity (Rood et al., 1980). These observations suggest that increased gibberellin levels in the developing pistils arrest stamen development (Dellaporta and Calderon-Urrea, 1994; Irish et al., 1994). They also indicate that mutation of the *ts2* gene does not directly cause the arrest of stamen development.

The *silkless1* (*sk1*) gene is necessary for the development of pistil primordia in the primary florets of the ears (Jones, 1925). In a *sk1* mutant, although the phenotype of the florets in the tassels is the same as wild-type plants, neither stamens nor pistils develop in the ear florets (Fig. 2). Thus, *sk1* mutation induces masculinization and leads to the production of sterile florets in the ears. In the *ts2sk1* double mutants, pistil primordia in the primary florets of the ears develop normally, similar to wild-type plants (Irish et al., 1994; Jones, 1934; Fig. 2). This suggests that pistil arrest in the primary florets of the ears in *sk1* mutant is caused by activation of the *TS2* gene. These results also indicate that the *SK1* gene might prevent the function of the *TS2* gene either directly or indirectly in the pistil primordia of the primary florets in the ears (Calderon-Urrea and Dellaporta, 1999).

On the basis of these genetic analyses, Calderon-Urrea and Dellaporta (1999) have proposed a model for pistil primordia arrest in maize. An outline of this model of sex determination in maize is summarized in Fig. 4. Three sex-differentiation phenotypes; primary and secondary florets in the tassels (Fig. 4A), primary florets in the ears (Fig. 4B), and secondary florets in the ears (Fig. 4C), can be considered in maize. In all florets (Fig. 4A–C), the *TS1* gene regulates the transcription of the *TS2* gene, either directly or indirectly.

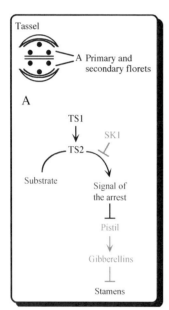
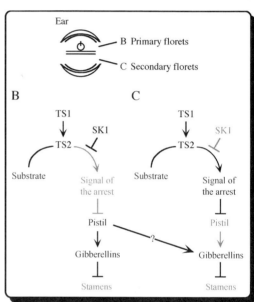

FIGURE 4. Outline of sex determination in maize. Three types of sex differentiation are considered here, primary and secondary florets in the tassels (A), primary florets in the ears (B), and secondary florets in the ears (C). In all florets (A–C), the *TS1* gene regulates transcription of the *TS2* gene either directly or indirectly. In the primary and secondary florets of the tassels (A) and the secondary florets of the ears (C), the TS2 protein induces the arrest of pistil primordia, presumably through the modification of an unknown substance related to steroids. On the contrary because SK1 prevents the action of TS2 in the primary florets of the ears (B), pistil primordia arrest does not occur. Thus, pistils develop only in the primary florets of the ears. In the primary florets of the ears (B), it is thought that the developing pistils produce enough gibberellins to arrest stamen primordia. Likewise, in the primary and secondary florets of the tassels (A), it is considered that low levels of gibberellins facilitate the development of stamens because of pistil primordia arrest. It is possible that higher levels of gibberellins that are produced by the developing pistils in the primary florets might affect the arrest of the stamen primordia in the secondary florets despite the spatial distance in the ears (C). TS1: TASSELSEED1; TS2: TASSELSEED2; SK1: SILKLESS1. (Modified from Calderon-Urrea and Dellaporta [1999]; © The Company of Biologists.)

In the primary and secondary florets of the tassels (Fig. 4A) and secondary florets of the ears (Fig. 4C), the TS2 protein induces an arrest of pistil primordia, presumably through the modification of an unknown substance that is speculated to be related to steroids. On the contrary, because SK1 prevent the action of TS2 in the primary florets of the ears (Fig. 4B), arrest of pistil primordia does not occur. Therefore, pistils develop only in the primary florets of the ears. As described previously, gibberellins are involved in sex differentiation of maize flowers. In the primary florets of the ears (Fig. 4B), it is thought that developing pistils produce enough gibberellins to arrest stamen primordia (Dellaporta and Calderon-Urrea, 1994; Irish et al., 1994). Likewise, in the primary and secondary florets of the tassels (Fig. 4A), it is assumed that the low levels of gibberellins, which occur due to the arrest of pistil primordial, facilitate the development of stamens. On the contrary, in the secondary florets of the ears (Fig. 4C) arrest of stamen primordia occurs in spite of the arrest of pistil primordia. Endogenous levels of gibberellins are reportedly 100-fold greater in the ears than in the tassels (Rood et al., 1980). Therefore, it is possible that high levels of gibberellins produced by developing pistils in the primary florets might affect the arrest of stamen primordia in the secondary florets despite there being spatial distance in the ears (Fig. 4C).

It has been suggested that in addition to gibberellins, cytokinin might also be involved in sex differentiation in maize plants (Young et al., 2004). Transgenic maize expressing the cytokinin-synthesizing isopentenyl transferase (IPT) enzyme has two functional florets in the ears reflecting the fact that its secondary floret pistils are rescued from abortion (Young et al., 2004). This finding indicated that cytokinin might prevent pistil abortion in the secondary florets of the ears and it determines the fate of pistils during the development of maize florets.

IV. DIFFERENTIATION OF FLOWER BUD AND SEX EXPRESSION IN CUCUMBER

One monoecious plant, cucumber, has been used as a model plant for the study of sex determination (Galun, 1961; Kubicki, 1969a,b,c; Shifriss, 1961). In addition to its basic monoecious sex phenotype, in which male and female flowers are produced on the same individual plant, cucumber can exhibit another four sex phenotypes, gynoecious, hermaphrodite, andromonoecious, and androecious (Malepszy and Niemirowicz-Szczytt, 1991). In principle, gynoecious cucumber plants bear only female flowers whereas hermaphrodite cucumber plants produce a substantial number of bisexual flowers, although in practice some male flowers appear in both plants. Andromonoecious cucumber bears bisexual and male flowers on the same plant whereas androecious plants bear only male flowers. Regardless of the sex phenotype of cucumber plants, male, female, or bisexual flowers are produced at each node along the stems. Generally, male flowers initially appear on the lower nodes followed by male and female flowers on the nodes of the middle region, and finally female flowers on the upper nodes in monoecious cucumber plants. Thus, sex expression in cucumber plants changes developmentally along the stems. Also, in flower buds destined to differentiate into male, female, or bisexual flowers, sexuality is not distinguishable at the early developmental stage (Fig. 5A–D). Specifically, in all cucumber flower buds, primordia of sepals, petals, stamens, and pistils are formed centripetally from the outside of the floral primordia (Atsmon and Galun, 1960; Fig. 5A–D). Following this bisexual stage (Fig. 5D), their sex is determined by the selective arrest of either pistil or stamen (Atsmon and Galun, 1960; Malepszy and Niemirowicz-Szczytt, 1991; Fig. 5E and F), continued pistil development and arrest of stamen development result in female flowers, whereas continued stamen development and arrest of pistil development result in male flowers. Therefore, vestiges of stamen primordia and of pistil primordia can be observed in mature female and mature male flowers, respectively (Fig. 5E and F). In the case of bisexual flowers, both pistils and stamens continue to develop. Thus, the processes of sex differentiation in maize and cucumber are similar in that both involve a common bisexual stage and the subsequent arrest of preformed sexual organs.

V. HORMONAL REGULATION OF SEX EXPRESSION IN CUCUMBER

A. ETHYLENE AS A PRIMARY PLANT HORMONE THAT REGULATES SEX EXPRESSION

Previous studies indicate that sex expression in cucumber plants is sensitive to modification by plant hormones (Atsmon and Galun, 1960; Durand and Durand, 1984; Frankel and Galun, 1977; Galun, 1961; Kubicki, 1969b; Shifriss, 1961; Shifriss and George, 1964; Takahashi *et al.*, 1983). Auxin, ethylene, abscisic acid, and cytokinin promote the formation of female flowers, whereas gibberellins promote the formation of male flowers (Galun, 1959b; Iwahori *et al.*, 1970; MacMurray and Miller, 1968; Peterson and Anhder, 1960; Saito and Takahashi, 1987; Takahashi and Suge, 1980, 1982). The enhancement of feminization by auxin possibly occurs through the induction of ethylene biosynthesis (Takahashi and Jaffe, 1984; Trebitsh *et al.*, 1987). Specifically, it has been reported that auxin regulates the expression of the genes that encode 1-aminocyclopropane-1-carboxylate (ACC) synthase, a key enzyme in the ethylene biosynthesis pathway (Arteca and Arteca, 1999; Botella *et al.*, 1992; Nakagawa *et al.*, 1991). Higher levels of ethylene were detected in gynoecious cucumber plants than in monoecious

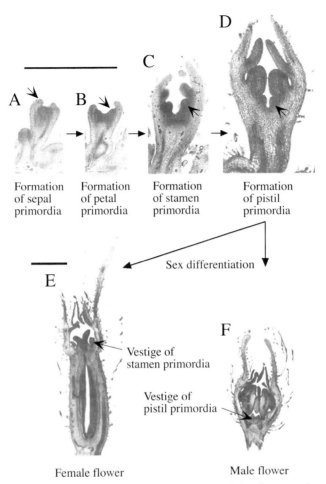

FIGURE 5. Formation of flower buds and sex differentiation in monoecious cucumber plants. Arrows indicate the primordia of sepals (A), petals (B), stamens (C), and pistils (D). (E) Mature female flower. (F) Mature male flower. All immature flower buds contain the primordia of both stamens and pistils at the early common bisexual stage (D), and sex differentiation occurs by selective arrest of the preformed sexual organ primordia. Thus, continued development of pistil primordia and arrest of stamen primordia result in the induction of female flowers (E), whereas continued development of stamen primordia and arrest of pistil primordia result in male flowers (F). Vestiges of stamen primordia and of pistil primordia can be found in female and male flowers, respectively (the bar indicates 500 μm).

ones, whereas gibberellin levels were lower in gynoecious cucumber plants than in monoecious ones (Atsmon et al., 1968; Rudich et al., 1972). Application of ethylene (or an ethylene-releasing agent) induces female flower expression, whereas gibberellins and inhibitors of ethylene biosynthesis or action promote male flower expression in cucumber (Atsmon and Tabbak,

1979; Galun, 1961; Malepszy and Niemirowicz-Szczytt, 1991; Rudich *et al.*, 1969). This suggests that ethylene is a primary plant hormone for the induction of femaleness, whereas gibberellins induce maleness. There are two hypotheses for the hormonal regulation of sex determination in unisexual plants. One is the two-hormone balance system, in which each sex is promoted by a different plant hormone (Chailakhan, 1979). The other is a single-hormone system, in which one plant hormone induces one sex while independently inhibiting the other sex (Yin and Quinn, 1992). The single-hormone system has been tested in cucumber plants by the application of ethylene, gibberellins, and their inhibitors (Yin and Quinn, 1995). The results of experiments in which combinations of the two plant hormones and their inhibitors were used, suggested that ethylene overrides the effects of gibberellins and acts more directly on sex expression in cucumber. Thus, the gaseous plant hormone, ethylene, is deemed to be a primary plant hormone regulating sex expression in cucumber plants, and as a consequence, molecular mechanisms of ethylene biosynthesis and action have been studied extensively in association with sex expression.

B. ETHYLENE BIOSYNTHESIS AND SEX EXPRESSION

The general pathway for ethylene biosynthesis in higher plants is described in Fig. 6. ACC synthase catalyzes the step from *S*-adenosyl methionine (SAM) to 1-aminocyclopropane-1-carboxylate, following which ACC oxidase catalyzes the final step in ethylene formation (Fig. 6). To determine which molecular components of the ethylene biosynthesis pathway regulate sex expression in cucumber, cDNAs of ACC synthase (*CS-ACSs*) and ACC oxidase (*CS-ACOs*) were isolated, and the relationship of their expression patterns to sex differentiation was characterized. It is known that a multigene family encodes ACC synthase and that expressions of these genes are regulated by auxin, extrinsic stimuli such as touch and injury, and developmental processes (Abel *et al.*, 1995; Liang *et al.*, 1992; Nakagawa *et al.*, 1991; Olson *et al.*, 1991, 1995; Rottmann *et al.*, 1991; Zarembinski and Theologis, 1993). Trebitsh *et al.* (1997) amplified the ACC synthase gene from cucumber genomic DNA by polymerase chain reaction using degenerate oligonucleotide primers and designated it *CS-ACS1*. When this *CS-ACS1* gene was used as a probe for Southern hybridization, the banding pattern between DNAs isolated from gynoecious and monoecious cucumber plants was virtually identical, except that an additional band was detected in gynoecious cucumber possessing the *F* gene relative to monoecious plants (Trebitsh *et al.*, 1997). The *F* gene is a partially dominant gene that controls femaleness and accelerates the production of female (or bisexual) flowers from the lower node (see Section VI and Table I). This additional band detected by the *CS-ACS1* probe was tightly linked to the *F* gene and Trebitsh *et al.* (1997) termed it *CS-ACS1G*. The same authors detected no difference in the accumulation

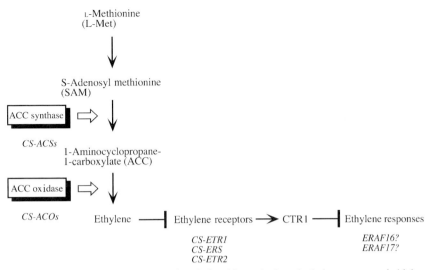

FIGURE 6. The general pathway for ethylene biosynthesis and ethylene responses in higher plants. 1-Aminocyclopropane-1-carboxylate (ACC) synthase catalyzes the step from S-adenosyl methionine (SAM) to ACC and ACC oxidase catalyzes the final step in ethylene formation. In cucumber plants, cDNAs of ACC synthase (*CS-ACSs*), ACC oxidase (*CS-ACOs*), and ethylene receptors (*CS-ETR1*, *CS-ERS*, *CS-ETR2*) have been isolated.

TABLE I. Major Genes Controlling Sex Expression in Cucumber Plants

Gene symbol	Character	References
F	Partially dominant gene that induces female flowers from the lower node of cucumber plants	Galun (1961)
Gy	Recessive gene that induces femaleness of cucumber plants	Kubicki (1974)
A	Recessive gene for a high degree of female flower expression. *A* gene is epistatic to *F* gene Androecious plants are produced only when both *A* and *F* genes are recessive (*aaff*)	Galun (1961)
M	This gene regulates the structure of flowers. Plants with the *M* form of the gene produce female flowers whereas plants with a recessive *m* gene (*mm*) produce bisexual flowers	Shifriss (1961)
In-F	Intensifier of female flower expression. This gene increases the degree of female flower expression of the plants with *F* gene	Kubicki (1969a)

of *CS-ACS1* mRNA between monoecious and gynoecious cucumber plants. In contrast, Kamachi *et al.* (2000) showed that the accumulation of *CS-ACS1* mRNA is more pronounced in gynoecious cucumber plants than in monoecious ones. It is unclear whether Kamachi *et al.* (2000) were actually detecting the accumulation of both *CS-ACS1* and *CS-ACS1G* mRNA or only *CS-ACS1* mRNA. Identifying the *CS-ACS1G* gene would possibly shed further light on *F* gene. Kamachi *et al.* (1997) isolated *CS-ACS2* cDNA and demonstrated that the accumulation of *CS-ACS2* mRNA was more pronounced in the shoot apices of gynoecious cucumber plants than in monoecious ones. Also, *CS-ACS2* mRNA accumulation coincided with the development of female flowers in monoecious and gynoecious cucumber plants (Kamachi *et al.*, 1997). In addition, the expression of the *CS-ACS2* gene in shoot apices correlates with the superiority or inferiority of the *F* gene when gynoecious, monoecious, and andromonoecious cucumber plants are compared (Yamasaki *et al.*, 2001). *In situ* hybridization revealed that *CS-ACS2* mRNA markedly accumulates in the flower bud pistil primordia of gynoecious cucumber plants (Yamasaki *et al.*, 2003b). In summary, these data strongly support the notion that the *CS-ACS2* genes are involved in the production of female flowers in cucumber plants.

ACC oxidase is also encoded by a multigene family in higher plants (Balague *et al.*, 1993; Holdsworth *et al.*, 1987; Kock *et al.*, 1991; Lasserre *et al.*, 1996; Spanu *et al.*, 1991). Kahana *et al.* (1999) isolated three types of cDNAs that encode ACC oxidase (*CS-ACO1*, *CS-ACO2*, and *CS-ACO3*) from cucumber plants and showed that the expression of *CS-ACO2* and *CS-ACO3* genes correlates with superiority or inferiority of the *F* gene in leaves. However, their expression is inversely correlated with superiority or inferiority of the *F* gene in the shoot apices and in the young flower buds, suggesting that the expression of *CS-ACO2* and *CS-ACO3* genes is inversely correlated with the production of female flowers in cucumber. Thus, genes that encode ACC synthase, rather than ACC oxidase, might function to regulate the aspects of ethylene biosynthesis that are necessary for feminization in cucumber plants.

C. ETHYLENE RESPONSES AND SEX EXPRESSION

After ethylene is biosynthesized, the responses it mediates are induced through an ethylene-signaling pathway (Fig. 6). Previous molecular and genetic studies in *Arabidopsis thaliana* resulted in the identification of five different putative ethylene receptor genes termed *Ethylene Response 1* (*ETR1*), *Ethylene Response Sensor 1* (*ERS1*), *ETR2*, *Ethylene Insensitive 4* (*EIN4*), and *ERS2* (Chang *et al.*, 1993; Hua *et al.*, 1995, 1998; Sakai *et al.*, 1998). In cucumber plants, ethylene-responsive genes, as well as ethylene receptor-related genes have been isolated and their expression patterns in relation to sex expression have been characterized.

In an effort to identify ethylene receptors involved in regulating sex expression in cucumber plants, Yamasaki et al. (2000) isolated three cDNAs of ethylene receptor-related genes, CS-ETR1, CS-ERS, and CS-ETR2. Of the three genes, CS-ERS and CS-ETR2 mRNA accumulated to a higher degree in the shoot apices of gynoecious cucumber plants than in those of monoecious ones. Also, ethylene exposure significantly increased CS-ERS and CS-ETR2 mRNA accumulation in monoecious and gynoecious cucumber plants but not in andromonoecious plants (Yamasaki et al., 2001). This provides the first compelling evidence that ethylene responses are reduced in andromonoecious cucumber plants compared with monoecious and gynoecious plants. This reduction in ethylene responsiveness could account for the ethylene-induced inhibition of stamen development that occurs in gynoecious and monoecious but not in andromonoecious cucumbers. These results provide the basis of a genetic model of sex expression in cucumber plants (see Section VI).

In cucumber plants, two ethylene-responsive genes were isolated using a differential display method and designated ethylene-responsive gene associated with the formation of female flowers 17 (ERAF17) (Ando et al., 2001) and ERAF16 (Ando and Sakai, 2002). The expression of ERAF17 and ERAF16 genes is increased when female flowers are induced by the application of ethephon, an ethylene-releasing agent. Sequence analysis revealed that the protein encoded by the ERAF17 gene contains a MADS-box domain. This is the first evidence that a MADS-box gene is induced by ethylene in cucumber plants. Phylogenetic analysis showed that the ERAF17 gene does not belong to any of the subfamilies of MADS-box genes that comprise the ABC model. The relationship of the ABC model to sex expression is described later in this article. The ERAF16 gene contains a motif that is conserved in genes encoding methyltransferase, but its functions as an enzyme and/or substrate are still unknown. ERAF17 mRNA was highly expressed in the sepals of mature female flowers, whereas the ERAF16 gene was expressed in opened flowers, stems, hypocotyls, and roots as well as in flower buds of gynoecious cucumber plants. Therefore, it is conceivable that these genes have pleiotropic functions and play roles in growth and development, including sex expression, in cucumber plants.

There have been several studies linking the molecular mechanisms of ethylene signaling with the sex differentiation of cucumber plants, with particular emphasis on expression analysis of the ethylene-responsive genes. Further functional analysis of ethylene-responsive genes may lead to the identification of genes that specifically regulate sex differentiation in cucumber plants.

VI. GENETIC REGULATION OF SEX EXPRESSION IN CUCUMBER

In cucumber plants, several genes that control sex expression have been characterized by different investigators since the 1960s (Table I). To date, three major genes, F, M, and A that regulate various sex phenotypes in cucumber have been identified and characterized (Kubicki 1969a,b,c; Pierce and Wehner 1990; Table I). The F gene is a partially dominant gene that controls femaleness and accelerates the production of female (or bisexual) flowers from the lower node. The M gene regulates the structure of flowers. Plants that have the M gene produce female flowers, but not bisexual flowers, whereas plants with recessive m genes (mm) bear bisexual flowers instead of female flowers. Monoecious plants are designated as M-ff. Similarly, gynoecious, hermaphrodite, and andromonoecious plants are described as M-F-, mmF-, and $mmff$, respectively. To characterize the F gene, gynoecious cucumber plants have been comprehensively compared with monoecious plants (George, 1971; Kamachi et al., 1997; Rudich et al., 1972; Trebitsh et al., 1987, 1997). The A gene acts downstream of the F gene and regulates the production of male flowers (Kubicki, 1969a,b). Since androecious plants are produced only when both A and F genes are recessive, these plants are described as $aaff$. In addition to F, M, and A, other genes that affect sex expression in cucumber plants have also been reported (Table I). These genes are thought to somehow influence F or A genes.

To clarify the roles of the F and M genes in regulating the sexual phenotype of cucumber cultivars, we previously analyzed ethylene production and ethylene responses in monoecious, gynoecious, and andromonoecious cucumber plants (Yamasaki et al., 2001; see Section V). On the basis of that study, we constructed a genetic model of the roles of the F and M genes and ethylene in the sex expression of cucumber plants (Fig. 7; Yamasaki et al., 2001). In constructing this genetic model, we focused on the development of stamens and pistils rather than on the expression of male, female, and bisexual flowers, which had previously been used to deduce the function of the F and M genes (Kubicki, 1969a,b,c; Pierce and Wehner, 1990). In this model, the product of the F gene regulates ethylene synthesis, and the downstream ethylene induces the development of pistils and reduces the development of stamens, thereby promoting femaleness (Fig. 7A). Also, this genetic model was the first to suggest that ethylene signals may influence the product of the M gene and thereby inhibit stamen development in cucumbers (Fig. 7A).

In the same genetic model, we also hypothesized that a recessive f gene does not necessarily affect all floral primordia because plants with an ff genotype (monoecious and andromonoecious) do not develop into androecious plants bearing only male flowers. Also, we proposed that a recessive m gene affects all floral primordia because plants with an mm genotype

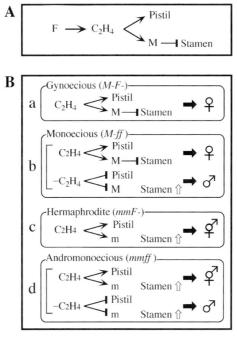

FIGURE 7. (A) A genetic model involving the F and M genes and ethylene in the sex expression of cucumbers (Yamasaki et al., 2001). The function of the F gene is to inhibit the development of stamens and to induce the development of pistils in some floral primordia. A recessive f gene does not affect all floral primordia. The function of the M gene is to inhibit the development of stamens in all floral primordia. A recessive m gene affects all floral primordia. (B) Sex phenotypes of cucumber plants explained by this genetic model. (a) Gynoecious (M-F-) cucumber plants produce only female flowers. (b) Monoecious (M-ff) cucumber plants produce male and female flowers. (c) Hermaphroditic (mmF-) cucumber plants produce only bisexual flowers. (d) Andromonoecious ($mmff$) cucumber plants produce bisexual and male flowers. If the genotype is F-, all floral primordia produce sufficient ethylene to induce femaleness. Otherwise, if the genotype is ff, not all floral primordia produce enough ethylene to induce femaleness. The product of the M gene mediates the inhibition of stamen development by ethylene. If the genotype is mm, then the ethylene signal is not transmitted, and stamen development is not inhibited.

(hermaphrodite and andromonoecious) always produce bisexual flowers and never female flowers. Based on our genetic model, all cucumber sex phenotypes can be accounted for as shown in Fig. 7B. In gynoecious cucumber plants, due to F-, it is likely that all floral primordia produce enough ethylene to induce femaleness, whereas stamen development is inhibited by ethylene due to M-. Thus, the induction of pistil development and the inhibition of stamen development occur in all floral primordia, leading to the formation of female flowers (Fig. 7Ba). In monoecious cucumber plants because of the ff genotype, not all floral primordia produce sufficient ethylene to induce femaleness, whereas stamen development is inhibited by ethylene, reflecting the *M-* genotype. Thus, in the subpopulation of floral primordia that produce enough ethylene for the induction of femaleness, pistil development and inhibition of the stamen development prompt the formation of female flowers. In the other floral primordia that do not produce enough ethylene to induce femaleness, neither induction of pistil development nor inhibition of the stamen development occurs, resulting in the appearance of male flowers. Consequently, monoecious cucumber plants produce both male and female flowers (Fig. 7Bb). In hermaphroditic cucumber plants, stamen development may have a reduced response to ethylene because of *mm*, which in turn leads to the constitutive development of stamens in all floral primordia. Because of their *F-* genotype, these plants may also produce sufficient ethylene to induce the development of pistils; thus, hermaphroditic cucumber plants produce many bisexual flowers (Fig. 7Bc). In andromonoecious cucumber plants, stamen development in all floral primordia exhibits a reduced response to ethylene because of the *mm* status, and this leads to constitutive stamen development. Moreover, because of the *ff* genotype, not all floral primordia produce enough ethylene to induce the development of pistils, and in those that do produce sufficient ethylene, induction of pistil development and constitutive stamen development lead to the formation of bisexual flowers. In the remaining floral primordia, inhibition of pistil development and constitutive stamen development lead to the formation of male flowers. Consequently, andromonoecious cucumber plants produce both bisexual and male flowers (Fig. 7Bd). Gynoecious and hermaphroditic plants sometimes produce male flowers at the lower node. In this case, it is thought that floral primordia do not produce enough ethylene to induce femaleness. Therefore, gynoecious plants bear male flowers via mechanisms similar to monoecious plants (Fig. 7Bb), while hermaphroditic plants bear male flowers as described for andromonoecious plants (Fig. 7Bd). Thus, our genetic model can reliably account for the sex phenotypes of gynoecious, monoecious, hermaphroditic, and andromonoecious cucumber plants.

VII. ENVIRONMENTAL REGULATION OF SEX EXPRESSION

In some monoecious plants, sex expression may be affected by environmental factors such as day-length and temperature. In maize, feminization of tassels can be induced by short-day and cool-night conditions (Heslop-Harrison, 1961). Rood *et al.* (1980) reported that the endogenous levels of gibberellins are 100-fold greater in the feminized tassels than in normal ones. Therefore, it is possible that short-day and cool-night conditions induce higher levels of gibberellins in tassels, thereby feminizing maize tassels. Also, in cucumber plants, sex expression may be influenced by environmental factors (Cantliffe, 1981; Fukushima *et al.*, 1968; Galun, 1961; Ito and Saito, 1957; Takahashi *et al.*, 1982). In general, short-day and cool-night conditions induce feminization in monoecious cucumber plants, whereas long-day and warm-night conditions promote maleness (Atsmon and Galun, 1962; Cantliffe, 1981; Fukushima *et al.*, 1968; Ito and Saito, 1957; Saito and Ito, 1964; Takahashi *et al.*, 1982). Although the mechanism of feminization under short-day conditions has not been fully elucidated, it is postulated that internal substances that induce feminization increase when leaves receive short-day signals. We have previously analyzed ethylene evolution and the expression of ACC synthase (*CS-ACS2* and *CS-ACS4*) genes and the ethylene receptor (*CS-ERS*) gene under short-day and long-day conditions in monoecious cucumber plants (Yamasaki *et al.*, 2003a). Ethylene evolution and expression of *CS-ACS2* and *CS-ACS4* were higher under short-day conditions than under long-day conditions, and exogenous application of ethylene increased the expression of *CS-ACS2* and *CS-ERS* in monoecious cucumber plants. Therefore, feminization of cucumber flowers under short-day conditions might be mediated by ethylene production through increased accumulation of *CS-ACS2*, *CS-ACS4*, and *CS-ERS* mRNA (Yamasaki *et al.*, 2003a). Thus, it is possible that ethylene is an internal factor responsible for feminization due to environmental cues in cucumber plants.

VIII. IDENTITY OF FLORAL ORGANS AND REGULATION OF SEX EXPRESSION

Genetic analysis of homeotic mutants in *A. thaliana* and *Antirrhinum majus* has yielded the ABC model that is the basic theorem for bisexual floral structure (Bowman *et al.*, 1991a,b; Coen and Meyerowitz, 1991; Meyerowitz *et al.*, 1991; Fig. 8A). The basic structure of well-studied bisexual flowers consists of four concentric whorls. From the outside inward, these are the sepals (whorl 1), petals (whorl 2), stamens (whorl 3), and finally the pistils (whorl 4). Genetic and molecular studies on the development of these flowers have shown that three distinct classes of floral homeotic genes control the identification of floral organs (Bowman *et al.*, 1991a; Carpenter and Coen, 1990; Coen and Meyerowitz, 1991; Schwarz-Sommer *et al.*, 1990); in whorl 1, class A genes control the development of sepals, whereas in whorl 2, the action of both class A and B genes leads to the formation of petals (Fig. 8A). In whorl 3, the combinatorial action of class B and C homeotic genes defines the formation of stamens, whereas only the class C genes define the identity of pistils in whorl 4 (Fig. 8A). Most of the class A, B, and C genes encode transcription factors belonging to the MADS-box family (Shore and Sharrocks, 1995; Theißen and Saedler, 1995). Since class A, B, and C genes have now been identified from a wide variety of plant species

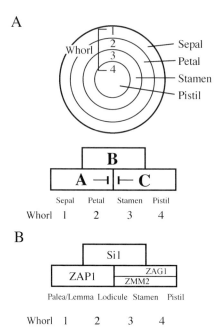

FIGURE 8. (A) Schematic diagram of the structure of bisexual flowers, four whorls (1, 2, 3, and 4), and three regions (A, B, and C) of homeotic gene action. In the whorl 1, action of the class A gene leads to the formation of sepals. In the whorl 2, action of the class A and B genes leads to the formation of petals. In the whorl 3, action of the class B and C genes leads to the formation of stamens. In the whorl 4, action of the class C gene leads to the formation of pistils. The function of class A and C genes are mutually antagonistic. (Modified from Coen and Meyerowitz [1991]; © Nature Publishing Group.) (B) The ABC model that is applied to monocot maize. C function is thought to be orchestrated by two genes, *Zag1* and *Zmm2*, both of which have partially redundant activities and distinct patterns of expression. B function is dictated by *Si1* and likely by *Zpi* which is the putative maize ortholog of *PI*. In maize, a combination of B and C function specifies stamens, a combination of B and A function (presumably dictated by *Zap1*) specifies lodicules, and C function specifies pistil development and determinacy of the floral meristem. Although no functional data yet exists, a presumed A function alone (ZAP1) is shown responsible for the formation of palea and lemmas. (Modified from Ambrose *et al.* [2000]; © Cell Press.)

including tomato, tobacco, and petunia, the generality of this ABC model has been established.

In monocot maize, several genes that specify floral organ identity have been cloned and analyzed (Ambrose *et al.*, 2000; Mena *et al.*, 1995, 1996; Schmidt *et al.*, 1993). As in dicot plants, maize flowers have stamens and pistils but lack obvious sepals and petals. Instead, organs specific to grass flowers, such as a pair of glumes, a lemma, a palea, and two lodicules, surround the stamens or pistils (Fig. 1). Analyses of the ABC genes of floral organs in maize indicate an evolutionary relationship between dicot and monocot floral organs as well as a relationship between floral organ identity and sex differentiation. Based on sequence similarity, the *Zea APETALA1* (*Zap1*) gene belongs to class A (Mena *et al.*, 1995); *silky1* (*si1*) belongs to class B (Ambrose *et al.*, 2000); and *Zea AGAMOUS 1* (*ZAG1*) and *Zea mays MADS2* (*ZMM2*) genes belong to class C (Schmidt *et al.*, 1993; Theißen and Saedler, 1995; Fig. 8B). In a *zag1* mutant, normal pollen-producing stamens and pistils are produced, and thus the identity of reproductive organs is largely unaffected (Mena *et al.*, 1996). This suggests a degree of redundancy in maize sex organ specification. Therefore, it is proposed that the class C gene functions in maize are orchestrated by two genes, *ZAG1* and *ZMM2* that have partially redundant activities and distinct patterns of expression (Mena *et al.*, 1996; Fig. 8B). Thus, it has been demonstrated that the ABC model is basically applicable to monocot and dicot plants.

The maize *si1* mutant exhibits male sterility with homeotic conversions of stamens into pistils and lodicules into palea/lemma-like structures (Fraser, 1933; Fig. 9). Also, the *si1zag1* double mutant has a phenotype in which normal glumes enclose reiterated palea/lemma-like structures (Ambrose *et al.*, 2000; Fig. 9). Although spikelets in the tassels and ears basically show a similar phenotype in the *si1zag1* double mutant, some older spikelets bear a fertile pistil (Ambrose *et al.*, 2000; Fig. 9). These results suggest that class B (*si1*) gene activity is conserved among monocot and dicot plants (Ambrose *et al.*, 2000; Fig. 8B). In addition, this study provides possible developmental evidence that lodicules are modified petals, whereas palea and lemma are modified sepals (Ambrose *et al.*, 2000; Fig. 8B). In the *si1* mutant, ears have a pistil in the centre and three additional pistil-like structures in the position

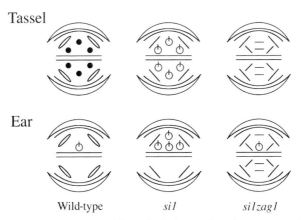

FIGURE 9. Schematic diagrams of florets in the homeotic maize mutants. *si1*, *silky1*; *zag1*, *Zea agamous1*. Symbols for floral parts are the same as those identified in Fig. 1. Lines (-) denote the palea/lemma-like organs in the primary and secondary florets of the *si1* and *si1zag1* mutants.

normally occupied by arrested stamens (Ambrose *et al.*, 2000; Fig. 9). This suggests that the arrest of stamen primordia in the ear depends on organ identity. In contrast, in the tassels, the *si1* mutation converts stamens to pistil-like structures, whereas it does not affect the arrest of pistil primordia in the centre, which is similar to wild-type plants (Ambrose *et al.*, 2000; Fig. 9). This indicates that arrest of pistil primordia in the tassels depends on whorl position rather than organ identity.

It was found that gibberellins promote the mRNA accumulation of floral homeotic genes, class B genes (*APETALA3* [*AP3*] and *PISTILLATA* [*PI*]) and class C gene (*AGAMOUS* [*AG*]), in *Arabidopsis* (Yu *et al.*, 2004). As described previously, gibberellins act to arrest the stamen primordia in the florets of the ears and play an important role in inducing femaleness in maize. Therefore, clarifying the effect of gibberellins on the mRNA accumulation of class B and C genes should help elucidate the relationship between the floral organ identity genes and sex expression.

Similar to maize, class B and C genes that define whorl 3 and 4 have been investigated in relation to sex differentiation in cucumber plants. Several MADS-box genes that have homology to class C genes, *AG* genes, have been isolated from cucumber plants (Filipecki *et al.*, 1997; Kater *et al.*, 1998; Perl-Treves *et al.*, 1998). Ectopic expression of *Cucumber MADS box gene 1* (*CUM1*) induced severe homeotic transformations of sepals into pistil-like structures and petals into stamens in transgenic petunia, which is similar to the effects of ectopic *AG* expression in *Arabidopsis* (Kater *et al.*, 1998). Over-expression of the other cucumber *AG* homolog, *CUM10*, resulted in petunia plants with partial transformations of the petals into anther-like structures, indicating that *CUM10* can also promote floral organ identity (Kater *et al.*, 1998).

Kater *et al.* (2001) produced cucumber mutants, similar to homeotic mutants of class A and C, by ectopically expressing or suppressing the *CUM1* gene and analyzed these mutants together with B function cucumber mutant, *green petals* (*gp*) (Fig. 10A and B). It is known that 15 amino acid residues are deleted in the coding region of the class B gene *CUM26* in the *gp* mutant (Kater *et al.*, 2001; Fig. 10A). The predicted transformations of whorls in class A, B, and C mutants are depicted in Fig. 10A, while the actual results of the conversion of the whorls are shown in Fig. 10B. In mutants ectopically expressing the *CUM1* gene (similar to class A mutant; Fig. 10A), sepals and petals were converted to pistils and stamens, respectively, in both male and female flowers (Fig. 10B). No organs were formed at whorl 4 in male flowers and at whorl 3 in female flowers (Fig. 10B). In a *gp* mutant (class B mutant; Fig. 10A), petals were converted to sepals in both male and female flowers (Fig. 10B). Also, stamens were converted to pistils, but pistils were not formed at whorl 4 in male flowers (Fig. 10B). In the female flowers, pistils were not formed at whorl 3 (Fig. 10B). In mutants in which the *CUM1* gene was suppressed (similar to class C mutant; Fig. 10A),

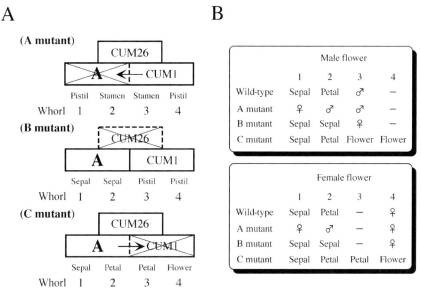

FIGURE 10. (A) Schematic representation of the homeotic mutant of the ABC model applied to cucumber plants. Ectopic expression of a class C gene, *CUM1*, leads to suppression of the class A function, resulting in a phenotype similar to that of a class A mutant. In this mutant, whorls 3 and 4 are normal but organs in whorls 1 and 2 are converted to pistils and stamens, respectively. In the cucumber B function mutant, *gp*, whorls 1 and 4 are normal but organs in whorls 2 and 3 are converted to sepals and pistils, respectively. Fifteen amino acid residues are deleted in the coding region of the class B gene *CUM26* in the *gp* mutant. C function mutants are produced by the suppression of the class C gene, *CUM1*. In the C function mutants, whorls 1 and 2 are normal, but organs in whorl 3 are converted to petals. Because class C genes are also essential in specifying meristem determinacy in the innermost whorl, indeterminate flowers are produced in whorl 4. (B) Outline of organ identities in the four floral whorls of male and female flowers of wild type and class A, B, and C cucumber plant mutants. "Flower" means indeterminate flowers. (Modified from Kater *et al.* [2001]; © American Society of Plant Biologists.)

indeterminate organs were produced at whorl 3 and 4 in male flowers and at whorl 4 in female flowers (Fig. 10B). Also, petals were formed at whorl 3 in female flowers (Fig. 10B). Studying these mutants prompted important concepts on the regulation of sex determination in cucumber plants. Namely, the arrest of either stamen or pistil development depends on their positions in the flower and is not associated with their sexual identity. Thus, the pathway that leads to the arrest of reproductive organs, sex differentiation, is independent of the pathway that defines the identity of the floral organs in cucumber plants. This idea is also supported by reports demonstrating that the expression of the *AG*-like genes, *CAG1*, *CAG2*, and *CAG3*, is not affected by the application of either ethephon, an ethylene-releasing agent, or gibberellins in cucumber plants (Perl-Treves *et al.*, 1998).

IX. PROGRAMMED CELL DEATH AND SEX DETERMINATION

Programmed cell death (PCD) is integral to development in both animals and plants. In animal development, PCD has been extensively studied and characterized at the cytological, biochemical, and molecular levels (Vaux and Korsmeyer, 1999). PCD involves specific biochemical and cytological events such as activation of caspases, chromatin condensation, and DNA fragmentation. Plants also undergo PCD in several situations such as hypersensitive defense responses against pathogens, the formation of xylem, and senescence (Vaux and Korsmeyer, 1999). In many vertebrate species, although sexuality between males and females is not distinguishable in early development, sexual dimorphism is mediated by cell death (Ellis *et al.*, 1991). In higher plants, cell death also occurs during the process of sexual differentiation, where it causes the selective abortion of either stamen or pistil primordia in some unisexual plants. It is also prevalent during the development of the male and female reproductive organs and gametes (Wu and Cheung, 2000).

In maize plants, sex differentiation involves abortion of pistil primordia in the tassel florets and arrest of stamen primordia in the ear florets. In both cases, cellular vacuolization and the loss of organelle and cytoplasmic integrity precede the abortion of pistil or stamen primordia (Cheng *et al.*, 1983). In addition, the tassels contain florets at various developmental stages, and the arrest of the pistil primordia in tassel florets is accompanied by nuclear degeneration (Calderon-Urrea and Dellaporta, 1999). This degeneration begins in subepidermal cells of the pistil primordia (Calderon-Urrea and Dellaporta, 1999), the same region in which *TS2* gene is highly expressed. Therefore, the localization patterns of *TS2* expression and nuclear degeneration are consistent with *TS2* playing a functional role in the arrest of pistil primordia in tassel florets and in the secondary florets of the ears (Calderon-Urrea and Dellaporta, 1999). Thus, the *TS2* gene is thought to induce PCD in the pistil primordia of the tassel florets.

In cucumber plants, morphological and cytological studies have revealed that the process of sex differentiation involves selective abortion of sexual organs (Bai *et al.*, 2004; Hao *et al.*, 2003). In the female flowers of monoecious cucumber, arrest of stamen development occurs mainly at the anther primordia and involves DNA damage, as evidenced by TdT-mediated dUTP nick-end labeling (TUNEL) assay and chromatin condensation (Hao *et al.*, 2003). Because the DNA damage and the chromatin condensation do not lead to cellular degeneration, there might exist a unique mechanism for the arrest of sexual organs, which is specific to cucumber plants. In the male flowers of cucumber, no TUNEL signal was detected in pistil primordial cells (Bai *et al.*, 2004). Therefore, Bai *et al.* (2004) concluded that PCD might not be involved in the arrest of pistil primordia in the male flowers. The same authors also revealed that arrest occurs only in portions of the sexual organs and all of the arrested portions, in both pistil and stamen primordial, are spore-bearing parts in cucumber.

The selective arrest of sexual organs is an outwardly visible event that mainly occurs downstream of the sex-differentiation process in maize and cucumber. In maize plants, it is thought that gibberellins induce the arrest of stamen primordia in the florets of the ears, but the precise mechanism of hormonal regulation is still unclear. In cucumber plants, Hao et al. (2003) reported that DNA damage that is involved in the arrest of stamen primordia in female flowers might be caused by anther-specific DNase activation. They also proposed that the anther-specific DNase might be activated by genetic, hormonal, or environmental signals. In this regard, *F* gene, ethylene, or short-day conditions are prime candidate signals. Thus, further investigation of the specific events involved in PCD and their association with genetic, hormonal, or environmental signals may lead to the identification of either universal or species-specific cell death mechanisms that determine the sexuality of maize and cucumber plants.

X. CONCLUSIONS

It is known that sexual differentiation of higher plants is controlled by sex chromosomes, genetic factors, plant hormones, and/or environmental factors (Dellaporta and Calderon-Urrea, 1993; Tanurdzic and Banks, 2004). Because higher plants ubiquitously have common plant hormones, it is predicted that the major processes underlying the biosynthesis and actions of plant hormones are reasonably conserved across higher plants. However, plant hormones specifically affect sex expression in some monoecious plants such as maize and cucumber. It is obvious that maize and cucumber have developed particular and different mechanisms for the hormonal regulation of sex expression. In maize plants, gibberellins act to arrest the stamen primordia in the primary and secondary florets of the ears, and low levels of gibberellins do not cause the arrest of pistil primordia in the primary florets of the ears. In principle, gibberellins do not affect the development of pistil primordial as evidenced by the analysis of *dwarf* (*d*) mutants (Fig. 2). On the contrary, in cucumber plants, ethylene acts on both the development of pistil primordia and the arrest of stamen primordia, which results in the induction of femaleness. Therefore, ethylene has opposing effects on the development of sexual organs, stamens and pistils. With regard to the hormonal regulation of sex expression, it is interesting that both plant hormones, gibberellins and ethylene, cause the arrest of stamen primordia in maize and cucumber, respectively. This indicates that ethylene- and gibberellins-signaling pathways mediate the arrest (by PCD) of stamen primordia in maize and cucumber. To further clarify the hormonal regulation of sex expression in maize and cucumber, therefore, the relationships between plant hormone signaling pathways and PCD should be analyzed.

In this review, we also presented models of sex expression in maize and cucumber (Figs. 4 and 7). These models have been constructed on the basis of analyses of mutants defective in sex expression in maize and cultivars with various sex phenotypes in cucumber. To dissect the molecular mechanism of sex expression in maize and cucumber, this is probably the best approach for identifying those factors responsible for plant hormone responses and ultimately for understanding the relationships between these factors. For example, it will be important to identify the *TS1* gene, *SK1* gene, and substrates of TS2 and to understand the role of gibberellins in arresting stamen primordia in maize. Also, identification of the *F*, *M*, and *A* genes and elucidation of the relationships between ethylene, the *M* gene, and the arrest of stamen primordia are a priority in cucumber plants. Further investigation of species-specific and common events in the sex expression of maize and cucumber may facilitate the understanding of the evolutionary aspects of sex differentiation in higher plants.

ACKNOWLEDGMENTS

This work was supported in part by a Grant-in-Aid (09460118) from the Ministry of Education, Culture, Sports, Science, and Technology of Japan and the "Ground Research Announcement for Space Utilization" promoted by NASDA and the Japan Space Forum to H.T. and the Research Fellowships of the Japan Society for the Promotion of Science for Young Scientists to S.Y.

REFERENCES

Abel, S., Nguyen, M. D., Chow, W., and Theologis, A. (1995). *ACS4*, a primary indoleacetic acid-responsive gene encoding 1-aminocyclopropane-1-carboxylate synthase in *Arabidopsis thaliana*. *J. Biol. Chem.* **270**, 19093–19099.

Ainsworth, C., Parker, J., and Buchanan-Wollaston, V. (1998). Sex determination in plants. *Curr. Top. Dev. Biol.* **38**, 167–223.

Ambrose, B. A., Lerner, D. R., Ciceri, P., Padilla, C. M., Yanofsky, M. F., and Schmidt, R. J. (2000). Molecular and genetic analysis of the *Silky1* gene reversal conservation in floral organ specification between eudicots and monocots. *Mol. Cell* **5**, 569–579.

Ando, S., and Sakai, S. (2002). Isolation of an ethylene-responsive gene (*ERAF16*) for a putative methyltransferase and correlation of *ERAF16* gene expression with female flower formation in cucumber plants (*Cucumis sativus*). *Physiol. Plant.* **116**, 213–222.

Ando, S., Sato, Y., Kamachi, S., and Sakai, S. (2001). Isolation of a MADS-box gene (*ERAF17*) and correlation of its expression with the induction of formation of female flowers by ethylene in cucumber plants (*Cucumis sativus* L.). *Planta* **213**, 943–952.

Arteca, J. M., and Arteca, R. N. (1999). A multi-responsive gene encoding 1-aminocyclopropane-1-carboxylate synthase (*ACS6*) in mature *Arabidopsis* leaves. *Plant Mol. Biol.* **39**, 209–219.

Atsmon, D., and Galun, E. (1960). A morphogenetic study of staminate, pistillate and hermaphrodite flowers in *Cucumis sativus* L. *Phytomorphology* **10**, 110–115.

Atsmon, D., and Galun, E. (1962). Physiology of sex in *Cucumis sativus* L. Leaf age patterns and sexual differentiation of floral buds. *Ann. Bot.* **26**, 137–146.

Atsmon, D., Lang, A., and Light, E. N. (1968). Contents and recovery of gibberellins in monoecious and gynoecious cucumber plants. *Plant Physiol.* **43**, 806–810.

Atsmon, D., and Tabbak, C. (1979). Comparative effects of gibberellin, silver nitrate, and aminoethoxyvinyl glycine on sexual tendency and ethylene evolution in the cucumber plant (*Cucumis sativus* L.). *Plant Cell Physiol.* **20**, 1547–1555.

Bai, S. L., Peng, Y. B., Cui, J. X., Gu, H. T., Xu, L. Y., Li, Y. Q., Xu, Z. H., and Bai, S. N. (2004). Developmental analysis reveal early arrests of the spore-bearing parts of reproductive organs in unisexual flowers of cucumber (*Cucumis sativus* L.). *Planta* **220**, 230–240.

Balague, C., Watson, C. F., Turner, A. J., Rouge, P., Picton, S., Pech, J. C., and Grierson, D. (1993). Isolation of a ripening and wound induced cDNA from *Cucumis melo* L. encoding a protein with homology to the ethylene-forming enzyme. *Eur. J. Biochem.* **212**, 27–34.

Bensen, R. J., Johal, G. S., Crane, V. C., Tossberg, J. T., Schnable, P. S., Meeley, R. B., and Briggs, S. P. (1995). Cloning and characterisation of the maize *An1* gene. *Plant Cell* **7**, 75–84.

Bonnett, O. T. (1940). Development of the staminate and pistillate inflorescences of sweet corn. *J. Agric. Res.* **60**, 25–37.

Bonnett, O. T. (1948). Ear and tassel development in maize. *Ann. Mo. Bot. Gard.* **35**, 269–287.

Botella, J. R., Arteca, J. M., Schlagnhaufer, C. D., Arteca, R. N., and Phillips, A. T. (1992). Identification and characterization of a full length cDNA encoding for an auxin-induced 1-aminocyclopropane-1-carboxylate synthase from etiolated mung bean hypocotyl segments and expression of its mRNA in response to indole-3-acetic acid. *Plant Mol. Biol.* **20**, 425–436.

Bowman, J. L., Drews, G. N., and Meyerowitz, E. M. (1991a). Expression of the *Arabidopsis* floral homeotic gene *AGAMOUS* is restricted to specific cell types late in flower development. *Plant Cell* **3**, 749–758.

Bowman, J. L., Smyth, D. R., and Meyerowitz, E. M. (1991b). Genetic interactions among floral homeotic genes of *Arabidopsis*. *Development* **112**, 1–20.

Calderon-Urrea, A., and Dellaporta, S. L. (1999). Cell death and cell protection genes determine the fate of pistils in maize. *Development* **126**, 435–441.

Cantliffe, D. J. (1981). Alteration of sex expression in cucumber due to changes in temperature, light intensity, and photoperiod. *J. Am. Soc. Hortic. Sci.* **106**, 133–136.

Carpenter, R., and Coen, E. S. (1990). Floral homeotic mutations produced by transposon-mutagenesis in *Antirrhinum majus*. *Genes Dev.* **4**, 1483–1493.

Chailakhan, M. K. (1979). Genetic and hormonal regulation of growth, flowering and sex expression in plants. *Am. J. Bot.* **66**, 717–736.

Chang, C., Kwok, S. F., Bleecker, A. B., and Meyerowitz, E. M. (1993). *Arabidopsis* ethylene response gene ETR1-similarity of product to two-component regulators. *Science* **262**, 539–544.

Cheng, P. C., Greyson, R. I., and Walden, D. B. (1983). Organ initiation and the development of unisexual flowers in the tassel and ear of *Zea mays*. *Am. J. Bot.* **70**, 450–462.

Coen, E. S., and Meyerowitz, E. M. (1991). The war of the whorls genetic interactions controlling flower development. *Nature* **353**, 31–37.

Dellaporta, S. L., and Calderon-Urrea, A. (1993). Sex determination in flowering plants. *Plant Cell* **5**, 1241–1251.

Dellaporta, S. L., and Calderon-Urrea, A. (1994). The sex determination process in maize. *Science* **266**, 1501–1505.

De Long, A., Calderon-Urrea, A., and Dellaporta, S. L. (1993). Sex determination gene *TASSELSEED2* of maize encodes a short-chain alcohol dehydrogenase required for stage-specific floral organ abortion. *Cell* **74**, 757–768.

Durand, R., and Durand, B. (1984). Sexual differentiation in higher plants. *Physiol. Plant.* **60**, 267–274.
Durand, R., and Durand, B. (1991). Sex determination and reproductive organ differentiation in *Mercuralis*. *Plant Sci.* **80**, 49–65.
Ellis, R. E., Yuan, J. Y., and Horvitz, H. R. (1991). Mechanisms and functions of cell death. *Annu. Rev. Cell Biol.* **7**, 663–698.
Emerson, R. A. (1932). The present status of maize genetics. *6th Int. Congr. Genet. Proc.* **1**, 141–152.
Filipecki, M. K., Sommer, H., and Malepszy, S. (1997). The MADS-box gene *CUS1* is expressed during cucumber somatic embryogenesis. *Plant Sci.* **125**, 63–74.
Frankel, R., and Galun, E. (1977). "Pollination Mechanisms, Reproduction, and Plant Breeding." Springer-Verlag, Heidelberg.
Fraser, A. C. (1933). Heritable characters of maize. XLIV-*silky ears*. *J. Hered.* **11**, 41–46.
Fujioka, S., Yamane, H., Spray, C. R., Gaskin, P., Mac Millan, J., Phinney, B. O., and Takahashi, N. (1988). Qualitative and quantitative analysis of gibberellins in vegetative shoots of normal, *dwarf-1*, *dwarf-2*, *dwarf-3*, and *dwarf-5* seedlings of *Zea mays* L. *Plant Physiol.* **88**, 1367–1372.
Fukushima, E., Matsuo, E., and Fujieda, K. (1968). Studies on the growth behaviour of cucumber, *Cucumis sativus* L. I. The types of sex expression and its sensitivity to various day length and temperature conditions. *J. Fac. Agric. Kyushu Univ.* **14**, 349–366.
Galun, E. (1959b). The role of auxins in the sex expression of the cucumber. *Physiol. Plant.* **12**, 48–61.
Galun, E. (1961). Study of the inheritance of sex expression in the cucumber: The interaction of major genes with modifying genetic and non-genetic factors. *Genetica* **32**, 134–163.
George, W. L. (1971). Influence of genetic background on sex conversion by 2-chloroethylphosphonic acid in monoecious cucumbers. *J. Am. Soc. Hortic. Sci.* **96**, 152–154.
Hao, Y. J., Wang, D. H., Peng, Y. B., Bai, S. L., Xu, L. Y., Li, Y. Q., Xu, Z. H., and Bai, S. N. (2003). DNA damage in the early primordial anther is closely correlated with stamen arrest in the female flower of cucumber (*Cucumis sativus* L.). *Planta* **217**, 888–895.
Heslop-Harrison, J. (1961). The experimental control of sexuality and inflorescence structure in *Zea mays* L. *Proc. Linn. Soc. Lond.* **172**, 108–123.
Holdsworth, M. J., Schuch, W., and Grieson, D. (1987). Nucleotide sequence of an ethylene-related gene from tomato. *Nucl. Acids Res.* **15**, 10600.
Hua, J., Chang, C., Sun, Q., and Meyerowitz, E. M. (1995). Ethylene insensitivity conferred by *Arabidopsis ERS* gene. *Science* **269**, 1712–1714.
Hua, J., Sakai, H., Nourizadeh, S., Chen, Q. G., Bleecker, A. B., Ecker, J. R., and Meyerowitz, E. M. (1998). *EIN4* and *ERS2* are members of the putative ethylene receptor gene family in *Arabidopsis*. *Plant Cell* **10**, 1321–1332.
Irish, E. E. (1996). Regulation of sex determination in maize. *Bioessays* **18**, 363–369.
Irish, E. E., Langdale, J. A., and Nelson, T. M. (1994). Interactions between *Tassel Seed* genes and other sex determining genes in maize. *Dev. Genet.* **15**, 155–171.
Ito, H., and Saito, T. (1957). Factors responsible for sex expression of Japanese cucumber. VI. Effects of the day length and night temperature, unsuitable for the pistillate flower formation, artificially controlled during the various stages of seedling development in the nursery bed. *J. Jpn. Soc. Hortic. Sci.* **26**, 1–8.
Iwahori, S., Lyons, J. M., and Smith, O. E. (1970). Sex expression in cucumber plants as affected by 2-chloroethylphosphonic acid, ethylene, and growth regulators. *Plant Physiol.* **46**, 412–415.
Jones, D. F. (1925). Heritable characters in maize. XXIII. Silkless. *J. Hered.* **16**, 339–341.
Jones, D. F. (1934). Unisexual maize plants and their bearing on sex differentiation in other plants and animals. *Genetics* **19**, 552–567.

Kahana, A., Silberstein, L., Kessler, N., Goldstein, R. S., and Perl-Treves, R. (1999). Expression of ACC oxidase genes differs among sex genotypes and sex phases in cucumber. *Plant Mol. Biol.* **41**, 517–528.

Kamachi, S., Mizusawa, H., Matsuura, S., and Sakai, S. (2000). Expression of two 1-aminocyclopropane-1-carboxylate synthase genes, *CS-ACS1* and *CS-ACS2*, correlated with sex phenotype in cucumber plants (*Cucumis sativus* L.). *Plant Biotechnol.* **17**, 69–74.

Kamachi, S., Sekimoto, H., Kondo, N., and Sakai, S. (1997). Cloning of a cDNA for a 1-aminocyclopropane-1-carboxylate synthase that is expressed during development of female flowers at the apices of *Cucumis sativus* L. *Plant Cell Physiol.* **38**, 1197–1206.

Kater, M. M., Colombo, L., Franken, J., Busscher, M., Masiero, S., Van Lookeren Campagne, M. M., and Angenent, G. C. (1998). Multiple *AGAMOUS* homologs from cucumber and petunia differ in their ability to induce reproductive organ fate. *Plant Cell* **10**, 171–182.

Kater, M. M., Franken, J., Carney, K. J., Colombo, L., and Angenent, G. C. (2001). Sex determination in the monoecious species cucumber is confined to specific floral whorls. *Plant Cell* **13**, 481–493.

Kellogg, E. A., and Birchler, J. A. (1993). Linking phylogeny and genetics: *Zea mays* as a tool for phylogenetic studies. *Syst. Biol.* **42**, 415–439.

Kock, M., Hamilton, A. J., and Grierson, D. (1991). *eth1*, a gene involved in ethylene synthesis in tomato. *Plant Mol. Biol.* **17**, 141–142.

Kubicki, B. (1969a). Investigations on sex determination in cucumber (*Cucumis sativus* L.). V. Genes controlling intensity of femaleness. *Genet. Pol.* **10**, 69–85.

Kubicki, B. (1969b). Investigations on sex determination in cucumber (*Cucumis sativus* L.). VI. Androecism. *Genet. Pol.* **10**, 87–98.

Kubicki, B. (1969c). Investigations on sex determination in cucumber (*Cucumis sativus* L.). VII. Andromonoecism and hermaphroditism. *Genet. Pol.* **10**, 101–120.

Kubicki, B. (1974). New sex types in cucumber and their uses in breeding work. *Proc. XIX Intl. Hortic. Congr.* **3**, 475–485.

Lasserre, E., Bouquin, F., Hernandez, J. A., Bull, J., Pech, J. C., and Balague, C. (1996). Structure and expression of three genes encoding ACC oxidase homologs from melon (*Cucumis melo*). *Mol. Gen. Genet.* **251**, 81–90.

Lebel-Hardenack, S., and Grant, S. R. (1997). Genetics of sex determination in flowering plants. *Trends Plant Sci.* **2**, 130–136.

Liang, X., Abel, S., Keller, J. A., Shen, N. F., and Theologis, A. (1992). The 1-aminocyclopropane-1-carboxylate synthase gene family of *Arabidopsis thaliana*. *Proc. Natl. Acad. Sci. USA* **89**, 11046–11050.

Louis, J. P. (1989). Genes for the regulation of sex-differentiation and male-fertility in *Mercurialis annua* L. *J. Hered.* **80**, 104–111.

Mac Murray, A. L., and Miller, C. M. (1968). Cucumber sex expression modified by 2-chloroethanephosphonic acid. *Science* **162**, 1397–1398.

Malepszy, S., and Niemirowicz-Szczytt, K. (1991). Sex determination in cucumber (*Cucumis sativus* L.) as a model system for molecular biology. *Plant Sci.* **80**, 39–47.

Mena, M., Ambrose, B. A., Meeley, R. B., Briggs, S. P., Yanofsky, M. F., and Schmidt, R. J. (1996). Diversification of C-function activity in maize flower development. *Science* **274**, 1537–1540.

Mena, M., Mandel, M. A., Lerner, D. R., Yanofsky, M. F., and Schmidt, R. J. (1995). A characterization of the MADS box gene family in maize. *Plant J.* **8**, 845–854.

Meyerowitz, E. M., Bowman, J. L., Brockman, L. L., Drews, G. N., Jack, T., Sieburth, L. E., and Weigel, D. (1991). A genetic and molecular model for flower development in *Arabidopsis thaliana*. *Dev.* **1**(Suppl.), 157–167.

Nakagawa, N., Mori, H., Yamazaki, K., and Imaseki, H. (1991). Cloning of a complementary DNA for auxin-induced 1-aminocyclopropane-1-carboxylate synthase and differential expression of the gene by auxin and wounding. *Plant Cell Physiol.* **32**, 1153–1163.

Olson, D. C., Oetiker, J. H., and Yang, S. F. (1995). Analysis of *LE-ACS3*, a 1-aminocyclopropane-1-carboxylic acid synthase gene expressed during flooding in the roots of tomato plants. *J. Biol. Chem.* **270**, 14056–14061.

Olson, D. C., White, J. A., Edelman, L., Harkins, R. N., and Kende, H. (1991). Differential expression of two genes for 1-aminocyclopropane-1-carboxylate synthase in tomato fruits. *Proc. Natl. Acad. Sci. USA* **88**, 5340–5344.

Peng, J., Richards, D. E., Hartley, N. M., Murphy, G. P., Devos, K. M., Flintham, J. E., Beales, J., Fish, L. J., Worland, A. J., Pelica, F., Sudhakar, D., Christou, P., Snape, J. W., Gale, M. D., and Harberd, N. P. (1999). 'Green revolution' genes encode mutant gibberellin response modulators. *Nature* **400**, 256–261.

Perl-Treves, R., Kahana, A., Rosenman, N., Xiang, Y., and Silberstein, L. (1998). Expression of multiple *AGAMOUS*-like genes in male and female flowers of cucumber (*Cucumis sativus* L.). *Plant Cell Physiol.* **39**, 701–710.

Peterson, C. E., and Anhder, L. D. (1960). Induction of staminate flowers on gynoecious cucumbers with gibberellin A_3. *Science* **131**, 1673–1674.

Pierce, L. K., and Wehner, T. C. (1990). Review of genes and linkage groups in cucumber. *Hortic. Sci.* **25**, 605–615.

Richards, D. E., King, K. E., Ait-ali, T., and Harberd, N. P. (2001). How gibberellin regulates plant growth and development: A molecular genetic analysis of gibberellin signaling. *Annu. Rev. Plant Physiol. Plant Mol. Biol.* **52**, 67–88.

Rood, S. B., Paris, R. P., and Major, D. J. (1980). Changes of endogenous gibberellin-like substances with sex reversal of the apical inflorescence of corn. *Plant Physiol.* **66**, 793–796.

Rottmann, W. H., Peter, G. F., Oeller, P. W., Keller, J. A., Shen, N. F., Nagy, B. P., Taylor, L. P., and Theologis, A. (1991). 1-Aminocyclopropane-1-carboxylate synthase in tomato is encoded by a multigene family whose transcription is induced during fruit and floral senescence. *J. Mol. Biol.* **222**, 937–961.

Rudich, J., Halevy, A. H., and Kedar, N. (1969). Increase of femaleness of three cucurbits by treatment with ethrel, an ethylene-releasing compound. *Planta* **86**, 69–76.

Rudich, J., Halevy, A. H., and Kedar, N. (1972). Ethylene evolution from cucumber plants as related to sex expression. *Plant Physiol.* **49**, 998–999.

Saito, T., and Ito, H. (1964). Factors responsible for the sex expression of the cucumber plant. XIV. Auxin and gibberellin content in the stem apex and the sex pattern of flowers. *Tohoku J. Agric. Res.* **14**, 227–239.

Saito, T., and Takahashi, H. (1987). Role of leaves in ethylene-induced femaleness of cucumber plants. *J. Jpn. Soc. Hortic. Sci.* **55**, 445–454.

Sakai, H., Hua, J., Chen, Q. G., Chang, C., Medrano, L. J., Bleecker, A. B., and Meyerowitz, E. M. (1998). *ETR2* is an *ETR1*-like gene involved in ethylene signaling in *Arabidopsis*. *Proc. Natl. Acad. Sci. USA* **95**, 5812–5817.

Schmidt, R. J., Veit, B., Mandel, M. A., Mena, M., Hake, S., and Yanofsky, M. F. (1993). Identification and molecular characterization of *ZAG1*, the maize homolog of the *Arabidopsis* floral homeotic gene *AGAMOUS*. *Plant Cell* **5**, 729–737.

Schwarz-Sommer, Z., Huijser, P., Nacken, W., Saedler, H., and Sommer, H. (1990). Genetic control of flower development: Homeotic genes in *Antirrhinum majus*. *Science* **250**, 931–936.

Shifriss, O. (1961). Sex control in cucumbers. *J. Hered.* **52**, 5–12.

Shifriss, O., and George, W. L., Jr. (1964). Sensitivity of female inbreds of *Cucumis sativus* to sex reversion by gibberellin. *Science* **143**, 1452–1453.

Shore, P., and Sharrocks, A. D. (1995). The MADS box family of transcription factors. *Eur. J. Biochem.* **229**, 1–13.

Spanu, P., Reinhardt, D., and Boller, T. (1991). Analysis and cloning of the ethylene forming enzyme from tomato by functional expression of its mRNA in *Xenopus laevis* oocytes. *EMBO J.* **10**, 2007–2013.

Spray, C. R., Phynney, B. O., Gaskin, P., Gilmour, S. J., and Mac Millan, J. (1984). Internode length in *Zea mays* L. The *dwarf-1* mutation controls the 3ß-hydroxylation of gibberellin A_{20} to gibberellin A_{20}. *Planta* **160,** 464–468.

Spray, C. R., Kobayashi, M., Suzuki, Y., Phinney, B. O., Gaskin, P., and Mac Millan, J. (1996). The *dwarf-1* (*d1*) mutant of *Zea mays* blocks steps in the gibberellin-biosynthetic pathway. *Proc. Natl. Acad. Sci. USA* **93,** 10515–10518.

Takahashi, H., and Jaffe, M. J. (1984). Further studies of auxin and ACC induced feminization in the cucumber plant using ethylene inhibitors. *Phyton* **44,** 81–86.

Takahashi, H., Saito, T., and Suge, H. (1982). Intergeneric translocation of floral stimulus across a graft in monoecious Cucurbitaceae with special reference to the sex expression of flowers. *Plant Cell Physiol.* **23,** 1–9.

Takahashi, H., Saito, T., and Suge, H. (1983). Separation of the effects of photoperiod and hormones on sex expression in cucumber. *Plant Cell Physiol.* **24,** 147–154.

Takahashi, H., and Suge, H. (1980). Sex expression in cucumber plants as affected by mechanical stress. *Plant Cell Physiol.* **21,** 303–310.

Takahashi, H., and Suge, H. (1982). Sex expression and ethylene production in cucumber plants as affected by 1-aminocyclopropane-1-carboxylic acid. *J. Jpn. Soc. Hortic. Sci.* **51,** 51–55.

Tanurdzic, M., and Banks, J. A. (2004). Sex-determining mechanisms in land plants. *Plant Cell* **16**(Suppl. 1), 61–71.

Theißen, G., and Saedler, H. (1995). MADS-box genes in plant ontogeny and phylogeny: Haeckel's 'biogenetic law' revisited. *Curr. Opin. Genet. Dev.* **5,** 628–639.

Trebitsh, T., Riov, J., and Rudich, J. (1987). Auxin, biosynthesis of ethylene and sex expression in cucumber (*Cucumis sativus*). *Plant Growth Regul.* **5,** 105–113.

Trebitsh, T., Staub, J. E., and O' Neill, S. D. (1997). Identification of a 1-aminocyclopropane-1-carboxylic acid synthase gene linked to the female (*F*) locus that enhances female sex expression in cucumber. *Plant Physiol.* **113,** 987–995.

Vaux, D. L., and Korsmeyer, S. J. (1999). Cell death in development. *Cell* **96,** 245–254.

Veit, B., Schmidt, R. J., Hake, S., and Yanofsky, M. (1993). Maize floral development: New genes and old mutants. *Plant Cell* **5,** 1205–1215.

Winkler, R. G., and Helentjaris, T. (1995). The maize *Dwarf3* gene encodes a cytochrome P450-mediated early step in gibberellin biosynthesis. *Plant Cell* **7,** 1307–1317.

Wu, H. M., and Cheung, A. Y. (2000). Programmed cell death in plant reproduction. *Plant Mol. Biol.* **44,** 267–281.

Yamasaki, S., Fujii, N., Matsuura, S., Mizusawa, H., and Takahashi, H. (2001). The *M* locus and ethylene-controlled sex determination in andromonoecious cucumber plants. *Plant Cell Physiol.* **42,** 608–619.

Yamasaki, S., Fujii, N., and Takahashi, H. (2000). The ethylene-regulated expression of *CS-ETR2* and *CS-ERS* genes in cucumber plants and their possible involvement with sex expression of flowers. *Plant Cell Physiol.* **41,** 608–616.

Yamasaki, S., Fujii, N., and Takahashi, H. (2003a). Photoperiodic regulation of *CS-ACS2*, *CS-ACS4*, and *CS-ERS* gene expression contributes to the femaleness of cucumber flowers through diurnal ethylene production under short-day conditions. *Plant Cell Env.* **26,** 537–546.

Yamasaki, S., Fujii, N., and Takahashi, H. (2003b). Characterization of ethylene effects on sex determination in cucumber plants. *Sex. Plant Reprod.* **16,** 103–111.

Yampolsky, C., and Yampolsky, H. (1922). Distribution of the sex forms in the phanerogamic flora. *Bibl. Genet.* **3,** 1–62.

Yin, T., and Quinn, J. A. (1992). A mechanistic model of one hormone regulating both sexes in flowering plants. *Bull. Torrey Bot. Club* **119,** 431–441.

Yin, T., and Quinn, J. A. (1995). Tests of a mechanistic model of one hormone regulating both sexes in *Cucumis sativus* (Cucurbitaceae). *Am. J. Bot.* **82,** 1537–1546.

Young, T. E., Giesler-Lee, J., and Gallie, D. R. (2004). Senescence-induced expression of cytokinin reverses pistil abortion during maize flower development. *Plant J.* **38**, 910–922.

Yu, H., Ito, T., Zhao, Y., Peng, J., Kumar, P., and Meyerowitz, E. M. (2004). Floral homeotic genes are targets of gibberellin signaling in flower development. *Proc. Natl. Acad. Sci. USA* **101**, 7827–7832.

Zarembinski, T. I., and Theologis, A. (1993). Anaerobiosis and plant growth hormones induce two genes encoding 1-aminocyclopropane-1-carboxylate synthase in rice (*Oryza sativa* L.). *Mol. Biol. Cell* **4**, 363–373.

FURTHER READINGS

Atal, C. K. (1959). Sex reversal in hemp by application of gibberellin. *Curr. Sci.* **28**, 408–409.

Byers, R. E., Baker, L. R., Sell, H. M., Herner, R. C., and Dilley, D. R. (1972). Ethylene: A natural regulator of sex expression of *Cucumis melo* L. *Proc. Natl. Acad. Sci. USA* **69**, 717–720.

Dauphin-Guerin, B., Teller, G., and Durand, B. (1980). Different endogenous cytokinins between male and female *Mercurialis annua* L. *Planta* **148**, 124–129.

Galun, E. (1959a). Effect of gibberellic acid and napthaleneacetic acid in sex expression and some morphological characters in the cucumber plant. *Phyton* **13**, 1–8.

Halevy, A. H., and Rudich, J. (1967). Modification of sex expression in muskmelon by treatment with growth retardant b-995. *Physiol. Plant.* **20**, 1052–1058.

Hamdi, S., Teller, G., and Louis, J. P. (1987). Master regulatory genes, auxin levels, and sexual organogenesis in the dioecious plant *Mercurialis annua*. *Plant Physiol.* **85**, 393–399.

Heslop-Harrison, J. (1956). Auxin and sexuality in *Cannabis sativa*. *Physiol. Plant.* **4**, 588–597.

Hua, J., and Meyerowitz, E. M. (1998). Ethylene responses are negatively regulated by a receptor gene family in *Arabidopsis thaliana*. *Cell* **94**, 261–271.

Lazarte, J. E., and Garrison, A. (1980). Sex modification in *Asparagus officinalis* L. *J. Am. Hortic. Sci.* **105**, 691–694.

Mohan Ram, H. Y., and Jaiswal, V. S. (1970). Induction of female flowers on male plants of *Cannabis sativa* by 2-chloroethane phosphonic acid. *Experientia* **26**, 214–216.

Weston, E. W. (1960). Changes in sex in the hop caused by plant growth substances. *Nature* **188**, 81–82.

4

Plant Peroxisomes

Shoji Mano and Mikio Nishimura

Department of Cell Biology, National Institute for Basic Biology
Okazaki 444-8585, Japan
Department of Molecular Biomechanics, School of Life Science
The Graduate University for Advanced Studies
Okazaki 444-8585, Japan

I. Introduction
II. Discovery, Structure, and Localization in Cells
III. Peroxisome Functions in Higher Plants
    A. Lipid Metabolism
    B. Photorespiration
    C. Biosynthesis of Plant Hormones
    D. Detoxification of Hydrogen Peroxide
IV. Transition of Peroxisome Function
V. Peroxisome Biogenesis
    A. Peroxisomal Targeting Signal and Protein Import
    B. Peroxins
    C. Genetic Approaches Using Mutants
VI. Challenge of Comprehensive Approach in the Postgenomic Era
    A. Transcriptome
    B. Proteome
VII. Few Final Considerations and Perspectives
    References

Peroxisomes, one of single membrane-bound organelles, are present ubiquitously in eukaryotic cells. They were originally identified as organelles for production of hydrogen peroxide, the degradation of its hydrogen peroxide, and metabolism of fatty acids, which are functions common to almost all the organisms. Meanwhile, photorespiration and assimilation of symbiotically induced nitrogen are plant-specific functions. Recent postgenetic approaches such as transcriptome and proteome showed that plant peroxisomes are differentiated in various tissues, and revealed that peroxisomes have more important roles in various metabolic processes including biosynthesis of plant hormones than we speculated. All peroxisomal proteins, including metabolic enzymes in the matrix, membrane proteins, and factors responsible for peroxisome biogenesis, are nuclear encoded, and are provided from the outside of peroxisomes. Peroxisome biogenesis, such as protein transport, division, and enlargement, requires various complicated steps and is one of the most intriguing topics. Analyses using peroxisome biogenesis mutants and the whole-scale sequencing projects among several organisms revealed the existence of essential factors responsible for peroxisome biogenesis such as peroxins. This review addresses a comprehensive issue relating to function and biogenesis of plant peroxisomes and *Arabidopsis* mutants that have been accelerating our understanding of peroxisomes *in planta*. © 2005 Elsevier Inc.

I. INTRODUCTION

Peroxisomes are single membrane-bound organelles that are found ubiquitously in eukaryotic cells including mammalian, yeasts, and plant cells. Peroxisomes are involved in various metabolic reactions in cells, such as fatty acid β-oxidation, photorespiration, scavenging of hydrogen peroxide, and biosynthesis of plant hormones, jasmonic acid (JA), and auxin. However, the functions differ with the species, developmental stage, and organ, even in the same organism. Based on their functions, plant peroxisomes are subcategorized into three groups, glyoxysomes, leaf peroxisomes, and unspecialized peroxisomes. Interestingly, each peroxisome can transform directly to another type of peroxisome.

In addition to cDNA cloning and characterization of peroxisomal genes, genetic analyses using mutants with defects in peroxisome function of various organisms have led to understanding the importance of correct maintenance of peroxisomes within cells. For example, the functional consequence of human peroxisome biogenesis disorders causes diseases such as Zellweger syndrome, neonatal adrenoleukodystrophy, infantile Refsum disease, and rhizomelic chondrodysplasia in mammals. Peroxisome-deficient mutants of yeasts are specifically unable to grow on fatty acids, methanol, or oleate as a sole carbon source. In higher plants, *Arabidopsis* mutants with defects in peroxisomal function or peroxisome biogenesis are unable to grow normally such as being unable to germinate or being sterile.

In this review, we briefly summarize the discovery, structure, and function of plant peroxisome, then review our current knowledge about peroxisome biogenesis and how they are regulated, and finally introduce the results that make use of the complete sequences of the *Arabidopsis* genome.

II. DISCOVERY, STRUCTURE, AND LOCALIZATION IN CELLS

In the early 1960s, peroxisomes were discovered as distinctive organelles about 0.2–1.5 μm in diameter, bounded by a single membrane, and with an amorphous or granular interior (Fig. 1). These organelles were initially given the name "microbody" based on the observation of animal tissues (Rhodin, 1958). Therefore, some textbooks use the word "microbody." However, they are now mainly called peroxisomes based on one of their important functions, peroxidase activity.

In the early stage of discovery, peroxisomes were believed to be only the site of production of hydrogen peroxide, the degradation of its hydrogen peroxide by catalase, and the conversion of fatty acids to carbohydrates. At the present time, the peroxisome has emerged as an essential organelle with a variety of roles in metabolism. In this regard, plant researches have contributed to the progress in the identification of the peroxisomal function.

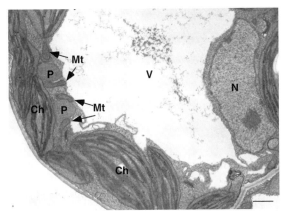

FIGURE 1. Electron microscopic observation of green cotyledon of *Arabidopsis*. P, peroxisome; Mt, mitochondrion; Ch, chloroplast; V, vacuole; N, nucleus.

In Beevers' laboratory, the mechanism for the conversion of fatty acids to sucrose has been clarified using castor bean endosperm (Beevers, 1961, 1979; Cooper and Beevers, 1969), and Tolbert (1971) discovered the glycolate pathway in photorespiration. In the field of medical science, the investigations of mammalian peroxisomes have provided essential information for clinical manifestation of genetic peroxisomal disorders.

Peroxisomes, unlike mitochondria and plastids, do not have their own genomes. Therefore, all proteins constituting peroxisomes are nuclear-encoded and transported to peroxisomes directly or indirectly. By a sucrose density gradient, peroxisomes are recovered at an equilibrium density of 1.22–1.25 g/cm^3. Interestingly, the shape, number, and size of peroxisomes vary with the species. In higher plants, they differ with the developmental stages or tissues even in the same plant (Mano *et al.*, 2002). Peroxisomes are motile organelles whose movement requires the interaction with microfilaments (Jedd and Chua, 2002; Mano *et al.*, 2002; Mathur *et al.*, 2002). In addition, it is clear that the position of peroxisomes and interaction with other organelles within cells are necessary for normal plant growth (Mano *et al.*, 2004).

III. PEROXISOME FUNCTIONS IN HIGHER PLANTS

Peroxisomes have a variety of functions, such as lipid metabolism, photorespiration, scavenging of hydrogen peroxide, biosynthesis of plant hormones, catabolism of branched chain amino acids, assimilation of symbiotically induced nitrogen, and so on. Among them, there are species-specific peroxisomal functions such as synthesis of cholesterol, bile acids, and plasmalogens in mammals; methanol oxidation in yeasts; or photorespiration, the biosynthesis of plant hormones, and the metabolism of ureides in plants. However, the scavenging of reactive oxygen species such as hydrogen peroxide is a function common to mammals, yeasts, and plants (Corpas *et al.*, 2001). As described in detail in Section VI, the existence of about 300 genes encoding peroxisomal proteins was anticipated based on the biological information from the sequencing project of the whole *Arabidopsis* genome (Kamada *et al.*, 2003). However, we have identified only a part of them, and information is lacking for most of them, suggesting that plant peroxisomes have a lot of unknown functions. In this first section, we focus first on lipid metabolism and photorespiration, both of which are important plant peroxisomal functions and have been characterized better than the other functions. Second, the function of peroxisomes in the biosynthesis of plant hormones, JA, and auxin, is reviewed. Finally, we describe the degradation of hydrogen peroxide, which is a function common to peroxisomes in all organisms.

A. LIPID METABOLISM

Oilseed plants, such as *Arabidopsis*, pumpkin, cucumber, and watermelon, reserve lipids as storage substrates. Immediately after germination, seedlings do not have the ability of photosynthesis to produce energy for growth. Instead, storage lipids are used for production of sucrose, which is called gluconeogenesis, until acquisition of ability of photosynthesis (Fig. 4). Storage lipids are stored as triacylglycerol in lipid bodies and are metabolized to fatty acids by the reaction of lipases. Fatty acids are released from lipid bodies and transported to peroxisomes where they are metabolized to succinate by fatty acid β-oxidation and the glyoxylate cycle in peroxisomes.

1. Fatty Acid β-Oxidation

Fatty acid β-oxidation, which releases two carbon units, is composed of several enzymes (Fig. 2A). Fatty acids are first activated to fatty acyl-CoA by fatty acyl-CoA synthetase. Fatty acyl-CoA is metabolized to acetyl-CoA, the end product, by four successive reactions. The first reaction catalyzed by acyl-CoA oxidase converts fatty acyl-CoA into *trans*-2-enoyl-CoA. The second and third reactions are catalyzed by a single enzyme,

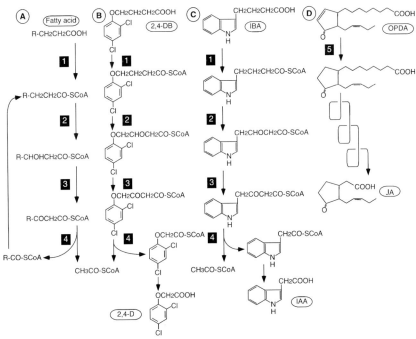

FIGURE 2. Conversion of fatty acids, 2,4-DB, IBA and OPDA in peroxisomes. Metabolisms of fatty acids (A), 2,4-DB (B), IBA (C), and OPDA (D). The enzymes involved in these pathways are: (1) acyl-CoA synthetase; (2) acyl-CoA oxidase; (3) the multifunctional protein; (4) 3-ketoacyl-CoA thiolase; (5) 12-oxophytodienoate reductase 3.

multifunctional protein, which possesses enoyl-CoA hydratase and β-hydroxyacyl-CoA dehydrogenase activities. This multifunctional protein converts *trans*-2-enoyl-CoA into 3-ketoacyl-CoA. The fourth reaction by 3-ketoacyl-CoA thiolase is the final step of β-oxidation and cleaves the acetyl group from the acyl-CoA. As a result, an acyl-CoA that is two carbons shorter is produced and this shorter acyl-CoA then re-enters the β-oxidation. Meanwhile, the produced acetyl-CoA is then used as a substrate for the glyoxylate cycle. It is noteworthy that there exist gene families in plant genomes. For example, *Arabidopsis* has several acyl-CoA oxidases that show different specificity for substrates (De Bellis *et al.*, 1999, 2000; Hayashi *et al.*, 1998a, 1999; Hooks *et al.*, 1996). Moreover, judging from the information from the *Arabidopsis* genome, there appeared to be other isoforms that have not been identified experimentally yet.

In mammalian cells, fatty acid β-oxidation is localized both in peroxisomes and mitochondria, whereas it is localized only in peroxisomes in higher plants. Because plant peroxisomes contain a system for degradation of short-chain fatty acids, storage lipids are degraded completely in peroxisomes (Hayashi *et al.*, 1999). However, mammalian peroxisomes are unable to metabolize short-chain fatty acids because they have no enzymes for the metabolism. Instead, short-chain fatty acids are transported from peroxisomes to mitochondria whose β-oxidation system is able to completely degrade them (Hashimoto, 1996; Lazarow, 1978).

The β-oxidation activity is induced during senescence as well as seed germination (Kato *et al.*, 1996). The β-oxidation during the senescence is thought to play a role in the turnover of lipids derived from membranes in various organelles, although the precise mechanism and function is unknown yet. Besides the high fatty acid β-oxidation activity during germination and senescence, a low activity is observed in different stages and different organs throughout the life cycle of the plant (Kamada *et al.*, 2003). As described later, there is growing evidence that β-oxidation is involved in the biosynthesis of plant hormones, JA, and auxin (Lange *et al.*, 2004; Sanders *et al.*, 2000; Stintzi and Browse, 2000; Zolman *et al.*, 2001a). Some β-oxidation activity might be required for biosynthesis of these plant hormones.

2. Glyoxylate Cycle

The glyoxylate cycle is a bypass for the decarboxylation steps of the citric acid cycle, allowing the net conversion of 2 mol of acetate into 1 mol of succinate, which was discovered in bacteria. In higher plants, the activities for this cycle are detected mainly in germinating seeds (Kornberg and Kerbs, 1957). The glyoxylate cycle in higher plants has a function to convert acetyl-CoA, which is provided from the fatty acid β-oxidation, into succinate, which is required for subsequent gluconeogenesis (Fig. 3). Consequently,

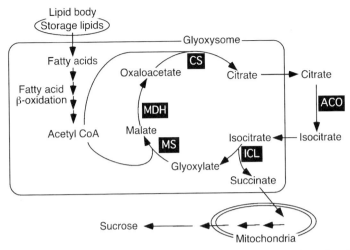

FIGURE 3. Glyoxylate cycle in glyoxysomes. The acetyl-CoA, which is produced by fatty acid β-oxidation, is used as a substrate for the glyoxylate cycle. Aconitase is not localized in the glyoxysome unlike the other four enzymes. Succinate is subsequently metabolized for production of sucrose. ICL, isocitrate lyase; MS, malate synthase; MDH, malate dehydrogenase; CS, citrate synthase; ACO, aconitase.

the glyoxylate cycle is considered essential for germinating seeds to obtain energy for growth until acquirement of the ability for photosynthesis, as in the fatty acid β-oxidation. The glyoxylate cycle consists of five enzymes: isocitrate lyase (ICL), malate synthase (MS), malate dehydrogenase (MDH), citrate synthase (CS), and aconitase (ACO). ICL and MS are peroxisome-specific enzymes and are used as marker enzymes, whereas, regarding MDH and CS, there are some homologous proteins that are localized in other organelles. ACO, however, is localized in the cytosol (Hayashi et al., 1995). This fact shows the necessity of shuttling intermediates across the peroxisomal membrane, but its mechanism is yet unknown.

It has been reported that an essential role of the glyoxylate cycle is to provide carbon after germination rather than during germination from the analyses using *Arabidopsis icl* and *ms* knockout mutants (see Section V; Cornah et al., 2004; Eastmond et al., 2000). Seeds of the *icl* and *ms* mutants could germinate even without sucrose. After germination, however, the seedling showed poor growth, especially in the dark, compared with the wild-type plant.

Glyoxylate cycle activity is detected at other stages such as developing seeds, senescent tissues, starved tissues, and pollen in addition to germinating seeds (Gut and Matile, 1988; Mano et al., 1996; Pistelli et al., 1995; Zhang et al., 1994). This suggests that the glyoxylate cycle has other

B. PHOTORESPIRATION

In photosynthetic tissues, such as green cotyledons and true leaves, peroxisomes (termed leaf peroxisomes) have a role in the glycolate pathway of photorespiration. Photorespiration involves the light-dependent uptake of O_2 and release of CO_2 during the metabolism of phosphoglycolate, the two-carbon by-product by the oxygenase activity of RubisCO. During photorespiration, up to 75% of the carbon diverted from the Calvin cycle as phosphoglycolate is returned to the cycle as 3-phosphoglycerate in a process involving metabolite flow through chloroplasts, mitochondria, and leaf peroxisomes (Fig. 4; Tolbert et al., 1968). Leaf peroxisomes contain hydroxypyruvate reductase (HPR; Greenler et al., 1989; Hayashi et al., 1996;

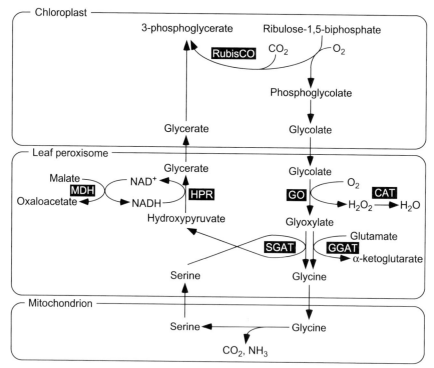

FIGURE 4. Photorespiratory pathway in leaf peroxisomes. RubisCO, ribulose-bisphosphate carboxylase/oxygenase; GO, glycolate oxidase; CAT, catalase; GGAT, glutamate: glutarate aminotransferase; SGAT, serine:glyoxylate aminotransferase; HPR, hydroxypyruvate reductase; MDH, malate dehydrogenase.

Mano *et al.*, 1997, 1999; Schwartz *et al.*, 1991; Titus and Becker, 1985), glycolate oxidase (GO; Ludt and Kindl, 1990; Tsugeki *et al.*, 1993; Volokita and Somerville, 1987), and several aminotransferases (Igarashi *et al.*, 2003; Liepman and Olsen, 2001) as enzymes for the photorespiration. Gene expression of these enzymes, and accumulation of mRNA and proteins are dramatically induced by exposure to light.

Several mutants with defects in the genes necessary for photorespiration have been isolated and characterized (see Section V). Most of these mutants show dwarfism in a normal atmosphere (36 Pa CO_2), whereas they grow normally as does the wild-type plant in an atmosphere containing high CO_2 (1000 Pa), under which the by-products are not produced because of the decrease of the oxygenase activity of RubisCO. Besides the leaf-peroxisomal enzymes described earlier, leaf peroxisomes require additional enzymes to maintain its homeostasis. For example, the conversion from glycolate to glyoxylate by GO generates hydrogen peroxide as a by-product (Fig. 4). Catalase (CAT) in the matrix (Bethards *et al.*, 1987; Esaka *et al.*, 1997; Frugoli *et al.*, 1996; Grotjohann *et al.*, 1997; Williamson and Scandalios, 1993) and ascorbate peroxidase (APX) on the membrane (Bunkelmann and Trelease, 1996; Yamaguchi *et al.*, 1995) act to detoxify this hydrogen peroxide immediately (see later for the difference between CAT and APX). Moreover, MDH is necessary for the electron shuttle system to regenerate NADH (Reumann *et al.*, 1994). As stated already, MDH is one of the glyoxylate cycle enzymes in watermelon (Gietl, 1990), cucumber (Kim and Smith, 1994), soybean (Guex *et al.*, 1995), and pumpkin (Kato *et al.*, 1998). In cucumber, MDH is encoded by a single gene so that glyoxysomal MDH has a function in leaf peroxisomes (Kim and Smith, 1994). However, unlike cucumber glyoxysomal MDH, another leaf-peroxisomal MDH homologue seems to act to regenerate NADH (Kamada *et al.*, 2003). *Arabidopsis* genome contains two MDH genes. One is specifically expressed in germinating seeds, whereas the other is expressed more in green cotyledons and leaves than in any other organ. It is, therefore, assumed that the former is involved in the glyoxylate cycle and the latter for regeneration of NADH in leaf peroxisomes.

C. BIOSYNTHESIS OF PLANT HORMONES

1. Auxin

Auxin is an essential plant hormone that influences various aspects of plant growth, such as vascular development, lateral root initiation, apical dominance, and gravitropism (Davis, 1995). Indole-3-acetic acid (IAA) is the most abundant endogenous auxin. Indole-3-butyric acid (IBA), which is used in agriculture because it can induce root formation, has been regarded as a synthetic auxin that functions through conversion to IAA. However,

IBA is often more effective than IAA for root initiation (Krieken et al., 1992), and IBA is more stable than IAA under various light and temperature conditions (Nissen and Sutter, 1990; Nördstrom et al., 1991). It is revealed that the function of IBA itself and the conversion to IAA are important *in vivo* (Epstein and Lavee, 1984; Krieken et al., 1992; Nordstrom et al., 1991), suggesting that a specific IBA to IAA ratio is essential for plant growth. This conversion from IBA to IAA is a step that is shortened by two carbons, namely β-oxidation, which is carried out in peroxisomes (Fig. 2C).

2,4-Dichlorophenoxyacetic acid (2,4-D), a synthetic auxin, is generated from the metabolism of 2,4-dichlorophenoxybutyric acid (2,4-DB) by β-oxidation, as in the conversion from IBA to IAA (Fig. 2B, Wain and Wightman, 1954). 2,4-D, but not 2,4-DB, inhibits root elongation of *Arabidopsis* at an early stage of seedling growth (Estelle and Somerville, 1987). Thus, although auxin is an essential hormone for plant growth, abnormal amounts of endogenous auxin or application of a synthetic auxin such as 2,4-D or the precursor 2,4-DB prevents normal plant growth. Thus, peroxisomal β-oxidation is an important function regulating the production of bioactive auxin in plant cells.

Since IBA, IAA, 2,4-DB, and 2,4-D inhibit root elongation in wild-type plants, 2,4-DB- and IBA-resistant mutants were screened using this phenomenon (Hayashi et al., 1998b; Zolman et al., 2000); it was anticipated that the β-oxidation–defective mutants would survive on a medium containing 2,4-DB or IBA because they could not metabolize them to 2,4-D and IAA, and such mutants would not be able to utilize storage oils. In fact, most but not all mutants were unable to germinate on a medium without sucrose (Fig. 6B). Based on Zolman's criteria, these mutants were subcategorized into four groups (classes 1–4) on the basis of their responses to IBA, IAA, 2,4-DB, auxin transport inhibitors, and the efficiency of the mobilization of storage lipids. Class 1 mutants are resistant to auxin and do not show inhibition of root elongation or promotion of lateral root initiation. In these mutants, seed storage lipids, such as eicosanoic acid, are metabolized more slowly compared to the wild-type plant, which reduces growth, suggesting that fatty acid β-oxidation is defective in these mutants. The *Arabidopsis* mutants of *ped* series, *pex5*, *chy1/drb5*, *pxa1*, and *pex6* are included in this class (Hayashi et al., 1998b, 2000, 2002; Lange et al., 2004; Zolman and Bartel, 2004; Zolman et al., 2000, 2001a,b). Other mutants can utilize storage lipids. Because class 2 mutants exhibit normal responses to IAA and 2,4-D, like the wild-type plant, these mutants are assumed to be defective in enzymes responsible for the conversion from IBA to IAA that is independent of the fatty acid β-oxidation of seed storage lipids. Alternatively, class 2 mutants may be defective in an IBA receptor. Class 3 and class 4 mutants are sensitive and resistant to another auxin 2,4-D, respectively. Class 3 mutants show the wild-type sensitivity to polar auxin transport inhibitors such as 1-napthylphythalamic acid (NPA), 9-hydroxyfluorene-9-carboxylic acid (HFCA), and 2,3,5-triiodobenzoic acid (TIBA), whereas class 4 mutants are resistant to these inhibitors. Therefore, class 4 mutants may be defective in IBA transport (Zolman et al., 2000). Some of these mutants, especially those in classes 2–4, were designated as *ibr* mutants and have been analyzed. *IBR* genes are expected to encode proteins essential for auxin function such as the recognition, signal transduction, response, or transport. Identification of *IBR* genes will provide useful information on the mechanism of auxin function and on how peroxisomes regulate the metabolism of the auxin precursor.

2. Jasmonic Acid

Jasmonic acid and its derivatives are lipid-derived compounds that affect a variety of developmental processes in plant growth such as carbon partitioning (Mason and Mullet, 1990), mechanotransduction (Weiler et al., 1993), and the maturation and release of pollen (McConn and Browse, 1996). Moreover, a predominant role of JA is in the activation of signal transduction pathways in response to insect predators, pathogen attacks, and wounding (Howe et al., 1996; McConn et al., 1997; Ryan and Farmer, 1991; Vijayan et al., 1998; Xu et al., 1994). JA and its derivatives are synthesized through multiple steps via chloroplasts and peroxisomes. α-Linolenic acid, which is formed by the reaction of lipases in plastids, is used as a substrate for the JA biosynthesis. 13-Hydroperoxylinolenic acid is synthesized from α-linolenic acid by the reaction of lipoxygenase, and then it is metabolized to (9S, 13S)-12-oxo-phytodienocid acid (OPDA) by the successive activities of allene oxide synthase and allene oxide cyclase. These reactions occur in plastids. OPDA is transported to peroxisomes and then metabolized to 3-oxo-2(2'[Z]-pentenyl)-cyclopentane-1-octanoic acid (OPC-8:0) by 12-oxophytodienoate reductase 3 (OPR3). Although there are four kinds of stereoisomers of OPDA (9S, 13S-OPDA; 9S, 13R-OPDA; 9R, 13R-OPDA, and 9R, 13S-OPDA), 9S, 13S-OPDA is the only biological precursor of JA, and is used as the substrate for the reduction by OPR3 (Schaller et al., 2000; Strassner et al., 2002). Lastly, three cycles of the β-oxidation shorten the OPC-8:0 side chain to yield OPC-2:0, namely JA (Fig. 2D). Thus, peroxisomes play an essential role in the final step of JA biosynthesis. In this pathway, the intermediate, OPDA, must be shuttled from plastids to peroxisomes. However, the mechanism about the transport of OPDA remains to be investigated. As stated in Section V, the *opr3* mutant is sterile (Sanders et al., 2000; Stintzi and Browse, 2000), as is the *coi1* mutant, which causes a defect in JA perception and the triple mutant, *fad3–2fad7–2fad8*, which lacks the hexadecatrienoic and linolenic fatty acid precursors of the jasmonic pathway (Feys et al., 1994; McConn and Browse, 1996). However, other JA response mutants, such as *jin* and *jar* in *Arabidopsis* and *def1* in tomato, are fertile (Berger et al., 1996; Howe et al., 1996; Staswick et al., 1992). In addition, all *Arabidopsis* mutants with defects in fatty acid β-oxidation do not always show sterility, suggesting that not all enzymes for fatty acid β-oxidation are commonly used in the processes between the metabolism of fatty acids to acetyl-CoA and the reduction of OPDA to JA.

D. DETOXIFICATION OF HYDROGEN PEROXIDE

Peroxisomes are sites for generating toxic reactive oxygen species (Corpas *et al.*, 2001). To degrade these toxic species, peroxisomes contain several enzymes, such as catalase (CAT; Bethards *et al.*, 1987; Esaka *et al.*, 1997; Frugoli *et al.*, 1996; Grotjohann *et al.*, 1997; Williamson and Scandalios, 1993), ascorbate peroxidase (APX; Bunkelmann and Trelease, 1996; Yamaguchi *et al.*, 1995), or superoxide dismutase (Río *et al.*, 1998), in the matrix or on the membrane. That is, a generation-degradation system for reactive oxygen species works all the time in peroxisomes. Hydrogen peroxide is the most typical reactive oxygen species in peroxisomes. As described in Section II, the peroxisome name was based on the existence of activities for peroxidases. CAT in the matrix has been considered to be mainly responsible for the degradation of hydrogen peroxide. In higher plants, however, APX on the membrane is also involved in the detoxification of hydrogen peroxide (Bunkelmann and Trelease, 1996; Yamaguchi *et al.*, 1995). Why are there different kinds of enzymes for scavenging hydrogen peroxide in peroxisomes? It is probably due to the difference in the affinity for hydrogen peroxide between CAT (Km = 0.047×10^3 to 1.1×10^3 mM) and APX (Km = 3×10^{-2} mM). CAT is unable to catalyze the low concentration of hydrogen peroxide, and instead, APX efficiently detoxifies the low concentration of hydrogen peroxide. Interestingly, it is apparent that reactive oxygen species have important roles in the signal transduction pathway as messenger molecules to communicate among organelles and/or among cells (Corpas *et al.*, 2001). It is, therefore, likely that the signal-generating function of plant peroxisomes is important.

IV. TRANSITION OF PEROXISOME FUNCTION

Functional transition of peroxisomes is the most interesting phenomenon in higher plants. Glyoxysomes, which have a role in lipid metabolism in etiolated tissues of germinating seeds, are transformed directly into leaf peroxisomes that are involved in photorespiration in photosynthetic tissues (Nishimura *et al.*, 1986; Titus and Becker, 1985). In the process of the transition from glyoxysomes to leaf peroxisomes, activities of glyoxysomal enzymes such as malate synthase and isocitrate lyase decreased, whereas activities of leaf-peroxisomal enzymes such as glycolate oxidase and hydroxypyruvate reductase increased (Hayashi *et al.*, 1998a, 1999; Kato *et al.*, 1995, 1996, 1998; Mano *et al.*, 1996; Tsugeki *et al.*, 1993). Moreover, the degradation and inhibition of translocation of glyoxysomal enzymes are regulated in underlying transition of peroxisomes. It should be noted that during the transition of peroxisomes novel peroxisomes do not appear and enzymes inside peroxisomes are replaced by new ones (Nishimura et al., 1986; Titus and Becker, 1985). This is called the single population hypothesis, which was initially proposed by Trelease et al. (1971). Although light probably acts as the major factor inducing this transition, the molecular mechanism remains to be understood. Interestingly, the reverse transition of leaf peroxisomes to glyoxysomes is observed in senescent tissues (Nishimura et al., 1993). In this case, glyoxysomal enzymes appear and leaf peroxisomal enzymes are degraded specifically. This reverse transition, like the transition from glyoxysomes to leaf peroxisome, occurs in the same peroxisomes, and is observed in flower petals (De Bellis and Nishimura, 1991). The search for proteases involved in the specific degradation of glyoxysomal or leaf-peroxisomal enzymes has not yet revealed such a unique protease.

V. PEROXISOME BIOGENESIS

Since peroxisomes do not possess a genome and protein-synthesizing machinery, all the proteins constituting peroxisomes are nuclear encoded, synthesized in the cytosol, and then imported to peroxisomes. Several pathways of protein transport to peroxisomes have been reported in higher plants (Johnson and Olsen, 2001; Mullen et al., 2001), as in mammals and yeasts. Peroxisomes proliferate by division of preexisting peroxisomes (Miyagishima et al., 1999). The analysis using peroxisome-defective mutants as described later revealed that the correct maintenance of peroxisomes within cells is essential for normal plant development, although the mechanism underlying peroxisome biogenesis, such as their assembly, differentiation, proliferation, and inheritance, has not been completely understood yet. In this section, we review the protein transport to peroxisomes and the peroxisome biogenesis factors, peroxin. Finally, several mutants with defects in peroxisomal function or peroxisome biogenesis are introduced.

A. PEROXISOMAL TARGETING SIGNAL AND PROTEIN IMPORT

Peroxisomal proteins in the matrix are transported into peroxisomes after synthesis on free polyribosomes in the cytosol. Most peroxisomal proteins have a targeting signal in their sequence. Although all of the targeting signals in the matrix proteins are not identified, there are at least two types of targeting signals. The peroxisomal targeting signal, termed PTS1, is localized at the C-terminus of the protein (Fig. 5A). PTS1 consists of Ser-Lys-Leu or derivations of this sequence, although the sequences show slight divergence

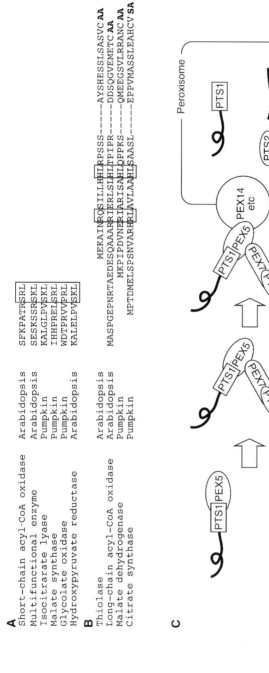

FIGURE 5. Transport of matrix proteins to peroxisomes. Representatives of targeting signals of PTS1 (A) and PTS2 (B). (A) Three carboxy-terminal amino acids that function as PTS1 are boxed. (B) Conserved amino acids that act as PTS2 are boxed and amino acids in bold letters indicate the N-terminus of mature protein after processing in the peroxisome. (C) PEX5 and PEX7 bind PTS1-containing or PTS2-containing proteins in the cytosol, respectively. The complexes of PEX7 with PTS2-containing proteins are transported via the interaction with PEX5. On the peroxisomal membrane, the receptors consisted of PEX14 with unidentified proteins exist.

among yeast, mammals, and higher plants (Gould et al., 1987, 1988, 1989, 1990; Hayashi et al., 1997). PTS2 was identified as a conserved N-terminal nonapeptide (Arg/Lys)-(Leu/Val/Ile)-x-x-x-x-x-(His/Qln)-(Leu/Ala) that occurs in the presequences of several enzymes such as 3-ketoacyl-CoA thiolase (Kato et al., 1996), citrate synthase (Kato et al., 1995), malate dehydrogenase (Kato et al., 1998), and long chain acyl-CoA oxidase in higher plants (Fig. 5B; Hayashi et al., 1998a; Kato et al., 1995, 1996, 1998). PTS1-containing proteins are synthesized as a mature form and are not proteolytically processed after the transport into peroxisomes, whereas PTS2-containing proteins are synthesized as larger precursors, and its presequence is cleaved after the import to peroxisomes (Fig. 5C). The protease(s), which recognize this presequence as a substrate, have not yet been identified. The fusion proteins consisting of reporter proteins, such as β-glucuronidase (GUS) or green fluorescent protein (GFP), with PTS1 or PTS2 directs non-peroxisomal proteins to peroxisomes in plant cells (Flynn et al., 1998; Hayashi et al., 1997; Kato et al., 1999; Mano et al., 1999, 2002; Mullen et al., 2001; Volokita, 1991).

It is intriguing that the various kinds of peroxisomes have similar machinery for protein transport because fusion proteins consisting of reporter proteins such as GUS or GFP with PTS1 or PTS2 are transported into all types of peroxisomes, glyoxysomes in etiolated tissues, leaf peroxisomes in greening tissues, and unspecialized peroxisomes in other tissues (Hayashi et al., 1997; Mano et al., 2002). As the components of protein transport to peroxisomes, PEXROXIN5 (PEX5, see next section for peroxin), and peroxin7 (PEX7) were identified as the PTS1 or PTS2 receptors, respectively, in yeasts (Leij et al., 1993; Marzioch et al., 1994), mammals (Fransen et al., 1995; Wiemer et al., 1995), and higher plants (Gurvitz et al., 2001; Kragler et al., 1998; Schumann et al., 1999a; Wimmer et al., 1998). These receptors recognize PTS1 or PTS2 in each protein, bind to them in the cytosol, and direct to the peroxisomal membrane. In addition to PEX5 and PEX7, other peroxins responsible for protein transport to peroxisomes have been identified and characterized in various organisms (Charlton and Lopez-Huertas, 2002; Fujiki, 2000; Holroyd and Erdmann, 2001).

Interestingly, even though PEX5 and PEX7 are present as components common to mammals, yeasts, and plants, the mechanism of protein transport using such components is slightly different. The PTS1 and PTS2 pathways are independent in yeasts because the PTS2 pathway is active in yeast mutants with defects of PTS1 pathway such as *pex5* and vice versa (Elgersma et al., 1998; Leij et al., 1993; Marzioch et al., 1994; McCollum et al., 1993; Rehling et al., 1996). However, the PTS2 pathway in mammalian cells is dependent on PTS1 pathway (Otera et al., 1998; Schliebs et al., 1999). Alternative splicing of the *PEX5* gene in humans and Chinese hamster ovary cells gave rise to two types of PEX5 proteins, termed long-form and short-form of PEX5s. The complex of PTS2-containing proteins with PEX7 can be transported to peroxisomes only after the binding with the long-form of

PEX5. The short-form of PEX5 is involved in only the PTS1 pathway, since there are two types of PTS1 pathways (Otera *et al.*, 1998; Schliebs *et al.*, 1999). However, higher plants use a mechanism different from that in yeasts and mammals (Nito *et al.*, 2002). In plant cells, the PTS2 pathway is completely dependent on the PTS1 pathway, as in mammalian cells. That is, after the interaction with PEX5, the complex of PEX7 with PTS2-containing proteins is transported to the machinery that consists of PEX14 and other unknown proteins on the peroxisomal membrane (Fig. 5C; Nito *et al.*, 2002). However, unlike mammalian PEX5, plant PEX5 protein exists in only one form.

Peroxisomes can accept proteins with a high-order structure (Flynn *et al.*, 1998; Kato *et al.*, 1999; Lee *et al.*, 1997). In the case of protein transport to mitochondria, plastids, or ER, the unfolding of proteins is required. Peroxisomes are able to accept the import of thiolase as a dimer in yeasts (Glover *et al.*, 1994), chloramphenicol acetyltransferase with PTS1 as a trimer in mammalian cultured cells (McNew and Goodman, 1996), and GUS with PTS2 as a tetramer in plants (Kato *et al.*, 1999). Additionally, in mammalian cultured cells, even 9-nm gold particles conjugated with PTS1 are translocated into peroxisomes when they are microinjected (Walton *et al.*, 1995). The existence of large pores, which show a porin-like character, has been suggested.

Peroxisomal membrane proteins such as APX, PMP22, PMP38, and PMP47 do not contain typical PTS1 or PTS2 sequences (Fukao *et al.*, 2001; McCammon *et al.*, 1994; Nito *et al.*, 2001; Pause *et al.*, 2000; Purdue and Lazarow, 1994). Although several peroxisomal membrane proteins have been cloned and the targeting sequences identified and termed mPTS, there are no consensus sequences like PTS1 and PTS2. Regarding the transport of peroxisomal membrane proteins, there seems to be at least two pathways to peroxisomes (Trelease, 2002). One is the direct transport from the cytosol to peroxisomes, like the PTS1- and PTS2-dependent pathways. The other is the indirect transport to peroxisomes via ER/ER-derived vesicles. The former is true of the transport of PMP22, PEX14, PMP47, and PMP70, whereas the latter is true of the transport of APX (Trelease, 2002). The receptor for mPTS in the cytosol, which is required for the direct transport, has not been characterized yet.

B. PEROXINS

Genes essential for peroxisome biogenesis were initially identified through the isolation of yeast mutants that were unable to grow on selective medium and the analyses of peroxisome biogenesis disorders in human, rat, and Chinese hamster ovary cells (Fujiki, 2000; Holroyd and Erdmann, 2001). Several acronyms such as PAS, PAF, PAY were initially used due to the different experimental systems used by several groups. Later, the nomenclature was unified to avoid confusion, and the peroxisome biogenesis factors were designated as peroxin, with PEX as the gene acronym (Distel *et al.*, 1996). To date, 32 *PEX* genes have been reported from various organisms. They are categorized into the following groups according to function: matrix protein import such as PEX2, PEX5, PEX7, PEX8, PEX9, PEX10, PEX12, PEX13, PEX14, PEX17, PEX18, PEX20, and PEX23; membrane protein import such as PEX3, PEX16, and PEX19; proliferation such as PEX11; early peroxisome biogenesis such as PEX1 and PEX6; and receptor recycling such as PEX4 and PEX22. However, the classification was according to the phenotype of the mutants with abnormal distribution of peroxisomal marker enzymes and/or unusual morphology in size, shape, and number. Therefore, the molecular mechanisms for most of the PEX functions, except for PEX5, PEX7, and PEX14 as stated earlier, remain to be elucidated. It should be noted that there are no organisms that have all *PEX* genes. Even in yeasts, the kinds of *PEX* genes in the genome depend on each strain. In *Arabidopsis*, several *PEX* genes were not found in the genome based on the database search for homologues. These results suggest that peroxisome biogenesis requires essential factors peculiar to each organism, although we cannot exclude the possibility that other factors, which do not contain high similarities, might have a function.

The *PEX* genes reported in higher plants are PEX1 (Lopez-Huertas *et al.*, 2000), PEX2 (Hu *et al.*, 2002), PEX5 (Gurvitz *et al.*, 2001; Kragler *et al.*, 1998; Nito *et al.*, 2002; Wimmer *et al.*, 1998), PEX6 (Zolman and Bartel, 2004), PEX7 (Nito *et al.*, 2002; Schumann *et al.*, 1999a), PEX14 (Hayashi *et al.*, 2000; Lopez-Huertas *et al.*, 1999), and PEX16 (Lin *et al.*, 1999, 2004). Most of them were basically characterized by the analyses using mutants (see section on phenotype of each mutant). In addition to the PEX genes described earlier, the existence of other *PEX* genes has been suggested based on the database search from the complete sequence of *Arabidopsis*. Attempts have been made to identify these genes and elucidate the peroxisomal function using T-DNA insertion lines and RNA interference method.

C. GENETIC APPROACHES USING MUTANTS

Much of the information on peroxisome function or peroxisome biogenesis has come from studies of mutants with defects of peroxisome function or biogenesis, especially *Arabidopsis* mutants. With regard to mutants with defects in peroxisome function or peroxisome biogenesis, there are more mutants in mammalian cultured cells and yeasts than in higher plants. However, the influence of each defect on the whole individuals is not clear because the data were derived from cultured cells and analysis in a multicellular organism is difficult. Peroxisomes function differently in the developmental stage of each organ so that it is necessary to understand the organ-/tissue-specific regulation of peroxisome biogenesis. Therefore, research in higher plants has the advantage of understanding the function at a multicellular level.

Several approaches have been taken to isolate mutants. The simplest approach is to utilize the change of the growth condition. For example, mutants with defects in photorespiration exhibit inhibited growth under a normal atmosphere (Somerville and Ogren, 1982). Moreover, a precursor of auxin, such as 2,4-DB and IBA, is metabolized by β-oxidation, and the metabolized product, 2,4-D and IAA, prevents normal plant growth in the wild-type plant, whereas mutants defective in β-oxidation can survive (Hayashi *et al.*, 1998b; Lange *et al.*, 2004; Zolman *et al.*, 2000). As a novel method, GFP was used for screening mutants (Mano *et al.*, 2004). Additionally, there are mutants that were selected based on other phenotypes, which had not been initially thought as defects of peroxisome function. *aim1* (Richmond and Bleecker, 1999), *opr3/delayed dehiscence1* (Sanders *et al.*, 2000; Stintzi and Browse, 2000), *pex2* (Hu *et al.*, 2002), and *pex16* (Lin *et al.*, 1999) mutants are cited as examples of such mutants. These results show that peroxisomes have more functions than expected. In addition, the large amount of biological information derived from the sequencing projects of whole plant genomes and the deposits of many EST clones and T-DNA insertion lines encouraged us not only to obtain the clone and its information about the homologous genes easily but also to take reverse-genetic approaches such as analysis of phenotypes in T-DNA insertion lines or RNA-interference lines (Cornah *et al.*, 2004; Eastmond *et al.*, 2000; Fulda *et al.*, 2004; Germain *et al.*, 2001; Igarashi *et al.*, 2003; Schumann *et al.*, 2003; Sparkes *et al.*, 2003). The next section is subcategorized into two issues: (1) mutants with defects in peroxisome function, mainly those defective in enzymes required for various metabolic pathways, and (2) mutants with defects in peroxisome biogenesis, including protein transport and regulation of size, shape, or number of peroxisomes.

1. Mutants with Defects in Peroxisome Function

a. lacs6 *and* lacs7 *Mutants*

Arabidopsis LACS6 and LACS7 have long-chain acyl-CoA synthetase activities and contain PTS (Fulda *et al.*, 2004). LACS6 has PTS2, whereas LACS7 has both PTS1 and PTS2. T-DNA insertion lines, termed *lacs6-1* and *lacs7-1*, were isolated and characterized (Fulda *et al.*, 2004). Each single mutant did not exhibit any differences in the timing or progress of germination, the lipid content and fatty acid composition of germinating seedlings, and the sensitivity to 2,4-DB compared with the wild-type plant. This suggests that LACS6 and LACS7 have substantially redundant functions. Therefore, the *lacs6-1lacs7-1* double mutant was generated by crossing each single mutant to abolish peroxisomal LACS activity completely. Unlike the single mutants and the wild type, the *lacs6-1lacs7-1* double mutant could germinate but stalled in their postgerminative growth, unless they were provided with exogenous sucrose, indicating that the fatty acid β-oxidation is not functional in the *lacs6-1lacs7-1* double mutant because of a lack of or a strong reduction of peroxisomal LACS activity. In the *lacs6-1lacs7-1* double mutant, the activity for utilization of triacylglycerol is markedly reduced and striking numbers of large lipid bodies are present even in 6-day-old cotyledonary cells in which lipid bodies are rarely observed in the wild-type plant. This result is consistent with the limited development during postgerminative growth. However, the peroxisomes in the *lacs6-1lacs7-1* double mutant are morphologically normal and the degree of growth inhibition on medium containing 2,4-DB in the *lacs6-1lacs7-1* double mutant is consistently similar to that in the wild-type plant. As mentioned later, the mutant with a defect in the JA biosynthesis displays male sterility (Sanders *et al.*, 2000; Stintzi and Browse, 2000). The *lacs6-1lacs7-1* double mutant is fully fertile so that the JA biosynthesis is active in this double mutant. These results suggest the presence of distinct CoA ligase responsible for the activation of OPC:8, the precursor of JA.

b. aim1 *Mutant*

Abnormal inflorescence meristem1 (*aim1*) mutant was isolated as a mutant exhibiting abnormal inflorescence and floral development from *Arabidopsis* T-DNA insertion pools (Richmond and Bleecker, 1999). In an *aim1* mutant, the vegetative phase of development is similar to that of the wild-type plant, although the rosette diameter and the morphology of leaves are affected only under short-day conditions. It is, however, after the transition to reproductive development that the *aim1* mutant phenotype becomes most pronounced. Although the basic pattern of inflorescence development is initially the same in the *aim1* mutant, as in the wild-type plant, it causes a wide range of abnormalities such as the defective production of floral structures, undifferentiation of tissues, production of abnormal buds, lack of internode elongation, and sterility. *AIM1* encodes protein showing significant similarity to the cucumber multifunctional protein (MFP) in the β-oxidation pathway (Preisig-Müller *et al.*, 1994). Recombinant AIM1 protein shows enoyl-CoA hydratase activity, although it has a higher affinity for short-chain acyl-CoAs, indicating a defect of the β-oxidation pathway in this mutant (Preisig-Müller *et al.*, 1994). In fact, the composition of total lipids, especially unsaturated 18-carbon fatty acids, of *aim1* mutant leaves was altered compared with the wild-type plant. In addition, an *aim1* mutant, like the β-oxidation–defective mutants, is resistant to 2,4-DB. These results confirm the decrease of activity for β-oxidation in this mutant. However, the *aim1* mutant, unlike the β-oxidation–defective mutants, is able to germinate on a medium without sucrose, indicating the presence of functional redundancy between AIM1 and the second MFP, *At*MFP2. The function of AIM1 protein is under investigation. Interestingly, the *Arabidopsis opr3/delayed dehiscence1* mutant, which is defective in the JA biosynthesis, also exhibit male sterility (see later; Sanders *et al.*, 2000; Stintzi and Browse, 2000). It is, therefore, plausible that AIM1 might be involved in the metabolism of fatty acids for the biosynthesis of fatty acid-derived signals but not for the gluconeogenesis.

c. ped1/kat2 *Mutant*

As described in Section III, 2,4-DB, a proherbicide, is metabolized to 2,4-D, which is an auxin by the action of β-oxidation in peroxisomes. 2,4-D but not 2,4-DB inhibits root elongation of *Arabidopsis* at an early stage of seedling growth (Estelle and Somerville, 1987). It was, therefore, assumed that application of 2,4-DB to the wild-type *Arabidopsis* would prevent root growth because 2,4-DB is converted to 2,4-D in peroxisomes, whereas mutants with defects in the peroxisomal β-oxidation show the resistance to 2,4-DB. On this assumption, a number of 2,4-DB-resistant mutants, termed *ped* (*peroxisome defective*), were screened (Hayashi *et al.*, 1998b). *ped* mutants showed sucrose dependency as well as the resistance to 2,4-DB, suggesting the defect of fatty acid β-oxidation (Fig. 6A and B). In one *ped* mutant, *ped1* has the mutation in the 3-ketoacyl-CoA thiolase (*KAT2*) gene. In the *Arabidopsis* genome, there are two other 3-ketoacyl-CoA thiolase genes, *KAT1*, *KAT5*, as well as other genes that encode proteins with thiolase activity. However, these gene products could not substitute for KAT2 activity. It should be noted that peroxisomes, especially glyoxysomes in etiolated tissues, in *ped1* mutants are two or three times greater in diameter than those in the wild-type plant and that these larger peroxisomes contained vesicle-like structures, which were probably generated because of the accumulation of intermediates by the fatty acid β-oxidation (Hayashi *et al.*, 1998b). The consecutive sections observed by electron microscopy suggested that the vesicle-like structures come from the lipid body and contain substances similar to those in lipid bodies and that fatty acids might be transported with vesicles at the glyoxysome-lipid body contact site (Hayashi *et al.*, 2001). In addition to *ped1* mutants, the T-DNA insertion line in *KAT2* gene was isolated and characterized (Germain *et al.*, 2001). Lipid bodies were observed in most *kat2* mutant cells, but not in the wild-type cells, in 5-day-old green seedlings. In fact, the amount of storage lipids remains unchanged in the *kat2* mutant during seed germination, whereas the amount in the wild type decreases rapidly. These results indicate that the metabolism in peroxisomes influences the metabolism in the lipid bodies.

d. icl *Mutant*

ICL is one of five enzymes required for the glyoxylate cycle, and it is used as a glyoxysomal marker protein. Two *icl* mutants (*icl-1* and *icl-2*) were isolated from an *Arabidopsis* population carrying the autonomous *En/Spn* transposable element from maize (Eastmond *et al.*, 2000). Surprisingly, based on their phenotypic analysis, seed germination and seedling growth in a favorable

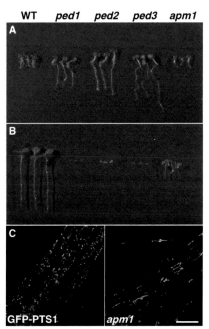

MANO AND NISHIMURA, FIGURE 6. Effects of 2,4-DB and sucrose on growth of mutants. Wild-type plants, *ped1*, *ped2*, *ped3* and *apm1* mutants were grown for 7 days on a medium containing 2,4-DB (A) or a medium without sucrose (B). Photographs were taken after the seedlings were removed from the media and rearranged on agar plate. The mutants of *ped* series are resistant to 2,4-DB and require exogenous sucrose for growth, whereas *apm1* mutant exhibits resistance to 2,4-DB and sucrose dependency partially. (C) In GFP-PTS1, peroxisomes normal in size and number are observed as spherical spots, whereas *apm1* mutants have slightly larger peroxisomes with long string-like tails. In addition, the number of peroxisomes in this mutant is decreased remarkably.

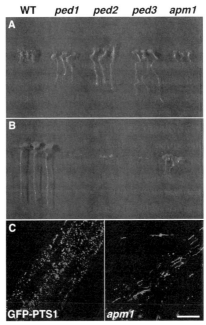

FIGURE 6. Effects of 2,4-DB and sucrose on growth of mutants. Wild-type plants, *ped1*, *ped2*, *ped3* and *apm1* mutants were grown for 7 days on a medium containing 2,4-DB (A) or a medium without sucrose (B). Photographs were taken after the seedlings were removed from the media and rearranged on agar plate. The mutants of *ped* series are resistant to 2,4-DB and require exogenous sucrose for growth, whereas *apm1* mutant exhibits resistance to 2,4-DB and sucrose dependency partially. (C) In GFP-PTS1, peroxisomes normal in size and number are observed as spherical spots, whereas *apm1* mutants have slightly larger peroxisomes with long string-like tails. In addition, the number of peroxisomes in this mutant is decreased remarkably. (See Color Insert.)

condition are indistinguishable between *icl* mutants and the wild-type plant. However, when light intensity was decreased or day length was shortened, the frequency of seedling establishment but not seed germination fell dramatically in *icl* mutants. This defect was rescued by supplying sugar exogenously. Although the total fatty acid content or composition was not altered in *icl* mutants, the breakdown of fatty acids was severely inhibited in these mutants. These results indicate that the gluconeogenesis from the storage lipids to sucrose by the glyoxylate cycle is essential for seedlinge stablishment, or postgerminative growth, rather than seed germination and that the breakdown process is controlled by metabolites (Eastmond *et al.*, 2000).

e. ms *Mutant*

Two *ms* mutants were also isolated by PCR screening from the *Arabidopsis* T-DNA collection (Cornah *et al.*, 2004). In contrast to *icl* mutants, *ms* mutants do not show a severe phenotype, although hypocotyl elongation and root development are inhibited in the dark. Interestingly, microarray analysis showed differences in gene expression between *icl* mutants and *ms* mutants. The gene expression in *ms* mutants differed very little from that of the wild-type plants, whereas expression of 397 genes was different in *icl* mutants compared with the wild-type plants. These results indicate that *icl* mutants but not *ms* mutants exhibit the features characteristic of carbohydrate-starvation and that *ms* mutants can generate more sugars compared with *icl* mutants. Actually, when seedlings of *ms* mutants were fed with ^{14}C-acetate, ^{14}C was incorporated to sugars from acetate, confirming the ability of gluconeogenesis from acetate in *ms* mutants. The rate of lipid breakdown and sugar accumulation in *ms* mutants is similar to that in the wild-type plant, indicating that *ms* mutants are capable of gluconeogenesis from acetate following the fatty acid β-oxidation. The pathway of MS-independent gluconeogenesis proposed by Cornah *et al.* (2004) is as follows. The glyoxylate would enter the photorespiratory gluconeogenic pathway that would be present in the tissues at the transition from germinated seeds to seedlings, where the glyoxylate is metabolized to glycine in the peroxisome and subsequently sugar is produced by the photorespiration. Metabolomic analysis showed an accumulation of serine and glycine in *ms* mutants, suggesting the presence of this new metabolic pathway that needs genetic and molecular characterization in detail (Cornah *et al.*, 2004).

f. sat *Mutant*

In the photorespiratory pathway, transamination and assimilation of ammonium are essential reactions. Several kinds of aminotransferases have been assumed: glutamate:glyoxylate aminotransferase (GGAT), alanine:glyoxylate aminotransferase (AGAT), alanine:2-oxoglutamate aminotransferase (AOAT), glutamate:pyruvate aminotransferase (GPAT), and serine:glyoxylate aminotransferase (SGAT). A *sat* mutant, which was isolated from the mutant pools with defects in the photorespiratory pathway in 1980 (Somerville and Ogren, 1980), was the first mutant with peroxisome function. This mutant grows poorly under a normal atmosphere but displays no apparent phenotype in a high-CO_2 condition, which suppresses photorespiration. The *sat* mutant is deficient in SGAT activity but not AGAT and GGAT activities. In addition, serine and glycine accumulate as end products of photosynthesis, mostly at the expense of starch and sucrose in a normal atmosphere. These results show that SGAT and AGAT/GGAT activities are probably distinct and SGAT is the essential enzyme in the photorespiration. Although the deficiency of SGAT activity had been detected, the *SAT* gene was not identified until 2001. AGT1, formerly AGT, encodes alanine:glyoxylate aminotransferase containing PTS1 (Liepman and Olsen, 2001). AGT1 has the highest specific activity with the SGAT reaction, although it also has AGAT activity at a lower rate, representing that the major role of AGT1 is

SGAT reaction rather than AGAT reaction. Sequencing analysis of the AGT1 loci in the *sat* mutant revealed the single-nucleotide substitution that causes the amino acid substitution from proline to leucine. The recombinant AGT1 mutant protein with the amino acid substitution from proline to leucine lacks any detectable SGAT and AGAT activities, demonstrating that the *SAT* gene encodes AGT1 (Liepman and Olsen, 2001).

g. aoat1 *Mutant*

GGAT gene, which had been assumed to play an important role for the transamination in peroxisomes, was not cloned until 2003. *Arabidopsis* has four kinds of *AOAT* genes (*AOAT1–4*), encoding alanine:2-oxoglutarate aminotransferase (AOAT). Among the four AOATs, AOAT1 and AOAT2 are localized in peroxisomes because they contain PTS1 at their C-termini. In fact, the fusion protein of AOAT1 with GFP is transported into peroxisomes. Characterization of an *aoat1* mutant, which was isolated from the T-DNA insertion lines, showed that AOAT1 had a major function in the transamination of alanine and 2-oxoglutarate in spite of the presence of AOAT2, AOAT3, and AOAT4 (Igarashi *et al.*, 2003). Interestingly, in an *aoat1* mutant, AOAT, AGAT, GPAT (glutamate:pyruvate aminotransferase), and GGAT activities are lower than those in the wild-type plant, whereas SGAT activity is indistinguishable, showing that AOAT exhibits activity with multiple donor:acceptor combinations and that AOAT, AGAT, and GPAT activities are distinct from SGAT activity. Moreover, based on the comparison of amino acid contents between the *aoat1* mutant and the wild-type plant, the concentration of glutamate, glutamine, and aspartate is higher and that of serine and glycine is lower in the *aoat1* mutant compared with the wild-type plant at a higher light intensity. At a lower light intensity, however, these differences were slight. These results show that AOAT1 catalyzed the peroxisomal GGAT reaction. The *aoat1* mutant had repressed growth with light green leaves under a normal condition. This defect was removed by the decrease of light intensity, the increase of CO_2 concentration, or application of sucrose exogenously, showing that the *aoat1* mutant is a photorespiration-deficient mutant (Igarashi *et al.*, 2003).

h. opr3/*Delayed* Dehiscence1

Jasmonic acid is generated from 3-oxo-2(2′[Z]-pentenyl)-cyclopentane-1-octanoic acid (OPC-8:0) as a substrate for subsequent three cycles of β-oxidation. This JA biosynthetic pathway has been believed to exist in the plastids and the cytosol. It has been clarified, however, that peroxisomes are the final sites of JA synthesis by the characterization of *Arabidopsis* mutants with a mutation in 12-oxophytodienoate reductase 3 (OPR3) and isolation of its homologue in tomato (Sanders *et al.*, 2000; Schaller *et al.*, 2000; Stintzi and Browse, 2000; Strassner *et al.*, 2002).

opr3/delayed dehiscence1 mutant was independently isolated by two groups (Sanders *et al.*, 2000; Stintzi and Browse, 2000). The most obvious phenotype in an *opr3/delayed dehiscence1* mutant is male sterility because of shortened anther filaments. In addition, this mutant shows the defect with dehisce of the anther locules at the time of flower opening and inviability of pollens. These phenotypes are alleviated by exogenous JA but not by OPDA. *OPR3* gene encodes 12-oxophytodienoate reductase, which is responsible for the conversion of OPDA to OPC-8:0. To date, three OPR isoenzymes (OPR1, OPR2, and OPR3) in *Arabidopsis* have been identified (Biesgen and Weiler, 1999; Schaller and Weiler, 1997a,b; Schaller *et al.*, 1998). Of three OPRs, only OPR3 is in the peroxisome (Strassner *et al.*, 2002). These results show that peroxisomes have an essential role in the JA biosynthesis. In spite of existence of OPR1 and OPR2, the *opr3/delayed dehiscence1* mutant shows the severe phenotype, suggesting that OPR3 is responsible for the major OPR activities.

i. chy1/dbr5 *Mutant*

Peroxisomes are involved in the conversion of IBA to IAA, one of the endogenous plant hormones, auxin. The *chy1* mutant was isolated as a mutant defective in response to IBA but not IAA (Zolman *et al.*, 2000). The peroxisomal β-oxidation is responsible for the conversion of IBA to IAA, and *chy1* mutants require exogenous sucrose, suggesting the defect of β-oxidation in *chy1* mutants. However, *chy1* mutants, like wild-type plants, respond to 2,4-DB, and thus the *chy1* mutation is specific to the conversion of IBA to IAA. *CHY1* gene encodes a protein 43% identical to a mammalian β-hydroxyisobutyryl-CoA hydrolase (HIBCH) that is responsible for valine catabolism by hydrolysis of HIBYL-CoA (Zolman *et al.*, 2001a). Interestingly, the mammalian HIBCH is localized in mitochondria (Shimomura *et al.*, 1994), whereas CHY1 protein was predicted to be peroxisomal because of the presence of PTS1. The mammalian HIBCH functionally complemented the *chy1* mutant when this protein was targeted to peroxisomes by the removal of the mitochondrial signal sequence and the addition of PTS1, showing that CHY1 is a peroxisomal HIBYL-CoA hydrolase (Zolman *et al.*, 2001a).

The decrease of CHY1 activity probably causes the accumulation of a toxic intermediate, methacrylyl-CoA, which is conjugated to free CoA and proteins containing cysteine. As a result, it causes the disturbance of β-oxidation because free CoA as a cofactor is removed and enzymes required for β-oxidation are inactivated. This phenomenon was confirmed using another *chy1* allele, *dbr5* (*2,4-DB resistant*), which was independently isolated based on the resistance to 2,4-DB (Lange *et al.*, 2004). In a *dbr5* mutant, the activity of 3-ketoacyl-CoA thiolase (KAT2), an enzyme for the β-oxidation, is reduced to approximately 30% of that in the wild-type plant, although the amount of KAT2 protein in this mutant is not different from that in the wild-type plant. This suggests that the decrease of KAT2 activity in a *chy1/dbr5* mutant is due to inhibition by methacrylyl-CoA, which is accumulated because of the defects in CHY1 (Lange *et al.*, 2004).

j. pxa1/ped3/cts *Mutant*

The *Arabidopsis pxa1* mutant was also originally isolated as a mutant exhibiting resistance to inhibition of root elongation by IBA (Zolman *et al.*, 2000), whereas the *ped3* mutant was isolated as a mutant showing resistance to 2,4-DB and requiring sucrose for germination (Hayashi *et al.*, 1998b). It has been revealed that *pxa1* and *ped3* are allelic (Hayashi *et al.*, 2002; Zolman *et al.*, 2001b). In addition, *COMATOSE* (*CTS*), whose mutants were originally isolated as mutants showing a decreased germination (Russell *et al.*, 2000), also corresponds to *PXA1/PED3* (Footitt *et al.*, 2002). *PXA1/PED3/CTS* encodes a peroxisomal ATP-binding cassette transporter (ABC transporter) (Footitt *et al.*, 2002; Hayashi *et al.*, 2002; Zolman *et al.*, 2001b). ABC transporters have been known to transport a variety of substrates from ions to polypeptides across various membranes by utilizing the energy of ATP hydrolysis. Peroxisomal ABC transporters, PMP70, ALDP, PMP70R, and ALDR have been identified in mammals (Kamijyo *et al.*, 1992; Lombard-Platet *et al.*, 1996; Mosser *et al.*, 1993; Shani *et al.*, 1997) and PXA1 and PXA2 in yeasts (Hettema *et al.*, 1996; Shani and Valle, 1996; Shani *et al.*, 1995; Swartzman *et al.*, 1996). They all belong to the half-size transporter class and have been shown to form homodimers and heterodimers, whereas PXA1/PED3/CTS is categorized to the full-size transporter. Of the peroxisomal ABC transporters, it has been reported that PXA2 in *Saccharomyces cerevisiae* can transport oleic acid as a CoA ester into peroxisomes (Verleur *et al.*, 1997). In addition, *pxa1* and *pxa2* mutants in yeasts showed the decrease of activity to long-chain fatty acids in fatty acid β-oxidation and the inability that utilized oleic acid as a carbon source (Hettema *et al.*, 1996; Shani and Valle, 1996; Shani *et al.*, 1995). In humans, ALDP is involved in X-linked adrenoleukodystrophy, one of peroxisomal disorders, and the mutation in ALDP results in the accumulation of long-chain fatty acids. In fact, *pxa1/ped3/cts* mutants also show decreased mobilization of storage oils and accumulated acyl-CoAs (Footitt *et al.*, 2002; Hayashi *et al.*, 2002).

Application of short-chain fatty acids, such as propionate or butylate, to the media increased the germination potential in a *cts* mutant, suggesting that the peroxisomal β-oxidation is functional in this mutant and that this mutant might have a defect in the transport of substrates, such as acyl-CoAs (Footitt *et al.*, 2002). Because *pxa1/ped3/cts* mutants show resistance to 2,4-DB and IBA with a short chain length and they accumulate long-chain and very long-chain fatty acids, the ABC transporter encoded by *PXA1/PED3/CTS* has a broad substrate specificity. In addition, in *cts* mutants, non-storage lipids, such as C20 fatty acids, are retained and the development of chloroplasts and vacuoles are abnormal (Footitt *et al.*, 2002). These results indicate that the carbon flux from storage oils is severely blocked.

2. Mutants with Defects in Peroxisome Biogenesis
a. ted3/pex2 *Mutant*

ted3 mutant was isolated as a dominant suppressor of *det1* mutant (Hu *et al.*, 2002). DET1 is a nuclear protein associated with chromatin, which probably regulates the expression of genes. Dark-grown *det1* mutants have short hypocotyls and opened cotyledons and developed chloroplasts, as does the wild-type plant in the light (Chory and Peto, 1990; Pepper *et al.*, 1994). In addition, light-dependently–regulated genes are expressed in a dark-grown *det1* mutant. Therefore, DET1 is assumed as a negative regulator for photomorphogenesis, although its precise mechanism is still unknown. Unexpectedly, it was revealed that the TED3 protein, whose mutation was able to suppress the phenotypes of a *det1* mutant and rescue the abnormal expression of many genes in a *det1* mutant, is homologous to PEX2 (Hu *et al.*, 2002). PEX2 is supposed to be responsible for peroxisome assembly and matrix protein import (Eitzen *et al.*, 1996; Liu *et al.*, 1996; Tsukamoto *et al.*, 1991). Although the fusion protein of PEX2/TED3 with GFP is transported to peroxisomes in *Arabidopsis*, the mechanism of how *pex2*/*ted3* alleviates the *det1* phenotype is unknown since PEX2/TED3 might not interact with DET1 directly because of the localization of DET1 in the nucleus (Hu *et al.*, 2002). Surprisingly, however, *pex2*/*ted3* partially suppressed another de-etiolated mutant, *cop1*. In addition, *cop1* mutants also exhibit some defects in sugar and IBA sensitivities, as are peroxisome-defective mutants (Hu *et al.*, 2002). Thus, the result that increased peroxisomal function can suppress the morphological and gene-expression defects caused by mutations in *DET1* and *COP1* shows that peroxisomes play an important role in the photomorphogenetic pathway negatively regulated by the DET1 and COP1 proteins. Although their mechanism is still unknown, signals are probably generated from peroxisomes that regulate nuclear gene expression through negative feedback. In fact, peroxisomes are thought to release signal molecules such as reactive oxygen species (Corpas *et al.*, 2001) and fatty acid-derived species (Weber, 2002), which might alter nuclear gene expression.

b. pex5 *Mutant*

PEX5 also has an essential role in the PTS1- and PTS2-dependent pathway in higher plants and mammals (Nito *et al.*, 2002; Otera *et al.*, 1998; Schliebs *et al.*, 1999). The *pex5* mutant is an *Arabidopsis* mutant showing resistance to IBA (Zolman *et al.*, 2000). PEX5 has two unique motifs, the TPR motif and Trp-x-x-(Phe/Tyr) repeat. The former is necessary for binding PTS1 in PTS1-containing proteins (Brocard *et al.*, 1994; Nito *et al.*, 2002). The latter is said to be essential for binding to PEX14, which acts as a initial docking site on the peroxisomal membrane (Nito *et al.*, 2002). Interestingly, this Trp-x-x-x-(Phe/Tyr) repeat is found in the amino acid sequences of PEX5 in various species but differs in number and spacing. In *Arabidopsis* and watermelon, PEX5 has nine repeats, whereas PEX5 from yeast has only two (Schliebs *et al.*, 1999). Based on the report by Nito *et al.* (2002), the number of repeats is responsible for the efficiency of the binding to PEX14. Why plants have more Trp-x-x-x-(Phe/Tyr) motif compared with other species is unknown. In higher plants, a large amount of metabolic enzymes have to be transported rapidly according to the developmental condition such as germination or peroxisome transition. In such cases, PEX5 in higher plants might require higher activity for protein import.

c. pex6 *Mutant*

An *Arabidopsis pex6* mutant was isolated based on the resistance to the inhibitory effects of IBA (Zolman and Bartel, 2004). Although the *pex6* mutation is partially rescued by exogenous sucrose, the most intriguing phenotype is that this mutation affects seedlings and adult plants. A *pex6* mutant exhibits a short root, siliques with fewer seeds and smaller and fewer rosettes (Zolman and Bartel, 2004). PEX6 belongs to the AAA family of ATPases and *PEX6* genes were isolated from yeast and CHO cells (Hashiguchi *et al.*, 2002; McCollum *et al.*, 1993; Titorenko and Rachubinski, 2000). Human *PEX6* cDNA could almost completely rescue the *pex6* phenotype in *Arabidopsis*, although sucrose dependency was not complemented (Zolman and Bartel, 2004). In an *Arabidopsis pex6* mutant, the fusion protein with GFP and PTS1 is transported to peroxisomes correctly, although peroxisomes were larger and fewer compared with the wild-type plant, suggesting that the protein transport, at least the PTS1-dependent pathway, is unaffected (Zolman and Bartel, 2004). *Pichia pastoris*, however, is defective in the import of proteins with PTS1 (McCollum *et al.*, 1993). An *Arabidopsis pex6* mutant showed decreased levels of the PTS1 receptor, PEX5. PEX6 is proposed to act in recycling PTS receptors from the peroxisome to the cytosol by the interaction with other AAA protein, PEX1, although the precise function of PEX6 protein is not clarified.

d. pex10 *Mutant*

The PEX10 protein is an integral membrane protein with a C_3H_4 RING finger motif and is thought to be involved in the transport of matrix proteins by interaction with PEX12, another RING finger protein, and the PTS1 receptor, PEX5 (Chang *et al.*, 1999; Okumoto *et al.*, 2000). *PEX10* was cloned from various species, including yeast (Kalish *et al.*, 1995), *Arabidopsis* (Schumann *et al.*, 1999b), and humans (Warren *et al.*, 1998).

In *Arabidopsis*, PEX10 protein is localized to peroxisomes and the *PEX10* gene is widely expressed in various plant tissues (Sparkes *et al.*, 2003). The

Arabidopsis pex10 mutant was identified from the pool of T-DNA insertion lines (Schumann *et al.*, 2003; Sparkes *et al.*, 2003). Interestingly, no viable homozygous mutant plants were obtained from self-heterozygous mutant plants, indicating lethality of the homozygous insertion mutants (Schumann *et al.*, 2003; Sparkes *et al.*, 2003). In fact, self-heterozygous plants produce abnormal seeds, which contained embryos at an earlier developmental stage than normal seeds in the same silique. Ultrastructural analyses of homozygous immature seeds revealed that normal biogenesis of peroxisomes, lipid bodies, and ER was disturbed and instead, the dysfunctional monolayer membranes, which might derive from the bilayer membrane of the ER, accumulated in the cytosol (Schumann *et al.*, 2003). These results suggest that the function of PEX10 protein might not be limited to peroxisome biogenesis and that it plays multiple roles in the formation of various organelle membranes.

e. ped2/pex14 *Mutant*

The *ped2* mutant was originally isolated as a mutant showing the resistance to 2,4-DB and sucrose dependency (Hayashi *et al.*, 1998b). PED2 protein is homologous to PEX14 on the peroxisomal membrane in yeasts and mammals (Albertini *et al.*, 1997; Brocard *et al.*, 1997; Komori *et al.*, 1997; Shimizu *et al.*, 1999). PEX14 is thought to be a key component of matrix protein transport machinery on the peroxisomal membrane. As mentioned earlier, PEX14 acts as the initial docking site on the peroxisomal membrane (Fig. 5C). Actually, the efficiency of both PTS1- and PTS2-dependent transport is decreased in a *ped2* mutant (Hayashi *et al.*, 2000). In this mutant, the morphology of the three kinds of peroxisomes, glyoxysomes, leaf peroxisomes, and unspecialized peroxisomes, is altered from round to shrunken shapes. These morphological changes might reflect the low content of matrix proteins in peroxisomes. The reduction of protein transport in a *ped2* mutant causes the decreased activities of photorespiration as well as fatty acid β-oxidation (Hayashi *et al.*, 2000). Therefore, the growth of a *ped2* mutant is complemented in a high CO_2 (1000 Pa) under low light (50 $\mu E/m^2/s$), where the activity of photorespiration is not required. PEX14 is known to interact with other PEX proteins such as PEX13 and PEX17, besides PEX5 and PEX7 in yeasts and mammals (Fujiki, 2000; Girzalsky *et al.*, 1999; Holroyd and Erdmann, 2001; Reguenga *et al.*, 2001; Urquhart *et al.*, 2000). As stated earlier, the mechanism of matrix protein transport to peroxisomes in yeasts and mammals slightly differs from that in plants. Therefore, we cannot exclude the possibility that different interaction(s) with PEX14 might contribute to the protein transport in higher plants.

f. sse1/pex16 *Mutant*

The *shrunken seed 1* (*sse1*) mutant was identified in an *Arabidopsis* T-DNA insertion line that exhibited the shrunken seed phenotype, whose seeds were not viable (Lin et al., 1999). In mature seeds of the wild-type *Arabidopsis*, oil bodies and protein bodies occupy over 90% of the cell volume, whereas *sse1* seeds contain no recognizable protein bodies and few oil bodies. Actually, storage oils are reduced to approximately 10–16% of the wild-type level and storage protein content is approximately 30–60% of the wild-type level. Instead, starch is increased over proteins and lipids as the major storage substrates in *sse1* seeds (Lin et al., 1999). Interestingly, the rate of fatty acid biosynthesis was also reduced, and the composition of fatty acids was altered in the developing embryo of *sse1*. The lack of normal peroxisomes, the defects in the greening process, development of embryo and thylakoid membrane assembly were partially removed by exogenous Tween 80 (Lin et al., 2004), suggesting that the deficiency of fatty acid synthesis causes various developmental defects in *Arabidopsis*. These results show that plant peroxisomes are involved in both degradation and synthesis of fatty acids and that their balance might be essential for normal plant growth.

The *SSE1* gene encodes the protein homologous to PEX16, which is thought to be required for the assembly and proliferation of peroxisomes (Eitzen et al., 1997). In fact, SSE1 partially complements the defects of *pex16* mutants in yeasts (Lin et al., 1999) and the GFP-SSE1 fusion protein was localized to peroxisomes (Lin et al., 2004). The *sse1* mutant does not have any normal peroxisomes and over-expression of *SSE1* cDNA causes the abnormal peroxisomal distribution such as aggregation of peroxisomes, suggesting that SSE1 functions as a peroxin in *Arabidopsis* (Lin et al., 2004). However, the function of SSE1/PEX16 in *Arabidopsis* might be different from that in yeasts because over-expression of PEX16 in the yeast *Yallowia lipolytica* results in fewer enlarged peroxisomes (Eitzen et al., 1997). As a whole, the fact that the defect in one peroxisomal gene, *PEX16*, alters the composition of organelles and it suggests that the unknown peroxisomal functions are generally important throughout the plant life cycle.

g. apm1/drp3a *Mutant*

apm (*a*berrant *p*eroxisome *m*orphology) mutants were isolated based on the pattern of GFP fluorescence (Mano et al., 2004). Seeds of transgenic *Arabidopsis* in which peroxisomes normal in size and number can be visualized with GFP (Mano et al., 2002) were mutagenized with ethylmethane sulfonate and M2 plants examined under a fluorescence microscope. The *apm* mutants were classified into mutants with long peroxisomes, giant peroxisomes, GFP fluorescence in the cytosol as well as in peroxisomes, and other distributions of peroxisomes.

In one of these mutants, *apm1*, the peroxisomes are long and reduced in number, apparently as a result of inhibition of division (Fig. 6C). Unexpectedly, the *apm1* mutation causes aberrant morphology of mitochondria. In *apm1* mutants, the growth of *Arabidopsis* is repressed, indicating that the changes in the morphology of peroxisomes and mitochondria reduce the efficiency of metabolism in these organelles because the cooperation of various organelles including peroxisomes and mitochondria is essential for the viability. *APM1* encodes DRP3A (dynamin-related protein *3A*), formerly ADL2A (*Arabidopsis d*ynamin-like *2A*) (Arimura and Tsutsumi, 2002). In fact, the transient expression analysis showed that DRP3A is associated with peroxisomes and mitochondria. These findings indicate that the same dynamin molecule is involved in peroxisomal and mitochondrial division in higher plants. Dynamins and dynamin-like proteins are high molecular GTPases. In agreement with our results, a dynamin-like protein 1 (DLP1), which has been identified as the dynamin responsible for mitochondrial division, was found to play a role in peroxisomal morphology in mammalian cultured cells (Koch *et al.*, 2003; Li and Gould, 2003).

VI. CHALLENGE OF COMPREHENSIVE APPROACH IN THE POSTGENOMIC ERA

The large-scale sequencing projects of the whole plant genome, such as *Arabidopsis* (*Arabidopsis* Genome Initiative, 2000) and rice (Goff *et al.*, 2002; Yu *et al.*, 2002), and the large number of deposits of EST clones in the public databases provide us an enormous amount of biological information. Moreover, the pools of transgenic *Arabidopsis* plants, which contain transposonable elements or T-DNA insertion in their genome, have been rapidly expanding so that it is easier to identify gene disruption mutants. By using this enormous amount of public information, we can obtain novel insight into the gene functions, which help us to understand comprehensively, but not individually, and the function of peroxisomes. In this section, we reviewed the literature on the global analysis of plant peroxisomes based on the transcriptome and proteome.

A. TRANSCRIPTOME

To survey genes related to peroxisomal function and biogenesis comprehensively, we constructed a peroxisome-specific DNA microarray based on the information of the whole genome sequence (Kamada *et al.*, 2003). As described in Section V, most matrix proteins have PTS1 or PTS2 as a targeting sequence to peroxisomes. Patterns of PTS1 and PTS2 that have been identified (Hayashi *et al.*, 1997; Kato *et al.*, 2000) were used to search databases. In addition to candidates extracted from the database, catalases, cytosolic aconitase, and peroxins, which are essential for peroxisomal function although they do not have obvious PTS1 or PTS2, were also arrayed. Unexpectedly, 256 gene candidates were identified in *Arabidopsis* genome (Kamada *et al.*, 2003). Among them, only 29 genes are reported to be functionally characterized as peroxisomal proteins in higher plans. Although further experimental confirmations is required, the existence of many PTS1- or PTS2-containing proteins indicates that peroxisomes might have more various functions than we have predicted.

Using this peroxisome-specific microarray, expression profiles of genes in various organs of *Arabidopsis* have been extensively investigated (Kamada *et al.*, 2003). Statistical analyses of these expression profiles revealed that peroxisomal genes could be divided into five groups. One group showed ubiquitous expression in all organs examined, suggesting that peroxisomes in this group have basic functions. This group contains genes encoding proteins for β-oxidation, degradation of hydrogen peroxide, catabolism of branched-chain amino acids, and some unknown functions, whereas the other four groups were classified as showing organ-specific expression in seedlings, cotyledons, roots, and in both cotyledons and leaves. Interestingly, it has been revealed that most genes for β-oxidation enzymes are multiple genes and that one set of genes is constantly expressed in all organs, whereas another set is only expressed in seedlings. These results suggest the more detailed differentiation of plant peroxisomes, although plant peroxisomes have been categorized into three groups of glyoxysomes, leaf peroxisomes, and unspecialized peroxisomes.

Although experimental studies such as an examination of the subcellular localization or the phenotype of peroxisomes in a mutant with a defect in each gene are required, the existence of a large number of candidates with PTS variants needs to be considered. Therefore, the prediction of the subcellular localization of unknown proteins is a challenge of the postgenomic era. New approaches have been undertaken to optimize the prediction of PTS1-targeted proteins (Emanuelsson *et al.*, 2003; Neuberger *et al.*, 2003a,b). However, since the consensus motifs of PTS1 and PTS2 in plants are slightly different from those in yeast and mammals, plant-specific algorithms should be developed. In fact, novel properties, such as the detection of accessory sequences with an auxiliary targeting function by sequence conservation, were found and analyzed (Reumann, 2004).

In yeasts, the analysis using oligonucleotide-based whole genome microarray identified 224 genes with expression profiles similar to those of genes encoding peroxisomal proteins and genes in peroxisome biogenesis (Smith *et al.*, 2002). In *Arabidopsis* and rice, the use of oligonucleotide-based microarrays, which cover most of the whole genome, has already been started. It is anticipated that new components responsible for peroxisomal function and/or peroxisome biogenesis will be identified, as in yeasts.

B. PROTEOME

To better understand the function and behavior of overall proteins in organelles of *Arabidopsis*, experimental proteomic analysis combined with

2D gel electrophoresis and mass spectrometry have been performed, and the results were verified by searching the public databases. These proteomic analyses of organelles in *Arabidopsis* have been reported for mitochondria from cell cultures (Millar et al., 2001) and green tissues (Kruft et al., 2001), lumenal and peripheral thylakoid (Peltier et al., 2002), and plasma membrane and ER of calli (Prime et al., 2000). Peroxisomes in *Arabidopsis* have not been considered to be suitable for a proteomic analysis because they are small and fragile single membrane organelles and because *Arabidopsis* is too small to isolate peroxisomes in large quantities. However, the method for isolation of glyoxysomes and leaf peroxisomes from *Arabidopsis* cotyledons in high purity was established (Fukao et al., 2002). Using these highly purified leaf peroxisomes as the material, 53 proteins were analyzed by matrix-associated laser desorption ionization time-of-flight mass spectroscopy (MALDI-TOF MS) and 29 proteins were identified (Fukao et al., 2002). The 29 proteins included five enzymes essential for photorespiration such as GO and HPR (Fig. 4) and four proteins for scavenging hydrogen peroxide such as CAT and APX. The other 20 proteins were identified as novel proteins, which had similarities to protein kinases, protein phosphatases, ubiquitin, and so on. This result suggests the existence of novel regulation by phosphorylation and/or dephosphorylation in plant peroxisomes. Of these identified proteins, some proteins had a typical PTS1 or PTS2 as a targeting signal, others did not have obvious PTS and others had the PTS1-like motif where one amino acid is substituted with another one in PTS1. This result revealed that PTS1 motifs have a little flexibility besides the identified PTS1 motifs (Hayashi et al., 1997) and that there is the possibility of another derivative of PTS1 motif.

From glyoxysomes, 19 proteins were identified including the peroxisomal enzymes responsible for fatty acid β-oxidation, glyoxylate cycle, and scavenging of hydrogen peroxide (Fukao et al., 2003). Of several novel proteins, one protein with similarity to protein kinase, designated as GPK1 (glyoxysomal protein kinase 1), was further investigated. It was found to be GPK1 having the Ala-Lys-Ile tripeptide at its C-termini, and it was localized on the peroxisomal membrane (Fukao et al., 2003). These data showed the existence of regulation by phosphorylation and/or dephosphorylation in glyoxysomes, like leaf peroxisomes. Thus, proteomic analyses provided us unexpected novel information comprehensively. In rat liver peroxisomes, magnetic beads conjugated with antibodies against peroxisomal membrane protein, PMP70, were used for purification of peroxisomes, and purified peroxisomes were applied for proteomic analysis (Kikuchi et al., 2004). Development of new approaches is expected to make proteomic analysis easier, and all protein profiles will be obtained using peroxisomes isolated from various organs or various developmental stages in the future.

VII. FEW FINAL CONSIDERATIONS AND PERSPECTIVES

Considerable progress has been made during the past decades in the research on plant peroxisomes. The peroxisome has emerged as an essential organelle with a variety of functions in the cell. These various peroxisomal functions are indispensable for maintenance of normal plant growth. However, much remains to be elucidated, although not described here because of limited space. An example is our limited knowledge on the origin of peroxisomes. It is apparent that mitochondria and plastids evolved from a-proteobacteria and cyanobacteria, respectively. Although there is no direct evidence, peroxisomes are thought to have evolved from an ancient symbiont (Duve, 1996) because peroxisomal proteins, like mitochondria and plastids, are translocated posttranslationally into organelles and because peroxisomes, like mitochondria and plastids, are multiplied by division (Miyagishima *et al.*, 1999). In addition, the same dynamin-related protein molecule is involved in both peroxisomal and mitochondrial division in *Arabidopsis* (Mano *et al.*, 2004) and mammalian cultured cells (Koch *et al.*, 2003; Li and Gould, 2003). In *Caenorhabditis elegans*, all peroxisomal proteins are transported only via PTS1 pathway because of lack of the PTS2-dependent pathway (Motley *et al.*, 2000). However, there is still an unanswered question: Why does *C. elegans* lack the PTS2-dependent pathway? Peroxisomes are gathering attention as the site for biosynthesis of biodegradable polymers, polyhydroxyalkanoate (PHA) that is naturally synthesized in bacteria (Arai *et al.*, 2002; Poirer, 2002) because plant peroxisomes can metabolize the short- and medium-chain acyl-CoAs so that these intermediates are used as substrates for synthesis of PHA. This approach provides insight into the plant peroxisome as a new genetic engineering tool. Further investigations using novel tools, such as transcriptome, proteome, and metabolome, should help us to understand the mechanism of the functions and biogenesis of peroxisomes in the future and provide useful knowledge on the peroxisomes in plants as a plastic factory.

REFERENCES

Albertini, M., Rehling, P., Erdmann, R., Girzalsky, W., Kiel, J. A. K. W., Veenhuis, M., and Kunau, W.-H. (1997). Pex14p, a peroxisomal membrane protein binding both receptors of the two PTS-dependent import pathways. *Cell* **89**, 83–92.

Arabidopsis Genome Initiative (2000). Analysis of the genome sequence of the flowering plant *Arabidopsis thaliana*. *Nature* **408**, 796–815.

Arai, Y., Nakashita, H., Suzuki, Y., Kobayashi, Y., Shimizu, T., Yasuda, M., Doi, Y., and Yamaguchi, I. (2002). Synthesis of a novel class of polyhydroxyalkanoates in *Arabidopsis* peroxisomes, and their use in monitoring short-chain-length intermediates of β-oxidation. *Plant Cell Physiol.* **43**, 555–562.

Arimura, S., and Tsutsumi, N. (2002). A dynamin-like protein (ADL2b), rather than FtsZ, is involved in *Arabidopsis* mitochondrial division. *Proc. Natl. Acad. Sci. USA* **99**, 5727–5731.

Beevers, H. (1961). Metabolic production of sucrose from fat. *Nature* **191**, 433–436.

Beevers, H. (1979). Microbodies in higher plants. *Annu. Rev. Plant Physiol.* **30**, 159–193.

Berger, S., Bell, E., and Mullet, J. E. (1996). Two methyl jasmonate-insensitive mutants show altered expression of *AtVsp* in response to methyl jasmonate and wounding. *Plant Physiol.* **111**, 525–531.

Bethards, L. A., Skadsen, R. W., and Scandalios, J. G. (1987). Isolation and characterization of a cDNA clone for the *Cat2* gene in maize and its homology with other catalases. *Proc. Natl. Acad. Sci. USA* **84**, 6830–6834.

Biesgen, C., and Weiler, E. W. (1999). Structure and regulation of OPR1 and OPR2, two closely related genes encoding 12-oxophytodienoic acid-10,11-reductases from *Arabidopsis thaliana*. *Planta* **208**, 155–165.

Brocard, C., Kragler, F., Simon, M. M., Schuster, T., and Hartig, A. (1994). The tetratricopeptide repeat-domain of the PAS10 protein of *Saccharomyces cerevisiae* is essential for binding the peroxisomal targeting signal -SKL. *Biochim. Biophys. Acta* **204**, 1016–1022.

Brocard, C., Lametschwandtner, G., Koudelka, R., and Hartig, A. (1997). Pex14p is a member of the protein linkage map of Pex5p. *EMBO J.* **16**, 5491–5500.

Bunkelmann, J. R., and Trelease, R. N. (1996). Ascorbate peroxidase: A prominent membrane protein in oilseed glyoxysomes. *Plant Physiol.* **110**, 589–598.

Chang, C.-C., Warren, D. S., Sacksteder, K. A., and Gould, S. J. (1999). PEX12 interacts with PEX5 and PEX10 and acts downstream of receptor docking in peroxisomal matrix protein import. *J. Cell Biol.* **147**, 761–773.

Charlton, W., and Lopez-Huertas, E. (2002). PEX genes in plant and other organisms. *In* "Plant Peroxisomes" (A. Baker and I. A. Graham, Eds.), pp. 385–426. Kluwer Academic Publishers, The Netherlands.

Chory, J., and Peto, C. (1990). Mutations in the *DET1* gene affect cell-type-specific expression of light-regulated genes and chloroplast development in *Arabidopsis*. *Proc. Natl. Acad. Sci. USA* **87**, 8776–8780.

Cooper, T. G., and Beevers, H. (1969). β-Oxidation in glyoxysomes from castor bean endosperm. *J. Biol. Chem.* **244**, 3514–3520.

Cornah, J. E., Germain, V., Ward, J. L., Beale, M. H., and Smith, S. M. (2004). Lipid utilization, gluconeogenesis and seedling growth in *Arabidopsis* mutants lacking the glyoxylate cycle enzyme malate synthase. *J. Biol. Chem.* **279**, 42916–42923.

Corpas, F. J., Barroso, J. B., and Río, L. A.d. (2001). Peroxisomes as a source of reactive oxygen species and nitric oxide signal molecules in plant cells. *Trends Plant Sci.* **6**, 145–150.

Davis, P. J. (1995). The plant hormone: Their nature, occurrence, and functions. *In* "Plant Hormones: Physiology, Biochemistry, and Molecular Biology," pp. 1–5. Kluwer Academic Publishers, The Netherlands.

De Bellis, L., Giuntini, P., Hayashi, H., Hayashi, M., and Nishimura, M. (1999). Purification and characterization of pumpkin long-chain acyl-CoA oxidase. *Physiol. Plant.* **106**, 170–176.

De Bellis, L., Gonzali, S., Alpi, A., Hayashi, H., Hayashi, M., and Nishimura, M. (2000). Purification and characterization of a novel pumpkin short-chain acyl-coenzyme A oxidase with structural similarity to acyl-coenzyme A dehydrogenase. *Plant Physiol.* **123**, 327–334.

De Bellis, L., and Nishimura, M. (1991). Development of enzymes of the glyoxylate cycle during senescence of pumpkin cotyledons. *Plant Cell Physiol.* **32**, 555–561.

Distel, B., Erdmann, R., Gould, S. J., Blobel, G., Crane, D. I., Cregg, J. M., Dodt, G., Fujiki, Y., Goodman, J. M., Just, W. W., Kiel, J. A. K. W., Kunau, W.-H., Lazarow, P. B., Mannaerts, G. P., Moser, H. W., Osumi, T., Tsukamoto, T., Valle, D., Klei, I. V. D., Veldhoven, P. P. V., and Veenhuis, M. (1996). A unified nomenclature for peroxisome biogenesis factors. *J. Cell Biol.* **135**, 1–3.

Duve, C. D. (1996). The birth of complex cells. *Sci. Am.* **274**, 38–45.
Eastmond, P. J., Germain, V., Lange, P. R., Bryce, J. H., Smith, S. M., and Graham, I. A. (2000). Postgerminative growth and lipid catabolism in oilseeds lacking the glyoxylate cycle. *Proc. Natl. Acad. Sci. USA* **97**, 5669–5674.
Eitzen, G. A., Szilard, R. K., and Rachubinski, R. A. (1997). Enlarged peroxisomes are present in oleic acid-grown *Yarrowia lipolytica* overexpressing the *PEX16* gene encoding an intraperoxisomal peripheral membrane peroxin. *J. Cell Biol.* **137**, 1265–1278.
Eitzen, G. A., Titorenko, V. I., Smith, J. F., Veenhuis, M., Szilard, R. K., and Rachuminski, R. A. (1996). The *Yarrowia lipolytica* gene *PAY5* encodes a peroxisomal integral membrane protein homologous to the mammalian peroxisome assembly factor PAF-1. *J. Biol. Chem.* **271**, 20300–20306.
Elgersma, Y., Elgersma-Hooisma, M., Wenzel, T., McCaffery, J. M., Farquhar, M. G., and Subramani, S. (1998). A mobile PTS2 receptor for peroxisomal protein import in *Pichia pastoris*. *J. Cell Biol.* **140**, 807–820.
Emanuelsson, O., Elofsson, A., Heijne, G.v., and Cristóbal, S. (2003). In silico prediction of the peroxisomal proteome in fungi, plants and animals. *J. Mol. Biol.* **330**, 443–456.
Epstein, E., and Lavee, S. (1984). Conversion of indole-3-butyric acid to indole-3-acetic acid by cutting of grapevine (*Vitis vinifera*) and olive (*Olea europea*). *Plant Cell Physiol.* **25**, 697–703.
Esaka, M., Yamada, N., Kitabayashi, M., Setoguchi, Y., Tsugeki, R., Kondo, M., and Nishimura, M. (1997). cDNA cloning and differential gene expression of three catalases in pumpkin. *Plant Mol. Biol.* **33**, 141–155.
Estelle, M. A., and Somerville, C. (1987). Auxin-resistant mutants of *Arabidopsis thaliana* with an altered morphology. *Mol. Gen. Genet.* **206**, 200–206.
Feys, B. J. F., Benedetti, C. E., Penfold, C. N., and Turner, J. G. (1994). *Arabidopsis* mutants selected for resistance to the phytotoxin coronatine are male sterile, insensitive to methyl jasmonate, and resistant to a bacterial pathogen. *Plant Cell* **6**, 751–759.
Flynn, C. R., Mullen, R. T., and Trelease, R. N. (1998). Mutational analysis of a type 2 peroxisomal targeting signal that is capable of directing oligomeric protein import into tobacco BY-2 glyoxysomes. *Plant J.* **16**, 709–720.
Footitt, S., Slocombe, S. P., Larner, V., Kurup, S., Wu, Y., Larson, T., Graham, I., Baker, A., and Holdsworth, M. (2002). Control of germination and lipid mobilization by *COMATOSE*, the *Arabidopsis* homologue of human ALDP. *EMBO J.* **21**, 2912–2922.
Fransen, M., Brees, C., Baumgart, E., Vanhooren, J. C. T., Baes, M., Mannaerts, G. P., and Veldhoven, P. P. V. (1995). Identification and characterization of the putative human peroxisomal C-terminal targeting signal import receptor. *J. Biol. Chem.* **270**, 7731–7736.
Frugoli, J. A., Zhong, H. H., Nuccio, M. L., McCourt, P., McPeek, M. A., Thomas, T. L., and McClung, C. R. (1996). Catalase is encoded by a multigene family in *Arabidopsis thaliana* (L.) Heynh. *Plant Physiol.* **112**, 327–336.
Fujiki, Y. (2000). Peroxisome biogenesis and peroxisome biogenesis disorders. *FEBS Lett.* **476**, 42–46.
Fukao, Y., Hayashi, M., Hara-Nishimura, I., and Nishimura, M. (2003). Novel glyoxysomal protein kinase, GPK1, identified by proteomic analysis of glyoxysomes in etiolated cotyledons of *Arabidopsis thaliana*. *Plant Cell Physiol.* **44**, 1002–1012.
Fukao, Y., Hayashi, M., and Nishimura, M. (2002). Proteomic analysis of leaf peroxisomal proteins in greening cotyledons of *Arabidopsis thaliana*. *Plant Cell Physiol.* **43**, 689–696.
Fukao, Y., Hayashi, Y., Mano, S., Hayashi, M., and Nishimura, M. (2001). Developmental analysis of a putative ATP/ADP carrier protein localized on glyoxysomal membranes during the peroxisome transition in pumpkin cotyledons. *Plant Cell Physiol.* **42**, 835–841.
Fulda, M., Schnurr, J., Abbadi, A., Heinz, E., and Browse, J. (2004). Peroxisomal acyl-CoA synthetase activity is essential for seedling development in *Arabidopsis thaliana*. *Plant Cell* **16**, 394–405.

Germain, V., Rylott, E. L., Larson, T. R., Sherson, S. M., Bechtold, N., Carde, J.-P., Bryce, J. H., Graham, I. A., and Smith, S. M. (2001). Requirement for 3-ketoacyl-CoA thiolase-2 in peroxisome development, fatty acid β-oxidation and breakdown of triacylglycerol in lipid bodies of *Arabidopsis* seedlings. *Plant J.* **28**, 1–12.

Gietl, C. (1990). Glyoxysomal malate dehydrogenase from watermelon is synthesized with an amino-terminal transit peptide. *Proc. Natl. Acad. Sci. USA* **87**, 5773–5777.

Girzalsky, W., Rehling, P., Stein, K., Kipper, J., Blank, L., Kunau, W.-H., and Erdmann, R. (1999). Involvement of Pex13p in Pex14p localization and peroxisomal targeting signal 2-dependent protein import into peroxisomes. *J. Cell Biol.* **144**, 1151–1162.

Glover, J. R., Andrews, D. W., and Rachubinski, R. A. (1994). *Saccharomyces cerevisiae* peroxisomal thiolase is imported as a dimer. *Proc. Natl. Acad. Sci. USA* **91**, 10541–10545.

Goff, S. A., Ricke, D., Lan, T.-H., Presting, G., Wang, R., Dunn, M., Glazebrook, J., Sessions, A., Oeller, P., Varma, H., Hadley, D., Hutchison, D., Martin, C., Katagiri, F., Lange, B. M., Moughamer, T., Xia, Y., Budworth, P., Zhong, J., Miguel, T., Paszkowski, U., Zhang, S., Colbert, M., Sun, W.-l., Chen, L., Cooper, B., Park, S., Wood, T. C., Mao, L., Quail, P., Wing, R., Dean, R., Yu, Y., Zharkikh, A., Shen, R., Sahasrabudhe, S., Thomas, A., Cannings, R., Gutin, A., Pruss, D., Reid, J., Tavtigian, S., Mitchell, J., Eldredge, G., Scholl, T., Miller, R. M., Bhatnagar, S., Adey, N., Rubano, T., Tusneem, N., Robinson, R., Feldhaus, J., Macalma, T., Oliphant, A., and Briggs, S. (2002). A draft sequence of the rice genome (*Oryza sativa* L. ssp. *japonica*). *Science* **296**, 92–100.

Gould, S. J., Keller, G.-A., Hosken, N., Wilkinson, J., and Subramani, S. (1989). A conserved tripeptide sorts proteins to peroxisomes. *J. Cell Biol.* **108**, 1657–1664.

Gould, S. J., Keller, G.-A., Schneider, M., Howell, S. H., Garrard, L. J., Goodman, J. M., Distel, B., Tabak, H., and Subramani, S. (1990). Peroxisomal protein import is conserved between yeast, plants, insects and mammals. *EMBO J.* **9**, 85–90.

Gould, S. J., Keller, G.-A., and Subramani, S. (1987). Identification of a peroxisomal targeting signal at the carboxy terminus of firefly luciferase. *J. Cell Biol.* **105**, 2323–2931.

Gould, S. J., Keller, G.-A., and Subramani, S. (1988). Identification of peroxisomal targeting signals located at the carboxy terminus of four peroxisomal proteins. *J. Cell Biol.* **107**, 897–905.

Greenler, J. M., Sloan, J. S., Schwartz, B. W., and Becker, W. M. (1989). Isolation, characterization and sequence analysis of a full-length cDNA clone encoding NADH-dependent hydroxypyruvate reductase from cucumber. *Plant Mol. Biol.* **13**, 139–150.

Grotjohann, N., Janning, A., and Eising, R. (1997). *In vitro* photoactivation of catalase isoforms from cotyledons of sunflower (*Helianthus annuus* L.). *Arch. Biochem. Biophys.* **346**, 208–218.

Guex, N., Henry, H., Flach, J., Richter, H., and Widmer, F. (1995). Glyoxysomal malate dehydrogenase and malate synthase from soybean cotyledons (*Glycine max* L.): Enzyme association, antibody production and cDNA cloning. *Planta* **197**, 369–375.

Gurvitz, A., Wabnegger, L., Langer, S., Hamilton, B., Ruis, H., and Hartig, A. (2001). The tetratricopeptide repeat domains of human, tobacco, and nematode PEX5 proteins are functionally interchangeable with the analogous native domain for peroxisomal import of PTS1-terminated proteins in yeast. *Mol. Gen. Genet.* **265**, 276–286.

Gut, H., and Matile, P. (1988). Apparent induction of key enzymes of the glyoxylic acid cycle in senescent barley leaves. *Planta* **176**, 548–550.

Hashiguchi, N., Kojidani, T., Imanaka, T., Haraguchi, T., Hiranaka, Y., Baumgart, E., Yokota, S., Tsukamoto, T., and Osumi, T. (2002). Peroxisomes are formed from complex membrane structures in *PEX6*-deficient CHO cells upon genetic complementation. *Mol. Biol. Cell* **13**, 711–722.

Hashimoto, T. (1996). Peroxisomal β-oxidation: Enzymology and molecular biology. *Annu. N.Y. Acad. Sci.* **804**, 86–98.

Hayashi, H., Bellis, L. D., Ciurli, A., Kondo, M., Hayashi, M., and Nishimura, M. (1999). A novel acyl-CoA oxidase can oxidize short-chain acyl-CoA in plant peroxisomes. *J. Biol. Chem.* **274,** 12715–12721.

Hayashi, H., Bellis, L. D., Yamaguchi, K., Kato, A., Hayashi, M., and Nishimura, M. (1998a). Molecular characterization of a glyoxysomal long chain acyl-CoA oxidase that is synthesized as a precursor of higher molecular mass in pumpkin. *J. Biol. Chem.* **273,** 8301–8307.

Hayashi, M., Aoki, M., Kondo, M., and Nishimura, M. (1997). Changes in targeting efficiencies of proteins to plant microbodies caused by amino acid substitutions in the carboxy-terminal tripeptide. *Plant Cell Physiol.* **38,** 759–768.

Hayashi, M., Bellis, L. D., Alpi, A., and Nishimura, M. (1995). Cytosolic aconitase participates in the glyoxylate cycle in etiolated pumpkin cotyledons. *Plant Cell Physiol.* **36,** 669–680.

Hayashi, M., Nito, K., Takei-Hoshi, R., Yagi, M., Kondo, M., Suenaga, A., Yamaya, T., and Nishimura, M. (2002). Ped3p is a peroxisomal ATP-binding cassette transporter that might supply substrates for fatty acid β-oxidation. *Plant Cell Physiol.* **43,** 1–11.

Hayashi, M., Nito, K., Toriyama-Kato, K., Kondo, M., Yamaya, T., and Nishimura, M. (2000). *At*Pex14p maintains peroxisomal functions by determining protein targeting to three kinds of plant peroxisomes. *EMBO J.* **19,** 5701–5710.

Hayashi, M., Toriyama, K., Kondo, M., and Nishimura, M. (1998b). 2,4-Dichlorophenoxy butyric acid-resistant mutants of *Arabidopsis* have defects in glyoxysomal fatty acid β-oxidation. *Plant Cell* **10,** 183–195.

Hayashi, M., Tsugeki, R., Kondo, M., Mori, H., and Nishimura, M. (1996). Pumpkin hydroxypyruvate reductases with and without a putative C-terminal signal for targeting to microbodies may be produced by alternative splicing. *Plant Mol. Biol.* **30,** 183–189.

Hayashi, M., Yagi, M., Nito, K., Kamada, T., and Nishimuma, M. (2005). Differential contribution of two peroxisomal protein receptors to the maintenance of peroxisomal functions in *Arabidopsis*. *J. Biol. Chem.* **280,** 14829–14835.

Hayashi, Y., Hayashi, M., Hayashi, H., Hara-Nishimura, I., and Nishimura, M. (2001). Direct interaction between glyoxysomes and lipid bodies in cotyledons of the *Arabidopsis thaliana ped1* mutant. *Protoplasma* **218.**

Hettema, E. H., van Roemund, C. W., Distel, B., van den Berg, M., Vilela, C., Rodrigues-Pousada, C., Wanders, R. J., and Tabak, H. F. (1996). The ABC transporter proteins Pat1 and Pat2 are required for import of long-chain fatty acids into peroxisomes of *Saccharomyces cerevisiae*. *EMBO J.* **15,** 3813–3822.

Holroyd, C., and Erdmann, R. (2001). Protein translocation machineries of peroxisomes. *FEBS Lett.* **501,** 6–10.

Hooks, M. A., Bode, K., and Couée, I. (1996). Higher-plant medium- and short-chain acyl-CoA oxidases: Identification, purification and characterization of two novel enzymes of eukaryotic peroxisomal β-oxidation. *Biochem. J.* **320,** 607–614.

Howe, G. A., Ligthner, J., Browse, J., and Ryan, C. A. (1996). An octadecanoid pathway mutant (JL5) of tomato is compromised in signaling for defense against insect attack. *Plant Cell* **8,** 2067–2077.

Hu, J., Aguirre, M., Peto, C., Alonso, J., Ecker, J., and Chory, J. (2002). A role for peroxisomes in photomorphogenesis and development of *Arabidopsis*. *Science* **297,** 405–409.

Igarashi, D., Miwa, T., Seki, M., Kobayashi, M., Kato, T., Tabata, S., and Shinozaki, K. (2003). Identification of photorespiratory *glutamate:glyoxylate aminotransferase (GGAT)* gene in *Arabidopsis*. *Plant J.* **33,** 975–987.

Jedd, G., and Chua, N.-H. (2002). Visualization of peroxisomes in living plant cells reveals acto-myosin-dependent cytoplasmic streaming and peroxisome budding. *Plant Cell Physiol.* **43,** 384–392.

Johnson, T. L., and Olsen, L. J. (2001). Building new models for peroxisome biogenesis. *Plant Physiol.* **127,** 731–739.

Kalish, J. E., Theda, C., Morrell, J. C., Berg, J. M., and Gould, S. J. (1995). Formation of the peroxisome lumen is abolished by loss of *Pichia pastoris* Pas7p, a zinc-binding integral membrane protein of the peroxisome. *Mol. Cell. Biol.* **15**, 6406–6419.

Kamada, T., Nito, K., Hayashi, H., Mano, S., Hayashi, M., and Nishimura, M. (2003). Functional differentiation of peroxisomes revealed by expression profiles of peroxisomal genes in *Arabidopsis thaliana*. *Plant Cell Physiol.* **44**, 1275–1289.

Kamijyo, K., Kamijyo, T., Ueno, I., Osumi, T., and Hashimoto, T. (1992). Nucleotide-sequence of the human 70 kDa peroxisomal membrane protein-a member of ATP-binding cassette transporters. *Biochim. Biophys. Acta* **1129**, 323–327.

Kato, A., Hayashi, M., Kondo, M., and Nishimura, M. (2000). Transport of peroxisomal proteins synthesized as large precursors in plants. *Cell Biochem. Biophys.* **32**, 269–275.

Kato, A., Hayashi, M., Mori, H., and Nishimura, M. (1995). Molecular characterization of a glyoxysomal citrate synthase that is synthesized as a precursor of higher molecular mass in pumpkin. *Plant Mol. Biol.* **27**, 377–390.

Kato, A., Hayashi, M., and Nishimura, M. (1999). Oligomeric proteins containing N-terminal targeting signals are imported into peroxisomes in transgenic *Arabidopsis*. *Plant Cell Physiol.* **40**, 586–591.

Kato, A., Hayashi, M., Takeuchi, Y., and Nishimura, M. (1996). cDNA cloning and expression of a gene for 3-ketoacyl-CoA thiolase in pumpkin cotyledons. *Plant Mol. Biol.* **31**, 843–852.

Kato, A., Takeda-Yoshikawa, Y., Hayashi, M., Kondo, M., Hara-Nishimura, I., and Nishimura, M. (1998). Glyoxysomal malate dehydrogenase in pumpkin: Cloning of a cDNA and functional analysis of its presequence. *Plant Cell Physiol.* **39**, 186–195.

Kikuchi, M., Hatano, N., Yokota, S., Shimozawa, N., Imanaka, T., and Taniguchi, H. (2004). Proteomic analysis of rat liver peroxisome. Presence of peroxisome-specific isozyme of Lon protease. *J. Biol. Chem.* **279**, 421–428.

Kim, D.-J., and Smith, S. M. (1994). Expression of a single gene encoding microbody NAD-malate dehydrogenase during glyoxysome and peroxisome development in cucumber. *Plant Mol. Biol.* **26**, 1833–1841.

Koch, A., Thiemann, M., Grabenbauer, M., Yoon, Y., McNiven, M. A., and Schrader, M. (2003). Dynamin-like protein 1 is involved in peroxisomal fission. *J. Biol. Chem.* **278**, 8597–8605.

Komori, M., Rasmussen, S. W., Kiel, J. A. K. W., Baerrends, R. J. S., Cregg, J. M., Klei, I. J. V. D., and Veenhuis, M. (1997). The *Hansenula polymorpha PEX14* gene encodes a novel peroxisomal membrane protein essential for peroxisome biogenesis. *EMBO J.* **16**, 44–53.

Kornberg, H. L., and Kerbs, H. A. (1957). Synthesis of cell constituents from C2-units by a modified tricarboxylic acid cycle. *Nature* **178**, 988–991.

Kragler, F., Lametschwandtner, G., Christmann, J., Hartig, A., and Harada, J. J. (1998). Identification and analysis of the plant peroxisomal targeting signal 1 receptor NtPEX5. *Proc. Natl. Acad. Sci. USA* **95**, 13336–13341.

Krieken, W. M. V. D., Breteler, H., and Visser, M. H. M. (1992). The effect of the conversion of indolebutyric acid into indoleacetic acid on root formation of microcuttings of *Malus*. *Plant Cell Physiol.* **33**, 709–713.

Kruft, V., Eubel, H., Jänsch, L., Werhahn, W., and Braun, H.-P. (2001). Proteomic approach to identify novel mitochondrial proteins in *Arabidopsis*. *Plant Physiol.* **127**, 1694–1710.

Lange, P. R., Eastmond, P. J., Madagan, K., and Graham, I. A. (2004). An *Arabidopsis* mutant disrupted in valine catabolism is also compromised in peroxisomal fatty acid β-oxidation. *FEBS Lett.* **571**, 147–153.

Lazarow, P. B. (1978). Rat liver peroxisomes catalyze the β oxidation of fatty acids. *J. Biol. Chem.* **253**, 1522–1528.

Lee, M. S., Mullen, R. T., and Trelease, R. N. (1997). Oilseed isocitrate lyases lacking their essential type 1 peroxisomal targeting signal are piggybacked to glyoxysomes. *Plant Cell* **9**, 185–197.

Leij, I. V. D., Franse, M. M., Elgersma, Y., Distel, B., and Tabak, H. F. (1993). PAS10 is a tetratricopeptide-repeat protein that is essential for the import of most matrix proteins into peroxisomes of *Saccharomyces cerevisiae*. *Proc. Natl. Acad. Sci. USA* **90**, 11782–11786.

Li, X., and Gould, S. J. (2003). The dynamin-like GTPase DLP1 is essential for peroxisome division and is recruited to peroxisomes in part PEX11. *J. Biol. Chem.* **278**, 17012–17020.

Liepman, A. H., and Olsen, L. J. (2001). Peroxisomal alanine:glyoxylate aminotransferase (AGT1) is a photorespiratory enzyme with multiple substrates in *Arabidopsis thaliana*. *Plant J.* **25**, 487–498.

Lin, Y., Cluette-Brown, J. E., and Goodman, H. M. (2004). The peroxisome deficient *Arabidopsis* mutant *sse1* exhibits impaired fatty acid synthesis. *Plant Physiol.* **135**, 814–827.

Lin, Y., Sun, L., Nguyen, L. V., Rachubinski, R. A., and Goodman, H. M. (1999). The Pex16p homolog SSE1 and storage organelle formation in *Arabidopsis* seeds. *Science* **284**, 328–330.

Liu, Y., Gu, K. L., and Dieckmann, C. L. (1996). Independent regulation of full-length and 5'-truncated PAS5 mRNAs in *Saccharomyces cerevisiae*. *Yeast* **12**, 135–143.

Lombard-Platet, G., Savary, S., Sarde, C.-O., Mandel, J.-L., and Chimini, G. (1996). A close relative of the adrenoleukodystrophy (*ALD*) gene codes for a peroxisomal protein with a specific expression pattern. *Proc. Natl. Acad. Sci. USA* **93**, 1265–1269.

Lopez-Huertas, E., Charlton, W. L., Johnson, B., Graham, I. A., and Baker, A. (2000). Stress induces peroxisome biogenesis genes. *EMBO J.* **19**, 6770–6777.

Lopez-Huertas, E., Oh, J., and Baker, A. (1999). Antibodies against Pex14p block ATP-dependent binding of matrix proteins to peroxisomes *in vitro*. *FEBS Lett.* **459**, 227–229.

Ludt, C., and Kindl, H. (1990). Characterization of a cDNA encoding *Lens culinaris* glycolate oxidase and developmental expression of glycolate oxidase mRNA in cotyledons and leaves. *Plant Physiol.* **94**, 1193–1198.

Mano, S., Hayashi, M., Kondo, M., and Nishimura, M. (1996). cDNA cloning and expression of a gene for isocitrate lyase in pumpkin cotyledons. *Plant Cell Physiol.* **37**, 941–948.

Mano, S., Hayashi, M., Kondo, M., and Nishimura, M. (1997). Hydroxypyruvate reductase with a carboxy-terminal targeting signal to microbodies is expressed in *Arabidopsis*. *Plant Cell Physiol.* **38**, 449–455.

Mano, S., Hayashi, M., and Nishimura, M. (1999). Light regulates alternative splicing of hydroxypyruvate reductase in pumpkin. *Plant J.* **17**, 309–320.

Mano, S., Nakamori, C., Hayashi, M., Kato, A., Kondo, M., and Nishimura, M. (2002). Distribution and characterization of peroxisomes in *Arabidopsis* by visualization with GFP: Dynamic morphology and actin-dependent movement. *Plant Cell Physiol.* **43**, 331–341.

Mano, S., Nakamori, C., Kondo, M., Hayashi, M., and Nishimura, M. (2004). An *Arabidopsis* dynamin-related protein, DRP3A, controls both peroxisomal and mitochondrial division. *Plant J.* **38**, 487–498.

Marzioch, M., Erdmann, R., Veenhuis, M., and Kunau, W.-H. (1994). PAS7 encodes a novel yeast member of the WD-40 protein family essential for import of 3-oxoacyl-CoA thiolase, a PTS2-containing protein, into peroxisomes. *EMBO J.* **13**, 4908–4918.

Mason, H. S., and Mullet, J. E. (1990). Expression of two soybean vegetative storage protein genes during development and in response to water deficit, wounding, and jasmonic acid. *Plant Cell* **2**, 569–579.

Mathur, J., Mathur, N., and Hülskamp, M. (2002). Simultaneous visualization of peroxisomes and cytoskeletal elements reveals actin and not microtubule-based peroxisome motility in plants. *Plant Physiol.* **128**, 1031–1045.

McCammon, M. T., McNew, J. A., Willy, P. J., and Goodman, J. M. (1994). An internal region of the peroxisomal membrane protein PMP47 is essential for sorting to peroxisomes. *J. Cell Biol.* **124**, 915–925.

McCollum, D., Monosov, E., and Subramani, S. (1993). The pas8 mutant of *Pichia pastoris* exhibits the peroxisomal protein import deficiencies of Zellweger syndrome cells—the PAS8 protein binds to the COOH-terminal tripeptide peroxisomal targeting signal, and is a member of the TPR protein family. *J. Cell Biol.* **121**, 761–774.

McConn, M., and Browse, J. (1996). The critical requirement for linolenic acid is pollen development, not photosynthesis, in an *Arabidopsis* mutant. *Plant Cell* **8**, 403–416.

McConn, M., Creelman, R. A., Bell, E., Mullet, J. E., and Browse, J. (1997). Jasmonate is essential for insect defense in *Arabidopsis*. *Proc. Natl. Acad. Sci. USA* **94**, 5473–5477.

McNew, J. A., and Goodman, J. M. (1996). The targeting and assembly of peroxisomal proteins: Some old rules do not apply. *Trends Biochem. Sci.* **21**, 54–58.

Millar, A. H., Sweetlove, L. J., Giegé, P., and Leaver, C. J. (2001). Analysis of the *Arabidopsis* mitochondrial proteome. *Plant Physiol.* **127**, 1711–1727.

Miyagishima, S.-y., Itoh, R., Toda, K., Kuroiwa, H., Nishimura, M., and Kuroiwa, T. (1999). Microbody proliferation and segregation in the single-microbody alga *Cyanidioschyzon merolae*. *Planta* **208**, 326–336.

Mosser, J., Douar, A. M., Sarde, C. O., Kioschis, P., Feil, R., Moser, H., Poustka, A. M. L. M. J., and Aubourg, P. (1993). Putative X-linked adrenoleukodystrophy gene shares unexpected homology with ABC transporters. *Nature* **361**, 726–730.

Motley, A. M., Hettema, E. H., Ketting, R., Platerk, R., and Tabak, H. F. (2000). *Caenorhabditis elegans* has a single pathway to target matrix proteins to peroxisomes. *EMBO reports* **1**, 40–46.

Mullen, R. T., Lisenbee, C. S., Flynn, C. R., and Trelease, R. N. (2001). Stable and transient expression of chimeric peroxisomal membrane proteins induces an independent "zippering" of peroxisomes and an endoplasmic reticulum subdomain. *Planta* **213**, 849–863.

Neuberger, G., Maurer-Stroh, S., Eisenhaber, B., Hartig, A., and Eisenhaber, F. (2003a). Motif refinement of the peroxisomal targeting signal 1 and evaluation of taxon-specific differences. *J. Mol. Biol.* **328**, 567–579.

Neuberger, G., Maurer-Stroh, S., Eisenhaber, B., Hartig, A., and Eisenhaber, F. (2003b). Prediction of peroxisomal targeting signal 1 containing proteins from amino acid sequence. *J. Mol. Biol.* **328**, 581–592.

Nishimura, M., Takeuchi, Y., Bellis, L. D., and Hara-Nishimura, I. (1993). Leaf peroxisomes are directly transformed to glyoxysomes during senescence of pumpkin cotyledons. *Protoplasma* **175**, 131–137.

Nishimura, M., Yamaguchi, J., Mori, H., Akazawa, T., and Yokota, S. (1986). Immunocytochemical analysis shows that glyoxysomes are directly transformed to leaf peroxisomes during greening of pumpkin cotyledon. *Plant Physiol.* **80**, 313–316.

Nissen, S. J., and Sutter, E. G. (1990). Stability of IAA and IBA in nutrient medium to several tissue culture procedures. *Hor. Sci.* **25**, 800–802.

Nito, K., Hayashi, M., and Nishimura, M. (2002). Direct interaction and determination of binding domains among peroxisomal import factors in *Arabidopsis thaliana*. *Plant Cell Physiol.* **43**, 355–366.

Nito, K., Yamaguchi, K., Kondo, M., Hayashi, M., and Nishimura, M. (2001). Pumpkin peroxisomal ascorbate peroxidase is localized on peroxisomal membranes and unknown membranous structures. *Plant Cell Physiol.* **42**, 20–27.

Nördstrom, A.-C., Jacobs, F. A., and Eliasson, L. (1991). Effect of exogenous indole-3-acetic acid and indole-3-butyric acid on internal levels of the respective auxins and their conjugation with aspartic acid during adventitious root formation in pea cuttings. *Plant Physiol.* **96**, 856–861.

Okumoto, K., Abe, I., and Fujiki, Y. (2000). Molecular anatomy of the peroxin Pex12p. Ring finger domain is essential for Pex12p function and interacts with the peroxisome-targeting signal type 1-receptor Pex5p and a ring peroxin, Pex10p. *J. Biochem.* **275**, 25700–25710.

Otera, H., Okumoto, K., Tateishi, K., Ikoma, Y., Matsuda, E., Nishimura, M., Tsukamoto, T., Osumi, T., Ohashi, K., Higuchi, O., and Fujiki, Y. (1998). Peroxisome targeting signal type

1 (PTS1) receptor is involved in import of both PTS1 and PTS2: Studies with *PEX5*-defective CHO cell mutants. *Mol. Cell. Biol.* **18**, 388–399.

Pause, B., Saffrich, R., Hunziker, A., Ansorge, W., and Just, W. W. (2000). Targeting of the 22 kDa integral peroxisomal membrane proteins. *FEBS Lett.* **471**, 23–28.

Peltier, J.-B., Emanuelsson, O., Kalume, D. E., Ytterberg, J., Friso, G., Rudella, A., Liberles, D. A., Söderberg, L., Roepstorff, P., Heijne, G. V., and Wijk, K. J. V. (2002). Central functions of the lumenal and peripheral thylakoid proteome of *Arabidopsis* determined by experimentation and genome-wide prediction. *Plant Cell* **14**, 211–236.

Pepper, A., Delaney, T., Washburn, T., Poole, D., and Chory, J. (1994). *DET1*, a negative regulator of light-mediated development and gene expression in *Arabidopsis*, encodes a novel nuclear-localized protein. *Cell* **78**, 109–116.

Pistelli, L., Bellis, L. D., and Alpi, A. (1995). Evidences of glyoxylate cycle in peroxisomes of senescent cotyledons. *Plant Sci.* **109**, 13–21.

Poirer, Y. (2002). Polyhydroxyalkanoate synthesis in plant peroxisomes. *In* "Plant peroxisome" (A. Baker and I. A. Graham, Eds.), pp. 465–496. Kluwer Academic Publishers, The Netherlands.

Preisig-Müller, R., Gühnemann-Schäfer, K., and Kindl, H. (1994). Domains of the tetrafunctional protein acting in glyoxysomal fatty acid β-oxidation. Demonstration of epimerase and isomerase activities on a peptide lacking hydratase activity. *J. Biochem.* **269**, 20475–20481.

Prime, T. A., Sherrier, D. J., Mahon, P., Packman, L. C., and Dupree, P. (2000). A proteomic analysis of organelles from *Arabidopsis thaliana*. *Electrophoresis* **21**, 3488–3499.

Purdue, P. E., and Lazarow, P. B. (1994). Peroxisomal biogenesis: Multiple pathways of protein import. *J. Biol. Chem.* **269**, 30065–30068.

Reguenga, C., Oliveira, M. E. M., Gouveia, A. M. M., Sa-Miranda, C., and Azevedo, J. E. (2001). Characterization of the mammalian peroxisomal import machinery. Pex2p, Pex5p, Pex12p and Pex14p are subunits of the same protein assembly. *J. Biol. Chem.* **276**.

Rehling, P., Marzioch, M., Niesen, F., Wittke, E., Veenhuis, M., and Kunau, W.-H. (1996). The import receptor for the peroxisomal targeting signal (PTS2) in *Saccharomyces cerevisiae* is encoded by the *PAS7* gene. *EMBO J.* **15**, 2901–2913.

Reumann, S. (2004). Specification of the peroxisome targeting signals type 1 and type 2 of plant peroxisomes by bioinformatics analyses. *Plant Physiol.* **135**, 783–800.

Reumann, S., Heupel, R., and Heldt, H. W. (1994). Compartmentation studies on spinach leaf peroxisomes; II. Evidence for the transfer of reductant from the cytosol to the peroxisomal compartment via a malate shuttle. *Planta* **193**, 167–173.

Rhodin, J. (1958). Anatomy of kidney tubules. *Int. Rev. Cytol.* **7**, 485–534.

Richmond, T. A., and Bleecker, A. B. (1999). A defect in β-oxidation causes abnormal inflorescence development in *Arabidopsis*. *Plant Cell* **11**, 1911–1923.

Río, L. A. D., Sandalio, L. M., Corpas, F. J., López-Huertas, E., Palma, J. M., and Pastori, G. M. (1998). Activated oxygen-mediated metabolic functions of leaf peroxisomes. *Physiol. Plant.* **104**, 673–680.

Russell, L., Larner, V., Kurup, S., Bougourd, S., and Holdsworth, M. (2000). The *Arabidopsis COMATOSE* locus regulates germination potential. *Development* **127**, 3759–3767.

Ryan, C. A., and Farmer, E. E. (1991). Oligosaccharide signals in plants: A current assessment. *Annu. Rev. Plant Physiol.* **42**, 651–674.

Sanders, P. M., Lee, P. Y., Biesgen, C., Boone, J. D., Beals, T. P., Weiler, E. W., and Goldberg, R. B. (2000). The *Arabidopsis DELAYED DEHISCENCE1* gene encodes an enzyme in the jasmonic acid synthesis pathway. *Plant Cell* **12**, 1041–1061.

Schaller, F., Biesgen, C., Müssig, C., Altmann, T., and Weiler, E. W. (2000). 12-oxophytodienoate reductase 3 (OPR3) is the isoenzyme involved in jasmonate biosynthesis. *Planta* **210**, 979–984.

Schaller, F., Henning, P., and Weiler, E. W. (1998). 12-Oxophytodienoate-10,11-reductase: Occurrence of two isoenzymes of different specificity against stereoisomers of 12-oxophytodienoic acid. *Plant Physiol.* **1998,** 1345–1351.

Schaller, F., and Weiler, E. W. (1997a). Enzymes of octadecanoid biosynthesis in plants 12-oxophytodienoate 10,11-reductase. *Eur. J. Biochem.* **245,** 294–299.

Schaller, F., and Weiler, E. W. (1997b). Molecular cloning and characterization of 12-oxophytodienoate reductase, an enzyme of the octadecanoid signaling pathway from *Arabidopsis thaliana*. Structural and functional relationship to yeast old yellow enzyme. *J. Biochem.* **272,** 28066–28072.

Schliebs, W., Saidowsky, J., Agianian, B., Dodt, G., Herberg, F. W., and Kunau, W.-H. (1999). Recombinant human peroxisomal targeting signal receptor PEX5. Structural basis for interaction of PEX5 with PEX14. *J. Biol. Chem.* **274,** 5666–5673.

Schumann, U., Gietl, C., and Schmid, M. (1999a). Sequence analysis of a cDNA encoding Pex7p, a peroxisomal targeting signal 2 receptor from *Arabidopsis*. *Plant Physiol.* **120,** 339.

Schumann, U., Gietl, C., and Schmid, M. (1999b). Sequence analysis of a cDNA encoding Pex10p, a zinc-binding peroxisomal integral membrane protein from *Arabidopsis*. *Plant Physiol.* **119,** 1147.

Schumann, U., Wanner, G., Veenhuis, M., Schmid, M., and Gietl, C. (2003). AthPEX10, a nuclear gene essential for peroxisome and storage organelle formation during *Arabidopsis* embryogenesis. *Proc. Natl. Acad. Sci. USA* **100,** 9626–9631.

Schwartz, B. W., Sloan, J. S., and Becker, W. M. (1991). Characterization of genes encoding hydroxypyruvate reductase in cucumber. *Plant Mol. Biol.* **17,** 941–947.

Shani, N., Jiminez-Sanchez, G., Steel, G., Dean, M., and Valle, D. (1997). Identification of a fourth half ABC transporter in the human peroxisomal membrane. *Hum. Mol. Genet.* **6,** 1925–1931.

Shani, N., and Valle, D. (1996). A *Saccharomyces cerevisiae* homolog of the human adrenoleukodystrophy transporter is a heterodimer of two half ATP-binding cassette transporters. *Proc. Natl. Acad. Sci. USA* **93,** 11901–11906.

Shani, N., Watkins, P. A., and Valle, D. (1995). *PXA1*, a possible *Saccharomyces cerevisiae* ortholog of the human adrenoleukodystrophy gene. *Proc. Natl. Acad. Sci. USA* **92,** 6012–6016.

Shimizu, N., Itoh, R., Hirono, Y., Otera, H., Ghaedi, K., Tateishi, K., Tamura, S., Okumoto, K., Harano, T., Mukai, S., and Fujiki, Y. (1999). The peroxin Pex14p, cDNA cloning by functional complementation on a chinese hamster ovary cell mutant, characterization, and functional analysis. *J. Biol. Chem.* **274,** 12593–12604.

Shimomura, Y., Murakami, T., Fujitsuka, N., Nakai, N., Sato, Y., Sugiyama, S., Shimomura, N., Irwin, J., Hawes, J. W., and Harris, R. A. (1994). Purification and partial characterization of 3-hydroxyisobutyryl-coenzyme A hydrolase of rat liver. *J. Biol. Chem.* **269,** 14248–14253.

Smith, J. J., Marelli, M., Christmas, R. H., Vizeacoumar, F. J., Dilworth, D. J., Ideker, T., Galitski, T., Dimitrov, K., Rachubinski, R. A., and Aitchison, J. D. (2002). Transcriptome profiling to identify genes involved in peroxisome assembly and function. *J. Cell Biol.* **158,** 259–271.

Somerville, C. R., and Ogren, W. L. (1980). Photorespiration mutants of *Arabidopsis thaliana* deficient in serine-glyoxylate aminotransferase activity. *Proc. Natl. Acad. Sci. USA* **77,** 2684–2687.

Somerville, C. R., and Ogren, W. L. (1982). Genetic modification of photorespiration. *Trends Biochem. Sci.* **7,** 171–174.

Sparkes, I. A., Brandizzi, F., Slocombe, S. P., El-Shami, M., Hawes, C., and Baker, A. (2003). An *Arabidopsis pex10* null mutant is embryo lethal, implicating peroxisomes in an essential role during plant embryogenesis. *Plant Physiol.* **133,** 1809–1819.

Staswick, P. E., Su, W., and Howell, S. H. (1992). Methyl jasmonate inhibition of root growth and induction of a leaf protein are decreased in an *Arabidopsis thaliana* mutant. *Proc. Natl. Acad. Sci. USA* **89**, 6837–3840.

Stintzi, A., and Browse, J. (2000). The *Arabidopsis* male-sterile mutant, *opr3*, lacks the 12-oxophytodienoic acid reductase required for jasmonate synthesis. *Proc. Natl. Acad. Sci. USA* **97**, 10625–10630.

Strassner, J., Schaller, F., Frick, U. B., Howe, G. A., Weiler, E. W., Amrhein, N., Macheroux, P., and Schaller, A. (2002). Characterization and cDNA-microarray expression analysis of 12-oxophytodienoate reductases reveals differential roles for octadecanoid biosynthesis in the local versus the systemic wound response. *Plant J.* **32**, 585–601.

Swartzman, E. E., Viswanathan, M. N., and Thorner, J. (1996). The *PAL1* gene product is a peroxisomal ATP-binding cassette transporter in the yeast *Saccharomyces cerevisiae*. *J. Cell Biol.* **4**, 549–563.

Titorenko, V. I., and Rachubinski, R. A. (2000). Peroxisomal membrane fusion requires two AAA family ATPases, Pex1p and Pex6p. *J. Cell Biol.* **150**, 881–886.

Titus, D. E., and Becker, W. M. (1985). Investigation of the glyoxysome-peroxisome transition in germinating cucumber cotyledons using double-label immunoelectron microscopy. *J. Cell Biol.* **101**, 1288–1299.

Tolbert, N. E. (1971). Microbodies-Peroxisomes and glyoxysomes. *Annu. Rev. Plant Physiol.* **21**, 45–74.

Tolbert, N. E., Oeser, A., Kisaki, T., Hageman, R. H., and Yamazaki, R. K. (1968). Peroxisomes from spinach leaves containing enzymes related to glycolate metabolism. *J. Biol. Chem.* **243**, 5179–5184.

Trelease, R. N. (2002). Peroxisomal biogenesis and acquisition of membrane proteins. In "Plant Peroxisome" (A. Baker and I. A. Graham, Eds.), pp. 305–337. Kluwer Academic Publishers, The Netherlands.

Trelease, R. N., Becker, W. M., Gruber, P. J., and Newcomb, E. H. (1971). Microbodies (glyoxysomes and peroxisomes) in cucumber cotyledons. *Plant Physiol.* **48**, 461–475.

Tsugeki, R., Hara-Nishimura, I., Mori, H., and Nishimura, M. (1993). Cloning and sequencing of cDNA for glycolate oxidase from pumpkin cotyledons and Northern blot analysis. *Plant Cell Physiol.* **34**, 51–57.

Tsukamoto, T., Miura, S., and Fujiki, Y. (1991). Restoration by a 35 K membrane protein of peroxisome assembly in a peroxisome-deficient mammalian cell mutant. *Nature* **350**, 77–81.

Urquhart, A. J., Kennedy, D., Gould, S. J., and Crane, D. I. (2000). Interaction of Pex5p, the type 1 peroxisome targeting signal receptor, with the peroxisomal membrane proteins Pex14p and Pex13p. *J. Biol. Chem.* **275**, 4127–4136.

Verleur, N., Hettema, E. H., Roermund, C. W. T. V., Tabak, H. F., and Wanders, R. J. A. (1997). Transport of activated fatty acids by the peroxisomal ATP-binding-cassette transporter Pxa2 in a semi-intact yeast cell system. *Eur. J. Biochem.* **249**, 657–661.

Vijayan, P., Shockey, J., Lévesque, C. A., Cook, R. J., and Browse, J. (1998). A role for jasmonate in pathogen defense of *Arabidopsis*. *Proc. Natl. Acad. Sci. USA* **95**, 7209–7214.

Volokita, M. (1991). The carboxy-terminal end of glycolate oxidase directs a foreign protein into tobacco leaf peroxisomes. *Plant J.* **1**, 361–366.

Volokita, M., and Somerville, C. R. (1987). The primary structure of spinach glycolate oxidase deduced from the DNA sequence of a cDNA clone. *J. Biochem.* **262**, 15825–15828.

Wain, R. L., and Wightman, F. (1954). The growth-regulating activity of certain β-substituted alkyl carboxylic acids in relation to their β-oxidation within the plant. *Proc. Roy. Soc. Lond. Biol. Sci.* **142**, 525–536.

Walton, P. E., Hill, P. E., and Subramani, S. (1995). Import of stably folded proteins into peroxisomes. *Mol. Biol. Cell* **6**, 675–683.

Warren, D. S., Morrell, J. C., Moser, H. W., Valle, D., and Gould, S. J. (1998). Identification of PEX10, the gene defective in complementation group 7 of the peroxisome-biogenesis disorders. *Am. J. Hum. Genet.* **63**, 347–359.

Weber, H. (2002). Fatty acid-derived signals in plants. *Trends Plant Sci.* **7**, 217–224.

Weiler, E. W., Albrecht, T., Groth, B., Xia, Z.-Q., Luxem, M., Lib, H., Andert, L., and Spengler, P. (1993). Evidence for the involvement of jasmonates and their octadecanoid precursors in the tendril coiling response of *Bryonia dioica*. *Phytochem.* **32**, 591–600.

Wiemer, E. A. C., Nuttley, W. M., Bertolaet, B. L., Li, X., Francke, U., Wheelock, M. J., Anne, U. K., Johnson, K. R., and Subramani, S. (1995). Human peroxisomal targeting signal-1 receptor restores peroxisomal protein import in cells from patients with fatal peroxisomal disorders. *J. Cell Biol.* **130**, 51–65.

Williamson, J. D., and Scandalios, J. G. (1993). Response of the maize catalases and superoxide dismutases to cercosporin-containing fungal extracts: The pattern of catalase response in scutella is stage specific. *Physiol. Plant.* **88**, 159–166.

Wimmer, C., Schmid, M., Veenhuis, M., and Gietl, C. (1998). The plant PTS1 receptor: Similarities and differences to its human and yeast counterparts. *Plant J.* **16**, 453–464.

Xu, Y., Chang, P.-F. L., Liu, D., Narasimhan, M. L., Raghothama, K. G., Hasegawa, P. M., and Bressan, R. A. (1994). Plant defense genes are synergistically induced by ethylene and methyl jasmonate. *Plant Cell* **6**, 1077–1085.

Yamaguchi, K., Mori, H., and Nishimura, M. (1995). A novel isoenzyme of ascorbate peroxidase localized on glyoxysomal and leaf peroxisomal membrane in pumpkin. *Plant Cell Physiol.* **36**, 1157–1162.

Yu, J., Hu, S., Wang, J., Wong, G. K.-S., Li, S., Liu, B., Deng, Y., Dai, L., Zhou, Y., Zhang, X., Cao, M., Liu, J., Sun, J., Tang, J., Chen, Y., Huang, X., Lin, W., Ye, C., Tong, W., Cong, L., Geng, J., Han, Y., Li, L., Li, W., Hu, G., Huang, X., Li, W., Li, J., Liu, Z., Li, L., Liu, J., Qi, Q., Liu, J., Li, L., Li, T., Wang, X., Lu, H., Wu, T., Zhu, M., Ni, P., Han, H., Dong, W., Ren, X., Feng, X., Cui, P., Li, X., Wang, H., Xu, X., Zhai, W., Xu, Z., Zhang, J., He, S., Zhang, J., Xu, J., Zhang, K., Zheng, X., Dong, J., Zeng, W., Tao, L., Ye, J., Tan, J., Ren, X., Chen, X., He, J., Liu, D., Tian, W., Tian, C., Xia, H., Bao, Q., Li, G., Gao, H., Cao, T., Wang, J., Zhao, W., Li, P., Chen, W., Wang, X., Zhang, Y., Hu, J., Wang, J., Liu, S., Yang, J., Zhang, G., Xiong, Y., Li, Z., Mao, L., Zhou, C., Zhu, Z., Chen, R., Hao, B., Zheng, W., Chen, S., Guo, W., Li, G., Liu, S., Tao, M., Wang, J., Zhu, L., Yuan, L., and Yang, H. (2002). A draft sequence of the rice genome (*Oryza sativa* L. ssp. *indica*). *Science* **296**, 79–92.

Zhang, J. Z., Laudencia-Chingcuanco, D. L., Comai, L., Li, M., and Harada, J. J. (1994). Isocitrate lyase and malate synthase genes from *Brassica napus* L. are active in pollen. *Plant Physiol.* **104**, 857–864.

Zolman, B. K., and Bartel, B. (2004). An *Arabidopsis* indole-3-butyric acid-response mutant defective in PEROXIN6, an apparent ATPase implicated in peroxisomal function. *Proc. Natl. Acad. Sci. USA* **101**, 1786–1791.

Zolman, B. K., Monroe-Augustus, M., Thompson, B., Hawes, J. W., Krukenberg, K. A., Matsuda, S. P. T., and Bartel, B. (2001a). *chy1*, an *Arabidopsis* mutant with impaired β-oxidation, is defective in a peroxisomal β-hydroxyisobutyryl-CoA hydrolase. *J. Biol. Chem.* **276**, 31037–31046.

Zolman, B. K., Silva, I. D., and Bartel, B. (2001b). The *Arabidopsis pxa1* mutant is defective in an ATP-binding cassette transporter-like protein required for peroxisomal fatty acid β-oxidation. *Plant Physiol.* **127**, 1266–1278.

Zolman, B. K., Yoder, A., and Bartel, B. (2000). Genetic analysis of indole-3-butyric acid responses in *Arabidopsis thaliana* reveals four mutant classes. *Genetics* **156**, 1323–1337.

5

Regulatory and Functional Interactions of Plant Growth Regulators and Plant Glutathione S-Transferases (GSTs)

Ann Moons

*National Research Council Canada, Biotechnology Research Institute
Montreal Canada, H4P 2R2*

I. Classification, Structure, and Subcellular Localization of Plant GSTs
II. Functions of Plant GSTs
   A. Plant GSTs Have Diverse GSH-Dependent Catalytic Functions
   B. Plant GSTs Can Act as Non-catalytic Carriers of Phytochemicals
   C. Emerging Alternative Functions of Plant GSTs
III. The Transcriptional Regulation of GST Gene Expression
   A. Role of Auxins in the Regulation of GST Gene Expression

    B. Role of ABA in the Regulation of GST Gene Expression
    C. Role of SA in Regulating GST Gene Expression
    D. Role of Jasmonates in the Regulation of GST Gene Expression During Biotic and Abiotic Stresses
    E. Role of Ethylene in the Regulation of GST Gene Expression During Biotic and Abiotic Stresses and Normal Development
    F. Role of NO in Regulating GST Gene Expression and Activity
    G. Integrated Roles of ROIs and Plant Growth Regulators on GST Gene Expression
    H. Integrated Effects of Antioxidants and Phytohormones on GST Gene Expression
IV. Potential Involvement of Plant Growth Regulators in Post-transcriptional Regulations of GST Gene Expression
    A. Alternative Splicing During GST Expression
    B. Differential Polyadenylation Processing During GST Expression
    C. Plant Growth Regulators May Post-transcriptionally Modify the Accumulation of Stress-response Proteins
V. Conclusions
References

Plant glutathione *S*-transferases (GSTs) are a heterogeneous superfamily of multifunctional proteins, grouped into six classes. The tau (GSTU) and phi (GSTF) class GSTs are the most represented ones and are plant-specific, whereas the smaller theta (GSTT) and zeta (GSTZ) classes are also found in animals. The lambda GSTs (GSTL) and the dehydroascorbate reductases (DHARs) are more distantly related.

  Plant GSTs perform a variety of pivotal catalytic and non-enzymatic functions in normal plant development and plant stress responses, roles that are only emerging. Catalytic functions include glutathione (GSH)-conjugation in the metabolic detoxification of herbicides and natural products. GSTs can also catalyze GSH-dependent peroxidase reactions that scavenge toxic organic hydroperoxides and protect from oxidative damage. GSTs can furthermore catalyze GSH-dependent isomerizations in endogenous metabolism, exhibit GSH-dependent thioltransferase safeguarding protein function from oxidative damage and DHAR activity functioning in redox homeostasis. Plant GSTs can also function as ligandins or binding proteins for phytohormones (i.e., auxins and cytokinins) or anthocyanins, thereby facilitating their distribution and transport. Finally, GSTs are also indirectly involved in the regulation of apoptosis and possibly also in stress signaling.

Plant GST genes exhibit a diversity of expression patterns during biotic and abiotic stresses. Stress-induced plant growth regulators (i.e., jasmonic acid [JA], salicylic acid [SA], ethylene [ETH], and nitric oxide [NO] differentially activate GST gene expression. It is becoming increasingly evident that unique combinations of multiple, often interactive signaling pathways from various phytohormones and reactive oxygen species or antioxidants render the distinct transcriptional activation patterns of individual GSTs during stress. Underestimated post-transcriptional regulations of individual GSTs are becoming increasingly evident and roles for phytohormones (i.e., ABA and JA) in these processes are being anticipated as well. Finally, indications are emerging that NO may regulate the activity of specific plant GSTs.

In this review, the current knowledge on the regulatory and functional interactions of phytohormones and plant GSTs are covered. We refer to a previous extensive review on plant GSTs (Marrs, 1996) for most earlier work. An introduction on the classification and roles of plant GSTs is included here, but these topics are more extensively discussed in other reviews (Dixon *et al.*, 2002a; Edwards *et al.*, 2000; Frova, 2003).

© 2005 Elsevier Inc.

I. CLASSIFICATION, STRUCTURE, AND SUBCELLULAR LOCALIZATION OF PLANT GSTs

The gluthathione transferases comprise a heterogeneous superfamily of proteins that possess a well-defined domain for glutathione-binding and catalysis at their active sites (Sheehan *et al.*, 2001). GSTs are ubiquitous in aerobic organisms and are encoded by large gene families in plants. The plant GSTs have been reclassified to be in tune with the mammalian GST phylogeny (Dixon *et al.*, 2002a; Edwards *et al.*, 2000). The current grouping into six plant GST classes (Fig. 1) is based on protein sequence similarity, active site residues, and gene organization (i.e., intron number and position). The majority of the plant GSTs belong to the tau (GSTU) and phi (GSTF) classes, which are plant-specific. The significantly smaller theta (GSTT) and zeta (GSTZ) classes have related GSTs in mammalians. The more distantly related lambda GSTs and the GST-like glutathione-dependent dehydroascorbate reductases have been included as well (Dixon *et al.*, 2002b). Table I presents the numbers of putative GST genes from the various classes for the model dicot *Arabidopsis thaliana* (Dixon *et al.*, 2002a) and the model monocot rice (*Oryza sativa* L.; Soranzo *et al.*, 2004), based on the genome sequences of these plants. The GST gene numbers for maize and soybean

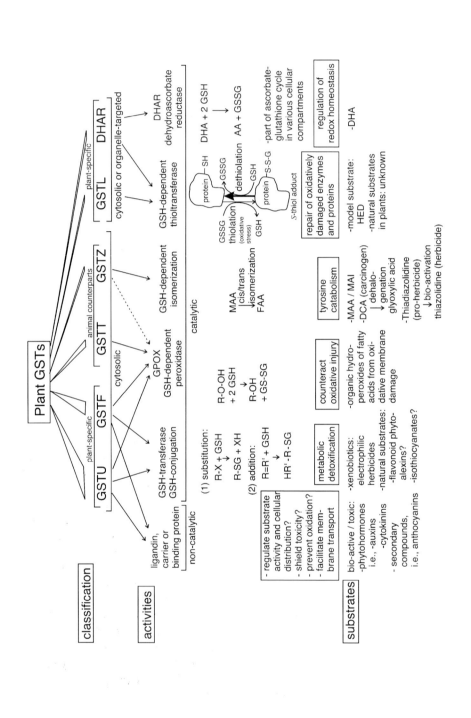

(Table I; McGonigle *et al.*, 2000) were derived from the analysis of expressed sequence tag (EST) databases and are most likely to be revised once the complete genome sequence of these plant species becomes available. GST genes are typically distributed in class-specific clusters on preferential chromosomes across the genomes of various organisms, including mammals, *Arabidopsis*, and rice. This distribution is indicative of a rapid evolution through gene duplication Dixon *et al.*, 2002a; Sheehan *et al.*, 2001; Soranzo *et al.*, 2004).

Most GSTs are active as homodimers or heterodimers of 23–30 kDa subunits, with the exception of the lambda GSTLs and the DHARs that function as monomers (Dixon *et al.*, 2002b). Each GST monomer typically consists of a conserved N-terminal, glutathione-specific binding domain (G-site), and a variable C-terminal domain (H-site) that binds the hydrophobic substrate. Tau-, phi-, zeta-, and theta-class plant GSTs have at the active site a conserved, catalytically essential serine that interacts with and activates the thiol group of GSH (Sheehan *et al.*, 2001). The GSTLs and DHARs contain instead a catalytically essential cysteine, which is involved in forming a mixed disulfide with GSH (Dixon *et al.*, 2002b). The three-dimensional structures of phi-class GST from maize and *Arabidopsis* and a zeta-class GST from *Arabidopsis*, which have been determined by X-ray crystallography revealed a high degree of three-dimensional structure conservation of plant GSTs, i.e., phi-class GSTs from maize and *Arabidopsis* and a zeta class GST from *Arabidopsis*, compared to GSTs from various other organisms, in spite of the primary sequence diversification (Dixon *et al.*, 2002a).

Most tau-, phi-, zeta-, and theta-class plant GSTs are cytosolic. Nuclear or apoplastic localizations have occasionally been reported for some GSTs within these four classes (Dixon *et al.*, 1998). By contrast, some GSTL and DHAR genes contain a distinct transit peptide for targeting to organelles, in particular chloroplasts, whereas other GSTLs and DHARs are predictably cytosolic (Table II; Dixon *et al.*, 2002b).

FIGURE 1. Overview of the classification, activities, and substrates of plant GSTs. The heterogeneous plant GST superfamily has been classified into the tau (GSTU) and phi (GSTF) GSTs that are abundant and plant-specific, the theta (GSTZ) and zeta (GSTZ) class GSTs that have mammalian counterparts, the more distantly related lambda GSTs (GSTL) and the GST-like dehydroascorbate reductases (DHARs). The arrow width indicates the relative representation of these classes in the *Arabidopsis* genome (see Table I for numbers). Plant GSTs can catalyze GSH conjugations, as well as a diversity of GSH-dependent reactions: Isomerizations, peroxidase (GPOX), thioltransferase, and dehydroascorbate reductase (DHAR) reactions. GST class subdivisions generally do not match specific functions. Moreover, individual GSTs can have multiple, substrate-dependent activities (see Table II for examples). Abbreviations: AA, ascorbate; DCA, dichloroacetic acid; DHA, dehydroascorbate; FAA, fumarylacetoacetate; HED, 2-hydroxyethyl disulfide; MAA, maleylacetoacetate; MAI, maleylacetone; ROOH and ROH, organic hydroperoxides and the corresponding monohydroxy alcohols; R–X and R = R′, electrophilic substrates; R–SG and HR′–R–SG, glutathione *S*-conjugates.

TABLE I. Reported Gene Numbers for the Tau-, Phi-, Theta-, Zeta-, Lambda-, and DHAR-Class GSTs From *Arabidopsis thaliana*, Rice, Maize, and Soybean[a]

Plant species	Total number of GST genes (typical exon–intron structure)	Tau-class GSTU genes (2 exons/ 1 intron)	Phi-class GSTF genes (3 exons/ 2 introns)	Theta-class GSTT genes (7 exons/ 6 introns)	Zeta-class GSTZ genes (10 exons/ 9 introns)	Lambda-class GSTL genes (9 exons/ 8 introns)	DHAR-class GST genes (3 exons/ 2 introns)	References
Arabidopsis thaliana L.	53	28	14	3	2	2	3 (and 1 pseudo)	Dixon et al. (2002a,b)
Rice (*Oryza sativa* L.)	59 (and 2 pseudo)	40	16	2	3	2	3	Soranzo et al. (2004)
Maize (*Zea mays*)	42	28	12	N/A	2	N/A	N/A	McGonigle et al. (2000)
Soybean (*Glycin max*)	25	20	4	N/A	1	N/A	N/A	McGonigle et al. (2000)

[a]The typical exon–intron structure of GST genes from each class are given. Additional database searches for rice indicated the existence of at least three *Os*DHAR and two *Os*GSTL genes with cDNA or protein accession numbers and chromosome (chrom.) locations: *Os*DHAR1 (AB037970, chrom. 5, Urano et al., 2000); *Os*DHAR2 (AK106013, chrom. 6); *Os*DHAR3 (AAQ01154, chrom. 3); *Os*GSTL1 (AAN64482, chrom. 3); and *Os*GSTL2 (XIG AF237487, AAF70831, chrom. 3).

N/A: Not available.

II. FUNCTIONS OF PLANT GSTs

Soluble GSTs (EC 2.5.1.18) are by definition a family of dimeric enzymes that catalyze the nucleophilic attack of the tripeptide glutathione on lipophilic compounds with electrophilic centers. In reality, however, GSTs exhibit an extraordinary diversity of functions (Fig. 1). Plant GSTs can use GSH as either co-substrate or as coenzyme and catalyze a variety of reactions including GSH conjugations, GSH-dependent isomerizations, GSH-dependent peroxidase reactions, GSH-dependent thioltransferase reactions, and GSH-dependent dehydroascorbate reductase reactions (Dixon *et al.*, 1998, 2002a,b; Edwards *et al.*, 2000; Frova, 2003). GST can also exert non-catalytic functions as binding or carrier proteins for phytochemicals between cellular compartments (Dixon *et al.*, 2002a; Edwards *et al.*, 2000; Sheehan *et al.*, 2001). In general, functional specifications do not match GST class subdivisions (Fig. 1). Moreover, individual GSTs can be multifunctional, having the ability to catalyze different reactions with structurally distinct substrates. Plant GSTs fulfill their versatile catalytic and non-catalytic functions during normal development as well as "on call" during diverse conditions of abiotic or biotic stress. However, the biological role of the various plant GST activities and their importance in plant development and stress survival remain poorly understood to date.

A. PLANT GSTs HAVE DIVERSE GSH-DEPENDENT CATALYTIC FUNCTIONS

1. Glutathione Transferase or GSH Conjugation Activities of Plant GSTs

The GST-catalyzed attachment of the tripeptide glutathione (γ-glutamyl-cysteinyl-glycine) (Fig. 1) to electrophilic and frequently hydrophobic, cytotoxic compounds of xenobiotic or natural origin mostly occurs through a nucleophilic substitution reaction (1) but is also accomplished by a nucleophilic addition reaction (2):

$$R-X + GSH \rightarrow R-SG + XH \qquad (1)$$

$$R=R' + GSH \rightarrow HR'-R-SG \qquad (2)$$

and results in the formation of non-toxic glutathione *S*-conjugates (Coleman *et al.*, 1997; Dixon *et al.*, 1998).

a. Herbicide Detoxification Through GST-Catalyzed GSH Conjugation

Historically, plant GSTs have received significant attention for their role in herbicide detoxification and selectivity. Phi-class GSTs (i.e., from maize and other cereals) and tau-class GSTs are able to conjugate GSH to a number of electrophilic herbicides, which results in the detoxification of these xenobiotics

TABLE II. GST Gene Expression in Response to Plant Growth Regulators, Biotic and Abiotic Stresses in *Arabidopsis thaliana* L.

GST class	New name (old names; Acc. no.)	Plant growth regulators					Biotic stresses		
		Induced (ind.)					Induced		
		Marked and rapid	Weak	Late	Not induced	Downregulated	Rapid, before PR-1	Late, same as PR-1	Not induced or downregulated
Phi	*At*GSTF2 (GST2, PM24.1; L11601, X75303)	ETH SA 2,4-D IAA NAA			MeJA		*Pst* (avir.)	*P. par.* (NOCO, vir.)	*P. par.* (EMWA, avir.)
	*At*GSTF3 (GST16; AF288181)				ETH SA	2,4-D (R) NAA (R)	*Pst* (avir.)		*Pst* (avir.)
	*At*GSTF6 (ERD11, GST1; D17672, L12057)	ETH SA MeJA 2,4-D NO			ABA BAP GA3		*Pst* (avir.)	*P. par.* (NOCO, vir.)	*P. par.* (EMWA, avir.)
	*At*GSTF8 (GST6; AF288176)	2,4-D SA NO			MeJA ETH				*P. par.* (NOCO, EMWA)
	*At*GSTF9 (GLUTTR; Y12295)				MeJA SA ETH	2,4-D			
	*At*GSTF10 (ERD13, D17673)	JA[h]	MeJA		ABA ACC ETH NAA BAP GA3	2,4-D SA			*P. par.* (NOCO, EMWA)

Family	Gene						
Theta	AtGSTT1 (GST10; AJ131580)			MeJA SA ETH	2,4-D	NAA (R)	
Zeta	AtGSTZ1 (GST18; AF288182, AJ278293)		MeJA 2,4-D	NAA (R) SA ETH		P. par. (EMWA, avir.)	P. par. (NOCO, vir.)
Tau	AtGSTU1 (GST19; AF288183)				2,4-D (R) NAA (R)		
	AtGSTU2 (GST20; AF288184)		2,4-D (R) NAA (R)				
	AtGSTU5 (AT103-1a; D44465)	2,4-D (R)		MeJA 2,4-D (L)	SA ETH		
	AtGSTU13 (GST12; AF288193)				MeJA SA ETH 2,4-D		
	AtGSTU19 (GST8; AJ012571)	NAA[h] BAP[h]	IAA[h] MeJA JA[h]	SA 2,4-D	ABA GA3 ACC ETH	P. par. (EMWA, avir.)	P. par. (NOCO, vir.)
Lambda	AtGSTL1 (AL162973)	2,4-D (R)		NAA (R)			
	AtGSTL2 (AL132970)				2,4-D (R) NAA (R)		
DHARs	AtDHAR1 (AC024609, AY039590)				2,4-D (R) NAA (R)		
	AtDHAR2 (AB026661, NM_106182)	2,4-D (R)		NAA (R)			
	AtDHAR3 (AC025814, AF301597)				2,4-D (R) NAA (R)		

(*Continues*)

163

TABLE II. (Continued)

GST class	New name (old names; Acc. no.)	Environmental stresses, chemicals, and xenobiotics			Organ or tissue specificity
		Induced		Not induced or downregulated	
		Marked and rapid	Weak or late		
Phi	AtGSTF2 (GST2, PM24.1; L11601, X75303)	GSH Cys DTT Cu HB (met, par, pri)		H_2O_2 HB (bxn)	Root–shoot transition zone and root distal elongation zone
	AtGSTF3 (GST16; AF288181)	GSH AA BSO BHP	HB (pro) HBS (d)	CDNB ETA	
	AtGSTF6 (ERD11, GST1; D17672, L12057)	D H_2O_2 HB (met, pri, pro)		HB (bxn)	
	AtGSTF8 (GST6; AF288176)	H_2O_2 HB (met)			Largest induction in roots
	AtGSTF9 (GLUTTR; Y12295)			H_2O_2 HB (met)	
	AtGSTF10 (ERD13, D17673)	D (R) W	D H_2O_2 C SO S HB (par)	HB (met)	D response: Stronger in root than shoot; D induction precedes peroxidase induction
Theta	AtGSTT1 (GST10; AJ131580)			H_2O_2 HB (met, f) W GSH AA BSO BHP HBS (d) CDNB ETA	
Zeta	AtGSTZ1 (GST18; AF288182, AJ278293)	AA BSO BHP HB (pri)	GSH	H_2O_2 HB (met, f) HBS (d) CDNB ETA	

Class	Gene				Notes
Tau	AtGSTU1 (GST19; AF288183)	CDNB	HB (f)	GSH AA BSO BHP HBS (d) ETA	
	AtGSTU2 (GST20; AF288184)	BSO	AA BHP CDNB HBS (d)	GSH HB (f) ETA	
	AtGSTU5 (AT103–1a; D44465)			H_2O_2 HB (met)	Auxin response is root-specific
	AtGSTU13 (GST12; AF288193)			H_2O_2 HB (met)	
	AtGSTU19 (GST8; AJ012571)	D (R)	D W H_2O_2 S HB (par)	C SO HB (met)	D response: Stronger in root than shoot; D induction simultaneous with peroxidase induction
Lambda	AtGSTL1 (AL162973)	GSH AA BSO BHP HBS (d)		HB (f) CDNB ETA	
	AtGSTL2 (AL132970)		GSH HB (f)	AA BSO BHP CDNB HBS (d) ETA	Chloroplast transit peptide
DHARs	AtDHAR1 (AC024609, AY039590)		GSH AA BSO BHP	CDNB HB (f) HBS (d) ETA	
	AtDHAR2 (AB026661, NM_106182)	BSO CDNB	AA HB (f) HBS (d) BHP ETA	GSH ETA	
	AtDHAR3 (AC025814, AF301597)		GSH HB (f)	AA BSO BHP CDNB HBS (d) ETA	Chloroplast or possibly mitochondrion transit peptide

(*Continues*)

TABLE II. (*Continued*)

		Catalytic activities on standard substrates			
GST class	New name (old names; Acc. no.)	GSH conjugating activities	Alternative GSH-dependent catalytic activities	Ligandin	References
Phi	AtGSTF2 (GST2, PM24.1; L11601, X75303)	CDNB > BITC	GPOX: LA-HPO > Cum-HPO	IAA	Zettl et al. (1993); Wagner et al. (2002); Lieberherr et al. (2003); Smith et al. (2003); Glombitza et al. (2004)
	AtGSTF3 (GST16; AF288181)				Dixon et al. (2002b); Lieberherr et al. (2003)
	AtGSTF6 (ERD11, GST1; D17672, L12057)				Kiyosue et al. (1993); Wagner et al. (2002); Lieberherr et al. (2003); Huang et al. (2002); Glombitza et al. (2004)
	AtGSTF8 (GST6; AF288176)	CDNB	GPOX: LA-HPO > Cum-HPO		Chen and Singh (1999); Wagner et al. (2002); Polverari et al. (2003)
	AtGSTF9 (GLUTTR; Y12295)	CDNB > BITC	GPOX: LA-HPO > Cum-HPO		Wagner et al. (2002)
	AtGSTF10 (ERD13, D17673)	CDNB > BITC	no GPOX		Kiyosue et al. (1993); Wagner et al. (2002); Bianchi et al. (2002)
Theta	AtGSTT1 (GST10; AJ131580)	pNBC > CDNB	high GPOX: LA-HPO = Cum-HPO		Dixon et al. (2002a,b); Wagner et al. (2002); Glombitza et al. (2004)
Zeta	AtGSTZ1 (GST18; AF288182, AJ278293)	None	Maleylacetone isomerase (MAI) GPOX: Cum-HPO		Dixon et al. (2000); Wagner et al. (2002); Dixon et al. (2002b)
Tau	AtGSTU1 (GST19; AF288183)				Dixon et al. (2002b)
	AtGSTU2 (GST20; AF288184)				Dixon et al. (2002b)
	AtGSTU5 (AT103–1a; D44465)	CDNB = BITC	no GPOX		Van der Kop et al. (1996); Wagner et al. (2002)

	AtGSTU13 (GST12; AF288193)	BITC CDNB	no GPOX	Wagner et al. (2002)
	AtGSTU19 (GST8; AJ012571)	CDNB	no GPOX	Bianchi et al. (2002); Wagner et al. (2002)
Lambda	AtGSTL1 (AL162973)	None	Thioltransferase no DHAR no GPOX	Dixon et al. (2002b)
	AtGSTL2 (AL132970)	None	Thioltransferase no DHAR no GPOX	Dixon et al. (2002b)
DHARs	AtDHAR1 (AC024609, AY039590)	None	DHAR thioltransferase no GPOX	Dixon et al. (2002b)
	AtDHAR2 (AB026661, NM_106182)	None	DHAR thioltransferase no GPOX	Dixon et al. (2002b)
	AtDHAR3 (AC025814, AF301597)	None	DHAR thioltransferase no GPOX	Dixon et al. (2002b)

Effects of exogenously applied abscisic acid (ABA), the cytokinin 6-benzylaminopurine (BAP), the natural auxin indole-3-acetic acid (IAA) and synthetic auxins 2,4-dichlorophenoxyacetic acid (2,4-D) and α-naphthalene acetic acid (NAA), the jasmonates, i.e., jasmonic acid (JA) and methyl jasmonate (MeJA), salicylic acid (SA), gibberellic acid (GA3) and ethylene (ETH) or its precursor 1-aminocyclopropane-1-carboxylic acid (ACC) and nitric oxide (NO) on the transcript levels of individual GST genes in the shoot or in the root (R) of Arabidopsis seedlings. Index (h) indicates the use of exceptionally high phytohormone concentrations. Phytohormone effects on GST gene expression in tobacco, carnation, and soybean have been summarized in Marrs (1996). Reported pathogen responses include inoculation with the avirulent Pseudomonas syringae pv. tomato (Pst) strain, the avirulent Peronospora parasitica (P. par.) isolate EMWA or the virulent P. par. isolate NOCO. Environmental stresses include cold (C), dehydration (D), wounding (W), metal stress imposed with Cu, osmotic stress imposed with sorbitol (SO) and salt stress (S). Xenobiotics include the herbicides (HB), fluorodifen (f), paraquat (p), metolachlor (met), bromoxynil octanoate (bxn) and the sulfonylurea herbicides primisulfuron (pri) and prosulfuron (pro) and the herbicide safener (HBS) and dichlormid (d). Other organic and inorganic chemicals used were the oxidants hydrogen peroxide (H_2O_2) and t-butyl hydroperoxide (BHP), the antioxidants ascorbic acid (AA) and glutathione (GSH), the GSH synthesis inhibitor L-buthionine-(SR)-sulfoximine (BSO), the sulphydryl compounds dithiothreitol (DTT) and cysteine, and 1% ethanol (ETA). Substrates used to assay GSH-conjugating activities were 1-chloro-2,4-dinitrobenzene (CDNB), benzyl isothiocyanate (BITC), 1,2-exoxy-3-(p-nitrophenoxy)propane (ENPP), and p-nitrobenzyl chloride (pNBG); substrates for GSH-dependent peroxidase (GPOX) activities were cumene hydroperoxide (Cum-HPO) and linoleic acid-13-hydroperoxide (LA-HPO); GSH-dependent thioltransferase activity was assayed using 2-hydroxyethyl disulfide (HED).

(Dixon et al., 1998, Edwards and Dixon, 2000; Kreuz et al., 1996). This herbicide detoxification process occurs at a significantly slower rate in weeds than in crops and accounts for herbicide selectivity. Herbicides susceptible for GSH conjugation include triazines, thiocarbamates, chloroacetanilides, sulphonylureas, and diphenylethers (Frova, 2003).

The metabolic detoxification of xenobiotics or foreign organic compounds is a well-known process in various organisms. It has beneficial roles in the detoxification of dietary toxins or carcinogenic organic pollutants, but it is notorious for its role in resistance to cancer chemotherapeutic drugs, insecticides, fungicides, and microbial antibiotics in mammals, insects, fungi, *Candida* sp., and bacteria, respectively. Metabolic detoxification of xenobiotics generally proceeds in three steps (Dixon et al., 1998; Edwards and Dixon, 2000; Frova, 2003; Kreuz et al., 1996; Sheehan et al., 2001).

1. *Phase I:* The xenobiotic is activated to facilitate further processing. Xenobiotics usually diffuse freely across the plasma membrane. Once in the cytosol, activation can occur through oxidation of the xenobiotic by oxidoreductases such as cytochrome P450 enzymes. Activation is generally not required prior to GSH conjugation for most herbicides that already contain electrophilic centers.
2. *Phase II:* Tagging of the (activated) xenobiotic through conjugation with highly water-soluble molecules. This increases solubility, limits movement across membranes, and allows identification in the subsequent phase. The prevalent tag in plants and animals is the tripeptide glutathione, which is added by GSTs. In plants, also sugars such as glucose and malonic acid are enzymatically added, whereas in animals, glucuronic acid is also added.

Herbicide safeners, which are chemicals used to protect monocotyledonous crops from herbicide injury, differentially increase the expression and activities of these GSTs and other herbicide detoxifying enzymes in cereals versus target weeds.

3. *Phase III:* Elimination of the tagged xenobiotic from the cytoplasm, which occurs through excretion in animals, vacuolar sequestration, and also apoplastic deposition in plants. The membrane transport of glutathione *S*-conjugates (GS-X) is mediated by specialized ATP-binding cassette (ABC) transporters that function as GS-X pumps. After import into the vacuole, xenobiotic conjugates are further metabolized into a range of sulfur-containing compounds.

ABC proteins typically consist of transmembrane-domain (TMD) and nucleotide-binding fold (NBF) modules and are encoded by significantly larger protein superfamilies in sessile plants than in animals (Sanchez-Fernandez et al., 2001). ABC transporters are directly energized by Mg-ATP instead of a proton gradient and can mobilize a broad range of structurally unrelated substrates across distinct cellular membranes (Martinoia *et al.*, 2002). Specific members of the multidrug resistance-related protein (MRP) subfamily of ABC transporters from *Arabidopsis* (i.e., *At*MRP1, *At*MRP2, and *At*MRP3) proved to function as GS-X pumps in the vacuolar sequestration of glutathionylated compounds of model substrates and herbicides and also transported malonylated chlorophyll catabolites and glucuronides (Martinoia *et al.*, 2002).

Since GSTs with GSH-conjugating activity evolved long before plants were exposed to herbicides or pollutants from anthropogenic origin, they are presumed to be *ad hoc* recruited for functioning in detoxification. The authentic roles of these GSTs in natural metabolism remain elusive since few endogenous substrates for GSH conjugation have been identified so far. This may be due to the observed instability of natural GSH conjugates (Edwards *et al.*, 2000) or could indicate alternative functionalities of these GSTs in natural plant metabolism.

b. Are Flavonoid Phytoalexins and Their Precursors Natural Substrates for GSH Conjugation in Plants?

Plant secondary metabolites, in particular those that function as phytoalexins, are considered potential natural substrates for GSH conjugation, although little evidence is available so far. Phytoalexins encompass per definition a wide variety of low molecular weight, structurally distinct, and often species-specific secondary metabolites that exert antibiotic activity and are *de novo* produced in plants upon infection with phytopathogenic organisms (Grayer and Kokubun, 2001). Phytoalexins are also produced in plants exposed to specific abiotic stresses (i.e., heavy metals or UV). Many of these secondary metabolites are toxic to the plant itself and would therefore require elimination from the cytosol (i.e., into the vacuole) or require export to the apoplast to exert their antifungal activities.

Glutathione-conjugation products have been detected for the chalcone isoliquirtigenin and the isoflavonoid phytoalexin medicarpin, the latter also proved to be transported into vacuoles (Li *et al.*, 1997). However, these GSH conjugates can form and break down spontaneously in a pH-dependent manner (Edwards *et al.*, 2000).

c. Isothiocyanates Can Be Substrates for GSH Conjugation in Plants

Glucosinolates, also called mustard oil glucosides, are a category of nitrogen-, sulfur-, and glucose-moiety containing secondary plant products that are mainly present in the order of the Capparales, including members of the *Brassica* species such as *Arabidopsis*. Upon tissue damage, glucosinolates are degraded by pre-existing thioglucosidases, denoted myrosinases, in reactive compounds that typically include isothiocyanates and nitriles (Wittstock and Halkier, 2002). These glucosinolate degradation products have important roles in defense against insects, herbivores, and probably pathogens in plants, whereas in mammalian food, effects can range from toxic to cancer-preventive.

The toxic isothiocyanates can function as substrates for GST-catalyzed GSH conjugation in plants. The *Arabidopsis* phi- and tau-class GSTs that were found to exhibit GSH-conjugating activity with the substrate benzyl isothiocyanate (BITC) (Table II) may conjugate isothiocyanates *in vivo* (Wagner *et al.*, 2002).

2. Plant GSTs with GPOX Activity Counteract Oxidative Injury Through Organic Hydroperoxide Scavenging

Many plant GSTs are multifunctional and possess additional GSH- dependent peroxidase (GPOX) activities (Dixon *et al.*, 1998, 2002a; Edwards *et al.*, 2000; Frova, 2003). These enzymes reduce organic hydroperoxides of fatty acids that are released following oxidative injury of membranes to the corresponding monohydroxy alcohols, using GSH as electron donor (Fig. 1):

$$R-O-OH + 2\ GSH \rightarrow R-OH + GS-SG$$

The scavenging of organic hydroperoxides prevents their degradation to cytotoxic aldehyde derivates.

The theta-class *Arabidopsis* GST, *At*GSTT1, typically exhibited high GPOX activity but no GSH-conjugating activity (Table II; Wagner *et al.*, 2002). Various phi- and tau-class GSTs of cereal crops (i.e., sorghum and wheat), monocot weeds such as blackgrass, and dicot weeds such as *A. thaliana* (Table II; Wagner *et al.*, 2002) were found to exhibit relatively high degrees of GPOX versus GSH-transferase activities with distinct substrate preferences. The over-expression of a phi-class GST with high GPOX activity conferred enhanced tolerance to salt and cold stress in tobacco seedlings (Roxas *et al.*, 1997). The constitutive expression of a phi-class GST with high GPOX activity, *Am*GSTF2, proved to correlate with tolerance to herbicide-induced oxidative injury in blackgrass populations, through preventing the accumulation of cytotoxic hydroperoxides (Cummins *et al.*, 1999). Plant GSTs with GPOX activity are now believed to contribute to defense against oxidative injury during various stresses, including oxidative stress, pathogen attack, herbicide treatment, and a prolonged or severe exposure to an array of abiotic stresses.

3. Plant GSTs with GSH-Dependent Isomerase Activities

Zeta GSTs of fungi and mammals catalyze the GSH-dependent *cis–trans* isomerization of maleylacetoacetate to fumarylacetoacetate (Fig. 1), a key step in the catabolism of tyrosine and phenylalanine (Fernandez-Canon and Penalva, 1998). Analogously, the *Arabidopsis* zeta-class GST (*At*GSTZ1) had no detectable GSH-conjugating and modest GPOX activity, but exhibited GSH-dependent maleylacetone isomerase (MAI) activity (Table I; Dixon *et al.*, 2000). Little is known concerning the regulation and roles of tyrosine catabolism in plants. However, the transcripts of two GSTZ genes of carnation accumulate in response to senescence and ethylene in flower petals (Meyer *et al.*, 1991), which can be consistent with a role in the degradation of aromatic amino acids. In addition, *At*GSTZ1 also catalyzed the GSH-dependent dehalogenation of the carcinogenic drinking water contaminant dichloroacetic acid (DCA) to glyoxylic acid, as found for human zeta GSTs (Fig. 1).

Plant GST isomerase activity has also been implicated in herbicide bioactivation. Not further specified plant GSTs were found to catalyze the isomerization of a thiadiazolidine proherbicide to the herbicide triazolidine (Fig. 1), thereby releasing its phytotoxicity (Edwards and Dixon, 2000; Edwards *et al.*, 2000).

4. Plant GSTs with GSH-Dependent Thioltransferase Activity and GSH-dependent Dehydroascorbate Reductase Activity

*At*GSTLs and *At*DHARs neither have GSH-conjugating activity nor GPOX activity but were found to exhibit GSH-dependent thioltransferase activities with a model substrate (Dixon *et al.*, 2002b; Fig. 1 and Table II). During oxidative stress, when cells show diminished GSH pools, some protein thiols are reversibly *S*-thiolated by oxidized nonprotein thiols (GSSG or oxidized cysteine) to form protein–thiol mixed disulfides (Fig. 1), which rapidly affects enzyme or protein activity. Thioltransferase activity is believed to protect thiols in enzymes and structural proteins from oxidative damage through reversing *S*-thiolation, thereby preserving their proper biological functions. In animal systems (i.e., eye lens) canonical thioltransferases have been found to catalyze the dethiolation of various glycolytic enzymes including GAPDH and antioxidant enzymes (Lou, 2003) probably in synergism with thioredoxins. As such thiolation can rapidly affect enzyme or protein activity during oxidative stress. It has been proposed that the GSH-dependent thioltransferase activity of plant GSTs can also participate in removing *S*-thiol protein adducts (Dixon *et al.*, 2002b), thereby complementing activities of proteins, such as thioredoxins, from which GSTs are believed to have evolved (Sheehan *et al.*, 2001). The substrates for dethiolation by the GST-catalyzed thioltransferase activity in plants remain unknown.

The *At*DHARs (Table II; Dixon *et al.*, 2002b) and *Os*DHAR1 (Urano *et al.*, 2000) further exhibited the classical dehydroascorbate reductase activity: GSH is used to reduce dehydroascorbate (DHA) back to ascorbate (AA, Vitamin C) (Fig. 1).

$$DHA + 2\ GSH \rightarrow AA + GSSG$$

In chloroplasts, DHARs were known to reduce the large amounts of ascorbate oxidized during hydrogen peroxide scavenging by ascorbate peroxidase (Mittler, 2002). The ascorbate–glutathione cycle (Fig. 2), which is

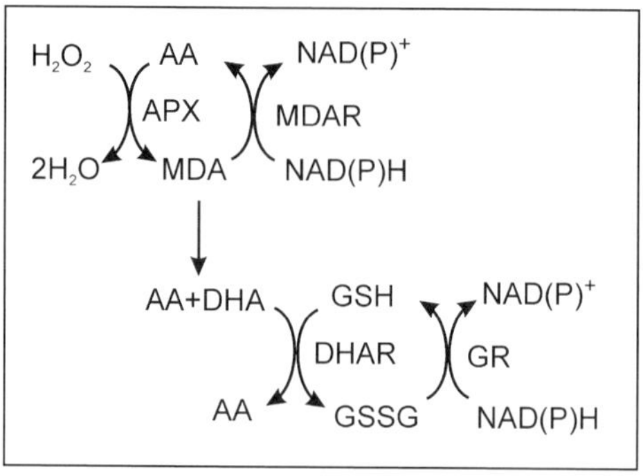

```
H₂O₂   AA      NAD(P)⁺
    ╲ ╱   ╲  ╱
    APX    MDAR
    ╱ ╲   ╱  ╲
2H₂O   MDA    NAD(P)H
         │
         ↓
AA+DHA    GSH     NAD(P)⁺
      ╲  ╱   ╲   ╱
      DHAR    GR
      ╱  ╲   ╱   ╲
   AA    GSSG    NAD(P)H
```

FIGURE 2. The ascorbate–glutathione cycle (with permission, adapted with modifications from Mittler, 2002). Basically, dehydroascorbate reductases (DHAR) are involved in regenerating ascorbic acid (AA) that was utilized during H_2O_2 scavenging by ascorbate peroxisase (APX). The cycle uses NAD(P)H as reducing power and has now been found in chloroplasts, cytosol, mitochondria, apoplast, and peroxisomes. Other abbreviations are MDAR: monohydroascorbate reductase; GR: glutathione reductase.

implicated in redox homeostasis, has now been found in almost all cellular compartments tested, including chloroplast, cytosol, mitochondrion, apoplast, and peroxisomes (Mittler, 2002). DHARs may analogously function in counteracting effects of large redox perturbations in the cytosol of plant cells, as suggested by the inducibility of cytosolic DHARs by oxidants as well as reducing agents (Table II; Dixon et al., 2002b).

B. PLANT GSTs CAN ACT AS NON-CATALYTIC CARRIERS OF PHYTOCHEMICALS

GSTs can function as noncatalytic binding proteins, also called carriers, ligandins, or cytoplasmic "escort" proteins for endogenous and exogenous, often bioactive substrates. Ligandin roles, which are well-established for mammalian GSTs (Sheehan et al., 2001), are also emerging in plant GSTs (Fig. 1). GSTs that function as binding proteins may play underestimated roles in facilitating the transport of secondary compounds and toxic intermediates of plant metabolism (Dixon et al., 2002a; Edwards et al., 2000).

1. GSTs Can Function as Anthocyanin Ligandins in Plants

The tau-class GST gene *Zm*GSTU4, also known as *bronze2*, and the phi-class GST gene *An9* represent the last genetically defined step in the anthocyanin biosynthesis pathways in maize and *Petunia hybrida*, respectively. Loss-of-function mutations of *Zm*GSTU4 caused a bronze color due to a cytosolic accumulation of anthocyanin that normally are deposited in the vacuole (Marrs, 1996). AN9 analogously proved essential for normal flavonoid biosynthesis and vacuolar transport (Alfenito et al., 1998). Since no

GST-catalyzed GSH conjugation with the nonelectrophilic flavonoids was found, *bronze2* and *An9* are now believed to function as cytosolic flavonoid-binding proteins (Mueller *et al.*, 2000). The way Bronze2 and AN9 would facilitate membrane transport while acting as carriers remains unclear, although an "indirect aid" as shuttle proteins has been hypothesized (Edwards *et al.*, 2000).

2. GSTs Can Function as Auxin or Cytokinin Ligandins, Thereby Affecting Phytohormone Homeostasis

Specific GSTs have been found to bind phytohormones (i.e., auxins) (Zettl *et al.*, 1994) or cytokinins (Gonneau *et al.*, 1998), albeit with a relatively low affinity. By acting as phytohormone-binding proteins or ligandins, these GSTs may affect the activity and distribution of these important plant growth regulators that control many aspects of plant development.

The *Arabidopsis* phi-class GST *At*GSTF2 has been implicated in auxin binding and auxin transport. *At*GSTF2 was isolated in a screen for auxin-binding proteins using photo-affinity labeling methods (Zettl *et al.*, 1994). *At*GSTF2 was found to bind the auxin phytohormone indole-3-acetic acid (IAA) and the synthetic auxin-transport inhibitor 1-*N*-naphthylphthalamic acid (NPA) (Zettl *et al.*, 1994), as well as the artificial auxin naphthalene acetic acid (NAA) and the endogenous flavonoids quercetin and kaempferol (Smith *et al.*, 2003). These flavonoids and NPA are believed to analogously block auxin efflux in *Arabidopsis* (Brown *et al.*, 2001). Consistently, competition assays indicated that NPA and the flavonoids bind to *At*GSTF2 at the same site, which is distinct from the IAA-binding site (Smith *et al.*, 2003). AtGSTF2 localizes near the plasma membrane of cells, where flavonoids also accumulate. The process by which the ligandin *At*GSTF2 would facilitate auxin-transport, thereby being regulated by flavonoids, remains unclear.

C. EMERGING ALTERNATIVE FUNCTIONS OF PLANT GSTs

1. Plant GSTs Can Indirectly Act as Regulators of Apoptosis in Plants

In plants, programmed cell death (PCD) is typically involved in the terminal differentiation of xylem vessels, the formation of necrotic lesions at infection sites during the "hypersensitive" defense response, and various plant organogenesis events. The study of apoptosis in plants has long been hampered because of the lack of a clear set of "core regulators" that are now beginning to be revealed (Lam, 2004). Since true structural orthologues for the BCL-2 protein family, with established roles in promoting or inhibiting apoptosis in mammals, proved not to be encoded in plants and yeast, genetic screens in yeast were used to search for functional plant homologues. Among others, tomato tau-class GSTs, with varying degrees of GPOX versus GSH-transferase activities, were found to indirectly suppress

Bax-lethality in yeast (Kampranis *et al.*, 2000). Co-expression of this tomato GST and Bax in yeast restored normal cellular GSH levels and preserved the membrane potential across the mitochondrial membrane, both of which had been disrupted by Bax-expression. Expression of these GSTs in yeast further provided tolerance to cell death caused by pro-oxidants including H_2O_2 and various hydroperoxides, in conjunction with the YAP1-activated oxidative stress response (Kilili *et al.*, 2004). These experiments suggested that the plant GSTs protected yeast cells from Bax-induced ROI-dependent cellular events, originating from the mitochondria and involved in PCD.

2. Plant GSTs May Play a Role in Stress-Signaling

A tau-class GST of parsley, *Pc*GST1, as well as GSH itself have been implicated in UV light signaling, which leads to transcriptional activation of genes involved in protective flavonoid biosynthesis (Loyall *et al.*, 2000). This was concluded from the observation that over-expression of *Pc*GST1 rapidly activated transcription from the UV-inducible chalcone synthase (CHS) promoter in transgenic cell lines. However, a direct role in signaling, as is known for specific mammalian GSTs, remains so far unproven for plant GSTs. The mammalian Pi-class GST, GSTp, can directly inhibit the activity of a Jun N-terminal kinase and thus contribute to the protection of cells against H_2O_2-induced cell death (Sheehan *et al.*, 2001).

III. THE TRANSCRIPTIONAL REGULATION OF GST GENE EXPRESSION

Plant GST genes usually are abundantly expressed and exhibit major transcriptional regulations. It is well documented that GST transcript levels can markedly increase in response to biotic and abiotic stresses, chemicals and xenobiotics. More specifically, GST gene inducing can include infection, biotic elicitors, ozone, UV, heavy metals, heat shock, cold, dehydration, salt stress, oxidative stress, wounding, senescence, ethanol, oxidants, antioxidants, herbicides, and herbicide safeners (Fig. 3). Studies in which the expression of multiple GST genes were assessed in parallel, have revealed an unexpected diversity in the transcriptional regulation of individual, often closely related genes within the six GST classes (McGonigle *et al.*, 2000; Wagner *et al.*, 2002). GST transcript levels can also vary during development (associated with cell division or senescence), or in various tissues or organs (root-, pollen-, or scutellum-specificity) that has rarely been studied in much detail.

Since many of the abiotic and biotic stresses that induce GST genes can generate oxidative stress, a generalized role for reactive oxygen intermediates has previously been proposed. However, GST genes since proved to exhibit a diversity of responses to a variety of molecules with potential signaling functions, including hydrogen peroxide, various plant growth regulators, and antioxidants. Particularly, the involvement of SA, JA, ethylene, auxins, and NO in regulating GST gene expression has become increasingly evident (Fig. 3). Other plant growth regulators, i.e., ABA, cytokinins, and gibberellic acids were not often tested but so far generally proved ineffective, while brassinosteroids—to our knowledge—have not been assayed yet. The current knowledge on GST expression patterns suggests that only a complex simultaneous employment of independent as well as interactive response pathways from various signals, including different plant growth regulators and ROI could have created the exceptional diversity that is evident now.

Our understanding on the biosynthesis, signal-transduction pathways, and nuclear events through which plant growth regulators control gene expression during development, biotic, or abiotic stresses has increased significantly over the past years. Research on plant GST gene expression has benefited from this progress, since various *Arabidopsis* mutants defective in phytohormone synthesis or signaling have been utilized. In the following sections, we subsequently discuss the roles of auxins, abscisic acid, salicylic acid, jasmonates (JAs), and ethylene, as well as the emerging role of nitric oxide, in regulating GST gene expression. Updates on the biosynthesis and signaling pathways of these plant growth regulators are included, while we refer to exciting reviews for further reading.

A. ROLE OF AUXINS IN THE REGULATION OF GST GENE EXPRESSION

1. Dose-Dependent Effects of Natural and Synthetic Auxins Acting as Plant Growth Regulators or Selective Herbicides

Natural auxins have been implicated in virtually every aspect of plant growth and development. In some tissues, auxins regulate cell elongation, while in others the hormone promotes cell division. Particularly well-known auxin effects include the control of apical dominance and lateral root initiation. Auxin activity is displayed by a chemically diverse group of naturally occurring compounds, predominantly indole-3-acetic acid, as well as synthetic compounds such as 2,4-dichlorophenoxyacetic acid (2,4-D) and naphthalene acetic acid.

Because of their importance in development, auxin concentrations are tightly regulated in plants, either through conjugation to various compounds or by degradation via multiple pathways. An endogenous or exogenous auxin overdose is phytotoxic and typically causes plant deformations, such as epinasty and growth inhibitions, followed by tissue necrosis and decay. Synthetic auxins, which are not rapidly inactivated in plants, display auxin activity at low concentrations but are phytotoxic at higher concentrations, and are therefore frequently used as selective herbicides. Effects of auxin overdose and auxin herbicides, such as quinclorac and 2,4-D, are complex

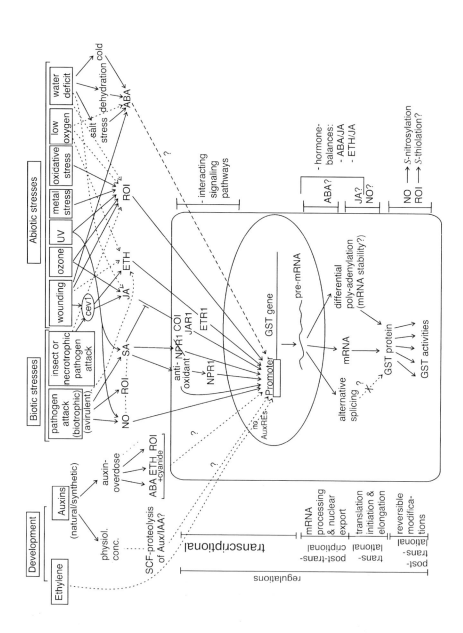

and pleiotropic, comprising among others the synthesis and effects of abscisic acid and ethylene and cyanide as its by-product (Grossmann, 2000) and most likely also the generation of ROI.

2. Auxin-Specific Responses Are Controlled by Ubiquitin-Mediated Proteolysis

Selective ubiquitin (Ub)-mediated proteolysis has emerged as a fundamental process in auxin responses, particularly in auxin-regulated gene expression (Kepinski and Leyser, 2002). Various *Arabidopsis* mutants impaired in auxin responses were found to encode crucial elements in a functional SCF (SKP1, CDC53p/CUL1 F-box) or E3 or Ub ligase complex, such as AXR1, an E1-like enzyme; ASK1, a SKP protein and TIR1, a F-box receptor protein involved in selectively recruiting a target protein for degradation.

FIGURE 3. GST genes exhibit a diversity of transcriptional regulations during various conditions of biotic and abiotic stresses. Plant growth regulators, i.e., auxins, salicylic acid (SA), ethylene (ETH), jasmonic acid (JA), nitric oxide (NO), as well as reactive oxygen intermediates (ROI) such as hydrogen peroxide (see Table I) also differentially affect the transcription of individual GST genes. It is now acknowledged that complex, stress-specific combinations of the signaling pathways of these phytohormones and ROI, render the unique diversity of the individual GST gene expression patterns. For each of the abiotic and biotic stress conditions, the plant growth regulators that are typically involved in transcription activation of stress-responsive marker genes are shown. Only some of these routes have so far been experimentally demonstrated for GST gene activation, while most others await testing. (The intensive cross-talk among the various phytohormone-signaling pathways is not shown in this figure.)

The pathogen-induced activation of GST gene expression has been characterized in detail. The rapid induction of two phi-class GST genes, *At*GSTF2 and *At*GSTF6, during avirulent plant–pathogen interactions proved SA-dependent, through NPR1-dependent and NPR1-independent pathways, as well as ETH-dependent and JA-independent, in a rather unorthodox combination of signaling events, while early signaling events remain elusive (Table III). By contrast, *At*GSTF2 induction by oxidative stress occurred through an ETH-independent pathway (Table III). Auxins are thought to be involved in the developmental regulation of *At*GSTF2. The auxin-signaling pathway that activates *At*GSTF2 gene expression remains unclarified, and no AuxREs were found in its promoter. All so far reported water-deficit conditions inducing GST expression suggest an involvement of ROI rather than ABA signaling.

In addition to the major transcriptional regulations, increasing evidence suggests posttranscriptional regulations of plant GST expression as well. Indications include alternative splicing, resulting in a noncoding transcript for *Bz-2*, and differential polyadenylation events. Stress-related plant growth regulators and their endogenous balances in particular, could also be involved in posttranscriptionally regulating the abundance of stress-response proteins such as GSTs for a number of reasons. ABA has been implicated in regulating pre-mRNA processing and nuclear export (Kuhn and Schroeder, 2003), JA in affecting protein abundance, possibly through effects on mRNA stability or translation initiation (Creelman and Mullet, 1997), and NO is thought to affect translation initiation and elongation (Lindermayr *et al.*, 2005).

A posttranslational modification of a tau-class GST of *Arabidopsis* through NO-mediated *S*-nitrosylation has been reported (Lindermayr *et al.*, 2005). Reversible NO-mediated *S*-nitrosylation and ROI-mediated *S*-thiolation can differentially affect the activity of specific microsomal and cytosolic GST isoforms in animals, which awaits to be demonstrated in plants.

3. Nuclear Events in the Effects of Auxins on Gene Expression

Auxin-regulated genes typically contain auxin-response promoter elements (AuxREs) with TGTCNC motif(s) (typically TGTCTC) in various contexts (Kepinski and Leyser, 2002). Members of the auxin-response factor (ARF) family of B3-type transcription factors bind specifically to these AuxREs, thereby generally activating auxin-regulated gene expression. Auxin responses involve members of a second protein family, the auxin/indoleacetic acids (Aux/IAAs) that are short-lived nuclear transcription regulators without DNA-binding domain. The Aux/IAAs can interact with the ARF transcription factors and thereby generally repress the transcription of auxin-inducible genes. Auxins were found to modify the SCF^{TIR1} complex through direct binding to TIR1, which promoted the selective recognition of Aux/IAA and condemned the transcription repressors for Ub-mediated proteolysis (Dharmasiri *et al.*, 2005).

4. Effects of Natural and Synthetic Auxins on GST Gene Expression in *Arabidopsis*

IAA and the synthetic auxins 2,4-D and NAA have frequently been reported to induce the expression of various GST genes in different plant species, including tobacco, *Arabidopsis*, and soybean (Marrs, 1996). Auxin responses are neither restricted nor characteristic to the members of the tau-class GST genes, as originally believed. In *Arabidopsis* seedlings, time-course spray applications of 1 mM 2,4-D caused a rapid increase in the abundance of transcripts of four phi-class GSTs (i.e., *At*GSTF2, *At*GSTF6, *At*GSTF10, and *At*GSTF8) and a delayed induction of the tau-class *At*GSTU19 and the zeta-class *At*GSTZ1 (Table II; Wagner *et al.*, 2002).

Auxin responses often proved root-specific to varying degrees. Natural and synthetic auxins markedly induced *At*GSTU5 in roots but not in leaves (Table II; Van der Kop *et al.*, 1996). *At*GSTF2 also exhibited a stronger auxin response in roots than in leaves (Smith *et al.*, 2003). In root cultures of *Arabidopsis* seedlings, a 100 μM 2,4-D for 24 h caused a transcript accumulation of *At*GSTL1, *At*DHAR2, *At*GSTU2, and *At*GSTZ1, which was particularly strong for *At*GSTL1 and *At*DHAR2, whereas the weaker auxin-analog NAA only induced *At*GSTU2 to some extent (Table II; Dixon *et al.*, 2002b). While many of these responses most likely are truly auxin-specific, others could be pleiotropic due to auxin overdose, which remains hard to discriminate at times.

Although auxins were the first phytohormones known to activate GST transcription, the pathway through which this occurs remains elusive. The AuxREs that are typically involved in auxin-regulated gene expression, have not been found in the promoters of auxin-responsive GST genes. The absence of these promoter elements may not necessarily exclude the involvement of all of the currently known elements in auxin-regulated gene expression. Typical auxin-signaling mutants can also affect auxin responses of genes without AuxREs (M. Estelle, personal communication). The employment of these mutants may help to clarify auxin-signaling pathways for GST gene expression.

Other cis-elements (i.e., the so-called *octopine synthase* [*ocs*] elements present in many GST gene promoters) had initially been associated with auxin-responsive GST gene expression (Marrs, 1996). However, *ocs*-like sequences soon proved to enhance the expression of GST genes in response to multiple stimuli including auxin-analogs, SA, JA, or hydrogen peroxide, depending on the gene studied. Some *ocs* elements now appear to contain a functional SA-response element, as discussed elsewhere in this review. Besides that finding, the role of most of the *ocs* elements in the pleiotropic activation of GST gene expression remains largely unclear to date.

Auxin-responsive GST gene expression may also require regulatory elements that are distinct from the AuxRE. Alternative promoter regions involved an auxin response gene expression are being studied. The auxin-inducible promoter of the *At*GSTF2 gene, which can function as an auxin-ligandin, neither contained a recognizable AuxRE nor an *ocs* element. Yet, auxin-responsive expression from the *At*GSTF2 promoter proved qualitatively similar to the expression from the DR5 promoter, an auxin-responsive marker gene (Smith *et al.*, 2003). Moreover, IAA-induced *At*GSTF2 transcripts accumulated at the root–shoot transition zone and in the root distal elongation zone, where endogenous auxins typically accumulate (Smith *et al.*, 2003). Transcription from the promoter of the *At*GSTF2 gene was not only activated by auxins, i.e., IAA (10 μM) and NAA (50 μM) but also induced by DTT, copper, and the herbicide paraquat, generating oxidative stress. It was proved that two regions of the *At*GSTF2 promoter, from -170 to -161 and from -350 to -170, were required for transcriptional activation by NAA as well as Cu, paraquat, and DTT. In addition, there were indications for differential signaling of auxin and Cu responses. *Cis*-elements for these auxins and oxidative stress responses remain to be identified.

Finally, a number of the observed auxin responses may result from auxin overdose, which can generate ETH, ABA, ROI, and possibly other stress signals, thereby potentially activating various combinations of stress-signaling pathways (Fig. 3).

B. ROLE OF ABA IN THE REGULATION OF GST GENE EXPRESSION

1. Roles of ABA, ABA Biosynthesis and its Regulation, and ABA Signaling

In developing seeds, ABA initiates embryo maturation, promotes desiccation tolerance, particularly through inducing the synthesis of *l*ate *e*mbryogenesis *a*bundant (LEA) proteins, and finally controls seed dormancy. When plants are exposed to dehydration, ABA triggers stomatal closure, thus preventing further water loss, and promotes dehydration tolerance through the collective induction of many proteins, including enzymes for osmolyte biosynthesis and LEA-like proteins, as has been reported extensively. In addition, ABA is involved in mediating aspects of responses to hypoxic stress, wounding as well as pathogen defense, which remains relatively less well documented.

In plants, ABA is synthesized from C_{40} carotenoid precursors through a well-known pathway, which has been corroborated through the characterization of various ABA-deficient (*aba*) mutants (Schwartz *et al.*, 2003). The dramatic increase of ABA levels in vegetative tissues exposed to drought and salt stress has now been attributed to transcriptional activations of several genes for ABA biosynthetic enzymes by these conditions (Xiong and Zhu, 2003).

The characterization of loss-of-function and gain-of-function ABA response mutants has led to the identification of various ABA-signaling elements active during seed germination, vegetative growth, or both. These ABA-signaling intermediates include among others, ABI1 and ABI2, which are two type-2C protein phosphatases, and ABI3, ABI4, and ABI5 which are B3, APETALA2, and basic-leucine zipper (bZIP) type transcription factors, respectively, ERA1, a farnesyl transferase β-subunit, a "protein kinase," and "an enzyme of phosphoinositide metabolism" (Finkelstein *et al.*, 2002). In addition, various ABA sensitive mutants proved to encode proteins with roles in RNA metabolism, such as HYL1, a dsRNA-binding protein; ABH1, an mRNA-cap-binding protein; and SAD1, an Sm-like small ribonucleoprotein (Kuhn and Schroeder, 2003). The involvement of posttranscriptional mRNA processing in ABA signal transduction remains mechanistically largely unclear.

2. Effects of ABA on GST Gene Expression

The involvement of ABA in transcriptional regulations has only been tested for a few GST genes, primarily those with known dehydration, high salt, or cold responses (Table II). ABA did not induce *At*GSTF6 and *At*GSTF10, also known as ERD11 and ERD13, whereas transcripts of these genes rapidly accumulated in dehydrated *Arabidopsis* plants (Table II; Kiyosue *et al.*, 1993). Exogenously applied ABA did neither induce *At*GSTU19, also known as GST8 (Table II; Kiyosue *et al.*, 1993). Moreover an endogenous ABA accumulation was not required for the dehydration-induced expression of *At*GSTU19 as evident from the use of the ABA biosynthesis mutant *aba 1–1* (Table III; Bianchi *et al.*, 2002). The delayed dehydration response of *At*GSTU19 coincided with the induction of a peroxidase, a marker for oxidative stress, which suggested a drought-associated response to oxidative damage. In rice roots, ABA caused a transcript accumulation of *Os*GSTU3, which was rapidly induced by salt stress (Moons, 2003). Conversely, ABA did not induce *Os*GSTU4, which exhibited a late salt stress induction, reminiscent

TABLE III. Analysis of the Role of Plant Growth Regulator Signaling Pathways in the Biotic and Abiotic Stress Responsive Transcriptional Regulation of GST Gene Expression using *Arabidopsis* Mutants

Arabidopsis GST gene	Arabidopsis mutant, defective in hormone biosynthesis or hormone signaling			Stress or hormone responsive GST gene expression			
	Mutant	Mutant type	Affected protein function	Stimulus	Response in mutant	Conclusion on signaling pathway	References
AtGSTU19 (GST8)	aba 1-1	ABA deficient	Zeaxanthin epoxidase (ZEP)	Dehydration	Unaffected	ABA independent	Bianchi et al. (2002)
AtGSTF2 and AtGSTF6	nahG transgene (bacterial salicylate hydroxylase)	SA non-accumulator	SA-inactivation by conversion to catechol	Pseudomonas syringae pv. tomato (Pst) (avirulent)	Both strongly affected	SA dependent	Lieberherr et al. (2003)
	npr1 (non-expressor of PR1)	SA signaling	NPR1: Redox-regulated transcriptional activator	Same	AtGSTF2 more negatively affected than AtGSTF6	Only partially dependent	
	cpr1 (constitutive expressor of PR1)	SA over-producer	Unknown	Same	Both constitutive in non-infected and upregulated in infected plants	SA dependent	
	jar1 (jasmonate resistant1)	JA signaling	JA-amino synthetase	Same	Inductions unaffected	Independent from JA activation by JA-Ile conjugation	
	etr1 (ethylene resistant1)	ETH signaling	ETR1 ETH receptor	Same	AtGSTF2 strongly affected; AtGSTF6 slightly affected	ETH perception dependent	
	nahG transgene	SA non-accumulator	SA conversion	ETH	Both negatively affected	Dependent on minimal SA levels	
AtGSTF2	etr1 (ethylene resistant1)	ETH signaling	ETR1 ETH receptor	NAA, GSH, Paraquat, Cu	Inductions unaffected	Independent of ETH perception	Smith et al. (2003)
	tt4 (transparent testa4)	Flavonoid deficient	Chalcone synthase (CHS)		Protein localization disrupted	Localization flavonoid-dependent	

of a secondary possibly oxidative damage response. High ABA concentrations induced a transcript accumulation of *Zm*GTU4 or *Bronze2*, which was also cold-responsive (Marrs and Walbot, 1997). In general, the current data provide little evidence for an involvement of ABA in the dehydration, salt, or cold stress response of most of these GST genes.

ABA can also play a role in wounding and pathogen responses and regulate the expression of defense-related genes (Moons *et al.*, 1997), which is often overlooked. A role for ABA in the wounding- or pathogen-responsive expression of individual *At*GST genes may be worth questioning.

C. ROLE OF SA IN REGULATING GST GENE EXPRESSION

1. SA: Role, Biosynthesis, and Signaling Network

Upon infection of a plant species with an incompatible or avirulent plant pathogen strain, gene-for-gene or avirulence signal/resistance protein (avr/R) recognition occurs (Ellis and Jones, 1998). The subsequent establishment of the hypersensitive response (HR) (i.e., local necrotic lesions that prevent further pathogen spread) triggers the accumulation of SA in both local and systemic tissue. SA is responsible for the establishment of the systemic SAR response, which provides long-lasting immunity against subsequent infections by a broad spectrum of pathogens, and is characterized by the systemic expression of a subset of pathogenesis-related (PR) genes. SA also accumulates during abiotic stress (i.e., exposure to UV light).

In plants, SA is most likely synthesized from phenylalanine through a phenylpropanoid pathway as well as from chorismate, a shikimate pathway product, as evident from the characterization of the SA-deficient mutant *SID2* (Shah, 2003). The characterized SA biosynthesis mutants now offer additional tools to assess the role of endogenous SA in gene expression studies, since the transgenic *Arabidopsis* lines that express bacterial salicylate hydroxylase (*nahG*) can exhibit pleiotropic effects.

2. Nuclear Events in SA-Mediated Gene Expression: NPR1-Dependent and NPR1-Independent Pathways and Long-Distance SA Signaling

The *Arabidopsis* NONEXPRESSOR OF PATHOGENESIS-RELATED GENES1 (NPR1) protein is a key regulator of salicyclic acid-responsive gene expression. NPR1 is a transcriptional activator which has no DNA-binding domain but containing an ankyrin-repeat protein–protein interaction domain and cys-residues involved in intermolecular disulfide bonds (Dong, 2001; Pieterse and Van Loon, 2004). SA mediates cellular redox changes, possibly an accumulation of antioxidants, which reduces inactive NPR1 oligomers to active NPR1 monomers and stimulates their translocation into the nucleus (Mou *et al.*, 2003). In the nucleus, NPR1 differentially interacts with members of the TGA family of basic-leucine zipper transcription factors, particularly TGA2 and TGA3. NPR1 also interacts with TGA1, after a SA-induced redox-mediated conformational changes of TGA1 (Despres *et al.*, 2003). The interaction with NPR1 enhances TGA binding to an a*s-1* promoter element containing the SA-responsive element TGACG, which is required for SA-induced PR gene expression (Lebel *et al.*, 1998). SA signaling can also proceed through a second pathway, which is SA-dependent but NPR1-independent, and exhibits cross-talk with JA and ethylene signaling (Dong, 2001; Shah, 2003).

A systemic accumulation of SA is required in both the NPR1-dependent and NPR1-independent pathways. Yet, the long-distance signal for both SA-response pathways is not SA itself but a so far unidentified, lipid-derived molecule. This became evident from the characterization of SA-response mutants including a desaturase, a glycerol-3-phosphate dehydrogenase, an apoplastic lipid-transfer protein, and a SA-binding lipase (Dong, 2001; Pieterse and Van Loon, 2004; Shah, 2003).

3. Effects of SA and SA Signaling on the Regulation of GST Gene Expression in *Arabidopsis* During Biotic and Abiotic Stresses

The effects of exogenously applied SA have been tested for nine GST genes in *Arabidopsis* leaves. SA caused a rapid transcript accumulation of the phi-class GSTs, *At*GSTF2, *At*GSTF6, and *At*GSTF8, which was particularly strong for *At*GSTF6, and a delayed induction of tau-class *At*GSTU19 (Table II; Wagner *et al.*, 2002; Glombitza *et al.*, 2004). A proteomic approach indicated among others a SA-induced accumulation of *At*DHAR2 and confirmed the SA-induced accumulation of the *At*GSTF8 protein in *Arabidopsis* cell cultures (Sappl *et al.*, 2004).

A functional SA-response element has been identified in the *At*GSTF8 promoter (Chen and Singh, 1999). The *At*GSTF8 promoter contains a 20 bp *octopine synthase* (*ocs*)-like enhancer element, with an exact copy of the salicylic acid response-element TGACG (Lebel *et al.*, 1998). Whereas TGACG occurs twice in the *as-1* CaMV promoter element, it is only occasionally present in the ocs-like elements, previously identified in the promoter of several GST genes (for sequences, see Marrs, 1996). Promoter mutations confirmed that the TGACG element in the *At*GSTF8 promoter was at least in part required for the induction of *At*GSTF8 by SA as well as hydrogen peroxide, particularly in roots. Another 50 bp promoter region was required for the SA and H_2O_2 responses in addition (Chen *et al.*, 1999). Conversely, the promoter of the *At*GSTF2, which is also SA-responsive, did not contain a recognizable *ocs* element.

Some GSTs are rapidly induced in systemic tissue in response to infection with an avirulent pathogen, markedly preceding the induction of SAR-related genes such as *PR1* (Table II; Alvares *et al.*, 1998). Inoculating *Arabidopsis* plants with an incompatible downy mildew isolate (*Peronospora parasitica* isolate EMWA) caused a rapid accumulation of *At*GSTU19 and *At*GSTF8 transcripts, while an avirulent strain of *Pseudomonas syringae* pv. *tomato* (*Pst*) caused a rapid accumulation of *At*GSTF2 and *At*GSTF6 transcripts, all of which preceded the induction of the *PR-1* gene (Table II; Lieberherr *et al.*, 2003; Wagner *et al.*, 2002). The rapid response suggested a protective role for these GSTs in pathogen defense and raised questions on the regulation of these GST genes compared to the rather well-established regulation of *PR-1* gene expression. The role of plant GSTs in pathogen defense remains largely hypothetical. It has been proposed that antioxidant activity of GSTs included the well-known role in hydroperoxide scavenging, could limit symptom development during compatible pathogen interactions, or restrict necrotic lesion development in non-compatible interactions (Alvares *et al.*, 1998). *At*GSTF2 and *At*GSTF8 do exhibit marked GPOX activity. *At*GSTU19 has no GPOX activity, but a role in oxidative stress protection has been suggested before (Table II; Bianchi *et al.*, 2002).

All four of the GST genes that were rapidly induced by avirulent pathogen infection were also markedly induced by exogenously applied SA (Table II). This prompted an investigation on the importance of endogenous SA and SA signaling in the induction of *At*GSTF6 and *At*GSTF2 by the avirulent pathogen (Lieberherr *et al.*, 2003). An endogenous accumulation of SA proved essential for the pathogen response of *At*GSTF2 and *At*GSTF6, as evident from the employment of *nahG* transgenics (Table III). However, the SA-dependent pathogen-induction of *At*GSTF2 and *At*GSTF6 was only partially dependent on NPR1, as evident from the limited impact of the *npr1* mutation versus the marked *nahG* effect, and the impact of an SA over-producer mutant (Table III). It was concluded that the pathogen response of *At*GSTF2 and *At*GSTF6 was regulated through SA-dependent, NPR1-dependent and NPR1-independent pathways (Table III), as is the case for PR1. However, a regulation through SA-dependent pathways, as had been found for *PR-1*, was insufficient to explain the rapid pathogen response, suggesting an involvement of other signals in addition (see further).

D. ROLE OF JASMONATES IN THE REGULATION OF GST GENE EXPRESSION DURING BIOTIC AND ABIOTIC STRESSES

1. Biosynthesis and Roles of the Jasmonate Family of Plant Growth Regulators

The Jasmonates represent a family of potent lipid signaling molecules or oxylipinses, which includes jasmonic acid, the JA intermediate 12-oxo-phytodienoic acid (OPDA), and JA derivates such as the volatile methyl ester, MeJA. In addition to the jasmonates, the role, the activities and biosynthetic pathways of other oxylipins are being characterized (Gidda *et al.*, 2003; Ponce de León *et al.*, 2002).

Jasmonic acids are synthesized from the membrane fatty acid precursor linolenic acid (LA) through the octadecanoic pathway that is well-characterized biochemically and genetically (Devoto and Turner, 2003; Liechti and Farmer, 2002; Turner et al., 2002). JAs are predominantly produced during plant wounding or stamen development. Whereas OPDA, JA, and MeJA all can elicit wound responses that mediate defense against insects or necrotrophic pathogens, JA is required for another development (Devoto and Turner, 2003). JAs can also be produced during diverse abiotic stresses including severe dehydration, salt and osmotic stress, and ozone exposure. JAs furthermore mediate protection from ozone stress and can regulate the biosynthesis of secondary metabolites (Devoto and Turner, 2003; Liechti and Farmer, 2002).

Early events in the wound-induced activation of JA biosynthesis include alterations in the cell wall, as evident from the characterization of a cellulose synthase encoded by the *constitutive expression of vegetative storage proteins 1* (CEV 1) mutant (Turner et al., 2002). The release of LA from plastid membranes is still believed to be the key regulatory step in wound-induced JA synthesis. No counterpart for the 18 amino acid polypeptide systemin from tomato, which triggers systemic JA responses, has so far been identified in *Arabidopsis*.

2. JA Signal Transduction and Nuclear Events in the Effects of JAs on Gene Expression

As for auxins, a ubiquitin-mediated proteolytic pathway proved a key regulator of JA responses. The *coronate insensitive1* (*coi1*) mutant, insensitive to coronate, a structural JA-conjugate analog, was found to encode an F-box protein that functions as a target receptor in an E3 ubiquitin ligase or SCF complex. COI1s can recruit a histone deacetylase indeed the small subunit of RUBISCO for ubiquitin-mediated degradation (Devoto and Turner, 2003). In addition, the JA-signaling mutant *jasmonate resistant1* (*jar1*) was found to encode a JA-amino synthetase that specifically conjugates isoleucine to JA, generating bioactive JA-ileu (Staswick and Tiryaki, 2004). Conjugation to the amino acid proved to activate JA, opposite to the effect of auxin-conjugations.

Jasmonates typically cause a massive transcriptional and posttranscriptional reprogramming, decreasing the expression of housekeeping genes and increasing the expression of defense genes. JA-induced proteins include proteinase inhibitors, thionins, plant defensins, a subset of the PR-proteins, JA biosynthetic enzymes, and enzymes involved in various secondary metabolite biosynthetic pathways.

ORCA 3, a rapidly JA-induced APETALA2 (AP-2) domain transcription factor, activates the transcription of JA-responsive genes for secondary metabolite enzymes in *Catharanthus roseus* (Turner et al., 2002) through binding to a JA and elicitor response promoter element (JERE). The

Arabidopsis homologue, ETHYLENE RESPONSE FACTOR1 (ERF1), which has binding specificity for the GCC-box, is analogously involved in regulating JA-responsive gene expression (Guo and Ecker, 2004).

3. Cross-Talk of JA Signaling with Auxin and SA Signaling

Ubiquitin-mediated proteolysis are key elements of both auxin and JA responses. While various elements of the SCF complex involved in auxin responses can be active in JA responses, the TIR1 and COI1 F-box proteins proved specific for auxin or JA responses, respectively.

Salicylic acid and JA generally activate different sets of plant defense genes. Moreover, activation of SAR generally suppresses JA signaling (Fig. 3), thereby prioritizing SA-dependent resistance over JA-dependent defenses. This antagonism is regulated through a substantial communication between both signaling networks. Wounding-activated mitogen-activated protein kinase (MAPK) pathways exert antagonistic effects on SA and JA signaling (Turner *et al.*, 2002). A cytosolic function of SA-activated NPR1 negatively affects JA-responsive gene expression, and a WRKY transcription factor activates SA-responsive genes and represses JA-inducible genes (Pieterse and Van Loon, 2004). Yet, JA is required for the SA-dependent NPR1-independent signaling pathway.

4. Effects of JAs and JA Signaling on GST Gene Expression in *Arabidopsis*

Responses to JAs vary significantly among "GTS" genes (Table II). Time–kinetic studies indicated that MeJA caused the rapid and marked transcript accumulation of *At*GSTF6, a weak induction of *At*GSTF10 and *At*GSTU19, and a late transcript accumulation of *At*GSTZ1 but did not induce or downregulate other phi- or tau-class GSTs (Table II; Wagner *et al.*, 2002). High JA concentrations (100 μM) caused a marked transcript accumulation of *At*GSTF10 and a weak induction of *At*GSTU19 (Table II; Bianchi *et al.*, 2002). The observed positive or negative JA/MeJA responses were antagonistic to SA responses for *At*GSTF2, *At*GSTF8, *At*GSTF10, and *At*GSTZ1.

*At*GSTF6 was markedly induced by JAs and SA, which is uncommon. An additional involvement of JA signaling in the rapid induction of *At*GSTF2 and *At*GSTF6 by the avirulent *Pseudomonas* strain *Pst* was therefore questioned (Lieberherr *et al.*, 2003). The avirulent pathogen response of both *At*GSTF2 and *At*GSTF6 proved unaffected by JA signaling through *jar1* (Table III), in spite of their differential responses to exogenous JA. Other JA response mutants were not tested.

Although direct evidence is lacking, indirect evidence suggests that JAs may be involved in signaling the abiotic stress-induced expression of *At*GSTF6 and *At*GSTF10. *At*GSTF6 (ERD11) and *At*GSTF10 (ERD13) were induced by 1 h dehydration, which is a rather severe stress (Table II; Kiyosue *et al.*, 1993). JAs can accumulate endogenously during salt dehydration or osmotic shock, particularly when stress conditions are drastic or long-term, presumably due to membrane damage and the release of LA precursors, and modulate the expression of sets of stress-responsive defense-related genes (Moons *et al.*, 1997). Wounding, which triggers JA biosynthesis, did markedly induce *At*GSTF10 (Table II; Bianchi *et al.*, 2002).

E. ROLE OF ETHYLENE IN THE REGULATION OF GST GENE EXPRESSION DURING BIOTIC AND ABIOTIC STRESSES AND NORMAL DEVELOPMENT

1. Ethylene Biosynthesis and Its Regulation in Plants

The gaseous plant hormone ethylene regulates plant responses to pathogen attack and abiotic stresses and controls various processes in plant development, including flower and leaf senescence, fruit ripening, and programmed cell death. Ethylene is produced in response to pathogen attack and abiotic stresses, such as wounding, hypoxia, ozone, metal stress, and extreme temperatures, and is tightly regulated during plant development.

Ethylene is synthesized in plants through a well-known two-step pathway (Wang *et al.*, 2002). In the first and rate-limiting step, the ethylene precursor, which is the cellular methyl group donor *S*-adenosyl-methionine, is converted to 1-aminocyclopropane-1-carboxylic acid (ACC) by the enzyme ACC synthase. The subsequent oxidation of ACC to ethylene is catalyzed by ACC oxidase. The control of ethylene biosynthesis occurs to a large extent through the differential transcriptional regulation of the various members of the ACC synthase multigene family by developmental and environmental signals (Wang *et al.*, 2002).

2. Ethylene Perception and Signal Transduction in *Arabidopsis*

The linear ethylene signaling pathway is generally well-known, thanks to the thorough characterization of various *Arabidopsis* ethylene response mutants (Guo and Ecker, 2004; Wang *et al.*, 2002). These mutants have been screened for based on the triple-response phenotype, which are ethylene-induced morphological changes in etiolated dicotyledonous seedlings.

Ethylene is perceived at the endoplasmic reticulum (ER) by membrane-associated receptors that are homologous to bacterial two-component histidine kinases (Guo and Ecker, 2004). There are five ethylene receptors in *Arabidopsis*, one of which is encoded by the *ethylene resistant1* (*etr1*) gene. The ethylene receptors negatively regulate the ETH response. In the absence of ethylene, the kinase domain of ETR1 directly interacts with a Raf-like serine/threonine (Ser/Thr) kinase, encoded by the *constitutive triple response1* (*ctr1*) gene. CTR1 also negatively regulates the downstream ETH response, through initiating a MAPK cascade. Upon ethylene binding to the

N-terminal transmembrane domain, for which a copper cofactor is required, the receptors are activated and CTR1 is released. CTR1 deactivation allows functioning of downstream positive regulators of the pathways including the integral membrane protein EIN2 (encoded by the *ethylene insensitive2* gene) and EIN3 (encoded by the *ethylene insensitive3* gene), the latter belonging to a family of six nuclear-localized EIN3-like (EIL1) transcription factors. EIN3 is regulated at the protein level, which may occur through ubiquitin-mediated proteolysis. The primary ethylene-signaling pathway, from ethylene perception to EIN3, is ethylene-specific.

3. Nuclear Events in the Effects of Ethylene on Gene Expression

A transcriptional cascade leads to the activation of ethylene-responsive gene expression in the nucleus (Guo and Ecker, 2004). Homodimers of the EIN3/EIL1 transcription factors bind to the primary ethylene response elements (PERE) in the promoters of genes for a second family of transcription factors, the ETHYLENE RESPONSE ELEMENT BINDING PROTEINS (EREBPs), including ETHYLENE RESPONSE FACTOR1 (ERF1), and activate their transcription. Members of the large, plant-specific EREBP family specifically bind to the GCC-box in the promoter of many ethylene-inducible genes that shape the ethylene response. However, only a few of the GCC-box binding EREBPs are regulated by ethylene, while others are involved in responses to wounding, cold, JA, high salt, drought, and other stresses. ERF1 is involved in JA- and ethylene-mediated gene regulation.

4. Cross-Talk of ETH and JA Signaling

ETH and JA induce distinct sets of genes that may act complementary during stress and defense responses. Among others, well-established JA/ETH signaling interactions include the *cev1* encoded cellulose synthase and ERF1. Alterations in the cell wall, apparently sensed through the cellulose synthase *cev1* gene, trigger early events in the activation of both JA and ETH biosynthesis (Devoto and Turner, 2003). Moreover, the transcription factor ERF1 represents a point of convergence among the ETH- and JA-response pathways.

5. Effects of Ethylene and Ethylene Signaling on GST Gene Expression in *Arabidopsis*

Effects of ethylene and its precursor ACC vary significantly among GST genes in a non class-specific manner. Exogenously applied ethylene caused a marked transcript accumulation of the phi-class *At*GSTF2 and *At*GSTF6 but did not induce four more *At*GSTFs, the *Arabidopsis* PR-1 gene nor *At*GSTT1, and ethylene did neither induce or down-regulate three tau-class *At*GSTUs (Table II; Glombitza *et al.*, 2004; Lieberherr *et al.*, 2003; Wagner *et al.*, 2002). Although it had previously been reported that ethylene markedly induced the expression of the senescence-related zeta-class *Dc*GSTZ1 and *Dc*GSTZ2 genes in carnation (Meyer *et al.*, 1991), ethylene did not induce *Arabidopsis At*GSTZ1 (Table II).

Remarkably, *At*GSTF2 and *At*GSTF6 were markedly induced by both ETH and SA, which is highly uncommon. A potential involvement of ETH in addition to SA signaling in the rapid induction of *At*GSTF2 and *At*GSTF6 by the avirulent *Pseudomonas* strain *Pst*, was therefore questioned (Lieberherr *et al.*, 2003). The pathogen-induced transcript accumulation of *At*GSTF2 and *At*GSTF6 correlated well with an endogenous increase in ethylene production, though dependence on an endogenous ETH accumulation remains unknown. The pathogen response of *At*GSTF2 proved also markedly dependent on ethylene perception through the ETR1 receptor, whereas the induction of *At*GSTF6 was much less ETR1-dependent (Table III). Remarkably, the ETH response of *At*GSTF2 and *At*GSTF6 proved dependent on an endogenous SA accumulation, as revealed in an *nahG* background (Table III). It was then demonstrated that the pathogen-induction of *At*GSTF2 and *At*GSTF6 required both ETH and SA signaling (Lieberherr *et al.*, 2003). Yet, the early phase of *At*GSTF6 induction remains unexplained for and must involve an additional unknown signaling pathway. Since synergistic effects of SA- and ETH-signaling pathways are very uncommon, it was suggested that the observed convergence of ETH and SA responses may occur downstream of the signaling pathways, possibly through an unusual combination of *cis*-elements in these GST gene promoters that have not been studied in much detail. ETH and SA signaling do interact synergistically in the SA-dependent NPR1-independent signaling pathway, through which *At*GSTF2 and *At*GSTF6 regulation partially occurs.

The role of ETH in the abiotic stress response of *At*GSTF2 has been investigated in addition (Smith *et al.*, 2003). *At*GSTF2 is markedly induced by a number of chemicals, metals, and xenobiotics, including NAA; the sulfhydryl compounds GSH, DTT, and cysteine; Cu, and herbicides, including paraquat, which can generate superoxide radicals. The induction of *At*GSTF2 by the auxin analog NAA, GSH, paraquat, and Cu proved independent of ETH perception through ETR1 (Table III). In summary, the current data suggest that *At*GSTF2, which diploids activities ranging from auxin-ligandin to GPOX and GSH-transferase (Table II), also exhibits a range of distinct regulations, i.e., auxin responses during normal development, ETH and SA regulations during pathogen infection, and ETH-independent regulations during chemical stresses.

In a previous work, the involvement of ethylene in the developmental, senescence-induced regulation of the zeta-class *Dc*GSTZ1 and *Dc*GSTZ2 genes in carnation petals has been demonstrated (Meyer *et al.*, 1991). An ERE had been identified in the *Dc*GSTZ1 promoter, as well as a corresponding EREBP from carnation (Marrs, 1996). A similar involvement of ETH in a developmental regulation of *At*GSTZ1 is unlikely, since no response to exogenous ETH was found (Table II).

F. ROLE OF NO IN REGULATING GST GENE EXPRESSION AND ACTIVITY

1. Roles of the Plant Growth Regulator NO

The reactive molecule NO has emerged as a crucial signal that modulates plant pathogen defense as well as diverse phytohormonal responses (Wendehenne *et al.*, 2004). NO is an important regulator of plant apoptosis, deploying PCD stimulating or inhibiting activities, depending on cross-communication with pro-oxidants or antioxidants. NO also modulates the regulation of stomatal closure.

During incompatible plant–pathogen interactions, NO cooperates with H_2O_2 to trigger hypersensitive cell death, activates the phenylpropanoid pathway, thereby stimulating phytoalexin and SA-synthesis, and facilitates SAR development in an SA-dependent manner (Romero-Puertas *et al.*, 2004; Van Camp *et al.*, 1998).

2. NO Biosynthesis in Plants

Nitric oxide rapidly peaks when plants are inoculated with avirulent bacteria or elicitors, concomitant with the oxidative burst, but NO does not accumulate in response to virulent phytopathogens (Romero-Puertas *et al.*, 2004). NO also accumulates in response to high temperatures, osmotic stress, and UV. High light does not cause an NO accumulation, while effects of wounding remain conflicting.

In plants, NO is generated by two unrelated nitric oxide synthase (NOS)-like enzymes that oxidize L-arginine to L-citrulline, and NO. iNOS, a variant of the P protein of the glycine decarboxylase complex, is inducible by avirulent plant–pathogen interactions and elicitors, whereas *At*NOS1 is activated by phytohormones including ABA (Romero-Puertas *et al.*, 2004; Wendehenne *et al.*, 2004). Moreover, nitrate reductase can catalyze the NAD(P)H-dependent reduction of nitrite to NO when photosynthesis is absent or blocked. NO can also form through the non-enzymatic reduction of nitrite, particularly in acidic and reducing conditions.

3. Effects of NO on Gene Expression

Microarray analysis indicated that nitric oxide affects the transcription of numerous plant genes (Huang *et al.*, 2002; Polverari *et al.*, 2003; Wendehenne *et al.*, 2004). NO rapidly induces genes for the first enzymes in the phenyl-propanoid pathway and flavonoid branch, i.e., phenylalanine ammonia lyase (PAL) and CHS, and upregulates PR genes. NO not only induces genes for proteins involved in PCD, but can also activate anti-apoptotic activities by inducing ROI scavenging or ROI avoiding and cyto-protective enzymes such as APX, catalases, superoxide dismutases (SOD), alternative oxidases (AOX), and GSTs. NO furthermore induces genes involved in JA and ETH biosynthesis and ETH signaling, indicating cross-talk with these pathways. The nuclear events in NO-mediated activation or suppression of gene expression have not yet been defined.

4. NO Signaling and NO-Specific Post-translational Modifications of Protein Function

Nitric oxide responses are signaled by cGMP and cyclic ADP ribose (cADPR), which stimulates Ca^{2+} release through intracellular ryanodine receptor channels (RYR), in a pathway that remains unknown.

Nitric oxide can furthermore exert major posttranslational regulations, as known in mammalian systems. NO can nitrosylate proteins through reacting with a thiol group from a single critical cysteine residue or with transition metal centers. S-nitrosylation can reversibly and specifically alter protein functions and enzyme activities. Using a proteomic approach, involving a biotin switch method, 63 and 50 targets for S-nitrosylation in plants have been identified in *Arabidopsis* cell cultures and leaves, respectively (Lindermayr *et al.*, 2005). The identified S-nitrosylated proteins included stress- and redox-related proteins, translation initiation and elongation factors, cytoskeleton proteins, metabolic enzymes, and photosynthetic proteins. NO is known to inhibit cytochrome *c* oxidase (COX) activity and respiration.

Considerable cross-talk is emerging among NO, ABA, auxin, SA, JA, and ROI signaling (Romero-Puertas *et al.*, 2004; Wendehenne *et al.*, 2004). NO cooperates with H_2O_2 in HR and coactivates the expression of PR genes in an SA-dependent manner. NO synthesis, along with cGMP and cADPR are required for ABA-induced stomatal closure through Ca^{2+} mobilization. NO is also involved in triggering auxin-mediated lateral root development.

Nitrosoglutathione (GSNO), a stabilized form of NO, has been suggested to function as intracellular and intercellular carrier for the NO signal in plants.

5. Effects of NO on GST Gene Expression in Plants

In soybean cell cultures, the NO donor sodium nitroprusside (SNP) caused a slow accumulation of a *gst* transcript, which was H_2O_2-dependent, in contrast to the rapid NO induction of *pal* and *chs*, which was H_2O_2-independent (Delledonne *et al.*, 1998). The rapid induction of this soybean *gst* gene by an avirulent *P. syringae* strain proved independent of NO, as had been concluded for ROI from the oxidative burst before. The primary signal for the early pathogen response of this GST gene remains elusive.

Microarray analysis has provided evidence for effects of NO on *Arabidopsis* GST gene expression. Treatment of *Arabidopsis* suspension cells with an NO donor induced a marked transcript accumulation of three GST ESTs (Huang et al., 2002) that upon BLAST analysis proved to correspond to AtGSTU7, AtGSTF6, and AtGSTF7 (Table II). In addition, treatment of Arabidopsis leaves with the NO donor sodium nitroprusside caused a rapid transcript accumulation of AtGSTF8 and a microsomal AtGST (Polverari et al., 2003). Both AtGSTF6 and AtGSTF8 are both markedly induced by NO, H_2O_2 and SA (Tables II, III), which suggests a potential involvement in HR, in addition to SAR during specific avirulent plant pathogen interactions. NO as well as alternative oxylipins, synthesized through dioxygenase pathways (Ponce de León et al., 2002), may now be worth assessing for early involvement in signaling of the particularly rapid avirulent pathogen response of AtGSTF6.

In addition, post-translational interactions of NO with GST proteins and activities are known in animal systems. S-nitrosylation is thought to differentially regulate the activity of microsomal versus cytosolic GSTs in rat liver. Treatment with S-nitrosoglutathione (GSNO) and other NO donors increased the activity of microsomal GSTs and partially inhibited the activities of cytosolic GSTs in these animal systems (Ji et al., 2002). Microsomal GSTs from rat liver are also activated through hydroperoxide-mediated S-thiolation, which could be reversed depending on the intracellular glutathione redox state (Dafre et al., 1996).

In plants, a proteomic strategy has identified AtGSTU19 in Arabidopsis cell cultures and leaves as a potential substrate for S-nitrosylation (Lindermayr et al., 2005). It remains to be determined whether AtGSTU19 (Table II) is activated or deactivated by nitrosylation. An activation would reinforce the proposed protective role of this GST isoform during oxidative stress (Bianchi et al., 2002) and pathogen attack (Wagner et al., 2002).

The human GST P1-1 has been suggested to act as an NO-carrier, when GSH-depletion occurs in the cell, in addition to its well-known detoxification activities (Lo Bello et al., 2001). GSTs that function as NO-carriers have not been identified in plants so far.

G. INTEGRATED ROLES OF ROIs AND PLANT GROWTH REGULATORS ON GST GENE EXPRESSION

1. Production, ROIs, and Scavenging of ROIs in Plants

Reactive oxygen intermediates, such as the superoxide radical (O_2^-), hydrogen peroxide (H_2O_2), and hydroxyl radicals (HO^-), are toxic by-products of the aerobic metabolism. ROIs are capable of unrestricted oxidation and destruction of various cellular compartments, in part by converting membrane fatty acids to toxic lipid peroxides. In addition, plants also actively produce ROIs, which act as signaling molecules that are either involved in stimulating programmed cell death or triggering ROI scavenging and other cyto-protective activities.

Hydrogen peroxide or superoxide radicals, which are rapidly converted to H_2O_2, are produced during photosynthesis in chloroplasts, electron transport in mitochondria, and photorespiration in peroxisomes. Pathogen attack and most abiotic stresses enhance the generation of H_2O_2 via these routes. Pathogen attack, wounding, UV, and ozone stress activate in addition the enzymatic H_2O_2 production through plasma membrane (PM)-localized NADPH oxidases, cell wall peroxidases, or enzymes such as xanthine, amine, or oxalate oxidases (Neill et al., 2002). The enzymatic H_2O_2 production is tightly regulated and participates in signaling PCD, pathogen defense, and abiotic stress responses. The well-known oxidative burst, which is the rapid enzymatic production and release of H_2O_2 into the apoplast during incompatible pathogen attack, minutes after avr/R recognition, participates in HR and SAR signaling.

Reactive oxygen intermediates are usually efficiently detoxified using antioxidants. Major ROI scavenging enzymes and pathways in plants include superoxide dismutases, with active isozymes in almost all cellular compartments, the water–water cycle in chloroplasts, ascorbate peroxidases (APX) within the ascorbate–glutathione cycle in various cellular compartments (Fig. 2), the glutathione peroxidase (GPX) cycle, and catalases (CAT) in peroxisomes (Mittler, 2002).

2. Effects of ROI on Gene Expression

Hydrogen peroxide signaling typically results in the activation of genes for ROI scavenging enzymes as well as proteins involved in PCD and the repression of genes involved in ROI-producing processes. However, ROI effects can differ during biotic and abiotic stress. ROIs also participate in activating systemic plant pathogen defense responses, wounding, and high light responses.

Nuclear events in H_2O_2-responsive gene expression in plants remain poorly understood. In bacteria and yeast, oxidative stress responses are triggered in part by the oxidative activation of the OxyR and Yap1 transcription factors, respectively, through the formation of intramolecular disulfide bridges. It remains unknown whether oxidative changes to transcription factors are required for oxidative stress responses in plants. Microarray analysis has indicated that ROI can affect the expression of a range of transcription factors that may all participate in modulating plant oxidative stress responses, including WRKY, EREBP, DREBA, myb, AP-1, Ocs/AS-1, and HSF (Mittler et al., 2002).

3. ROI Signal Transduction and Cross-Talk with Phytohormone Signaling

Unlike for yeast, the molecule through which H_2O_2 is perceived remains unknown in plants. Known elements in H_2O_2 signaling include Rho-like small G proteins (Rops), calmodulin, NADPH oxidases that are activated by Ca^{2+} binding, and MAPK cascade components (Neill et al., 2002).

Increasing evidence indicates an integrated role of ROIs in various phytohormone-signaling pathways (Neill et al., 2002). During incompatible plant–pathogen interactions, H_2O_2, SA, and NO cooperate in repressing ROI scavenging and stimulating PCD for HR development, and, in conjunction with a lipid-derived molecule, further cooperate in long-distance signaling of SAR. H_2O_2 and NO also participate in the regulation of ABA-induced stomatal closure. H_2O_2 is further involved in specific auxin responses such as gravitropism.

4. Integrated Effects of ROIs and Plant Growth Regulators on GST Gene Expression in *Arabidopsis*

Reactive oxygen intermediates are important regulators of GST gene expression in various organisms. However, the role of ROI in directly activating GST gene expression has proven less general than what was originally believed. *Arabidopsis* GST genes exhibit a range of different responses to exogenous hydrogen peroxide, with only a few of the tested GST genes being markedly induced (Wagner et al., 2002). H_2O_2 caused a marked transcript accumulation of *At*GSTF6 and *At*GSTF8, a weak induction of *At*GSTF10 and *At*GSTU19 but had no effect or downregulated the expression of two other phi-, two tau-, a theta-, and a zeta-class *At*GST (Table II). The survey suggests that H_2O_2, NO, and SA could act synergistically in the induction of *At*GSTF6 and *At*GSTF8 expression by avirulent pathogens. The herbicide, paraquat, which can generate superoxide radicals, also proved to directly induce the expression of a selected number of individual GST genes. Paraquat caused a marked transcript accumulation of *At*GSTF2 and a weak induction of *At*GSTF10 and *At*GSTU19. A possible involvement of ROI signaling in the auxin response of *At*GSTF2 may be worth investigating.

H. INTEGRATED EFFECTS OF ANTIOXIDANTS AND PHYTOHORMONES ON GST GENE EXPRESSION

1. Effects of Redox Perturbations and Antioxidants in Plants

The antioxidants, AA and GSH, are crucial for the plants defense against oxidative stress. ROI scavenging through GPX and APX therefore involves AA and GSH regenerating cycles, usually using NAD(P)H as reducing power (Mittler, 2002). Abiotic and biotic stress conditions that stimulate ROI production can temporarily cause large redox perturbations in various cellular compartments. In addition, SA is known to mediate redox perturbation during incompatible pathogen interactions. Upon R gene recognition, the bifacial oxidative burst is followed by SA-mediated reducing conditions, probably through the accumulation of antioxidants that remain to be identified (Mou et al., 2003). The oxidized versus reduced ratio of antioxidants such as GSH and AA has frequently been suggested to possibly serve as a signal for plant stress responses. Although a variety of plant genes are markedly induced by antioxidants and sulfhydryl compounds, the knowledge on redox sensing and signaling in plants remains sketchy. An exception is the activation of the transcriptional regulator NPR1 and the transcription factor TGA1, through conformational changes mediated by SA-induced reducing conditions, as discussed earlier. More transcriptional regulations involving reducing conditions may remain to be discovered in plants.

2. Integrated Effects of Antioxidants and Phytohormones on GST Gene Expression in *Arabidopsis*

Salicylic acid, GSH, and the sulfhydryl compounds Cys and DTT caused a marked induction of *At*GSTF2 (Table II). This suggested an involvement of NPR1 in the pathogen response of *At*GSTF2, which has been confirmed through mutant expression studies (Table III). However, the antioxidants GSH and AA also induced a marked transcript accumulation of *At*GSTF3 and *At*GSTZ1, which are not SA-responsive (Table II) and therefore unlikely involve NPR1 signaling. Genes from the *At*GSTL and *At*DHARs classes are particularly sensitive to redox perturbations (Table II) but SA responses have not been studied for these GST classes.

IV. POTENTIAL INVOLVEMENT OF PLANT GROWTH REGULATORS IN POST-TRANSCRIPTIONAL REGULATIONS OF GST GENE EXPRESSION

There are ample indications that GST expression is not only transcriptionally but also post-transcriptionally regulated (Fig. 3), which has received little attention so far.

There is evidence for alternative splicing and differential polyadenylation of GST transcripts, and differences between GST transcript and protein levels have been reported (Smith *et al.*, 2004). Phytohormones could be involved in differential post-transcriptional regulations of the abundance of stress-response proteins such as GSTs.

A. ALTERNATIVE SPLICING DURING GST EXPRESSION

Differential splicing may significantly contribute to the expression regulation for some GSTs during stress responses as well as during normal development (Fig. 3). For the tau-class *ZmGSTU4* or *bronze2* gene, both spliced and unspliced transcripts occur in maize leaves. *ZmGSTU4* transcripts that contained the unspliced single intron were particularly abundant during Cd stress, while negligible during cold, ABA, or auxin responses

(Marrs and Walbot, 1997). Cd stress was found to promote the activation of an alternative mRNA start site. The encoded truncated protein was not detected, indicating that the alternative transcripts were most likely noncoding (Pairoba and Walbot, 2003). Alternative splicing events have also been detected for the tau-class *AtGSTU9* and *AtGSTU17* genes and the phi-class *At*GSTF4 gene (Wagner *et al.*, 2002) but have not been studied further.

B. DIFFERENTIAL POLYADENYLATION PROCESSING DURING GST EXPRESSION

Differential polyadenylation processing can affect mRNA stability and mRNA half-life (Fig. 3). Polyadenylation can occur at two distinct sites in the petunia *An9* gene (Alfenito *et al.*, 1998). Polyadenylation was found to occur at three sites each preceded by a potential polyadenylation signal sequence in the rice *osgstu4* gene (Moons, 2003). The impact of these differential RNA processing events remains unknown.

C. PLANT GROWTH REGULATORS MAY POST-TRANSCRIPTIONALLY MODIFY THE ACCUMULATION OF STRESS-RESPONSE PROTEINS

Several lines of evidence suggest that plant growth regulators may exert important post-transcriptional effects on gene expression that can help to rapidly fine-tune stress responses (Fig. 3). Evidence is mounting for regulatory effects of ABA on pre-mRNA processing events, including splicing, 3′ processing, mRNA stability, and nuclear export, while target transcripts and mechanisms often remain to be elucidated (Kuhn and Schroeder, 2003). JAs also exert important post-transcriptional regulations, which are thought to include effects on RNA processing and translation initiation, that differentially affect the abundance of defense versus housekeeping proteins (Creelman and Mullet, 1997).

In addition, effects of NO on mRNA translation are now suggested in addition, based on the identification of translation initiation and elongation factors as targets for *S*-nitrosylation (Lindermayr *et al.*, 2005).

Earlier work has indicated the importance of the endogenous balance of stress-related hormones on post-transcriptional regulations of stress-response gene expression, a route that remains under-investigated (Fig. 3). SA, MeJA, and ethylene caused a transcript but no protein accumulation for a tobacco taumatin-like *PR-5* gene, whereas various combinations of these phytohormones did cause a protein accumulation (Xu *et al.*, 1994). ABA induced a marked transcript but no protein accumulation for a rice *PR-10* gene and other defense-related genes, JA induced a transcript and marked protein accumulation, and combinations of both hormones had differential effects (Moons *et al.*, 1997). The endogenous ABA versus JA balance was suggested to antagonistically and post-transcriptionally regulate the levels of proteins from water deficit and defense responses. The mechanisms explaining these observations remain unknown. Finally, post-translational mechanisms, such as Ub-mediated proteolysis could also fine tune the level of stress-response proteins, depending on phytohormonal signals, which remains unknown.

Assessing the expression of stress-response genes such as GSTs only at the transcript level may leave these potentially important post-transcriptional regulations undetected. Various interesting proteomic approaches to study GST expression have been introduced (Sappl *et al.*, 2004).

V. CONCLUSIONS

The completion of *Arabidopsis* and rice genome sequencing projects have rendered an unprecedented insight into the composition of plant GST gene families, revealing similarities and clear differences to mammalian GST family inventories. Guided by the knowledge on the function of mammalian GSTs, various similar catalytic and non-catalytic activities have been demonstrated for plant GSTs. In spite of that progress, the versatile roles of plant GSTs during normal development, biotic and abiotic stresses still remain largely unclarified. Classic functional genetic approaches will most likely be hampered by the redundancy of multiple GSTs for specific functions. Instead, small molecule approaches to modulate and reveal GST functions may be more useful.

Global approaches have indicated an unprecedented diversity of GST gene expression patterns. Microarray transcription profiling has enabled us to study these exceptionally large gene families, provided that the experimental design safeguards gene-specificity. Important progress has been made in understanding the involvement of SA, ethylene, and JA in signaling the transcriptional activation of GST gene expression during pathogen defense. Much has been owed to the availability of various well-characterized phytohormone response mutants of *Arabidopsis* and the current exponential increase in the understanding of phytohormone signaling, although these tools remain underutilized. Conversely, the role of JA, SA, and ethylene in the abiotic stress-responsive GST gene expression has received less attention so far. The signaling pathways through which auxins activate GST gene transcription remain elusive so far. Lately, NO is emerging as a novel signal in regulating GST expression and possibly also controlling specific GST activities. A major challenge remains understanding the interaction of the various stress-related plant growth regulator signaling pathways and ROI in generating the unique diversity of individual GST gene expression patterns. Finally, transcriptional profiling can leave potentially important post-transcriptional regulations unnoticed, so that effects of distinct plant growth regulators may incorrectly appear redundant. An increased use of proteomic approaches is therefore advisable.

ACKNOWLEDGMENTS

I would like to gratefully acknowledge Dr. R. Edwards for his insightful comments as well as Dr. G. Gheysen for critical reading of the manuscript, Dr. M. Whiteway for helpful comments and Dr. M. Estelle for interesting discussion points. I also would like to acknowledge partial financial support from the Rockefeller Foundation through grants 2000 FS 184 and 2004 FS 035.

REFERENCES

Alfenito, M. R., Souer, E., Goodman, C. D., Buell, R., Mol, J., Koes, R., and Walbot, V. (1998). Functional complementation of the anthocyanin sequestration in the vacuole by widely divergent glutathione S-transferases. *Plant Cell* **10,** 1135–1149.

Alvares, M. E., Pennell, R. I., Meijer, P.-J., Ishikawa, A., Dixon, R. A., and Lamb, C. (1998). Reactive oxygen intermediates mediate a systemic signal network in the establishment of plant immunity. *Cell* **92,** 773–784.

Bianchi, M. W., Roux, C., and Vartanian, N. (2002). Drought regulation of *GST8*, encoding the *Arabidopsis* homologue of ParC/Nt107 glutathione transferase/peroxidase. *Physiol. Plant* **116,** 96–105.

Brown, D. E., Rashotte, A. M., Murphy, A. S., Normanly, J., Tague, B. W., Peer, W. A., Taiz, L., and Muday, G. K. (2001). Flavonoids act as negative regulators of auxin transport in vivo in *Arabidopsis*. *Plant Physiol.* **126,** 524–535.

Chen, W., and Singh, K. B. (1999). The auxin, hydrogen peroxide and salicylic acid induced expression of the *Arabidopsis* GST6 promoter is mediated in part by an ocs element. *Plant J.* **19,** 667–677.

Creelman, R. A., and Mullet, J. E. (1997). Biosynthesis and action of jasmonates in plants. *Annu. Rev. Plant Physiol. Plant Mol. Biol.* **48,** 355–381.

Coleman, J. O. D., Blake-Kalff, M. M. A., and Davies, T. G. E. (1997). Detoxification of xenobiotics by plants: Chemical modification and vacuolar compartmentation. *Trends Plant Sci.* **2,** 144–151.

Cummins, I., Cole, D. J., and Edwards, R. (1999). A role for glutathione transferases functioning as glutathione peroxidases in resistance to multiple herbicides in black-grass. *Plant J.* **18,** 285–292.

Dafre, A. L., Sies, H., and Akerboom, T. (1996). Protein S-thiolation and regulation of microsomal glutathione transferase activity by the glutathione redox couple. *Arch. Biochem. Biophys.* **332,** 288–294.

Delledonne, M., Xia, Y., Dixon, R. A., and Lamb, C. (1998). Nitric oxide functions as a signal in plant disease resistance. *Nature* **394,** 585–588.

Despres, C., Chubak, C., Rochon, A., Clark, R., Bethune, T., Desveaux, D., and Fobert, P. R. (2003). The *Arabidopsis* NPR1 disease resistance protein is a novel cofactor that confers redox regulation of DNA binding activity to the basic domain/leucine zipper transcription factor TGA1. *Plant Cell* **15,** 2181–2191.

Devoto, A., and Turner, J. G. (2003). Regulation of jasmonate-mediated plant responses in *Arabidopsis*. *Ann. Bot.* **92,** 329–337.

Dharmasiri, N., Dharmasiri, S., and Estelle, M. (2005). The F-box protein TIR1 is an auxin receptor. *Nature* **435,** 441–445.

Dixon, D. P., Cole, D. J., and Edwards, R. (2000). Characterization of a zeta class glutathione transferase from *Arabidopsis thaliana* with a putative role in tyrosine catabolism. *Arch. Biochem. Biophys.* **384**, 407–412.

Dixon, D. P., Cummins, I., Cole, D. J., and Edwards, R. (1998). Glutathione-mediated detoxification systems in plants. *Curr. Opin. Plant Biol.* **1**, 258–266.

Dixon, D. P., Lapthorn, A., and Edwards, R. (2002a). Plant glutathione transferases. *Genome Biol.* **3**, 3004.1–3004.10.

Dixon, D. P., Davis, B. G., and Edwards, E. (2002b). Functional divergence in the glutathione transferase superfamily in plants. *J. Biol. Chem.* **277**, 30859–30869.

Dong, X. (2001). Genetic dissection of systemic acquired resistance. *Curr. Opin. Plant Biol.* **4**, 309–314.

Edwards, R., and Dixon, D. P. (2000). The role of glutathione transferases in herbicide metabolism. *In* "Herbicides and Their Mechanisms of Action" (A. H. Cobb and R. C. Kirkwood, Eds.), pp. 38–71. Sheffield Academic Press, Sheffield, England.

Edwards, R., Dixon, D. P., and Walbot, V. (2000). Plant glutathione *S*-transferases: Enzymes with multiple functions in sickness and in health. *Trends Plant Sci.* **5**, 193–198.

Ellis, J., and Jones, D. (1998). Structure and function of proteins controlling strain-specific pathogen resistance in plants. *Curr. Opin. Plant Biol.* **1**, 288–293.

Fernandez-Canon, J. M., and Penalva, M. A. (1998). Characterisation of a fungal maleylacetoacetate isomerase gene and identification of its human homologue. *J. Biol. Chem.* **273**, 329–337.

Finkelstein, R. R., Gampala, S. S. L., and Rock, C. D. (2002). Abscisic acid signaling in seeds and seedlings. *Plant Cell* **14**(Suppl.), S15–S45.

Frova, C. (2003). The plant glutathione transferase gene family: Genomic structure, functions, expression and evolution. *Physiol. Plant* **119**, 469–479.

Gidda, K. S., Miersch, O., Levitin, A., Schmidt, J., Wasternack, C., and Varin, L. (2003). Biochemical and molecular characterization of a hydroxyljasmonate sulfotransferase from *Arabidopsis thaliana*. *J. Biol. Chem.* **278**, 17895–17900.

Glombitza, S., Dubuis, P.-H., Thulke, O., Welzl, G., Bovet, L., Gotz, M., Affenzeller, M., Geist, B., Hehn, A., Asnaghi, C., Ernst, D., Seidlitz, H. K., Gundlach, H., Mayer, K. F., Martinoia, E., Werck-Reichhart, D., Mauch, F., and Schaffner, A. R. (2004). Crosstalk and differential response to abiotic and biotic stressors reflected at the transcriptional level of effector genes from secondary metabolism. *Plant Mol. Biol.* **54**, 817–835.

Gonneau, J., Mornet, R., and Laloue, M. (1998). A *Nicotaina plumbaginifolia* protein labeled with an azido cytokinin agonist is a glutathione *S*-transferase. *Physiol. Plant* **103**, 114–124.

Grayer, R. J., and Kokubun, T. (2001). Plant–fungal interactions: The search for photoalexins and other antifungal compounds from higher plants. *Phytochemistry* **56**, 253–263.

Grossmann, K. (2000). The mode of action of quinclorac: A case study of a new auxin-type herbicide. *In* "Herbicides and Their Mechanisms of Action" (A. H. Cobb and R. C. Kirkwood, Eds.), pp. 181–214. Sheffield Academic Press, Sheffield, England.

Guo, H., and Ecker, J. R. (2004). The ethylene signaling pathways: New insights. *Curr. Opin. Plant Biol.* **7**, 40–49.

Huang, X., von Rad, U., and Durner, J. (2002). Nitric oxide induces transcriptional activation of the nitric oxide-tolerant alternative oxidase in *Arabidopsis* suspension cells. *Planta* **215**, 914–923.

Ji, Y., Toader, V., and Bennett, B. M. (2002). Regulation of microsomal and cytosolic glutathione *S*-transferase activities by *S*-nitrosylation. *Biochem. Pharmacol.* **63**, 1397–1404.

Kampranis, S. C., Damianova, R., Atallah, M., Toby, G., Kondi, G., Tsichlis, P. N., and Makris, A. M. (2000). A novel plant glutathione *S*-transferase/peroxidase suppresses Bax lethality in yeast. *J. Biol. Chem.* **275**, 29207–29216.

Kepinski, S., and Leyser, O. (2002). Ubiquitination and auxin signaling: A degrading story. *Plant Cell* **14**(Suppl.), S81–S95.

Kilili, K. G., Atanassova, N., Vardanyan, A., Clatot, N., Al-Sabarna, K., Kanellopoulos, P. N., Makris, A. M., and Kampranis, S. C. (2004). Differential roles of tau class glutathione S-transferases in oxidative stress. *J. Biol. Chem.* **279,** 24540–24551.

Kiyosue, T., Yamaguchi-Shinozaki, K., and Shinozaki, K. (1993). Characterization of two cDNAs (ERD11 and ERD13) for dehydration-inducible genes that encode putative glutathione S-transferases in *Arabidopsis thaliana* L. *FEBS Lett.* **335,** 189–192.

Kreuz, K., Tommasini, R., and Martinoia, E. (1996). Old enzymes for a new job (herbicide detoxification in plants). *Plant Physiol.* **111,** 349–353.

Kuhn, J. M., and Schroeder, J. I. (2003). Impacts of altered RNA metabolism on abscisic acid signaling. *Curr. Opin. Plant Biol.* **6,** 463–469.

Lam, E. (2004). Controlled cell death, plant survival and development. *Nat. Rev. Mol. Cell Biol.* **5,** 305–315.

Lebel, E., Heifetz, P., Thorne, L., Ukness, S., Ryals, J., and Ward, E. (1998). Functional analysis of regulatory sequences controlling PR-1 gene expression in *Arabidopsis*. *Plant J.* **16,** 223–233.

Li, Z.-S., Alfenito, M., Rea, P. A., Walbot, V., and Dixon, R. A. (1997). Vacuolar uptake of the phytoalexin medicarpin by the glutathione conjugate pump. *Phytochemistry* **45,** 689–693.

Lieberherr, D., Wagner, U., Dubuis, P.-H., Metraux, J.-P., and Mauch, F. (2003). The rapid induction of glutathione S-transferases *At*GSTF2 and *At*GSTF6 by avirulent *Pseudomonas syringae* is the result of combined salicylic acid and ethylene signaling. *Plant Cell Physiol.* **44,** 750–757.

Liechti, R., and Farmer, E. F. (2002). The jasmonate pathway. *Science* **296,** 1649–1650.

Lindermayr, C., Saalbach, G., and Durner, J. (2005). Proteomic identification of S-nitrosylated proteins in *Arabidopsis*. *Plant Physiol.* **137,** 921–930.

Lo Bello, M., Nuccetelli, M., Caccuri, A. M., Stella, L., Parker, M. W., Rossjohn, J., McKinstry, W. J., Mozzi, A. F., Federici, G., Polizio, F., Pedersen, J. Z., and Ricci, G. (2001). Human glutathione transferase P1-1 and nitric oxide carriers; a new role for an old enzyme. *J. Biol. Chem.* **276,** 42138–42145.

Lou, M. F. (2003). Redox regulation in the lens. *Prog. Retin. Eye Res.* **22,** 657–682.

Loyall, L., Uchida, K., Braun, S., Furuya, M., and Frohmeyer, H. (2000). Glutathione and a UV light-induced glutathione S-transferase are involved in signaling to chalcone synthase in cell cultures. *Plant Cell* **12,** 1939–1950.

Marrs, K. A. (1996). The functions and regulation of glutathione S-transferases in plants. *Annu. Rev. Plant Physiol. Plant Mol. Biol.* **47,** 127–158.

Marrs, K. A., and Walbot, V. (1997). Expression and RNA splicing of the maize glutathione S-transferase Bronze2 gene is regulated by cadmium and other stresses. *Plant Physiol.* **113,** 93–102.

Martinoia, E., Klein, M., Geisler, M., Bovet, L., Forestier, C., Kolukisaoglu, U., Muller-Rober, B., and Schulz, B. (2002). Multifunctionalilty of plant ABC transporters—more than just detoxifiers. *Planta* **214,** 345–355.

McGonigle, B., Keeler, S. J., Lau, S.-M. C., Koeppe, M. K., and O' Keefe, D. P. (2000). A genomic approach to the comprehensive analysis of the glutathione S-transferase gene family in soybean and maize. *Plant Physiol.* **124,** 1105–1120.

Meyer, R. C., Goldsborough, P. B., and Woodson, W. R. (1991). An ethylene-responsive flower senescence-related gene from carnation encodes a protein homologous to glutathione S-transferases. *Plant Mol. Biol.* **17,** 277–281.

Mittler, R. (2002). Oxidative stress, antioxidants and stress tolerance. *Trends Plant Sci.* **7,** 405–410.

Moons, A. (2003). *Osgstu3* and *osgstu4*, encoding tau class glutathione S-transferases, are heavy metal- and hypoxic stress-induced and differentially salt stress-responsive in rice roots. *FEBS Lett.* **553,** 427–432.

Moons, A., Prinsen, E., Bauw, G., and Van Montagu, M. (1997). Antagonistic effects of abscisic acid and jasmonates on salt stress-inducible transcripts in rice roots. *Plant Cell* **9**, 2243–2259.

Mou, Z., Fan, W., and Dong, X. (2003). Inducers of plant systemic acquired resistance regulate NPR1 function through redox changes. *Cell* **113**, 935–944.

Mueller, L. A., Goodman, C. D., Silady, R. A., and Walbot, V. (2000). AN9, a petunia glutathione *S*-transferase required for anthocyanin sequestration, is a flavonoid-binding protein. *Plant Physiol.* **123**, 1561–1570.

Neill, S., Desikan, R., and Hancock, J. (2002). Hydrogen peroxide signalling. *Curr. Opin. Plant Biol.* **5**, 388–395.

Pairoba, C. F., and Walbot, V. (2003). Post-transcriptional regulation of expression of the *bronze-2* gene of *Zea mays* L. *Plant Mol. Biol.* **53**, 75–86.

Pieterse, C. M. J., and Van Loon, L. C. (2004). NPR1: The spider in the web of induced resistance signaling pathways. *Curr. Opin. Plant Biol.* **7**, 456–464.

Polverari, A., Molesini, B., Pezzotti, M., Buonaurio, R., Marte, M., and Delledonne, M. (2003). Nitric oxide-mediated transcriptional changes in *Arabidopsis thaliana*. *Mol. Plant Microbe Interact* **16**, 1094–1105.

Ponce de León, I., Sanz, A., Hamberg, M., and Castresana, C. (2002). Involvement of the *Arabidopsis* α-DOX1 fatty acid dioxygenase in protection against oxidative stress and cell death. *Plant J.* **29**, 61–72.

Romero-Puertas, M. C., Perazzolli, M., Zago, E. D., and Delledonne, M. (2004). Nitric oxide signalling functions in plant–pathogen interactions. *Cell. Microbiol.* **6**, 795–803.

Roxas, V. P., Smith, R. K., Jr., Allen, E. R., and Allen, R. D. (1997). Overexpression of glutathione *S*-transferase/glutathione peroxidase enhances the growth of transgenic tobacco seedlings during stress. *Nat. Biotechnol.* **15**, 988–991.

Sanchez-Fernandez, R., Davies, T. G. E., Coleman, J. O. D., and Rea, P. A. (2001). The *Arabidopsis thaliana* ABC protein superfamily, a complete inventory. *J. Biol. Chem.* **276**, 30231–30244.

Sappl, P. G., Oñate-Sánchez, L., Singh, K. B., and Millar, A. H. (2004). Proteomic analysis of glutathione *S*-transferases of *Arabidopsis thaliana* reveals differential salicylic acid-induced expression of the plant-specific phi and tau classes. *Plant Mol. Biol.* **54**, 205–219.

Schwartz, S. H., Qin, X., and Zeevaart, J. A. D. (2003). Elucidation of the indirect pathway of abscisic acid biosynthesis by mutants, genes and enzymes. *Plant Physiol.* **131**, 1591–1601.

Shah, J. (2003). The salicylic acid loop in plant defense. *Curr. Opin. Plant Biol.* **6**, 365–371.

Sheehan, D., Meade, G., Foley, V. M., and Dowd, C. A. (2001). Structure, function and evolution of glutathione transferases: Implications for classification of non-mammalian members of an ancient enzyme superfamily. *Biochem. J.* **360**, 1–16.

Smith, A. P., De Ridder, B. P., Guo, W.-J., Seeley, E. H., Regnier, F. E., and Goldsbrough, P. B. (2004). Proteomic analysis of *Arabidopsis* glutathione *S*-transferases from Benoxacor- and copper-treated seedlings. *J. Biol. Chem.* **279**, 26098–26104.

Smith, A. P., Nourizadeh, S. D., Peer, W. A., Xu, J., Bandyopadhyay, A., Murphy, A. S., and Goldsbrough, P. B. (2003). *Arabidopsis AtGSTF2* is regulated by ethylene and auxin, and encodes a glutathione *S*-transferase that interacts with flavonoids. *Plant J.* **36**, 433–442.

Soranzo, N., Sari Gorla, M., Mizzi, L., De Toma, G., and Frova, C. (2004). Organisation and structural evolution of the rice glutathione *S*-transferase gene family. *Mol. Genet. Genomics* **271**, 511–521.

Staswick, P. E., and Tiryaki, I. (2004). The oxylipin signal jasmonic acid is activated by an enzyme that conjugates it to isoleucine in *Arabidopsis*. *Plant Cell* **16**, 2117–2127.

Turner, J. G., Ellis, C., and Devoto, A. (2002). The jasmonate signal pathway. *Plant Cell* **14** (Suppl.), S153–S164.

Urano, J., Nakagawa, T., Maki, Y., Masumura, T., Tanaka, K., Murata, N., and Ushimaru, T. (2000). Molecular cloning and characterization of a rice dehydroascorbate reductase. *FEBS Lett.* **466**, 107–111.

Van Camp, W., Van Montagu, M., and Inze, D. (1998). H$_2$O$_2$ and NO: Redox signals in disease resistance. *Trends Plant. Sci.* **3**, 330–334.

Van der Kop, D. A. M., Schuyer, M., Scheres, B., van der Zaal, B. J., and Hooykaas, P. J. J. (1996). Isolation and characterization of an auxin-inducible glutathione *S*-transferase gene of *Arabidopsis thaliana*. *Plant Mol. Biol.* **30**, 839–844.

Wagner, U., Edwards, R., Dixon, D. P., and Mauch, F. (2002). Probing the diversity of the *Arabidopsis glutathione S*-transferase gene family. *Plant Mol. Biol.* **49**, 515–532.

Wang, K. L.-C., Li, H., and Ecker, J. R. (2002). Ethylene biosynthesis and signaling networks. *Plant Cell* **14**(Suppl.), S131–S151.

Wendehenne, D., Durner, J., and Klessig, D. F. (2004). Nitric oxide: A new player in plant signaling and defense. *Curr. Opin. Plant Biol.* **7**, 449–455.

Wittstock, U., and Halkier, B. A. (2002). Glucosinolate research in the *Arabidopsis* era. *Trends Plant Sci.* **7**, 263–270.

Xiong, L., and Zhu, J.-K. (2003). Regulation of abscisic acid biosynthesis. *Plant Physiol.* **133**, 29–36.

Xu, Y., Chang, P.-F. L., Liu, D., Narasimhan, L., Raghothama, K. G., Hasegawa, P. M., and Bressan, R. A. (1994). Plant defense genes are synergistically induced by ethylene and methyl jasmonate. *Plant Cell* **6**, 1077–1085.

Zettl, R., Schell, J., and Palme, K. (1994). Photoaffinity labeling of *Arabidopsis thaliana* plasma membrane vesicles by 5-azido-[7-^{3}H]indole-3-acetic acid: Identification of a glutathione *S*-transferase. *Proc. Natl. Acad. Sci. USA* **91**, 689–693.

6

Auxins

Catherine Perrot-Rechenmann* AND Richard M. Napier[†]

*ISV-CNRS, 91198 Gif-sur-Yvette, Cedex, France
[†]Warwick HRI, University of Warwick, Wellesbourne
Warwick, CV35 9EF, United Kingdom

I. Introduction
II. Generation of IAA
   A. Auxin Conjugates
   B. De Novo *Synthesis*
   C. Deactivation
   D. Homeostasis
   E. Overproduction of IAA
III. Cell Division and Tissue Culture
   A. Cell Division
   B. Secondary Meristems
IV. Transport of IAA: Introduction
   A. Uptake Carriers
   B. The *Efflux Complex and the Importance of Vesicle Cycling*
V. Embryonic Patterning
VI. Vascular Patterning
VII. Auxin Perception, Receptors, and Signaling
   A. Perception and Receptors
   B. Signaling

VIII. Organ Patterning
  A. *Phyllotaxis*
  B. *Root Initiation and Gravitropism*
IX. Apical Dominance and Branching
X. Tropisms and Epinasty
XI. Adventitious Rooting and Wound Responses
XII. Fruit Growth
XIII. Herbicides
  References

Auxin is a multifactorial phytohormone that is required for cell division. Fine gradients determine points of developmental change in time and space. It is associated intimately with the axiality of plant growth, and increasing doses lead to cell expansion or inhibition of cell expansion in different tissues. From embryonic patterning to fruit dehiscence every plant process has some involvement with auxin as a hormonal signal, including responses to wounding. Moreover, synthetic auxins have widespread uses as agrochemicals, particularly as selective herbicides. Despite the importance of auxin as a plant signal the pathways of its biosynthesis are still not clear. Much more is known about auxin perception and the mechanisms through which gene transcription is regulated. One receptor has been identified, and protein crystallography data has explained its auxin-binding capacity, but this is likely to control only a subset of auxin-mediated responses. Little is known of the signal transduction intermediates. A second receptor has been nominated and may be involved in controlling auxin-mediated gene transcription. A complex set of proteins comprising signalosome and proteasome contribute to the regulation of sets of transcription factors to confer regulation by derepression. A set of auxin transport proteins has been described with associated regulatory interactors, and these account for polar auxin flow and the control of auxin movements across cells, tissues, and around the plant. The gradients these transport systems build regulate the responses of growth and differentiation, including the plant's response to gravity. These areas are described and discussed by relating the physiology of the whole plant to the details of genetic and protein activities.
© 2005 Elsevier Inc.

I. INTRODUCTION

Auxins are phytohormones, mobile signaling molecules actively transported along local gradients and throughout the plant to coordinate growth and drive responses to environmental signals. Auxin biologists consider the formative work of Charles and Francis Darwin on phototropic bending in grass coleoptiles in the 1880s as the beginning of their science. Much has been learned since. Fritz Went and others in the early twentieth century collected and applied diffusates to coleoptiles (Chronicled by Pennazio, 2002). In 1934, the chemistry and crystallography of Kogl and colleagues eventually led to the identification of indole-3-acetic acid (IAA) as the active compound. More recent advances on the molecular and genetic mechanisms of auxin synthesis, transport, recognition, and action are summarized here.

The archetypal and naturally predominant auxin in all plants is IAA. In seeds and seedlings it is synthesized primarily from storage conjugates in the endosperm, but just a few days after germination *de novo* synthesis starts and from then on it seems that young leaves close to the shoot apex are the primary, but not the sole source of IAA (Ljung *et al.*, 2002). Some plant species synthesize additional active auxin molecules such as 4-chloroindole-3-acetic acid, phenylacetic acid, and indole-3-butyric acid (IBA), but it is likely that IBA is converted to IAA by β-oxidation in peroxisomes (Bartel *et al.*, 2001).

Auxin is an absolute requirement for plant cell division, as well as to many other developmental and environmental responses that are mediated through auxin movements. Of these, a good number are commercially relevant, such as branching, rooting, and fruiting, but the biggest market for synthetic auxins is as selective herbicides.

In spite of the importance of auxins for plant growth and development, there are still surprising gaps in our knowledge of their biosynthesis and mechanisms of action. The biggest advances in the field have been in the elucidation of components of the polar auxin transport system and the piecing-together of the enzyme complexes mediating auxin-controlled gene expression, or more accurately, derepression of transcription through regulated proteolysis of transcriptional repressors. Each of these areas will be covered at a level of detail appropriate for the information available.

II. GENERATION OF IAA

A. AUXIN CONJUGATES

In germinating seeds, IAA is produced from the breakdown of stored forms of the hormone, conjugates of amino acids, proteins, and sugars. Broadly, monocotyledonous plants (cereals and grasses) accumulate sugar conjugates; dicotyledonous plants accumulate amino acid conjugates, although this reflects only the most abundant storage products. Hydrolysis of these conjugates during germination precedes or coincides with the start of root extension. In cereal seeds, large fluxes of IAA released from conjugates stored in the kernal mediate extension of the coleoptile, an observation used extensively as the basis of what has been a widely used auxin bioassay. In intact seedlings, the auxin moves in the vascular strands from seed to the coleoptile tip, from where it is distributed back down to the growth-limiting epidermal cells by the polar auxin transport system (see later) (Bialek and Cohen, 1992; Jones, 1990; Reinecke and Bandurski, 1987) as well as to the other parts of the seedling.

The storage endosperm of seeds is not the only site of IAA conjugate synthesis. Conjugate synthesis is developmentally regulated and can be switched on at any time by exogenous application of auxins. In addition to the major storage forms noted earlier, it appears that a variety of both glucoside esters and amide conjugates can be made in vegetative tissues of all plants. In *Arabidopsis* leaves, for example, IAA is conjugated to the aspartic acid amide, several other amino acid amides as well as the glucose ester (Östin *et al.*, 1998). A UDPG-dependent IAA glucosyltransferase has been cloned from *Arabidopsis* and its activities described in some detail (Jackson *et al.*, 2002), extending earlier biochemistry recorded for a similar enzyme from maize endosperm (Reinecke and Bandurski, 1987). Whether amide or glucose conjugated, the immediate reaction products do not accumulate greatly but are further metabolized. The ester conjugate 1-*O*-IAA-glucose is metabolized into indole-3-acetyl-*myo*-inositol and glucosides of this, while amide conjugates build up as protein and polypeptide conjugates to become the major bound IAA pool. The pathways remain unclear (Bialek and Cohen, 1992). A family of amidohydrolases has been described from *Arabidopsis* that are active at releasing free IAA from amide conjugates (Bartel and Fink, 1995; Davies *et al.*, 1999; Rampey *et al.*, 2004). These proteins are targeted to the endoplasmic reticulum, although it remains unclear whether or not their substrates accumulate here. The protein structure of one member of this family has been determined (PDB file 1XMB).

Conjugates are likely to be moved into storage compartments such as the plant vacuole and, possibly, the endoplasmic reticulum. A member of the ATP-binding cassette (ABC) transporter proteins (*At*MRP5) has been shown to carry conjugates, and when it is defective, it leads to reduced root growth and an increase in lateral root formation (Gaedeke *et al.*, 2001), phenotypes indicative of local increases in auxin concentration. The authors suggest that by failing to remove conjugates from the root cytoplasm, the *Atmrp5* mutant becomes auxin rich.

B. *DE NOVO* SYNTHESIS

The flow of free IAA from stored conjugates in germinating seeds is unusual, and these reserves soon run low to be replaced by *de novo* synthesis. There are parallel, degenerate biosynthetic pathways referred to as the tryptophan-dependent and tryptophan-independent pathways. Both derive from anthranilic acid, but the latter branches from the tryptophan pathway at indole or indole-3-glycerol phosphate. A few days after germination, the tryptophan-independent pathway is activated. The intermediates on this pathway have not been identified even though it seems likely that this pathway remains the constitutive biosynthetic source of IAA throughout the rest of a plant's life. However, when a plant is stressed, such as by wounding, or during major developmental events such as seed filling (Glawischnig *et al.*, 2000) and at least in pine (*Pinus sylvestris*) seedlings (Ljung *et al.*, 2001b) early after germination, then IAA synthesis is stepped up by switching on the tryptophan-dependent pathways.

Tryptophan is the precursor for a set of parallel IAA biosynthetic pathways, none of which are fully mapped or understood at the genetic or protein levels (Bartel *et al.*, 2001; Ljung *et al.*, 2002). An *Arabidopsis* mutant named YUCCA accumulates more IAA than the wild-type and exhibits characteristic auxin-enriched phenotypes such as elongated hypocotyls, epinastic leaves, and increased apical dominance (Zhao *et al.*, 2001). The YUCCA protein is a cytosolic flavin monooxygenase. Neither the tryptophan decarboxylase needed to produce the substrate for YUCCA nor the oxidase(s) required for the rest of the pathway to IAA have yet been identified, although the *Arabidopsis* genome does contain candidates for each. There is also evidence for a pathway through indole-3-pyruvate (Bartel *et al.*, 2001; Ljung *et al.*, 2002). In the Brassicaceae (which includes *Arabidopsis*), a plastidic pathway using a set of cytochrome P450s and a C-S lyase (Mikkelsen *et al.*, 2004) has been identified through the analysis of a set of auxin-enriched mutants such as the superroot lines *sur1* and *sur2*. However, despite the apparent importance of this pathway suggested by the *Arabidopsis* mutants, it is likely that the plastidic pathway relates primarily to the synthesis of glucosinolates and other secondary metabolites and remains restricted to this plant family. All parts of young, growing plants appear to participate in IAA synthesis, although this synthesis is tightly controlled in order to maintain homeostasis (Ljung *et al.*, 2001a). Very young leaves, less than 0.5 mm long in *Arabidopsis*, produce large amounts of IAA. These early high concentrations and high synthetic capacity help drive leaf cell division. As the leaf expands towards its final size, synthesis and concentrations fall. Concentrations of free IAA were found to be 100-fold lower in the expanded leaf lamina than in the youngest leaves (Ljung *et al.*, 2001a). In fully grown *Arabidopsis* plants, the highest concentrations of IAA are found in expanding fruiting bodies, the siliques (Müller *et al.*, 2002). It is likely that much of this is synthesized by seeds during embryo development (see later). The inflorescence stalk contains more IAA than most other parts of the plant, but it is not clear if this is synthesized *in situ* or if it is in transit.

C. DEACTIVATION

Control of auxin action can be mediated by removing IAA as well as by synthesis. Auxins are removed by conjugation into inactive storage compounds (see earlier section) and by oxidation. The primary oxidation product of IAA in *Arabidopsis* and other plants is likely to be 2-oxoindoleacetic acid (OxIAA), although quantities of oxidized conjugates, OxIAA-aspartate, and -*O*-glucoside are also found (Östin *et al.*, 1998). Free IAA is metabolized by decarboxylation by a variety of plant peroxidases *in vitro* and in crude extracts, but the quantities of such products from intact tissues are generally very low suggesting that these are less important catabolic pathways *in vivo*.

D. HOMEOSTASIS

The balance of synthesis, breakdown, conjugation, and transport is regulated rigorously to give auxin homeostasis. Some of the fine measurements of IAA concentration have illustrated that leaf expansion, for example, is reduced by both increases and decreases of auxin. Clearly, changes in auxin concentration are important as plants respond to stimuli, but homeostasis is critical for both optimal development and to keep the system primed for stimulatory responses. At present, the feedback mechanisms for homeostasis are not well characterized. However, it is clear that auxin transport (see later) plays a crucial role in both establishing and perturbing homeostasis.

E. OVERPRODUCTION OF IAA

The auxin IAA is made by certain microorganisms as well as by plants, but it appears that only plants can control auxin concentrations to give auxin homeostasis. Bacteria, such as the pathogens *Pseudomonas syringiae* and *Agrobacterium tumefaciens*, synthesize copious amounts of auxin from tryptophan, making use of a tryptophan-dependent pathway not exploited by plants (Costacurta and Vanderleyden, 1995). Tryptophan monooxygenase converts tryptophan to indoleacetamide, which is converted by a hydrolase to IAA. Such auxin contributes to the pathogenic responses induced in plant hosts such as in the crown gall tumors caused by agrobacterial infections. Likewise, there is evidence to suggest that symbiotic mycorrhizal fungi produce auxin and that this auxin then contributes to the changing morphology of the root system in these associations.

The bacterial enzymes have been used in experiments to increase IAA concentrations in plants through genetic engineering and transformation. Generally, the measured concentrations rise only a few-fold, although this is sufficient to give the characteristic phenotypes noted earlier of epinasty, increased apical dominance, and hypocotyl extension (Gaudin *et al.*, 1994; Sitbon *et al.*, 1992; van der Graaff *et al.*, 2003). When expression is directed to the reproductive tissues, parthenocarpy is induced, which is a useful agricultural trait (Rotino *et al.*, 1997).

III. CELL DIVISION AND TISSUE CULTURE

A. CELL DIVISION

One of the most profound actions of auxin on plants is the control of cell division. Auxin is a permissive signal for cell division, providing the necessary competence to enter into the cell cycle. The specific target of auxin action is unknown. A primary event in the stimulation of cell division is the cytokinin-mediated initiation of cyclin D transcription (Cockcroft et al., 2000). The newly synthesized cyclin D associates with a cyclin-dependent kinase (CDK) to create an active complex at the G1-to-S transition. The CDK-a/cyclin D complex leads to phosphorylation of the retinoblastoma tumor suppressor protein (Rb). Evidence suggests that auxin increases expression of CDK-a, but activation of this kinase may require cytokinin (Hemerly et al., 1993; John et al., 1993). Inactivation of Rb in late G1 provokes release of the transcription factor E2F, and genes controlled by E2F factors are therefore activated, driving cells into DNA replication and committing them to the cell cycle. Stimulatory effects of auxins and cytokinins on cell proliferation have been reported for most cell types, and both are associated with progression through G1-S and G2-M checkpoints (den Boer and Murray, 2000). Cells can be arrested either in G1 or in G2 phase after auxin deprivation (Planchais et al., 1997). Generally, higher auxin concentrations stimulate cell division, and low auxin concentrations drive cell elongation (orientated growth), cell enlargement (with loss of polarity), and cell differentiation (Winicur et al., 1998; Zazimalova et al., 1995). There is some evidence of plant MAPkinases controlling feedback loops (Mockaitis and Howell, 2000). More substantive evidence suggests that plant homologues of both the α and β subunits of heterotrimeric G proteins play a role in the control of auxin-regulated cell division (Ullah et al., 2001). Regulation by auxin is indirect (not coupled directly to the G-proteins), with G_β acting as a negative regulator of division (Ullah et al., 2003).

The requirement for plant growth substances, such as auxin and cytokinin, for the growth of cell cultures, and for the reentry of quiescent cells into the division cycle has been known for many years. Among the most striking examples of this critical control over division is in cell and tissue culture where auxin must be included in almost all media. Some measure of cell cycle synchrony can be established in cell suspension cultures by auxin starvation and re-addition (Nishida et al., 1992). Furthermore, auxin and cytokinin concentrations can be manipulated to promote proliferative callus growth, regeneration of vegetative tissue, or root induction. Elevated auxin/cytokinin ratio is favorable for rooting, whereas a reduced ratio facilitates buds regeneration. Horticultural practices, such as cutting and clonal micropropagation, are directly based on these characteristics.

In plants, most cell divisions take place in meristems that are active throughout a plant's life. Most newly formed cells escape the cell cycle and become quiescent, elongate, and then enter into programs of differentiation. However, many plant cells retain the capacity to dedifferentiate and to reenter into the cell cycle when exposed to exogenous auxin application. Such cell cycle reactivation of differentiated cells is a normal process during plant growth. Lateral and adventitious root formations are well-documented examples of such developmental processes (e.g., Himanen et al., 2002). In lateral roots, primordial auxin carriers facilitate auxin redistribution and give rise to local elevated concentrations of auxin, stimulating pericycle cell division (Bhalerao et al., 2002; Casimiro et al., 2001; Marchant et al., 2002).

B. SECONDARY MERISTEMS

Secondary meristems are important for the generation of many plant tissues. Among the most prolific is the cambium that gives rise to secondary xylem and phloem and the tissues of the phellum (bark). Auxin concentrations have been measured in thin, sequential sections across the cambial region of pine trees, from xylem (wood) to bark (Uggla et al., 1996, 1998). The highest concentration corresponded with cambial cells, and it was suggested that this zone of high auxin conveys positional information. In this way, auxin is acting as a morphogen, and the steepness of the declining gradient of auxin across cells adjacent to the cambium determines the radial width of differentiating zones of division and expansion as well as patterning (Bhalerao and Bennett, 2003). As described earlier for cell division (high) and cell expansion (low), there are concentration thresholds that combine with other positional stimuli to mediate the length of time cambial derivatives spend dividing, differentiating, or laying down secondary walls. Much secondary growth is governed by auxin concentrations in and delivery from the cambium. However, it is also clear that IAA is not the only signal determining division and differentiation around the cambium (Hellgren et al., 2004).

IV. TRANSPORT OF IAA: INTRODUCTION

All plant hormones are moved around the plant in the vasculature where, once loaded, they move passively as passenger molecules. Auxin is special in that plants also have a polar transport system to move IAA vectorally. Polar auxin transport requires energy and contributes to many of the important auxin-dependent responses described in earlier sections on cell division and cambial development and later sections on apical dominance and gravitropism. Transport not only mediates control by delivering stimulatory or inhibitory doses of auxin but also appears intimately involved with the orientation of division (Petrasek *et al.*, 2002) and polarity (Friml *et al.*, 2003). Indeed, there have been suggestions that axial growth in the plant kingdom may have co-evolved with polar auxin transport (Poli *et al.*, 2003).

The consequence of polar auxin transport is that certain cells (or tissues) receive extra auxin, an auxin stimulus. The donor cells might or might not become depleted of auxin, depending on whether or not they are replenished by new synthesis or by influx from other cells. Polar transport functions at the cellular level and is effected by the combined activities of both auxin influx and efflux carrier proteins. In vascular plants there is transcriptional control of both classes of auxin transport protein in many tissues, including the cambium (Schrader *et al.*, 2003), the embryo, and root primordia (Blilou *et al.*, 2005; Furutani *et al.*, 2004).

The speed of polar auxin transport is generally measured at about 10–20 mm/h. Under circumstances in which auxin might be moved through few cells to effect a gradient for differential growth (such as for gravitropism in roots or coleoptiles) and for small plants, such a transport rate seems sustainable. However, the passage of an auxin stimulus from the apex of a tree to its roots presents a challenge of a different magnitude. Observations have been made of a wave of auxin release traveling in advance of and more rapidly than net polar auxin transport (Wodzicki *et al.*, 1987). The mechanisms for such a wave are still to be identified, but merit further research.

It is not correct to surmise that all auxin movement is through the polar transport system. Considerable quantities of free IAA and of auxin conjugates are carried in the vasculature, particularly the phloem (Baker, 2000; Cambridge and Morris, 1996; Else *et al.*, 2004). Indeed, the relatively high pH in the phloem acts to concentrate free IAA, and unloading from the phloem will require the action of efflux proteins and uptake carrier proteins in adjacent cells, although the mechanisms regulating unloading sites remain to be identified. Many of the auxin-driven responses considered in a later section fail if auxin is absent or in excess, or if polar auxin transport is defective. Certainly, hormone delivery is as important as the hormone itself. As a result, auxin physiology has benefited from the use of drugs that act specifically to inhibit polar auxin transport. Of these, naphthylphthalamic acid (NPA) is most response-specific and has been used widely. Triiodobenzoic acid (TIBA) has also been used but has additional activities. Whereas these compounds are synthetic, the flavonoids are natural secondary metabolites of plants and some can act to regulate polar auxin transport *in vivo* (Brown *et al.*, 2001). 1-Naphthoxyacetic acid (1-NOA) has been identified as a useful tool to inhibit uptake (Parry *et al.*, 2001), and its use on stem apical meristems has contributed to dramatic new understanding of auxin flow and action during organogenesis at the shoot tip (see later section; Stieger *et al.*, 2002).

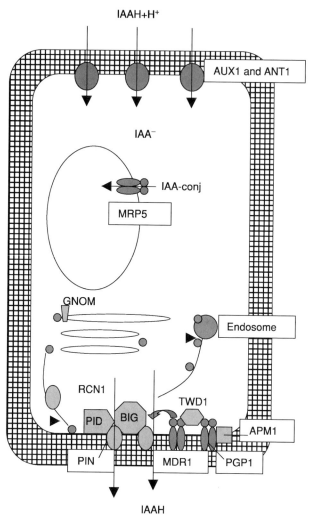

FIGURE 1. The dynamics of auxin transport. A plant cell encased by the plant cell wall is indicated. Auxin is moved vectorally and this is indicated here as downwards. Auxin (indole-3-acetic acid, IAA) is imported into the cell (at the *top*) through influx carriers, notably AUX1 and the less specific ANT1. Influx is cotransport with protons. Inside the cell IAA dissociates to the IAA⁻ anion and is subject to conjugation as well as export through an efflux complex. Conjugates (and possibly free IAA) are probably compartmentalized through carrier proteins like MRP5, of which there are likely to be a number carrying different conjugates. Energy is required to pump indole-3-acetic acid out of the cell and a number of proteins contribute to this activity (although the exact associations suggested here remain a model). A necessary component of the efflux complex is one or more PIN proteins and it seems likely that the kinase PID and phosphatase RCN1 and BIG are all involved with regulation of PIN activity, although it is not possible to tell if they regulate directly or control passage of PINs to the plasma membrane. The PIN proteins are rapidly recycled away from the plasma membrane to the plant

A. UPTAKE CARRIERS

Candidate proteins and their cognate genes have been described for both uptake and efflux, for the carriers themselves and for associated regulatory components. All have been identified from screening mutant populations and the application of molecular genetics. The influx carrier gene, now known as *AUX1*, was isolated from seedlings insensitive to the auxin 2,4-dichlorophenoxyacetic acid (2,4-D) (Bennett *et al.*, 1996), and the protein has been mapped in some detail (Swarup *et al.*, 2004). The AUX1 protein is a member of the amino acid-proton cotransporter protein superfamily (in a clade known as the amino acid/auxin permease group). It is an integral membrane protein spanning the lipid bilayer with 10 or 11 helices joined by short loops of more hydrophilic amino acids. *Arabidopsis* and other plant species tend to have small gene families of AUX1-like proteins (Schnabel and Frugoli, 2004). In addition, a member of the aromatic and neutral amino acid transporter family (AtANT1) has been found to carry IAA in heterologous expression studies using yeast (Chen *et al.*, 2001) and seems likely to contribute to overall auxin influx.

B. THE EFFLUX COMPLEX AND THE IMPORTANCE OF VESICLE CYCLING

The first efflux carrier gene was identified from plants with an extreme flower morphology. Pinlike organs formed with no development of flower parts. This morphology gave rise to the name *PIN* for the gene family, and there are eight *PIN* genes in *Arabidopsis* (Galweiler *et al.*, 1998; Paponov *et al.*, 2005). Individual members of the *PIN* gene family are expressed in defined sets of cells or tissues (Friml *et al.*, 2002; Galweiler *et al.*, 1998), but family members appear to be expressed in most parts of the plant. The proteins themselves sit in the plasma membrane and span the membrane 10 times but are not related to *AUX1*. They are localized asymmetrically around the cell so that efflux activity induces polar auxin pumping (Galweiler *et al.*, 1998). However, as will be discussed later, since the discovery of the PIN proteins a growing set of proteins has been described that contributes to and controls the presence of active efflux complexes at the cell surface (Fig. 1).

endosomal system. New PIN proteins are secreted through a GNOM-regulated ER–Golgi–vesicle pathway. Additionally, MDR1 may regulate the PIN complex and transport IAA directly. The twisted protein (TWD1) interacts with a cytosolic domain of MDR1 and also with PGP1, which in turn interacts with APM1. The polar organization of AUX1 and PIN proteins has been observed in several tissues and PIN proteins can be relocated to, or differentially enriched at, different parts of the cell surface by appropriate stimuli, such as gravity. Asymmetrical distribution of the other proteins named here has not yet been established. See text for more details.

A principal target of NPA has been found to be a member of the multidrug-resistant (MDR) class of transport proteins (Noh et al., 2001). A class of plasma membrane-bound regulatory amidopeptidases (Murphy et al., 2002) has also been identified as NPA-binding sites, but the PIN proteins are not direct targets of NPA. If NPA is an effective auxin polar transport inhibitor and yet does not bind to PINs, it is clear that each of the NPA targets plays a role in auxin efflux and that this is a complex, multicomponent carrier system.

The *Arabidopsis* MDR1 (AtMDR1) protein might transport auxin itself because it has also been found to interact with a plasma membrane-anchored FKBP-like immunophilin known as TWISTED DWARF1 (TWDI) (Geisler et al., 2003). The peptidyl prolyly isomerase activity of this plant immunophilin is absent and it appears that the PPI-like domain is the site of interaction with the N-terminal domains of both *At*MDR1 and a similar transporter AtPGP1. Double mutants of these MDRs give a polar auxin-transport–deficient phenotype similar to *TWD1* mutants, supporting the suggestion that there is a functional interaction between these proteins during auxin efflux. A further protein shown to contribute to polar auxin efflux is BIG (also known as *At*TIR3), a member of the calossin protein family (Gil et al., 2001). As the name suggests, the calossins are very large proteins and membrane-associated with multiple functional domains, and they appear to be involved in regulation of the efflux complex at the plasma membrane.

Two other proteins are known to regulate polar auxin transport. The product of the *AtRCN1* gene is a protein phosphatase A2, and this protein appears to act as a negative regulator of auxin efflux, giving a similar phenotype as NPA (Garbers et al., 1996). Loss of function in the *PID* gene (also *PINOID*) gives a less extreme pinlike phenotype, and these mutants are polar auxin-transport–impaired. The PID protein is a serine-threonine kinase; its expression is closely controlled and dependent on PIN1 expression and can be induced by adding exogenous auxin (Benjamins et al., 2001; Furutani et al., 2004). PID appears to be a positive regulator of auxin efflux.

Correct regulation of auxin efflux is clearly vital for normal plant development. Regulation is conferred at several levels: by transcriptional control of components of the transport complex (Schrader et al., 2004), by post-translational control of efflux activity (Benjamins et al., 2001), and by the rapid cycling of active PIN efflux carriers around plasma membrane–endosomal vesicle pathways. Certainly, the PIN proteins cycle rapidly, and this cycling is sensitive to pharmaceutical intervention by Brefeldin A (BFA) (Muday et al., 2003). The site for BFA action is GNOM, an ADP-ribosylation factor–GTP exchange factor protein of the family known to regulate vesicle trafficking (Geldner et al., 2003). *GNOM* mutant plants display a host of characteristics consistent with perturbation of the auxin signaling system, including disorganized or apolar development and the failure of cells to organize an asymmetrical distribution of PIN proteins during embryo development (Steinmann et al., 1999). These observations and earlier physiological experiments using BFA and other vesicle-active drugs (Morris and Robinson, 1998) illustrated the key role played in the correct delivery and cycling of PIN (and other) proteins to plant cell plasma membranes (Muday et al., 2003). Nevertheless, with the exception of GNOM, the exact roles for all the proteins associated with auxin efflux remains unclear. Even the role of PIN proteins as the efflux carrier remains questionable, although PINs are certainly needed for the efflux complex to work.

V. EMBRYONIC PATTERNING

The phenotypes of *PIN*, *PID*, and *GNOM* mutants have been mentioned earlier and indicate the key role polar auxin transport plays in establishment of the plant body axis in young embryos. A *pid pin1* double mutant displays a total absence of bilateral symmetry, for example (Furutani et al., 2004). Auxin transport inhibitors also prevent the establishment of bilateral symmetry in embryos (Benjamins et al., 2001; Fischer and Neuhaus, 1996; Friml et al., 2003; Geldner et al., 2003; Steinmann et al., 1999).

Auxin responses during embryogenesis have also been studied using other morphogenic *Arabidopsis* mutants. One known as *MONOPTEROS* shows the same embryonic failure to establish a polar body axis as seen after NPA treatment. The *MONOPTEROS* gene codes for an auxin-regulated transcription factor and is one of a family of transcription factors known collectively as the auxin response factor (ARF) genes. These are considered in a later section. It appears that local auxin gradients generated by efflux proteins are a common module for plant organ formation (Benkova et al., 2003).

VI. VASCULAR PATTERNING

In the same way the auxin transport inhibitors cause defects in embryonic patterning, they have dramatic effects on the development of the vasculature in developing tissues. For example, addition of NPA to developing cotyledons or leaves leads to confinement of the vasculature to the leaf margins (Mattsson et al., 1999). Auxin determines sites of vascular differentiation and promotes vascular connectivity (Aloni et al., 2003; Avsian-Kretchmer et al., 2002; Berleth et al., 2000; Sieburth, 1999). The zone of cell division in leaves lies at the margins, and auxin synthesis in very young leaves is extremely active, as noted earlier. It seems likely that auxin from the margins induces vascular development. The resulting auxin gradient from edge to the centre of the leaf lays down the direction for early patterning. As the vasculature forms, it generates an auxin sink as the new phloem exports IAA, reinforcing local gradients. Both before and during vascular development polar transport provides directionality.

The pattern of primary and secondary vascular strands is set very early in leaf development, and once set is not changed by auxin stimuli (Berleth et al., 2000). In addition to synthesis at the leaf margins it has been suggested that specialized leaf cells in the lamina, such as glandular hydathodes, provide auxin and help determine tertiary patterning (Aloni et al., 2003). It can be seen that combining these various axes and gradients of auxin, along with other stimuli, gives a complex set of developmental cues that result in diverse, yet reproducible, venation patterns.

VII. AUXIN PERCEPTION, RECEPTORS, AND SIGNALING

A. PERCEPTION AND RECEPTORS

Many techniques have been used in experiments to identify auxin receptors, and many candidates have been examined in detail (Napier et al., 2002). Photoactive and radiolabeled auxins have been shown to bind to a diverse set of proteins, affinity purification has identified others, and genetic screens have yielded a vast amount of information on signaling intermediates and on the transcriptional regulation of auxin-induced genes. There is still scope for the discovery of novel receptors.

The first candidate auxin receptor was identified in the early 1970s in membrane fractions of *Zea* coleoptiles (Hertel et al., 1972). It was later purified and characterized as a protein that bound active auxins and is referred to as auxin-binding protein 1 (ABP1) (Löbler and Klambt, 1985). Sequencing showed it had no homologues (Inohara et al., 1989). It is a soluble protein targeted to the endoplasmic reticulum. It is also ubiquitous throughout the plant kingdom, including a close homologue in the green alga *Chlamydomonas* (Woo et al., 2002).

Early experiments to test for the function of ABP1 used electrophysiology to record ion fluxes across the plasma membrane. These tests showed that ABP1 was necessary for auxin to stimulate proton efflux and membrane hyperpolarization (Barbier-Brygoo et al., 1991; Rück et al., 1993). However, they also required ABP1 to be present on the outer face of the plasma membrane, not just in the ER, and both immunological and biochemical techniques were used to show that a small population of ABP1 can reach the cell surface (Bauly et al., 2000; Diekmann et al., 1995). Transformation to give plants that over-express ABP1 confers a phenotype in which auxin-induced cell expansion is dependent on ABP1 (Jones et al., 1998). A homozygous null mutant (ABP1 knockout) was found to be embryo lethal (Chen et al., 2001). These observations are consistent with ABP1 being an auxin receptor, but many auxin-regulated events remain unlinked to ABP1 activity, most notably auxin-regulated gene activation. Consequently, ABP1 is unlikely to be the only auxin receptor, and its activities as a receptor have long been recognized to lie in a set of cell elongation responses mediated at the plasma membrane. A new receptor candidate has just been identified as the TIR1 protein, a component of the plant proteasomal targeting machinery; see later section (Dharmasiri et al., 2005; Kepinski and Leyser, 2005). The TIR1 protein is an F-box domain protein with a large set of leucine-rich repeat domains, and, if confirmed, this would be the first example of an F-box domain protein acting as a hormone receptor.

The three-dimensional crystal structure of ABP1 with and without auxin bound has been determined at high resolution and it gives the first high resolution structural detail of a plant hormone recognition site (Woo et al., 2002). There has been a history of binding site models for the auxin receptor and all their crucial details are met by the structure of ABP1 (Napier, 2001). This class of auxin-binding proteins is distinct from any of the other plant hormone receptors, all of which have orthologues in the repertoire of animal and microbial receptor families (Napier, 2004). ABP1 is a member of the cupin protein superfamily, which contains proteins of varied function diversified from a core antiparallel β-barrel fold (Dunwell et al., 2000).

B. SIGNALING

There is a gap in our knowledge about signaling intermediates involved in the expected biochemical cascade from ABP1 activation to effectors. The data have suggested that the outcome of ABP1 action is the activation of the plasma membrane proton ATPase, giving rise to acidification of the cell wall space and allowing cell expansion (Rück et al., 1993). Other ion transport activities have also been implicated (Bauly et al., 2000), and a series of models have been suggested (Macdonald, 1997).

1. Auxin-Mediated Gene Transcription

In contrast, a great deal is known about the proteins that mediate auxin-regulated transcriptional activity (Leyser, 2002). If TIR1 is the receptor for the control of auxin-mediated gene transcription (Dharmasiri et al., 2005; Kepinski and Leyser, 2005), no further intermediates may be required. However, TIR1 functions as part of a large complex of proteins, all of which appear to be required for correct auxin activity because loss-of-function mutants give auxin-insensitive phenotypes.

The first auxin-insensitive mutant for which the defective gene was cloned and sequenced was a mutant known as *axr1* (Leyser et al., 1993). The sequence data suggested the gene product coded for an orthologue of a ubiquitin-activating enzyme E1, and therefore, the AXR1 protein was likely to be involved in protein degradation. Many other components of the ubiquitination pathway have been shown to be involved in auxin-mediated transcriptional control (Dharmasiri and Estelle, 2004). Of these, most insight followed from the cloning of *TIR1*, the sequence of which showed it to be an F-box protein (with multiple C-terminal leucine-rich repeats) and likely to contribute to a ubiquitination E3 complex (Ruegger et al., 1998). The E3 complex linked to auxin signaling is known as the SCF^{TIR1} complex after the exemplar complex in yeast (Skp protein-Cullin-F-Box protein complex) and comprises Skp1/ASK1, Cullin1/Cdc53, Rbx1/ROC1/Hrt1, and the F-box protein TIR1 (Gray et al., 1999). Over-expression of either *AtTIR1* or *AtASK1* promoted auxin responses and auxin-dependent gene expression; mutations decreased auxin responses, suggesting that the complex was limiting responsiveness and that the principal targets of SCF^{TIR1} are negative regulators of auxin action.

Arabidopsis CUL1 requires not only its SCF partners but is also regulated by covalent modification with a ubiquitin-like protein known as RUB1 (del Pozo et al., 1998). RUB1 is necessary for auxin action and appears to function as a parallel pathway to ubiquitation. It is on this pathway that AXR1 works (along with heterodimer partner *At*ECR1) with an E2 (ubiquitin-conjugating enzyme) known as RCE (RUB-conjugating enzyme), and thus, although AXR1 first suggested the involvement of ubiquitination pathways, its involvement is through RUB1 regulation of SCF^{TIR1} rather than directly through addition of ubiquitin to substrates. Once the substrates are polyubiquitinated they pass to the 26S proteasome where they are degraded and ubiquitin is recycled. However, there is an intermediary even on this pathway, the COP9 signalosome (Schwecheimer et al., 2002). A cartoon illustrates the key elements of SCF^{TIR1} regulation (Fig. 2).

2. Aux/IAA Proteins, the Substrates of SCF^{TIR1}

During auxin signaling, the targets of proteasomal degradation are auxin-regulated transcription factors, in particular a family of transcriptional repressors known as the Aux/IAA proteins (Tiwari et al., 2001; Zenser et al., 2001). This has been the latest and most incisive step in understanding auxin signaling and control, although the protein family itself has been studied over many years. Early biochemical studies discovered several families of protein strongly upregulated by addition of exogenous auxin (Napier and Venis, 1995). Once cloned and sequenced, it became clear that many of these were transcription factors. Since these were examined further in different species, they gave rise to the *Aux/IAA* gene family (Abel and Theologis, 1996; Abel et al., 1994). The Aux/IAA proteins are nuclear targeted and are turned over very rapidly, which might be anticipated of regulatory elements. They are induced as some of the first transcription products after an auxin stimulus (detected within just a few minutes) and

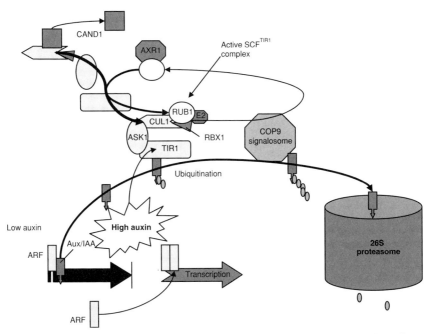

FIGURE 2. Auxin-regulated gene expression. The thick arrows at the foot of the figure represent auxin-regulated genes. At low auxin levels, a heterodimer of an ARF and an Aux/IAA protein represses transcription. Addition of auxin leads to the dissociation of the Aux/IAA protein, likely assembly of an ARF homodimer and the acceleration of transcription. The Aux/IAA protein binds to TIR1 in the SCF^{TIR1} complex where it is ubiquitinated, targeting it for the COP9 signalosome and degradation in the 26S proteasome. The SCF^{TIR1} complex is assembled when CUL1 is modified by the addition of RUB1, a ubiquitin-like peptide. This addition requires the actions of the ubiquitin (RUB1)-activating enzyme AXR1 and an E2 ubiquitin (RUB1) ligase as well as RBX1 which remains bound to CUL1. Once CUL1 is activated by RUB1 it assembles with ASK1 and TIR1. The COP9 signalosome plays a role in removing RUB1 and recycling the CUL1 complex. TIR1 is likely to be a site of auxin perception, and therefore, signaling intermediates may not be necessary in this hormone receptor system.

have short half-lives once translated (Oeller and Theologis, 1995). The proteins themselves comprise a set of four highly conserved domains, domains I and II contribute to their 'instability' and domain II in particular is sufficient and necessary for binding to TIR1 (Gray et al., 2001; Ramos et al., 2001). Mutations in domain II lead to stabilization of the proteins and confer enhanced auxin activity in these plants (e.g., Rouse et al., 1998). Domains III and IV are involved in dimerization of Aux/IAA proteins, particularly the formation of heterodimers with auxin-response factor (ARF) proteins, and it is through the formation and breaking of these that transcription is controlled (Reed, 2001). The heterodimer pair ARF-Aux/IAA binds to auxin-response elements (ARE) in the promoter regions of auxin-regulated genes and repress transcription. The addition of auxin promotes the dissociation of the dimer and the Aux/IAA proteins are ubiquitinated and degraded. A second ARF then forms a homodimer on the promoter and permits transcription (Fig. 2). As will be seen later, the Aux/IAA proteins form a family of 29 members in *Arabidopsis* (http://datf. cbi.pku.edu.cn/browsefamily.php?familyname=AUX/IAA) and mediate a huge array of auxin-driven responses.

3. The ARF Proteins

The ARF family of transcription factors has been identified as the regulatory partners of Aux/IAA proteins. The description arose from analysis of the auxin-response elements of promoters rapidly induced by auxin (e.g., Ulmasov *et al.*, 1995), rapidly followed by identification of ARF1 as a transcription factor that binds to AREs (Ulmasov *et al.*, 1997). Thereafter, a large family of ARF proteins was identified, and there is a family of 23 in *Arabidopsis*. The generic structure of ARF proteins is of an N-terminal DNA-binding domain, a transcriptional activator domain, and then a pair of dimerization domains homologous to those of Aux/IAA proteins at the C-terminus (Guilfoyle and Hagen, 2001). The defective gene in the embryo morphology mutant *MONOPTEROS*, mentioned earlier, is a member of the ARF gene family (Hardtke and Berleth, 1998).

4. Transcriptional Control

The current model for auxin-regulated gene expression is that at low auxin concentrations, mixed ARF-Aux/IAA dimers bind tightly to the DNA repressing transcription. A rise in auxin concentration leads to the dissociation of the dimers releasing the Aux/IAA protein, allowing transcription to proceed. The ARF remains bound to the promoter DNA and may associate with another ARF allowing transcription to proceed. The dissociated Aux/IAA proteins become phosphorylated and substrates for rapid degradation by the ubiquitin-proteosome pathway. A fall in auxin concentration slows protolysis and newly transcribed Aux/IAAs reassemble with the ARFs to repress transcription again.

Given the number of ARFs and Aux/IAAs and the large number of different, mixed dimers that can be made, it can be seen that this system can account for the great variety of different responses controlled by auxin, even if redundancy is reduced by the requirement that any pair needs to be expressed in the same cell before they may interact to give a phenotype. Linking each auxin-mediated response to a defined pair of transcriptional regulators will take considerable effort.

The list of genes regulated by auxin has been growing since the mid-1980s. Many of these genes were anticipated from physiological observations such as the genes encoding the ethylene biosynthetic enzymes ACC synthase and ACC oxidase. Microarray analysis of the *Arabidopsis* genome's responses to a set of active auxins has illustrated the full size of the auxin-induced transcriptome (Pufky et al., 2003).

VIII. ORGAN PATTERNING

A. PHYLLOTAXIS

Evidence is accumulating to show that auxin and its polar transport determine the apical–basal pattern of development in some floral tissues such as the gynoecium (the role of auxin in fruit development is discussed later) (Nemhauser et al., 2000). Perhaps the most widely observed outcome of auxin-induced patterning is in phyllotaxis, the regular arrangement of leaves on a stem. Once again, polar auxin transport inhibitors and *Arabidopsis* mutants defective in polar transport like *pin1-1* have helped illustrate that accumulation of IAA in cells at the side of the shoot apical meristem initiates organogenesis and determines the position of the next leaf primordium. Some delicate and beautiful experiments have been described by a group in Switzerland (Kuhlemeier and Reinhardt, 2001). Microapplications of transport inhibitors or IAA to tomato shoot apices have shown that for leaves and flowers, auxin application determines primordium development in the radial dimension (Reinhardt et al., 2000). Primordium development always arises at a fixed distance from the summit of the apical meristem in the apical–basal dimension. The cellular and subcellular distributions of the efflux proteins PIN and AUX1 in the surface layers of cells towards the tip of the shoot apical meristem were shown to correspond to the subsequent differentiation of leaf primordial (Reinhardt et al., 2003).

B. ROOT INITIATION AND GRAVITROPISM

A similar picture can be drawn for the origination of lateral root primordia. Auxin transport into distinct tissues of the root drives lateral primordium development (Blilou et al., 2005; Casimiro et al., 2001). Specific PIN proteins then contribute to root patterning, particularly AtPIN4 (Friml et al., 2002), although the AUX1 protein is again an essential contributor. Indeed, it was from a root phenotype of tolerance to the auxin 2,4-D, fewer root laterals, and agravitropism that mutants in *AUX1* were identified.

The centre of auxin concentration in roots is at the root tip, just behind the cap around the quiescent centre and columella cells (Bhalerao et al., 2002). From here both AUX1 and PIN proteins work to produce defined gradients along the root epidermal layers stretching back into the elongation zone (Ottenschlager et al., 2003). These control not just lateral rooting but also gravitropism. Several PIN genes have been identified that contribute to the gravitropic response, and AtPIN3 has been shown to relocate within columella cells from a uniform, peripheral location to a lateral location on the plasma membrane in response to a changing gravity vector (Friml et al., 2002). The movement appears to be actin-dependent. AtPIN2 protein is distributed in cortical and epidermal cells and may deliver auxin back along the root from the gravity-sensitive columella (Müller et al., 1998).

IX. APICAL DOMINANCE AND BRANCHING

In addition to controlling the initiation of leaf primordia at shoot apices, it has been recognized for many years that auxin transported back from the shoot apex controls the outgrowth of side branches from axillary buds. Gardeners and horticulturalists make use of this apical dominance to control plant structure by pruning. By removal of the stem apex the source of auxin is removed and its inhibitory action on axillary buds is released, leading to side branch outgrowth and bushier plants. Consistent with these utilitarian observations, classic experiments by Thimann and Skoog showed that if auxin is applied to the pruned apical stump, bud outgrowth is inhibited (see Davies, 1995). Experimental plants generated with low auxin levels are bushier (e.g., Jackson et al., 2002). However, apically derived auxin does not enter the axillary bud and it is unclear how the signal is transferred. Root-derived cytokinins are also involved in controlling bud break and other root–shoot signals are likely to be added to the list of controls before long.

Apical dominance is active in roots as well as stems, but for roots the contribution of additional specific signals is less clear. Root tip-derived auxin certainly inhibits lateral root initiation close to the apex; however, lateral root development is controlled by auxin at several stages. Further up the root, behind the elongation zone of primary roots, lateral primordia lie formed, but remain dormant. These primordia will not grow out unless the tip is removed (apical dominance) or until a second, shoot-derived, polar flux of auxin arrives. Not long after emergence a lateral root becomes auxin-autonomous (Casimiro et al., 2001).

X. TROPISMS AND EPINASTY

Plants respond to the vectoral signals, light, and gravity by differential growth. Auxin gradients, generated by lateral (polar) auxin transport, have been implicated, and the experiments of Cholodny and Went are common schoolroom lessons (Davies, 1995). Blocks of agar were used to collect endogenous signals from two sides of gravistimulated seedling stems. The blocks were then applied to unstimulated seedlings and bending was observed. Auxin applied in similar blocks was shown to give the same differential growth.

In stems, greater auxin concentrations enhance elongation growth, in roots auxin inhibits growth. Auxin measurements have shown that the side of stems expanding more rapidly was found to contain a little more IAA, although the gradient across the stem is generally only of the order of twofold to threefold (Harrison and Pickard, 1989). Auxin reporter genes have shown expression patterns corresponding to anticipated auxin gradients (Li *et al.*, 1991; Muday, 2001), but even so it is argued that the magnitude of the gradient (measured or assumed) is insufficient to support the considerable differences in elongation rate across the same tissues. The Cholodny–Went hypothesis remains controversial. Recent data on the stimulus-induced redistribution of specific members of the PIN family in roots (e.g., PIN3), data from auxin measurements and observations from auxin-response mutants all support the hypothesis. Other data, such as kinetic measurements of gravistimulated roots and experiments in which seedling shoots are bathed in excess IAA that are still found to be able to respond by differential growth (Edelmann, 2001), argue against such a straightforward hypothesis. Either way, it is clear that auxin is a requirement for differential and tropic growth, and many auxin sensitive mutants have agravitropic root phenotypes (e.g., *aux1* lines, Bennett *et al.*, 1996). It might not be the only signal involved.

A similar phenomenon to tropic growth is epinasty, petiole bending induced by a range of nondirectional stimuli. The contribution of auxin in epinasty is seen most clearly in plants treated with auxinic herbicides (see later). The top sides of the petioles extend excessively leading to curvature turning the face of the leaf away from light. Other hormones, in particular ethylene, are also involved in epinasty.

XI. ADVENTITIOUS ROOTING AND WOUND RESPONSES

Adventitious roots form from stem tissues, generally as a result of damage or removal of the primary root system. Many plants do produce some adventitious roots naturally, and they add valuable lateral support to tall stems in cereals, for example. However, addition of auxin to cut or damaged stems often induces a strong adventitious rooting response, and horticultural industries relying on clonal propagation make good use of this response for ornamentals, trees, flowers, and general garden plants. Synthetic auxins IBA and 1-NAA are both sold widely in hormone rooting powders and dips.

For lateral root primordia to form, cells of the root xylem pericycle divide and differentiate. For adventitious rooting a similar series of events is likely to take place in equivalent cells in the stem vasculature, but the exact cells giving rise to the first division are less certain. New root primordia form at the side of the procambium, or within callus tissue that forms at the cut or wounded surface. The initiation of root primordia in stem tissues requires a redifferentiation response, and, although auxin promotes this, the molecular or genetic mechanisms are unclear.

There are mutants of *Arabidopsis* that show precocious rooting along the seedling hypocotyl, *superroot 1* and *2* (*sur1* and *sur2*), for example (Boerjan *et al.*, 1995). In both, auxin levels are elevated due to distinct effects on auxin biosynthetic pathways.

Wounding plant tissues gives rise to rapid cell division and callus formation. Callus is undifferentiated and apolar in its growth, but not only can new, polar organs arise from it (like adventitious roots) but adjacent cells can redifferentiate to reform vascular connectivity (Davies, 1995). The process is slow, commencing many days after damage and callus growth, but xylem elements start to form with a polar orientation and line up to reestablish the vasculature as fully formed, thickened xylem elements die (Berleth *et al.*, 2000). Auxin and polar auxin transport are essential for this response and it is likely that this polar redifferentiation occurs along a reforming apical–basal auxin gradient. This is not a general event; this redifferentiation occurs only in a narrow band of cells that starts and finishes at damaged vascular strands. A similar process is likely to take place during graft development so that vascular continuity is regained.

XII. FRUIT GROWTH

Auxin plays a vital role in all stages of reproductive growth. The role of auxin in gynoecium development and embryo polarity have been recorded earlier, and some of the highest auxin concentrations have been found to be in developing fruit (Müller *et al.*, 2002). In the early 1950s, Nitsch showed that achene (seed) removal from strawberry receptacles inhibited receptacle enlargement (Pennazio, 2002). Replacing the achene with a supply of auxin maintained fruit growth. Cessation of auxin supply also led to ripening in this nonclimacteric fruit.

In fruit as well as in cereal seed-heads, it is clear that the developing embryo and seed are rich auxin sources and that this auxin mediates many developmental responses in adjacent tissues. Indeed, parthenocarpy can be induced in many species by auxin application, and this has been used to generate seed-free fruit by both topical application and by genetic engineering (Donzella *et al.*, 2000; Ficcadenti *et al.*, 1999). The bacterial IAA biosynthetic gene *iaaM* was transferred into tomato, aubergine, and other fruit under the control of a carpel-specific promoter. This gave rise to localized overproduction of IAA, induced seed-free fruit of marketable quality with the added advantage that, without the need for bees, production was no longer seasonal. Some apple and cherry crops are sprayed with auxin (mixed with other hormones) at flowering to induce fruit set.

Crops of fruit and nuts, particularly citrus, are sprayed with auxins to improve fruit size by promoting "June drop", but more widely, orchards are sprayed after the drop to increase fruit size and hexose content and to delay fruit abscission (Agusti et al., 2002). Addition of auxin delays the development of the abscission zone (Davies, 1995). Exocarp and seeds are rich in IAA, but, as in strawberries, it is likely that as the testa matures around the seed it forms an impervious barrier blocking auxin release and allowing abscission to proceed. Treatment of orchards to delay fruit abscission is becoming increasingly common to facilitate automated harvesting, thus maximizing the yield of ripe fruit collected in the minimum number of passes by harvesters. Leaf abscission is also initiated as auxin transport from the leaf declines (Morris, 1993).

XIII. HERBICIDES

The largest commercial exploitation of any plant hormone has been the use of synthetic auxins as selective herbicides. Around the world, a number of populations of auxinic herbicide-tolerant weeds have arisen, an indication of the long time over which applications have been made (Hall et al., 1996). Synthetic auxins such as 2,4-D acid were developed in the 1940s and use has been extensive, particularly because of their selectivity. A much higher activity is generally found against dicotyledonous plants than against cereals, and this had led to widespread use to control broad-leaved weeds in cereal crops, sugarcane, and turf. Auxins were the active defoliants in "Agent Orange," and phenoxy auxins are still used to control brushwood.

The mechanisms of herbicidal action and the basis of selectivity between monocots and dicots are unclear. Induction of a massive synthesis of ethylene has been suggested to be the mechanism of activity, at least in part because a by-product of ethylene overproduction is cyanide (Grossmann, 2003). However, recent work shows that inhibition of either ethylene synthesis or ethylene receptors does not prevent herbicidal activity. Therefore, although induction of the ethylene-synthesizing enzymes is a recognized effect of auxin application (Hansen and Grossmann, 2000; Kim et al., 1992), ethylene remains a symptom of herbicidal auxin and is not the sole mediator.

Auxinic herbicides also induce ABA synthesis and this might be a consequence of elevated ethylene synthesis. Ethylene and ABA act antagonistically in growth control. Elevated ABA will shut stomata and the consequent block on photosynthesis might be the source of excessive hydrogen peroxide production, as recorded in auxinic herbicide-treated plants (Grossmann et al., 2001). In addition to the damage done by peroxide itself, such reactive oxygen species will induce defense and senescence responses, giving rise to necrosis and death. Saturating auxin clearly gives rise to a number of damaging symptoms, but genetic data from resistant biotypes, as well as the observations of monocot/dicot selectivity, suggest that there is a single target site for these herbicides, and this seems likely to be the auxin receptor. As more becomes known about the receptor, the understanding of plant development and auxins as agrochemicals will also grow.

ACKNOWLEDGMENTS

Funding has been provided by BBSRC and EC contract HPRN-CT-2002-00334.

REFERENCES

Abel, S., and Theologis, A. (1996). Early genes and auxin action. *Plant Physiol.* **111**, 9–17.
Abel, S., Oeller, P. W., and Thoeologis, A. (1994). Early auxin-induced genes encode short-lived nuclear proteins. *Proc. Natl. Acad. Sci. USA* **91**, 326–330.
Agusti, M., Zaragoza, S., Iglesias, D. J., Almela, V., Primo-Millo, E., and Talon, M. (2002). The synthetic auxin 3,5,6-TPA stimulates carbohydrate accumulation and growth in citrus fruit. *J. Plant Growth Regul.* **36**, 141–147.
Aloni, R., Schwalm, K., Langhans, M., and Ullrich, C. I. (2003). Gradual shifts in sites of free-auxin production during leaf primordium development and their role in vascular differentiation and leaf morphogenesis in *Arabidopsis*. *Planta* **216**, 841–853.
Avsian-Kretchmer, O., Cheng, J.-C., Chen, L., Moctezuma, E., and Sung, Z. R. (2002). Indole acetic acid distribution coincides with vascular differentiation pattern during *Arabidopsis* leaf ontogeny. *Plant Physiol.* **130**, 199–209.
Baker, D. A. (2000). Long-distance vascular transport of endogenous hormones in plants and their role in source:sink regulation. *Isr. J. Plant Sci.* **48**, 199–203.
Barbier-Brygoo, H., Ephritikhine, G., Klambt, D., Maurel, C., Palme, K., Schell, J., and Guern, J. (1991). Perception of the auxin signal at the plasma membrane of tobacco mesophyll protoplasts. *Plant J.* **1**, 83–93.
Bartel, B., and Fink, G. R. (1995). ILR1, an amidohydrolase that releases active indole-3-acetic acid from conjugates. *Science* **268**, 1745–1748.
Bartel, B., LeClere, S., Magidin, M., and Zolman, B. K. (2001). Inputs to the active indole-3-acetic acid pool: *de novo* synthesis, conjugate hydrolysis, and indole-3-butyric acid β-oxidation. *J. Plant Growth Regul.* **20**, 198–216.
Bauly, J., Sealy, I. M., Macdonald, H., Brearley, J., Droge, S., Hillmer, S., Robinson, D. G., Venis, M. A., Blatt, M. R., Lazarus, C. M., and Napier, R. M. (2000). Overexpression of auxin-binding protein enhances the sensitivity of guard cells to auxin. *Plant Physiol.* **124**, 1229–1338.
Benjamins, R., Quint, A., Weijers, D., Hooykaas, P., and Offringa, R. (2001). The PINOID protein kinase regulates organ development in *Arabidopsis* by enhancing polar auxin transport. *Development* **128**, 4057–4067.
Benkova, E., Michniewicz, M., Sauer, M., Teichmann, T., Seifertova, D., Jurgens, G., and Friml, J. (2003). Local, efflux-dependent auxin gradients as a common module for plant organ formation. *Cell* **115**, 591–602.
Bennett, M. J., Marchant, A., Green, H. G., May, S. T., Ward, S. P., Millner, P. A., Walker, A. R., Schulz, B., and Feldmann, K. A. (1996). *Arabidopsis* AUX1 gene: A permease-like regulator of root gravitropism. *Science* **273**, 948–950.

Berleth, T., Mattsson, J., and Hardtke, C. S. (2000). Vascular continuity and auxin signals. *Trends Plant Sci.* **5**(9), 387–393.

Bhalerao, R., and Bennett, M. J. (2003). The case for morphogens in plants. *Nat. Cell Biol.* **5**(11), 939–943.

Bhalerao, R. P., Eklof, J., Ljung, K., Marchant, A., Bennett, M., and Sandberg, G. (2002). Shoot-derived auxin is essential for early lateral root emergence in *Arabidopsis* seedlings. *Plant J.* **29**(3), 325–332.

Bialek, L., and Cohen, J. D. (1992). Amide-linked indoleacetic acid conjugates may control levels of indoleacetic acid in germinating seedlings of *Phaseolus vulgaris*. *Plant Physiol.* **100**, 2002–2007.

Blilou, I., Xu, J., Wildwater, M., Willemsen, V., Paponov, I., Friml, J., Heidstra, R., Aida, M., Palme, K., and Scheres, B. (2005). The PIN auxin efflux facilitator network controls growth and patterning in *Arabidopsis* roots. *Nature* **433**, 39–44.

Boerjan, W., Cervera, M. T., Delarue, M., Beeckman, T., Dewitte, W., Bellini, C., Caboche, M., Vanonckelen, H., Vanmontagu, M., and Inze, D. (1995). Superroot, a recessive mutation in *Arabidopsis*, confers auxin overproduction. *Plant Cell* **7**, 1405–1419.

Brown, D. A., Rashotte, A. M., Murphy, A. S., Normanly, J., Tague, B. W., Peer, W. A., Taiz, L., and Muday, G. K. (2001). Flavonoids act as negative regulators of auxin transport *in vivo* in *Arabidopsis*. *Plant Physiol.* **126**, 524–535.

Cambridge, A. P., and Morris, D. A. (1996). Transfer of exogenous auxin from the phloem to the polar auxin transport pathway in pea (*Pisum sativum* L.). *Planta* **199**, 583–588.

Casimiro, I., Marchant, A., Bhalerao, R. P., Beeckman, T., Dhooge, S., Swarup, R., Graham, N., Inze, D., Sandberg, G., Casero, P. J., and Bennett, M. (2001). Auxin transport promotes *Arabidopsis* lateral root initiation. *Plant Cell* **13**(4), 843–852.

Chen, J.-G., Ullah, H., Young, J. C., Sussman, M. R., and Jones, A. M. (2001). ABP1 is required for organised cell elongation and division in *Arabidopsis* embryogenesis. *Genes Dev.* **15**, 902–911.

Cockcroft, C. E., den Boer, B. G., Healy, J. M., and Murray, J. A. (2000). Cyclin D control of growth rate in plants. *Nature* **405**(6786), 575–579.

Costacurta, A., and Vanderleyden, J. (1995). Synthesis of phytohormones by plant-associated bacteria. *Crit. Rev. Microbiol.* **21**, 1–18.

Davies, P. J. (Ed.) (1995). "Plant Hormones: Physiology, Biochemistry and Molecular Biology." Kluwer Academic Publishers, Dordrecht, The Netherlands.

Davies, R. T., Goetz, D. H., Lasswell, J., Anderson, M. N., and Bartel, B. (1999). *IAR3* encodes an auxin conjugate hydrolase from *Arabidopsis*. *Plant Cell* **11**, 365–376.

del Pozo, J. C., Timpte, C., Tan, S., Callis, J., and Estelle, M. (1998). The ubiquitin-related protein RUB1 and auxin response in *Arabidopsis*. *Science* **280**, 1760–1763.

den Boer, B. G. W., and Murray, J. A. H. (2000). Triggering the cell cycle in plants. *Trends Cell Biol.* **10**(6), 245–250.

Dharmasiri, N., and Estelle, M. (2004). Auxin signaling and regulated protein degradation. *Trends Plant. Sci.* **9**, 302–308.

Dharmasiri, N., Dharmasiri, S., and Estelle, M. (2005). The F-box protein TIR1 is an auxin receptor. *Nature* **435**, 441–445.

Diekmann, W., Venis, M. A., and Robinson, D. G. (1995). Auxins induce clustering of the auxin-binding protein at the surface of maize coleoptile protoplasts. *Proc. Natl. Acad. Sci. USA* **92**, 3425–3429.

Donzella, G., Spena, A., and Rotino, G. L. (2000). Transgenic parthenocarpic eggplants: Superior germplasm for increased winter production. *Mol. Breed.* **6**, 79–86.

Dunwell, J. M., Khuri, S., and Gane, P. J. (2000). Microbial relatives of the seed storage proteins of higher plants: Conservation of structure and diversification of function during evolution of cupin superfamily. *Microbiol. Mol. Biol. Rev.* **64**, 153–179.

Edelmann, H. G. (2001). Lateral redistribution of auxin is not the means for gravitropic differential growth of coleoptiles: A new model. *Physiol. Plantarum* **112**(1), 119–126.
Else, M. A., Stankiewicz-Davies, A. P., Crisp, C. M., and Atkinson, C. J. (2004). The role of polar auxin transport through pedicels of *Prunus avium* L. in relation to fruit development and retention. *J. Exp. Bot.* **55**, 2099–2109.
Ficcadenti, N., Sestili, S., Pandolfini, T., Cirillo, C., Rotino, G. L., and Spena, A. (1999). Genetic engineering of parthenocarpic fruit development in tomato. *Mol. Breed.* **5**, 463–470.
Fischer, C., and Neuhaus, G. (1996). Influence of auxin on the establishment of bilateral symmetry in monocots. *Plant J.* **9**, 659–669.
Friml, J., Benkova, E., Blilou, I., Wisniewska, J., Hamann, T., Ljung, K., Woody, S., Sandberg, G., Scheres, B., Jurgens, G., and Palme, K. (2002). AtPIN4 mediates sink-driven auxin gradients and root patterning in *Arabidopsis. Cell* **108**, 661–673.
Friml, J., Vieten, A., Sauer, M., Weijers, D., Schwarz, H., Hamann, T., Offringa, R., and Jurgens, G. (2003). Efflux-dependent auxin gradients establish the apical-basal axis of *Arabidopsis. Nature* **426**, 147–153.
Furutani, M., Vernoux, T., Traas, J., Kato, T., Tasaka, M., and Aida, M. (2004). PIN-FORMED1 and PINOID regulate boundary formation and cotyledon development in *Arabidopsis* embryogenesis. *Development* **131**, 5021–5030.
Gaedeke, N., Klein, M., Kolukisaoglu, U., Forestier, C., Muller, A., Ansorge, M., Becker, D., Mamnun, Y., Kuchler, K., Schulz, B., Mueller-Roeber, B., and Martinoia, E. (2001). The *Arabidopsis thaliana* ABC transporter AtMRP5 controls root development and stomata movement. *EMBO J.* **20**, 1875–1887.
Galweiler, L., Guan, C. H., Muller, A., Wisman, E., Mendgen, K., Yephremov, A., and Palme, K. (1998). Regulation of polar auxin transport by AtPIN1 in *Arabidopsis* vascular tissue. *Science* **282**, 2226–2230.
Garbers, C., DeLong, A., Deruere, J., Bernasconi, P., and Soll, D. (1996). A mutation in protein phosphatase A2 regulatory subunit A affects auxin transport in *Arabidopsis. EMBO J.* **15**, 2115–2124.
Gaudin, V., Vrain, T., and Jouanin, L. (1994). Bacterial genes modifying hormonal balances in plants. *Plant Physiol. Biochem.* **32**, 11–29.
Geisler, M., Kolukisaoglu, H. U., Bouchard, R., Billion, K., Berger, J., Saal, B., Frangne, N., Koncz-Kalman, Z., Koncz, C., Dudler, R., Blakeslee, J. J., Murphy, A. S., Martinoia, E., and Schulz, B. (2003). TWISTED DWARF1, a unique plasma membrane-anchored immunophilin-like protein, interacts with *Arabidopsis* multidrug resistance-like transporters AtPGP1 and AtPGP19. *Mol. Biol. Cell* **14**, 4238–4249.
Geldner, N., Anders, N., Wolters, H., Keicher, J., Kornberger, W., Muller, P., Delbarre, A., Ueda, T., Nakano, A., and Jurgens, G. (2003). The *Arabidopsis* GNOM ARF-GEF mediates endosomal recycling, auxin transport, and auxin-dependent plant growth. *Cell* **112**, 219–230.
Gil, P., Dewey, E., Friml, J., Zhao, Y., Snowden, K. C., Putterill, J., Palme, K., Estelle, M., and Chory, J. (2001). BIG: A calossin-like protein required for polar auxin transport in *Arabidopsis. Genes Dev.* **15**, 1985–1997.
Glawischnig, E., Tomas, A., Eisenreich, W., Spiteller, P., Bacher, A., and Gierl, A. (2000). Auxin biosynthesis in maize kernels. *Plant Physiol.* **123**, 1109–1119.
Gray, W. M., Kepinski, S., Rouse, D., Leyser, O., and Estelle, M. (2001). Auxin regulates SCF^{TIR1}-dependent degradation of AUX/IAA proteins. *Nature* **414**, 271–276.
Gray, W. M., del Pozo, J. C., Walker, L., Hobbie, L., Riseeuw, E., Banks, T., Crosby, W. L., Yang, M., Ma, H., and Estelle, M. (1999). Identification of an SCF ubiquitin ligase complex required for auxin response in *Arabidopsis thaliana. Genes Dev.* **13**, 1678–1691.
Grossmann, K. (2003). Mediation of herbicide effects by hormone interactions. *J. Plant Growth Regul.* **22**, 109–122.

Grossmann, K., Kwiatkowski, J., and Tresch, S. (2001). Auxin herbicides induce H_2O_2 overproduction and tissue damage in cleavers (*Galium aparine* L.). *J. Exp. Bot.* **52**, 1811–1816.
Guilfoyle, T. J., and Hagen, G. (2001). Auxin response factors. *J. Plant Growth Regul.* **20**, 281–291.
Hall, J. C., Webb, S. R., and Deshpande, S. (1996). An overview of auxinic herbicide resistance: Wild mustard (*Sinapis arvensis* L.) as a case study. *ACS Symp. Ser.* **645**, 28–43.
Hansen, H., and Grossmann, K. (2000). Auxin-induced ethylene triggers abscisic acid biosynthesis and growth inhibition. *Plant Physiol.* **124**, 1437–1448.
Hardtke, C. S., and Berleth, T. (1998). The *Arabidopsis* gene MONOPTEROS encodes a transcription factor mediating embryo axis formation and vascular development. *EMBO J.* **17**, 1405–1411.
Harrison, M. A., and Pickard, B. G. (1989). Auxin asymmetry during gravitropism by tomato hypocotyls. *Plant Physiol.* **89**, 652–657.
Hellgren, J. M., Olofsson, K., and Sundberg, B. (2004). Patterns of auxin distribution during gravitational induction of reaction wood in polar and pine. *Plant Physiol.* **135**, 1–9.
Hemerly, A. S., Ferreira, P., de Almeida Engler, J., Van Montagu, M., Engler, G., and Inze, D. (1993). cdc2a expression in *Arabidopsis* is linked with competence for cell division. *Plant Cell* **5**(12), 1711–1723.
Hertel, R., Thompson, K., and Russo, V. E. A. (1972). *In vitro* auxin binding to particulate cell fractions from corn coleoptiles. *Planta* **107**, 325–340.
Himanen, K., Boucheron, E., Vanneste, S., Engler, J. D., Inze, D., and Beeckman, T. (2002). Auxin-mediated cell cycle activation during early lateral root initiation. *Plant Cell* **14**, 2339–2351.
Inohara, N., Shimomura, S., Fukuui, T., and FuTai, M. (1989). Auxin-binding protein located in the endoplasmic reticulum of maize shoots. *Proc. Natl. Acad. Sci. USA* **83**, 3654–3658.
Jackson, R. G., Kowalczyk, M., Li, Y., Higgins, G., Ross, J., Sandberg, G., and Bowles, D. J. (2002). Over-expression of an *Arabidopsis* gene encoding a glucosyltransferase of indole-3-acetic acid: Phenotypic characterisation of transgenic lines. *Plant J.* **32**, 573–583.
John, P. C. L., Zhang, K., Dong, C., Diederich, L., and Wightman, F. (1993). $p34^{cdc2}$ related proteins in control of the cell cycle progression, the switch between division and differentiation in tissue development, and stimulation of division by auxin and cytokinin. *Aust. J. Plant Physiol.* **20**, 503–526.
Jones, A. M. (1990). Location of transported auxin in etiolated maize shoots using 5-azidoindole-3-acetic acid. *Plant Physiol.* **93**, 1154–1161.
Jones, A. M., Im, K. H., Savka, M. A., Wu, M. J., Dewitt, N. G., Shillito, R., and Binns, A. N. (1998). Auxin-dependent cell expansion mediated by overexpression of auxin-binding protein 1. *Science* **282**, 1114–1117.
Kepinski, S., and Leyser, O. (2005). The *Arabidopsis* TIR1 protein is an auxin receptor. *Nature* **435**, 446–451.
Kim, W. T., Silverstone, A., Yip, W. K., Dong, J. G., and Yang, S. F. (1992). Induction of 1-aminocyclopropane-1-carboxylate synthase messenger-RNA by auxin in mung bean hypocotyls and cultured apple shoots. *Plant Physiol.* **98**, 465–471.
Kuhlemeier, C., and Reinhardt, D. (2001). Auxin and phyllotaxis. *Trends Plant. Sci.* **6**(5), 187–189.
Leyser, O. (2002). Molecular genetics of auxin signaling. *Ann. Rev. Plant. Biol.* **53**, 377–398.
Leyser, H. M. O., Lincoln, C. A., Timpte, C., Lammer, D., Turner, J., and Estelle, M. (1993). *Arabidopsis* auxin resistance gene AXR1 encodes a protein related to ubiquitin-activating enzyme E1. *Nature* **364**, 161–164.
Li, Y., Hagen, G., and Guilfoyle, T. J. (1991). An auxin-responsive promoter is differentially induced by auxin gradients during tropisms. *Plant Cell* **3**, 1167–1175.

Ljung, K., Bhalerao, R. P., and Sandberg, G. (2001a). Sites and homeostatic control of auxin biosynthesis in *Arabidopsis* during vegetative growth. *Plant J.* **28**, 465–474.

Ljung, K., Hull, A. K., Kowalczyk, M., Marchant, A., Celenza, J., Cohen, J. D., and Sandberg, G. (2002). Biosynthesis, conjugation, catabolism and homeostasis of indole-3-acetic acid in *Arabidopsis thaliana*. *Plant Mol. Biol.* **49**, 249–272.

Ljung, K., Östin, A., Lioussanne, L., and Sandberg, G. (2001b). Developmental regulation of indole-3acetic acid turnover in Scots pine seedlings. *Plant Physiol.* **125**, 464–475.

Löbler, M., and Klambt, D. (1985). Auxin-binding protein from coleoptile membranes from corn (*Zea mays* L.). I. Purification by immunological methods and characterisation. *J. Biol. Chem.* **260**, 9848–9853.

Macdonald, H. (1997). Auxin perception and signal transduction. *Physiol. Plantarum* **100**, 423–430.

Marchant, A., Bhalerao, R., Casimiro, I., Eklof, J., Casero, P. J., Bennett, M., and Sandberg, G. (2002). AUX1 promotes lateral root formation by facilitating indole-3-acetic acid distribution between sink and source tissues in the *Arabidopsis* seedling. *Plant Cell* **14**(3), 589–597.

Mattsson, J., Sung, Z. R., and Berleth, T. (1999). Responses of plant vascular systems to auxin transport inhibition. *Development* **126**, 2979–2991.

Mikkelsen, M. D., Naur, P., and Halkier, B. A. (2004). *Arabidopsis* mutants in the C-S lyase of glucosinolate biosynthesis establish a critical role for indole-3-acetaldoxime in auxin homeostasis. *Plant J.* **37**, 770–777.

Mockaitis, K., and Howell, S. H. (2000). Auxin induces mitogenic activated protein kinase (MAPK) activation in roots of *Arabidopsis* seedlings. *Plant J.* **24**, 785–796.

Morris, D. A. (1993). The role of auxin in the apical regulation of leaf abscission in cotton (*Gossypium hirsutum* L.). *J. Exp. Bot.* **44**, 807–814.

Morris, D. A., and Robinson, J. S. (1998). Targeting of auxin carriers to the plasma membrane: Differential effects of brefeldin A on the traffic of auxin uptake and efflux carriers. *Planta* **205**, 606–612.

Muday, G. K. (2001). Auxins and tropisms. *J. Plant Growth Regul.* **20**, 226–243.

Muday, G. K., Peer, W. A., and Murphy, A. S. (2003). Vesicular cycling mechanisms that control auxin transport polarity. *Trends Plant. Sci.* **8**, 301–304.

Müller, A., Duchting, P., and Weiler, E. W. (2002). A multiplex GC-MS/MS technique for the sensitive and quantitative single-run analysis of acidic phytohormones and related compounds, and its application to *Arabidopsis thaliana*. *Planta* **216**, 44–56.

Müller, A., Guan, C., Galweiler, L., Tanzer, P., Huijser, P., Marchant, A., Parry, G., Bennett, M., Wisman, E., and Palme, K. (1998). AtPIN2 defines a locus of *Arabidopsis* for root gravitropism control. *EMBO J.* **17**, 6903–6911.

Murphy, A. S., Hoogner, K. R., Peer, W. A., and Taiz, L. (2002). Identification, purification, and molecular cloning of N-1-naphthylphthalmic acid-binding plasma membrane-associated aminopeptidases from *Arabidopsis*. *Plant Physiol.* **128**, 935–950.

Napier, R. M. (2001). Models of auxin binding. *J. Plant Growth Regul.* **20**, 244–254.

Napier, R. M. (2004). Plant hormone binding sites. *Ann. Bot.* **93**, 227–233.

Napier, R. M., and Venis, M. A. (1995). Auxin action and auxin-binding proteins. *New Phytologist* **129**, 167–201.

Napier, R. M., David, K. M., and Perrot-Rechenmann, C. (2002). A short history of auxin-binding proteins. *Plant Mol. Biol.* **49**, 339–348.

Nemhauser, J. L., Feldman, L. J., and Zambryski, P. C. (2000). Auxin and ETTIN in *Arabidopsis* gynoecium morphogenesis. *Development* **127**, 3877–3888.

Nishida, T., Ohnishi, N., Kodama, H., and Komamine, A. (1992). Establishment of synchrony by starvation and readdition of auxin in suspension cultures of *Catharanthus roseus* cells. *Plant Cell Tissue Organ Cult.* **28**, 37–43.

Noh, B., Murphy, A. S., and Spalding, E. P. (2001). Multidrug resistance-like genes of *Arabidopsis* required for auxin transport and auxin-mediated development. *Plant Cell* **13**, 2441–2454.

Oeller, P. W., and Theologis, A. (1995). Induction kinetics of the nuclear proteins encoded by the early indoleacetic acid-inducible genes, *PsIAA4/5* and *PsIAA6* in pea (*Pisum sativum* L.). *Plant J.* **7**, 37–48.

Ottenschlager, I., Wolff, P., Wolverton, C., Bhalerao, R., Sandberg, G., Ishikawa, H., Evans, M., and Palme, K. (2003). Gravity-regulated differential auxin transport from columella to lateral root cap cells. *Proc. Natl. Acad. Sci. USA* **100**, 2987–2991.

Östin, A., Kowalyczk, M., Bhalerao, R. P., and Sandberg, G. (1998). Metabolism of indole-3-acetic acid in *Arabidopsis*. *Plant Physiol.* **118**, 285–296.

Paponov, L. A., Teale, W., Trebar, M., Blilou, I., and Palme, K. (2005). The PIN auxin efflux facilitators: Evolutionary and functional perspectives. *Trends Plant. Sci.* in press.

Parry, G., Delbarre, A., Marchant, A., Swarup, R., Napier, R. M., Perrot-Rechenmann, C., and Bennett, M. J. (2001). Novel auxin transport inhibitors phenocopy the auxin influx carrier mutation aux1. *Plant J.* **25**, 399–406.

Pennazio, S. (2002). The discovery of the chemical nature of the plant hormone auxin. Rivista di Biologia. *Biology Forum* **95**, 289–307.

Petrasek, J., Elckner, M., Morris, D. A., and Zazimalova, E. (2002). Auxin efflux carrier activity and auxin accumulation regulate cell division and polarity in tobacco cells. *Planta* **216**, 302–308.

Planchais, S., Glab, N., Trehin, C., Perennes, C., Bureau, J. M., Meijer, L., and Bergounioux, C. (1997). Roscovitine, a novel cyclin-dependent kinase inhibitor, characterizes restriction point and G2/M transition in tobacco BY-2 cell suspension. *Plant J.* **12**(1), 191–202.

Poli, D. B., Jacobs, M., and Cooke, T. J. (2003). Auxin regulation of axial growth in bryophyte sporophytes: Its potential significance for the evolution of early land plants. *Am. J. Bot.* **90**, 1405–1415.

Pufky, J., Qui, Y., Rao, M. V., Hurban, P., and Jones, A. M. (2003). The auxin-induced transcriptome for etiolated *Arabidopsis* seedlings using a structure/function approach. *Funct. Integr. Genomics* **3**, 135–143.

Ramos, J. A., Zenser, N., Leyser, O., and Callis, J. (2001). Rapid degradation of auxin/indoleacetic acid proteins requires conserved amino acids of domain II and is proteasome dependent. *Plant Cell* **13**, 2349–2360.

Rampey, R. A., LeClere, S., Kowalczyk, M., Ljung, K., Sandberg, G., and Bartel, B. (2004). A family of auxin-conjugate hydrolases that contributes to free indole-3-acetic acid levels during *Arabidopsis* germination. *Plant Physiol.* **135**, 978–988.

Reed, J. W. (2001). Roles and activities of Aux/IAA proteins in *Arabidopsis*. *Trends Plant. Sci.* **6**, 420–425.

Reinecke, D. M., and Bandurski, R. S. (1987). Auxin biosynthesis and metabolism. *In* "Plant Growth and Development" (P. J. Davies, Ed.), pp. 24–42. Martinus Nijhoff, Dordrecht, The Netherlands.

Reinhardt, D., Mandel, T., and Kuhlemeier, C. (2000). Auxin regulates the initiation and radial position of plant lateral organs. *Plant Cell* **12**, 507–518.

Reinhardt, D., Pesce, E. R., Stieger, P., Mandel, T., Baltensperger, K., Bennett, M., Traas, J., Friml, J., and Kuhlemeier, C. (2003). Regulation of phyllotaxis by polar auxin transport. *Nature* **426**(6964), 255–260.

Rotino, G. L., Perri, E., Zottini, M., Sommer, H., and Spena, A. (1997). Genetic engineering of parthenocarpic plants. *Nature Biotech.* **15**, 1398–1401.

Rouse, D., Mackay, P., Stirnberg, P., Estelle, M., and Leyser, O. (1998). Changes in auxin response from mutations in an Aux/IAA gene. *Science* **279**, 1371–1373.

Rück, A., Palme, K., Venis, M. A., Napier, R. M., and Felle, R. H. (1993). Patch-clamp analysis establishes a role for an auxin-binding protein in the auxin stimulation of plasma membrane current in *Zea mays* protoplasts. *Plant J.* **4**, 41–46.

Ruegger, M., Dewey, E., Gray, W. M., Hobie, M., Turner, J., and Estelle, M. (1998). The TIR1 protein of *Arabidopsis* functions in auxin response and is related to human SKP2 and yeast grr1p. *Genes Dev.* **12**, 198–207.

Schnabel, E. L., and Frugoli, J. F. (2004). The PIN and LAX families of auxin transport genes in *Medicago truncatula*. *Mol. Genet. Genomics* **272**, 420–432.

Schrader, J., Baba, K., May, S. T., Palme, K., Bennett, M., Bhalerao, R. P., and Sandberg, G. (2003). Polar auxin transport in the wood-forming tissues of hybrid aspen is under simultaneous control of developmental and environmental signals. *Proc. Natl. Acad. Sci. USA* **100**(17), 10096–10101.

Schrader, J., Moyle, R., Bhalerao, R., Hertzberg, M., Lundeberg, J., Nilsson, P., and Bhalerao, R. P. (2004). Cambial meristem dormancy in trees involves extensive remodelling of the transcriptome. *Plant J.* **40**, 173–187.

Schwechheimer, C., Serino, G., and Deng, X. W. (2002). Multiple ubiquitin ligase-mediated processes require COP9 signalosome and AXR1 function. *Plant Cell* **14**, 2553–2563.

Sieburth, L. (1999). Auxin is required for leaf vein pattern in *Arabidopsis*. *Plant Physiol.* **121**, 1179–1190.

Sitbon, F., Hennion, S., Sundberg, B., Little, C. H. A., Olsson, O., and Sandberg, G. (1992). Transgenic tobacco plants coexpressing the agrobacterium-tumefaciens-IAAM and IAAH genes display altered growth and indoleacetic-acid metabolism. *Plant Physiol.* **99**(3), 1062–1069.

Steinmann, T., Geldner, N., Grebe, M., Mangold, S., Jackson, C. L., Paris, S., Galweiler, L., Palme, K., and Jurgens, G. (1999). Coordinated polar localization of auxin efflux carrier *PIN1* by GNOM ARF GEF. *Science* **286**, 316–318.

Stieger, P. A., Reinhardt, D., and Kuhlemeier, C. (2002). The auxin influx carrier is essential for correct leaf positioning. *Plant J.* **32**, 509–517.

Swarup, R., Kargul, J., Marchant, A., Zadik, D., Rahman, A., Mills, R., Yemm, A., May, S., Williams, L., Millner, P., Tsurumi, S., Moore, I., Napier, R., Kerr, I. D., and Bennett, M. J. (2004). Structure-function analysis of the presumptive *Arabidopsis* auxin permease AUX1. *Plant Cell* **16**, 3069–3083.

Tiwari, S. B., Wang, X.-J., Hagen, G., and Guilfoyle, T. J. (2001). AUX/IAA proteins are active repressors, and their stability and activity are modulated by auxin. *Plant Cell* **13**, 2809–2822.

Uggla, C., Mellerowicz, E. J., and Sundberg, B. (1998). Indole-3-acetic acid controls cambial growth in Scots pine by positional signalling. *Plant Physiol.* **117**, 113–121.

Uggla, C., Moritz, T., Sandberg, G., and Sundberg, B. (1996). Auxin is a positional signal in pattern formation in plants. *Proc. Natl. Acad. Sci. USA* **93**, 9282–9286.

Ullah, H., Chen, J.-G., Temple, B., Boyes, D. C., Alonso, J. M., Davis, K. R., Ecker, J. R., and Jones, A. M. (2003). The β-subunit of the *Arabidopsis* G protein negatively regulates auxin-induced cell division and affects multiple developmental processes. *Plant Cell* **15**, 393–409.

Ullah, H., Chen, J.-G., Young, J. C., Im, K.-H., Sussman, M. R., and Jones, A. M. (2001). Modulation of cell proliferation by heterotrimeric G protein in *Arabidopsis*. *Science* **292**, 2066–2069.

Ulmasov, T., Hagen, G., and Guilfoyle, T. J. (1997). ARF1, a transcription factor that binds to auxin response elements. *Science* **276**, 1865–1868.

Ulmasov, T., Liu, Z.-B., Hagen, G., and Guilfoyle, T. J. (1995). Composite structure of auxin response elements. *Plant Cell* **7**, 1611–1623.

van der Graaff, E., Boot, K., Granbom, R., Sandberg, G., and Hooykaas, P. J. J. (2003). Increased endogenous auxin production in *Arabidopsis thaliana* causes both earlier described and novel auxin-related phenotypes. *J. Plant Growth Regul.* **22,** 240–252.

Winicur, Z. M., Zhang, G. F., and Staehelin, L. A. (1998). Auxin deprivation induces synchronous Golgi differentiation in suspension-cultured tobacco BY-2 cells. *Plant Physiol.* **117**(2), 501–513.

Wodzicki, T. J., Abe, H., Wodzicki, A. B., Pharis, R. P., and Cohen, J. D. (1987). Investigations on the nature of the auxin-wave in the cambial region of pine stems. *Plant Physiol.* **84,** 135–143.

Woo, E.-J., Marshall, J., Bauly, J., Chen, J.-G., Venis, M. A., Napier, R. M., and Pickersgill, R. W. (2002). Crystal structure of auxin-binding protein 1 in complex with auxin. *EMBO J.* **21,** 2877–2885.

Zazimalova, E., Opatrny, Z., and Brezinova, A. (1995). The effect of auxin starvation on the growth of auxin-dependent tobacco cell culture—dymanics of auxin binding activity and endogenous free IAA content. *J. Exp. Bot.* **46,** 1205–1213.

Zenser, N., Ellsmore, A., Leasure, C., and Callis, J. (2001). Auxin modulates the degradation rate of Aux/IAA proteins. *Proc. Natl. Acad. Sci. USA* **98,** 11795–11800.

Zhao, Y., Christensen, S. K., Fankhauser, C., Cashman, J. R., Cohen, J. D., Weigel, D., and Chory, J. (2001). A role for flavin monooxygenase-like enzymes in auxin biosynthesis. *Science* **291,** 306–309.

FURTHER READING

Geldner, N. (2004). The plant endosomal system—its structure and role in signal transduction and plant development. *Planta* **219,** 547–560.

7

REGULATORY NETWORKS OF THE PHYTOHORMONE ABSCISIC ACID

ZHEN XIE, PAUL RUAS, AND QINGXI J. SHEN

Department of Biological Sciences, University of Nevada, Las Vegas, Nevada 89154

I. Introduction
II. Signaling Pathways
    A. Receptors
    B. G-Proteins
    C. Secondary Messengers
    D. Phosphatases and Kinases
    E. Transcriptional Regulation
    F. Posttranscriptional Regulation
III. Cross-talk of ABA and GA
IV. Conclusions
    References

Structurally similar to retinoic acid (RA), the phytohormone abscisic acid (ABA) controls many developmental and physiological processes via complicated signaling networks that are composed of receptors, secondary messengers, protein kinase/phosphatase cascades, transcription factors, and chromatin-remodeling factors. In addition, ABA signaling is further modulated by mRNA maturation and stability, microRNA (miRNA) levels, nuclear speckling, and protein degradation.

This chapter highlights the identified regulators of ABA signaling and reports their homologues in dicotyledonous and monocotyledonous plants. © 2005 Elsevier Inc.

I. INTRODUCTION

Abscisic acid plays a variety of roles in plant development, bud and seed dormancy, germination, cell division and movement, leaf senescence and abscission, and cellular response to environmental stresses (Leung and Giraudat, 1998; Rohde et al., 2000; Zhu, 2002). It is ubiquitous in lower and higher plants and has also been found in algae (Hirsch et al., 1989), fungi (Yamamoto et al., 2000), and even mammalian brain tissue (Le Page-Degivry et al., 1986). Abscisic acid and RA are similar in several aspects: (1) the structure of ABA, a 15-carbon sesquiterpenoid carboxylic acid, is very similar to RA (Fig. 1); (2) both ABA and RA are synthesized ultimately from β-carotene (provitamine A); and (3) only certain geometric isomers

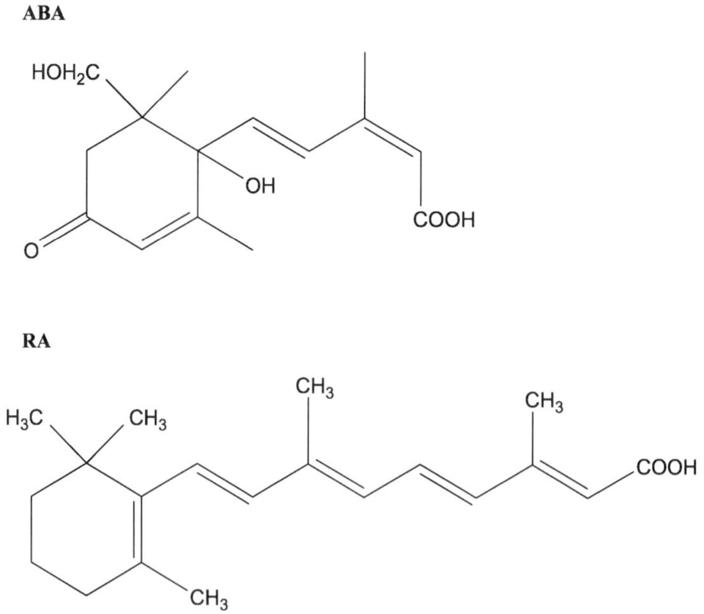

FIGURE 1. Abscisic acid (ABA) is structurally similar to retinoic acid (RA). Shown here are the geometric isomers of biologically active ABA (C2-*cis*, C4-*trans* isomer) and RA (all *trans*). Another geometric isomer of RA, 9-*cis* RA is also biologically active.

are biologically active. Retinoic acid is active in two forms: all *trans* RA and 9-*cis* RA. For ABA, the C2-*cis*, C4-*trans* isomer, but not the C2-*trans*, C4-*trans* isomer, is biologically active (Milborrow, 1978). However, the mechanisms of cellular response to RA and ABA are quite different. Retinoic acid is perceived by an intracellular receptor that belongs to the nuclear receptor superfamily. The RA receptor (RAR) forms a heterodimer with a common nuclear receptor monomer, RXR, that is located exclusively in the nucleus. In the absence of ligand, the heterodimer represses transcription of promoters that contain the cognate RA response elements by directing histone deacetylation at nearby nucleosomes. Binding of RA to RAR results in a dramatic conformational change of RAR that can still heterodimerize with RXR. However, in the ligand-bound conformation, the heterodimeric nuclear receptors direct hyperacetylation of histones in nearby nucleosomes to reverse the effects of the ligand-free heterodimer. The ligand-binding domain of nuclear receptors also binds mediators and stimulates the assembly of transcriptional pre-initiation complexes (Lodish *et al.*, 2004). In contrast, the response of plant cells to ABA involves a signal network containing receptors, secondary messengers, protein kinases and phosphatases, chromatin-remodeling proteins, transcriptional regulators, RNA-binding proteins, and protein degradation complexes (Chinnusamy *et al.*, 2004; Fan *et al.*, 2004; Finkelstein and Rock, 2001; Hare *et al.*, 2003; Himmelbach *et al.*, 2003; Kuhn and Schroeder, 2003; Lovegrove and Hooley, 2000; Ritchie *et al.*, 2002; Rock, 2000; Schroeder *et al.*, 2001).

Aleurone cells, suspension cells, protoplasts, and mutants/transgenic plants of several species have been used to address the complicated ABA-signaling networks for guard cell movement and other aspects of stress responses, seed germination, and growth of vegetative tissues. However, it is believed that ABA-signaling networks are conserved among higher plant species; information derived from several plant species has been used to compile a network map of ABA signaling (Finkelstein and Rock, 2001; Himmelbach *et al.*, 2003). Assessment of the universality of ABA-signaling mechanisms is greatly facilitated by the availability of the genome sequences and full-length cDNA sequences of *Arabidopsis* (Seki *et al.*, 2002b; The *Arabidopsis* Initiative, 2000) and rice (Goff *et al.*, 2002; Kikuchi *et al.*, 2003; Yu *et al.*, 2002). In this review, we summarize the advances in ABA-signaling research and report the closest (lowest *E*-value) rice homologues of known ABA-signaling regulators. We identified these homologues by BLAST searching against a comprehensive rice peptide database that contains the sequences downloaded from NCBI (http://www.ncbi.nlm.nih.gov) and TIGR (http://www.tigr.org/) and those deduced from the longest open reading frame (ORF) of rice full-length cDNA sequences (Kikuchi *et al.*, 2003).

II. SIGNALING PATHWAYS

A. RECEPTORS

The site and nature of ABA perception were addressed in barley aleurone cells and guard cells of several plant species. Externally applied but not microinjected, ABA could repress gibberellin (GA)-induced α-amylase expression in aleurone protoplasts, suggesting an extracellular perception of ABA (Gilroy and Jones, 1994). This notion is supported by two studies using ABA-protein conjugates that cannot enter the cell, yet are able to regulate ion channel activity (Jeannette *et al.*, 1999) and gene expression (Jeannette *et al.*, 1999; Schultz and Quatrano, 1997). It is also supported by a study in *Commelina* guard cells (Anderson *et al.*, 1994). In contrast, introduction of ABA into the cytoplasm by microinjection (Schwartz *et al.*, 1994) or a patch-clamp electrode (Allan *et al.*, 1994) triggered or maintained stomatal closure arguing for intracellular perceiving sites. Other approaches taken to identify ABA receptors (Desikan *et al.*, 1999; Leyman *et al.*, 1999, 2000; Sutton *et al.*, 2000; Yamazaki *et al.*, 2003) have resulted in several leads. One promising receptor candidate is ABAP1 (Fig. 2) that is located in membrane fractions of ABA-treated barley aleurone cells. It is capable of specifically yet reversibly binding to ABA at a capacity of 0.8 mol of ABA mol^{-1} protein with a K_d of 2.8×10^{-8} M, and it is present in diverse monocotyledonous and dicotyledonous species (Razem *et al.*, 2004). Another candidate is GCR1, a putative G-protein–coupled receptor identified in *Arabidopsis* (Pandey and Assmann, 2004) that can directly interact with GPA1, the α-subunit of G-proteins. The *Arabidopsis gcr1* knockout mutant is more sensitive to ABA and more tolerant to drought stress due to reduced rates of water loss. These data suggest that GCR1 may function as a negative regulator of ABA signaling (Pandey and Assmann, 2004). The closest rice homologues of ABAP1, GPA1, and GCR1 are shown in Table I.

B. G-PROTEINS

As mentioned previously, heterotrimeric G-proteins are involved in the transduction of ABA signal in *Arabidopsis* (Pandey and Assmann, 2004; Wang *et al.*, 2001). In cereal aleurone cells, the activation of a plasma-membrane–bound ABA-inducible phospholipase D (PLD) is essential for ABA response (Ritchie and Gilroy, 2000). This process is GTP-dependent; addition of GTPγS transiently stimulates PLD in an ABA-independent manner, whereas treatment with GDPβS or pertussis toxin blocks the PLD activation by ABA. These data suggest the involvement of G-protein activity in the ABA response of barley (Ritchie and Gilroy, 2000). Monomeric G-proteins also regulate ABA responses (Lemichez *et al.*, 2001; Yang, 2002). ROP10, a plasma-membrane–associated small GTPase, appears to negatively

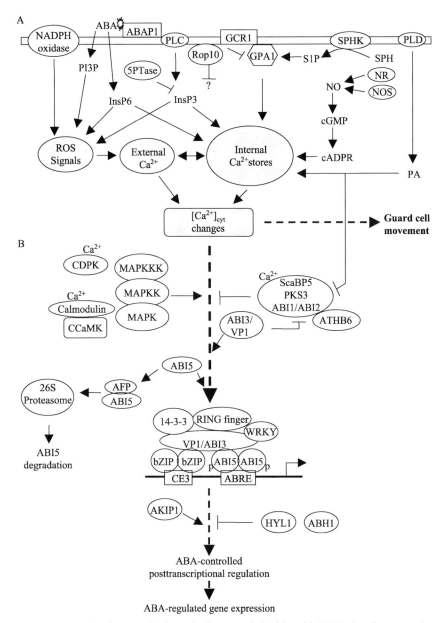

FIGURE 2. An integrated schematic diagram of abscisic acid (ABA)-signaling networks. The model is not comprehensive and does not address tissue specificity. Most relationships in (A) are derived from studies in guard cells while those in (B) are derived from studies in seeds or seedlings. Abscisic acid signaling is perceived by ABAP1 and GCR1 that interacts with GPA1 and functions as a negative regulator. Abscisic acid-induced sphingosine kinase (SPHK) converts sphingosine (SPH) into sphingosine-1-phosphate (S1P), which promotes ABA

regulate ABA responses in seed germination and seedling growth of *Arabidopsis* (Yang, 2002). The recruitment of ROP10 to the plasma membrane requires a functional farnesylation site. However, farnesylation of ROP10 appears to be independent of *ERA1*, that encodes the β-subunit of farnesyl transferase (Cutler *et al.*, 1996) because ROP10 localization is only weakly affected in the *era1* mutant (Zheng *et al.*, 2002). The closest rice homologues of this group of proteins are shown in Table I.

C. SECONDARY MESSENGERS

The primary intracellular messenger of ABA responses is Ca^{2+} which also mediates the signaling of other hormones. However, the specificity of Ca^{2+} signaling is thought to be determined by the magnitude, timing, spatial distribution, and frequency of its change. Abscisic acid activates the vacuolar H^+ATPase (Barkla *et al.*, 1999) and regulates the influx of Ca^{2+} across the plasma membrane through ABA-activated channels (Hamilton *et al.*, 2000; Schroeder and Hagiwara, 1990). In addition, the concentration of Ca^{2+} in cytosol ($[Ca^{2+}]_{cyt}$) is further modulated by other secondary messengers including inositol 1,4,5 triphosphate (InsP3), phosphatidic acid (PA), myo-inositol hexakisphosphate (InsP6), sphingosine-1-phosphate (S1P), hydrogen peroxide (H_2O_2), nitric oxide (NO), cyclic ADP ribose (cADPR), and cyclic guanosine monophosphate (cGMP) (Himmelbach *et al.*, 2003; Leckie *et al.*, 1998; Wu *et al.*, 1997). Abscisic acid enhances the activities of phospholipase C (PLC), PLD, and ADPR cyclase to produce InsP3 (Sanchez and Chua, 2001), PA (Ritchie and Gilroy, 2000), and cADPR (Sanchez *et al.*, 2004), respectively. Overexpression of inositol 5-phosphatase, an enzyme that hydrolyzes InsP3 and inositol 1,3,4,5-tetrakisphosphate, results in

---

responses by mobilizing internal Ca^{2+} stores. In addition, other secondary messengers including cyclic ADP ribose (cADPR), inositol 1,4,5 triphosphate (InsP3), myo-inositol hexakisphosphate (InsP6), phosphatidic acid (PA), and NO, control $[Ca^{2+}]_{cyt}$ by releasing Ca^{2+} from internal storages. Reactive oxygen species (ROS), which is produced by NADPH-oxidase or promoted by secondary messengers (PI3P, InsP6, and InsP3), enhances $[Ca^{2+}]_{cyt}$ by activating Ca^{2+} channels on the plasma membrane. $[Ca^{2+}]_{cyt}$ changes ultimately control guard cell movement and gene expression. Ca^{2+} signaling is negatively regulated by a protein complex including protein phosphatases (ABI1 or ABI2), a protein kinase (CIPK15/PKS3), a Ca^{2+}-binding protein (CBL/ScaBP5), and a homeodomain leucine zipper protein (ATHB6). In turn, ABI1/ABI2 is repressed by both the secondary messenger (PA) and ABI3/VP1. This pathway is positively regulated by a MAP kinase cascade, calcium-dependent protein kinases (CDPKs), and probably Ca^{2+}-calmodulin-dependent protein kinases II (CCaMKs). Abscisic acid-response promoter complex is composed of an ACGT-box and a coupling element. ABI3/VP1 and 14-3-3 increase the binding affinity of ABI5 to ABRE and CE elements, forming a transcriptional complex that likely includes a ring zinc finger and WRKY proteins. ABI5 binding protein (AFP) promotes the degradation of ABI5 by 26S proteasomes. Three RNA-binding proteins, one functioning as a positive regulator (AKIP1) and two as negative regulators (HYL1 and ABH1) modulate ABA signaling at the posttranscriptional level.

TABLE I. Genes Involved in Early ABA Signaling Events and Production of Secondary Messenger

Gene[a]	Accession[b]	Length[c]	Gene product	Mutation	Material	Response[d]	Reference	Rice homologue accession[e]	Length[f]	E-value[g]
HvABAP1	AAF97846	472	ABA-binding protein		Aleurone cells	Induced	Razem et al. (2004)	**AAT72462**	738	0
AtERA1	NP_198844	482	Farnesyl transferase, beta-subunit	era1, wiggum	Seed/ guard cells	Oversensitive	Cutler et al. (1996); Pei et al. (1998)	5147.m00126	478	e−132
AtGCR1	AAN15633	326	G protein-coupled receptors	gcr1-3, gcr1-4	Root/ guard cells	Oversensitive	Pandey and Assmann (2004)	6498.m00149	321	e−120
AtGPA1	AAA32805	383	Heterotrimeric GTP-binding (G) protein	gpa1	Guard cells	Insensitive	Wang et al. (2001)	4351.m00157	390	e−166
At5PTase1	AAD10828	590	Inositol 5-phosphatase	Ecotypic expression	Guard cells	Insensitive	Burnette et al. (2003)	4384.m00157	676	e−152
zAtFRY1	AAC49263	353	Inositol polyphosphate-1-phosphatase	fry1, hos2	Seed/ seedling	Oversensitive	Xiong et al. (2001a)	7124.m00171	441	e−126
AtIP5PII	NP_849402	613	Ins(1,4,5)P$_3$ 5-phosphatase	Ecotypic expression	Seed/ seedling	Insensitive	Sanchez and Chua (2001)	4384.m00157	676	e−124

(Continues)

TABLE 1. (Continued)

Gene[a]	Accession[b]	Length[c]	Gene product	Mutation	Material	Response[d]	Reference	Rice homologue accession[e]	Length[f]	E-value[g]
AtrbohD	NP_199602	921	NADPH oxidase catalytic subunit genes	atrbohD/F	Seed/guard cells/root	Insensitive	Kwak et al. (2003)	7173.m00220	941	0
AtrbohF	NP_564821	944	NADPH oxidase catalytic subunit genes	atrbohD/F	Seed/guard cells/root	Insensitive	Kwak et al. (2003)	NP_916447	943	0
AtNIA1	CAA31786	393	Nitrate reductase	nia1/nia2	Guard cells	Insensitive	Desikan et al. (2002)	4982.m00153	681	e−153
AtNIA2	AAK56261	917	Nitrate reductase	nia1/nia2	Guard cells	Insensitive	Desikan et al. (2002)	BAD09558	916	0
AtNOS1	AAU95423	561	Nitric oxide synthase	Atnos1	Seedling/guard cells	Insensitive	Quo et al. (2003)	2463.m00107	547	0
AtPLC1	BAA07547	561	Phospholipase C1	Antisense inhibition	Seed/seedling	Insensitive	Sanchez and Chua (2001)	AAS90683	598	e−153

Gene	Accession[e]	Length[c]	Description	Mutation/expression[d]	Tissue	Phenotype	Reference	Rice accession[b]	Rice length[f]	E-value[g]
AtPLDa1	Q38882	810	Phospholipase D	PLDalpha1	Guard cells	Insensitive	Zhang et al. (2004a)	**BAA11136**	812	0
AtROP2	Q38919	195	Rho-type small GTPase	Dominant negative	Seed	Oversensitive	Li et al. (2001)	**AAF28764**	197	e−101
AtROP10	Q9SU67	208	Rop subfamily of Rho GTPases	*rop10*	Seed/root	Oversensitive	Zheng et al. (2002)	3857.m00131	215	e−100
AtSYP61	AAK40222	245	SNARE superfamily of proteins. SNAP receptor	*osm1*	Guard cells	Insensitive	Zhu et al. (2002)	**NP_914267**	270	5.00 e−74
AtSphK	BAB07787	763	Sphingosine kinase	SphKs inhibitor	Guard cells	Insensitive	Coursol et al. (2003)	903.m00132	757	0

Peptide sequences of the genes listed in the first column were used to search against the comprehensive rice peptide database for rice homologues.

[a] Gene names with the abbreviated names of the species: At, *Arabidopsis thaliana*; Cp, *Craterostigma plantagineum*; Hv, *Hordeum vulgare*; Lt, *Larrea tridentata*; Os, *Oryza sativa*; Pv, *Phaseolus vulgaris*; Ta, *Triticum aestivum*; Vf, *Vicia faba*; Zm, *Zea mays*.
[b] Genbank accession numbers for the peptide sequences.
[c] Lengths of the peptide sequences.
[d] Phenotypes of the mutations or expression of the reporter genes driven by the ABA-responsive promoters.
[e] Accession numbers.
[f] Lengths of the homologous rice peptide sequences.
[g] E-values of the blast analyses. Accession numbers in bold represent sequences from NCBI; those in regular font represent sequences from TIGR; and those in italic represent sequences from translated full-length cDNAs.

hyposensitivity of guard cells to ABA (Burnette et al., 2003). InsP6 promotes the releases of Ca^{2+} from endomembrane compartments such as the vacuole (Lemtiri-Chlieh et al., 2003). Sphingosine-1-phosphate, which is converted from the long-chain amine alcohol (sphingosine) by ABA-induced activation of sphingosine kinase (Coursol et al., 2003), acts at trimeric G-protein GPA1 (Coursol et al., 2003) and its receptor GCR1 (Pandey and Assmann, 2004) to mobilize calcium (Ng et al., 2001). Reactive oxygen species (ROS), such as H_2O_2 produced by a membrane-bound NADPH-oxidase (Kwak et al., 2003), and NO resulting from the activities of nitrate reductase (Desikan et al., 2002) and a glycine decarboxylase complex (Chandok et al., 2003) also serve as secondary messengers in ABA signaling. Mutations in the *Arabidopsis* nitrate reductase apoprotein genes, *NIA1* and *NIA2* (Desikan et al., 2002), or the NO synthase gene, *AtNOS1* (Guo et al., 2003), diminish NO synthesis and impair stomatal closure in response to ABA, although stomatal opening is still inhibited by ABA. Cyclic ADP ribose and cGMP are required for the induction of ABA response by NO, suggesting that NO acts upstream of these two secondary messengers (Desikan et al., 2002). It was shown that a new inositol phosphate, phosphatidylinositol 3-phophate (PI3P), might act upstream of ROS in ABA signaling because treatments with phosphatidylinositol 3-kinase inhibitors impair ABA-induced stomatal closure in *Vicia faba* (Jung et al., 2002), and inhibition can be partially rescued by applying H_2O_2 (Park et al., 2003). These messengers control $[Ca^{2+}]_{cyt}$ by releasing Ca^{2+} from the internal storage sites (such as vacuoles and the ER), producing Ca^{2+} oscillations (Allen et al., 2001) that serve as a primary regulator of ABA signaling to control the movement of guard cells for the closing and opening of stomata (Fan et al., 2004).

The calcium oscillations regulated by these secondary messengers also control ABA-regulated gene expressions in other cell types (Chen et al., 1997; Sheen, 1996; Wu et al., 1997). Indeed, inactivation of an inositol polyphosphate 1-phosphatase that is capable of dephosphorylating InsP3 results in oversensitivity to ABA in seed germination and postembryonic development (Xiong et al., 2001b). Double mutation of the NADPH-oxidase catalytic subunit genes *AtrbohD* and *AtrbohF*, impairs ABA-induced ROS production and increases in $[Ca^{2+}]_{cyt}$, thereby interfering with ABA-induced stomatal closing and ABA-inhibition of seed germination and root elongation (Kwak et al., 2003). The closest rice homologues of the enzymes producing these secondary messengers are shown in Table I.

D. PHOSPHATASES AND KINASES

Mutation studies suggest that several *Arabidopsis* protein phosphatases 2C, such as ABI1 and ABI2, function as negative regulators of ABA signaling (Himmelbach et al., 2003; Ianzano et al., 2004; Leonhardt et al., 2004b; Merlot et al., 2001). Electrophysiological studies indicate that *abi1-1* and *abi2-1* mutations disrupt ABA activation of calcium channels (Murata *et al.*, 2001) and reduce ABA-induced cytosolic calcium increases in guard cells (Allen *et al.*, 1999), suggesting these two phosphatases act upstream of $[Ca^{2+}]_{cyt}$. However, other studies suggest they act downstream of cADPR (Sanchez *et al.*, 2004; Wu *et al.*, 2003) and NO (Desikan *et al.*, 2002). The activities of protein phosphatases are modulated by secondary messengers (PA and Ca^{2+}) and protein kinases. Phosphatidic acid binds to and inhibits ABI1 activity (Zhang *et al.*, 2004a). ABI2 and ABI1 physically interact with PKS3 (or its homologue CIPK3), a Ser/Thr protein kinase. This kinase is also associated with the calcineurin B-like Ca^{2+} binding protein, SCaBP5 (or its homologue CBL), forming a complex that negatively controls ABA sensitivity (Guo *et al.*, 2002; Kim *et al.*, 2003). Another calcium sensor (CBL9) functions as a negative regulator of ABA signaling and biosynthesis (Pandey *et al.*, 2004). In contrast, the protein phosphatase 2A encoded by *RCN1* functions as a positive regulator of ABA signaling (Kwak *et al.*, 2002).

Protein kinases also can function as positive regulators of ABA signaling. Calcium-dependent protein kinases (CDPKs) contain a protein kinase domain and a carboxyl-terminal calmodulin-like structure that directly binds calcium (Cheng *et al.*, 2002). Two *Arabidopsis* CDPKs (AtCPK10 and AtCPK30) activate an ABA-inducible barley promoter in the absence of the hormone (Cheng *et al.*, 2002). Abscisic acid and H_2O_2 activate the *Arabidopsis* mitogen-activated protein kinase, ANP1, which initiates a phosphorylation cascade involving two mitogen-activated protein kinases (MAPK), AtMPK3, and AtMPK6 (Kovtun *et al.*, 2000). Overexpression of *AtMAPK3* increases ABA sensitivity while inhibition of MAPK activity by inhibitor PD98059 decreases ABA sensitivity (Lu *et al.*, 2002). Sucrose nonfermenting1-related protein kinases function as activators of ABA signaling in rice (Kobayashi *et al.*, 2004) and wheat (Johnson *et al.*, 2002). There are several other protein kinase genes whose expressions are induced or whose protein activities are activated by ABA. However, it is unknown how they are involved in ABA responses (Finkelstein and Rock, 2001). The closest homologues of these regulators in rice are shown in Table II.

E. TRANSCRIPTIONAL REGULATION

1. *Cis*-Acting Elements

The ABA-response elements include those with an ACGT-core (G-box/ABRE, /ACGT-box), a CGT-core (CE3-like) or a GCC-core (Motif I-like), Sph/RY sequences (CATGCA(TG)), DRE (CCGA(C/G)), MYC and MYB binding sites (ACACGCATGTG and YAAC(G/T)G, respectively), and coupling elements. Most of these elements are defined in transient expression systems, including protoplasts, suspension cells, and aleurone cells (Rock, 2000; Shen and Ho, 1997). The cereal aleurone layers are composed of

TABLE II. Kinases and Phosphatases Involved in ABA Signaling

Gene[a]	Accession[b]	Length[c]	Gene product	Mutation	Material	Response[d]	Reference	Rice homologue accession[e]	Length[f]	E-value[g]
AtCDPK1	BAA04829	493	Calcium-dependent protein kinases	Ecotypic expression	Leaf protoplast	Upregulation	Sheen (1996)	**AAT81734**	574	0
AtCDPK1a	EAA19816	251	Calcium-dependent protein kinases	Ecotypic expression	Leaf protoplast	Upregulation	Sheen (1996)	8220.m00104	328	4.00 e−33
AtPKS3	AAK26842	421	Protein kinase	RNA interference	Seed/seedling/guard cells	Oversensitive	Quo et al. (2002)	4351.m00164	456	e−151
AtCIPK3	NP_850095	382	Ser/Thr protein kinase	cipk3	Seed	Oversensitive	Kim et al. (2003)	6635.m00186	454	e−173
HvPKABA1	BAB61736	342	Ser/Thr protein kinase	Over-expression	Aleurone cells	Suppression	Gómez-Cadenas et al. (1999)	8364.m00150	342	e−179
AtSCaBP5	AAC26008	213	Ca(2+) binding protein	RNA interference	Seed/seedling/guard cells	Oversensitive	Quo et al. (2003)	6848.m00127	225	2.00 e−74

Gene	Accession[b]	Length[c]	Protein family	Mutant allele[d]	Tissue	Phenotype	Reference	Rice homologue[f]	Length	E-value[g]
AtCBL9	AAL10301	213	Calcineurin B-like proteins	*cbl9*	Seed/seedling	Oversensitive	Pandey et al. (2004)	6848.m00127	225	5.00e−74
AtRCN1	AAC49255	588	Protein phosphatases 2A	*rcn1*	Seed/guard cells	Insensitive	Kwak et al. (2002)	5660.m00128	587	0
AtABI1	NP_194338	434	Protein phosphatase 2C	*abi1-1*	Seedling/guard cells	Insensitive	Leung et al. (1994); Wu et al. (2003)	**XP_463364**	467	e−109
AtABI2	O04719	423	Protein phosphatase 2C	*abi2-1*	Guard cells	Insensitive	Finkelstein (1993)	**XP_463364**	467	e−112
AtP2C-HA	AAG51849	511	Protein phosphatase 2C	*atp2c-ha*	Seed/guard cells	Oversensitive	Leonhardt et al. (2004a,b)	**XP_463364**	467	e−115
AtPP2CA	BAA07287	399	Protein phosphatases 2C	Antisense inhibition	Seed/seedling	Oversensitive	Tahtiharju and Palva (2001)	3955.m00148	416	3.00e−97

Peptide sequences of the genes listed in the first column were used to search against the comprehensive rice peptide database for rice homologues.
[a] Gene names with the abbreviated names of the species: *At*, *Arabidopsis thaliana*; *Cp*, *Craterostigma plantagineum*; *Hv*, *Hordeum vulgare*; *Lt*, *Larrea tridentata*; *Os*, *Oryza sativa*; *Pv*, *Phaseolus vulgaris*; *Ta*, *Triticum aestivum*; *Vf*, *Vicia faba*; *Zm*, *Zea mays*.
[b] Genbank accession numbers for the peptide sequences.
[c] Lengths of the peptide sequences.
[d] Phenotypes of the mutations or expression of the reporter genes driven by the ABA-responsive promoters.
[e] Accession numbers.
[f] Lengths of the homologous rice peptide sequences.
[g] E-values of the blast analyses. Accession numbers in bold represent sequences from NCBI; those in regular font represent sequences from TIGR; and those in italic represent sequences from translated full-length cDNAs.

uniform, synchronized, and highly differentiated cells that can be easily prepared in large quantity within a short period of time (Bethke *et al.*, 1997). None of the *cis*-acting elements described earlier can function alone (Hobo *et al.*, 1999a; Narusaka *et al.*, 2003). Instead, they form ABA-response promoter complexes called ABRC (Shen and Ho, 1995, 1997, 1998; Shen *et al.*, 1996, 2001, 2004). For two barley genes, each ABRC consists of an ACGT core containing element (ACGT-box) and a coupling element (CE1 or CE3), forming two different ABRCs called ABRC1 and ABRC3 (Fig. 2). These two promoter complexes are different in the sequences of the coupling elements, the orientation constraints of the coupling elements, and the distances between an ACGT-box and a CE (Shen *et al.*, 2004). Extensive deletion and point mutation analyses suggest that the ACGT element requires the sequence 5'-ACGTGGC-3' and the elements CE1 and CE3 require the sequences CCACC and GCGTGTC, respectively. It is suggested that the ACGT-box and CE3 are functionally equivalent because the OsTRAB1/ABI5 binds to both the ACGT-box and CE3 element *in vitro* (Hobo *et al.*, 1999a). However, data indicate that the coupling between an ACGT-box and a CE, or between two ACGT-boxes is essential for a high level of ABA induction; two copies of CE3 are much less active (Shen *et al.*, 2004). Furthermore, a partially purified nuclear extract from barley embryos has specific binding activity for the ACGT-box present in ABRC3. It recognizes the wild-type version of the ABRC3 and two copies of the ACGT-box but possesses low affinity for two copies of the coupling element CE3, suggesting that it is likely a bZIP protein that is different from ABI5 binds to the CE3 element *in vivo* (Casaretto and Ho, 2003; Shen *et al.*, 2004). An ACGT-box can form other types of ABRCs by coupling with elements such as DRE (Narusaka *et al.*, 2003).

2. *Trans*-Acting Factors

Several transcription factors have been well documented to mediate ABA signaling. The ABI5-type bZIP proteins from *Arabidopsis*, sunflower, wheat, barley, and rice bind as dimers to the ACGT-box or CE3 to activate the promoters (Finkelstein and Rock, 2001). ABI5 is upregulated by ABA through an increase in the transcript level as well as the stability of the protein. AP2-type proteins from maize and barley, ZmABI4, HvDRF1, ZmDBF1, ZmDBF2, DREB1s/CBFs, and DREB2s (Narusaka *et al.*, 2003), interact with CE1 or its related C-rich motifs including DRE (Himmelbach *et al.*, 2003; Xue and Loveridge, 2004). AtMYC2 and AtMYB2 bind to MYC and MYB recognition sites, respectively, and function as activators of ABA signaling (Abe *et al.*, 2003). Abscisic acid-inducible NAC activator proteins were also found to interact with the MYC site (Fujita *et al.*, 2004; Tran *et al.*, 2004).

The activities of ABI5 and its orthologues/homologues (Table III) are modified by some kinases (Johnson *et al.*, 2002; Kagaya *et al.*, 2002; Lu *et al.*, 2002), VP1 (Casaretto and Ho, 2003; Suzuki *et al.*, 2003), FUS3, and LEC1 (Finkelstein and Rock, 2001). Phosphorylation in the nucleus of preexisting AtABI5/OsTRAB1/TaABF is found to be the nearly terminal event of ABA response (Johnson *et al.*, 2002; Kagaya *et al.*, 2002; Lopez-Molina *et al.*, 2001). VP1 has a coactivation/repression domain at the N-terminus and three basic domains (B1, B2, and B3) at the C-terminus. The N-terminal domain is necessary for activating the ABA pathway and repressing the gibberellin (GA) pathway (Suzuki *et al.*, 2003). The C-terminal B3 domain is shown to bind specifically to the Sph1/RY element, although the full-length VP1 does not bind to DNA (Suzuki *et al.*, 1997). The B1 and B2 domains are likely to be involved in nuclear localization and interaction with ABI5, WRKY, 14-3-3, ring (C3HC3-type) zinc finger proteins, and RNA polymerase II subunit RPB5 to potentiate ABA-inducible gene expression (Hobo *et al.*, 1999b; Jones *et al.*, 2000; Kurup *et al.*, 2000; Nakamura *et al.*, 2001; Schultz *et al.*, 1998; Zou *et al.*, 2004).

The activity/assembly of the transcription complex for ABA signaling appears to be modulated by at least four classes of transcriptional repressors. The first class of repressors are bZIP proteins that negatively regulate ABA-induced gene expression by sequestering bZIP activators or competing with bZIP activators for binding to the ACGT-box. For example, two rice bZIP proteins (OsZIP-2a and OsZIP-2b) do not bind to ABRE by themselves. However, they heterodimerize via the leucine zipper with EmBP-1 (Table IV) and prevent it from binding to the ACGT-box (Nantel and Quatrano, 1996). In contrast, ROM2 binds to the ACGT-box but functions as a repressor (Chern *et al.*, 1996). The second class of repressors are protein phosphatases. In addition to ABI1 and ABI2 protein phosphatases 2C described earlier, the C-terminal domain phosphatase-like protein, AtCPL3, also functions as a repressor of ABA signaling. AtCPL3 specifically downregulates ABA-responsive gene expression possibly by contacting and dephosphorylating the carboxyl-terminal domain (CTD) of the RNA polymerase II, thereby blocking transcription initiation (Koiwa *et al.*, 2002). The third class of repressors are homeodomain proteins that bind to the *cis*-acting element, CAATTATTA; ATHB6 physically interacts with ABI1 and acts downstream of ABI-1 in mediating ABA signaling (Himmelbach *et al.*, 2002). The fourth class of repressors are WRKY proteins; of the 77 published *OsWRKY* genes (Zhang *et al.*, 2004b) at least two function as repressors of ABA signaling in aleurone cells (Z. Xie and Q. Shen, unpublished data).

Inactivation of repressors hence is essential for ABA signaling. Indeed, prior to becoming part of the transcription complex, VP1/ABI3 appears to play two additional roles: to inactivate ABI1 and ABI2 protein phosphatases (Suzuki *et al.*, 2003) and to modify the chromatin structure (Li *et al.*, 1999). Two other B3 proteins, LEC2 and FUS3, might participate in the chromatin-remodeling process (Luerssen *et al.*, 1998; Stone *et al.*, 2001). In addition,

TABLE III. Transcriptional Regulators Involved in ABA Signaling

Gene[a]	Accession[b]	Length[c]	Gene product	Mutation	Material	Response[d]	Binding site	Reference	Rice homologue accession[e]	Length[f]	E-value[g]
AtABI4	AAF18736	328	AP2 domain protein	abi4, gin6, isi3, san5, sis5, sun6	Seed/seedling	Insensitive		Finkelstein et al. (1998)	3353.m00161	318	2.00e−23
ZmABI4	AAM95247	248	AP2 domain protein	Over expression in abi4	Seed	Insensitive	CE1	Niu et al. (2002)	3353.m00161	318	1.00e−23
AtABI3	NP_189108	720	B3 domain protein	abi3	Seed	Insensitive	ABRE, RY/G-box	Ezcurra et al. (2000)	**BAA04066**	728	3.00e−95
AtFUS3	AAC35246	310	B3 domain protein	ML1::FUS3-GFP	Seed	Oversensitive		Gazzarrini et al. (2004)	2880.m00122	728	2.00e−31
OsVP1	BAA04066	728	B3 domain protein	Over expression	Protoplast	Upregulation		Hattori et al. (1995)			
HvVP1	AAO06117	394	B3 domain protein	Over expression	Aleurone cells	Upregulation	RY/Sph	Casaretto and Ho (2003)	2880.m00122	728	e−133
PvALF	T10864	700	B3 domain protein	Over expression	Cotyledon	Upregulation	RY/Sph	Bobb et al. (1997)	2880.m00122	728	2.00e−84
ZmVP1	CAA04889	449	B3 domain protein	vp1	Seed	Insensitive	ABRE	McCarty et al. (1991); Schultz et al. (1998)	2880.m00121	704	e−138

AtMYC2	Q39204	623	bHLH	atmyc2	Seed	Insensitive	MYC site	Abe et al. (2003)	749.m00139	688	e−146
AtIMB1	AAO22056	386	Bromodomain proteins	imb1	Seedling	Oversensitive		Duque and Chua (2003)	3028.m00190	344	5.00e−87
AtABF3	BAD43614	454	bZIP protein	Over expression	Seed/seedling	Oversensitive	ABRE	Choi et al. (2000); Kang et al. (2002)	**BAD17130**	357	1.00e−64
AtABF4	AAF27182	431	bZIP protein	Over expression	Seed/seedling	Oversensitive	ABRE	Choi et al. (2000); Kang et al. (2002)	3486.m00122	644	1.00e−50
AtABI5	AAD21438	442	bZip protein	abi5	Seed	Insensitive	ABRE	Carles et al. (2002); Finkelstein and Lynch (2000)	5149.m00135	388	4.00e−61
HvABI5	AAO06115	353	bZip protein	Over expression	Aleurone cells	Upregulation	ABRE, CE3	Casaretto and Ho (2003)	**BAD38293**	376	e−142
OsTRAB1	XP_482899	318	bZip protein	Over expression	Protoplast	Upregulation	ABRE	Hobo et al. (1999a)			

(*Continues*)

TABLE III. (Continued)

Gene[a]	Accession[b]	Length[c]	Gene product	Mutation	Material	Response[d]	Binding site	Reference	Rice homologue accession[e]	Length[f]	E-value[g]
OsZip-1a	AAC49556	390	bZip protein	Over expression	Protoplast	Heterodimerized, upregulation	ABRE	Nantel and Quatrano (1996)			
OsZip-2a	AAC49557	124	bZip protein	Over expression	Protoplast		ABRE	Nantel and Quatrano (1996)			
PvROM2	T10985	424	bZip protein	Over expression	Cotyledon	Downregulation	ABRE	Chern et al. (1996)	AK065440	608	4.00e−60
TaEmBP-1	P25032	354	bZip protein	Over expression	Protoplast	Upregulation	ABRE	Guiltinan et al. (1990); Hill et al. (1996)	BAC83673	423	3.00e−73
AtCPL1	NP_193898	995	C-terminal domain phosphatase-like	cpl1, fry2	Seed/seedling	Insensitive(seed)/oversensitive (seedling)		Koiwa et al. (2002); Xiong et al. (2002)	BAD25346	940	0
AtCPL3	NP_180912	1190	C-terminal domain phosphatase-like	cpl3	Seedling	Upregulation		Koiwa et al. (2002)	7174.m00156	1272	0

Gene	Accession	Length	Protein	Homolog	Expression	Tissue	ABA response	Cis-element	Reference	Rice homolog accession	Rice length	E-value
AtLEC1	NP_173616	238	HAP3 subunit of CCAAT protein	lec1		Seed	Insensitive	CCAAT	Brocard-Gifford et al. (2003)	2534.m00206	254	4.00e−43
AtHB5	P46667	312	Homeodomain protein		Ecotypic expression	Seed/seedling	Oversensitive	AT-rich	Johannesson et al. (2001, 2003)	7458.m00131	269	4.00e−40
AtHB6	AAD41726	291	Homeodomain protein		Over expression	Seed/guard cells	Insensitive	AT-rich	Himmelbach et al. (2002)	**BAD22271**	277	1.00e−39
AtMYB2	BAA03534	273	MYB protein		Overexpression	Seed	Oversensitive	MYB site	Abe et al. (2003)	8346.m00103	329	4.00e−62
AtRD26	NP_849457	164	NAC protein		Over expression	Seedling	Oversensitive		Fujita et al. (2004)	2885.m00175	303	1.00e−05
LtWRKY21	AY792618	314	WRKY protein		Over expression	Seed	Upregulation		Zou et al. (2004)	**DAA05093**	374	2.00e−52
AtMARD1	AAK92226	263	Zinc-finger protein	mard1		Seed	Insensitive		He and Gan (2004)	*AK066202*	429	3.00e−29

[a] Gene names with the abbreviated names of the species: At, *Arabidopsis thaliana*; Cp, *Craterostigma plantagineum*; Hv, *Hordeum vulgare*; Lt, *Larrea tridentata*; Os, *Oryza sativa*; Pv, *Phaseolus vulgaris*; Ta, *Triticum aestivum*; Vf, *Vicia faba*; Zm, *Zea mays*.
[b] Genbank accession numbers for the peptide sequences.
[c] Lengths of the peptide sequences.
[d] Phenotypes of the mutations or expression of the reporter genes driven by the ABA-responsive promoters.
[e] Accession numbers.
[f] Lengths of homologous rice peptide sequences.
[g] E-values of the BLAST analyses. Accession numbers in bold represent sequences from NCBI; those in regular font represent sequences from TIGR; and those in italic represent sequences from translated full-length cDNAs.

TABLE IV. Genes Modulating ABA Responses at Posttranscriptional Regulation

Gene[a]	Accession[b]	Length[c]	Gene product	Mutation	Material	Response[d]	Reference	Rice homologue accession[e]	Length[f]	E-value[g]
AtAFP	AAF67775	335	ABI five binding protein	afp-1, afp-2	Seed/seedling	Oversensitive	Lopez-Molina et al. (2003)	6877.m00115	307	9.00e−41
AtHYL1	AAG49890	419	Double-stranded RNA binding protein	hyl1	Seed/root	Oversensitive	Lu and Fedoroff (2000)	AK103543	577	5.00e−58
AtABH1	NP_565356	848	mRNA CAP binding protein	abh1	Seed/guard cells	Oversensitive	Hugouvieux et al. (2001)	**AAG54079**	910	0
AtSAD1	AAK61592	88	U6-related Sm-like small ribonucleoprotein	sad1	Seed/root	Oversensitive	Xiong et al. (2001b)	AK059190	218	2.00e−40

Gene	Accession	Length	Protein type	Mutant	Tissue	ABA response	Reference	Rice homolog	Rice length	E-value
AtPRL1	NP_193325	486	WD40 domain protein	*prl1*	Seedling	Oversensitive	Nemeth et al. (1998)	6133.m00106	472	0
AtOST1	CAC87047	362	Ser/Thr protein kinase	*ost1-1, ost1-2, srk2e*	Guard cells	Insensitive	Mustilli et al. (2002); Yoshida et al. (2002)	8370.m00178	362	e−176
VfAAPK	AAF27340	349	Ser/Thr protein kinase	Dominant negative	Guard cells	Insensitive	Li et al. (2002)	8370.m00178	362	e−161
VfAKIP1	AAM73765	515	RNA-binding protein		Guard cells		Li et al. (2002)	2729.m00136	490	e−100

[a] Gene names with the abbreviated names of the species: *At*, *Arabidopsis thaliana*; *Cp*, *Craterostigma plantagineum*; *Hv*, *Hordeum vulgare*; *Lt*, *Larrea tridentata*; *Os*, *Oryza sativa*; *Pv*, *Phaseolus vulgaris*; *Ta*, *Triticum aestivum*; *Vf*, *Vicia faba*; *Zm*, *Zea mays*.
[b] Genbank accession numbers for the peptide sequences.
[c] Lengths of the peptide sequences.
[d] Phenotypes of the mutations or expression of the reporter genes driven by the ABA-responsive promoters.
[e] Accession numbers.
[f] Lengths of homologous rice peptide sequences.
[g] E-values of the BLAST analyses. Accession numbers in bold represent sequences from NCBI; those in regular font represent sequences from TIGR; and those in italic represent sequences from translated full-length cDNAs.

LEC1 encodes a transcription factor homologous to CCAAT box-binding factor HAP3 subunit (Lotan *et al.*, 1998). Transcription of *LEC1*, *LEC2*, and *FUS3* genes is repressed by *PKL* that encodes a CHD3-chromatin-remodeling factor. Hence, ABI3, FUS3, LEC2, and PKL might work together to control the remodeling of chromatin structure prior to the binding of transcriptional activators such as ABI4 and ABI5 to promoters. Although *ABI3*, *LEC1*, and *FUS3* all interact with *ABI4* and *ABI5* genetically (Brocard-Gifford *et al.*, 2003), only ABI3/VP1 has been shown to directly interact with ABI5/OsTRAB1 (Hobo *et al.*, 1999b; Nakamura *et al.*, 2001). Furthermore, only *VP1* and *LEC1*, but not *LEC2* and *FUS3*, have been implicated in ABA signaling.

Although VP1 binds to the Sph/RY element (Suzuki *et al.*, 1997) to activate the *C1* promoter in the absence of ABA (Kao *et al.*, 1996), VP1 also can enhance the transcription of the ABRC-containing promoters that lacks an Sph/RY element. This has been well demonstrated by over-expression studies in barley aleurone cells (Shen *et al.*, 1996) and rice protoplasts (Gampala *et al.*, 2002; Hobo *et al.*, 1999b) and by double-stranded RNA interference experiments in barley aleurone cells (Casaretto and Ho, 2003). As many as 70 VP1-dependent ABA-activated genes have been found in a transcriptional profiling study with transgenic *Arabidopsis* carrying 35S promoter::VP1 in an *abi3* null mutant background (Suzuki *et al.*, 2003). However, VP1 does not always function as an agonist of ABA responses. In fact, 49 *Arabidopsis* ABA-inducible genes are repressed by VP1 and nine ABA-repressed genes are enhanced by VP1 (Suzuki *et al.*, 2003). The closest homologues of these regulators in rice are shown in Table III.

F. POSTTRANSCRIPTIONAL REGULATION

Abscisic acid regulation is also exerted at the posttranscriptional level (Fig. 2, Table IV). Abscisic acid induces the expression of several RNA-binding proteins, including: (1) the maize glycine-rich protein MA16 that preferentially interacts with uridine-rich and guanosine-rich RNA fragments (Freire and Pages, 1995); (2) AtABH1 and AtCBP20 that form a dimeric *Arabidopsis* mRNA cap-binding complex (Hugouvieux *et al.*, 2002); (3) AtSAD1 that is similar to multifunctional Sm-like small nuclear ribonucleoproteins; and (4) the dsRNA-binding protein HYL1, mutations in which lead to enhanced levels of ABI5 and MAPK (Lu and Fedoroff, 2000). Except for MA16, whose function remains unknown, these RNA-binding proteins function as negative regulators of ABA signaling. Another RNA-binding protein, AKIP1, is a substrate of the protein kinase AAPK. Phosphorylated AKIP1 interacts with the mRNA that encodes a dehydrin, a protein implicated in cell protection under stress conditions (Li *et al.*, 2002).

It is unknown how these RNA-binding proteins function in regulating ABA responses. However, homologues of ABH1, SAD1, and AKIP1 have been reported to be components of RNA spliceosomes and exporting machinery. In addition, ABA enhances the partitioning of AKIP1 and HYL1 into subnuclear foci that are reminiscent of nuclear speckles (Han *et al.*, 2004; Li *et al.*, 2002). Finally, the levels of several miRNAs are reduced in the *hyl1* ABA hypersensitive mutant, suggesting that HYL1 protein is part of a nuclear macromolecular complex that is involved in miRNA-mediated gene regulation (Han *et al.*, 2004).

Protein degradation is also part of ABA signaling (Hare *et al.*, 2003). A nuclear-localized ABA-regulated protein AFP that physically interacts with ABI5 as shown by a yeast two-hybrid assay and co-immunoprecipitation, functions as a negative regulator of ABA signaling (Lopez-Molina *et al.*, 2003). Proteasome inhibitor studies show that ABI5 stability is regulated by ABA through ubiquitin-related events. Both AFP and ABI5 are co-localized in nuclear bodies that also contain COP1, a RING-finger–containing protein and WD40-repeat–containing protein that functions as a key repressor of seedling de-etiolation (Ang *et al.*, 1998). COP1 possesses autoubiquitination activity (E3) *in vitro* and can ubiquitinate, hence can likely promote the degradation of MYB-type transcription factors (Seo *et al.*, 2003). Although COP1 has not been shown to mediate ABA signaling, the mutation of another WD-40 protein, PRL1 (Table IV), results in oversensitivity to ABA (Nemeth *et al.*, 1998), suggesting that PRL1 is a repressor of ABA signaling. Phosphorylation of the ABI5 stabilizes the protein probably by blocking its AFP-promoted degradation by the 26S proteasome (Lopez-Molina *et al.*, 2003). These data suggest that AFP and PRL1 modulate ABA signaling by promoting degradation of transcriptional activators.

Data suggest that some *cis*-acting elements can be bound by both repressors and activators. Removing the repressors by a hormone-promoted and 26S-proteasome–mediated process facilitates binding of activators to the *cis*-acting elements, thereby enhancing transcription (Zhang *et al.*, 2004b). It remains to be determined whether AFP is also involved in the degradation of repressors of ABA signaling.

III. CROSS-TALK OF ABA AND GA

Increasing evidence suggests the connections of ABA, ethylene, sugar, and auxin synthesis and signaling (Fedoroff, 2002). However, the best-known interaction is the ABA and GA cross-talk in controlling seed germination. Abscisic acid downregulates many genes, especially those upregulated by GA. This effect is so drastic that ABA completely blocks GA-induced seed germination (Lovegrove and Hooley, 2000). In cereal aleurone tissue, GA induces and ABA suppresses the expression of α-amylases that are essential for the utilization of starch stored in the endosperm. The cross-talk of GA and ABA signaling is mediated by secondary messengers. For example, application of PA to barley aleurone inhibits α-amylase production and induces an ABA-inducible amylase inhibitor and RAB (response to ABA) protein expression, mimicking the effect of ABA (Ritchie and Gilroy, 1998). The ABA inhibition also involves kinases. For example, although the ABA-induced protein kinase, PKABA1, has little activity on regulating the expression of ABA-inducible *HVA1* and *HVA22* genes, it almost completely suppresses the GA-induced expression of α-amylase and protease genes (Gómez-Cadenas *et al.*, 1999, 2001; Zentella *et al.*, 2002; Zhang *et al.*, 2004b). Because GA induction and ABA suppression of the α-amylase gene expression in barley aleurone cells appear to be dependent on the same set of *cis*-acting elements in the amylase promoter (Lanahan *et al.*, 1992), an intriguing question is at which site the ABA suppression on the GA-signaling pathway is exerted. Data indicate that PKABA1 acts upstream from the formation of functional GAMyb (a transcriptional activator of GA signaling) but downstream from the site of action of the Slender (a negative regulator of GA signaling) (Gómez-Cadenas *et al.*, 2001). However, there are more pathways mediating the suppression of GA signaling by PKABA1 because *PKABA1* RNA interference does not hamper the inhibitory effect of ABA on the expression of α-amylase (Zentella *et al.*, 2002). Indeed, two ABA-inducible OsWRKY proteins (Z. Xie and Q. Shen, unpublished data) also block GA signaling. Whether they represent components of the PKABA1-independent ABA-suppression pathway remains to be studied.

IV. CONCLUSIONS

Our understanding of ABA signaling has been dramatically improved in the past years with the studies in several dicotyledonous and monocotyledonous plants. Orthologues of a dozen reported ABA-signaling regulators have been found in these two great classes of angiosperms. In addition, for the 53 regulators that are reported only in dicotyledonous plants (mainly *Arabidopsis*), we have found their homologues in rice although the homology for 10 (19%) of these genes is quite low, with the *E*-values higher than e−50 (Tables I–IV). These data suggest that ABA-signaling networks might be highly conserved in several dicotyledonous and monocotyledonous plants. However, we should be cautious in reaching such a conclusion because of the following reasons: (1) the 10 ABA-signaling genes that share a low homology with those in rice, might be unique to dicotyledonous plants. (2) Conserved protein sequences do not necessarily mean conserved functions; experiments need to be carried out to study whether the proteins listed in the tables are truly the orthologues of the ABA-signaling regulators. (3) Even if they are indeed orthologues, their expression patterns upon ABA expression might be completely different, even reversed, as reported from the study of key regulator genes controlling photoperiodism in *Arabidopsis* (a long-day plant) and rice (a short-day plant) (Hayama et al., 2003). This question can be addressed in future by comparing the transcriptional profiling data of *Arabidopsis* (Duque and Chua, 2003; Leonhardt et al., 2004b; Seki et al., 2002a) and rice (Rabbani et al., 2003; Yazaki et al., 2004). (4) The signaling network might be regulated differently in different tissues. Example are certain mutations that only affect one or two aspects of ABA-regulated processes (Tables I–IV); ABA has opposite effects on $[Ca^{2+}]_{cyt}$ in the aleurone and guard cell (Ritchie et al., 2002); and cGMP treatments elucidate ABA responses in guard cells but not aleurone cells (Penson et al., 1996).

Now the challenge is to address the functions of the rapidly growing number of ABA-regulated genes and analyze their promoters experimentally after bioinformatics studies. Transgenic plants (over-expression and RNAi) and mutants (chemically-induced, T-DNA-induced, and transposon-induced) will continue to play important roles in helping to address gene functions. However, transient expression systems, especially the naturally synchronized aleurone cells, will remain extremely valuable for the dissection of promoter structures, definition of protein motifs, and determination of gene interactions.

ACKNOWLEDGMENTS

The authors are grateful to Drs. Andrew Andres, Frank van Breukelen, Jose Casaretto, Deborah Hoshizaki, and Chris Ross for helpful comments on this manuscript.

REFERENCES

Abe, H., Urao, T., Ito, T., Seki, M., Shinozaki, K., and Yamaguchi-Shinozaki, K. (2003). *Arabidopsis* AtMYC2 (bHLH) and AtMYB2 (MYB) function as transcriptional activators in abscisic acid signaling. *Plant Cell* **15**, 63–78.

Allan, A. C., Fricker, M. D., Ward, J. L., Beale, M. H., and Trewavas, A. J. (1994). Two transduction pathways mediate rapid effects of abscisic acid in *Commelina* guard cells. *Plant Cell* **6**, 1319–1328.

Allen, G. J., Kuchitsu, K., Chu, S. P., Murata, Y., and Schroeder, J. I. (1999). *Arabidopsis* abi1-1 and abi2-1 phosphatase mutations reduce abscisic acid-induced cytoplasmic calcium rises in guard cells. *Plant Cell* **11**, 1785–1798.

Allen, G. J., Chu, S. P., Harrington, C. L., Schumacher, K., Hoffmann, T., Tang, Y. Y., Grill, E., and Schroeder, J. I. (2001). A defined range of guard cell calcium oscillation parameters encodes stomatal movements. *Nature* **411**, 1053–1057.

Anderson, B. E., Ward, J. M., and Schroeder, J. I. (1994). Evidence for an extracellular reception site for abscisic acid in *Commelina* guard cells. *Plant Physiol.* **104**, 1177–1183.

Ang, L. H., Chattopadhyay, S., Wei, N., Oyama, T., Okada, K., Batschauer, A., and Deng, X. W. (1998). Molecular interaction between COP1 and HY5 defines a regulatory switch for light control of *Arabidopsis* development. *Mol. Cell* **1**, 213–222.

Barkla, B. J., Vera-Estrella, R., Maldonado-Gama, M., and Pantoja, O. (1999). Abscisic acid induction of vacuolar H$^+$-ATPase activity in *Mesembryanthemum crystallinum* is developmentally regulated. *Plant Physiol.* **120**, 811–820.

Bethke, P. C., Schuurink, R., Jones, R. J., and Hooley, R. (1997). Hormonal signalling in cereal aleurone. *J. Exp. Bot.* **48**, 1337–1356.

Bobb, A. J., Chern, M. S., and Bustos, M. M. (1997). Conserved RY-repeats mediate transactivation of seed-specific promoters by the developmental regulator PvALF. *Nucl. Acids Res.* **25**, 641–647.

Brocard-Gifford, I. M., Lynch, T. J., and Finkelstein, R. R. (2003). Regulatory networks in seeds integrating developmental, abscisic acid, sugar, and light signaling. *Plant Physiol.* **131**, 78–92.

Burnette, R. N., Gunesekera, B. M., and Gillaspy, G. E. (2003). An *Arabidopsis* inositol 5-phosphatase gain-of-function alters abscisic acid signaling. *Plant Physiol.* **132**, 1011–1019.

Carles, C., Bies-Etheve, N., Aspart, L., Leon-Kloosterziel, K. M., Koornneef, M., Echeverria, M., and Delseny, M. (2002). Regulation of *Arabidopsis thaliana* Em genes: Role of ABI5. *Plant J.* **30**, 373–383.

Casaretto, J., and Ho, T. H. (2003). The transcription factors HvABI5 and HvVP1 are required for the abscisic acid induction of gene expression in barley aleurone cells. *Plant Cell* **15**, 271–284.

Chandok, M. R., Ytterberg, A. J., van Wijk, K. J., and Klessig, D. F. (2003). The pathogen-inducible nitric oxide synthase (iNOS) in plants is a variant of the P protein of the glycine decarboxylase complex. *Cell* **113**, 469–482.

Chen, X., Chang, M., Wang, B., and Wu, B. (1997). Cloning of a Ca^{2+}-ATPase gene and the role of cytosolic Ca^{2+} in the gibberellin-dependent signaling pathway in aleurone cells. *Plant J.* **11**, 363–371.

Cheng, S. H., Willmann, M. R., Chen, H. C., and Sheen, J. (2002). Calcium signaling through protein kinases. The *Arabidopsis* calcium-dependent protein kinase gene family. *Plant Physiol.* **129**, 469–485.

Chern, M. S., Bobb, A. J., and Bustos, M. M. (1996). The regulator of *MAT2* (ROM2) protein binds to early maturation promoters and represses PvALF-activated transcription. *Plant Cell* **8**, 305–321.

Chinnusamy, V., Schumaker, K., and Zhu, J. K. (2004). Molecular genetic perspectives on cross-talk and specificity in abiotic stress signalling in plants. *J. Exp. Bot.* **55**, 225–236.

Choi, H., Hong, J., Ha, J., Kang, J., and Kim, S. Y. (2000). ABFs, a family of ABA-responsive element binding factors. *J. Biol. Chem.* **275**, 1723–1730.

Coursol, S., Fan, L. M., Le Stunff, H., Spiegel, S., Gilroy, S., and Assmann, S. M. (2003). Sphingolipid signalling in *Arabidopsis* guard cells involves heterotrimeric G proteins. *Nature* **423**, 651–654.

Cutler, S., Ghassemian, M., Bonetta, D., Cooney, S., and McCourt, P. (1996). A protein farnesyl transferase involved in abscisic acid signal transduction in *Arabidopsis*. *Science* **273**, 1239–1241.

Desikan, R., Hagenbeek, D., Neill, S. J., and Rock, C. D. (1999). Flow cytometry and surface plasmon resonance analyses demonstrate that the monoclonal antibody JIM19 interacts with a rice cell surface component involved in abscisic acid signalling in protoplasts. *FEBS Lett.* **456**, 257–262.

Desikan, R., Griffiths, R., Hancock, J., and Neill, S. (2002). A new role for an old enzyme: Nitrate reductase-mediated nitric oxide generation is required for abscisic acid-induced stomatal closure in *Arabidopsis thaliana*. *Proc. Natl. Acad. Sci. USA* **99**, 16314–16318.

Duque, P., and Chua, N. H. (2003). IMB1, a bromodomain protein induced during seed imbibition, regulates ABA- and phyA-mediated responses of germination in *Arabidopsis*. *Plant J.* **35**, 787–799.

Ezcurra, I., Wycliffe, P., Nehlin, L., Ellerstrom, M., and Rask, L. (2000). Transactivation of the *Brassica napus* napin promoter by ABI3 requires interaction of the conserved B2 and B3 domains of ABI3 with different *cis*-elements: B2 mediates activation through an ABRE, whereas B3 interacts with an RY/G-box. *Plant J.* **24,** 57–66.

Fan, L. M., Zhao, Z., and Assmann, S. M. (2004). Guard cells: A dynamic signaling model. *Curr. Opin. Plant Biol.* **7,** 537–546.

Fedoroff, N. V. (2002). Cross-talk in abscisic acid signaling. *Sci. STKE* **2002,** RE10.

Finkelstein, R. R. (1993). Abscisic acid-insensitive mutations provide evidence for stage-specific signal pathways regulating expression of an *Arabidopsis* late embryogenesis-abundant (lea) gene. *Mol. Gen. Genet.* **238,** 401–408.

Finkelstein, R. R., Wang, M. L., Lynch, T. J., Rao, S., and Goodman, H. M. (1998). The *Arabidopsis* abscisic acid response locus ABI4 encodes an APETALA 2 domain protein. *Plant Cell* **10,** 1043–1054.

Finkelstein, R. R., and Lynch, T. J. (2000). The *Arabidopsis* abscisic acid response gene ABI5 encodes a basic leucine zipper transcription factor. *Plant Cell* **12,** 599–609.

Finkelstein, R. R., and Rock, C. D. (2001). Abscisic acid biosynthesis and response. *In* "The *Arabidopsis* Book" (C. Somerville and E. Meyerowitz, Eds.). American Society of Plant Biologists, Rockville, MDhttp://www.aspb.org/publications/arabidopsis/.

Freire, M. A., and Pages, M. (1995). Functional characteristics of the maize RNA-binding protein MA16. *Plant Mol. Biol.* **29,** 797–807.

Fujita, M., Fujita, Y., Maruyama, K., Seki, M., Hiratsu, K., Ohme-Takagi, M., Tran, L. S., Yamaguchi-Shinozaki, K., and Shinozaki, K. (2004). A dehydration-induced NAC protein, RD26, is involved in a novel ABA-dependent stress-signaling pathway. *Plant J.* **39,** 863–876.

Gampala, S. S., Finkelstein, R. R., Sun, S. S., and Rock, C. D. (2002). ABI5 interacts with abscisic acid signaling effectors in rice protoplasts. *J. Biol. Chem.* **277,** 1689–1694.

Gazzarrini, S., Tsuchiya, Y., Lumba, S., Okamoto, M., and McCourt, P. (2004). The transcription factor FUSCA3 controls developmental timing in *Arabidopsis* through the hormones gibberellin and abscisic acid. *Dev. Cell* **7,** 373–385.

Gilroy, S., and Jones, R. L. (1994). Perception of gibberellin and abscisic acid at the external face of the plasma membrane of barley (*Hordeum vulgare* L.) aleurone protoplasts. *Plant Physiol.* **104,** 1185–1192.

Goff, S. A., Ricke, D., Lan, T. H., Presting, G., Wang, R., Dunn, M., Glazebrook, J., Sessions, A., Oeller, P., Varma, H., Hadley, D., Hutchison, D., Martin, C., Katagiri, F., Lange, B. M., Moughamer, T., Xia, Y., Budworth, P., Zhong, J., Miguel, T., Paszkowski, U., Zhang, S., Colbert, M., Sun, W. L., Chen, L., Cooper, B., Park, S., Wood, T. C., Mao, L., Quail, P., Wing, R., Dean, R., Yu, Y., Zharkikh, A., Shen, R., Sahasrabudhe, S., Thomas, A., Cannings, R., Gutin, A., Pruss, D., Reid, J., Tavtigian, S., Mitchell, J., Eldredge, G., Scholl, T., Miller, R. M., Bhatnagar, S., Adey, N., Rubano, T., Tusneem, N., Robinson, R., Feldhaus, J., Macalma, T., Oliphant, A., and Briggs, S. (2002). A draft sequence of the rice genome (*Oryza sativa* L. ssp. *japonica*). *Science* **296,** 92–100.

Gómez-Cadenas, A., Verhey, S. D., Holappa, L. D., Shen, Q., Ho, T. H. D., and Walker-Simmons, M. K. (1999). An abscisic acid-induced protein kinase, PKABA1, mediates abscisic acid-suppressed gene expression in barley aleurone layers. *Proc. Natl. Acad. Sci. USA* **96,** 1767–1772.

Gómez-Cadenas, A., Zentella, R., Walker-Simmons, M. K., and Ho, T. H. D. (2001). Gibberellin/abscisic acid antagonism in barley aleurone cells: Site of action of the protein kinase PKABA1 in relation to gibberellin signaling molecules. *Plant Cell* **13,** 667–679.

Guiltinan, M. J., Marcotte, W. R., Jr., and Quatrano, R. S. (1990). A plant leucine zipper protein that recognizes an abscisic acid response element. *Science* **250,** 267–271.

Guo, F. Q., Okamoto, M., and Crawford, N. M. (2003). Identification of a plant nitric oxide synthase gene involved in hormonal signaling. *Science* **302,** 100–103.

Guo, Y., Xiong, L., Song, C. P., Gong, D., Halfter, U., and Zhu, J. K. (2002). A calcium sensor and its interacting protein kinase are global regulators of abscisic acid signaling in *Arabidopsis*. *Dev. Cell* **3**, 233–244.

Hamilton, D. W., Hills, A., Kohler, B., and Blatt, M. R. (2000). Ca^{2+} channels at the plasma membrane of stomatal guard cells are activated by hyperpolarization and abscisic acid. *Proc. Natl. Acad. Sci. USA* **97**, 4967–4972.

Han, M. H., Goud, S., Song, L., and Fedoroff, N. (2004). The *Arabidopsis* double-stranded RNA-binding protein HYL1 plays a role in microRNA-mediated gene regulation. *Proc. Natl. Acad. Sci. USA* **101**, 1093–1098.

Hare, P. D., Seo, H. S., Yang, J. Y., and Chua, N. H. (2003). Modulation of sensitivity and selectivity in plant signaling by proteasomal destabilization. *Curr. Opin. Plant Biol.* **6**, 453–462.

Hattori, T., Terada, T., and Hamasuna, S. (1995). Regulation of the Osem gene by abscisic acid and the transcriptional activator VP1: Analysis of *cis*-acting promoter elements required for regulation by abscisic acid and VP1. *Plant J.* **7**, 913–925.

Hayama, R., Yokoi, S., Tamaki, S., Yano, M., and Shimamoto, K. (2003). Adaptation of photoperiodic control pathways produces short-day flowering in rice. *Nature* **422**, 719–722.

He, Y., and Gan, S. (2004). A novel zinc-finger protein with a proline-rich domain mediates ABA-regulated seed dormancy in *Arabidopsis*. *Plant Mol. Biol.* **54**, 1–9.

Hill, A., Nantel, A., Rock, C. D., and Quatrano, R. S. (1996). A conserved domain of the viviparous-1 gene product enhances the DNA binding activity of the bZIP protein EmBP-1 and other transcription factors. *J. Biol. Chem.* **271**, 3366–3374.

Himmelbach, A., Hoffmann, T., Leube, M., Hohener, B., and Grill, E. (2002). Homeodomain protein ATHB6 is a target of the protein phosphatase ABI1 and regulates hormone responses in *Arabidopsis*. *EMBO J.* **21**, 3029–3038.

Himmelbach, A., Yang, Y., and Grill, E. (2003). Relay and control of abscisic acid signaling. *Curr. Opin. Plant Biol.* **6**, 470–479.

Hirsch, R., Hartung, W., and Gimmler, H. (1989). Abscisic acid content in algae under stress. *Bot. Acta* **102**, 326–334.

Hobo, T., Asada, M., Kowyama, Y., and Hattori, T. (1999a). ACGT-containing abscisic acid response element (ABRE) and coupling element 3 (CE3) are functionally equivalent. *Plant J.* **19**, 679–689.

Hobo, T., Kowyama, Y., and Hattori, T. (1999b). A bZIP factor, TRAB1, interacts with VP1 and mediates abscisic acid-induced transcription. *Proc. Natl. Acad. Sci. USA* **96**, 15348–15353.

Hugouvieux, V., Kwak, J. M., and Schroeder, J. I. (2001). An mRNA cap binding protein, ABH1, modulates early abscisic acid signal transduction in *Arabidopsis*. *Cell* **106**, 477–487.

Hugouvieux, V., Murata, Y., Young, J. J., Kwak, J. M., Mackesy, D. Z., and Schroeder, J. I. (2002). Localization, ion channel regulation, and genetic interactions during abscisic acid signaling of the nuclear mRNA cap-binding protein, ABH1. *Plant Physiol.* **130**, 1276–1287.

Ianzano, L., Young, E. J., Zhao, X. C., Chan, E. M., Rodriguez, M. T., Torrado, M. V., Scherer, S. W., and Minassian, B. A. (2004). Loss of function of the cytoplasmic isoform of the protein laforin (EPM2A) causes Lafora progressive myoclonus epilepsy. *Hum. Mutat.* **23**, 170–176.

Jeannette, E., Rona, J. P., Bardat, F., Cornel, D., Sotta, B., and Miginiac, E. (1999). Induction of RAB18 gene expression and activation of K^+ outward rectifying channels depend on an extracellular perception of ABA in *Arabidopsis thaliana* suspension cells. *Plant J.* **18**, 13–22.

Johannesson, H., Wang, Y., and Engstrom, P. (2001). DNA-binding and dimerization preferences of *Arabidopsis* homeodomain-leucine zipper transcription factors *in vitro*. *Plant Mol. Biol.* **45**, 63–73.

Johannesson, H., Wang, Y., Hanson, J., and Engstrom, P. (2003). The *Arabidopsis thaliana* homeobox gene ATHB5 is a potential regulator of abscisic acid responsiveness in developing seedlings. *Plant Mol. Biol.* **51,** 719–729.

Johnson, R. R., Wagner, R. L., Verhey, S. D., and Walker-Simmons, M. K. (2002). The abscisic acid-responsive kinase PKABA1 interacts with a seed-specific abscisic acid response element-binding factor, TaABF, and phosphorylates TaABF peptide sequences. *Plant Physiol.* **130,** 837–846.

Jones, H. D., Kurup, S., Peters, N. C., and Holdsworth, M. J. (2000). Identification and analysis of proteins that interact with the *Avena fatua* homologue of the maize transcription factor VIVIPAROUS 1. *Plant J.* **21,** 133–142.

Jung, J. Y., Kim, Y. W., Kwak, J. M., Hwang, J. U., Young, J., Schroeder, J. I., Hwang, I., and Lee, Y. (2002). Phosphatidylinositol 3- and 4-phosphate are required for normal stomatal movements. *Plant Cell* **14,** 2399–2412.

Kagaya, Y., Hobo, T., Murata, M., Ban, A., and Hattori, T. (2002). Abscisic acid-induced transcription is mediated by phosphorylation of an abscisic acid response element binding factor, TRAB1. *Plant Cell* **14,** 3177–3189.

Kang, J. Y., Choi, H. I., Im, M. Y., and Kim, S. Y. (2002). *Arabidopsis* basic leucine zipper proteins that mediate stress-responsive abscisic acid signaling. *Plant Cell* **14,** 343–357.

Kao, C. Y., Cocciolone, S. M., Vasil, I. K., and McCarty, D. R. (1996). Localization and interaction of the *cis*-acting elements for abscisic acid, VIVIPAROUS1, and light activation of the C1 gene of maize. *Plant Cell* **8,** 1171–1179.

Kikuchi, S., Satoh, K., Nagata, T., Kawagashira, N., Doi, K., Kishimoto, N., Yazaki, J., Ishikawa, M., Yamada, H., Ooka, H., Hotta, I., Kojima, K., Namiki, T., Ohneda, E., Yahagi, W., Suzuki, K., Li, C. J., Ohtsuki, K., Shishiki, T., Otomo, Y., Murakami, K., Iida, Y., Sugano, S., Fujimura, T., Suzuki, Y., Tsunoda, Y., Kurosaki, T., Kodama, T., Masuda, H., Kobayashi, M., Xie, Q., Lu, M., Narikawa, R., Sugiyama, A., Mizuno, K., Yokomizo, S., Niikura, J., Ikeda, R., Ishibiki, J., Kawamata, M., Yoshimura, A., Miura, J., Kusumegi, T., Oka, M., Ryu, R., Ueda, M., Matsubara, K., Kawai, J., Carninci, P., Adachi, J., Aizawa, K., Arakawa, T., Fukuda, S., Hara, A., Hashidume, W., Hayatsu, N., Imotani, K., Ishii, Y., Itoh, M., Kagawa, I., Kondo, S., Konno, H., Miyazaki, A., Osato, N., Ota, Y., Saito, R., Sasaki, D., Sato, K., Shibata, K., Shinagawa, A., Shiraki, T., Yoshino, M., and Hayashizaki, Y. (2003). Collection, mapping, and annotation of over 28,000 cDNA clones from japonica rice. *Science* **301,** 376–379.

Kim, K. N., Cheong, Y. H., Grant, J. J., Pandey, G. K., and Luan, S. (2003). CIPK3, a calcium sensor-associated protein kinase that regulates abscisic acid and cold signal transduction in *Arabidopsis*. *Plant Cell* **15,** 411–423.

Kobayashi, Y., Yamamoto, S., Minami, H., Kagaya, Y., and Hattori, T. (2004). Differential activation of the rice sucrose nonfermenting1-related protein kinase2 family by hyperosmotic stress and abscisic acid. *Plant Cell* **16,** 1163–1177.

Koiwa, H., Barb, A. W., Xiong, L., Li, F., McCully, M. G., Lee, B. H., Sokolchik, I., Zhu, J., Gong, Z., Reddy, M., Sharkhuu, A., Manabe, Y., Yokoi, S., Zhu, J. K., Bressan, R. A., and Hasegawa, P. M. (2002). C-terminal domain phosphatase-like family members (AtCPLs) differentially regulate *Arabidopsis thaliana* abiotic stress signaling, growth, and development. *Proc. Natl. Acad. Sci. USA* **99,** 10893–10898.

Kovtun, Y., Chiu, W. L., Tena, G., and Sheen, J. (2000). Functional analysis of oxidative stress-activated mitogen-activated protein kinase cascade in plants. *Proc. Natl. Acad. Sci. USA* **97,** 2940–2945.

Kuhn, J. M., and Schroeder, J. I. (2003). Impacts of altered RNA metabolism on abscisic acid signaling. *Curr. Opin. Plant Biol.* **6,** 463–469.

Kurup, S., Jones, H. D., and Holdsworth, M. J. (2000). Interactions of the developmental regulator ABI3 with proteins identified from developing *Arabidopsis* seeds. *Plant J.* **21,** 143–155.

Kwak, J. M., Moon, J. H., Murata, Y., Kuchitsu, K., Leonhardt, N., De Long, A., and Schroeder, J. I. (2002). Disruption of a guard cell-expressed protein phosphatase 2A regulatory subunit, RCN1, confers abscisic acid insensitivity in *Arabidopsis*. *Plant Cell* **14,** 2849–2861.

Kwak, J. M., Mori, I. C., Pei, Z. M., Leonhardt, N., Torres, M. A., Dangl, J. L., Bloom, R. E., Bodde, S., Jones, J. D., and Schroeder, J. I. (2003). NADPH oxidase AtrbohD and AtrbohF genes function in ROS-dependent ABA signaling in *Arabidopsis*. *EMBO J.* **22,** 2623–2633.

Lanahan, M. B., Ho, T. H. D., Rogers, S. W., and Rogers, J. C. (1992). A gibberellin response complex in cereal alpha-amylase gene promoters. *Plant Cell* **4,** 203–211.

Le Page-Degivry, M.-T., Bidard, J.-N., Rouvier, E., Bulard, C.,and, and Lazdunski, M. (1986). Presence of abscisic acid, a phytohormone, in the mammalian brain. *Proc. Natl. Acad. Sci. USA* **83,** 1155–1158.

Leckie, C. P., McAinsh, M. R., Allen, G. J., Sanders, D., and Hetherington, A. M. (1998). Abscisic acid-induced stomatal closure mediated by cyclic ADP-ribose. *Proc. Natl. Acad. Sci. USA* **95,** 15837–15842.

Lemichez, E., Wu, Y., Sanchez, J. P., Mettouchi, A., Mathur, J., and Chua, N. H. (2001). Inactivation of AtRac1 by abscisic acid is essential for stomatal closure. *Genes Dev.* **15,** 1808–1816.

Lemtiri-Chlieh, F., Mac Robbie, E. A. C., Webb, A. A. R., Manison, N. F., Brownlee, C., Skepper, J. N., Chen, J., Prestwich, G. D., and Brearley, C. A. (2003). Inositol hexakisphosphate mobilizes an endomembrane store of calcium in guard cells. *Proc. Natl. Acad. Sci. USA* **100,** 10091–10095.

Leonhardt, N., Kwak, J. M., Robert, N., Waner, D., Leonhardt, G., and Schroeder, J. I. (2004a). Microarray expression analyses of *Arabidopsis* guard cells and isolation of a recessive abscisic acid hypersensitive protein phosphatase 2C mutant. *Plant Cell* **16,** 596–615.

Leonhardt, N., Kwak, J. M., Robert, N., Waner, D., Leonhardt, G., and Schroeder, J. I. (2004b). Microarray expression analyses of *Arabidopsis* guard cells and isolation of a recessive abscisic acid hypersensitive protein phosphatase 2C mutant. *Plant Cell* **14,** 596–615.

Leung, J., Bouvier-Durand, M., Morris, P. C., Guerrier, D., Chefdor, F., and Giraudat, J. (1994). *Arabidopsis* ABA response gene *ABI1*: Features of a calcium-modulated protein phosphatase. *Science* **264,** 1448–1452.

Leung, J., and Giraudat, J. (1998). Abscisic acid signal transduction. *Annu. Rev. Plant Physiol. Plant Mol. Biol.* **49,** 199–222.

Leyman, B., Geelen, D., Quintero, F. J., and Blatt, M. R. (1999). A tobacco syntaxin with a role in hormonal control of guard cell ion channels. *Science* **283,** 537–540.

Leyman, B., Geelen, D., and Blatt, M. R. (2000). Localization and control of expression of Nt-Syr1, a tobacco SNARE protein. *Plant J.* **24,** 369–381.

Li, G., Bishop, K. J., Chandrasekharan, M. B., and Hall, T. C. (1999). Beta-phaseolin gene activation is a two-step process: PvALF-facilitated chromatin modification followed by abscisic acid-mediated gene activation. *Proc. Natl. Acad. Sci. USA* **96,** 7104–7109.

Li, H., Shen, J. J., Zheng, Z. L., Lin, Y., and Yang, Z. (2001). The Rop GTPase switch controls multiple developmental processes in *Arabidopsis*. *Plant Physiol.* **126,** 670–684.

Li, J., Kinoshita, T., Pandey, S., Ng, C. K., Gygi, S. P., Shimazaki, K., and Assmann, S. M. (2002). Modulation of an RNA-binding protein by abscisic-acid-activated protein kinase. *Nature* **418,** 793–797.

Lodish, H., Berk, A., Matsudaira, P., Kaiser, C. A., Krieger, M., Scott, M. P., Zipursky, S. L., and Darnell, J. (2004). Regulation of transcription factor activity. *In* "Molecular Cell Biology," pp. 481–484. W. H. Freeman and Company, New York.

Lopez-Molina, L., Mongrand, S., and Chua, N. H. (2001). A postgermination developmental arrest checkpoint is mediated by abscisic acid and requires the ABI5 transcription factor in *Arabidopsis*. *Proc. Natl. Acad. Sci. USA* **98,** 4782–4787.

Lopez-Molina, L., Mongrand, S., Kinoshita, N., and Chua, N. H. (2003). AFP is a novel negative regulator of ABA signaling that promotes ABI5 protein degradation. *Genes Dev.* **17**, 410–418.

Lotan, T., Ohto, M., Yee, K. M., West, M. A., Lo, R., Kwong, R. W., Yamagishi, K., Fischer, R. L., Goldberg, R. B., and Harada, J. J. (1998). *Arabidopsis* LEAFY COTYLEDON1 is sufficient to induce embryo development in vegetative cells. *Cell* **93**, 1195–1205.

Lovegrove, A., and Hooley, R. (2000). Gibberellin and abscisic acid signalling in aleurone. *Trends Plant Sci.* **5**, 102–110.

Lu, C., and Fedoroff, N. (2000). A mutation in the *Arabidopsis* HYL1 gene encoding a dsRNA binding protein affects responses to abscisic acid, auxin, and cytokinin. *Plant Cell* **12**, 2351–2366.

Lu, C., Han, M. H., Guevara-Garcia, A., and Fedoroff, N. V. (2002). Mitogen-activated protein kinase signaling in postgermination arrest of development by abscisic acid. *Proc. Natl. Acad. Sci. USA* **99**, 15812–15817.

Luerssen, H., Kirik, V., Herrmann, P., and Misera, S. (1998). FUSCA3 encodes a protein with a conserved VP1/ABI3-like B3 domain which is of functional importance for the regulation of seed maturation in *Arabidopsis thaliana*. *Plant J.* **15**, 755–764.

McCarty, D. R., Hattori, T., Carson, C. B., Vasil, V., Lazar, M., and Vasil, I. K. (1991). The Viviparous-1 developmental gene of maize encodes a novel transcriptional activator. *Cell* **66**, 895–905.

Merlot, S., Gosti, F., Guerrier, D., Vavasseur, A., and Giraudat, J. (2001). The ABI1 and ABI2 protein phosphatases 2C act in a negative feedback regulatory loop of the abscisic acid signalling pathway. *Plant J.* **25**, 295–303.

Milborrow, B. V. (1978). Phytohormones and related compounds: A comprehensive treatise. *In* "The Biochemistry of Phytohormones and Related Compounds" (D. Letham, P. Goodwin, and T. Higgins, Eds.), pp. 295–347. Elsevier, North-Holland.

Murata, Y., Pei, Z. M., Mori, I. C., and Schroeder, J. (2001). Abscisic acid activation of plasma membrane Ca(2+) channels in guard cells requires cytosolic NAD(P)H and is differentially disrupted upstream and downstream of reactive oxygen species production in abi1-1 and abi2-1 protein phosphatase 2C mutants. *Plant Cell* **13**, 2513–2523.

Mustilli, A. C., Merlot, S., Vavasseur, A., Fenzi, F., and Giraudat, J. (2002). *Arabidopsis* OST1 protein kinase mediates the regulation of stomatal aperture by abscisic acid and acts upstream of reactive oxygen species production. *Plant Cell* **14**, 3089–3099.

Nakamura, S., Lynch, T. J., and Finkelstein, R. R. (2001). Physical interactions between ABA response loci of *Arabidopsis*. *Plant J.* **26**, 627–635.

Nantel, A., and Quatrano, R. S. (1996). Characterization of three rice basic/leucine zipper factors, including two inhibitors of EmBP-1 DNA binding activity. *J. Biol. Chem.* **271**, 31296–31305.

Narusaka, Y., Nakashima, K., Shinwari, Z. K., Sakuma, Y., Furihata, T., Abe, H., Narusaka, M., Shinozaki, K., and Yamaguchi-Shinozaki, K. (2003). Interaction between two cis-acting elements, ABRE and DRE, in ABA-dependent expression of *Arabidopsis* rd29A gene in response to dehydration and high-salinity stresses. *Plant J.* **34**, 137–148.

Nemeth, K., Salchert, K., Putnoky, P., Bhalerao, R., Koncz-Kalman, Z., Stankovic-Stangeland, B., Bako, L., Mathur, J., Okresz, L., Stabel, S., Geigenberger, P., Stitt, M., Redei, G. P., Schell, J., and Koncz, C. (1998). Pleiotropic control of glucose and hormone responses by PRL1, a nuclear WD protein, in *Arabidopsis*. *Genes Dev.* **12**, 3059–3073.

Ng, C. K., Carr, K., McAinsh, M. R., Powell, B., and Hetherington, A. M. (2001). Drought-induced guard cell signal transduction involves sphingosine-1-phosphate. *Nature* **410**, 596–599.

Niu, X., Helentjaris, T., and Bate, N. J. (2002). Maize ABI4 binds coupling element1 in abscisic acid and sugar response genes. *Plant Cell* **14**, 2565–2575.

Pandey, G. K., Cheong, Y. H., Kim, K. N., Grant, J. J., Li, L., Hung, W., D' Angelo, C., Weinl, S., Kudla, J., and Luan, S. (2004). The calcium sensor calcineurin B-like 9 modulates abscisic acid sensitivity and biosynthesis in *Arabidopsis*. *Plant Cell* **16**, 1912–1924.

Pandey, S., and Assmann, S. M. (2004). The *Arabidopsis* putative G protein-coupled receptor GCR1 interacts with the G protein alpha subunit GPA1 and regulates abscisic acid signaling. *Plant Cell* **16**, 1616–1632.

Park, K. Y., Jung, J. Y., Park, J., Hwang, J. U., Kim, Y. W., Hwang, I., and Lee, Y. (2003). A role for phosphatidylinositol 3-phosphate in abscisic acid-induced reactive oxygen species generation in guard cells. *Plant Physiol.* **132**, 92–98.

Pei, Z. M., Ghassemian, M., Kwak, C. M., McCourt, P., and Schroeder, J. I. (1998). Role of farnesyltransferase in ABA regulation of guard cell anion channels and plant water loss. *Science* **282**, 287–290.

Penson, S. P., Schuurink, R. C., Fath, A., Gubler, F., Jacobsen, J. V., and Jones, R. L. (1996). cGMP is required for gibberellic acid-induced gene expression in barley aleurone. *Plant Cell* **8**, 2325–2333.

Rabbani, M. A., Maruyama, K., Abe, H., Khan, M. A., Katsura, K., Ito, Y., Yoshiwara, K., Seki, M., Shinozaki, K., and Yamaguchi-Shinozaki, K. (2003). Monitoring expression profiles of rice genes under cold, drought, and high-salinity stresses and abscisic acid application using cDNA microarray and RNA gel-blot analyses. *Plant Physiol.* **133**, 1755–1767.

Razem, F. A., Luo, M., Liu, J. H., Abrams, S. R., and Hill, R. D. (2004). Purification and characterization of a barley aleurone abscisic acid-binding protein. *J. Biol. Chem.* **279**, 9922–9929.

Ritchie, S., and Gilroy, S. (1998). Abscisic acid signal transduction in the barley aleurone is mediated by phospholipase D activity. *Proc. Natl. Acad. Sci. USA* **95**, 2697–2702.

Ritchie, S., and Gilroy, S. (2000). Abscisic acid stimulation of phospholipase D in the barley aleurone is G-protein-mediated and localized to the plasma membrane. *Plant Physiol.* **124**, 693–702.

Ritchie, S. M., Swanson, S. J., and Gilroy, S. (2002). From common signalling components to cell specific responses: Insights from the cereal aleurone. *Physiol. Plant* **115**, 342–351.

Rock, C. D. (2000). Pathways to abscisic acid-regulated gene expression. *New Phytol.* **148**, 357–396.

Rohde, A., Kurup, S., and Holdswoth, M. (2000). ABI3 emerges from the seed. *Trends Plant Sci.* **5**, 418–419.

Sanchez, J. P., and Chua, N. H. (2001). *Arabidopsis* PLC1 is required for secondary responses to abscisic acid signals. *Plant Cell* **13**, 1143–1154.

Sanchez, J. P., Duque, P., and Chua, N. H. (2004). ABA activates ADPR cyclase and cADPR induces a subset of ABA-responsive genes in *Arabidopsis*. *Plant J.* **38**, 381–395.

Schroeder, J. I., and Hagiwara, S. (1990). Repetitive increases in cytosolic Ca^{2+} of guard cells by abscisic acid activation of nonselective Ca^{2+} permeable channels. *Proc. Natl. Acad. Sci. USA* **87**, 9305–9309.

Schroeder, J. I., Allen, G. J., Hugouvieux, V., Kwak, J. M., and Waner, D. (2001). Guard cell signal transduction. *Annu. Rev. Plant Physiol. Plant Mol. Biol.* **52**, 627–658.

Schultz, T. F., and Quatrano, R. S. (1997). Evidence for surface perception of abscisic acid by rice suspension cells as assayed by Em gene expression. *Plant Sci.* **130**, 63–71.

Schultz, T. F., Medina, J., Hill, A., and Quatrano, R. S. (1998). 14-3-3 proteins are part of an abscisic acid-VIVIPAROUS1 (VP1) response complex in the Em promoter and interact with VP1 and EmBP1. *Plant Cell* **10**, 837–847.

Schwartz, A., Wu, W. H., Tucker, E. B., and Assmann, S. M. (1994). Inhibition of inward K^+ channels and stomatal response by abscisic acid: An intracellular locus of phytohormone action. *Proc. Natl. Acad. Sci. USA* **91**, 4019–4023.

Seki, M., Narusaka, M., Ishida, J., Nanjo, T., Fujita, M., Oono, Y., Kamiya, A., Nakajima, M., Enju, A., Sakurai, T., Satou, M., Akiyama, K., Taji, T., Yamaguchi-Shinozaki, K., Carninci, P., Kawai, J., Hayashizaki, Y., and Shinozaki, K. (2002a). Monitoring the expression profiles of 7000 Arabidopsis genes under drought, cold and high-salinity stresses using a full-length cDNA microarray. *Plant J.* **31,** 279–292.

Seki, M., Narusaka, M., Kamiya, A., Ishida, J., Satou, M., Sakurai, T., Nakajima, M., Enju, A., Akiyama, K., Oono, Y., Muramatsu, M., Hayashizaki, Y., Kawai, J., Carninci, P., Itoh, M., Ishii, Y., Arakawa, T., Shibata, K., Shinagawa, A., and Shinozaki, K. (2002b). Functional annotation of a full-length *Arabidopsis* cDNA collection. *Science* **296,** 141–145.

Seo, H. S., Yang, J. Y., Ishikawa, M., Bolle, C., Ballesteros, M. L., and Chua, N. H. (2003). LAF1 ubiquitination by COP1 controls photomorphogenesis and is stimulated by SPA1. *Nature* **423,** 995–999.

Sheen, J. (1996). Ca^{2+}-dependent protein kinases and stress signal transduction in plants. *Science* **274,** 1900–1902.

Shen, Q., and Ho, T. H. D. (1995). Functional dissection of an abscisic acid (ABA)-inducible gene reveals two independent ABA-responsive complexes each containing a G-box and a novel *cis*-acting element. *Plant Cell* **7,** 295–307.

Shen, Q., Zhang, P., and Ho, T. H. D. (1996). Modular nature of abscisic acid (ABA) response complexes: Composite promoter units that are necessary and sufficient for ABA induction of gene expression in barley. *Plant Cell* **8,** 1107–1119.

Shen, Q., and Ho, T. H. D. (1997). Promoter switches specific for abscisic acid (ABA)-induced gene expression in cereals. *Physiol. Plant.* **101,** 653–664.

Shen, Q., and Ho, T. H. D. (1998). Abscisic acid- and stress-induced promoter swithes in the control of gene expression. *In* "Inducible Gene Expression in Plants" (P. H. S. Reynolds, Ed.), pp. 187–218. CABI Publishing, Wallingford.

Shen, Q., Gómez-Cadenas, A., Zhang, P., Walker-Simmons, M. K., Sheen, J., and Ho, T. H. (2001). Dissection of abscisic acid signal transduction pathways in barley aleurone layers. *Plant Mol. Biol.* **47,** 437–448.

Shen, Q. J., Casaretto, J. A., Zhang, P., and Ho, T. H. (2004). Functional definition of ABA-response complexes: The promoter units necessary and sufficient for ABA induction of gene expression in barley (*Hordeum vulgare* L.). *Plant Mol. Biol.* **54,** 111–124.

Stone, S. L., Kwong, L. W., Yee, K. M., Pelletier, J., Lepiniec, L., Fischer, R. L., Goldberg, R. B., and Harada, J. J. (2001). LEAFY COTYLEDON2 encodes a B3 domain transcription factor that induces embryo development. *Proc. Natl. Acad. Sci. USA* **98,** 11806–11811.

Sutton, F., Paul, S. S., Wang, X. Q., and Assmann, S. M. (2000). Distinct abscisic acid signaling pathways for modulation of guard cell versus mesophyll cell potassium channels revealed by expression studies in *Xenopus laevis* oocytes. *Plant Physiol.* **124,** 223–230.

Suzuki, M., Kao, C. Y., and McCarty, D. R. (1997). The conserved B3 domain of VIVIPAROUS1 has a cooperative DNA binding activity. *Plant Cell* **9,** 799–807.

Suzuki, M., Ketterling, M. G., Li, Q. B., and McCarty, D. R. (2003). Viviparous1 alters global gene expression patterns through regulation of abscisic acid signaling. *Plant Physiol.* **132,** 1664–1677.

Tahtiharju, S., and Palva, T. (2001). Antisense inhibition of protein phosphatase 2C accelerates cold acclimation in *Arabidopsis thaliana*. *Plant J.* **26,** 461–470.

The *Arabidopsis*, Initiative (2000). Analysis of the genome sequence of the flowering plant *Arabidopsis thaliana*. *Nature* **408,** 796–815.

Tran, L. S., Nakashima, K., Sakuma, Y., Simpson, S. D., Fujita, Y., Maruyama, K., Fujita, M., Seki, M., Shinozaki, K., and Yamaguchi-Shinozaki, K. (2004). Isolation and functional analysis of *Arabidopsis* stress-inducible NAC transcription factors that bind to a drought-responsive *cis*-element in the early responsive to dehydration stress 1 promoter. *Plant Cell* **16,** 2481–2498.

Wang, X. Q., Ullah, H., Jones, A. M., and Assmann, S. M. (2001). G protein regulation of ion channels and abscisic acid signaling in *Arabidopsis* guard cells. *Science* **292**, 2070–2072.

Wu, Y., Kuzma, J., Marechal, E., Graeff, R., Lee, H. C., Foster, R., and Chua, N. H. (1997). Abscisic acid signaling through cyclic ADP-ribose in plants. *Science* **278**, 2126–2130.

Wu, Y., Sanchez, J. P., Lopez-Molina, L., Himmelbach, A., Grill, E., and Chua, N. H. (2003). The abi1-1 mutation blocks ABA signaling downstream of cADPR action. *Plant J.* **34**, 307–315.

Xiong, L., Gong, Z., Rock, C. D., Subramanian, S., Guo, Y., Xu, W., Galbraith, D., and Zhu, J. K. (2001a). Modulation of abscisic acid signal transduction and biosynthesis by an Sm-like protein in *Arabidopsis*. *Dev. Cell* **1**, 771–781.

Xiong, L., Lee, B., Ishitani, M., Lee, H., Zhang, C., and Zhu, J. K. (2001b). FIERY1 encoding an inositol polyphosphate 1-phosphatase is a negative regulator of abscisic acid and stress signaling in *Arabidopsis*. *Genes Dev.* **15**, 1971–1984.

Xiong, L., Lee, H., Ishitani, M., Tanaka, Y., Stevenson, B., Koiwa, H., Bressan, R. A., Hasegawa, P. M., and Zhu, J. K. (2002). Repression of stress-responsive genes by FIERY2, a novel transcriptional regulator in *Arabidopsis*. *Proc. Natl. Acad. Sci. USA* **99**, 10899–10904.

Xue, G. P., and Loveridge, C. W. (2004). HvDRF1 is involved in abscisic acid-mediated gene regulation in barley and produces two forms of AP2 transcriptional activators, interacting preferably with a CT-rich element. *Plant J.* **37**, 326–339.

Yamamoto, H., Inomata, M., Tsuchiya, S., Nakamura, M., Uchiyama, T., and Oritani, T. (2000). Early biosynthetic pathway to abscisic acid in *Cercospora cruenta*. *Biosci. Biotechnol. Biochem.* **64**, 2075–2082.

Yamazaki, D., Yoshida, S., Asami, T., and Kuchitsu, K. (2003). Visualization of abscisic acid-perception sites on the plasma membrane of stomatal guard cells. *Plant J.* **35**, 129–139.

Yang, Z. (2002). Small GTPases: Versatile signaling switches in plants. *Plant Cell* **14**(Suppl.), S375–S388.

Yazaki, J., Shimatani, Z., Hashimoto, A., Nagata, Y., Fujii, F., Kojima, K., Suzuki, K., Taya, T., Tonouchi, M., Nelson, C., Nakagawa, A., Otomo, Y., Murakami, K., Matsubara, K., Kawai, J., Carninci, P., Hayashizaki, Y., and Kikuchi, S. (2004). Transcriptional profiling of genes responsive to abscisic acid and gibberellin in rice: Phenotyping and comparative analysis between rice and *Arabidopsis*. *Physiol. Genomics* **17**, 87–100.

Yoshida, R., Hobo, T., Ichimura, K., Mizoguchi, T., Takahashi, F., Aronso, J., Ecker, J. R., and Shinozaki, K. (2002). ABA-activated SnRK2 protein kinase is required for dehydration stress signaling in *Arabidopsis*. *Plant Cell Physiol.* **43**, 1473–1483.

Yu, J., Hu, S., Wang, J., Wong, G. K., Li, S., Liu, B., Deng, Y., Dai, L., Zhou, Y., Zhang, X., Cao, M., Liu, J., Sun, J., Tang, J., Chen, Y., Huang, X., Lin, W., Ye, C., Tong, W., Cong, L., Geng, J., Han, Y., Li, L., Li, W., Hu, G., Li, J., Liu, Z., Qi, Q., Li, T., Wang, X., Lu, H., Wu, T., Zhu, M., Ni, P., Han, H., Dong, W., Ren, X., Feng, X., Cui, P., Li, X., Wang, H., Xu, X., Zhai, W., Xu, Z., Zhang, J., He, S., Xu, J., Zhang, K., Zheng, X., Dong, J., Zeng, W., Tao, L., Ye, J., Tan, J., Chen, X., He, J., Liu, D., Tian, W., Tian, C., Xia, H., Bao, Q., Li, G., Gao, H., Cao, T., Zhao, W., Li, P., Chen, W., Zhang, Y., Hu, J., Liu, S., Yang, J., Zhang, G., Xiong, Y., Li, Z., Mao, L., Zhou, C., Zhu, Z., Chen, R., Hao, B., Zheng, W., Chen, S., Guo, W., Tao, M., Zhu, L., Yuan, L., and Yang, H. (2002). A draft sequence of the rice genome (*Oryza sativa* L. ssp. *indica*). *Science* **296**, 79–92.

Zentella, R., Yamauchi, D., and Ho, T. H. D. (2002). Molecular dissection of the gibberellin/abscisic acid signaling pathways by transiently expressed RNA interference in barley aleurone cells. *Plant Cell* **14**, 2289–2301.

Zhang, W., Qin, C., Zhao, J., and Wang, X. (2004a). Phospholipase D alpha 1-derived phosphatidic acid interacts with ABI1 phosphatase 2C and regulates abscisic acid signaling. *Proc. Natl. Acad. Sci. USA* **101**, 9508–9513.

Zhang, Z. L., Xie, Z., Zou, X., Casaretto, J., Ho, T. H., and Shen, Q. J. (2004b). A rice WRKY gene encodes a transcriptional repressor of the gibberellin signaling pathway in aleurone cells. *Plant Physiol.* **134,** 1500–1513.

Zheng, Z. L., Nafisi, M., Tam, A., Li, H., Crowell, D. N., Chary, S. N., Schroeder, J. I., Shen, J., and Yang, Z. (2002). Plasma membrane-associated ROP10 small GTPase is a specific negative regulator of abscisic acid responses in *Arabidopsis*. *Plant Cell* **14,** 2787–2797.

Zhu, J., Gong, Z., Zhang, C., Song, C. P., Damsz, B., Inan, G., Koiwa, H., Zhu, J. K., Hasegawa, P. M., and Bressan, R. A. (2002). OSM1/SYP61: A syntaxin protein in *Arabidopsis* controls abscisic acid-mediated and non-abscisic acid-mediated responses to abiotic stress. *Plant Cell* **14,** 3009–3028.

Zhu, J. K. (2002). Salt and drought stress signal transduction in plants. *Annu. Rev. Plant Biol.* **53,** 247–273.

Zou, X., Seemann, J. R., Neuman, D., and Shen, Q. J. (2004). A *WRKY* gene from creosote bush encodes an activator of the ABA signaling pathway. *J. Biol. Chem.* , 10.1074/jbc.M408536200.

8

Cytokinin Biosynthesis and Regulation

Hitoshi Sakakibara

Plant Science Center, RIKEN 1-7-22 Suehiro, Tsurumi, Yokohama 230–0045, Japan

I. Introduction
II. Structural Variation of Cytokinin
    A. *Structural Variations*
    B. *Biological Activity and Stability*
III. Biosynthesis of Cytokinin
    A. *Basic Scheme of Cytokinin Biosynthesis*
    B. *Agrobacterium*
    C. *Slime Mold*
    D. *Higher Plants*
    E. *Trans-Zeatin Biosynthesis in Higher Plants*
IV. Regulation of Cytokinin Biosynthesis in Higher Plants
    A. *Regulation by Plant Hormones*
    B. *Regulation by Nitrogen Sources*
    References

Most natural cytokinins (CKs) are adenine derivatives that carry an isoprene-derived side chain at the N^6-terminus. Structural variation at the isoprenoid side chain alters their biological activity and stability. The first step of *de novo* synthesis of CKs is catalyzed by adenosine phosphate-isopentenyltransferase (IPT), which produces isopentenyl-adenine nucleotide. In higher plants, *trans*-zeatin (tZ), a major CK, is formed by subsequent hydroxylation, which is catalyzed by a cytochrome P450 monooxygenase (P450), CYP735A1 or CYP735A2. Biochemical characterization of IPTs revealed that the substrate specificities differ between *Agrobacterium* and higher plants. *Agrobacterium* IPTs have the ability to produce tZ-type species directly by use of hydroxymethylbutenyl diphosphate as the side chain donor. Analyses of expression patterns of genes for CK metabolic enzymes suggest that CK biosynthesis and homeostasis are finely controlled by internal and external environmental factors such as phytohormones and inorganic nitrogen sources. This regulatory system appears important in linking nutrient signals and morphogenetic responses. © 2005 Elsevier Inc.

I. INTRODUCTION

Cytokinins are involved in the regulation of various processes of plant growth and development, such as germination, shoot differentiation, and leaf senescence, while interacting with other phytohormones (Mok, 1994). After the identification of kinetin, a CK (Fig. 1; Miller *et al.*, 1955), extensive chemical analyses have identified several adenine derivatives and phenylurea derivatives as CKs. To date, natural CKs are adenine derivatives that carry either an isoprene-derived side chain or an aromatic side chain at the N^6-terminus (Fig. 1; Mok and Mok, 2001; Strnad, 1997); they are conventionally called isoprenoid CKs or aromatic CKs, respectively. Since isoprenoid CKs are more abundant in plants than are aromatic CKs, most studies have focused on isoprenoid CK synthesis and metabolism. Phenylurea-type species (Fig. 1) have strong CK activity (Mok *et al.*, 1982; Shudo, 1994), but there is no evidence that any of them occur naturally in plants (Mok and Mok, 2001).

Thanks to the accomplishment of the *Arabidopsis thaliana* genome project (Arabidopsis Genome Initiative, 2000) and the development of sensitive technologies for plant hormone analysis, several breakthroughs in CK biosynthesis, metabolism, and signal transduction were achieved during the last decade. Findings tell us that CK biosynthesis pathways in plants and bacteria evolved independently in terms of substrate specificity and that CK biosynthesis and homeostasis are finely controlled by factors such as other phytohormones and inorganic nitrogen sources. After a brief overview of CK structure and activity, we focus on genes and the way they regulate the biosynthesis of isoprenoid CKs. Excellent reviews on CK signaling that complement our paper are available in the literature (Aoyama and Oka, 2003; Grefen and Harter, 2004; Hutchison and Kieber, 2002; Kakimoto, 2003; Mizuno, 2004).

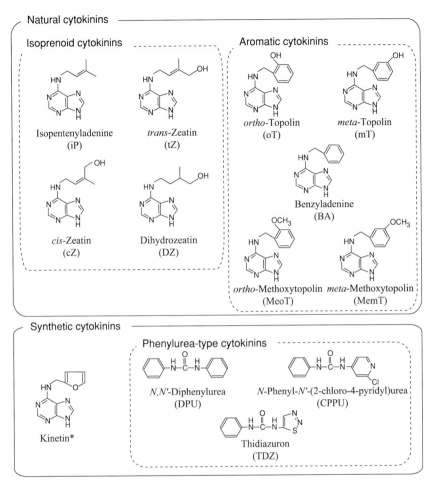

FIGURE 1. Structures of representative active CK species; only trivial names are shown. Commonly used abbreviations in parentheses. *Kinetin was reported to be present in human urine (Barciszewski et al., 2000) but not in plants. Thus, it is categorized as a synthetic CK in this review.

II. STRUCTURAL VARIATION OF CYTOKININ

A. STRUCTURAL VARIATIONS

The side chains of natural isoprenoid CKs in plants vary in the presence or the absence of a hydroxyl group at the end of the prenyl chain and the stereoisomeric position (Mok and Mok, 2001; Skoog and Armstrong, 1970). Common natural derivatives are isopentenyladenine (iP), tZ, cis-zeatin (cZ), and dihydrozeatin (DZ) (Fig. 1). Several additional synthetic variants also showed CK activity in bioassays (Iwamura, 1994; Skoog and Armstrong, 1970).

Aromatic CKs were identified in plant species including poplar (Strnad, 1997) and *Arabidopsis* (Tarkowska *et al.*, 2003). They include benzyladenine and its hydroxylated derivatives, such as *meta*-topolins and *ortho*-topolins (Fig. 1). It is not clear to date whether aromatic CKs are common in plants or not.

In most cases, natural CKs are also present in plant tissues as the corresponding nucleosides, nucleotides, and glycosides (Fig. 2). Thus, common or similar metabolic pathways can be expected to be involved in their interconversion regardless of the side chain variations. Due to the structural similarity between CKs and adenine, the ribosylation and phosphoribosylation reactions of the adenine moiety have been suggested to be shared with the purine metabolic pathway (i.e., salvage pathway; Chen, 1997; Mok and Mok, 2001). The enzymes for the N-glucosylation of the adenine moiety (Hou *et al.*, 2004) and those for the hydroxylation (Takei *et al.*, 2004b), O-glucosylation, and O-xylosylation of the side chain (Martin *et al.*, 1999a,b, 2001) are specific for particular CK substrates.

B. BIOLOGICAL ACTIVITY AND STABILITY

While classic bioassays have clarified the general structural features of active CKs, the identification of CK receptors enabled us to understand the ligand specificity in more detail. Assays using heterologous expression

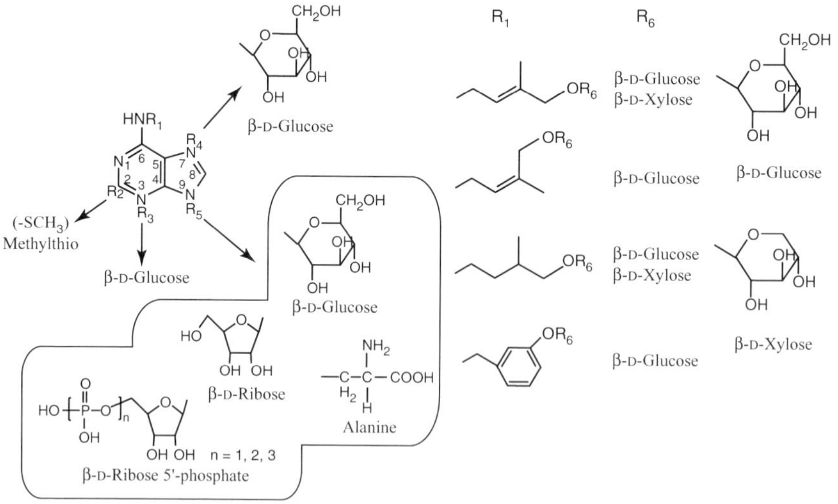

FIGURE 2. Cytokinin conjugates with sugars, sugar-phosphates, and others.

systems of CK receptors in yeasts (Inoue *et al.*, 2001; Suzuki *et al.*, 2001) and *Escherichia coli* (Spichal *et al.*, 2004; Suzuki *et al.*, 2001; Yamada *et al.*, 2001; Yonekura-Sakakibara *et al.*, 2004) demonstrated that free-base forms are the primary ligands, whereas the sugar conjugates (i.e., nucleosides, nucleotides, and glycosides) are less active or inactive. In some previous studies, doses of conjugates required to evoke typical CK responses seemed comparable to the equivalent doses of the free bases (Letham *et al.*, 1983; Schmitz and Skoog, 1972; Spiess, 1975; Takei *et al.*, 2001b), but in most cases this probably was caused by the corresponding free base being released by deconjugation during the experiments. In the free-base species tZ, iP, and cZ, tZ exhibited the highest and cZ the lowest activity in bioassays (Kaminek *et al.*, 1979). However, in the *E. coli* expression system, some of the CK receptors from maize responded to cZ with a similar sensitivity as to tZ (e.g., ZmHK1), while others responded to tZ riboside as well as to tZ (e.g., ZmHK2) (Yonekura-Sakakibara *et al.*, 2004). Cytokinin receptors in higher plants are encoded by a small gene family (Inoue *et al.*, 2001; Suzuki *et al.*, 2001; Ueguchi *et al.*, 2001; Yonekura-Sakakibara *et al.*, 2004). Studies suggest that each receptor has a different ligand preference (Spichal *et al.*, 2004; Yamada *et al.*, 2001; Yonekura-Sakakibara *et al.*, 2004). Structural variation of side chains therefore might enable the transduction of different physiological messages to regulate plant development.

Cytokinin levels *in planta* are determined by the balance of *de novo* synthesis and degradation. Cytokinin oxidase/dehydrogenases (CKX) catalyze the degradation step (Armstrong, 1994; Galuszka *et al.*, 2001; Schmülling *et al.*, 2003). *In vitro* studies showed that CKXs have different affinities for each of their substrates. Some CKXs have a higher affinity for iP than tZ (Bilyeu *et al.*, 2001), while others prefer tZ (Galuszka *et al.*, 2004). *cis*-Zeatin is generally less efficiently processed by CKXs than iP and tZ (Bilyeu *et al.*, 2001). Since CKXs recognize the double bond of the isoprenoid side chain, DZ is resistant to CKX (Armstrong, 1994). Sugar conjugates, such as *O*-glucosides of tZ and cZ, are also less affected by CKXs (Armstrong, 1994). Thus, turnover rates of different CKs *in vivo* vary, which may be important for the different physiological roles of the CK species.

III. BIOSYNTHESIS OF CYTOKININ

A. BASIC SCHEME OF CYTOKININ BIOSYNTHESIS

The first step in the biosynthesis of isoprenoid CKs is *N*-prenylation of adenosine 5′-phosphates (AMP, ADP, or ATP) with dimethylallyl diphosphate (DMAPP) or hydroxymethylbutenyl diphosphate (HMBDP), which is catalyzed by adenosine phosphates-isopentenyltransferase (IPT; EC

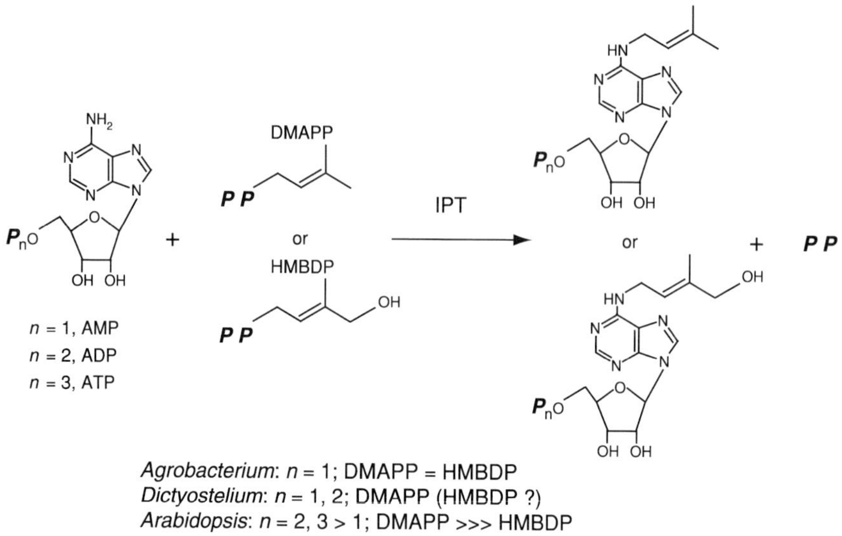

FIGURE 3. Primary reaction of CK biosynthesis in bacteria, slime molds, and higher plants. Substrate specificity of IPT in each organism is summarized at the bottom. *PP*, diphospholic acid. Other abbreviations as defined in the text.

2.5.1.27; Fig. 3). Hydroxymethylbutenyl diphosphate is a metabolic intermediate of the methylerythritol phosphate (MEP) pathway that occurs in bacteria and plastids (Hecht et al., 2001). Dimethylallyl diphosphate is synthesized *via* the MEP pathway and the mevalonate (MVA) pathway that is commonly found in the cytosol of eukaryotes (Lichtenthaler, 1999; Rohmer, 1999). When DMAPP is used as a substrate, the primary product is iP nucleotide. The tZ nucleotide is formed when IPT utilizes HMBDP. Biochemical studies revealed that IPT substrate specificities differ between bacteria, slime mold, and higher plants (Kakimoto, 2001; Sakakibara, 2004).

On the other hand, tRNA prenylation also contributes, at least to some extent, to CK production. Some of the tRNA species with anticodons complementary to codons beginning with uridine, such as tRNA[Leu] and tRNA[Ser], carry a prenylated adenosine adjacent to the anticodon. This decoration increases the fidelity of translation. When the tRNA is degraded, the prenylated adenosine is released as a CK (Fig. 4). The first step of the pathway leading to CKs is catalyzed by tRNA-isopentenyltransferase (tRNA-IPT; EC 2.5.1.8). Historically, tRNA-degradation had been the first source of CK that was identified (Skoog et al., 1966). However, calculations of turnover rates of tRNA suggested that the metabolic flow of tRNA-derived CKs was not the major pathway in ordinary plants (Klämbt, 1992). Since the prenyl-moiety of the tRNA contains a *cis*-hydroxylated group (Vreman et al., 1978), tRNA-degradation is a source of cZ-type CKs. Studies

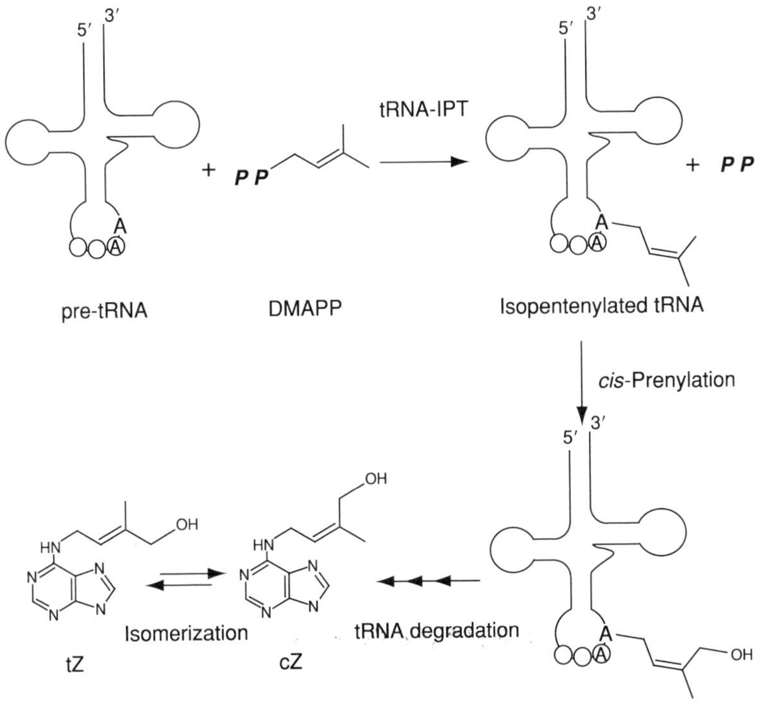

FIGURE 4. Proposed pathway for tRNA-mediated cZ biosynthesis. pre-tRNA, precursor tRNA; *PP*, diphospholic acid. Other abbreviations as defined in the text. *cis*-Zeatin can be reversibly converted to tZ by CK *cis–trans* isomerases (Bassil *et al.*, 1993), the genes of which remain to be identified.

demonstrated that a large fraction of the cZ side chains are derived from the MVA pathway, whereas side chains of iP and tZ predominantly originate from the MEP pathway (Kasahara *et al.*, 2004). This suggests that plants can modulate the levels of cZ-type species independently.

B. AGROBACTERIUM

Cytokinins are produced not only in plants but also in some microorganisms such as *Agrobacterium tumefaciens*. In *A. tumefaciens*, *Tmr*, an IPT gene, is located in the T-DNA region of the Ti-plasmid, a large extrachromosomal element (Van Larebeke *et al.*, 1974). The gene is integrated into the nuclear genome of the host plant after infection. *Tmr* functions in host plant cells to produce CKs that support tumorigenesis. *Tmr* has similar affinities for DMAPP and HMBDP (Sakakibara *et al.*, 2005) and a specific and high affinity for AMP (Blackwell and Horgan, 1993). Thus, the primary product of the enzymatic reaction *in vitro* is iP riboside 5′-monophosphate (iPRMP) or tZ riboside 5′-monophosphate. In contrast to *in vitro* studies, crown galls and transgenic plants expressing *Tmr* predominantly contain tZ-type species (Åstot *et al.*, 2000; Faiss *et al.*, 1997; Stuchbury *et al.*, 1979). We have shown that *Tmr* has the potential of being imported into the stroma of the host plastids, where it utilizes HMBDP to produce tZ-type CKs directly (Sakakibara *et al.*, 2005).

Tzs is another IPT gene found in the *vir*-region of nopaline-type Ti-plasmids that promotes T-DNA transfer efficiency (John and Amasino, 1988; Powell *et al.*, 1988). The enzyme utilizes HMBDP as an isoprene donor as well as DMAPP (Krall *et al.*, 2002).

C. SLIME MOLD

Dictyostelium discoideum, a slime mold, produces discadenine, 3-(3-amino-3-carboxypropyl)-N^6-Δ^2-isopentenyladenine, an inhibitor of spore germination (Abe *et al.*, 1976). The first step of discadenine biosynthesis is prenylation of AMP that is catalyzed by IPT. *D. discoideum* IPT utilizes AMP or ADP as a prenyl side-chain acceptor but not ATP or cyclic AMP (Ihara *et al.*, 1984; Taya *et al.*, 1978). Substrate preferences of the *D. discoideum* IPT for isoprenoid donors have not yet been well characterized.

D. HIGHER PLANTS

Higher plant IPT genes have been found in *Arabidopsis* (Kakimoto, 2001; Takei *et al.*, 2001a), petunia (Zubko *et al.*, 2002), and hop (Sakano *et al.*, 2004). In *Arabidopsis*, seven IPT genes (*AtIPT1* and *AtIPT3–AtIPT8*) have been identified as CK biosynthesis genes (Kakimoto, 2001; Sun *et al.*, 2003; Takei *et al.*, 2001a). Biochemical studies showed that plant IPTs prefer ADP or ATP to AMP as prenyl acceptors, resulting in the production of iP riboside 5′-diphosphate (iPRDP) or iP riboside 5′-triphosphate (iPRTP), respectively (Kakimoto, 2001; Sakakibara, 2004; Takei *et al.*, 2003b). Although some *Arabidopsis* IPTs could utilize HMBDP as a prenyl donor *in vitro*, the affinity was quite low and there is little evidence that tZ-type species are formed via the direct reaction *in vivo* (Sakakibara *et al.*, 2005; Takei *et al.*, 2003a).

In *Arabidopsis*, IPTs are differentially distributed to plastids, mitochondria, and the cytosol. AtIPT1, AtIPT3, AtIPT5, and AtIPT8 locate in the plastids and produce CKs with side chain precursors derived from the MEP pathway (Kasahara *et al.*, 2004). On the other hand, lovastatin, an inhibitor of the MVA pathway, was reported to significantly decrease CK accumulation in tobacco BY-2 cells (Laureys *et al.*, 1998, 1999). In *Arabidopsis*, AtIPT4 and AtIPT7 are localized in the cytosol and mitochondria, respectively

(Kasahara *et al.*, 2004). Thus, the MVA pathway seems to provide DMAPP for CK synthesis in some situations.

Analyses of spatial expression patterns of *AtIPTs* using their promoter:: reporter genes suggested that the synthesis of CKs was restricted to specific tissues and cells in roots and aerial organs (Miyawaki *et al.*, 2004; Takei *et al.*, 2004a). For instance, *AtIPT3* and *AtIPT5* are expressed in phloem companion cells and lateral root primordia, respectively (Miyawaki *et al.*, 2004; Takei *et al.*, 2004a). Despite a frameshift mutation occurring in some ecotypes (Kakimoto, 2001; K. Takei, unpublished results), the transcript of *AtIPT6* is generally accumulated in siliques (Miyawaki *et al.*, 2004; Takei *et al.*, 2004a).

E. *TRANS*-ZEATIN BIOSYNTHESIS IN HIGHER PLANTS

In higher plants, there are two possible pathways for tZ biosynthesis, the iPRMP-dependent and the iPRMP-independent one (Fig. 5; Åstot *et al.*, 2000; Nordström *et al.*, 2004; Takei *et al.*, 2004b). In the iPRMP-dependent pathway, tZ synthesis generally was thought to be catalyzed by a P450, although this enzymatic activity had been demonstrated only in a microsomal fraction of cauliflower (Chen and Leisner, 1984). The genes *CYP735A1*

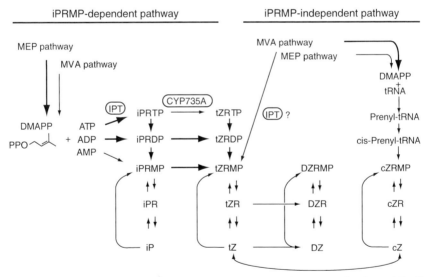

FIGURE 5. Current model of the isoprenoid CK metabolic pathway in plants. iPR, iP riboside; tZRTP, tZ riboside 5'-triphosphate; tZRDP, tZ riboside 5'-diphosphate; tZRMP, tZ riboside 5'-monophosphate; tZR, tZ riboside; DZR, DZ riboside; cZR, cZ riboside. Other abbreviations as defined in the text. Interconversions between nucleotides, nucleosides, and free bases (e.g., iPRMP, iPR, and iP) are catalyzed by enzymes of the purine salvage pathway.

and *CYP735A2* were shown to encode CK hydroxylases in *Arabidopsis* (Takei *et al.*, 2004b). Interestingly, both enzymes utilize iP-nucleotides but not the nucleoside and free-base forms. Comparison of the specificity constants (k_{cat}/K_m) for iP-nucleotides suggested that CYP735As preferentially use iPRMP or iPRDP rather than iPRTP (Takei *et al.*, 2004b). Although the physiological role of the CK nucleotides remains to be clarified, the nucleotide-specific hydroxylation indicates that they form a metabolic pool for side chain modifications. If the free-base cytokinins iP and tZ serve distinct physiological functions, the metabolic compartmentalization of the corresponding nucleotides may be important to maintain the physiological division of tasks at the free-base level.

In the iPRMP-independent pathway, tZ riboside 5′-phosphates are assumed to be produced directly by IPT using an unknown hydroxylated sidechain precursor (Åstot *et al.*, 2000). This precursor probably is derived from the MVA pathway because mevastatin, an inhibitor of that pathway, reduces the rate of tZ biosynthesis (Åstot *et al.*, 2000). Although HMBDP is the best candidate for the side-chain precursor, it is an intermediate of the MEP pathway (Hecht *et al.*, 2001) and plant IPTs hardly use HMBDP (Sakakibara *et al.*, 2005; Takei *et al.*, 2003a). Thus, the biochemical nature of the iPRMP-independent pathway remains obscure.

IV. REGULATION OF CYTOKININ BIOSYNTHESIS IN HIGHER PLANTS

A. REGULATION BY PLANT HORMONES

The expression of *IPT*, *CKX*, and *CYP735A*s is coordinately or differentially regulated by phytohormones. Cytokinins, auxin, and abscisic acid (ABA) appear to be the major factors controlling the CK accumulation level. In *Arabidopsis*, for instance, the accumulation of the transcripts of *AtIPT5* and *AtIPT7* is promoted by auxin, whereas the transcript levels of *AtIPT1*, *AtPT3*, *AtIPT5*, and *AtIPT7* are negatively regulated by CK (Miyawaki *et al.*, 2004). In pea, some of the expression of IPT genes is negatively regulated by auxin in the stem (H. Mori, personal communication). On the other hand, the expression of both *CYP735A*s is upregulated by CKs in roots but downregulated by auxin or ABA (Takei *et al.*, 2004b). Genes for CKX in maize are upregulated by CK or ABA (Brugiere *et al.*, 2003). These regulation patterns suggest that the products of the genes mentioned are involved as antagonists in the regulation of the cellular CK level and of the balance between iP and tZ that interact with auxin and/or ABA (Fig. 6). The interdependent regulation of phytohormones might provide the basis for the variable morphogenetic responses of plants to environmental factors. For example, in the control of outgrowth and dormancy of

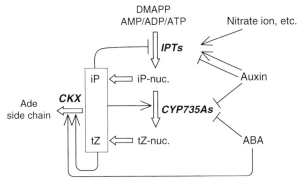

FIGURE 6. Schematic representation of the regulation of genes for CK metabolizing enzymes. Ade, adenine; iP-nuc., isopentenyladenine nucleotides; tZ-nuc., *trans*-zeatin nucleotides. Other abbreviations as defined in the text. Thick arrows indicate metabolic flow. Lines ending with arrowheads show positive regulation, whereas lines ending in cross-lines indicate negative regulation.

axillary buds, the mutual regulation of auxin, ABA, and CKs has been proposed to play a central role (Shimizu-Sato and Mori, 2001).

B. REGULATION BY NITROGEN SOURCES

Inorganic nitrogen is a crucial, often limiting factor for plant growth and development. In most natural soils, NO_3^- is the major form of inorganic nitrogen. NO_3^- functions not only as a nutrient but also as a primary signal to activate the expression of the assimilatory and related genes (Stitt, 1999; Wang *et al.*, 2000). Studies have revealed a close relationship between NO_3^- and CK biosynthesis. The expression profiles of *AtIPTs* in response to inorganic nitrogen sources showed that *AtIPT3* and *AtIPT5* are regulated differentially by nitrogen availability (Miyawaki *et al.*, 2004; Takei *et al.*, 2004a): *AtIPT3* rapidly and specifically responded to NO_3^- under nitrogen-limited conditions, while *AtIPT5* responded to both NO_3^- and NH_4^+ under long-term treatment (Takei *et al.*, 2004a). This dual response system regulating CK biosynthesis might be important in allowing plants to cope with changes in nitrogen status. It is well-known that NO_3^- induces accumulation of CKs in roots and shoots (Samuelson and Larsson, 1993; Takei *et al.*, 2001b, 2004a). In a *Ds* transposon-insertion mutant of *AtIPT3*, NO_3^--dependent CK accumulation was greatly reduced (Takei *et al.*, 2004a). These lines of evidence strongly suggest that *AtIPT3* is a key determinant of CK biosynthesis in response to rapid changes in the availability of NO_3^-.

*CYP735A*s are expressed in roots and stem but the expression is upregulated by CK specifically in roots (Takei *et al.*, 2004b). Thus, tZ biosynthesis

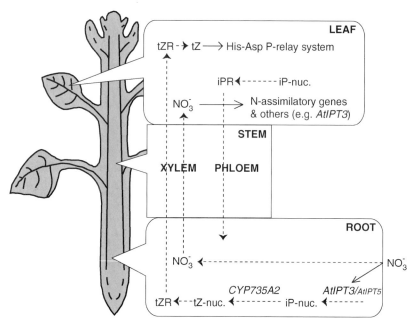

FIGURE 7. Nitrogen-dependent regulation of CK biosynthesis. Solid lines with arrowheads indicate positive regulation of gene expression. Broken lines with arrowheads show metabolic flow or translocation of CKs. iPR, iP riboside; tZR, tZ riboside; iP-nuc., isopentenyladenine nucleotides; tZ-nuc., *trans*-zeatin nucleotides. Other abbreviations as defined in the text.

in roots should also be regulated by NO_3^-. Cytokinin has been proposed as a long-distance signal in communicating nitrogen availability (Fig. 7; Simpson *et al.*, 1982; Takei *et al.*, 2002) because the translocation rate of tZ riboside via xylem vessels is controlled by NO_3^- in the root medium (Takei *et al.*, 2001b). While CKs play an important role as a local signal to regulate plant development, root-to-shoot signaling by tZ-type species in the transpiration stream must be essential for the integration of nutrient signals on the whole-plant level. On the other hand, leaf exudates contained mainly iP-type species (Corbesier *et al.*, 2003). Thus, *trans*-hydroxylation may be important for the compartmentalization of the CK species and may control the direction of translocation.

ACKNOWLEDGMENTS

This study was supported by the Ministry of Education, Culture, Sports, Science, and Technology and the Ministry of Agriculture, Forestry, and Fisheries, Japan.

REFERENCES

Abe, H., Uchiyama, M., Tanaka, Y., and Saito, H. (1976). Structure of discadenine, a spore germination inhibitor from cellular slime-mold, *Dictyostelium discoideum*. *Tetrahedron Lett.* **42,** 3807–3810.

Aoyama, T., and Oka, A. (2003). Cytokinin signal transduction in plant cells. *J. Plant Res.* **116,** 221–231.

Arabidopsis Genome, Initiative (2000). Analysis of the genome sequence of the flowering plant *Arabidopsis thaliana*. *Nature* **408,** 796–815.

Armstrong, D. J. (1994). Cytokinin oxidase and the regulation of cytokinin degradation. *In* "Cytokinins: Chemistry, Activity, and Function" (D. W. S. Mok and M. C. Mok, Eds.), pp. 139–154. CRC Press, Boca Raton, Florida.

Åstot, C., Dolezal, K., Nordström, A., Wang, Q., Kunkel, T., Moritz, T., Chua, N.-H., and Sandberg, G. (2000). An alternative cytokinin biosynthesis pathway. *Proc. Natl. Acad. Sci. USA* **97,** 14778–14783.

Barciszewski, J., Mielcarek, M., Stobiecki, M., Siboska, G., and Clark, B. F. (2000). Identification of 6-furfuryladenine (kinetin) in human urine. *Biochem. Biophys. Res. Commun.* **279,** 69–73.

Bassil, N. V., Mok, D., and Mok, M. C. (1993). Partial purification of a *cis–trans*-isomerase of zeatin from immature seed of *Phaseolus vulgaris* L. *Plant Physiol.* **102,** 867–872.

Bilyeu, K. D., Cole, J. L., Laskey, J. G., Riekhof, W. R., Esparza, T. J., Kramer, M. D., and Morris, R. O. (2001). Molecular and biochemical characterization of a cytokinin oxidase from maize. *Plant Physiol.* **125,** 378–386.

Blackwell, J. R., and Horgan, R. (1993). Cloned *Agrobacterium tumefaciens ipt1* gene product, DMAPP:AMP isopentenyl transferase. *Phytochemistry* **34,** 1477–1481.

Brugiere, N., Jiao, S., Hantke, S., Zinselmeier, C., Roessler, J. A., Niu, X., Jones, R. J., and Habben, J. E. (2003). Cytokinin oxidase gene expression in maize is localized to the vasculature, and is induced by cytokinins, abscisic acid, and abiotic stress. *Plant Physiol.* **132,** 1228–1240.

Chen, C.-M., and Leisner, S. M. (1984). Modification of cytokinins by cauliflower microsomal enzymes. *Plant Physiol.* **75,** 442–446.

Chen, C.-M. (1997). Cytokinin biosynthesis and interconversion. *Physiol. Plant.* **101,** 665–673.

Corbesier, L., Prinsen, E., Jacqmard, A., Lejeune, P., Van Onckelen, H., Perilleux, C., and Bernier, G. (2003). Cytokinin levels in leaves, leaf exudate and shoot apical meristem of *Arabidopsis thaliana* during floral transition. *J. Exp. Bot.* **54,** 2511–2517.

Faiss, M., Zalubilová, J., Strnad, M., and Schmülling, T. (1997). Conditional transgenic expression of the ipt gene indicates a function for cytokinins in paracrine signaling in whole tobacco plants. *Plant J.* **12,** 401–415.

Galuszka, P., Frebort, I., Sebela, M., Sauer, P., Jacobsen, S., and Pec, P. (2001). Cytokinin oxidase or dehydrogenase? Mechanism of cytokinin degradation in cereals. *Eur. J. Biochem.* **268,** 450–461.

Galuszka, P., Frebortova, J., Werner, T., Yamada, M., Strnad, M., Schmülling, T., and Frebort, I. (2004). Cytokinin oxidase/dehydrogenase genes in barley and wheat. *Eur. J. Biochem.* **271,** 3990–4002.

Grefen, C., and Harter, K. (2004). Plant two-component systems: Principles, functions, complexity and cross talk. *Planta* **219,** 733–742.

Hecht, S., Eisenreich, W., Adam, P., Amslinger, S., Kis, K., Bacher, A., Arigoni, D., and Rohdich, F. (2001). Studies on the nonmevalonate pathway to terpenes: The role of the GcpE (IspG) protein. *Proc. Natl. Acad. Sci. USA* **98,** 14837–14842.

Hou, B., Lim, E.-K., Higgins, G. S., and Bowles, D. J. (2004). *N*-glucosylation of cytokinins by glycosyltransferases of *Arabidopsis thaliana*. *J. Biol. Chem.* **279,** 47822–47832.

Hutchison, C. E., and Kieber, J. J. (2002). Cytokinin signaling in *Arabidopsis*. *Plant Cell* **14**, S47–S59.

Ihara, M., Taya, Y., Nishimura, S., and Tanaka, Y. (1984). Purification and some properties of delta 2-isopentenylpyrophosphate: 5′AMP delta 2-isopentenyltransferase from the cellular slime mold *Dictyostelium discoideum*. *Arch. Biochem. Biophys.* **230**, 652–660.

Inoue, T., Higuchi, M., Hashimoto, Y., Seki, M., Kobayashi, M., Kato, T., Tabata, S., Shinozaki, K., and Kakimoto, T. (2001). Identification of CRE1 as a cytokinin receptor from *Arabidopsis*. *Nature* **409**, 1060–1063.

Iwamura, H. (1994). Cytokinin antagonists: Synthesis and biological activity. *In* "Cytokinins: Chemistry, Activity, and Function" (D. W. S. Mok and M. C. Mok, Eds.), pp. 43–55. CRC Press, Boca Raton, Florida.

John, M. C., and Amasino, R. M. (1988). Expression of an *Agrobacterium* Ti plasmid gene involved in cytokinin biosynthesis is regulated by virulence loci and induced by plant phenolic compounds. *J. Bacteriol.* **170**, 790–795.

Kakimoto, T. (2001). Identification of plant cytokinin biosynthetic enzymes as dimethylallyl diphosphate: ATP/ADP isopentenyltransferases. *Plant Cell Physiol.* **42**, 677–685.

Kakimoto, T. (2003). Perception and signal transduction of cytokinins. *Ann. Rev. Plant Biol.* **54**, 605–627.

Kaminek, M., Paces, V., Corse, J., and Challice, J. S. (1979). Effect of stereospecific hydroxylation of N^6-(delta-2-isopentenyl)adenosine on cytokinin activity. *Planta* **145**, 239–243.

Kasahara, H., Takei, K., Ueda, N., Hishiyama, S., Yamaya, T., Kamiya, Y., Yamaguchi, S., and Sakakibara, H. (2004). Distinct isoprenoid origins of *cis*- and *trans*-zeatin biosyntheses in *Arabidopsis*. *J. Biol. Chem.* **279**, 14049–14054.

Klämbt, D. (1992). The biogenesis of cytokinins in higher plants: Our present knowledge. *In* "Physiology and Biochemistry of Cytokinins in Plants" (M. Kaminek, D. W. S. Mok, and E. Zazímalová, Eds.), pp. 25–27. SPB Academic Publishing, The Hague.

Krall, L., Raschke, M., Zenk, M. H., and Baron, C. (2002). The Tzs protein from *Agrobacterium tumefaciens* C58 produces zeatin riboside 5′-phosphate from 4-hydroxy-3-methyl-2-(*E*)-butenyl diphosphate and AMP. *FEBS Lett.* **527**, 315–318.

Laureys, F., Dewitte, W., Witters, E., Van Montagu, M., Inze, D., and Van Onckelen, H. (1998). Zeatin is indispensable for the G2-M transition in tobacco BY-2 cells. *FEBS Lett.* **426**, 29–32.

Laureys, F., Smets, R., Lenjou, M., Van Bockstaele, D., Inze, D., and Van Onckelen, H. (1999). A low content in zeatin type cytokinins is not restrictive for the occurrence of G1/S transition in tobacco BY-2 cells. *FEBS Lett.* **460**, 123–128.

Letham, D. S., Palni, L. M. S., Tao, G.-Q., Gollnow, B. I., and Bates, C. M. (1983). Regulators of cell division in plant tissues XXIX. The activities of cytokinin glucosides and alanine conjugates in cytokinin bioassays. *J. Plant Growth Regul.* **2**, 103–115.

Lichtenthaler, H. K. (1999). The 1-deoxy-D-xylulose-5-phosphate pathway of isoprenoid biosynthesis in plants. *Ann. Rev. Plant Physiol. Plant Mol. Biol.* **50**, 47–65.

Martin, R. C., Mok, M. C., Habben, J. E., and Mok, D. W. (2001). A maize cytokinin gene encoding an *O*-glucosyltransferase specific to *cis*-zeatin. *Proc. Natl. Acad. Sci. USA* **98**, 5922–5926.

Martin, R. C., Mok, M. C., and Mok, D. W. (1999a). A gene encoding the cytokinin enzyme zeatin *O*-xylosyltransferase of *Phaseolus vulgaris*. *Plant Physiol.* **120**, 553–558.

Martin, R. C., Mok, M. C., and Mok, D. W. (1999b). Isolation of a cytokinin gene, *ZOG1*, encoding zeatin *O*-glucosyltransferase from *Phaseolus lunatus*. *Proc. Natl. Acad. Sci. USA* **96**, 284–289.

Miller, C. O., Skoog, F., Saltza, v.N. H., and Strong, M. (1955). Kinetin, a cell division factor from deoxyribonucleic acid. *J. Am. Chem. Soc.* **77**, 1329–1334.

Miyawaki, K., Matsumoto-Kitano, M., and Kakimoto, T. (2004). Expression of cytokinin biosynthetic isopentenyltransferase genes in *Arabidopsis*: Tissue specificity and regulation by auxin, cytokinin, and nitrate. *Plant J.* **37,** 128–138.

Mizuno, T. (2004). Plant response regulators implicated in signal transduction and circadian rhythm. *Curr. Opin. Plant Biol.* **7,** 499–505.

Mok, D. W., and Mok, M. C. (2001). Cytokinin metabolism and action. *Annu. Rev. Plant Physiol. Plant Mol. Biol.* **52,** 89–118.

Mok, M. C. (1994). Cytokinins and plant development—an overview. In "Cytokinins: Chemistry, Activity, and Function" (D. W. S. Mok and M. C. Mok, Eds.), pp. 155–166. CRC Press, Boca Raton, Florida.

Mok, M. C., Mok, D. W. S., Armstrong, D. J., Shudo, K., Isogai, Y., and Okamoto, T. (1982). Cytokinin activity of N-phenyl-N'-1,2,3-thiadiazol-5-ylurea (thidiazuron). *Phytochemistry* **21,** 1509–1511.

Nordström, A., Tarkowski, P., Tarkowska, D., Norbaek, R., Åstot, C., Dolezal, K., and Sandberg, G. (2004). Auxin regulation of cytokinin biosynthesis in *Arabidopsis thaliana*: A factor of potential importance for auxin-cytokinin-regulated development. *Proc. Natl. Acad. Sci. USA* **101,** 8039–8044.

Powell, G. K., Hommes, N. G., Kuo, J., Castle, L. A., and Morris, R. O. (1988). Inducible expression of cytokinin biosynthesis in *Agrobacterium tumefaciens* by plant phenolics. *Mol. Plant Microbe. Interact.* **1,** 235–242.

Rohmer, M. (1999). The discovery of a mevalonate-independent pathway for isoprenoid biosynthesis in bacteria, algae and higher plants. *Nat. Prod. Rep.* **16,** 565–574.

Sakakibara, H. (2004). Cytokinin biosynthesis and metabolism. In "Plant Hormones: Biosynthesis, Signal Transduction, Action!" (P. J. Davies, Ed.), pp. 95–114. Springer, Dordrecht.

Sakakibara, H., Kasahara, H., Ueda, N., Kojima, M., Takei, K., Hishiyama, S., Asami, T., Okada, K., Kamiya, Y., Yamaya, T., and Yamaguchi, S. (2005). *Agrobacterium tumefaciens* increases cytokinin production in plastids by modifying the biosynthetic pathway in the host plant. *Proc. Natl. Acad. Sci. USA* **102,** 9972–9977.

Sakano, Y., Okada, Y., Matsunaga, A., Suwama, T., Kaneko, T., Ito, K., Noguchi, H., and Abe, I. (2004). Molecular cloning, expression, and characterization of adenylate isopentenyltransferase from hop (*Humulus lupulus* L.). *Phytochemistry* **65,** 2439–2446.

Samuelson, M. E., and Larsson, C.-M. (1993). Nitrate regulation of zeatin riboside levels in barley roots: Effects of inhibitors of N assimilation and comparison with ammonium. *Plant Sci.* **93,** 77–84.

Schmitz, R. Y., and Skoog, F. (1972). Cytokinins: Synthesis and biological activity of geometric and position isomers of zeatin. *Plant Physiol.* **50,** 702–705.

Schmülling, T., Werner, T., Riefler, M., Krupkova, E., Bartrina, Y., and Manns, I. (2003). Structure and function of cytokinin oxidase/dehydrogenase genes of maize, rice, *Arabidopsis* and other species. *J. Plant Res.* **116,** 241–252.

Shimizu-Sato, S., and Mori, H. (2001). Control of outgrowth and dormancy in axillary buds. *Plant Physiol.* **127,** 1405–1413.

Shudo, K. (1994). Chemistry of phenylurea cytokinins. In "Cytokinins: Chemistry, Activity, and Function" (D. W. S. Mok and M. C. Mok, Eds.), pp. 35–42. CRC Press, Boca Raton, Florida.

Simpson, R. J., Lambers, H., and Dalling, M. J. (1982). Kinetin application to roots and its effect on uptake, translocation and distribution of nitrogen in wheat (*Triticum aestivum*) grown with a split root system. *Physiol. Plant.* **56,** 430–435.

Skoog, F., and Armstrong, D. J. (1970). Cytokinins. *Ann. Rev. Plant Physiol.* **21,** 359–384.

Skoog, F., Armstrong, D. J., Cherayil, J. D., Hampel, A. E., and Bock, R. M. (1966). Cytokinin activity: Localization in transfer RNA preparations. *Science* **154,** 1354–1356.

Spichal, L., Rakova, N. Y., Riefler, M., Mizuno, T., Romanov, G. A., Strnad, M., and Schmulling, T. (2004). Two cytokinin receptors of *Arabidopsis thaliana*, CRE1/AHK4 and AHK3, differ in their ligand specificity in a bacterial assay. *Plant Cell Physiol.* **45**, 1299–1305.

Spiess, L. D. (1975). Comparative activity of isomers of zeatin and ribosyl-zeatin on *Funaria hygrometrica*. *Plant Physiol.* **55**, 583–585.

Stitt, M. (1999). Nitrate regulation of metabolism and growth. *Curr. Opin. Plant Biol.* **2**, 178–186.

Strnad, M. (1997). The aromatic cytokinins. *Physiol. Plant.* **101**, 674–688.

Stuchbury, T., Palni, L. M. S., Horgan, R., and Wareing, P. F. (1979). The biosynthesis of cytokinins in crown-gall tissue of *Vinca rosea*. *Planta* **147**, 97–102.

Sun, J., Niu, Q. W., Tarkowski, P., Zheng, B., Tarkowska, D., Sandberg, G., Chua, N. H., and Zuo, J. (2003). The *Arabidopsis* AtIPT8/PGA22 gene encodes an isopentenyl transferase that is involved in *de novo* cytokinin biosynthesis. *Plant Physiol.* **131**, 167–176.

Suzuki, T., Miwa, K., Ishikawa, K., Yamada, H., Aiba, H., and Mizuno, T. (2001). The *Arabidopsis* sensor His-kinase, AHK4, can respond to cytokinins. *Plant Cell Physiol.* **42**, 107–113.

Takei, K., Dekishima, Y., Eguchi, T., Yamaya, T., and Sakakibara, H. (2003a). A new method for enzymatic preparation of isopentenyladenine-type and *trans*-zeatin-type cytokinins with radioisotope-labeling. *J. Plant Res.* **116**, 259–263.

Takei, K., Sakakibara, H., and Sugiyama, T. (2001a). Identification of genes encoding adenylate isopentenyltransferase, a cytokinin biosynthesis enzyme, in *Arabidopsis thaliana*. *J. Biol. Chem.* **276**, 26405–26410.

Takei, K., Sakakibara, H., Taniguchi, M., and Sugiyama, T. (2001b). Nitrogen-dependent accumulation of cytokinins in root and the translocation to leaf: Implication of cytokinin species that induces gene expression of maize response regulator. *Plant Cell Physiol.* **42**, 85–93.

Takei, K., Takahashi, T., Sugiyama, T., Yamaya, T., and Sakakibara, H. (2002). Multiple routes communicating nitrogen availability from roots to shoots: A signal transduction pathway mediated by cytokinin. *J. Exp. Bot.* **53**, 971–977.

Takei, K., Ueda, N., Aoki, K., Kuromori, T., Hirayama, T., Shinozaki, K., Yamaya, T., and Sakakibara, H. (2004a). *AtIPT3* is a key determinant of nitrate-dependent cytokinin biosynthesis in *Arabidopsis*. *Plant Cell Physiol.* **45**, 1053–1062.

Takei, K., Yamaya, T., and Sakakibara, H. (2003b). A method for separation and determination of cytokinin nucleotides from plant tissues. *J. Plant Res.* **116**, 265–269.

Takei, K., Yamaya, T., and Sakakibara, H. (2004b). *Arabidopsis CYP735A1* and *CYP735A2* encode cytokinin hydroxylases that catalyze the biosynthesis of *trans*-zeatin. *J. Biol. Chem.* **279**, 41866–41872.

Tarkowska, D., Dolezal, K., Tarkowski, P., Åstot, C., Holub, J., Fuksova, K., Schmülling, T., Sandberg, G., and Strnad, M. (2003). Identification of new aromatic cytokinins in *Arabidopsis thaliana* and *Populus x canadensis* leaves by LC-(+)ESI-MS and capillary liquid chromatography/frit-fast atom bombardment mass spectrometry. *Physiol. Plant.* **117**, 579–590.

Taya, Y., Tanaka, Y., and Nishimura, S. (1978). $5'$-AMP is a direct precursor of cytokinin in *Dictyostelium discoideum*. *Nature* **271**, 545–547.

Ueguchi, C., Koizumi, H., Suzuki, T., and Mizuno, T. (2001). Novel family of sensor histidine kinase genes in *Arabidopsis thaliana*. *Plant Cell Physiol.* **42**, 231–235.

Van Larebeke, N., Engler, G., Holsters, M., Van den Elsacker, S., Zaenen, I., Schilperoort, R. A., and Schell, J. (1974). Large plasmid in *Agrobacterium tumefaciens* essential for crown gall-inducing ability. *Nature* **252**, 169–170.

Vreman, H. J., Thomas, R., and Corse, J. (1978). Cytokinins in tRNA obtained from *Spinacia oleracea* L. leaves and isolated chloroplasts. *Plant Physiol.* **61**, 296–306.

Wang, R., Guegler, K., La Brie, S. T., and Crawford, N. M. (2000). Genomic analysis of a nutrient response in *Arabidopsis* reveals diverse expression patterns and novel metabolic and potential regulatory genes induced by nitrate. *Plant Cell* **12,** 1491–1509.

Yamada, H., Suzuki, T., Terada, K., Takei, K., Ishikawa, K., Miwa, K., Yamashino, T., and Mizuno, T. (2001). The *Arabidopsis* AHK4 histidine kinase is a cytokinin-binding receptor that transduces cytokinin signals across the membrane. *Plant Cell Physiol.* **42,** 1017–1023.

Yonekura-Sakakibara, K., Kojima, M., Yamaya, T., and Sakakibara, H. (2004). Molecular characterization of cytokinin-responsive histidine kinases in maize: Differential ligand preferences and response to *cis*-zeatin. *Plant Physiol.* **134,** 1654–1661.

Zubko, E., Adams, C. J., Machaekova, I., Malbeck, J., Scollan, C., and Meyer, P. (2002). Activation tagging identifies a gene from *Petunia hybrida* responsible for the production of active cytokinins in plants. *Plant J.* **29,** 797–808.

9

GIBBERELLIN METABOLISM AND SIGNALING

STEPHEN G. THOMAS,* IVO RIEU,* AND
CAMILLE M. STEBER[†]

*IACR Rothamsted Research, CPI Division, Harpenden, Hertfordshire,
AL5 2JQ, United Kingdom
[†]USDA-ARS and Department of Crop and Soil Science,
Washington State University Pullman, Washington 99164

I. Introduction
  A. Historical Perspective
  B. Gibberellins and Plant Development
II. Gibberellin Biosynthesis
  A. Introduction
  B. Involvement of MVA and MEP Pathways in Gibberellin Biosynthesis
  C. ent-Copalyl-Diphosphate Synthase
  D. ent-Kaurene Synthase
  E. ent-Kaurene Oxidase
  F. ent-Kaurenoic Acid Oxidase
  G. Gibberellin 13β-Hydroxylase
  H. Gibberellin 20-Oxidase
  I. Gibberellin 3-Oxidase
  J. Gibberellin 2-Oxidase and Gibberellin Inactivation
  K. Feedback and Feedforward Regulation of Gibberellin Metabolism
  L. Regulation of GA Metabolism by Light

III. Gibberellin Signal Transduction
   A. DELLA Proteins in Gibberellin Signaling
   B. Control of DELLA Protein Accumulation by E3 Ubiquitin Ligases
   C. Negative Regulation of Gibberellin Response
   D. Positive Regulation of Gibberellin Response
   E. Gibberellin-Response Genes
   F. Model for Gibberellin Signaling
IV. Cross-talk with Other Hormone-Signaling Pathways
   A. Gibberellin and Abscisic Acid Signaling
   B. Gibberellin and Brassinosteroid Signaling
   C. Gibberellin and Auxin Signaling
V. Perspectives
   References

Gibberellins (GAs) are a family of plant hormones controlling many aspects of plant growth and development including stem elongation, germination, and the transition from vegetative growth to flowering. Cloning of the genes encoding GA biosynthetic and inactivating enzymes has led to numerous insights into the developmental regulation of GA hormone accumulation that is subject to both positive and negative feedback regulation. Genetic and biochemical analysis of GA-signaling genes has revealed that posttranslational regulation of DELLA protein accumulation is a key control point in GA response. The highly conserved DELLA proteins are a family of negative regulators of GA signaling that appear subject to GA-stimulated degradation through the ubiquitin-26S proteasome pathway. This review discusses the regulation of GA hormone accumulation and signaling in the context of its role in plant growth and development. © 2005 Elsevier Inc.

I. INTRODUCTION

A. HISTORICAL PERSPECTIVE

Unlike mammals, plants have evolved to be very plastic in their development. Every plant cell is ostensibly a "stem cell" capable of giving rise to a wide array of developmental fates in response to signals from plant hormones, also referred to as phytohormones. Also, unlike mammals, plants do not have clearly defined source and target organs for hormone signals.

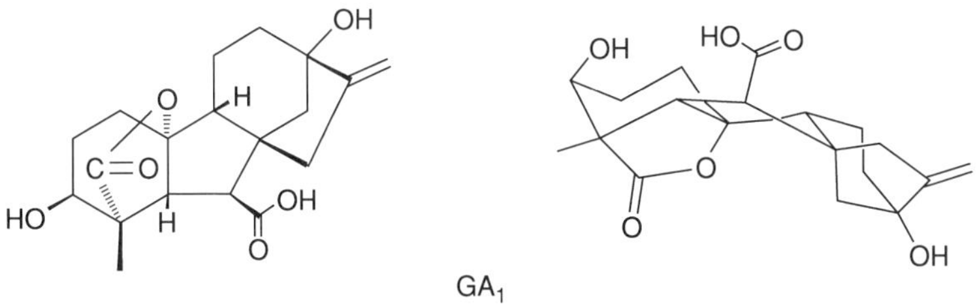

FIGURE 1. Example of the molecular structure of a gibberellin, GA_1, presented in 2D and 3D view. One-hundred and thirty-six naturally occurring GAs have been found in plants and fungi so far.

This has complicated the study of plant hormones. Numerous advances have been made in understanding the regulation of plant hormone accumulation, transport, and signaling through genetic, biochemical, and physiological approaches. This review is focused on the plant hormone gibberellin.

Gibberellins are a large family of tetracyclic diterpene plant hormones characterized by the *ent*-gibberellane ring system (Fig. 1). Gibberellins have been shown to promote many facets of plant growth and development including germination and stem elongation, and in most species transition to flowering, pollen tube elongation, and seed development (Olszewski *et al.*, 2002; Sun and Gubler, 2004). Every hormone signal transduction pathway is composed of two essential components, the control of hormone accumulation and reception of the hormone signal. This chapter will: (1) briefly review the history of GA research and the role of GA in regulating plant growth and development; (2) review the control of GA hormone accumulation through gene regulation; (3) review GA signal reception in the context of its role in plant growth and development; and (4) review the interaction of GA signaling with other hormone-signaling pathways.

Gibberellins were the first plant hormone identified (Phinney, 1983; Tamura, 1991). Ironically, the discovery of gibberellin by the Japanese scientist Eiichi Kurosawa in 1926 was based on its synthesis by the fungus *Gibberella fujikuroi*, the causative agent of *bakanae* disease in rice. The "foolish seedlings" infected by *bakanae* disease grew excessively tall and spindly. The rare infected seedlings that survived produced poor seed set. Kurosawa demonstrated that the fungal pathogen infecting these plants synthesized a chemical that could stimulate shoot elongation in rice and other grasses (Kurosawa, 1926). The structure of this chemical, gibberellin A_3 or GA_3, was proposed in 1956 and revised in 1961. The occurrence of gibberellins in higher plant species was discovered in the mid-1950s. This discovery marked the beginning of research on the role of GA in plant growth and development.

Since their discovery, over 136 GAs have been identified in plants and fungi; however, only a small fraction of these are biologically active in plants (Olszewski et al., 2002). Each unique GA has a number ranging from GA_1 to GA_{136}. Gibberellins are divided into two classes based on the number of carbon atoms, C20-GAs and C19-GAs, in which C20 has been replaced by a gamma-lactone ring. The synthesis of bioactive GAs is essentially a three-step process involving: (1) the formation of *ent*-kaurene in the proplastid, (2) the formation of $GA_{12/53}$ in the ER, and (3) the formation of active GA in the cytoplasm by successive oxidation steps. In most plant species, GA_1 or GA_4 are the bioactive GA. GA_1 and GA_4 are formed by similar pathways differing only in early 13-hydroxylation in the case of GA_1.

B. GIBBERELLINS AND PLANT DEVELOPMENT

The role of GA in plant growth and development has been elucidated through the physiological characterization of GA biosynthesis and signaling mutants and the characterization of GA-responsive genes. This section deals with the role of GA in seed development and germination, plant growth and elongation, flowering, and meristem cell identity.

1. GA in Seed Development and Germination

Our understanding of GA in seed development and germination is based on mutants or tissues with reduced accumulation of GAs (Bentsink and Koornneef, 2002; Ni and Bradford, 1993; Singh et al., 2002). For example, the *ga1*, *ga2*, and *ga3* mutants of *Arabidopsis* were isolated in an elegant screen for GA-dependent germination by Koornneef and van der Veen (1980). These mutants cause marked reduction in endogenous GA and are unable to germinate unless GA is applied externally. While seeds are an excellent source of GA, the failure to synthesize GA in these mutants does not completely block seed development (Bentsink and Koornneef, 2002). Thus, it was originally thought that GA is not required for seed development. However, physiological characterization of *Arabidopsis* plants constitutively expressing the GA catalytic enzyme *GAox2* revealed that reduced accumulation of GA in seed leads to increased probability of seed abortion (Singh et al., 2002). This suggests that GA is actually required in seed development. Moreover, reduced GA accumulation leads to reduced seed set by interfering with pollen tube elongation and silique expansion (Singh et al., 2002; Swain et al., 2004). How does GA stimulate germination? Germination and seedling growth require the production of hydrolytic enzymes to weaken the seed coat, mobilize seed nutrient storage reserves, stimulate plant embryo expansion and hypocotyl elongation, and activate the embryo meristem to produce new shoots and roots (Bewley and Black, 1994). Gibberellin has been implicated in all of these processes.

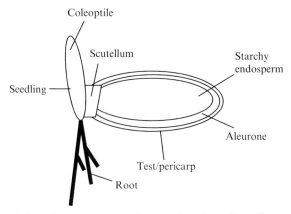

FIGURE 2. Schematic of a germinating cereal grain. Gibberellin produced in the germinating embryo stimulates production of α-amylase and other hydrolytic enzymes in the aleurone layer. These enzymes break down starch in the endosperm providing nutrition for the emerging seedling.

The germination process is considered complete when any part of the plant embryo emerges from the seed (Bewley and Black, 1994). Initial studies in tomato and muskmelon suggested that the decision to germinate results from the balance between the internal pressure of an expanding embryo and the external restraint of the endosperm cap or seed coat (Groot and Karssen, 1987; Ni and Bradford, 1993). Gibberellin-induced hydrolytic enzymes such as endo-[β]-mannase are apparently needed to weaken the endosperm cap in these species (Still and Bradford, 1997).

Gibberellin stimulation of seed nutrient storage mobilization is best illustrated by the cereal aleurone system (Jacobsen et al., 1995). Gibberellin synthesized by the plant embryo stimulates secretion of the hydrolytic enzymes including α-amylase by the aleurone layer. Aleurone-derived hydrolases diffuse to the adjacent endosperm where they degrade starch for use by the embryo (Fig. 2). Because the aleurone layer itself secretes no GA, it can be isolated and used to assay α-amylase secretion in response to hormone (Bush and Jones, 1988; Varner et al., 1965). α-Amylase is arguably the best characterized GA-responsive gene. Measurement of α-amylase enzyme activity and mRNA accumulation has been used to identify GA-responsive promoter elements and transcription factors (Sun and Gubler, 2004).

2. Gibberellin Stimulation of Growth and Elongation

Gibberellin stimulation of plant stem elongation was the basis for the hormone's discovery and remains a reliable assay for GA response. Research suggests that GA stimulates stem elongation through stimulation of cell elongation and cell division (Huttly and Phillips, 1995). Gibberellin treatment causes microtubules to reorientate so as to encourage axial elongation (Shibaoka, 1994). It is thought that GA promotes cell elongation by induction of enzymes that promote cell wall loosening and expansion such as xyloglucan endotransglycosylase/hydrolase (XET or XTH), expansins, and pectin methylesterase (PME). Xyloglucan endotransglycosylases split cell wall xyloglucan polymers endolytically and then rejoin the free ends with another xyloglucan chain (Campbell and Braam, 1999). Xyloglucan endotransglycosylase activity has been associated with expanding regions and shown to be GA-induced in *Arabidopsis*, lettuce, and pea (Kauschmann *et al.*, *et al.*, 2003; Potter and Fry, 1993). Expansins disrupt hydrogen bonding in the cell wall and appear to be GA-induced in *Arabidopsis* and rice (Cosgrove, 2000; Lee and Kende, 2001; Ogawa *et al.*, 2003). Pectin methylesterase is thought to induce stem elongation by loosening the cell wall via pectin modification and is GA-induced in *Arabidopsis* (Ogawa *et al.*, 2003). Gibberellin was first shown to stimulate growth through induction of the cell cycle in rapidly growing deepwater rice (Sauter *et al.*, 1995). In rice, GA induces expression of the cyclin *cycA1;1* and the cyclin-dependent kinase *cdc2Os-3* in the G2/M phase transition (Fabian *et al.*, 2000). Microarray analysis in *Arabidopsis* has demonstrated GA induction of genes involved in the G1/S transition including *cyclinD*, *MCM*, and *replication protein A* (Ogawa *et al.*, 2003). Further research on the mechanism of GA induction of these genes and their exact mode of action is needed.

3. Gibberellins and Flowering

In most species, the transition to floral development is stimulated by gibberellins (Sun and Gubler, 2004). However, gibberellins are not the sole factor in determining transition to flowering. In *Arabidopsis*, a facilitative long-day (LD) plant, transition to flowering is controlled by the integration of signals from the GA pathway, the autonomous pathway, the vernalization pathway, and the light-dependent pathway (Komeda, 2004). It is clear that gibberellins are required for transition to flowering in short days (SD, 8-h light) because the strong GA biosynthesis mutant *ga1-3* cannot transition to flowering without application of GA under these conditions (Wilson *et al.*, 1992). The failure of *ga1-3* to flower under SD appears to be due to reduced expression of the *LEAFY* (*LFY*) gene (Blazquez *et al.*, 1998). The fact that the *ga1-3* mutant causes poor development of floral organs including petals and stamen shows that GA is also involved in the stimulation of floral development. Gibberellin has also been shown to induce expression of floral homeotic genes *APETELA3*, *PISTILLATA*, and *AGAMOUS* (Yu *et al.*, 2004).

Studies on *Lolium temulentum* have suggested that GA is an inducer of flowering or "florigen" in LD-responsive grasses (King and Evans, 2003). In *Lolium*, GA_1, GA_3, and GA_4 are more active for stem elongation, whereas

GA_5 and GA_6 are more active in triggering transition to flowering. It has been proposed that GA_5 and GA_6 are more active in the floral meristem because they have greater resistance to the expression of the GA catabolic enzyme GA2ox early in floral induction.

4. Gibberellin in Shoot Apical Meristem Development

Studies in *Arabidopsis* have indicated an emerging role for GA in shoot apical meristem (SAM) cell identity (Hay *et al.*, 2002, 2004). The SAM is a reservoir of undifferentiated cells that gives rise to the aerial leaves and stems of higher plants. Knotted-like homeobox (KNOX) transcription factors appear to control meristem versus leaf cell identity. The KNOX gene *SHOOTMERISTEMLESS* (*STM*) has been shown to prevent expression of the GA biosynthesis gene *GA20ox1* in the SAM (see Section II). The fact that ectopic GA signaling is detrimental to meristem maintenance suggests that GA signaling is antagonistic to meristem cell identity and may be involved in the transition from meristem to leaf cell fate.

II. GIBBERELLIN BIOSYNTHESIS

A. INTRODUCTION

In the last 50 years our understanding of GA metabolism has been advanced by using a variety of experimental systems, including most notably the characterization of GA metabolic enzymes and the reactions they catalyze using cell-free systems derived from immature seeds of *Cucurbita maxima* (pumpkin), *Pisum sativum* (pea), and *Phaseolus vulgaris* (bean) (Graebe, 1987). In several cases, expression of these enzymes in these immature seeds has served as a basis for cloning of their respective genes. Over the last two decades, *Arabidopsis* has become an experimental system of choice for studying GA metabolism. The power of *Arabidopsis* molecular genetic analyses was illustrated when the GA biosynthesis mutants *ga1*, *ga2*, *ga3*, *ga4*, and *ga5* served as a basis for cloning several of the biosynthetic genes (Koornneef and van der Veen, 1980). The characterizations of these genes are rapidly uncovering the complex regulatory mechanisms controlling GA metabolism. In addition, the *Arabidopsis* and rice genome sequences, together with convenient transformation procedures, have greatly improved our understanding of GA metabolism and the role of these phytohormones in regulating plant growth and development. The following sections describing GA metabolism will focus on advancements, including the discovery and regulation of GA biosynthetic and catabolic genes. The primary aim of this section is to review the steps in GA metabolism that are exclusive to this class of compounds.

B. INVOLVEMENT OF MVA AND MEP PATHWAYS IN GIBBERELLIN BIOSYNTHESIS

There are some excellent reviews that comprehensively describe the earlier steps in terpenoid biosynthesis (Goodwin, 1965; Rodriguez-Concepcion and Boronat, 2002; Sponsel, 2001). Although we will not discuss these steps in detail, it is necessary to mention some important findings that are relevant to GA biosynthesis.

Geranylgeranyl diphosphate (GGPP) is the precursor isoprenoid necessary for the synthesis of many terpenoid compounds, including GAs. The initial step in isoprenoid biosynthesis is the condensation of isopentenyl diphosphate (IPP) and dimethylallyl diphosphate (DMAPP). For many years, it was believed that the IPP destined for isoprenoid biosynthesis in plants was synthesized exclusively via the mevalonic acid (MVA) pathway. The incorporation of ^{14}C MVA into *ent*-kaurene in cell-free systems provided some initial support for this pathway in the biosynthesis of GAs. It is now known that another route for IPP biosynthesis, the plastidic methylerythritol 4-phosphate (MEP) pathway, exists in plants (Rodriguez-Concepcion and Boronat, 2002). Kasahara *et al.* (2002), has directly addressed the contribution of the MEP and MVA pathways to GA biosynthesis. Using ^{13}C feeding studies of *Arabidopsis* plants blocked in either of these pathways, they demonstrated that the MEP pathway has a predominant role in the biosynthesis of GAs, but it appears that the MVA pathway also contributes under certain conditions. Further studies are necessary to uncover the regulation of these two pathways controlling the production of IPP destined for isoprenoid biosynthesis.

C. *ENT*-COPALYL-DIPHOSPHATE SYNTHASE

The first committed step of GA biosynthesis is the cyclization of GGPP producing *ent*-copalyl diphosphate (CPP) (Fig. 3). In plants, this reaction is catalyzed by *ent*-copalyl-diphosphate synthase (CPS), a diterpene cyclase. The potential of *Arabidopsis* genetic analyses to identify genes encoding GA biosynthetic enzymes was illustrated when Sun and coworkers elegantly cloned the *GA1* gene using genomic subtraction and demonstrated that it encodes a functional CPS enzyme, AtCPS (Sun and Kamiya, 1994; Sun *et al.*, 1992). The authors subsequently provided evidence that AtCPS is localized in the plastids as a processed form (Sun and Kamiya, 1994). This is consistent with biochemical studies demonstrating CPS activity in the proplastids of several plant species (Aach *et al.*, 1995; Simcox *et al.*, 1975). Based on sequence homologies, there appears to be a single gene encoding a CPS enzyme in *Arabidopsis* (Hedden and Phillips, 2000); although it is interesting to note that *AtCPS* null mutants have detectable levels of GAs (Silverstone *et al.*, 2001; Zeevaart and Talon, 1992). This supports the existence of

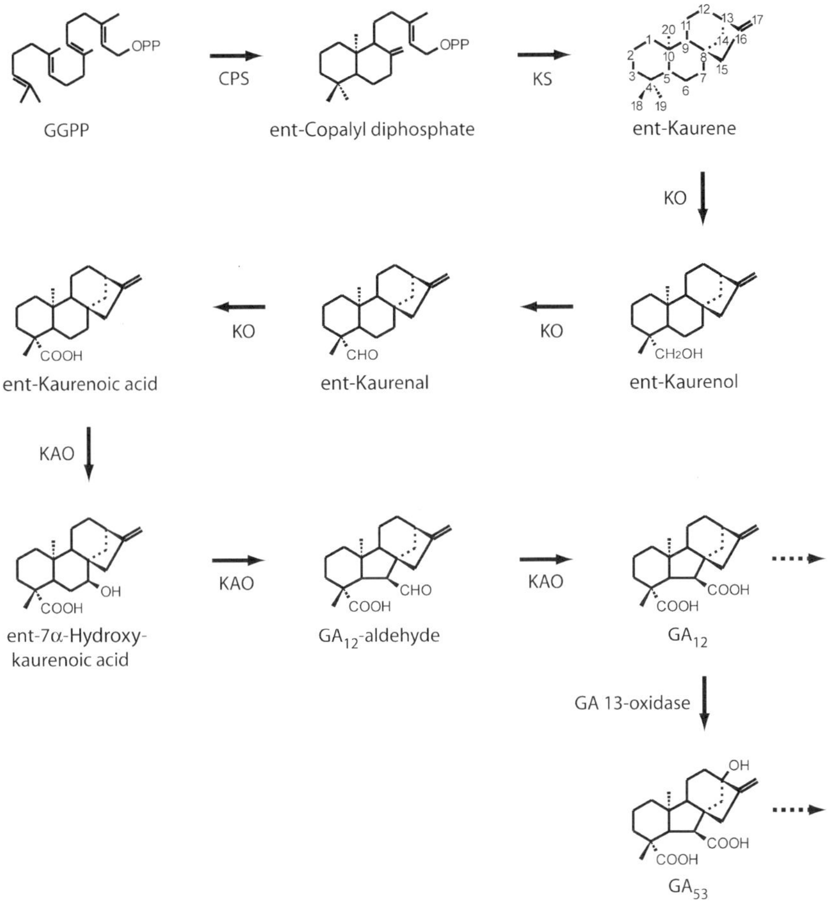

FIGURE 3. Early GA biosynthetic pathway showing conversions from geranylgeranyl diphosphate (GGPP) to GA_{12} and GA_{53}. Numbering of the C-atoms is shown for *ent*-kaurene.

another pathway capable of producing CPP or *ent*-kaurene. A study suggests that rice also contains a single gene encoding a CPS enzyme, *OsCPS1* (Sakamoto *et al.*, 2004). Null alleles of *OsCPS1* produce plants with a severe GA-deficient dwarf character reminiscent of the *Arabidopsis ga1* loss-of-function mutants.

The identity of the CPS-encoding genes has allowed the characterization of their spatial and temporal expression patterns with a view to determine the precise cellular sites of GA biosynthesis. In *Arabidopsis*, *AtCPS* demonstrates highly specific developmental and cell-specific expression patterns. Highest levels of promoter activity are localized to actively growing regions, consistent with GAs having a growth promoting role (Silverstone *et al.*, 1997b).

Interestingly, *AtCPS* expression is also observed in vascular tissue of expanded leaves, suggesting that these may be a source for GAs to be transported to a responsive tissue. Subsequent deletion analysis of the *AtCPS* promoter has identified *cis*-regulatory elements necessary for their tissue-specific expression (Chang and Sun, 2002).

D. *ENT*-KAURENE SYNTHASE

The formation of *ent*-kaurene from CPP is catalyzed by another diterpene cyclase, *ent*-kaurene synthase (KS). This enzyme catalyzes the cyclization reaction needed to produce the characteristic tetracyclic backbone of GAs (Figs. 3 and 4). A gene encoding KS was first isolated from pumpkin (Yamaguchi *et al.*, 1996). The presence of high levels of KS activity in the developing cotyledons of immature pumpkin seeds allowed purification of the enzyme to homogeneity and amino acid sequencing. A cDNA clone was subsequently identified using a degenerate PCR strategy and demonstrated to encode a functional KS enzyme. This work led to the isolation of a gene encoding a KS from *Arabidopsis* (*AtKS*). The authors demonstrated that the *GA2* locus encodes AtKS (Yamaguchi *et al.*, 1998a).

Biochemical studies suggest that the KS enzymes are localized in proplastids (Aach *et al.*, 1995, 1997). This is supported by the presence of a putative N-terminal transit peptide in both CmKS and AtKS likely to direct targeting to the plastid (Yamaguchi *et al.*, 1996, 1998a). Furthermore, fusion of the first 100 amino acids of AtKS to GFP (TPKS-GFP) demonstrated plastid localization in transiently transformed tobacco epidermal cells (Helliwell *et al.*, 2001b). In the same study, the TPKS-GFP fusion protein was imported into isolated pea chloroplasts. The potential co-localization of CPS and KS raises the possibility that they may form a plastidic complex involved in *ent*-kaurene production.

The *Arabidopsis* genome appears to contain a single *AtKS* gene (Hedden and Phillips, 2000). This is consistent with the severity of the loss-of-function *ga2-1* allele that closely resembles the extreme dwarf *ga1* null mutants (Koornneef and van der Veen, 1980; Yamaguchi *et al.*, 1998a). Interestingly, there are differences in the expression profiles of *AtCPS* and *AtKS*, with *AtCPS* demonstrating a more localized pattern (Silverstone *et al.*, 1997b; Yamaguchi *et al.*, 1998a). In addition, differences in the expression profiles of *CmCPS1/2* and *CmKS* genes were also observed in pumpkin (Smith *et al.*, 1998; Yamaguchi *et al.*, 1996). It is conceivable that the more localized expression pattern of the *CPS* genes is indicative of the CPS enzymes catalyzing the rate-limiting step in the production of *ent*-kaurene. This is supported by studies showing that transgenic *Arabidopsis* plants over-expressing AtCPS have elevated *ent*-kaurene levels, whereas plants over-expressing AtKS have wild-type levels (Fleet *et al.*, 2003).

E. ENT-KAURENE OXIDASE

The biosynthesis of $GA_{12/53}$ from *ent*-kaurene is catalyzed by cytochrome-P450-dependent monooxygenase enzymes. The first of these steps is catalyzed by *ent*-kaurene oxidase (KO), a multifunctional enzyme that catalyzes the successive oxidation at the C-19 position (Fig. 3), producing *ent*-kaurenoic acid (KA) (Helliwell *et al.*, 1999; Swain *et al.*, 1997). Studies by Helliwell *et al.* (1998, 1999, 2001a) have proved instrumental in improving our understanding of the cytochrome-P450 monooxygenases involved in GA biosynthesis. This work initially involved the confirmation that the *ga3* mutants were deficient in KO activity. They subsequently confirmed that the *GA3* locus encoded a cytochrome-P450 monooxygenase that was capable of converting *ent*-kaurene to *ent*-kaurenoic acid when it was heterologously expressed in yeast (Helliwell *et al.*, 1998, 1999). This gene was designated as *AtKO* and appears to be present as a single copy in the *Arabidopsis* genome. RNAse protection analysis of *AtKO* gene expression demonstrated developmental regulation, with highest levels of transcripts in young seedlings, elongating stems, and inflorescences (Helliwell *et al.*, 1998). Gibberellin treatment did not affect the levels of *AtKO* mRNA.

Rice contains five KO-like genes (OsKOL1–5) that are arranged in tandem as a cluster of genes on chromosome 6 (Itoh *et al.*, 2004; Sakamoto *et al.*, 2004). One of these genes, OsKOL2, has been shown to correspond to the *D35* loci. Null mutations at *D35* produce a severe GA-deficient phenotype that is probably blocked at the GA biosynthetic step of *ent*-kaurene oxidation (Itoh *et al.*, 2004; Ogawa *et al.*, 1996). A weak allele of *D35*, $d35^{Tan-Ginbozu}$, produces a rice plant with a semidwarf character (Itoh *et al.*, 2004). The introduction of this allele in the 1950s, producing the Tan-Ginbozu cultivar, led to dramatic increases in rice crop yields. This is one of many examples where mutations affecting GA biosynthesis or response have been instrumental in producing crops with improved agronomic traits.

The cytochrome-P450-dependent monooxygenases involved in GA biosynthesis have generally been considered as being localized to the endoplasmic reticulum (ER). This is based on studies showing that the enzymatic activity co-purifies with a microsomal fraction (Graebe, 1979). The availability of the *AtKO* gene has provided the opportunity to investigate the localization of these enzymes using more sensitive cell biology-based approaches. Interestingly, Helliwell *et al.* (2001b) found that an AtKO-GFP fusion protein was localized to the outer plastid membrane of transiently transformed tobacco epidermal cells. They have hypothesized that AtKO provides a link between the plastid and ER located steps of the GA biosynthetic pathway (Helliwell *et al.*, 2001b). In a somewhat conflicting study by Yamaguchi and coworkers, aimed at understanding the localization of enzymes involved in the biosynthesis of GAs in germinating *Arabidopsis* seeds, it was found that *AtCPS* and *AtKO* display distinctly different cell-specific expression patterns (Yamaguchi et al., 2001). Based on these studies, they proposed that intercellular transport of GA intermediates, possibly *ent*-kaurene, is occurring between the provasculature and the cortex/endodermis. Further studies aimed at detecting the localization of the endogenous proteins will be necessary to establish the precise subcellular distribution of these enzymes.

Repression of shoot growth (RSG) was identified, rather fortuitously, in a screen designed to isolate *trans*-acting factors that bind to an auxin-responsive *cis*-regulatory element in tobacco (Fukazawa et al., 2000). It was demonstrated that RSG did not bind the auxin-responsive element but instead bound to the *AtKO* promoter *in vitro*. Furthermore, expression of a dominant-negative form of RSG in transgenic tobacco produced a GA-responsive dwarf phenotype with lower levels of bioactive GAs and reduced expression of the *AtKO* homologue. Studies have shown that GA signaling promotes RSG disappearance from the nucleus through its binding to a cytoplasmic 14-3-3 protein (Igarashi et al., 2001; Ishida et al., 2004). The interaction of RSG with the 14-3-3 protein appears to be dependent on phosphorylation of a serine residue. The authors propose a model in which RSG is negatively regulated by GAs and has a role in the maintenance of GA levels (Ishida et al., 2004). Further studies are necessary to confirm whether RSG is a direct regulator of GA biosynthesis.

F. *ENT*-KAURENOIC ACID OXIDASE

The conversion of *ent*-kaurenoic acid to GA_{12} is catalyzed by another cytochrome-P450 monooxygenase, *ent*-kaurenoic acid oxidase (KAO). The multifunctional KAO enzyme oxidizes the C-7 of *ent*-kaurenoic acid to produce *ent*-7α-hydroxy-kaurenoic acid, which is then oxidized by this enzyme on C-6 to form GA_{12}-aldehyde. Finally, KAO oxidizes GA_{12}-aldehyde on C-7 to produce GA_{12} (Fig. 3). Gibberellin-deficient mutants blocked at this step in the biosynthetic pathway have not been identified in *Arabidopsis*. In contrast, the barley *grd5* and pea *na* mutants, both of which display a GA-responsive dwarf character, demonstrate reduced KAO activity (Helliwell et al., 2001a; Ingram and Reid, 1987). The maize *dwarf3* (*d3*) mutants have a similar GA-deficient phenotype. Although the precise GA biosynthetic step blocked in the *d3* mutants was unknown, the identity of the *D3* gene proved instrumental in the identification of a *KAO* gene. The *D3* gene was cloned using a transposon tagging strategy and demonstrated to encode a cytochrome-P450–dependent monooxygenase (Winkler and Helentjaris, 1995) belonging to the CYP88A subfamily (Helliwell et al., 2001a). Helliwell and coworkers isolated a *Grd5* cDNA clone based on its homology to *D3* and confirmed that it encoded a cytochrome-P450 monooxygenase, also belonging to the CYP88A subfamily (Helliwell et al., 2001a). Furthermore, they identified two *Arabidopsis* genes encoding CYP88A enzymes. Using a

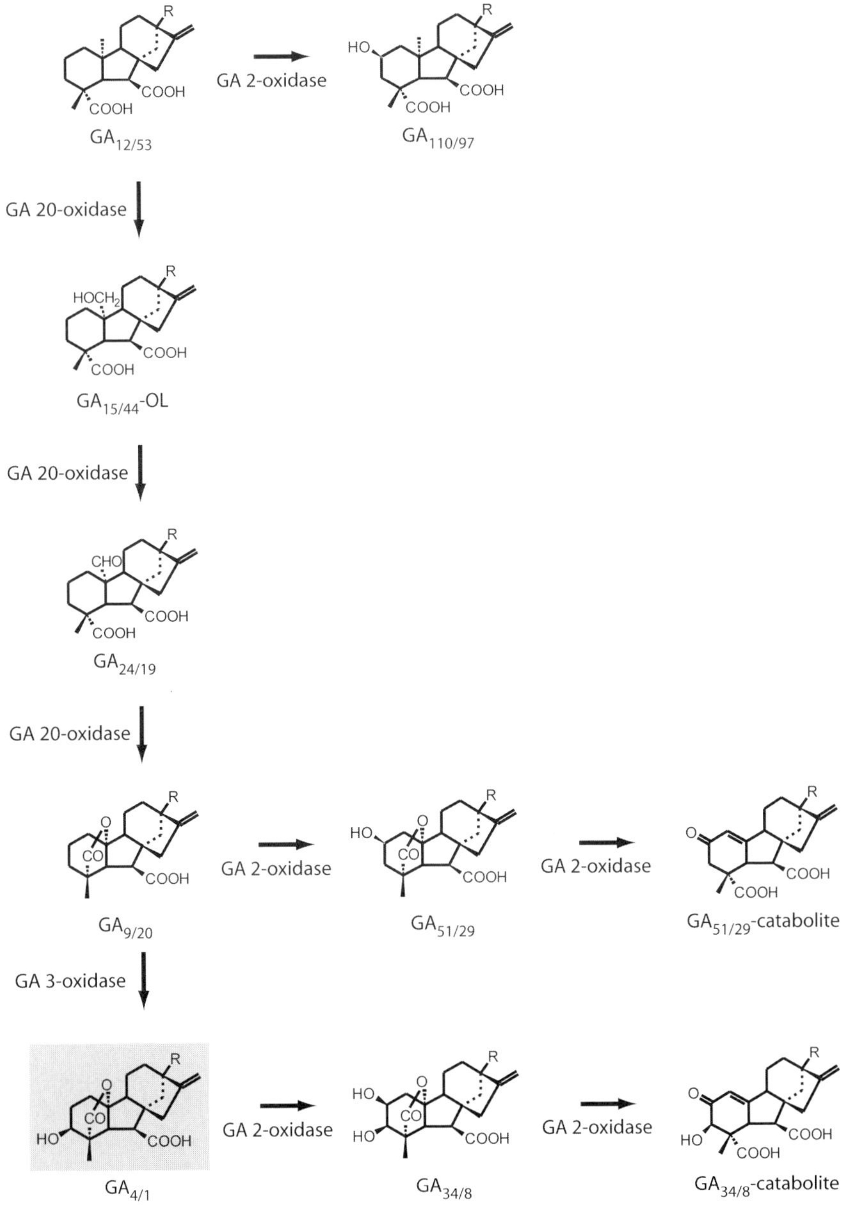

FIGURE 4. Late GA biosynthetic and catabolic pathways. The bioactive GAs, GA_4 and GA_1, are synthesized from GA_{12} and GA_{53}, respectively. Subscripted numbers before the slash indicate the non-13-hydroxylated GA (R=H) and after the slash indicate the 13-hydroxylated equivalent (R=OH).

yeast heterologous expression system developed for testing the functionality of *AtKO*, it was confirmed that the barley and the two *Arabidopsis* CYP88A enzymes catalyzed the three steps of GA biosynthesis from KA to GA_{12} (Helliwell *et al.*, 2001a). They were subsequently designated as *ent*-kaurenoic acid oxidases. It is likely that other CYP88A enzymes, including *D3*, encode KAO enzymes. Interestingly, a novel gene encoding a 2-oxoglutarate–dependent dioxygenase (2-ODD) enzyme (GA 7-oxidase), which catalyzes the single-step conversion of GA_{12}-aldehyde to GA_{12}, was identified from pumpkin (Lange, 1997). The significance of this class of enzymes is unknown as they have not been identified in other plant species.

The presence of two *AtKAO* genes is in contrast to those encoding earlier steps in the GA biosynthetic pathway. It is likely that this functional redundancy explains why no *Arabidopsis* mutants blocked at this step have been identified. Although the AtKAO genes exhibit similar expression patterns (Helliwell *et al.*, 2001a), characterization of knockout mutants is needed to determine whether they have specific roles in regulating plant development. In pea, there are two KAO genes, *PsKAO1* and *PsKAO2*, that do appear to have distinct developmental roles (Davidson *et al.*, 2003). The pea *NA* gene encodes PsKAO1, and *na* mutants exhibit severe GA-deficient phenotypes but normal seed development. These characteristics are potentially explained by the differential expression pattern of the *PsKAO* genes; *PsKAO1* is expressed ubiquitously in the plant whereas *PsKAO2* is only expressed in the developing seeds (Davidson *et al.*, 2003). The role of *PsKAO2* in seed development remains to be resolved.

In some plants, including *Thlaspi arvense*, GAs have an important role in mediating vernalization (or cold)-induced bolting and flowering (Metzger and Dusbabek, 1991). It has been proposed that thermoinduction stimulates GA biosynthesis and the resulting GA accumulation promotes stem elongation. In *Thlaspi*, the site of perception of cold is the shoot apex (Hazebroek and Metzger, 1990) where the levels of KA were dramatically reduced following vernalization (Hazebroek *et al.*, 1993). This suggests that KAO is the primary step in GA metabolism regulated by vernalization in this species. The identification and characterization of GA biosynthetic genes in *T. arvense* should help to determine how vernalization regulates GA metabolism.

G. GIBBERELLIN 13β-HYDROXYLASE

In many plants, including most monocots and pea, GA_1 is the predominant bioactive GA, illustrating the importance of 13-hydroxylation in the biosynthetic pathway. At present, the exact point in GA biosynthesis at which 13-hydroxylation occurs is still not entirely clear. Gibberellin-feeding experiments in pea suggest that this reaction occurs early in the pathway, with both GA_{12} and GA_{12}-aldehyde proving to be good substrates (Kamiya and Graebe, 1983). Biochemical studies suggest that this class of enzymes is predominantly cytochrome-P450–dependent monooxygenases (Grosselindemann et al., 1992; Hedden et al., 1984; Kamiya and Graebe, 1983), although a soluble enzyme activity was detected in cell-free extracts from spinach leaves (Gilmour et al., 1986). It is tempting to speculate that GA_{12} is the endogenous substrate as this is produced by other microsomal located cytochrome-P450 monooxygenases. There are no characterized GA 13-hydroxylase mutants, and a gene encoding this enzyme has not been identified in plants. A better understanding of the 13-hydroxylase enzymes awaits the cloning of these elusive genes.

H. GIBBERELLIN 20-OXIDASE

The final steps in the metabolism of bioactive GAs are catalyzed by 2-oxoglutarate–dependent dioxygenases (Hedden and Phillips, 2000). These enzymes are believed to be soluble and cytoplasmic. The GA 20-oxidase catalyzes the penultimate step in the biosynthesis of bioactive GAs, a stage that involves the oxidation of C-20 to an aldehyde followed by the removal of this C atom and the formation of a lactone (Hedden and Phillips, 2000). Some plants contain a GA 20-oxidase enzyme with different properties. For example, in spinach a GA_{44}-oxidase activity that converts the lactone, rather than the free alcohol form of this GA, has been identified (Gilmour et al., 1986; Ward et al., 1997).

Over the last decade, our understanding of GA 20-oxidation has improved dramatically since Lange and coworkers identified the first GA 20-oxidase gene (Lange et al., 1994). Their strategy aimed at cloning this gene involved purifying a GA 20-oxidase enzyme from immature pumpkin seeds, a tissue extremely rich in GA metabolic enzymes (Lange, 1994). Antibodies raised against a peptide sequence contained within the purified GA 20-oxidase were subsequently used to isolate a corresponding cDNA clone by expression screening (Lange, 1994). The recombinant pumpkin GA 20-oxidase (CmGA20ox1) expressed from this cDNA clone was confirmed as a multifunctional enzyme capable of converting GA_{12} to GA_9 (Lange et al., 1994). Surprisingly, the predominant reaction catalyzed by Cm20ox1 was the complete oxidation of the carbon-20 to the carboxylic acid, rather than its loss. The C-20 tricarboxylic acid GAs produced by Cm20ox1 are essentially biologically inactive. This raises the question: what functional role does it play in the development of pumpkin seeds?

The identity of *Cm20ox1* led directly to the isolation of three GA 20-oxidase genes from *Arabidopsis* and the confirmation that one of these corresponds to the *GA5* locus, *AtGA20ox1* (Phillips et al., 1995; Xu et al., 1995). In *Arabidopsis*, it is now apparent that there are five putative GA 20-oxidase genes (Hedden et al., 2001). Three of these genes, *AtGA20ox1, 2, and 3*, have been confirmed to encode functional enzymes that predominantly metabolize GA_{12} to GA_9 (Phillips et al., 1995; Xu et al., 1995). In rice, the recessive *semidwarf1* (*sd1*) mutations have been instrumental in producing higher yielding dwarf varieties that are more resistant to environmental damage (Hedden, 2003). Studies from three independent labs have demonstrated that the *SD1* locus encodes a GA 20-oxidase, OsGA20ox2 (Monna et al., 2002; Sasaki et al., 2002; Spielmeyer et al., 2002). On the basis of the rice genome sequence, it appears there are four GA 20-oxidase genes in rice (Sakamoto et al., 2004). Further work is necessary to confirm the functional role of the three other putative OsGA20ox genes.

The importance of GA 20-oxidation as a key regulatory step within the biosynthetic pathway of most plants is demonstrated by the finding that the GA 20-oxidase catalyzes a rate-limiting step. This is clearly illustrated in *Arabidopsis*, where GA_{24} and GA_{19} have been shown to accumulate in stems (Coles et al., 1999; Talon et al., 1990a). Furthermore, it was demonstrated that GA 20-oxidase over-expression in transgenic *Arabidopsis* leads to elevated levels of bioactive GAs and a corresponding GA overdose phenotype compared to wild-type plants (Coles et al., 1999; Huang et al., 1998). In contrast, over-expression of enzymes catalyzing earlier steps in the GA biosynthetic pathway do not have this effect (Fleet et al., 2003). It is, therefore, likely that GA 20-oxidase activity provides an important step in the regulation of bioactive GA levels and the subsequent developmental programs these control. This is supported by the observations that transcript levels of GA 20-oxidase genes demonstrate tight spatial and developmental regulation.

The presence of GA 20-oxidase multigene families in higher plants raises the possibility that certain members have roles in regulating specific developmental programs. This hypothesis is supported by studies that show distinct spatial and development expression profiles for individual genes (Carrera et al., 1999; Garcia-Martinez et al., 1997; Phillips et al., 1995; Rebers et al., 1999). For example, in *Arabidopsis*, *AtGA20ox2* is expressed predominantly in flowers and siliques, whereas *AtGA20ox3* expression is exclusively found in siliques (Phillips et al., 1995). In contrast, *AtGA20ox1* is expressed predominantly in the stem, providing a possible explanation for the semidwarf character of the *ga5* mutant (Phillips et al., 1995; Xu et al., 1995). The identification of loss-of-function mutations in other GA 20-oxidase genes should help to uncover specific roles for these family members.

The *KNOX* genes are involved in maintenance of the meristem (Hake et al., 2004). There is evidence to suggest that KNOX proteins achieve this, in part, by controlling GA levels through the regulation of their biosynthesis. Tanaka-Ueguchi and coworkers demonstrated that over-expression of the *NTH15* KNOX gene in tobacco produced a GA-responsive dwarf phenotype caused, in part, by the reduced expression of a GA 20-oxidase gene, *Ntc12* (Tanaka-Ueguchi et al., 1998). They propose that NTH15 directly represses *Ntc12* to maintain the indeterminate state of cells in the SAM.

At the periphery of the meristem, *NTH15* expression is suppressed, allowing GA 20-oxidase expression and subsequent determination of cell fate. Similarly, in *Arabidopsis* the KNOX gene *STM* is involved in repressing *AtGA20ox1* expression in the meristem (Hay *et al.*, 2002).

I. GIBBERELLIN 3-OXIDASE

Growth active GAs are hydroxylated at the C-3β position (Fig. 4). The 2-ODD enzyme responsible for this modification is a GA 3-oxidase (Hedden and Phillips, 2000). The *Arabidopsis ga4* mutants are GA-responsive semidwarf plants (Chiang *et al.*, 1995; Koornneef and van der Veen, 1980) that contain reduced levels of 3β-hydroxy GAs, together with increased levels of GA_{19}, GA_{20}, and GA_9. These observations suggested that *GA4* may encode a 3β-hydroxylase (Talon *et al.*, 1990a). This was subsequently confirmed when the *GA4* locus was identified by T-DNA tagging (Chiang *et al.*, 1995) and the recombinant GA4 enzyme demonstrated to convert GA_9 to GA_4 (Williams *et al.*, 1998). To prevent confusion, the *GA4* gene has been renamed *AtGA3ox1*, following the nomenclature suggested by Coles *et al.* (1999).

Gregor Mendel's pioneering experiments using garden peas to investigate the transmission of hereditary elements are widely accepted as the foundation of genetics. In these studies he followed seven pairs of traits, including stem length (Le) (Mendel, 1865). The *le* mutations are recessive and produce GA-responsive dwarf plants (Brian and Hemming, 1955; Mendel, 1865). Analysis of endogenous GA levels in the *le* mutant (Potts *et al.*, 1982) and the finding that these plants were unable to convert GA_{20} to GA_1 (Ingram *et al.*, 1984) suggested that Le is involved in GA biosynthesis at the step of 3β-hydroxylation. The identity of *AtGA3ox1* directly led to the isolation and characterization of the *Le* gene by two independent groups (Lester *et al.*, 1997; Martin *et al.*, 1997). Both of these groups confirmed that *Le* encodes a functional 3β-hydroxylase, whereas the *le* mutant form exhibited reduced activity, when expressed in *Escherichia coli*. The reduction in activity was associated with an alanine to threonine substitution in the predicted amino acid sequence of the enzyme near its proposed active site.

In addition to *Arabidopsis* and pea, other plant species containing mutations that affect GA biosynthesis at the 3β-hydroxylation step have been identified (Fujioka *et al.*, 1988a; Ross, 1994). In general, all of the GA 3-oxidase loss-of-function mutants have a semidwarf phenotype, in contrast to the severe dwarf phenotype of GA auxotrophs blocked at earlier steps in the pathway. The most likely explanation for this observation is the functional redundancy of GA 3-oxidase genes. For example, there are at least four GA 3-oxidase genes in *Arabidopsis*, whereas rice contains two genes (Phillips and Hedden, 2000, Sakamoto *et al.*, 2004). There is also evidence to indicate that different GA 3-oxidase genes have specific roles regulating plant development. For example, two of the *Arabidopsis* GA 3-oxidase genes, *AtGA3ox1* and *AtGA3ox2* (formerly *GA4H*), display differential spatial and temporal expression patterns (Yamaguchi et al., 1998b). *AtGA3ox1* was expressed in all growing tissues tested, whereas *AtGA3ox2* was predominantly expressed in germinating seeds and young seedlings but not in other tissues (Yamaguchi et al., 1998b). Similarly, *OsGA3ox2* expression was detected in all aerial portions of the rice plant; in contrast, *OsGA3ox1* was exclusively expressed in floral tissue (Kaneko et al., 2003). A more detailed analysis of *AtGA3ox1* and *AtGA3ox2* mRNA transcripts in germinating seeds found that both were predominantly expressed in the cortex and endodermis of the embryo axes (Yamaguchi et al., 2001). These observations suggest that GA production occurs in GA-responsive cells.

Considering the importance of the GA 3-oxidases in producing bioactive GAs, it is not surprising that expression of the respective genes is tightly regulated, by both developmental and environmental stimuli. Work is underway to uncover the complex environmental regulation of these genes, most notably with respect to germination of *Arabidopsis* seeds (Ogawa et al., 2003; Yamaguchi et al., 1998b, 2001; Yamauchi et al., 2004). These studies have demonstrated that expression of *AtGA3ox1* is regulated by light, bioactive GAs, and temperature (regulation by light and GAs will be discussed later). The treatment of imbibed *Arabidopsis* seeds to low temperatures (stratification) is known to promote germination. Stratification has also been implicated in increasing the GA levels (Derkx et al., 1994). These observations suggest that cold treatment promotes germination of *Arabidopsis* seeds by stimulating GA biosynthesis. Studies have clearly demonstrated that stratification produces an increase in the levels of *AtGA3ox1* transcripts, which is directly responsible for the increase in bioactive GA_4 levels that promote germination (Yamauchi et al., 2004). These elegant studies provide a benchmark for future studies investigating the regulation of GA metabolism. They demonstrate the potential for integration of genomics, genetic analysis, and biochemical studies to improve our understanding of the role of GAs in regulating plant development.

J. GIBBERELLIN 2-OXIDASE AND GIBBERELLIN INACTIVATION

The amount of bioactive GAs is determined by both the rate of GA biosynthesis and inactivation. Inactivation can be achieved by glucosyl conjugation or by 2β-hydroxylation, the relative contributions of the two pathways being unknown (Schneider and Schliemann, 1994). A clear physiological role of GA conjugation, however, has not been shown, whereas the importance of 2β-hydroxylation in regulating bioactive GA content is well established. Gibberellin 2β-hydroxylase activity is abundant in seeds during the later stages of maturation, particularly in legume seeds that accumulate large amounts of 2β-hydroxylated GAs (Albone *et al.*, 1984; Durley *et al.*, 1971; Frydman *et al.*, 1974). Indeed, GA$_8$, the first 2β-hydroxy GA to be identified was extracted from seeds of runner bean (*Phaseolus coccineus*) (MacMillan *et al.*, 1962). In certain species, including legumes, further metabolism of 2β-hydroxy GAs occurs to form the so-called catabolites, in which C-2 is oxidized to a ketone and the lactone is opened with the formation of a double bond between C-10 and an adjacent C atom (Albone *et al.*, 1984; Sponsel and MacMillan, 1980). Biochemical characterization of the proteins responsible for 2β-hydroxylation showed they belong to the soluble 2-oxoglutarate–dependent dioxygenases (Griggs *et al.*, 1991).

A gene encoding for GA 2-oxidase was first identified in runner bean by screening an embryo-cDNA expression library for 2β-hydroxylase activity (Thomas *et al.*, 1999) and studies using a similar approach with seed-cDNA libraries led to the identification of two GA 2-oxidase genes from pea (*P. sativum* L.; Lester *et al.*, 1999; Martin *et al.*, 1999). Five *Arabidopsis* GA 2-oxidase genes have since been identified based on sequence homology and their identity has been confirmed by activity assays (Hedden and Phillips, 2000; Thomas *et al.*, 1999; Wang *et al.*, 2004). Two more *Arabidopsis* proteins capable of GA 2β-hydroxylation were identified using an activation tagging screen for dwarf mutants (Schomburg *et al.*, 2003). Interestingly, these two proteins, AtGA2ox7 and AtGA2ox8, are more related to GA 20-oxidases than to the other GA 2-oxidases. Evidence that all these proteins function in GA inactivation *in vivo* comes from experiments in which over-expression in *Arabidopsis* resulted in dwarfed plants (Schomburg *et al.*, 2003; Thomas, Phillips, and Hedden, 2000, unpublished data; Wang *et al.*, 2004). Similar results have been obtained with GA 2-oxidases from poplar and rice (Busov *et al.*, 2003; Sakamoto *et al.*, 2001).

Detailed characterization of the enzymatic activities of GA 2-oxidases from various plants has shown that they can convert a range of GAs. Most of the enzymes tested show activity towards the bioactive GA$_{1/4}$ and their non-3-hydroxylated precursors GA$_{20/9}$, although there are differences in the preferred substrate (e.g., Lester *et al.*, 1999; Thomas *et al.*, 1999). A subset of the enzymes is capable of further oxidation to a ketone at C-2. AtGA2ox7 and AtGA2ox8 are somewhat exceptional, in that they are specific for C-20 GAs (Schomburg *et al.*, 2003).

Because of the highly similar activities of the various GA 2-oxidases, any functional diversity between the family members may be expected to lie in differential expression patterns. Support for this comes from studies in pea, where *PsGA2ox1* is highly expressed in maturing seed and *PsGA2ox2* preferentially in the shoot (Lester *et al.*, 1999; Martin *et al.*, 1999). This differentiation may partly explain the strong block in the conversion of GA$_{20}$ to GA$_{29}$ observed in seed of the *sln* mutant that carries a point mutation in *PsGA2ox1* (Lester *et al.*, 1999; Martin *et al.*, 1999; Ross *et al.*, 1995). The elongated shoot phenotype of this mutant is due to enhanced elongation of the first internodes only and appears to arise from transport of GA_{20} from the seed into the young shoot after germination (Reid et al., 1992; Ross et al., 1993).

A very specific expression pattern has been reported for *OsGA2ox1* in rice (Sakamoto et al., 2001). mRNA from this gene was observed in a ring around the vegetative shoot apical meristem, at the bases of the youngest leaf primordia. After phase transition to the inflorescence stage, however, expression was drastically reduced. This prompted the authors to speculate on a role of GA 2-oxidases in floral transition, a hypothesis further elaborated by King and Evans (2003) to account for the effects of various applied GAs on floral transition in *L. temulentum*. However, this interesting hypothesis still awaits testing using knockout mutants.

Due to their rather recent discovery, little is known about the regulation of the GA 2-oxidase genes. Using a chromatin immunoprecipitation approach, Wang et al. (2002) isolated a portion of the *AtGA2ox6* promoter. They convincingly showed that *AtGA2ox6* is a direct target of AGL15 and is transcriptionally activated during embryogenesis (Wang et al., 2004). The function of *AtGA2ox6* expression during embryogenesis is not yet fully clear, but it seems to contribute to seed dormancy.

K. FEEDBACK AND FEEDFORWARD REGULATION OF GIBBERELLIN METABOLISM

In plants, a homeostatic regulatory mechanism exists whereby biologically active GAs control their own levels through the processes of feedback and feedforward regulation of GA metabolism. Evidence for this level of regulation originally came from studies in which GA levels were compared between GA-response mutants and the respective wild-type controls (Hedden and Croker, 1992). The GA-insensitive dwarf *rht3* and *d8* mutants in wheat and maize, respectively, were found to contain highly elevated levels of the bioactive GA_1, whereas the levels of GA_{19} were lower, compared to wild-type seedlings (Appleford and Lenton, 1991; Fujioka et al., 1988b). Similar observations were made in the GA-insensitive *gai-1* mutant in *Arabidopsis* (Talon et al., 1990b). These studies suggested that GA 20-oxidation was increased in these GA-response mutants and hence under feedback control. Hedden and Croker subsequently demonstrated that the maize *d1* mutant, which is defective in 3-oxidation, has high levels of GA_{20} but reduced levels of GA_{53} and GA_{19} compared to wild-type plants (Hedden and Croker, 1992). The subsequent application of bioactive GA to *d1* restored the levels of these GAs close to those of wild-type plants, providing strong support for feedback regulation of GA 20-oxidation in maize.

The identity of genes encoding GA biosynthetic enzymes has provided further clues to the control of GA metabolism by feedback and feedforward regulation. It was found that the *Arabidopsis ga4-1* mutant accumulated high levels of the *ga4* transcripts compared to wild-type plants (Chiang *et al.*, 1995). Treating the *ga4-1* plants with GA dramatically reduced the *ga4* transcript levels, indicating that the expression of *AtGA3ox1* is under feedback control by bioactive GAs. Further evidence for the GA 20-oxidation step being under feedback control was also provided by the demonstration that the expression of *AtGA20ox1*, *AtGA20ox2*, and *AtGA20ox3* genes were reduced by exogenous applications of GA (Phillips *et al.*, 1995; Xu *et al.*, 1995). It is now apparent that feedback regulation of most GA 20-oxidase and GA 3-oxidase genes is conserved in higher plants. Although, it is interesting to note that the expression of *AtGA3ox2* is apparently not under feedback control (Yamaguchi *et al.*, 1998b). Currently, there is no evidence to suggest that earlier steps in the GA biosynthetic pathway are controlled by feedback regulation. In contrast to the GA-induced downregulation of GA biosynthetic 2-ODD genes, the expression of the inactivating *Arabidopsis* GA 2-oxidase genes, *AtGA2ox1* and *AtGA2ox2*, is upregulated by GA treatment of *ga1-2* plants (Thomas *et al.*, 1999). A similar effect on the expression of *PsGA2ox1* and *PsGA2ox2* genes was observed in pea (Elliott *et al.*, 2001). In this case, levels of the *PsGA2ox1/2* transcripts were elevated in the WT background compared to the *ls* and *na* mutants. Interestingly, in this study there was no evidence of feedforward regulation based on the endogenous levels of 2β-hydroxy GAs. The authors suggest that other uncharacterized 2-oxidase activities could account for this anomaly. These two studies suggest that bioactive GAs regulate their own levels by adjusting inactivation through a feedforward controlling mechanism. It will be necessary to confirm the biological significance of feedforward regulation. The isolation of GA 2-oxidase loss-of-function mutants in *Arabidopsis* should provide help in these studies.

The studies showing that feedback regulation is perturbed in GA-insensitive response mutants support a direct role for the GA-response pathway in controlling this process. In *Arabidopsis*, the GA-induced decrease in expression of *AtGA3ox1* is currently one of the earliest markers of GA-responsive gene expression (Ogawa *et al.*, 2003; Thomas and Sun, unpublished). Changes in *AtGA3ox1* and *AtGA20ox1* expression levels are observed only 30 min after treating the *ga1-3* mutant with bioactive GAs (Thomas and Sun, unpublished). It is not clear whether these are primary responses to GA signaling because studies using cycloheximide demonstrate that protein synthesis is necessary for GA-mediated feedback regulation of *AtGA20ox1* expression (Bouquin *et al.*, 2001).

L. REGULATION OF GA METABOLISM BY LIGHT

The intrinsic ability of plants to respond to their environmental conditions is clearly essential for them to survive and reproduce. Light quantity, quality, and photoperiod are certainly the most important of these factors, and it is therefore not surprising that they regulate all aspects of plant growth and development. The role of light in controlling plant developmental processes has been studied in great detail. It has emerged that in some of these cases, light exerts its effect by causing changes in the concentration and/or sensitivity to GAs (Kamiya and Garcia-Martinez, 1999; Olszewski et al., 2002; Yamaguchi and Kamiya, 2000). The most well-characterized examples that we will discuss further include seed germination, de-etiolation, photoperiodic control of flowering, and tuberization in potato.

The germination of *Arabidopsis* seeds has an absolute requirement for both GAs and red light. There is strong evidence to suggest that the response of seeds to light is mediated by an increase in GA biosynthesis (Yamaguchi and Kamiya, 2000). In addition, red light was shown to increase the sensitivity of the seed to the concentration of GA required for germination (Hilhorst and Karssen, 1988). Studies of germination in *Arabidopsis* (Yamaguchi et al., 1998b) and lettuce (Toyomasu et al., 1993, 1998) have provided strong evidence that phytochrome upregulates GA biosynthesis by promoting GA 3-oxidation. In *Arabidopsis*, red light was demonstrated to upregulate the expression of both *AtGA3ox1* and *AtGA3ox2* within 1 h of treatment (Yamaguchi et al., 1998b). Interestingly, the red light induction of these two genes appears to be mediated by different phytochrome (PHY) light receptors.

Bioactive GAs are required for establishing etiolated growth and repressing photomorphogenesis. This was illustrated in a study demonstrating that reductions in *Arabidopsis* GA levels partially derepress photomorphogenesis in the dark (Alabadi et al., 2004). Similarly, in pea, the GA-deficient mutant *na* exhibited a dramatic photomorphogenic phenotype when grown in the dark.

Upon exposure to light, dark-grown seedlings demonstrate a dramatic change in phenotype that is known as de-etiolation. These changes include a significant reduction in stem elongation, which coincides with decreased levels of bioactive GAs in peas (Ait-Ali et al., 1999; O'Neill et al., 2000; Reid et al., 2002). A study by Reid and coworkers demonstrated that this reduction mediated by red and blue light is most likely caused by a rapid (within 30 min) downregulation in the expression levels of *PsGA3ox1* and by an upregulation of *PsGA2ox1* (Reid et al., 2002). They went on to confirm that the light-induced reduction in GA levels is mediated through phytochrome A and a blue light receptor. After the initial decline in GA_1 levels following 8-h exposure of light, there is a subsequent increase over the next 16 h, which results in plants that have similar GA_1 concentrations to those grown exclusively in the dark (O'Neill et al., 2000; Reid et al., 2002). There is a direct correlation between the recovery in GA_1 levels and increases in the expression levels of *PsGA20ox1* and *PsGA3ox1*, presumably due to feedback regulation of GA biosynthesis (Reid et al., 2002). The continued inhibition of stem elongation by light appears to be attributed to reduced responsiveness to GA in the light-grown plants compared to those grown in the dark (O'Neill et al., 2000; Reid, 1988).

Tuberization of *Solanum tuberosum* (potato) occurs when the plants are exposed to a SD photoperiod (Jackson, 1999). There is strong evidence to suggest that the photoperiodic control of tuberization is mediated, in part, by GAs. More specifically, GAs appear to inhibit tuberization in long days (LD). This is illustrated by the observation that exogenous applications of GAs can inhibit or delay tuberization under inductive SD photoperiods (Jackson and Prat, 1996). In contrast, reducing the levels of GAs promotes tuberization under noninducing LD (Jackson and Prat, 1996; Vandenberg et al., 1995). Furthermore, endogenous levels of GA_1 were reduced in stolons and leaves of plants induced to tuberize compared to those grown under noninductive conditions (Xu et al., 1998). It has been demonstrated that the potato leaves are the principal site of photoperiod perception (Ewing and Wareing, 1978). A role for PHYB in inhibiting tuberization in LD has been suggested by the findings that transgenic potato plants with reduced PHYB levels will tuberize in both SD and LD (Jackson and Prat, 1996). The identification of potato genes encoding GA biosynthetic enzymes has provided important tools to help understand the role of GAs in the photoperiodic control of tuberization. Using a degenerate PCR-based approach, Carrera and coworkers cloned three GA 20-oxidase genes from potato that displayed differential tissue-specific expression profiles (Carrera et al., 1999). One of these genes, *StGA20ox1*, was expressed at relatively high levels in leaves and exhibited photoperiodic regulation of transcript levels (Carrera et al., 1999, 2000; Jackson et al., 2000). The photoperiodic transcriptional regulation of *StGA20ox1* appears to be controlled by PHYB, along with an unidentified blue light receptor (Jackson et al., 2000). The role of *StGA20ox1* in tuberization was investigated by producing transgenic potato plants expressing sense or antisense copies of this gene (Carrera et al., 2000). Although the over-expression of *StGA20ox1* did not prevent tuberization under SD, it did result in plants that required a longer duration of SD photoperiod to tuberize compared to control plants. Conversely, the *StGA20ox1* antisense lines tuberized earlier than the controls and showed increased tuber yields. This study supports a role for StGA20ox1 in tuberization, although it indicates that other factors are also necessary for LD inhibition of this process.

In *Arabidopsis* and spinach (*Spinacia oleracea*), *GA20ox* genes are also subject to transcriptional regulation by LD photoperiods (Wu et al., 1996; Xu et al., 1997). Expression of *AtGA20ox1* is enhanced by exposure to LDs that promotes rapid stem elongation and flowering. Spinach has an absolute requirement for LD photoperiods to initiate bolting and flowering. The increase in GA levels, which is a necessary requirement for LD-induced bolting in spinach, is directly attributable to increased transcription of the *SoGA20ox1* gene (Lee and Zeevaart, 2002; Wu et al., 1996). It was also found that expression of *SoGA2ox1* was repressed by LDs (Lee and Zeevaart, 2002). This suggests that the LD-induced increases in bioactive GA levels may also be maintained by a reduction in the rate of their inactivation.

III. GIBBERELLIN SIGNAL TRANSDUCTION

Much has been learned about GA-signal transduction using a combination of genetic, physiological, and biochemical analyses. Regulatory elements of the GA-signal transduction pathway have been identified using: (1) screens for mutants with altered GA sensitivity, (2) identification of transcriptional regulators of the GA-responsive genes, and (3) methods for identifying differentially expressed genes. Such approaches have recovered both positive and negative regulators of GA response that have been the subject of several reviews (Jacobsen *et al.*, 1995; Olszewski *et al.*, 2002; Sun and Gubler, 2004).

Mutant analysis is often used to determine the role of a gene in a signaling pathway. The hallmark of a GA-insensitive mutant is that it shares all or a subset of the phenotypes of a GA biosynthesis mutant, but cannot be rescued by hormone application. This failure to be rescued by GA indicates that plants are unable to perceive the GA signal. Gibberellin-insensitive mutants may show poor germination or increased seed dormancy, growth as a dark green dwarf, delayed flowering, and reduced fertility. Conversely, mutants with a constitutive GA response have phenotypes expected in plants subject to a GA overdose, such as increased plant height and internode length, slender stems, parthenocarpy, and a reduced requirement for GA in germination. Table I summarizes the GA-response genes identified to date.

A. DELLA PROTEINS IN GIBBERELLIN SIGNALING

The current model of GA signaling is centered on the control of DELLA protein accumulation (see model in Fig. 5A). DELLA proteins are negative regulators of GA response subject to GA-stimulated disappearance (Itoh *et al.*, 2003). Loss of DELLA gene function results in a recessive constitutive GA-response phenotype. Such mutants can be tall and slender with a reduced requirement for GA in stem elongation and transition to flowering. Gain-of-function mutations in DELLA genes have the opposite effect resulting in a semidominant GA-insensitive semi-dwarf phenotype and increased sensitivity to GA biosynthesis inhibitors (Dill *et al.*, 2001; Peng *et al.*, 1997). The DELLA proteins in a number of species have been shown to disappear following GA application including: (1) *Oryza sativa SLENDER RICE1* (rice *OsSLR1*; Itoh *et al.*, 2002); (2) *Hordeum vulgare SLENDER1* (barley *HvSLN1*; Gubler *et al.*, 2002); and (3) *Arabidopsis thaliana REPRESSOR OF gal-3* (*RGA*; Silverstone

TABLE I. GA Signaling Genes

Gene	Isolated in	Phenotypes	Encodes
Positive regulators			
D1	Rice	GA-insensitive dwarf	α-Subunit of heterotrimeric G-protein
GAMYB	Barley, rice	Activator of α-amylase	Myb transcription factor
GID1	Rice	Recessive GA-insensitive dwarf	Serine hydrolase
GID2	Rice	GA-insensitive dwarf, poor fertility, overproduces SLR1 protein	F-box protein, homologous to SLY1
GSE1	Barley	Recessive GA-insensitive dwarf, SLN1 protein overproduced	Unknown
PHOR1	Potato	Antisense gives a GA-insensitve dwarf, over-expression gives increased internode length	U-box protein with Armadillo repeats, a potential component of an E3 Ub ligase
PKL	Arabidopsis	Recessive dark green semidwarf, GA overproduction, embryonic root in mature plant	Chromatin remodeling factor
SLY1	Arabidopsis	GA-insensitive dwarf, increased seed dormancy, poor fertility, overproduces RGA protein	F-box protein, homologous to GID2
SNE	Arabidopsis	Over-expression suppresses sly1 dwarf	F-box protein, homologous to SLY1
Negative regulators			
GAI	Arabidopsis	Semidominant semidwarf, also recessive increased internode length, partly redundant with RGA	DELLA subfamily of GRAS family of putative transcription factors
RGA	Arabidopsis	Recessive increased internode length, reduced requirement for GA in germination	DELLA
RGL1, RGL2, RGL3	Arabidopsis	RGL1 is involved in germination and stature, RGL2 is specific to germination	DELLA
RSG	Tobacco	Dominant-negative dwarf, reduced GA_1	bZIP transcription factor
SHI	Arabidopsis	Over-expression leads to dwarf stature	Ring finger protein
SLN1	Barley	Recessive increased internode length	DELLA
SLR1	Rice	Recessive increased internode length	DELLA
SPY	Arabidopsis, barley	Recessive increased internode length, parthenocarpy, reduced requirement for GA in germination	O-Glc-NAc transferase

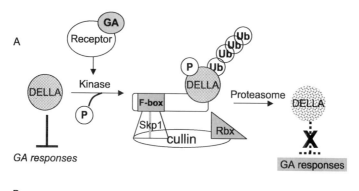

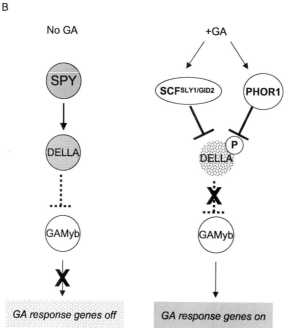

FIGURE 5. Gibberellin signaling in plants. (A) Regulation of DELLA proteins by the ubiquitin-proteasome pathway is mediated by GA-dependent phosphorylation. In the absence of GA, DELLA inhibits GA responses. Gibberellin-binding by the GA receptor stimulates a kinase to phosphorylate the DELLA protein. The phosphorylated DELLA is recognized by the SCF$^{SLY1/GID2}$ E3 ubiquitin ligase complex (F-box protein, Skp1 homologue, cullin, and ring finger protein Rbx). The SCF complex catalyzes the transfer of ubiquitin from Rbx to the target protein. Formation of a polyubiquitin chain targets the DELLA for degradation by the 26S proteasome. (B) Genetic model for GA signaling. In the absence of GA, DELLA proteins inhibit expression of GA-responsive genes either directly or indirectly through inhibition of transcription factors like GAMYB. SPINDLY may negatively regulate GA response by stabilizing the DELLA protein by O-Glc-NAc modification. In the presence of GA, DELLA is negatively regulated by the SCF$^{SLY1/GID2}$ and possibly by the U-box protein PHOR1. DELLA destruction allows activation of GA-responsive gene expression possibly via GAMYB or other transcription factor.

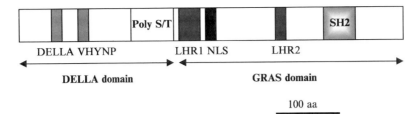

FIGURE 6. DELLA protein structure. The DELLA protein family consists of a number of conserved domains. This schematic is drawn approximately to scale based on an alignment of *Arabidopsis* and rice DELLA proteins. The DELLA protein domain consists of two conserved elements, DELLA and VHYNP. Deletions within this domain lead to loss of GA regulation. The GRAS superfamily domain contains two LHR and one SH2-like domain. These domains are found in STAT transcription factors of metazoans.

et al., 2001), *GA-INSENSITIVE* (*GAI*; Dill *et al.*, 2004; Fu *et al.*, 2004), and *RGA-LIKE2* (*RGL2*; Tyler *et al.*, 2004). Thus, the model is that GA induces GA responses like stem elongation by triggering the destruction of the DELLA protein inhibiting stem elongation (Dill and Sun, 2001; King *et al.*, 2001).

DELLA proteins consist of a DELLA domain required for GA regulation and a GRAS domain required for function (Fig. 6). The DELLA genes are members of the GRAS (GAI-RGA and Scarecrow) family of putative transcription factors (Pysh *et al.*, 1999). The C-terminal GRAS domain contains sequences similar to those found in metazoan signal transducers and activators of transcription (STAT) factors including two leucine heptad repeats (LHR) and an SH2-like domain (Peng *et al.*, 1999). GAI-RGA and Scarecrow proteins contain a variable N-terminal domain. The N-terminal domain of the DELLA subfamily is defined by the consensus "DELLA" and "VHYNP" amino acid sequences. Deletions, N-terminal truncations, and amino acid substitutions within the DELLA domain have been shown to result in a semidominant GA-insensitive dwarf phenotype (Boss and Thomas, 2002; Dill *et al.*, 2001; Peng *et al.*, 1999). Thus, the DELLA domain is required for GA regulation. The DELLA proteins are nuclear localized. A consensus nuclear localization sequence is located within the GRAS domain (Fig. 6). Domain analysis of the rice DELLA protein OsSLR1 was performed by over-expressing *SLR1* constructs containing domain deletions. In spite of the fact that LHR1 domain deletion (termed LZ; Itoh *et al.*, 2002) does not disappear when treated with GA, it results in no phenotype. In contrast, deletion of sequences on the C-terminal side of the NLS results in a dominant-negative tall/slender phenotype. Itoh and coworkers suggest the C-terminal domain is required for function while the LHR1 domain is required for homodimerization. Failure to form homodimer makes the LHR1 deletion both inactive and unregulated, whereas the dominant-negative phenotype of the C-terminal deletion results from dimerization of the truncated protein with wild-type protein via the LHR1 domain. Leucine heptad repeat1-dependent homodimerization was detected by 2-hybrid analysis. Further studies are needed to establish whether SLR1 forms a homodimer in plants.

DELLA gene function is conserved in a wide range of plant species. A single DELLA gene has been functionally defined in monocot species barley (*SLN1*; Gubler *et al.*, 2002), rice (*SLR1*; Ikeda *et al.*, 2001), and maize (*dwarf8* or *d8*; Peng *et al.*, 1999). There are three known DELLA genes in hexaploid wheat, *Rht-A1*, *Rht-B1*, and *Rht-D1* (Peng *et al.*, 1999). It has been demonstrated that mutations in the DELLA domain of *Rht-B1* and *Rht-D1* resulted in the semidominant GA-insensitive semidwarf varieties that were the basis of the 20% increase in yield called the "Green Revolution" in the 1960s and 1970s (Allan, 1986; Peng *et al.*, 1999). These semidwarf mutations appear to increase yield by: (1) making plants with shorter and stronger stems that are resistant to falling over, and (2) causing the plant to put more energy into producing grain than into biomass. Two DELLA genes have been identified in Hawaiian Silversword and a single DELLA gene has been characterized in wine grape (Boss and Thomas, 2002; Remington and Purugganan, 2002). There are five DELLA genes in the dicot species *A. thaliana* (Itoh *et al.*, 2003). It is not yet known why this dicot species has evolved so many copies of this gene family. However, it is known that the five *Arabidopsis* genes serve partly overlapping functions. *RGA* and *GAI* have been shown to act redundantly in repressing stem elongation, transition to flowering, and the juvenile-to-adult phase transition (Dill and Sun, 2001; King *et al.*, 2001). *RGA* and *RGL1* have the strongest role in the transition to flowering (Cheng *et al.*, 2004; Tyler *et al.*, 2004; Yu *et al.*, 2004). *RGL2* is the main DELLA regulating seed germination, but also appears to act in the regulation of flower development (Lee *et al.*, 2002; Tyler *et al.*, 2004; Yu *et al.*, 2004). The combination of knockouts in *RGA* and *RGL2* is sufficient to restore normal flower development in *ga1-3*. While the function of *RGL3* is not yet known, its transcript appears mainly in young plant tissues (Tyler *et al.*, 2004).

B. CONTROL OF DELLA PROTEIN ACCUMULATION BY E3 UBIQUITIN LIGASES

Growing evidence suggests that GA targets the DELLA proteins for destruction via the ubiquitin-26S proteasome pathway. Supporting evidence comes from the study of a conserved F-box protein of a *S*kp1, *C*ullin or Cdc53, F-box (SCF) E3 ubiquitin ligase in rice and in *Arabidopsis* GA signaling. *S*kp1, *C*ullin or Cdc53, F-box complexes are one form of E3 ubiquitin ligase previously defined in yeast and animals (Itoh et al., 2003). The crystal structure of SCF^{Skp2} has been solved and was used as a basis for the model structure in Fig. 5A (Zheng *et al.*, 2002). The F-box protein binds to a specific substrate at its C-terminus that typically contains a consensus protein–protein interaction domain such as leucine rich repeats (LRR), WD repeats, or kelch repeats. The N-terminus contains an F-box domain for Skp1 binding. Skp1 tethers the F-box protein to the N-terminus of cullin, the backbone of the complex. Cullin binds a RING-H2 motif subunit (Rbx1/Hrt1/Roc1) like Rbx1 at the C-terminus. The RING-H2 motif protein binds to the E2-conjugating enzyme. The E3 catalyzes the transfer of ubiquitin from the cysteine of E2 to a lysine residue on the substrate. Addition of four or more ubiquitin moieties to the substrate protein targets it for destruction by the 26S proteasome. The presence of 694 F-box proteins in the *A. thaliana* genome points to their important role in plant signal transduction. The ubiquitin-proteasome has become a recurrent theme in plant hormone signaling as E3 ubiquitin ligases act in auxin, jasmonic acid, ethylene, abscisic acid (ABA), and gibberellin signaling.

The F-box genes rice *GA-INSENSITIVE DWARF2* (*OsGID2*) and *Arabidopsis SLEEPY1* (*AtSLY1*) appear to be positive regulators of GA response because they are negative regulators of the DELLA negative regulators of GA response (Fig. 5B). This model is supported both by genetic and biochemical evidence. Recessive mutations in *sly1* and *gid2* result in a recessive GA-insensitive phenotype. Double mutant analysis showed that the *sly1-10* and *gid2-1* dwarf phenotype was suppressed by knockout mutations in DELLA genes, indicating that the DELLA genes act downstream of GID2/SLY1 in GA signaling (Fig. 5B). Moreover, recessive mutations in *OsGID2* and in *AtSLY1* result in high-level accumulation of DELLA proteins even in the presence of GA (Dill *et al.*, 2004; Fu *et al.*, 2004; McGinnis *et al.*, 2003; Sasaki *et al.*, 2003; Tyler *et al.*, 2004). These results suggested that the GA signal causes SCF$^{GID2/SLY1}$ to target the DELLA proteins for destruction by ubiquitylation. Further evidence for this model include: (1) these F-box proteins have been shown to interact with DELLA proteins using yeast two-hybrid, GST pull down assay, and co-immunoprecipitation (Dill *et al.*, 2004; Fu *et al.*, 2004; Gomi *et al.*, 2004); (2) DELLA protein accumulates in a ubiquitylated form in wild-type plants, but not in *gid2* mutants (Sasaki *et al.*, 2003); and (3) 26S proteasome inhibitors cause the DELLA protein HvSLN1 to accumulate at elevated levels (Fu *et al.*, 2002).

How does GA signal to SCF$^{SLY1/GID2}$ to ubiquitylate the DELLA proteins and target them for destruction? In yeast and in mammals, SCF complexes often ubiquitylate their substrate when the substrate is phosphorylated. It appears that phosphorylation of the DELLA protein is at least one signal that stimulates their ubiquitylation by the SCF$^{SLY1/GID2}$ complex (Fig. 5A). The DELLA OsSLR1 accumulates in a phosphorylated form in the *gid2* mutant. In addition, only the phosphorylated form of OsSLR1 interacts with the OsGID2 protein (Gomi *et al.*, 2004). Similarly, AtSLY1 interacts more strongly with the phosphorylated form of the gai-1 protein, the form of GAI that contains the 17 amino acid deletion of the

DELLA domain (Fu *et al.*, 2004). Thus, it will be important to define the DELLA phosphorylation sites and to identify the kinase responsible for DELLA protein phosphorylation.

Is *AtSLY1/OsGID2* the only F-box protein acting in GA signaling? Evidence suggests that the homologue of *SLY1* in *Arabidopsis*, *SNEEZY* (*SNE*) may act redundantly with *SLY1* in GA signaling. Over-expression of *SNE* suppresses the *sly1–10* phenotype (Fu *et al.*, 2004; Strader *et al.*, 2004). In addition, it appears that the C-terminal truncations encoded by *sly1-2* and *sly1-10* mutant alleles can interfere with wild-type *SNE* function (Strader *et al.*, 2004). Thus, it is possible that SLY1 and SNE interact in GA signaling. The *SNE* gene is conserved in plant species ranging from grape to rice (Strader *et al.*, 2004).

C. NEGATIVE REGULATION OF GIBBERELLIN RESPONSE

This section summarizes additional genes that have been identified as negative regulators of GA response.

1. *SHI*

Over-expression of the *SHORT INTERNODES* (*SHI*) gene leads to a semidwarf GA-insensitive phenotype in *Arabidopsis*. The *SHI* gene is a member of a multigene family whose predicted protein sequence has homology to RING fingers that mediate protein–protein interactions in ubiquitylation and in transcription (Fridborg *et al.*, 1999, 2001). Epistasis studies may shed light on the position of *SHI* in the GA-signaling pathway.

2. *SPY* and *SEC*

Recessive mutations in *SPINDLY* (*SPY*) were isolated in *Arabidopsis* based on resistance to the inhibitory effect of the GA biosynthesis inhibitor paclobutrazol on germination (Jacobsen and Olszewski, 1993) and on the ability to suppress the GA biosynthesis mutant *ga1-3* (Silverstone *et al.*, 1997a). Loss of *SPY* function results in a GA-overdose phenotype including increased internode length, parthenocarpy, and increased resistance to the GA biosynthetic inhibitor paclobutrazol both vegetatively and in germination. The SPY homologue of barley has also been shown to negatively regulate GA response in the aleurone (Robertson *et al.*, 1998). *SPINDLY* encodes an O-linked-β-N-acetylglucosamine transferase (OGT; Thornton *et al.*, 1999). O-linked-β-N-acetylglucosamine transferases catalyze post-translational modification of Ser/Thr residues by addition of a single O-linked β-N-acetylglucosamine. Evidence from animal systems suggests that OGTs can regulate transcription factors by multiple mechanisms, including competition with kinases for modification of protein phosphorylation sites (Vosseller *et al.*, 2002). *SPINDLY* has a single homologue in

Arabidopsis, *SECRET AGENT* (*SEC*). Genetic data indicate that *SPY* and *SEC* agent are required for plant viability as the double mutant is defective in gametogenesis and embryogenesis (Hartweck *et al.*, 2002). Since DELLA protein destruction is apparently induced by DELLA phosphorylation, it will be important to determine whether *SPY* negatively regulates GA signaling by stabilizing DELLA proteins through competition for phosphorylation sites (Fig. 5B). It is known that the dwarf phenotype of the gain-of-function mutation *gai-1* requires *SPY* function (Swain *et al.*, 2001; Tseng *et al.*, 2001). The possibility that SPY may directly affect DELLA protein activity needs to be investigated.

D. POSITIVE REGULATION OF GIBBERELLIN RESPONSE

This section reviews additional genes that have been identified as positive regulators of GA response in plants.

1. *D1*

The *d1* mutant of rice has a recessive GA-insensitive dwarf phenotype. The *DWARF1* (*D1*) gene encodes the α-subunit of a heterotrimeric G-protein (Ueguchi-Tanaka *et al.*, 2000). Epistasis analysis suggests that *D1* acts upstream of the DELLA gene *OsSLR1* to positively regulate GA signaling. Heterotrimeric G-proteins in yeast and other systems can act in conjunction with a G-protein–coupled receptor. More research is needed to determine if D1 may play a similar role in rice GA signaling. The notion that the heteromeric G-protein plays a role in GA signaling is supported by pharmacological studies in oat aleurone (Jones *et al.*, 1998). G-protein α-subunit (*GPA1*) is the single α-subunit of heterotrimeric G-proteins found in the *Arabidopsis* genome (Jones and Assman, 2004). While T-DNA disruption of *gpa1* does cause reduced response to GA in germination, it does not cause reduced plant height. Thus, the heterotrimeric GA protein may have different roles in GA signaling in different plant species.

2. *GAMYB*

GAMYB is a GA-regulated transcription factor first isolated as a positive regulator of α-amylase in the barley aleurone system (Cercos *et al.*, 1999; Gubler *et al.*, 1995, 1999) and subsequently found to regulate anther development (Murray *et al.*, 2003). GAMYB has been shown to act by directly binding to the GA-response element (GARE) promoter element (Sun and Gubler, 2004). Three transposon insertions have been identified in *GAMYB* of rice (Kaneko *et al.*, 2004). As expected, these mutants produced no α-amylase in the endosperm. These mutants show no change in vegetative growth or in the timing of floral induction. However, upon induction of flowering they show reduced internode length, reduced number of spiklets per panicle, and varying degrees of floral defects, including pale shrunken sterile anthers, whitened lemma, malformed palea, and malformed pistils.

The dicot *Arabidopsis* contains three homologues of barley and rice *GAMYB*, AtMYB33, AtMYB65, and AtMYB101. Each of these three homologues is able to induce α-amylase expression when expressed in barley (Gocal *et al.*, 2001). Expression of the AtMYB33 transcript, the closest homologue to *GAMYB* of barley and rice, is induced by GA and LD in the shoot apex. AtMYB33 appears to mediate GA induction of flowering because it is able to bind the GARE of the *LEAFY* gene promoter (Gocal *et al.*, 2001). Based on microarray analysis, 20% of the GA-inducible genes of *Arabidopsis* contain a consensus GARE element in the promoter region, suggesting that *GAMYB* may regulate additional GA-response genes (Ogawa *et al.*, 2003). AtMYB33 transcript is negatively regulated by a microRNA, miR159 (Achard *et al.*, 2004). Accumulation of miR159 is positively regulated by GA and negatively regulated by DELLA proteins.

3. GID1

Recessive mutations in *GA-INSENSITIVE DWARF1* (*GID1*) result in a GA-insensitive dwarf phenotype. The predicted GID1 protein is a member of the serine hydrolase family that includes esterases, lipases, and proteases. Epistasis analysis indicates that *GID1* acts upstream of the DELLA protein *OsSLR1*. The SLR1 protein accumulates at high levels in *gid1* mutants suggesting that *GID1* is involved in control of SLR1 protein degradation (Gomi and Matsuoka, 2003).

4. GSE1

Recessive mutations in *GA-SENSITIVITY1* (*GSE1*) of barley result in a GA-insensitive dwarf phenotype (Chandler and Robertson, 1999). While the gene remains uncloned, studies indicate that *GSE1* is required for the GA-stimulated disappearance of the DELLA protein SLN1 (Gubler *et al.*, 2002). It will be interesting to learn whether *GSE1* is a unique gene or whether it encodes the barley homologue of rice genes *GID1* or *GID2*.

5. PHOR1

PHOTOPERIOD-RESPONSIVE1 (*PHOR1*) is a GA-signaling gene identified in potato based on its role in promoting tuberization (Amador *et al.*, 2001). Tuberization of wild potato plants is induced under SDs (8 h of light) and not under LDs (16 h of light). The tuberization process under SD appears to be due, in part, to inhibition of GA signaling (Garcia-Martinez and Gil, 2001). *PHOTOPERIOD-RESPONSIVE1* was recovered using RT-PCR differential display to identify genes expressed during SD-induced tuberization. Antisense expression of *PHOR1* results in a GA-insensitive semi-dwarf phenotype, whereas over-expression of PHOR1 results in enhanced

GA response. In addition, the observation that a PHOR1-GFP translation fusion protein shows GA-dependent nuclear localization supports the view that *PHOR1* is involved in GA signaling. The predicted PHOR1 protein encodes a U-box protein with armadillo repeats (Amador et al., 2001). Evidence suggests that U-box proteins may act independently as E3 ubiquitin ligases (Hatakeyama and Nakayama, 2003). It will be interesting to see whether future research supports a role for *PHOR1* in negatively regulating DELLA proteins via the ubiquitin-proteasome pathway (Monte et al., 2003).

6. *PKL*

The possible role for GA in the transition from embryo to adult development is highlighted by studies of the recessive *pickle* (*pkl*) mutant of *Arabidopsis*. Originally identified based on its tendency to retain embryonic characteristics upon germination, it was subsequently suggested that *PICKLE* (*PKL*) is a positive regulator of GA response (Ogas et al., 1997). This recessive mutation imparts a partially GA-insensitive semidwarf phenotype, reduced response to GA in hypocotyl elongation assays, and enhancement of its embryo-like phenotype when treated with GA biosynthesis inhibitor uniconazole-P (Henderson et al., 2004). While the *pkl* mutant results in overaccumulation of bioactive GAs, a hallmark of GA-insensitive mutants, it does not result in overproduction of *GA3ox1* or of *GA20ox1*. Thus, unlike GA-insensitive mutants *gai-1* and *sly1-10*, the *pkl* mutant does not stimulate positive feedback control of these GA biosynthetic genes. It will be interesting to learn if *pkl* alters expression of the GA inactivating enzyme *GA2ox*. The *PKL* gene encodes a CHD3 protein, a chromatin remodeling factor found throughout eukaryotes that acts as a developmentally regulated repressor of transcription (Dean Rider et al., 2003; Ogas et al., 1999). The model proposed is that *PKL* is a hormone-responsive negative regulator of embryo-specific gene transcription (Henderson et al., 2004). In this case, GA stimulates the transition from embryo to adult developmental state both via a *PKL*-dependent and *PKL*-independent pathway. This raises the intriguing possibility that GA is needed in germination, in part, to signal for the transition to adult development. If this is true, one expects GA biosynthesis mutants to retain some embryonic characteristics after germination. Evidence supporting this model includes: (1) *PKL* appears to be a negative regulator of master regulators of embryonic identity genes *FUSCA3* (*FUS3*), and *LEAFY COTYLEDONS1* and *LEAFY COTYLEDONS2* (*LEC1* and *LEC2*) (Dean Rider et al., 2003; Ogas et al., 1999); (2) *pkl* mutants accumulate seed storage compounds in roots including triacylglycerol, seed storage proteins, and phytate (Rider et al., 2004); (3) GA is able to suppress embryonic characteristics in the *pkl* mutant (Henderson et al., 2004); and (4) GA appears to destabilize a FUS3-GFP fusion protein (Gazzarrini et al., 2004). Mutations in *PKL* have also been identified as enhancers of *crabsclaw* (*crc*) based on ectopic production of ovules on carples, suggesting that *PKL* may be a general inhibitor of indeterminacy (Hay et al., 2004).

E. GIBBERELLIN-RESPONSE GENES

The final targets of GA signaling are the GA-response genes responsible for the effects of the hormone. Known GA-response genes include: (1) hydrolytic genes acting in germination such as genes encoding α-amylase, endo-[beta]-mannase, and β-1,3-glucanase (Jacobsen et al., 1995; Ni and Bradford, 1993; Wu et al., 2001); (2) cell cycle and cell wall loosening enzymes involved in stem elongation such as cyclins, CDKs, XETs, and expansins (Cosgrove, 2000; Ogawa et al., 2003; Sauter, 1997); and (3) genes involved in induction of flowering and floral development such as *LEAFY*, *APETELA3*, *PISTILLATA*, and *AGAMOUS* (Gocal et al., 2001; Yu et al., 2004).

The precise mechanism by which DELLA genes control expression of GA-response genes is still unknown. However, it is known that the DELLA gene *HvSLN1* of barley is required for repression of *GAMYB* transcription (Fig. 5B, Gubler et al., 2002). *GAMYB* family members positively control expression of α-amylase and of the flowering gene *LEAFY* by direct binding to a GARE promoter element (Gocal et al., 1999, 2001; Gubler et al., 1999; Rogers et al., 1992). GAMYB is known to regulate both α-amylase in germination and LEAFY expression in flowering. Future research will need to determine if *GAMYB* or related genes participate in regulation of other GA-response genes, including those involved in stem elongation or feedback regulation of GA biosynthesis (Fig. 5B).

F. MODEL FOR GIBBERELLIN SIGNALING

Figure 5B shows a current model for control of GA-responsive gene expression in plants. In the absence of GA, DELLA proteins inhibit expression of GA-responsive genes. *GAMYB* is known to induce expression of GA-responsive genes such as those encoding α-amylase and *AtLEAFY*. It will be important to determine whether DELLAs inhibit GA-response gene expression directly or indirectly through inhibition of *GAMYB* or other transcription factors. *SPINDLY* negatively regulates GA response, possibly by stabilizing the DELLA protein by O-Glc-NAc modification. In the presence of GA, SCF$^{SLY1/GID2}$ and possibly also the U-box protein PHOR1 target the DELLA protein for destruction via the ubiquitin-proteasome pathway. This relieves DELLA repression, allowing GAMYB or other transcription factors to induce expression of GA-response genes. It appears that GA targets the DELLA protein for destruction by phosphorylation (Fig. 5A). In this case, the unidentified GA receptor causes activation of a kinase that phosphorylates the DELLA protein. The phosphorylated

DELLA is recognized by the SCF$^{SLY1/GID2}$ E3 ubiquitin ligase. Polyubiquitylation of DELLA by SCF$^{SLY1/GID2}$ targets the DELLA protein for destruction by the 26S proteasome. Degradation of DELLA allows activation of GA-responsive gene expression possibly via GAMYB.

IV. CROSS-TALK WITH OTHER HORMONE-SIGNALING PATHWAYS

The regulation of specific developmental processes is controlled by multiple plant hormones. It is therefore not surprising to find the existence of multiple levels of cross-talk between these phytohormone-signaling pathways. Cross-talk between hormone-signaling pathways is seen both in the control of hormone accumulation and in control of hormone sensitivity.

A. GIBBERELLIN AND ABSCISIC ACID SIGNALING

The antagonism between GA and ABA in the control of seed germination is a well-characterized interaction between two plant hormone-signaling pathways (Koornneef et al., 2002). Studies in *Arabidopsis* show that ABA biosynthesis is transiently induced during embryo maturation and is needed for the embryo to achieve dormancy and dessication tolerance (Karssen et al., 1983). Gibberellin is needed to break seed dormancy and induce germination. Many studies have shown that mutations in ABA and GA biosynthesis and signaling pathways alter response to the other hormone in germination. One can think of this as a tug-of-war over germination with the ABA players pulling for seed dormancy and the GA players pulling for germination. For example, mutations that reduce ABA biosynthesis or sensitivity suppress the requirement for GA in germination (Karssen and Lacka, 1986; Léon-Kloosterziel et al., 1996; Nambara et al., 1991; Steber et al., 1998). This failure to respond or synthesize ABA alleviates the requirement for GA in germination because the seeds never become dormant in the first place. Conversely, the GA-insensitive mutants in *SLY1* result in increased seed dormancy and increased sensitivity to ABA in germination while GA-hypersensitive mutations in *SPY* cause slight ABA-insensitivity in germination (Steber et al., 1998; Strader et al., 2004; Swain et al., 2001).

Abscisic acid and GA may negatively regulate the other hormone-signaling pathway at multiple levels including: (1) hormone biosynthesis, (2) hormone signaling, and (3) transcriptional control. Gibberellin treatment has been shown to reduce accumulation of ABA in dark-germinating lettuce seeds after a pulse of far-red light (Toyomasu et al., 1994). Further research is needed to examine the effects of GA and ABA on one another's biosynthesis. Gibberellin and ABA have been shown to differentially regulate the transcription of genes in a number of plant systems. The cereal aleurone system has eloquently demonstrated the ability of ABA to block GA induction of α-amylase at the level of transcription (Jacobsen *et al.*, 1995). This may occur, in part, via the ABA-induced protein kinase *PKABA1*, as transient over-expression of *PKABA1* represses the GA-induced genes *GAMYB* and α-amylase (Gomez-Cadenas *et al.*, 2001; Zentella *et al.*, 2002). (B) In tomato, ABA induces and GA represses expression of the sugar-sensing gene *LeSNF4* (Bradford *et al.*, 2003). Finally, microarray analysis in *Arabidopsis* has shown that many GA-downregulated genes have ABA response elements (ABRE) in their promoters (Ogawa *et al.*, 2003). The downregulation of these genes in GA-treated *ga1-3* did not appear to correlate with reduced endogenous ABA suggesting that GA is downregulating ABA signaling. Further research is needed to precisely determine how these hormones negatively regulate one another's signaling cascades.

B. GIBBERELLIN AND BRASSINOSTEROID SIGNALING

Cross-talk has been seen between GA and brassinosteroid (BR) signaling during seed germination and hypocotyl elongation. Brassinosteroid partially rescues seed germination and elongation of dark-grown hypocotyls in the *Arabidopsis* GA biosynthesis mutant *ga1-3* and in GA-insensitive mutant *sly1-2* (Steber and McCourt, 2001). Gibberellin does not, however, rescue hypocotyl elongation of dark-grown BR biosynthesis mutant *det2-1*. Thus, BR appears to be able to bypass GA signaling in these processes, but GA cannot bypass BR in hypocotyl elongation. Work on tobacco indicates that GA and BR promote germination by distinct mechanisms (Leubner-Metzger, 2001). Gibberellin and light appear to act in a common pathway to release photodormancy and to induce expression of the hydrolytic enzyme β-1,3-glucanase in the endosperm. In contrast, BR could not overcome photodormancy or induce β-1,3-glucanase. However, both BR and GA could stimulate germination of ABA-inhibited seeds and accelerate the germination of non-photodormant seeds. Leubner-Metzger proposes that BR stimulates germination solely through stimulation of hypocotyl elongation. This would suggest that BR acts in parallel with, rather than downstream of GA signaling to stimulate germination and hypocotyl elongation of *ga1-3* in *Arabidopsis*.

Further research is needed to understand the interaction between GA-, BR-, and ABA-signaling pathways in germination. One possibility is that GA and BR may regulate one another's biosynthesis. Interestingly, Bouquin and coworkers found that whereas GA negatively regulates the GA biosynthesis gene *AtGA20ox1*, BR positively regulates *AtGA20ox1* (Bouquin *et al.*, 2001). Thus, BR may act, in part, by stimulating GA biosynthesis. This does not fully explain the interaction between GA and BR because BR is able to rescue the germination of *ga1-3*, a mutant blocked upstream of *AtGA-20ox1* in GA biosynthesis (Steber and McCourt, 2001). However, the fact that a mutation in the BR receptor *AtBRI1* resulted in increased expression of *AtGA20ox1* suggest that plants may induce GA biosynthesis in response to reduced flux in the BR-signaling pathway. Future research will need to determine whether the converse is true.

Research in *Arabidopsis* suggests that the heterotrimeric GTP-binding protein (G-protein) and putative G-protein–coupled receptor may be involved in GA and BR signaling in germination (Chen *et al.*, 2004; Ullah *et al.*, 2002). The *Arabidopsis* genome contains one prototypical *GPA1*, one G-protein β-subunit (*AGB1*), and two G-protein γ-subunits (*AGG1* and *AGG2*) (Jones and Assmann, 2004). One putative G-protein–coupled receptor (*GCR1*) containing a predicted seven-transmembrane domain has been identified in *Arabidopsis*. T-DNA insertional mutations in *GPA1* and in *GCR1* result in reduced response to GA and BR in germination (Chen *et al.*, 2004; Ullah *et al.*, 2002). Ullah *et al.* (2002) proposed that BR may potentiate GA signaling in *Arabidopsis* via *GPA1*.

C. GIBBERELLIN AND AUXIN SIGNALING

In pea, elegant experiments studying both the shoot apex regulation of stem elongation and seed regulation of pericarp growth have provided many insights into the interaction of GA and auxin (O'Neill and Ross, 2002; Ozga *et al.*, 2003; Ross *et al.*, 2001; van Huizen *et al.*, 1997).

Removal of the pea shoot apex inhibits stem elongation because the growth promoting IAA source has been removed. Ross *et al.* (2000) have demonstrated that auxin exerts this effect on stem growth by increasing bioactive GA. This is achieved by promoting expression of the *PsGA3ox1* gene, whereas the levels of *PsGA2ox1/2* transcripts were suppressed (O'Neill and Ross, 2002). Similarly, in the case of seed-stimulated pericarp growth, it has been demonstrated that auxin (4-Cl-IAA) and the presence of seeds promotes pericarp growth by upregulating expression of *PsGA3ox1* (Ozga *et al.*, 2003). The effect of auxin on GA metabolism and promotion of stem growth appears conserved in monocots. Wolbang and coworkers confirmed that auxin from the developing inflorescence of barley plants is required for bioactive GA production and subsequent growth in the stem (Wolbang *et al.*, 2004). A barley GA 3-oxidase gene, *HvGA3ox2*, is implicated in this response to auxin. Interestingly, a study in tobacco indicates that auxin promotes a different GA biosynthetic step, GA 20-oxidation (Wolbang and Ross, 2001). These studies have demonstrated that auxin is likely transported to its site of action where it stimulates the biosynthesis of GAs, which in turn promote growth. Further work is necessary to understand the molecular basis of this cross-talk.

Considering the role of GA signaling in regulating the expression levels of 2-ODD genes, one possible explanation for the auxin-mediated regulation of GA metabolism is that this hormone directly modulates the GA-signal transduction pathway. It is this effect on GA signaling that leads to changes in expression of GA metabolic genes. This model is supported by a study of root growth in *Arabidopsis*. Fu and Harberd (2003) demonstrated that the shoot apex-derived auxin controls root elongation by modulating the GA-response pathway. More specifically, auxin was shown to affect GA-regulated root growth by modifying the stability of the DELLA protein, RGA. The same group has also demonstrated that ethylene can affect GA-regulated root and hypocotyl growth by a similar process (Achard *et al.*, 2003). In view of the role of SCF E3 ubiquitin ligases in these three hormone-signaling pathways, it is tempting to speculate that these complexes may provide the molecular link to this hormone cross-talk. Biochemical and proteomics approaches should help to provide answers to these questions.

V. PERSPECTIVES

We have seen that mutations affecting GA biosynthesis and response have been essential for improving yields in many agronomically important crops. Although the molecular basis of several of the mutations has been revealed, in most cases, we still have little understanding of how they confer these beneficial traits. In contrast to GA metabolism, our knowledge of GA signaling and the downstream processes that promote GA-responsive growth is rather limited. To further our understanding, it is crucial that we identify the respective components of these processes. It will then be possible to fully understand the developmental and environmental factors that regulate GA metabolism, signaling, and responsive components. Furthermore, the precise spatial and temporal localization patterns can be determined, leading to an understanding of the relationships between these components and their roles in mediating GA-responsive growth. We believe that this understanding will, in part, lead to a second "Green Revolution" in the not too distant future.

REFERENCES

Aach, H., Bose, G., and Graebe, J. E. (1995). *ent*-Kaurene biosynthesis in a bell-free system from wheat (*Triticum aestivum* L.) seedlings and the localization of *ent*-kaurene synthetase in plastids of three species. *Planta* **197**, 333–342.

Aach, H., Bode, H., Robinson, D. G., and Graebe, J. E. (1997). *ent*-Kaurene synthase is located in proplastids of meristematic shoot tissues. *Planta* **202**, 211–219.

Achard, P., Vriezen, W. H., Van Der, Straeten, D., and Harberd, N. P. (2003). Ethylene regulates *Arabidopsis* development via the modulation of DELLA protein growth repressor function. *Plant Cell* **15**, 2816–2825.

Achard, P., Herr, A., Baulcombe, D. C., and Harberd, N. P. (2004). Modulation of floral development by a gibberellin-regulated microRNA. *Development* **131**, 3357–3365.

Ait-Ali, T., Frances, S., Weller, J. L., Reid, J. B., Kendrick, R. E., and Kamiya, Y. (1999). Regulation of gibberellin 20-oxidase and gibberellin 3beta-hydroxylase transcript accumulation during de-etiolation of pea seedlings. *Plant Physiol.* **121,** 783–791.

Alabadi, D., Gil, J., Blazquez, M. A., and Garcia-Martinez, J. L. (2004). Gibberellins repress photomorphogenesis in darkness. *Plant Physiol.* **134,** 1050–1057.

Albone, K. S., Gaskin, P., Mac Millan, J., and Sponsel, V. M. (1984). Identification and localization of gibberellins in maturing seeds of the cucurbit *Sechium edule*, and a comparison between this cucurbit and the legume *Phaseolus coccineus*. *Planta* **162,** 560–565.

Allan, R. E. (1986). Agronomic comparison among wheat lines nearly isogenic for three reduced-height genes. *Crop Sci.* **26,** 707–710.

Amador, V., Monte, E., Garcia-Martinez, J. L., and Prat, S. (2001). Gibberellins signal nuclear import of PHOR1, a photoperiod-responsive protein with homology to *Drosophila armadillo*. *Cell* **106,** 343–354.

Appleford, N. E. J., and Lenton, J. R. (1991). Gibberellins and leaf expansion in near-sogenic wheat lines containing *rht1* and *rht3* dwarfing alleles. *Planta* **183,** 229–236.

Bentsink, L., and Koornneef, M. (2002). Seed dormancy and germination. In "The *Arabidopsis* Book" (C. R. Somerville and E. M. Meyerowitz, Eds.). doi: 10.1199/tab.0050. American Society of Plant Biologist, Rockville, MD.

Bewley, J. D., and Black, M. (1994). "Seeds: Physiology of Development and Germination." Plenum Press, New York.

Blazquez, M. A., Green, R., Nilsson, O., Sussman, M. R., and Weigel, D. (1998). Gibberellins promote flowering of *Arabidopsis* by activating the LEAFY promoter. *Plant Cell* **10,** 791–800.

Boss, P. K., and Thomas, M. R. (2002). Association of dwarfism and floral induction with a grape 'green revolution' mutation. *Nature* **416,** 847–850.

Bouquin, T., Meier, C., Foster, R., Nielsen, M. E., and Mundy, J. (2001). Control of specific gene expression by gibberellin and brassinosteroid. *Plant Physiol.* **127,** 450–458.

Bradford, K. J., Downie, A. B., Gee, O. H., Alvarado, V., Yang, H., and Dahal, P. (2003). Abscisic acid and gibberellin differentially regulate expression of genes of the SNF1-related kinase complex in tomato seeds. *Plant Physiol.* **132,** 1560–1576.

Brian, P. W., and Hemming, H. G. (1955). The effect of gibberellic acid on shoot growth of pea seedlings. *Physiol. Plantarum* **8,** 669–681.

Bush, D., and Jones, R. (1988). Cytoplasmic calcium and amylase secretion from barley aleurone protoplasts. *Eur. J. Cell Biol.* **46,** 466–469.

Busov, V. B., Meilan, R., Pearce, D. W., Ma, C. P., Rood, S. B., and Strauss, S. H. (2003). Activation tagging of a dominant gibberellin catabolism gene (GA 2-oxidase) from poplar that regulates tree stature. *Plant Physiol.* **132,** 1283–1291.

Campbell, P., and Braam, J. (1999). *In vitro* activities of four xyloglucan endotransglycosylases from *Arabidopsis*. *Plant J.* **18,** 371–382.

Carrera, E., Jackson, S. D., and Prat, S. (1999). Feedback control and diurnal regulation of gibberellin 20-oxidase transcript levels in potato. *Plant Physiol.* **119,** 765–774.

Carrera, E., Bou, J., Garcia-Martinez, J. L., and Prat, S. (2000). Changes in GA 20-oxidase gene expression strongly affect stem length, tuber induction and tuber yield of potato plants. *Plant J.* **22,** 247–256.

Cercos, M., Gomez-Cadenas, A., and Ho, T. H. (1999). Hormonal regulation of a cysteine proteinase gene, EPB-1, in barley aleurone layers: *cis*- and *trans*-acting elements involved in the co-ordinated gene expression regulated by gibberellins and abscisic acid. *Plant J.* **19,** 107–118.

Chandler, P. M., and Robertson, M. (1999). Gibberellin dose-response curves and the characterization of dwarf mutants of barley. *Plant Physiol.* **120,** 623–632.

Chang, C. W., and Sun, T. P. (2002). Characterization of *cis*-regulatory regions responsible for developmental regulation of the gibberellin biosynthetic gene GA1 in *Arabidopsis thaliana*. *Plant Mol. Biol.* **49,** 579–589.

Chen, J. G., Pandey, S., Huang, J., Alonso, J. M., Ecker, J. R., Assmann, S. M., and Jones, A. M. (2004). GCR1 can act independently of heterotrimeric G-protein in response to brassinosteroids and gibberellins in *Arabidopsis* seed germination. *Plant Physiol.* **135**, 907–915.

Cheng, H., Qin, L., Lee, S., Fu, X., Richards, D. E., Cao, D., Luo, D., Harberd, N. P., and Peng, J. (2004). Gibberellin regulates *Arabidopsis* floral development via suppression of DELLA protein function. *Development* **131**, 1055–1064.

Chiang, H. H., Hwang, I., and Goodman, H. M. (1995). Isolation of the *Arabidopsis GA4* locus. *Plant Cell* **7**, 195–201.

Coles, J. P., Phillips, A. L., Croker, S. J., Garcia-Lepe, R., Lewis, M. J., and Hedden, P. (1999). Modification of gibberellin production and plant development in *Arabidopsis* by sense and antisense expression of gibberellin 20-oxidase genes. *Plant J.* **17**, 547–556.

Cosgrove, D. J. (2000). Loosening of plant cell walls by expansins. *Nature* **407**, 321–326.

Davidson, S. E., Elliott, R. C., Helliwell, C. A., Poole, A. T., and Reid, J. B. (2003). The pea gene NA encodes ent-kaurenoic acid oxidase. *Plant Physiol.* **131**, 335–344.

Dean Rider, S., Jr., Henderson, J. T., Jerome, R. E., Edenberg, H. J., Romero-Severson, J., and Ogas, J. (2003). Coordinate repression of regulators of embryonic identity by PICKLE during germination in *Arabidopsis*. *Plant J.* **35**, 33–43.

Derkx, M. P. M., Vermeer, E., and Karssen, C. M. (1994). Gibberellins in seeds of *Arabidopsis thaliana*—biological activities, identification and effects of light and chilling on endogenous levels. *Plant Growth Regul.* **15**, 223–234.

Dill, A., and Sun, T. (2001). Synergistic derepression of gibberellin signaling by removing RGA and GAI function in *Arabidopsis thaliana*. *Genetics* **159**, 777–785.

Dill, A., Jung, H. S., and Sun, T. P. (2001). The DELLA motif is essential for gibberellin-induced degradation of RGA. *Proc. Natl. Acad. Sci. USA* **98**, 14162–14167.

Dill, A., Thomas, S. G., Hu, J., Steber, C. M., and Sun, T. P. (2004). The *Arabidopsis* F-box protein SLEEPY1 targets gibberellin signaling repressors for gibberellin-induced degradation. *Plant Cell* **16**, 1392–1405.

Durley, R. C., MacMillan, J., and Pryce, R. J. (1971). Investigation of gibberellins and other growth substances in the seed of *Phaseolus multiflorus* and *Phaseolus vulgaris* by gas chromatography and gas chromatography-mass spectrometry. *Phytochemistry* **10**, 1891–1908.

Elliott, R. C., Ross, J. J., Smith, J. L., Lester, D. R., and Reid, J. B. (2001). Feed-forward regulation of gibberellin deactivation in pea. *J. Plant Growth Regul.* **20**, 87–94.

Ewing, E. E., and Wareing, P. F. (1978). Shoot, stolon and tuber formation on potato (*Solanum tuberosum* L.) cuttings in response to photoperiod. *Plant Physiol.* **61**, 348–353.

Fabian, T., Lorbiecke, R., Umeda, M., and Sauter, M. (2000). The cell cycle genes cycA1;1 and cdc2Os-3 are coordinately regulated by gibberellin in planta. *Planta* **211**, 376–383.

Fleet, C. M., Yamaguchi, S., Hanada, A., Kawaide, H., David, C. J., Kamiya, Y., and Sun, T. P. (2003). Overexpression of AtCPS and AtKS in *Arabidopsis* confers increased *ent*-kaurene production but no increase in bioactive gibberellins. *Plant Physiol.* **132**, 830–839.

Fridborg, I., Kuusk, S., Moritz, T., and Sundberg, E. (1999). The *Arabidopsis* dwarf mutant shi exhibits reduced gibberellin responses conferred by overexpression of a new putative zinc finger protein. *Plant Cell* **11**, 1019–1032.

Fridborg, I., Kuusk, S., Robertson, M., and Sundberg, E. (2001). The *Arabidopsis* protein SHI represses gibberellin responses in *Arabidopsis* and barley. *Plant Physiol.* **127**, 937–948.

Frydman, V. M., Gaskin, P., and Mac Millan, J. (1974). Qualitative and quantitative analyses of gibberellins throughout seed maturation in *Pisum sativum* cv. Progress No. 9. *Planta* **118**, 123–132.

Fu, X., Richards, D. E., Ait-Ali, T., Hynes, L. W., Ougham, H., Peng, J., and Harberd, N. P. (2002). Gibberellin-mediated proteasome-dependent degradation of the barley DELLA protein SLN1 repressor. *Plant Cell* **14**, 3191–3200.

Fu, X., Richards, D. E., Fleck, B., Xie, D., Burton, N., and Harberd, N. P. (2004). The *Arabidopsis* mutant sleepy1gar2-1 protein promotes plant growth by increasing the affinity

of the SCFSLY1 E3 ubiquitin ligase for DELLA protein substrates. *Plant Cell* **16,** 1406–1418.
Fu, X. D., and Harberd, N. P. (2003). Auxin promotes *Arabidopsis* root growth by modulating gibberellin response. *Nature* **421,** 740–743.
Fujioka, S., Yamane, H., Spray, C. R., Gaskin, P., Macmillan, J., Phinney, B. O., and Takahashi, N. (1988a). Qualitative and quantitative-analyses of gibberellins in vegetative shoots of normal, Dwarf-1, Dwarf-2, Dwarf-3, and Dwarf-5 seedlings of *Zea mays* L. *Plant Physiol.* **88,** 1367–1372.
Fujioka, S., Yamane, H., Spray, C. R., Katsumi, M., Phinney, B. O., Gaskin, P., Macmillan, J., and Takahashi, N. (1988b). The dominant non-gibberellin-responding dwarf mutant (*d8*) of maize accumulates native gibberellins. *Proc. Natl. Acad. Sci. USA* **85,** 9031–9035.
Fukazawa, J., Sakai, T., Ishida, S., Yamaguchi, I., Kamiya, Y., and Takahashi, Y. (2000). Repression of shoot growth, a bZIP transcriptional activator, regulates cell elongation by controlling the level of gibberellins. *Plant Cell* **12,** 901–915.
Garcia-Martinez, J. L., and Gil, J. (2001). Light regulation of gibberellin biosynthesis and mode of action. *J Plant Growth Regul.* **20,** 354–368.
Garcia-Martinez, J. L., Lopez Diaz, I., Sanchez Beltran, M. J., Phillips, A. L., Ward, D. A., Gaskin, P., and Hedden, P. (1997). Isolation and transcript analysis of gibberellin 20-oxidase genes in pea and bean in relation to fruit development. *Plant Mol. Biol.* **33,** 1073–1084.
Gazzarrini, S., Tsuchiya, Y., Lumba, S., Okamoto, M., and McCourt, P. (2004). The transcription factor FUSCA3 controls developmental timing in *Arabidopsis* through the hormones gibberellin and abscisic acid. *Dev. Cell* **7,** 373–385.
Gilmour, S. J., Zeevaart, J. A. D., Schwenen, L., and Graebe, J. E. (1986). Gibberellin metabolism in cell-free-extracts from spinach leaves in relation to photoperiod. *Plant Physiol.* **82,** 190–195.
Gocal, G. F., Poole, A. T., Gubler, F., Watts, R. J., Blundell, C., and King, R. W. (1999). Long-day up-regulation of a GAMYB gene during *Lolium temulentum* inflorescence formation. *Plant Physiol.* **119,** 1271–1278.
Gocal, G. F., Sheldon, C. C., Gubler, F., Moritz, T., Bagnall, D. J., MacMillan, C. P., Li, S. F., Parish, R. W., Dennis, E. S., Weigel, D., and King, R. W. (2001). GAMYB-like genes, flowering, and gibberellin signaling in *Arabidopsis*. *Plant Physiol.* **127,** 1682–1693.
Gomez-Cadenas, A., Zentella, R., Walker-Simmons, M. K., and Ho, T. H. (2001). Gibberellin/abscisic acid antagonism in barley aleurone cells: Site of action of the protein kinase PKABA1 in relation to gibberellin signaling molecules. *Plant Cell* **13,** 667–679.
Gomi, K., and Matsuoka, M. (2003). Gibberellin signaling pathway. *Curr. Opin. Plant Biol.* **6,** 489–493.
Gomi, K., Sasaki, A., Itoh, H., Ueguchi-Tanaka, M., Ashikari, M., Kitano, H., and Matsuoka, M. (2004). GID2, an F-box subunit of the SCF E3 complex, specifically interacts with phosphorylated SLR1 protein and regulates the gibberellin-dependent degradation of SLR1 in rice. *Plant J.* **37,** 626–634.
Goodwin, T. W. (1965). Regulation of terpenoid biosynthesis in higher plants. *In* "Biosynthetic Pathways in Higher Plants" (J. B. Pridham and T. Swain, Eds.), pp. 57–71. Academic Press, London.
Graebe, J. E. (1979). *In* "Proceedings of the 10th International Conference on Plant Growth Substances", pp. 180–187.
Graebe, J. E. (1987). Gibberellin biosynthesis and control. *Annu. Rev. Plant Physiol. Plant Mol. Biol.* **38,** 419–465.
Griggs, D. L., Hedden, P., and Lazarus, C. M. (1991). Partial-purification of 2 gibberellin 2b-hydroxylases from cotyledons of *Phaseolus vulgaris*. *Phytochemistry* **30,** 2507–2512.
Groot, S. P. C., and Karssen, C. M. (1987). Gibberellins regulate seed germination in tomato by endosperm weakening: A study with gibberellin-deficient mutants. *Planta* **171,** 525–531.

Grosselindemann, E., Lewis, M. J., Hedden, P., and Graebe, J. E. (1992). Gibberellin biosynthesis from gibberellin A12-aldehyde in a cell-free system from germinating barley (*Hordeum vulgare* L., Cv Himalaya) embryos. *Planta* **188**, 252–257.

Gubler, F., Kalla, R., Roberts, J. K., and Jacobsen, J. V. (1995). Gibberellin-regulated expression of a myb gene in barley aleurone cells: Evidence for Myb transactivation of a high-pI alpha-amylase gene promoter. *Plant Cell* **7**, 1879–1891.

Gubler, F., Chandler, P. M., White, R. G., Llewellyn, D. J., and Jacobsen, J. V. (2002). Gibberellin signaling in barley aleurone cells. Control of SLN1 and GAMYB expression. *Plant Physiol.* **129**, 191–200.

Gubler, F., Raventos, D., Keys, M., Watts, R., Mundy, J., and Jacobsen, J. V. (1999). Target genes and regulatory domains of the GAMYB transcriptional activator in cereal aleurone. *Plant J.* **17**, 1–9.

Hake, S., Smith, H. M., Holtan, H., Magnini, E., Mele, G., and Ramirez, J. (2004). The role of knox genes in plant development. *Annu. Rev. Cell. Dev. Biol.* **20**, 125–151.

Hartweck, L. M., Scott, C. L., and Olszewski, N. E. (2002). Two O-linked N-acetylglucosamine transferase genes of *Arabidopsis thaliana* L. Heynh. Have overlapping functions necessary for gamete and seed development. *Genetics* **161**, 1279–1291.

Hatakeyama, S., and Nakayama, K. I. (2003). U-box proteins as a new family of ubiquitin ligases. *Biochem. Biophys. Res. Commun.* **302**, 635–645.

Hay, A., Craft, J., and Tsiantis, M. (2004). Plant hormones and homeoboxes: Bridging the gap? *Bioessays* **26**, 395–404.

Hay, A., Kaur, H., Phillips, A., Hedden, P., Hake, S., and Tsiantis, M. (2002). The gibberellin pathway mediates KNOTTED1-type homeobox function in plants with different body plans. *Curr. Biol.* **12**, 1557–1565.

Hazebroek, J. P., and Metzger, J. D. (1990). Thermoinductive regulation of gibberellin metabolism in *Thlaspi arvense* L. 1. Metabolism of [H-2]-ent-kaurenoic acid and [C-14] gibberellin-A_{12}-aldehyde. *Plant Physiol.* **94**, 157–165.

Hazebroek, J. P., Metzger, J. D., and Mansager, E. R. (1993). Thermoinductive regulation of gibberellin metabolism in *Thlaspi arvense* L. 2. Cold induction of enzymes in gibberellin biosynthesis. *Plant Physiol.* **102**, 547–552.

Hedden, P. (2003). The genes of the Green Revolution. *Trends Genet.* **19**, 5–9.

Hedden, P., and Croker, S. J. (1992). Regulation of gibberellin biosynthesis in maize seedlings. In "Progress in Plant Growth Regulation: Proceedings of the 14th International Conference on Plant Growth Substances" (D. Vreugdenhil, Ed.), pp. 534–544. Kluwer, Dordrecht.

Hedden, P., and Phillips, A. L. (2000). Gibberellin metabolism: New insights revealed by the genes. *Trends Plant Sci.* **5**, 523–530.

Hedden, P., Graebe, J. E., Beale, M. H., Gaskin, P., and Macmillan, J. (1984). The biosynthesis of 12-alpha-hydroxylated gibberellins in a cell-free system from *Cucurbita maxima* endosperm. *Phytochemistry* **23**, 569–574.

Hedden, P., Phillips, A. L., Rojas, M. C., Carrera, E., and Tudzynski, B. (2001). Gibberellin biosynthesis in plants and fungi: A case of convergent evolution? *J. Plant Growth Regul.* **20**, 319–331.

Helliwell, C. A., Poole, A., Peacock, W. J., and Dennis, E. S. (1999). *Arabidopsis* ent-kaurene oxidase catalyzes three steps of gibberellin biosynthesis. *Plant Physiol.* **119**, 507–510.

Helliwell, C. A., Chandler, P. M., Poole, A., Dennis, E. S., and Peacock, W. J. (2001a). The CYP88A cytochrome P450, ent-kaurenoic acid oxidase, catalyzes three steps of the gibberellin biosynthesis pathway. *Proc. Natl. Acad. Sci. USA* **98**, 2065–2070.

Helliwell, C. A., Sullivan, J. A., Mould, R. M., Gray, J. C., Peacock, W. J., and Dennis, E. S. (2001b). A plastid envelope location of *Arabidopsis* ent-kaurene oxidase links the plastid and endoplasmic reticulum steps of the gibberellin biosynthesis pathway. *Plant J.* **28**, 201–208.

Helliwell, C. A., Sheldon, C. C., Olive, M. R., Walker, A. R., Zeevaart, J. A., Peacock, W. J., and Dennis, E. S. (1998). Cloning of the *Arabidopsis* ent-kaurene oxidase gene GA3. *Proc. Natl. Acad. Sci. USA* **95**, 9019–9024.

Henderson, J. T., Li, H. C., Rider, S. D., Mordhorst, A. P., Romero-Severson, J., Cheng, J. C., Robey, J., Sung, Z. R., de Vries, S. C., and Ogas, J. (2004). PICKLE acts throughout the plant to repress expression of embryonic traits and may play a role in gibberellin-dependent responses. *Plant Physiol.* **134**, 995–1005.

Hilhorst, H. W. M., and Karssen, C. M. (1988). Dual effect of light on the gibberellin-stimulated and nitrate-stimulated seed-germination of *Sisymbrium officinale* and *Arabidopsis thaliana*. *Plant Physiol.* **86**, 591–597.

Huang, S., Raman, A. S., Ream, J. E., Fujiwara, H., Cerny, R. E., and Brown, S. M. (1998). Overexpression of 20-oxidase confers a gibberellin-overproduction phenotype in *Arabidopsis*. *Plant Physiol.* **118**, 773–781.

Huttly, A. K., and Phillips, A. L. (1995). Gibberellin-regulated plant genes. *Physiol. Plantarum* **95**, 310–317.

Igarashi, D., Ishida, S., Fukazawa, J., and Takahashi, Y. (2001). 14-3-3 proteins regulate intracellular localization of the bZIP transcriptional activator RSG. *Plant Cell* **13**, 2483–2497.

Ikeda, A., Ueguchi-Tanaka, M., Sonoda, Y., Kitano, H., Koshioka, M., Futsuhara, Y., Matsuoka, M., and Yamaguchi, J. (2001). Slender rice, a constitutive gibberellin response mutant, is caused by a null mutation of the SLR1 gene, an ortholog of the height-regulating gene GAI/RGA/RHT/D8. *Plant Cell* **13**, 999–1010.

Ingram, T. J., Reid, J. B., Murfet, I. C., Gaskin, P., Willis, C. L., and Mac Millan, J. (1984). Internode length in *Pisum*. The *Le* gene controls the 3β-hydroxylation of gibberellin A$_{20}$ to gibberellin A$_1$. *Planta* **160**, 455–463.

Ingram, T. J., and Reid, J. B. (1987). Internode length in *Pisum* 1. Gene *na* may block gibberellin synthesis between *ent-7-alpha*-hydroxykaurenoic acid and gibberellin A$_{12}$-aldehyde. *Plant Physiol.* **83**, 1048–1053.

Ishida, S., Fukazawa, J., Yuasa, T., and Takahashi, Y. (2004). Involvement of 14-3-3 signaling protein binding in the functional regulation of the transcriptional activator REPRESSION OF SHOOT GROWTH by gibberellins. *Plant Cell* **16**, 2641–2651.

Itoh, H., Matsuoka, M., and Steber, C. M. (2003). A role for the ubiquitin-26S-proteasome pathway in gibberellin signaling. *Trends Plant Sci.* **8**, 492–497.

Itoh, H., Ueguchi-Tanaka, M., Sato, Y., Ashikari, M., and Matsuoka, M. (2002). The gibberellin signaling pathway is regulated by the appearance and disappearance of SLENDER RICE1 in nuclei. *Plant Cell* **14**, 57–70.

Itoh, H., Tatsumi, T., Sakamoto, T., Otomo, K., Toyomasu, T., Kitano, H., Ashikari, M., Ichihara, S., and Matsuoka, M. (2004). A rice semi-dwarf gene, Tan-Ginbozu (D35), encodes the gibberellin biosynthesis enzyme, ent-kaurene oxidase. *Plant Mol. Biol.* **54**, 533–547.

Jackson, S. D. (1999). Multiple signaling pathways control tuber induction in potato. *Plant Physiol.* **119**, 1–8.

Jackson, S. D., and Prat, S. (1996). Control of tuberisation in potato by gibberellins and phytochrome-B. *Physiol. Plantarum* **98**, 407–412.

Jackson, S. D., James, P. E., Carrera, E., Prat, S., and Thomas, B. (2000). Regulation of transcript levels of a potato gibberellin 20-oxidase gene by light and phytochrome B. *Plant Physiol.* **124**, 423–430.

Jacobsen, J. V., Gubler, F., and Chandler, P. M. (1995). Gibberellin action in germinating cereal grains. *In* "Plant Hormones: Physiology, Biochemistry and Molecular Biology" (P. J. Davies, Ed.), pp. 246–271. Kluwer, Dordrecht.

Jacobsen, S. E., and Olszewski, N. E. (1993). Mutations at the SPINDLY locus of *Arabidopsis* alter gibberellin signal transduction. *Plant Cell* **5**, 887–896.

Jones, A. M., and Assmann, S. M. (2004). Plants: The latest model system for G-protein research. *EMBO Rep.* **5,** 572–578.

Jones, H. D., Smith, S. J., Desikan, R., Plakidou-Dymock, S., Lovegrove, A., and Hooley, R. (1998). Heterotrimeric G proteins are implicated in gibberellin induction of α-amylase gene expression in wild oat aleurone. *Plant Cell* **10,** 245–254.

Kamiya, Y., and Graebe, J. E. (1983). The biosynthesis of all major pea gibberellins in a cell-free system from *Pisum sativum*. *Phytochemistry* **22,** 681–689.

Kamiya, Y., and Garcia-Martinez, J. L. (1999). Regulation of gibberellin biosynthesis by light. *Curr. Opin. Plant Biol.* **2,** 398–403.

Kaneko, M., Itoh, H., Inukai, Y., Sakamoto, T., Ueguchi-Tanaka, M., Ashikari, M., and Matsuoka, M. (2003). Where do gibberellin biosynthesis and gibberellin signaling occur in rice plants? *Plant J.* **35,** 105–115.

Kaneko, M., Inukai, Y., Ueguchi-Tanaka, M., Itoh, H., Izawa, T., Kobayashi, Y., Hattori, T., Miyao, A., Hirochika, H., Ashikari, M., and Matsuoka, M. (2004). Loss-of-function mutations of the rice GAMYB gene impair alpha-amylase expression in aleurone and flower development. *Plant Cell* **16,** 33–44.

Karssen, C. M., and Lacka, E. (1986). A revision of the hormone balance theory of seed dormancy: Studies on gibberellin and/or abscisic acid-deficient mutants of *Arabidopsis thaliana*. In "Plant Growth Substances 1985" (M. Bopp, Ed.), pp. 315–323. Springer-Verlag, Heidelberg.

Karssen, C. M., Brinkhorst-van der Swan, D. L. C., Breekland, A. E., and Koornneef, M. (1983). Induction of dormancy during seed development by endogenous abscisic acid: Studies on abscisic acid deficient genotypes of *Arabidopsis thaliana* (L.) Heynh. *Planta* **157,** 158–165.

Kasahara, H., Hanada, A., Kuzuyama, T., Takagi, M., Kamiya, Y., and Yamaguchi, S. (2002). Contribution of the mevalonate and methylerythritol phosphate pathways to the biosynthesis of gibberellins in *Arabidopsis*. *J. Biol. Chem.* **277,** 45188–45194.

Kauschmann, A., Jessop, A., Koncz, C., Szekeres, M., Willmitzer, L., and Altmann, T. (1996). Genetic evidence for an essential role of brassinosteroids in plant development. *Plant J.* **9,** 701–713.

King, K. E., Moritz, T., and Harberd, N. P. (2001). Gibberellins are not required for normal stem growth in *Arabidopsis thaliana* in the absence of GAI and RGA. *Genetics* **159,** 767–776.

King, R. W., and Evans, L. T. (2003). Gibberellins and flowering of grasses and cereals: Prizing open the lid of the "florigen" black box. *Annu. Rev. Plant Biol.* **54,** 307–328.

Komeda, Y. (2004). Genetic regulation of time to flower in *Arabidopsis thaliana*. *Annu. Rev. Plant Biol.* **55,** 521–535.

Koornneef, M., and van der Veen, J. H. (1980). Induction and analysis of gibberellin sensitive mutants in *Arabidopsis thaliana* (L.) Heynh. *Theor. Appl. Genet.* **58,** 257–263.

Koornneef, M., Bentsink, L., and Hilhorst, H. (2002). Seed dormancy and germination. *Curr. Opin. Plant Biol.* **5,** 33–36.

Kurosawa, E. (1926). Experimental studies on the nature of the substance secreted by the "bakanae" fungus. *Nat. Hist. Soc. Formosa* **16,** 213–227.

Lange, T. (1994). Purification and partial amino-acid sequence of gibberellin 20-oxidase from *Cucurbita maxima* L. endosperm. *Planta* **195,** 108–115.

Lange, T. (1997). Cloning gibberellin dioxygenase genes from pumpkin endosperm by heterologous expression of enzyme activities in *Escherichia coli*. *Proc. Natl. Acad. Sci. USA* **94,** 6553–6558.

Lange, T., Hedden, P., and Graebe, J. E. (1994). Expression cloning of a gibberellin 20-oxidase, a multifunctional enzyme involved in gibberellin biosynthesis. *Proc. Natl. Acad. Sci. USA* **91,** 8552–8556.

Lee, D. J., and Zeevaart, J. A. (2002). Differential regulation of RNA levels of gibberellin dioxygenases by photoperiod in spinach. *Plant Physiol.* **130,** 2085–2094.

Lee, S., Cheng, H., King, K. E., Wang, W., He, Y., Hussain, A., Lo, J., Harberd, N. P., and Peng, J. (2002). Gibberellin regulates *Arabidopsis* seed germination via RGL2, a GAI/RGA-like gene whose expression is up-regulated following imbibition. *Genes Dev.* **16**, 646–658.

Lee, Y., and Kende, H. (2001). Expression of beta-expansins is correlated with internodal elongation in deepwater rice. *Plant Physiol.* **127**, 645–654.

Léon-Kloosterziel, K. M., Alvarez Gil, M., Ruijs, G. J., Jacobsen, S. E., Olszewski, N. E., Schwartz, S. H., Zeevaart, J. A. D., and Koornneef, M. (1996). Isolation and characterization of abscisic acid-deficient *Arabidopsis* mutants at two new loci. *Plant J.* **10**, 655–661.

Lester, D. R., Ross, J. J., Davies, P. J., and Reid, J. B. (1997). Mendel's stem length gene (Le) encodes a gibberellin 3 beta-hydroxylase. *Plant Cell* **9**, 1435–1443.

Lester, D. R., Ross, J. J., Smith, J. J., Elliott, R. C., and Reid, J. B. (1999). Gibberellin 2-oxidation and the SLN gene of *Pisum sativum*. *Plant J.* **19**, 65–73.

Leubner-Metzger, G. (2001). Brassinosteroids and gibberellins promote tobacco seed germination by distinct pathways. *Planta* **213**, 758–763.

MacMillan, J., Seaton, J. C., and Suter, P. J. (1962). Plant hormones-II: Isolation and structures of gibberellin A_6 and gibberellin A_8. *Tetrahedron* **18**, 349–355.

Martin, D. N., Proebsting, W. M., and Hedden, P. (1997). Mendel's dwarfing gene: cDNAs from the Le alleles and function of the expressed proteins. *Proc. Natl. Acad. Sci. USA* **94**, 8907–8911.

Martin, D. N., Proebsting, W. M., and Hedden, P. (1999). The *SLN* gene of pea encodes a GA 2-oxidase. *Plant Physiol.* **121**, 775–781.

McGinnis, K. M., Thomas, S. G., Soule, J. D., Strader, L. C., Zale, J. M., Sun, T. P., and Steber, C. M. (2003). The *Arabidopsis* SLEEPY1 gene encodes a putative F-box subunit of an SCF E3 ubiquitin ligase. *Plant Cell* **15**, 1120–1130.

Mendel, G. (1865). Versuche über Pflanzen-Hybriden. *Verh. Naturfosch. Ver. Brünn* **4**, 3–47.

Metzger, J. D., and Dusbabek, K. (1991). Determination of the cellular mechanisms regulating thermoinduced stem growth in *Thlaspi arvense* L. *Plant Physiol.* **97**, 630–637.

Monna, L., Kitazawa, N., Yoshino, R., Suzuki, J., Masuda, H., Maehara, Y., Tanji, M., Sato, M., Nasu, S., and Minobe, Y. (2002). Positional cloning of rice semidwarfing gene, sd-1: Rice "green revolution gene" encodes a mutant enzyme involved in gibberellin synthesis. *DNA Res.* **9**, 11–17.

Monte, E., Amador, V., Russo, E., Martínez-García, J., and Prat, S. (2003). PHOR1: A U-box GA signaling component with a role in proteasome degradation? *J. Plant Growth Regul.* **22**, 152–162.

Murray, F., Kalla, R., Jacobsen, J., and Gubler, F. (2003). A role for HvGAMYB in anther development. *Plant J.* **33**, 481–491.

Nambara, E., Akazawa, T., and McCourt, P. (1991). Effects of the gibberellin biosynthetic inhibitor uniconazol on mutants of *Arabidopsis*. *Plant Physiol.* **97**, 736–738.

Ni, B. R., and Bradford, K. J. (1993). Germination and dormancy of abscisic acid- and gibberellin-deficient mutant tomato (*Lycopersicon esculentum*) seeds (sensitivity of germination to abscisic acid, gibberellin, and water potential). *Plant Physiol.* **101**, 607–617.

Ogas, J., Cheng, J. C., Sung, Z. R., and Somerville, C. (1997). Cellular differentiation regulated by gibberellin in the *Arabidopsis thaliana* pickle mutant. *Science* **277**, 91–94.

Ogas, J., Kaufmann, S., Henderson, J., and Somerville, C. (1999). PICKLE is a CHD3 chromatin-remodeling factor that regulates the transition from embryonic to vegetative development in *Arabidopsis*. *Proc. Natl. Acad. Sci. USA* **96**, 13839–13844.

Ogawa, M., Hanada, A., Yamauchi, Y., Kuwahara, A., Kamiya, Y., and Yamaguchi, S. (2003). Gibberellin biosynthesis and response during *Arabidopsis* seed germination. *Plant Cell* **15**, 1591–1604.

Ogawa, S., Toyomasu, T., Yamane, H., Murofushi, N., Ikeda, R., Morimoto, Y., Nishimura, Y., and Omori, T. (1996). A step in the biosynthesis of gibberellins that is controlled by the mutation in the semidwarf rice cultivar Tan-Ginbozu. *Plant Cell Physiol.* **37**, 363–368.

Olszewski, N., Sun, T. P., and Gubler, F. (2002). Gibberellin signaling: Biosynthesis, catabolism, and response pathways. *Plant Cell* **14**(Suppl.), S61–S80.

O' Neill, D. P., and Ross, J. J. (2002). Auxin regulation of the gibberellin pathway in pea. *Plant Physiol.* **130**, 1974–1982.

O' Neill, D. P., Ross, J. J., and Reid, J. B. (2000). Changes in gibberellin A(1) levels and response during de-etiolation of pea seedlings. *Plant Physiol.* **124**, 805–812.

Ozga, J. A., Yu, J., and Reinecke, D. M. (2003). Pollination-, development-, and auxin-specific regulation of gibberellin 3beta-hydroxylase gene expression in pea fruit and seeds. *Plant Physiol.* **131**, 1137–1146.

Peng, J., Carol, P., Richards, D. E., King, K. E., Cowling, R. J., Murphy, G. P., and Harberd, N. P. (1997). The *Arabidopsis GAI* gene defines a signaling pathway that negatively regulates gibberellin responses. *Genes Dev.* **11**, 3194–3205.

Peng, J., Richards, D. E., Hartley, N. M., Murphy, G. P., Devos, K. M., Flintham, J. E., Beales, J., Fish, L. J., Worland, A. J., Pelica, F., Sudhakar, D., Christou, P., Snape, J. W., Gale, M. D., and Harberd, N. P. (1999). 'Green revolution' genes encode mutant gibberellin response modulators. *Nature* **400**, 256–261.

Phillips, A. L., Ward, D. A., Uknes, S., Appleford, N. E., Lange, T., Huttly, A. K., Gaskin, P., Graebe, J. E., and Hedden, P. (1995). Isolation and expression of three gibberellin 20-oxidase cDNA clones from *Arabidopsis*. *Plant Physiol.* **108**, 1049–1057.

Phinney, B. O. (1983). The history of gibberellins. *In* "The Chemistry and Physiology of Gibberellins" (A. Crozier, Ed.), pp. 19–52. Praeger Press, New York, NY.

Potter, I., and Fry, S. C. (1993). Xyloglucan endotransglycosylase activity in pea internodes. Effects of applied gibberellic acid. *Plant Physiol.* **103**, 235–241.

Potts, W. C., Reid, J. B., and Murfet, I. C. (1982). Internode length in *Pisum*. I. The effect of the *Le/le* gene difference on endogenous gibberellin-like substances. *Physiol. Plant.* **55**, 323–328.

Pysh, L. D., Wysocka-Diller, J. W., Camilleri, C., Bouchez, D., and Benfey, P. N. (1999). The GRAS gene family in *Arabidopsis*: Sequence characterization and basic expression analysis of the SCARECROW-LIKE genes. *Plant J.* **18**, 111–119.

Rebers, M., Kaneta, T., Kawaide, H., Yamaguchi, S., Yang, Y. Y., Imai, R., Sekimoto, H., and Kamiya, Y. (1999). Regulation of gibberellin biosynthesis genes during flower and early fruit development of tomato. *Plant J.* **17**, 241–250.

Reid, J. (1988). Internode length in *Pisum*: Comparison of genotypes in the light and dark. *Physiol. Plant.* **74**, 83–88.

Reid, J. B., Ross, J. J., and Swain, S. M. (1992). Internode length in *Pisum*—a new, slender mutant with elevated levels of C_{19} gibberellins. *Planta* **188**, 462–467.

Reid, J. B., Botwright, N. A., Smith, J. J., O' Neill, D. P., and Kerckhoffs, L. H. (2002). Control of gibberellin levels and gene expression during de-etiolation in pea. *Plant Physiol.* **128**, 734–741.

Remington, D. L., and Purugganan, M. D. (2002). GAI homologues in the Hawaiian silversword alliance (Asteraceae-Madiinae): Molecular evolution of growth regulators in a rapidly diversifying plant lineage. *Mol. Biol. Evol.* **19**, 1563–1574.

Rider, S. D., Jr., Hemm, M. R., Hostetler, H. A., Li, H. C., Chapple, C., and Ogas, J. (2004). Metabolic profiling of the *Arabidopsis* pkl mutant reveals selective derepression of embryonic traits. *Planta* **219**, 489–499.

Robertson, M., Swain, S. M., Chandler, P. M., and Olszewski, N. E. (1998). Identification of a negative regulator of gibberellin action, HvSPY, in barley. *Plant Cell* **10**, 995–1007.

Rodriguez-Concepcion, M., and Boronat, A. (2002). Elucidation of the methylerythritol phosphate pathway for isoprenoid biosynthesis in bacteria and plastids. A metabolic milestone achieved through genomics. *Plant Physiol.* **130**, 1079–1089.

Rogers, J. C., Lanahan, M. B., Rogers, S. W., and Mundy, J. (1992). The gibberellin response element: A DNA sequence in cereal alpha-amylase gene promoters that mediates GA and ABA effects. *In* "Progress in Plant Growth Regulation" (C. M. Karssen, L. C. Van Loon, and D. Vreugdenhil, Eds.), pp. 136–146. Kluwer, Boston.

Ross, J. J., Reid, J. B., and Swain, S. M. (1993). Control of stem elongation by gibberellin A_1: Evidence from genetic studies including the slender mutant *sln*. *Aust. J. Plant Physiol.* **20**, 585–599.

Ross, J. J. (1994). Recent advances in the study of gibberellin mutants. *J. Plant Growth Regul.* **15**, 193–206.

Ross, J. J., O' Neill, D. P., Smith, J. J., Kerckhoffs, L. H., and Elliott, R. C. (2000). Evidence that auxin promotes gibberellin A1 biosynthesis in pea. *Plant J.* **21**, 547–552.

Ross, J. J., O' Neill, D. P., Wolbang, C. M., Symons, G. M., and Reid, J. B. (2001). Auxin–gibberellin interactions and their role in plant growth. *J. Plant Growth Regul.* **20**, 336–353.

Ross, J. J., Reid, J. B., Swain, S. M., Hasan, O., Poole, A. T., Hedden, P., and Willis, C. L. (1995). Genetic regulation of gibberellin deactivation in *Pisum*. *Plant J.* **7**, 513–523.

Sakamoto, T., Kobayashi, M., Itoh, H., Tagiri, A., Kayano, T., Tanaka, H., Iwahori, S., and Matsuoka, M. (2001). Expression of a gibberellin 2-oxidase gene around the shoot apex is related to phase transition in rice. *Plant Physiol.* **125**, 1508–1516.

Sakamoto, T., Miura, K., Itoh, H., Tatsumi, T., Ueguchi-Tanaka, M., Ishiyama, K., Kobayashi, M., Agrawal, G. K., Takeda, S., Abe, K., Miyao, A., Hirochika, H., Kitano, H., Ashikari, M., and Matsuoka, M. (2004). An overview of gibberellin metabolism enzyme genes and their related mutants in rice. *Plant Physiol.* **134**, 1642–1653.

Sasaki, A., Ashikari, M., Ueguchi-Tanaka, M., Itoh, H., Nishimura, A., Swapan, D., Ishiyama, K., Saito, T., Kobayashi, M., Khush, G. S., Kitano, H., and Matsuoka, M. (2002). Green revolution: A mutant gibberellin-synthesis gene in rice. *Nature* **416**, 701–702.

Sasaki, A., Itoh, H., Gomi, K., Ueguchi-Tanaka, M., Ishiyama, K., Kobayashi, M., Jeong, D. H., An, G., Kitano, H., Ashikari, M., and Matsuoka, M. (2003). Accumulation of phosphorylated repressor for gibberellin signaling in an F-box mutant. *Science* **299**, 1896–1898.

Sauter, M. (1997). Differential expression of a CAK (cdc2-activating kinase)-like protein kinase, cyclins and cdc2 genes from rice during the cell cycle and in response to gibberellin. *Plant J.* **11**, 181–190.

Sauter, M., Mekhedov, S. L., and Kende, H. (1995). Gibberellin promotes histone H1 kinase activity and the expression of cdc2 and cyclin genes during the induction of rapid growth in deepwater rice internodes. *Plant J.* **7**, 623–632.

Schneider, G., and Schliemann, W. (1994). Gibberellin conjugates—an overview. *J. Plant Growth Regul.* **15**, 247–260.

Schomburg, F. M., Bizzell, C. M., Lee, D. J., Zeevaart, J. A. D., and Amasino, R. M. (2003). Overexpression of a novel class of gibberellin 2-oxidases decreases gibberellin levels and creates dwarf plants. *Plant Cell* **15**, 151–163.

Shibaoka, H. (1994). Plant hormone-induced changes in the orientation of cortical microtubules. *Annu. Rev. Plant Physiol. Plant Mol. Biol.* **45**, 527–544.

Silverstone, A. L., Mak, P. Y. A., Martinez, E. C., and Sun, T. P. (1997a). The new RGA locus encodes a negative regulator of gibberellin response in *Arabidopsis thaliana*. *Genetics* **146**, 1087–1099.

Silverstone, A. L., Chang, C., Krol, E., and Sun, T. P. (1997b). Developmental regulation of the gibberellin biosynthetic gene GA1 in *Arabidopsis thaliana*. *Plant J.* **12**, 9–19.

Silverstone, A. L., Jung, H. S., Dill, A., Kawaide, H., Kamiya, Y., and Sun, T. P. (2001). Repressing a repressor: Gibberellin-induced rapid reduction of the RGA protein in *Arabidopsis*. *Plant Cell* **13**, 1555–1566.

Simcox, P. D., Dennis, D. T., and West, C. A. (1975). Kaurene synthetase from plastids of developing plant tissues. *Biochem. Biophys. Res. Commun.* **66**, 166–172.

Singh, D. P., Jermakow, A. M., and Swain, S. M. (2002). Gibberellins are required for seed development and pollen tube growth in *Arabidopsis*. *Plant Cell* **14**, 3133–3147.

Smith, M. W., Yamaguchi, S., Ait-Ali, T., and Kamiya, Y. (1998). The first step of gibberellin biosynthesis in pumpkin is catalyzed by at least two copalyl diphosphate synthases encoded by differentially regulated genes. *Plant Physiol.* **118**, 1411–1419.

Spielmeyer, W., Ellis, M. H., and Chandler, P. M. (2002). Semidwarf (sd-1), "green revolution" rice, contains a defective gibberellin 20-oxidase gene. *Proc. Natl. Acad. Sci. USA* **99**, 9043–9048.

Sponsel, V. M. (2001). The deoxyxylulose phosphate pathway for the biosynthesis of plastidic isoprenoids: Early days in our understanding of the early stages of gibberellin biosynthesis. *J. Plant Growth Regul.* **20**, 332–345.

Sponsel, V. M., and MacMillan, J. (1980). Metabolism of [$^{13}C_1$] gibberellin A_{29} to [$^{13}C_1$] gibberellin catabolite in maturing seeds of *Pisum sativum* cv. Progress No. 9. *Planta* **150**, 46–52.

Steber, C. M., and McCourt, P. (2001). A role for brassinosteroids in germination in *Arabidopsis*. *Plant Physiol.* **125**, 763–769.

Steber, C. M., Cooney, S. E., and McCourt, P. (1998). Isolation of the GA-response mutant sly1 as a suppressor of ABI1–1 in *Arabidopsis thaliana*. *Genetics* **149**, 509–521.

Still, D. W., and Bradford, K. J. (1997). Endo-[beta]-mannanase activity from individual tomato endosperm caps and radicle tips in relation to germination rates. *Plant Physiol.* **113**, 21–29.

Strader, L. C., Ritchie, S., Soule, J. D., McGinnis, K. M., and Steber, C. M. (2004). Recessive-interfering mutations in the gibberellin signaling gene SLEEPY1 are rescued by overexpression of its homologue, SNEEZY. *Proc. Natl. Acad. Sci. USA* **101**, 12771–12776.

Sun, T. P., and Gubler, F. (2004). Molecular mechanism of gibberellin signaling in plants. *Annu. Rev. Plant Biol.* **55**, 197–223.

Sun, T. P., and Kamiya, Y. (1994). The *Arabidopsis GA1* locus encodes the cyclase *ent*-kaurene synthetase A of gibberellin biosynthesis. *Plant Cell* **6**, 1509–1518.

Sun, T. P., Goodman, H. M., and Ausubel, F. M. (1992). Cloning the *Arabidopsis GA1* locus by genomic subtraction. *Plant Cell* **4**, 119–128.

Swain, S. M., Reid, J. B., and Kamiya, Y. (1997). Gibberellins are required for embryo growth and seed development in pea. *Plant J.* **12**, 1329–1338.

Swain, S. M., Tseng, T. S., and Olszewski, N. E. (2001). Altered expression of SPINDLY affects gibberellin response and plant development. *Plant Physiol.* **126**, 1174–1185.

Swain, S. M., Muller, A. J., and Singh, D. P. (2004). The gar2 and rga alleles increase the growth of gibberellin-deficient pollen tubes in *Arabidopsis*. *Plant Physiol.* **134**, 694–705.

Talon, M., Koornneef, M., and Zeevaart, J. A. (1990a). Endogenous gibberellins in *Arabidopsis thaliana* and possible steps blocked in the biosynthetic pathways of the semidwarf ga4 and ga5 mutants. *Proc. Natl. Acad. Sci. USA* **87**, 7983–7987.

Talon, M., Koornneef, M., and Zeevaart, J. A. D. (1990b). Accumulation of C_{19}-gibberellins in the gibberellin-insensitive dwarf mutant *GAI* of *Arabidopsis thaliana* (L.) Heynh. *Planta* **182**, 501–505.

Tamura, S. (1991). Historical aspects of gibberellins. *In* "Gibberellins" (N. Takahashi, B. O. Phinney, and J. MacMillan, Eds.), pp. 1–8. Springer-Verlag, New York.

Tanaka-Ueguchi, M., Itoh, H., Oyama, N., Koshioka, M., and Matsuoka, M. (1998). Overexpression of a tobacco homeobox gene, NTH15, decreases the expression of a gibberellin biosynthetic gene encoding GA 20-oxidase. *Plant J.* **15**, 391–400.

Thomas, S. G., Phillips, A. L., and Hedden, P. (1999). Molecular cloning and functional expression of gibberellin 2-oxidases, multifunctional enzymes involved in gibberellin deactivation. *Proc. Natl. Acad. Sci. USA* **96**, 4698–4703.

Thornton, T. M., Swain, S. M., and Olszewski, N. E. (1999). Gibberellin signal transduction presents the SPY who O-GlcNAc'd me. *Trends Plant Sci.* **4**, 424–428.

Toyomasu, T., Tsuji, H., Yamane, H., Nakayama, M., Yamaguchi, I., Murofushi, N., Takahashi, N., and Inoue, Y. (1993). Light effects on endogenous levels of gibberellins in photoblastic lettuce seeds. *J. Plant Growth Regul.* **12**, 85–90.

Toyomasu, T., Yamane, H., Murofushi, N., and Inoue, Y. (1994). Effects of exogenously applied gibberellin and red-light on the endogenous levels of abscisic acid in photoblastic lettuce seeds. *Plant Cell Physiol.* **35**, 127–129.

Toyomasu, T., Kawaide, H., Mitsuhashi, W., Inoue, Y., and Kamiya, Y. (1998). Phytochrome regulates gibberellin biosynthesis during germination of photoblastic lettuce seeds. *Plant Physiol.* **118**, 1517–1523.

Tseng, T. S., Swain, S. M., and Olszewski, N. E. (2001). Ectopic expression of the tetratricopeptide repeat domain of SPINDLY causes defects in gibberellin response. *Plant Physiol.* **126**, 1250–1258.

Tyler, L., Thomas, S. G., Hu, J., Dill, A., Alonso, J. M., Ecker, J. R., and Sun, T. P. (2004). DELLA proteins and gibberellin-regulated seed germination and floral development in *Arabidopsis*. *Plant Physiol.* **135**, 1008–1019.

Ueguchi-Tanaka, M., Fujisawa, Y., Kobayashi, M., Ashikari, M., Iwasaki, Y., Kitano, H., and Matsuoka, M. (2000). Rice dwarf mutant d1, which is defective in the alpha subunit of the heterotrimeric G protein, affects gibberellin signal transduction. *Proc. Natl. Acad. Sci. USA* **97**, 11638–11643.

Ullah, H., Chen, J. G., Wang, S., and Jones, A. M. (2002). Role of a heterotrimeric G protein in regulation of *Arabidopsis* seed germination. *Plant Physiol.* **129**, 897–907.

van Huizen, R., Ozga, J. A., and Reinecke, D. M. (1997). Seed and hormonal regulation of gibberellin 20-oxidase expression in pea pericarp. *Plant Physiol.* **115**, 123–128.

Vandenberg, J. H., Simko, I., Davies, P. J., Ewing, E. E., and Halinska, A. (1995). Morphology and [C-14] gibberellin A(12) metabolism in wild-type and dwarf *Solanum tuberosum* spp Andigena grown under long and short photoperiods. *J. Plant Physiol.* **146**, 467–473.

Varner, J. E., Chandra, G. R., and Chrispeels, M. J. (1965). Gibberellic acid-controlled synthesis of alpha-amylase in barley endosperm. *J. Cell Physiol.* **66**(Suppl. 1), 55–67.

Vosseller, K., Sakabe, K., Wells, L., and Hart, G. W. (2002). Diverse regulation of protein function by O-GlcNAc: A nuclear and cytoplasmic carbohydrate post-translational modification. *Curr. Opin. Chem. Biol.* **6**, 851–857.

Wang, H., Tang, W. N., Zhu, C., and Perry, S. E. (2002). A chromatin immunoprecipitation (ChIP) approach to isolate genes regulated by AGL15, a MADS domain protein that preferentially accumulates in embryos. *Plant J.* **32**, 831–843.

Wang, H., Caruso, L. V., Downie, A. B., and Perry, S. E. (2004). The embryo MADS domain protein AGAMOUS-Like 15 directly regulates expression of a gene encoding an enzyme involved in gibberellin metabolism. *Plant Cell* **16**, 1206–1219.

Ward, J. L., Jackson, G. J., Beale, M. H., Gaskin, P., Hedden, P., Mander, L. N., Phillips, A. L., Seto, H., Talon, M., Willis, C. L., Wilson, T. M., and Zeevaart, J. A. D. (1997). Stereochemistry of the oxidation of gibberellin 20-alcohols, GA(15) and GA(44), to 20-aldehydes by gibberellin 20-oxidases. *Chem. Commun.* **1**, 13–14.

Williams, J., Phillips, A. L., Gaskin, P., and Hedden, P. (1998). Function and substrate specificity of the gibberellin 3b-hydroxylase encoded by the *Arabidopsis GA4* gene. *Plant Physiol.* **117**, 559–563.

Wilson, R. N., Heckman, J. W., and Somerville, C. R. (1992). Gibberellin is required for flowering in *Arabidopsis thaliana* under short days. *Plant Physiol.* **100**, 403–408.

Winkler, R. G., and Helentjaris, T. (1995). The maize Dwarf3 gene encodes a cytochrome P450-mediated early step in gibberellin biosynthesis. *Plant Cell* **7**, 1307–1317.

Wolbang, C. M., and Ross, J. J. (2001). Auxin promotes gibberellin biosynthesis in decapitated tobacco plants. *Planta* **214**, 153–157.
Wolbang, C. M., Chandler, P. M., Smith, J. J., and Ross, J. J. (2004). Auxin from the developing inflorescence is required for the biosynthesis of active gibberellins in barley stems. *Plant Physiol.* **134**, 769–776.
Wu, C. T., Leubner-Metzger, G., Meins, F., Jr., and Bradford, K. J. (2001). Class I beta-1,3-glucanase and chitinase are expressed in the micropylar endosperm of tomato seeds prior to radicle emergence. *Plant Physiol.* **126**, 1299–1313.
Wu, K., Li, L., Gage, D. A., and Zeevaart, J. A. (1996). Molecular cloning and photoperiod-regulated expression of gibberellin 20-oxidase from the long-day plant spinach. *Plant Physiol.* **110**, 547–554.
Xu, X., van Lammeren, A. A., Vermeer, E., and Vreugdenhil, D. (1998). The role of gibberellin, abscisic acid, and sucrose in the regulation of potato tuber formation *in vitro*. *Plant Physiol.* **117**, 575–584.
Xu, Y. L., Gage, D. A., and Zeevaart, J. A. (1997). Gibberellins and stem growth in *Arabidopsis thaliana*. Effects of photoperiod on expression of the GA4 and GA5 loci. *Plant Physiol.* **114**, 1471–1476.
Xu, Y. L., Li, L., Wu, K., Peeters, A. J., Gage, D. A., and Zeevaart, J. A. (1995). The GA5 locus of *Arabidopsis thaliana* encodes a multifunctional gibberellin 20-oxidase: Molecular cloning and functional expression. *Proc. Natl. Acad. Sci. USA* **92**, 6640–6644.
Yamaguchi, S., and Kamiya, Y. (2000). Gibberellin biosynthesis: Its regulation by endogenous and environmental signals. *Plant Cell Physiol.* **41**, 251–257.
Yamaguchi, S., Kamiya, Y., and Sun, T. (2001). Distinct cell-specific expression patterns of early and late gibberellin biosynthetic genes during *Arabidopsis* seed germination. *Plant J.* **28**, 443–453.
Yamaguchi, S., Saito, T., Abe, H., Yamane, H., Murofushi, N., and Kamiya, Y. (1996). Molecular cloning and characterization of a cDNA encoding the gibberellin biosynthetic enzyme *ent*-kaurene synthase B from pumpkin (*Cucurbita maxima* L.). *Plant J.* **10**, 203–213.
Yamaguchi, S., Sun, T., Kawaide, H., and Kamiya, Y. (1998a). The GA2 locus of *Arabidopsis thaliana* encodes ent-kaurene synthase of gibberellin biosynthesis. *Plant Physiol.* **116**, 1271–1278.
Yamaguchi, S., Smith, M. W., Brown, R. G. S., Kamiya, Y., and Sun, T. P. (1998b). Phytochrome regulation and differential expression of gibberellin 3 beta-hydroxylase genes in germinating *Arabidopsis* seeds. *Plant Cell* **10**, 2115–2126.
Yamauchi, Y., Ogawa, M., Kuwahara, A., Hanada, A., Kamiya, Y., and Yamaguchi, S. (2004). Activation of gibberellin biosynthesis and response pathways by low temperature during imbibition of *Arabidopsis thaliana* seeds. *Plant Cell* **16**, 367–378.
Yu, H., Ito, T., Zhao, Y., Peng, J., Kumar, P., and Meyerowitz, E. M. (2004). Floral homeotic genes are targets of gibberellin signaling in flower development. *Proc. Natl. Acad. Sci. USA* **101**, 7827–7832.
Zeevaart, J. A. D., and Talon, M. (1992). Gibberellin mutants in *Arabidopsis thaliana*. *In* "Progress in Plant Growth Regulation: Proceedings of the 14th International Conference on Plant Growth Substances" (D. Vreugdenhil, Ed.), pp. 34–42. Kluwer, Dordrecht.
Zentella, R., Yamauchi, D., and Ho, T. H. (2002). Molecular dissection of the gibberellin/abscisic acid signaling pathways by transiently expressed RNA interference in barley aleurone cells. *Plant Cell* **14**, 2289–2301.
Zheng, N., Schulman, B. A., Song, L., Miller, J. J., Jeffrey, P. D., Wang, P., Chu, C., Koepp, D. M., Elledge, S. J., Pagano, M., Conaway, R. C., Conaway, J. W., Harper, J. W., and Pavletich, N. P. (2002). Structure of the Cul1-Rbx1-Skp1-F boxSkp2 SCF ubiquitin ligase complex. *Nature* **416**, 703–709.

10

NITRIC OXIDE SIGNALING IN PLANTS

ALLAN D. SHAPIRO

Biotechnology Program, Florida Gulf Coast University, Fort Myers Florida 33965-6565

    I. Introduction
   II. Nitric Oxide Biosynthesis
      A. Detection and Quantitation of NO in Plants
      B. Pathways and Enzymes
  III. NO Metabolism and Transport
      A. Autoxidation will not Limit Intracellular Diffusion of NO but Might Limit Paracrine Effects in Some Cases
      B. Does NO Undergo Catalyzed Oxidative or Reductive Catabolism in Plants?
      C. Does Superoxide Limit NO Accumulation?
      D. Evidence for Additional Reactions of NO With Reactive Oxygen
      E. S-Nitrosylation Reactions are not Likely to be a Major Fate of Plant NO Except Perhaps During Hypoxia
      F. Metalloenzymes are Likely to be the Major NO Targets in Plants
  IV. NO Function in Plants
      A. Caveats and Potential Artefacts in Assessing NO Function
      B. NO-induced Gene Expression

C. NO Regulates Mitochondrial and
   Chloroplast Functions
D. NO and Iron Homeostasis
E. NO in Plant Development
F. NO in Responses to the Abiotic Environment
G. NO in Response to Pathogens
V. Conclusions
   References

Plants have four nitric oxide synthase (NOS) enzymes. NOS1 appears mitochondrial, and inducible nitric oxide synthase (iNOS) chloroplastic. Distinct peroxisomal and apoplastic NOS enzymes are predicted. Nitrite-dependent NO synthesis is catalyzed by cytoplasmic nitrate reductase or a root plasma membrane enzyme, or occurs nonenzymatically. Nitric oxide undergoes both catalyzed and uncatalyzed oxidation. However, there is no evidence of reaction with superoxide, and *S*-nitrosylation reactions are unlikely except during hypoxia. The only proven direct targets of NO in plants are metalloenzymes and one metal complex. Nitric oxide inhibits apoplastic catalases/ascorbate peroxidases in some species but may stimulate these enzymes in others. Plants also have the NO response pathway involving cGMP, cADPR, and release of calcium from internal stores. Other known targets include chloroplast and mitochondrial electron transport. Nitric oxide suppresses Fenton chemistry by interacting with ferryl ion, preventing generation of hydroxyl radicals. Functions of NO in plant development, response to biotic and abiotic stressors, iron homeostasis, and regulation of respiration and photosynthesis may all be ascribed to interaction with one of these targets. Nitric oxide function in drought/abscisic acid (ABA)-induction of stomatal closure requires nitrate reductase and NOS1. Nitric oxide synthase1 likely functions to produce sufficient NO to inhibit photosynthetic electron transport, allowing nitrite accumulation. Nitric oxide is produced during the hypersensitive response outside cells undergoing programmed cell death immediately prior to loss of plasma membrane integrity. A plasma membrane lipid-derived signal likely activates apoplastic NOS. Nitric oxide diffuses within the apoplast and signals neighboring cells via hydrogen peroxide (H_2O_2)-dependent induction of salicylic acid biosynthesis. Response to wounding appears to involve the same NOS and direct targets. © 2005 Elsevier Inc.

I. INTRODUCTION

"I'm afraid that if you look at a thing long enough, it loses all of its meaning." Although Andy Warhol was not likely referring to the study of plant hormones, this caution certainly applies. Effects of hormones that initially appear dramatic and straightforward have later been revealed to be subtle and complex. Qualitative dependence on hormone dosage, site of action, kinetics of synthesis, metabolism and transport, and interactions with other regulators is more the rule than the exception. By this criterion, NO is emerging as a typical plant hormone and signaling molecule. It has been implicated in a bewildering array of processes related to plant development, function, and response to the abiotic and biotic environment. Yet, NO does not seem to take center stage in any of these dramas. Its role is more often supporting, as a regulator or embedded in highly complicated networks. As such, initial superficial explanations must be replaced by highly nuanced descriptions to avoid losing the "meaning" of NO in plant biology.

This nuanced view is the goal to which this review article strives. Attention is paid first to the details of NO biosynthesis and fate. These considerations then underlie a critical review of the literature on the roles of NO in plants. Finally, a speculative synthesis of this literature is attempted out of which NO emerges as a ubiquitous character actor rather than the "star." Nonetheless, by playing similar parts on different stages, NO contributes in important ways to the integration of higher-order plant functions.

II. NITRIC OXIDE BIOSYNTHESIS

A. DETECTION AND QUANTITATION OF NO IN PLANTS

Establishing functions for NO requires methods of measuring its production *in vivo* and/or in tissue samples. The simple, high-throughput assays most frequently used by mammalian researchers are usually inappropriate. The Griess reaction for the NO catabolic products nitrite and nitrate (Schmidt and Kelm, 1996) cannot be used because of the high nitrate levels in plants. Assay via spectroscopic changes induced in hemoglobin is only appropriate with supernatants from plant tissue culture cells (Clarke *et al.*, 2000; Delledonne *et al.*, 1998) or with extracts from which compounds that will interfere with light absorbance readings have been removed (Orozco-Cardenas and Ryan, 2002). The most frequently used methods in plant biology are semiquantitative *in vivo* methods based on fluorescent reporter dyes, electron paramagnetic spectroscopy (EPR) methods, and methods that detect emission of gaseous NO. These and other approaches are described in this section.

1. *In Vivo* Methods Based on Reporter Dyes

Semiquantitative measurement of NO with subcellular resolution has been achieved using fluorescence microscopy. The most commonly used dye, 4,5-diaminofluorescein diacetate (DAF-DA) reacts with autoxidation products of NO to form a fluorescent triazole derivative (Kojima *et al.*, 1998). This approach has been used extensively with plants (Desikan *et al.*, 2002; Foissner *et al.*, 2000; Garcês *et al.*, 2001; Gould *et al.*, 2003; Huang *et al.*, 2004; Jih *et al.*, 2003; Neill *et al.*, 2002; Pagnussat *et al.*, 2003; Pedroso *et al.*, 2000a; Prado *et al.*, 2004; Tada *et al.*, 2004; Xin *et al.*, 2003).

Unfortunately, this dye fluoresces very poorly at even mildly acidic pH. As such, some results using DAF-DA *in planta* may have been misleading in that NO signals in the apoplast (as well as in vacuolar and other acidic compartments) would have been grossly underestimated. The pH of the *Arabidopsis* leaf apoplast was measured using fluorescence-based methods (Hoffmann and Kosegarten, 1995) to be 5.7 ± 0.1 (A. D. Shapiro and K. J. Czymmek, unpublished data). Apoplastic pH measurements from a wide variety of other plants generally varied between 5.0 and 6.5 (Grignon and Sentenac, 1991). However, not all studies using DAF-DA were truly *in planta*. Apoplastic NO was noted when DAF-DA was used with epidermal peels immersed in a neutral pH buffers (Corpas *et al.*, 2004; Gould *et al.*, 2003; Huang *et al.*, 2004). Whether the use of a nonphysiological pH affected NO production was not determined.

Derivatives of fluorescein are available that are much less sensitive to pH (Itoh *et al.*, 2000; Kojima *et al.*, 1999). 4-Amino-5-methylamino-2',7'-difluorofluorescein (DAF-FM) diacetate was used in studies of the *Arabidopsis* hypersensitive response (HR) that showed the initial site of accumulation was in the apoplast (Zhang *et al.*, 2003). 4-Amino-5-methylamino-2',7'-difluorofluorescein was also used to document exogenous nitrite-dependent NO secretion by barley aleurone tissue into the highly acidic incubation media (Bethke *et al.*, 2004a). In the former study, propidium iodide was used for simultaneous detection of programmed cell death (PCD) events. The fluorescence spectrum for this dye is largely nonoverlapping with that of DAF-FM triazole. Although use in plants has not yet been reported, pH-independent NO-reporters that are rhodamine derivatives are also available (Kojima *et al.*, 2001). These dyes should allow simultaneous monitoring of other physiological events that can be reported with fluorescein derivatives or green fluorescent protein. 4-Amino-5-methylamino-2',7'-difluorofluorescein diacetate has also been used in conjunction with a multi-well plate reader in kinetic assays to generate semiquantitative measurements of NO production rates (Zeidler *et al.*, 2004).

2. EPR Spectroscopy

Much of the work detecting emission of NO from plant samples has employed electron paramagnetic resonance (EPR) spectroscopy (Caro and Puntarulo, 1999; Corpas *et al.*, 2004; Dordos *et al.*, 2004; Guo *et al.*, 2004; Huang *et al.*, 2004). Although NO has an unpaired electron, the ground state does not generate an EPR signal because of coupling between spin and orbital angular momentum (Singel and Lancaster, 1996). Spin traps are employed that react with NO to quench angular momentum and consequently generate stable signals. Vanin *et al.* (2004) described water soluble and lipid soluble spin traps to detect NO in both aqueous and membrane compartments of leaf slices.

However, unless corrections are made for endogenous superoxide levels, EPR methods will grossly underestimate NO production. Superoxide will oxidize the complexes responsible for the NO signal into EPR-silent species. Use of a superoxide scavenger revealed up to18-fold underestimation of the basal level of NO release from *Arabidopsis* or bean leaves (Vanin *et al.*, 2004). As such, EPR results must be interpreted with caution even when taken as semiquantitative if changes in superoxide levels are expected during an experiment.

3. Detection of Gaseous NO Emitted from Plant Samples

The first reports of biological NO production documented emission of gaseous NO by plants under specific extreme stress conditions, with certain chemical treatments or following vacuum infiltration of high levels of nitrate (Dean and Harper, 1986; Harper, 1981; Klepper, 1979, 1990, 1991). It was thought at that time that NO was not emitted by healthy plants growing under "standard" conditions (Harper, 1981). However, when more sensitive methodology was employed, constitutive NO emission was detected from a wide variety of plants (Wildt *et al.*, 1997). This study used ozone-induced chemiluminescence (Hampl *et al.*, 1996) to detect NO. This technique is based on the highly specific reaction of NO with ozone to form NO_2 in an excited state. The spontaneous decay to the ground state releases light that is detected using photomultipliers. The technique requires NO to be flushed out of solution with an inert gas and thus has a working detection limit in the low nanomolar range (Feelisch and Stamler, 1996b).

Highly sensitive detection has also been achieved using laser photoacoustic spectroscopy (Leshem, 2000). In this technique, gas samples in a closed volume are subjected to infrared light. Absorption of an infrared photon excites sample gases that emit heat upon relaxation. The temperature variation is accompanied by pressure variation that creates sound waves. These sound waves are detected with a sensitive microphone. The microphone signal is sensitive to absorption strength (allowing identification of trace gases) as well as trace gas concentration. This technique was used to confirm increases in NO emission in ripening fruit and concomitant decreases in ethylene emission using a noninvasive technique that did not wound the plants (Leshem and Pinchasov, 2000). The specialized facility required for this work is a disadvantage for routine use in plant biology.

Conrath *et al.* (2004) described measurement of NO emitted from intact, nitrate-fed *Arabidopsis* plants and detached, fungicide-treated tobacco leaves. This plant material was placed in an enclosed chamber and emitted gases were passed into a benchtop mass spectrometer through a restriction capillary. Nitric oxide emission was also detected from pathogen-infected plant tissue cultures. In this case, emitted gases from 10-ml samples were passed through a gas-permeable membrane into the mass spectrometer. Mass spectrometry (MS) methods are more quantitative than ozone-induced chemiluminescence or EPR methods with these applications (albeit less sensitive), and the instrumentation required is the least expensive and most readily available of all the methods presented. In addition, as MS methods can be used in conjunction with isotope labeling, this approach will have advantages for NO biosynthesis studies. Gas chromatography–mass spectrometry (GC–MS) methods have also been used to detect NO emission from *Arabidopsis* resulting from hypoxia or abrupt shift from light to darkness (Perazzolli *et al.*, 2004).

4. Other Methods

Nitric oxide–sensitive electrodes have been used for *in situ* detection of NO (Leshem and Haramaty, 1996; Leshem *et al.*, 1998; Malinski and Czuchajowski, 1996). These methods are suitable for detection of NO in plant cell cultures (Delledonne *et al.*, 2001). However, intact plant material is usually impaled with the electrode, potentially confounding the observed physiology with a wound response. If the probe is instead held in a closed atmosphere near the plant surface, the sensitivity of the method is decreased by several orders of magnitude (Leshem, 2000). Nitric oxide has also been detected in supernatants of cultured *Agave pacifica* cells by *in situ* derivization with 2,3-diaminonaphthalene, HPLC separation of cell filtrate, and fluorescence detection (Wada *et al.*, 2002).

B. PATHWAYS AND ENZYMES

1. Plant Nitric Oxide Synthases (NOSs) Catalyze Production of Nitric Oxide (NO) from Arginine

In mammals, most NO synthesis is catalyzed by nitric oxide synthase (NOS) enzymes that produce NO from arginine with citrulline as a coproduct. Although no close homologues of these genes have been found in plants, they contain enzymes that catalyze this reaction. Evidence indicates there are at least four distinct NOS enzymes in plants (see Table I). An enzyme activity that is induced by avirulent (gene-for-gene resistance response-eliciting) viral or bacterial infection of tobacco, tomato, or *Arabidopsis* was purified from infected tobacco and shown to encode a variant form of the glycine decarboxylase complex P protein (Chandok *et al.*, 2003, 2004).[1] This

---

[1] While this manuscript was in press, both cited papers (Chandok *et al.*, 2003, 2004) were retracted by the authors (Klessig, D. F., Ytterberg, A. J., and van Wijk, K. J. (2004). *Cell* **119**, 445; Klessig, D. F., Martin, G. B., and Ekengren, S. K. (2004). *Proc. Natl. Acad. Sci. USA* **101**, 16081.). The first author of the papers did not approve of either retraction.

TABLE I. Plant Enzymes that Catalyze NO Synthesis

Enzyme	Likely subcellular location	Substrate	Proven cofactors and co-substrates	Inhibitors	Mutants/VIGS
iNOS	Chloroplast	Arginine	Pyridoxal phosphate, Ca^{++}/calmodulin, $THFglu_1$, FAD, FMN, heme, NADPH, O_2[a]	Mammalian NOS inhibitors, methotrexate, carboxymethoxylamine, aminoacetonitrile	VIGS for loss of function in tomato plants
AtNOS1 (NOS1)	Mitochondria	Arginine	Ca^{++}/calmodulin, NADPH, O_2[a]	Mammalian NOS inhibitors	*Arabidopsis nos1* mutant
pNOS	Peroxisome	Arginine	Ca^{++}/calmodulin, BH_4 ($THFglu_1$?), FAD, FMN, heme, NADPH, O_2[a]	Mammalian NOS inhibitors, anti-mouse iNOS antibody	None, gene not cloned
aNOS	Leaf apoplast (plasma membrane or cell wall)	Arginine	Unknown	Mammalian NOS inhibitors	None, gene not cloned
NR	Cytoplasm	Nitrite	NADH, molybdenum cofactor, FAD, heme[b]	Cyanide, azide, anti-NR IgG	*Arabidopsis nia1-1/nia2-5* double mutant
NI-NOR	Root plasma membrane	Nitrite	Unknown	Unknown	None, gene not cloned

[a] Participation of O_2 as a cosubstrate is assumed based on similarities to the mammalian enzymes.
[b] Participation of molybdenum cofactor, FAD and heme is established for nitrate reduction but not yet known to be essential in NO formation.

NOS activity (iNOS) was indeed inhibited by the glycine decarboxylase inhibitors carboxymethoxylamine and aminoacetonitrile. Carboxymethoxylamine inhibits glycine decarboxylase activity by interacting with its cofactor pyridoxal phosphate. A binding motif for pyridoxal phosphate was found in the iNOS gene and pyridoxal phosphate addition modestly stimulated iNOS activity. However, in most other respects, the enzymatic activity strongly resembled that of mammalian NOSs. Spectroscopic studies provided evidence for binding of heme and flavins. Enzyme stimulation/inhibition studies provided evidence for requirement of the other cofactors and co-substrates utilized by mammalian NOS enzymes.

Glycine decarboxylase is a mitochondrial enzyme. However, indirect evidence suggests that iNOS is a chloroplast protein. Tobacco NO biosynthesis elicited by the *Phytophthora*-derived compound cryptogein is known to occur in the chloroplast (Foissner *et al.*, 2000). This NO synthesis is inhibited by the glycine decarboxylase inhibitors (Lamotte *et al.*, 2004), implicating iNOS. Sequence analysis was also used to argue against a mitochondrial localization (Chandok *et al.*, 2003).

On the other hand, there is excellent evidence for NO function in the mitochondria (see Section III.C). A different enzyme has emerged as the leading candidate to be the mitochondrial NOS. *AtNOS1* (*NOS1*) is the *Arabidopsis* member of a ubiquitous family of related NOS genes that include a conserved domain with homology to GTPases (Zemojtel *et al.*, 2004). To date, NOS activity has been verified only in the snail and *Arabidopsis* enzymes, and GTPase activity studies have not been reported for any family member. Sequence analysis suggested a mitochondrial location for *NOS1*, although this has yet to be verified experimentally. Analysis of a knockout mutant indicated that it is responsible for 75% of the basal NOS activity in *Arabidopsis* leaf extracts and most of the ABA-induced NOS activity in roots. The activity of the protein produced by heterologous expression in *Escherichia coli* was dependent upon NADPH and Ca^{2+}/calmodulin and inhibited by a mammalian NOS inhibitor. However, it was not stimulated by tetrahydrobiopterin (BH_4), flavin adenine dinucleotide (FAD), riboflavin 5'-phosphate (FMN), or heme. Determining whether these cofactors are truly unnecessary or were merely tightly bound to the enzyme or present in the crude protein fraction used for these experiments requires further experimentation.

Nitric oxide synthase1 was shown to be required for vigorous growth, full fertility, and ABA responses via analysis of the *Atnos1* (later renamed *nos1*) knockout mutant (Guo *et al.*, 2003). Nitric oxide synthase1 was also implicated in the regulation of flowering time (He *et al.*, 2004). Subcellular localization of NO production was not assessed in these studies. However, the *NOS1*-dependent production of NO by lipopolysaccharide-stimulated *Arabidopsis* cells was imaged using DAF-FM diacetate (Zeidler *et al.*, 2004). The earliest production of NO was in a cellular region that would have included mitochondria. Although a cytoplasmic, peroxisomal, or endomembrane location could not be ruled out, initial staining was clearly absent in chloroplasts, vacuole, nucleus, and extracellular space. Co-localization studies with mitochondria-specific dyes in this system could be highly informative.

Peroxisomes also contain a distinct NOS activity. Subcellular fractionation of pea seedling extracts indicated that most of the pea basal NOS activity was in this organelle (Barroso et al., 1999). This activity (here termed "pNOS" for peroxisomal NOS) was not inhibited by glycine decarboxylase inhibitors and is thus distinct from iNOS (Corpas et al., 2004). The activity did require all of the cofactors and co-substrates required by the mammalian NOS enzymes and was inhibited by inhibitors of mammalian NOSs (Barroso et al., 1999). The presence of NO in peroxisomes was confirmed by EPR spectroscopy as well as dye-based fluorometric analysis (Corpas et al., 2004). The functions of NO in this organelle are unknown; however, given the known functions of this organelle (Corpas et al., 2001) a regulatory role in the glyoxylate cycle, photorespiration, or oxidative catabolism is conceivable. Nitric acid levels were shown to decrease with leaf senescence, in parallel with increases in the activity of a glyoxylate cycle enzyme (Corpas et al., 2004).

A fourth distinct NOS (here termed "aNOS" for apoplastic NOS) is responsible for NO synthesis in the apoplast. During the *Arabidopsis* HR, NO synthesis was first detected at punctate foci at the cell surface (Zhang et al., 2003). This NO synthesis was blocked by an NOS inhibitor that is an arginine substrate analog. As NO synthesis immediately preceded onset of plasma membrane permeability to a dye, it is likely that NOS may be activated by lipid signals derived from enzymatic breakdown of plasma membrane lipids. In a different study (Huang et al., 2004), wound-induced NO production was noted in the *Arabidopsis* apoplast. Nitric oxide was also detected in this study in the cytoplasm. However, time course studies were not performed that would have identified the initial site of synthesis. As apoplastic NO produced during the HR diffuses into the cytoplasm (Zhang et al., 2003), it is likely that the same may happen with wound-induced NO production. Wound-induced NO production was partially inhibited by an NOS inhibitor in one study (Huang et al., 2004) and more fully inhibited by a higher concentration of the inhibitor in a different study (Garcês et al., 2001).

Although more complicated scenarios are certainly possible, the most parsimonious prediction from these studies is that a distinct apoplastic (cell wall-localized or plasma membrane-localized with the enzymatic activity being extracellular) NOS exists that is activated by lipid signals produced by cellular damage that occurs as a consequence of the HR or severe wounding. The alternative hypothesis for aNOS activation that the cellular damage releases cofactors or substrates required for NOS activity is highly unlikely. All known NOS cofactors are not significantly smaller in size than propidium iodide, and NO synthesis was detected prior to onset of plasma membrane permeability to this dye (Zhang et al., 2003). The arginine substrate was significantly smaller than propidium iodide and could thus conceivably have been released from cells immediately prior to aNOS activation. However, infiltrating *Arabidopsis* leaves with up to 10mM arginine did not cause NO accumulation as assessed by microscopy using the DAF-FM dye (A. D. Shapiro and K. J. Czymmek, unpublished data).

2. Nitrate Reductases Catalyze Production of NO from Nitrite

Plants synthesize NO not only from arginine but also from nitrite (Yamasaki, 2000). The best-characterized pathway is catalyzed by the cytoplasmic enzyme nitrate reductase. The main function of nitrate reductase is in nitrate assimilation, in which it catalyzes reduction of nitrate to nitrite, which is subsequently reduced to ammonia in the chloroplasts. Nitrite is toxic and does not accumulate under most conditions. However, when nitrite does accumulate, nitrate reductase catalyzes one electron reduction of nitrite using NAD(P)H as an electron donor. It is not yet known whether the cofactors required for nitrate reduction (molybdenum cofactor, heme, and FAD; Meyer and Stitt, 2001) also function in NO synthesis.

Nitrite accumulates under conditions for which rates of nitrite reduction are limiting. As the nitrite reductase reaction uses reduced ferredoxin as an electron donor, herbicides that interfere with photosynthetic electron transport will cause nitrite accumulation that leads to NO production. As import of nitrite into chloroplasts requires ΔpH across the chloroplast envelope, uncouplers will also lead to accumulation of nitrite and NO synthesis. Nitrite accumulation and consequent NO production also occurs during hypoxia in the dark with nitrate-fertilized plants (Rockel *et al.*, 2002). In the absence of oxygen, nitrate acts as an alternative electron acceptor from NADH, regenerating NAD^+ for glycolysis (Polyakova and Vartapetian, 2003). As such, high levels of nitrite accumulate because of the lack of photosynthetic electron transport in the dark. These observations suggest a possible role for NO in response to hypoxic stresses, such as growth in waterlogged soils. A transient burst of nitrite accumulation and NO production was also noted following abrupt turning off of the lights. Nitrite accumulated because cessation of photosynthetic electron transport was more rapid than darkness-induced downregulation of nitrate reductase activity (Kaiser *et al.*, 2002). Antisense tobacco plants under-expressing nitrite reductase also showed enhanced nitrite accumulation and NO production (Morot-Gaudry-Talarmain *et al.*, 2002), as did transgenic tobacco plants expressing a copy of nitrate reductase that lacked the site for phosphorylation implicated in downregulation of nitrate reductase activity that accompanies the light-to-dark transition (Lea *et al.*, 2004).

Although these studies established that nitrate reductase could catalyze NO synthesis, the physiological relevance was questionable. The availability of an *Arabidopsis* mutant with severely reduced levels of both nitrate reductase gene products (Wilkinson and Crawford, 1993) was critical to establishing the relevance of this pathway. Use of this double mutant (*nia1-1/nia2-5*) uncovered an important role for this enzyme in the production of NO implicated in ABA-induced stomatal closure (Desikan *et al.*, 2002). Further roles for this pathway may emerge.

3. A Root Plasma Membrane Enzyme Catalyzes NO Production from Nitrite

Nitric oxide synthesis from nitrite may also occur in the root apoplast. An enzyme activity was identified in tobacco root plasma membrane preparations that could catalyze this reaction when provided with an electron donor (Stöhr *et al.*, 2001). As nitrite, not arginine, is the substrate for NO synthesis, this enzyme is distinct from aNOS. This root enzyme, plasma membrane-bound nitrite: NO-reductase (NI-NOR), was insensitive to cyanide and could not be inhibited by antibodies that blocked function of the succinate-dependent, plasma membrane-bound nitrate reductase (PM-NR). The enzyme activity was therefore ascribed to a distinct protein. However, it was speculated that the two enzymes might work together with the latter catalyzing nitrite production from nitrate and the former catalyzing NO synthesis. A role in sensing soil nitrate levels and transmitting this information to the cytoplasm via NO production was proposed. Purification of this protein has not yet been reported, and as such, the corresponding gene is not yet identified. The physiological reductant is also not yet known.

4. Can NO be Produced from Nitrite Nonenzymatically?

Nitric oxide can also be produced as a decomposition product of nitrous acid (Yamasaki, 2000). However, protonation of nitrite to form nitrous acid is unlikely to occur in most physiological situations in plants, as the pK_a is ~3.2. Nonetheless, it has been argued that this reaction might be significant in germinating cereal grain (Bethke *et al.*, 2004a). Barley aleurone layers do acidify the apoplast to pH 3–4. Although soil nitrite levels are generally very low, nitrite can accumulate when certain mixtures of ammonia and nitrate are used in fertilizer. With high levels of ammonia, elevated soil pH will inhibit growth of nitrite-oxidizing bacteria. With nitrite oxidation inhibited, both nitrifying and denitrifying bacteria contribute to the accumulation of soil nitrite (Burns *et al.*, 1996). Nitrite could enter the seed from the soil following rupture of the seed coat on germination. The elevated pH in bulk soil may be overcome near the germinating grain by the strong proton pumping activity of the grain aleurone tissue. Alternatively, it is conceivable that nitrite would accumulate in nitrate-fed grain tissues because of depletion of the pool of reducing equivalents required for nitrite reduction. The high levels of H_2O_2 resulting from glyoxosomal lipid catabolism (Fath *et al.*, 2002) may provide the oxidant required for such a process.

One additional requirement for this pathway to proceed at a physiologically relevant rate is a reductant (Yamasaki, 2000). Abscisic acid-treated or GA-treated aleurone layers were shown to secrete reductant(s) into the incubation media (Bethke *et al.*, 2004a). No ascorbate and very little glutathione were detected in this media. However, high levels of phenolics were found in this tissue. In the same study, a model phenolic compound (catechin) was able to stimulate synthesis of NO from nitrite added to ABA-treated or GA-treated barley aleurone layers in a pH-dependent fashion. Although this study proved the feasibility of nonenzymatic synthesis of NO under these conditions, the importance of this pathway in comparison with NO synthesis catalyzed by the nitrate-inducible nitrate reductase known to be present in this tissue (Ferrari and Varner, 1970) has not yet been determined.

III. NO METABOLISM AND TRANSPORT

A. AUTOXIDATION WILL NOT LIMIT INTRACELLULAR DIFFUSION OF NO BUT MIGHT LIMIT PARACRINE EFFECTS IN SOME CASES

The biological fate of NO is partly a consequence of its physical properties. Even though NO is often described as a gas, it is better regarded as an aqueous solution in most instances of biological relevance. Equilibration of pure NO gas with water (1atm pressure, 25 °C) yields a 1.9-mM solution (Lancaster, 2000). As this level is vastly in excess of that seen in living organisms, biological NO will act as a gas only at air–water interfaces (although the possibility of NO also acting as a gas in interplant communication or soil microbe-to-plant communication has not been adequately explored).

In the absence of other potential reactants, NO will undergo autoxidation according to the following equation:

$$2NO + O_2 \rightarrow 2NO_2 \tag{1}$$

The reaction proceeds with kinetics that are second order with respect to NO and first order with respect to molecular oxygen. This rate law was shown to apply in air, aqueous solution, or nonpolar solvents. The trimolecular rate constants were shown to be similar in aqueous solution (6×10^6 M^{-2} s^{-1}; Wink et al., 1993) and in the nonpolar solvent carbon tetrachloride (2.3×10^6 M^{-2} s^{-1}; Nottingham and Sutter, 1986). The rate constant in air was ~300-fold lower (7.1×10^3 M^{-2} s^{-1}; Olbregts, 1985). However, the higher concentration of molecular oxygen in pure air (41 mM; Tsukahara et al., 1999) as compared with air-saturated aqueous solution (0.25 mM; Lancaster, 2000) will largely eliminate the discrepancy in rates.

In biological systems, autoxidation will occur primarily in the lipid phase of membranes. Nitric oxide is ~ninefold more soluble in nonpolar organic solvents than in water and molecular oxygen is ~threefold more soluble. Given the trimolecular nature of the reaction, it was calculated that autoxidation would proceed ~240-fold faster in the membranes than in aqueous phases. Experiments using phospholipid vesicles or detergents qualitatively validated these calculations *in vitro* (Lancaster, 2000). Because of these considerations, NO levels could be dramatically higher within the cellular compartment in which it is synthesized than outside of a membrane-delimited barrier. These considerations could also suggest that specificity of action will be obtained from NO added to an experimental system only when it is synthesized *in situ* by targeting an NOS to a specific cellular compartment as opposed to when it is added as a gas or when an NO donor is injected into the apoplast.

However, these conclusions are not necessarily correct if diffusion is very fast relative to autoxidation. The rate of autoxidation places a maximum limit on the extent of diffusion of NO from a point source. At limiting concentrations of NO (when $[O_2] \gg [NO]$), the first half-life for the oxidation reaction can be calculated according to the following equation (Wink *et al.*, 1996):

$$t_{1/2} = \frac{1}{k \times [O_2] \times [NO]} \tag{2}$$

In mammals, NO levels of 0.1–5 μM have been reported, except in activated neutrophils and macrophages where levels have been reported in excess of 10 μM (Wink *et al.*, 1996). A typical calculation using Eq. (1) assuming 200 μM O_2 and 1 μM NO predicted a first half-life for NO of ~830 s. However, the concentration of O_2 in mammalian cells is usually taken to be of a much lower value (20–50 μM in the cytoplasm, 1–5 μM in mitochondria; Takehara *et al.*, 1995). Using a value of 3 μM would yield a first half-life in animal mitochondria of 15.4 h. By contrast, in plant leaves, assuming the air-saturated water value for oxygen concentration (0.25 mM) to be a good approximation and maintaining the assumption of 1 μM NO, the first half-life becomes ~670 s.

The extent of diffusion of NO within these spans of time can also be calculated. The average distance a single molecule travels from a point source depends upon the diffusion constant (D) and time according to the Einstein–Smolochowski equation:

$$\Delta x = \sqrt{2 \times D \times t} \tag{3}$$

The diffusion constant for NO in water was calculated to be ~3400 $\mu m^2 \, s^{-1}$, a value that was consistent with experimental measurements (Lancaster, 2000). Using this value, average displacement was plotted with respect to time in Fig. 1. Within 670 s (first half-life for NO autoxidation calculated earlier), the average displacement of an NO molecule will be ~2.1 mm. This value indicates that autoxidation would not limit NO action either within plant cells or locally between plant cells. By contrast, if one was to assume a concentration of NO to be ~100 μM (the upper end of values reported for high-output

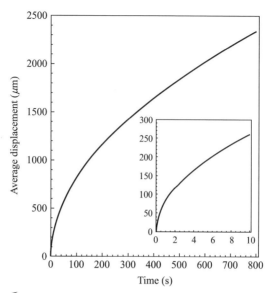

FIGURE 1. Nitric oxide diffusion calculated from the Einstein–Smolochowski equation. Average displacement of an NO molecule from a point source as a function of time is plotted assuming 1 μM NO (large plot) or 100 μM NO (inset).

production from stimulated macrophages and neutrophils; Akaike and Maeda, 2000), a rate of autoxidation of 15 μMs^{-1} can be calculated from the rate law. As shown in the inset to Fig. 1, although NO autoxidation would be much faster, diffusion of NO throughout a 200-μm-diameter plant cell would nonetheless be possible. Thus, autoxidation does not limit NO action within plant cells. However, paracrine actions of NO would be limited by autoxidation if NO production were comparable to the highest reported mammalian values rather than comparable to most typical values.

B. DOES NO UNDERGO CATALYZED OXIDATIVE OR REDUCTIVE CATABOLISM IN PLANTS?

Although autoxidation of NO is slow, catalyzed oxidation has been reported. Members of the globin family (Joshi *et al.*, 2002; Poole and Hughes, 2000) as well as the unrelated mammalian flavoprotein dihydrolipoamide dehydrogenase (Igamberdiev *et al.*, 2004a) are known to catalyze the oxidation of NO to nitrate. With prokaryotic and eukaryotic microbial flavohemoglobins, molecular oxygen and NAD(P)H are co-substrates for this NO dioxygenase activity (Gardner *et al.*, 1998; Poole and Hughes, 2000). Mutants in the corresponding genes in *E. coli* and *Salmonella typhimurium* were hypersensitive to growth inhibition by NO, NO donors, or paraquat (Crawford and Goldberg, 1998; Gardner *et al.*, 1998; Membrillo-Hernández *et al.*, 1999).

Plant hemoglobin enzymes (phytoglobins) appear to play similar roles. These enzymes are found in the nucleus and cytoplasm (Seregélyes et al., 2000), and they resemble the microbial enzymes by criteria of cofactor requirements, kinetics, and response to inhibitors (Igamberdiev et al., 2004b; Seregélyes et al., 2004). An alfalfa gene (*MHb1*) that was expressed exclusively in the roots was strongly induced by hypoxia (Seregélyes et al., 2000). Hypoxia was shown to elicit NO synthesis in alfalfa root culture cell lines and maize cell lines, with transgenic expression of a barley phytoglobin reducing NO emission and antisense-induced under-expression of *MHb1* increasing emission (Dordos et al., 2003, 2004). Cultured alfalfa root extracts displayed NAD(P)H-dependent NO metabolism whose rate was increased by transgenic expression of barley phytoglobin and decreased by antisense-induced under-expression of *MHb1* (Igamberdiev et al., 2004b). A similar activity was enhanced in extracts of tobacco seedlings expressing the *MHb1* transgene (Seregélyes et al., 2004). Overexpression of *Arabidopsis AHb1* reduced and antisense-mediated under-expression of *AHb1* increased emission of gaseous NO immediately following abrupt shift of light-adapted plants to darkness (Perazzolli et al., 2004).

Based on these results, it can be concluded that phytoglobin-catalyzed oxidation is one fate of NO in plants, at least in certain tissues/cells/subcellular locations under certain conditions. Catalyzed NO oxidation appears to play an important role in the detoxification of NO produced during hypoxic stress (Perazzolli et al., 2004). It has also been suggested that catalase (in the presence of H_2O_2) can catalyze NO oxidation based on catalase-dependent increases in NO oxidation rates that were too large to explain merely by increases in the concentration of O_2 (Brown, 1995).

Oxidation is not the only possible metabolic fate of NO. Denitrifying bacteria and some fungi also reductively catabolize NO (Averill, 1996). Nitric oxide is an obligatory intermediate in the bacterial conversion of nitrate to N_2. The bacterial enzymes are membrane-bound dimers consisting of a cytochrome *b* and a cytochrome *c*. The fungal (*Fusarium* spp.) NO reductases are cytochrome P450 enzymes that use NADH as a reductant. In all cases, the immediate product is nitrous oxide (N_2O). In fungi, N_2O is excreted, and the overall process appears to serve a detoxification function. Nitrous oxide emission has been detected from a wide variety of aseptically-grown plants (Hakata et al., 2003). It has not yet been determined whether the enzymes catalyzing NO reduction are of plant or endophytic microbial origin.

C. DOES SUPEROXIDE LIMIT NO ACCUMULATION?

Oxidation is only one of the possible fates of NO. The reaction with superoxide is one of the fastest known. The bimolecular rate constant (1×10^{10} M^{-1} s^{-1}) is fivefold higher than values typically reported for interaction of superoxide with superoxide dismutase (Radi et al., 2000). The product of this reaction, peroxynitrite, reacts readily with carbon dioxide to give an adduct ($ONOOCO_2^-$) (Poole and Hughes, 2000). This adduct can decompose to yield nitrogen dioxide (NO_2) and the carbonate radical ion. The former will ultimately lead to nitration reactions (see Section III.E for mechanism). The latter is highly reactive in one-electron oxidations and is largely responsible for the contribution of peroxynitrite to the pathology of numerous human diseases as well as peroxynitrite's antimicrobial activity. Alternatively, the adduct can decompose to release NO_2^+ that is itself a nitrating agent.

Whether this fate occurs in plants will depend upon the relative concentration of NO and superoxide dismutase. As these values are difficult to determine, investigations have centered on peroxynitrite. Detection of peroxynitrite decomposition products can not be used as evidence for peroxynitrite formation because they (nitrate at acid pH and additionally nitrite plus molecular oxygen at alkaline pH; Radi *et al.*, 2000) are not uniquely diagnostic. Tyrosine nitration is accepted as evidence of peroxynitrite formation in mammals. However, nitration of tyrosine as well as other phenolics can also occur in plants with nitrite (rather than peroxynitrite) as a substrate in the presence of peroxidases and H_2O_2 (Sakihama *et al.*, 2003).

As peroxynitrite concentrations as low as 1 μM have been shown to cause death of mammalian cells, addition of exogenous peroxynitrite to plant cell cultures or plant leaves was attempted to establish plausibility of a role for peroxynitrite. Addition of up to 1mM peroxynitrite did not cause death of soybean cells (Delledonne *et al.*, 2001). Slower release of peroxynitrite from an NO donor (up to 5 mM SIN-1) was also ineffective (Delledonne *et al.*, 2001). *Arabidopsis* leaves showed damage only when very high concentrations of these compounds were used (>2 mM SIN-1 or 1 mM peroxynitrite were required for consistent necrotic responses; Alamillo and García-Olmedo, 2001).

Urate, a known scavenger of peroxynitrite, was also used in an attempt to demonstrate a functional role for plant peroxynitrite (Alamillo and García-Olmedo, 2001). Urate reduced the extent of *Arabidopsis* hypersensitive cell death seen in response to *Pseudomonas syringae* carrying the avirulence gene *avrRpm1*. However, this effect was highly dependent upon experimental conditions, with a twofold difference in the level of inoculum abolishing the effects of urate. The same study showed that urate caused a necrotic response when co-infiltrated with virulent *P. syringae*. By contrast, urate did not affect the HR to *Xanthomonas campestris* pv. *campestris*. Perhaps tellingly, urate was shown to affect *P. syringae* growth in culture. Moreover, it was not established whether the effects of urate were caused by peroxynitrite scavenging. Urate is also known to have concentration-dependent prooxidant or antioxidant functions in mammals that result from its ability to reduce copper (Filipe *et al.*, 2002).

It has been suggested that nitrate reductase can catalyze production of superoxide (Barber and Kay, 1996) and peroxynitrite (Yamasaki and

Sakihama, 2000) *in vitro*. However, the dye whose fluorescence was taken as evidence of peroxynitrite formation also reacts with H_2O_2 (Wolfe *et al.*, 2000) and the appropriate controls (e.g., effects of superoxide dismutase and catalase) were not performed.

As such, conclusive evidence for peroxynitrite formation or function in plants is still lacking. Experiments with mutant plants lacking superoxide dismutase activity (currently unavailable) might be more definitive. However, it would be prudent at present to avoid assuming direct reaction with superoxide to be a major fate of NO in plant biology.

D. EVIDENCE FOR ADDITIONAL REACTIONS OF NO WITH REACTIVE OXYGEN

Evidence for additional reactions of NO with reactive oxygen has been presented. Much of the early literature on NO autoxidation reported much faster rates for the trimolecular reaction with molecular oxygen than referenced previously. It was later established that these rates of reaction occurred only near industrial areas or along transportation corridors and reflected the participation of atmospheric ozone (Leshem, 2000). This reaction must be taken into account in laboratory studies that expose plants to NO in gaseous form.

Nitric oxide also reacts with lipid hydroperoxide radicals with rapid kinetics (bimolecular rate constants of $\sim 1 \times 10^9$ $M^{-1}s^{-1}$; Wink *et al.*, 1993). In this capacity, NO functions as a chain reaction-terminating antioxidant. It is possible that NO plays this role in plant biology (e.g., in limiting the extent of peroxidation of plasma membrane lipids during the HR). In support of this hypothesis, treatment of potato leaf pieces or isolated chloroplasts with NO donors reduced lipid peroxidation caused by treatment with the herbicide diquat (Beligni and Lamattina, 2002). Diquat functions as an artificial photosystem I electron acceptor that subsequently reduces oxygen to superoxide. Dismutation followed by Fenton chemistry can produce hydroxyl radical that can initiate lipid peroxidation.

However, the locus of action of NO in this case may not have been the lipid hydroperoxide radicals. Nitric oxide has been suggested to play a role in preventing Fenton chemistry (Sharpe *et al.*, 2003). The simultaneous presence of Fe^{2+} and H_2O_2 is known to result in the *in vitro* production of hydroxyl radicals. Hydroxyl radicals are one of the most reactive and thus damaging of reactive oxygen species. The mechanism of hydroxyl radical generation (Fenton chemistry) involves oxidation of iron to the ferryl, $[Fe=O]^{2+}$, species as the first step. Nitric oxide is known to reduce this species, resulting in the production of NO_2 and Fe^{2+}. In support of the relevance of this role for NO, diquat-induced hydroxyl radical production was shown to be reduced to near basal levels by pretreatment with the NO donor sodium nitroprusside (SNP) (Beligni and Lamattina, 2002).

E. S-NITROSYLATION REACTIONS ARE NOT LIKELY TO BE A MAJOR FATE OF PLANT NO EXCEPT PERHAPS DURING HYPOXIA

The preceding discussion suggested that chemical reaction with molecular or reactive oxygen is not the major fate of NO in plants. Another documented fate of NO in biology is in the nitrosylation of thiol compounds. The most prevalent thiol in plants is glutathione. As yet, there is no direct proof that S-nitrosyl glutathione (GSNO) occurs in plants (Sakamoto et al., 2002). However, GSNO formation has been shown to be biologically significant in mammals. Protein S-nitrosylation, which occurs as a consequence of transnitrosylation between GSNO and cysteine moieties, is a common mammalian regulatory posttranslational modification (Martinez-Ruiz and Lamas, 2004). By contrast, S-nitrosylation of plant proteins has not been documented (Huber and Hardin, 2004) except for the S-nitrosylation that occurs as part of the NO dioxygenase mechanism of *Arabidopsis* AHb1 (Perazzolli et al., 2004).

It is possible that these interkingdom differences are a direct consequence of the action of the NOS cofactor BH_4. In the absence of BH_4, mammalian NOS will catalyze synthesis of superoxide. Tetrahydrobiopterin binds anticompetitively to NOS dimers (Gorren et al., 1996). As such, the subunit that binds BH_4 will catalyze NO synthesis, while the other subunit will catalyze simultaneous superoxide synthesis. In the presence of reduced glutathione, the product seen from simultaneous generation of NO and superoxide was GSNO (not peroxynitrite that is seen only in the absence of glutathione) (Mayer et al., 1998). In the presence of trace levels of Cu^+, GSNO will subsequently break down to release NO. Cu^+-specific chelators were used in the same study (Cu^{2+}-specific chelators had no effect) to demonstrate that GSNO formation and breakdown was essential to the stimulation of soluble guanylate cyclase resulting either from NOS action or from exogenous application of NO plus superoxide to perfused rat heart.

By contrast, current evidence suggests that physiological GSNO formation in plants is unlikely. The existence of BH_4 in plants has not been established (Hanson, 2004). Nitric oxide synthase1 does not require BH_4 for activity (Guo et al., 2003) and monoglutamylated tetrahydrofolate (which is found in plants; Hanson and Gregory, 2002) can substitute for tetrahydrobiopterin with iNOS (Chandok et al., 2004). The native molecular weight of iNOS does suggest it exists as a dimer; however, the EC_{50} for BH_4 activation of the *Arabidopsis* or tobacco enzyme was 32-fold or 25-fold higher, respectively, than for the mammalian enzyme. EC_{50} determinations for all other substrates and cofactors were remarkably similar to those for the mammalian enzymes (Chandok et al., 2003). As such, it is not clear whether the known plant NOS enzymes can coproduce NO and superoxide.

In the absence of this mechanism, generation of GSNO would require autoxidation. Under aerobic conditions, NO does not react directly with glutathione (Hogg et al., 1996). Nitric oxide first undergoes autoxidation to form NO_2 (nitrogen dioxide) that subsequently reacts with NO to form N_2O_3 (dinitrogen trioxide). This species partially dissociates into [$^+ON \cdot \cdot NO_2^-$], allowing nucleophilic attack of a free sulfhydryl (or other nucleophile) on the nitrosonium moiety, releasing nitrite. Transnitrosylation of proteins proceeds analogously with cysteine sulfhydryl moieties acting as the nucleophiles (Martinez-Ruiz and Lamas, 2004). However, as the rate-limiting step for aerobic GSNO formation would be autoxidation (which is very slow; see Section III.A), GSNO formation and protein *S*-nitrosylation are unlikely to be major occurrences in plants.

Nonetheless, the presence of plant homologs of an enzyme known to function in GSNO catabolism have been presented as indirect evidence for GSNO in plants (Sakamoto *et al.*, 2002). These enzymes are likely cytoplasmic in that no organellar targeting sequences were noted. The genes are expressed constitutively at high levels, and the *Arabidopsis* and tobacco genes were upregulated by salicylic acid treatment and downregulated by wounding (*Arabidopsis*) or jasmonate treatment (tobacco) (Díaz *et al.*, 2003). Like their counterparts in other kingdoms, these enzymes converted GSNO to ammonia and oxidized glutathione in an NADH-dependent reaction. Deleting the GSNO reductase homolog from yeast or mice abolished GSNO reductase activity and led to increased accumulation of GSNO and nitrosylated proteins (Liu *et al.*, 2001). However, the existence of a GSNO reductase in plants cannot be taken as evidence for GSNO because these enzymes also possess formaldehyde dehydrogenase activity that may be their major function in plants (Achkor *et al.*, 2003).

Nevertheless, nitrosylation of at least one protein family can occur under hypoxia (Perazzolli *et al.*, 2004). Stoichiometric *S*-nitrosylation of AHb1 was documented *in vitro* as a partial reaction in the absence of NADPH. *S*-nitrosylation was also seen *in vivo* under conditions of hypoxia or AHb1 overexpression. Whether transnitrosylation of other proteins or that of glutathione occurs in plants under these conditions is unknown. *De novo* GSNO formation is not likely to occur during hypoxia because *in vitro* studies have shown that the products of the anaerobic reaction of NO and glutathione are glutathione disulfide, N_2O, and water, rather than GSNO (Hogg *et al.*, 1996). Moreover, this reaction is very slow ($t_{1/2} = 24$ min with physiological concentration of glutathione).

F. METALLOENZYMES ARE LIKELY TO BE THE MAJOR NO TARGETS IN PLANTS

The only known protein targets for NO proven relevant in plant biology are metalloproteins. Nitric oxide can bind to most transition metals. Nonetheless, with the exception of cytochrome oxidase, with which initial NO interactions are likely to be with a copper ion (Torres *et al.*, 1998), plant metalloprotein interactions with NO of proven significance all involve iron. The chemistry of these interactions was the subject of an excellent review

(Cooper, 1999) and as such only the most relevant details for plant signaling are summarized here. Binding to metalloproteins can be very rapid, with typical bimolecular association rate constants of $\sim 10^8$ $M^{-1} s^{-1}$. As NO binds more tightly in the absence of a *trans* ligand coordinating Fe^{2+}, its binding can displace a *trans* ligand. This displacement can induce conformational changes in enzymes. In the case of the prototypical metalloprotein NO target guanylate cyclase, a histidine residue that ligates the iron from the *trans* position is displaced. The resulting conformational change increases cyclase activity. Nitric oxide dissociation can be quite slow from some metalloproteins (e.g., 1.8×10^{-5} s^{-1} measured off-rate constant from hemoglobin in the "R" state). For enzymes that are reversibly modulated by NO, the protein microenvironment weakens the NO–iron bond, leading to faster off-rates. Dissociation of NO from the mammalian guanylate cyclase is slow (off-rate constant of 7×10^{-4} s^{-1}); however, this rate is markedly increased by the conformational changes that occur upon rebinding of GTP and Mg^{2+} (to 0.05 s^{-1}).

These off-rates dictate that NO–hemoprotein interactions leading to enzyme inhibition or activation generally will be with the Fe^{2+} form rather than the Fe^{3+} form of iron. With model Fe^{3+}-containing compounds, off-rates are typically 10^4- to 10^8-fold higher than with related Fe^{2+}-containing compounds. The resulting weak binding usually precludes enzyme inhibition/activation. However, catalase is an exception. Nitric oxide binds to mammalian Fe^{3+}-catalase with a K_D of 0.5 μM (the measured apparent K_i for enzyme inhibition was comparable). This concentration is of potential biological relevance.

Nitric oxide binding to globin family members under aerobic conditions leads to catalyzed oxidation with a transient peroxynitrite intermediate breaking down to release nitrate. The iron is oxidized in the process and must be re-reduced to the Fe^{2+} form by a methemoglobin reductase activity. This metabolism of NO is required to avoid irreversible inhibition of protein function because of the slow off-rates for NO.

IV. NO FUNCTION IN PLANTS

A. CAVEATS AND POTENTIAL ARTEFACTS IN ASSESSING NO FUNCTION

1. The Method of NO Delivery Affects Experimental Results

Much of the research implicating NO in various aspects of plant biology relied upon delivery of exogenous NO to the plants. Prior to examining claims of NO function, it is prudent to consider the limitations and potential artefacts associated with these studies. Nitric oxide can be provided either as a gas or from "donor compounds." Potential artefacts associated with compartmentalization of NO delivery/action and of the reaction of gaseous NO with ambient ozone are considered in Sections II.A and II.D. This section will focus on the various donor compounds.

With regard to potential artefacts, the gauntlet was laid down with the publication of one recent study (Murgia *et al.*, 2004a). Four "NO donors" were compared. In virtually all experiments performed, sodium nitroprusside, which was referred to as a NO^+ (nitrosonium) donor, differed markedly in its action from the compounds stated to release NO. The three NO-releasing compounds examined were *S*-nitroso-*N*-acetylpenicillamine (SNAP), *S*-nitrosoglutathione (GSNO), and 3,3bis(aminoethyl)-1-hydroxy-2-oxo-1-triazene (NOC-18).

Sodium nitroprusside elicited and the other compounds did not elicit the following NO-mediated responses on *Arabidopsis* and tobacco cell cultures: induction of ferritin transcript accumulation, induction of PCD when presented in combination with H_2O_2, and inhibition of ascorbate peroxidase and catalase activities. Additionally, levels of cellular ascorbate and glutathione were lower, and the pools were more highly oxidized in cultures undergoing PCD in response to SNP plus a H_2O_2 generating system than in response to H_2O_2 alone. The other NO donors either were without effect or had the opposite effects in a few cases. Rates of NO release were not likely to be responsible for these differences as the rate was lowest with SNP and comparison with levels of the other compounds leading to the same rate of release did not change the results. Sodium nitroprusside decomposition products were also shown not to be responsible for these differences via experiments with SNP analogs that do not release NO. However, this concern may exist for other studies. In one study, the effects of SNP on tobacco seedling growth were partially mimicked by an SNP preparation treated with sufficient light to cause SNP decomposition (Seregélyes *et al.*, 2003).

The differences between the donor compounds were striking and incontrovertible. Nonetheless, the explanation for these differences requires further examination. It was speculated that the capacity for direct nitrosylation of targets due to NO^+ "release" was the key difference between SNP and the other NO donors (Murgia *et al.*, 2004a). The SNP ion, $[Fe(CN)_5(NO)]^{2-}$, although commonly referred to as an NO^+ donor, does not spontaneously release NO^+ (Poole and Hughes, 2000). The nitrosonium ion is coordinated to iron. Nitric oxide is released from the iron in one of three ways. Ambient light causes photolysis of the bond between iron and NO (Singh *et al.*, 1995) according to the following equation:

$$[Fe^{2+}(CN)_5NO]^{2-} \xrightarrow[H_2O]{h\nu} [Fe^{3+}(CN)_5H_2O]^{2-} + NO \qquad (4)$$

Nitric oxide is the species that is released, not NO^+. Additionally, in biological systems, NO can be released following reduction, loss of one of the cyanide moieties that coordinate the iron, and subsequent nucleophilic attack. In mammalian systems, the nucleophile is thought to be an as-yet-unidentified membrane-bound thiol. The release is known to be catalyzed by a membrane-bound enzyme and utilizes either NADH or NADPH as a cofactor (Feelisch and Stamler, 1996a). Glutathione is unlikely to be involved as its depletion did not prevent NO release (Rochelle et al., 1994).

Although the physiological nucleophile has not been identified in any system, in plants, it is also clearly not glutathione. S-nitrosyl glutathione (and other nitrosylated thiols) can decompose spontaneously to release NO (Poole and Hughes, 2000; Singh et al., 1996). This reaction is either caused by ambient light (Eq. 5) or catalyzed by transition metals (Eq. 6):

$$GSNO \xrightarrow{h\nu} GS^{\bullet} + {}^{\bullet}NO \quad (5)$$

$$GSNO + Cu^+ + H^+ \rightarrow GSH + {}^{\bullet}NO + Cu^{2+} \quad (6)$$

With reaction (5), two molecules of the thyl radical coproduct react to form the disulfide. With reaction (6), the reduced glutathione coproduct is oxidized to the disulfide, regenerating the Cu^+ catalyst.

Most importantly, either reaction leads to NO release. In the comparison of donor compounds study, GSNO was compared directly with SNP that also releases NO. The differences seen thus strongly suggest that a thiolate nucleophile other than glutathione was responsible for NO release from SNP. Differences in internalization and/or intracellular transport of this unidentified S-nitrosothiol product and GSNO, resulting in differences in the site of NO delivery, may explain the differences seen with cell cultures. With regards to intact plants, the unknown rates of transport of SNP and GSNO in the vasculature present a further complication. As such, it is not clear which chemical species was directly responsible for the seemingly NO-induced effects in leaves seen upon application of SNP to roots in two studies (Capone et al., 2004; Parani et al., 2004).

To confuse matters even further, both SNP and nitrosylated thiols including GSNO and SNAP can participate in transnitrosylation reactions (formally a donation of NO^+), and the nitrosylated thiols are capable of heterolytic decomposition to release nitroxyl ion (NO^-) (Arnelle and Stamler, 1995; Feelisch, 1998). Nitroxyl ion subsequently reacts with two equivalents of thiol to form a disulfide and hydroxylamine (NH_2OH).

To summarize, when a donor compound is added to plants, in most cases, it is completely unclear what form of NO is "released," what the rate of release will be, or in which subcellular compartment (or sometimes even which part of the plant) the release will occur. However, it is known that seemingly minor differences in experimental conditions can dramatically affect these parameters. These considerations may explain differences between "NO donors." They could account for differences seen between research groups in the effects of the "NO donors" on iron metabolism (Graziano *et al.*, 2002; Murgia *et al.*, 2004a) or other assays (Murgia *et al.*, 2004a) as slight differences in plants, reagents, light, buffer trace metal contamination, or experimental protocols. They could also explain why NO gas has been shown to elicit biological effects similar to SNP despite not being a "nitrosonium donor" (Graziano *et al.*, 2002; Huang *et al.*, 2002b). Because of these considerations, any study relying exclusively upon exogenous application of NO to establish physiological relevance must be questioned. Fortunately, loss-of-function studies have provided complementary evidence in many cases.

2. High Levels of Exogenous NO are Cytotoxic

A second caveat to bear in mind with regards to studies of NO function is its cytotoxicity. SNP or SNAP (10 μM) was sufficient to elicit PCD in cultured sweet orange (*Citrus sinensis*) cells (Saviani *et al.*, 2002). Significant increases in cell death were seen upon exposure of haploid Pacific yew (*Taxus brevifolia*) cultures derived from female gametophytes to even 1 μM SNP (Pedroso *et al.*, 2000b). *Arabidopsis* cells were somewhat less sensitive, requiring 50 μM SNP to show a statistically significant increase in cell death (Clarke *et al.*, 2000). Soybean cultured cells showed no increases in cell death upon exposure to 500 μM SNP (Delledonne *et al.*, 1998). This huge species-specific variation necessitates control experiments in the system of choice under the precise conditions of assay prior to ascribing an effect of exogenously supplied NO to something other than induced cell death.

3. Problems with Interpreting NO Loss-of-Function Studies

Globin family members have been exploited as tools to investigate NO functions in plants. However, interpretation of NO loss-of-function experiments performed with transgenic plants and plant cell lines expressing microbial (Frey *et al.*, 2004; Zeier *et al.*, 2004) or plant hemoglobins (Dordos *et al.*, 2003; Perazzolli *et al.*, 2004; Seregélyes *et al.*, 2003) may not be straightforward. The *E. coli* flavohemoglobin is also known to synthesize superoxide during NADH oxidation (Membrillo-Hernández *et al.*, 1996). If phytoglobins also synthesize superoxide, it might explain the increased basal superoxide levels and increased superoxide accumulation during a nonhost HR to *P. syringae* pv. *maculicola* displayed by tobacco plants expressing alfalfa *MHb1* (Seregélyes *et al.*, 2003).

Another possible complication results from the observation of efficient alkylhydroperoxide reductase activity of the *E. coli* flavohemoglobin (Bonamore *et al.*, 2003). It was speculated in this paper that a major function of all microbial flavohemoglobins is in the repair of oxidatively damaged membrane phospholipids. The HR is accompanied by damage to membrane lipids (Keppler and Novacky, 1986, 1987), and activation of both NO production (Zhang *et al.*, 2003) and reactive oxygen accumulation in adjacent living cells (Shapiro and Zhang, 2001; Zhang *et al.*, 2004) appear to require this membrane damage (see Sections II.B.1, IV.G.1, and IV.G.2.a). No effects of overexpressing *AHb1* were seen on the *Arabidopsis* HR or accompanying NO production (Perazzolli *et al.*, 2004) arguing against such a role for plant globins. However, this result may be inconclusive in that initial NO production is extracellular (Zhang *et al.*, 2003), the introduced reporter dye was initially extracellular, and transgenic AHb1 expression was presumably intracellular.

Recombinant *Arabidopsis* globins were also shown to possess peroxidase activity with diverse substrates including nitrite (Sakamoto *et al.*, 2004). Nitrite oxidation to nitrogen dioxide (NO_2) was catalyzed by three different *Arabidopsis* globins. As NO_2 reacts to form a nitrating agent (see Section III. E), tyrosine nitration presents a further complication of studies with the transgenic lines under conditions where nitrite accumulation is expected. Superoxide dismutase-insensitive (and thus peroxynitrite-independent) tyrosine nitrating activity was documented for these proteins.

As such, it is far from clear that the NO dioxygenase activity is the primary explanation for phenotypes seen with the transgenic lines. Additional complications can result from perturbations in NADH/NAD and NADPH/NADP ratios (Igamberdiev *et al.*, 2004b). However, transgenic plant lines expressing the *Vitreoscilla* hemoglobin (which lacks the FAD-binding domain with the associated functions provided by a separate protein in the native bacteria; Frey *et al.*, 2002) did lack one potential complication seen with some other transgenic lines. Resistance to paraquat, which induces oxidative stress, was not seen in tobacco plants expressing this hemoglobin and was likely a function of the FAD-binding domain of the flavohemoglobins (Frey *et al.*, 2004). The lack of this domain in plant globins would argue that transgenic lines based on them would also be free from this potential complication. Further research using these transgenic lines is described in Section IV.G.2.a.

B. NO-INDUCED GENE EXPRESSION

Transcript profiling has become common in plant biology research. Application of NO to plants does result in expression changes for a large number of genes (Huang *et al.*, 2002a,b; Parani *et al.*, 2004; Polverari *et al.*, 2003; Zeidler *et al.*, 2004). However, identification of "NO-induced genes" is far from straightforward. As described in earlier section, there are distinct enzymes catalyzing NO synthesis in the chloroplasts, mitochondria, cytoplasm, peroxisomes, and apoplast. In many cases, it appears that the function is within the compartment of synthesis. When the lack of evidence for *S*-nitrosylation in plants is also considered, it can be predicted that most NO-regulation of gene expression will be highly indirect. Moreover, in the best-characterized examples, NO interaction with its direct target leads to numerous downstream signals that can affect gene expression. For example, in the case of the HR (see Section IV.G.2.a), H_2O_2 (Desikan et al., 2001; Perazzolli et al., 2004), salicylic acid (Tao et al., 2003; Ward et al., 1991), changes to ascorbate levels (de Gara et al., 2003; Pastori et al., 2003), and signals directly associated with cell death events (Zhang and Shapiro, 2002) all regulate gene expression in primary responding and/or surrounding cells. One group has in fact argued that much of the observed "NO-regulated" gene expression could be attributed solely to change in salicylic acid levels (Huang et al., 2002a). Disentangling truly NO-regulated genes from genes more properly considered regulated by these secondary processes would require sophisticated studies involving appropriate loss-of-function mutants/transgenic lines.

Moreover, NO effects are likely to vary based upon concentration. As such, one group chose to treat with two different doses of an NO donor (Parani et al., 2004). However, prolonged exposure to NO will yield artefacts associated with NO-induction of stomatal closure (which would lead to gene regulation related to changes in water relations and gas exchange). To avoid these effects, a different group chose to induce with a high concentration (5000 ppm) of NO gas only for a short period of time (1 min) (Huang et al., 2002a).

Controls addressing all of these issues would be necessary to identify true NO-regulated genes. No transcript profiling study to date has been this comprehensive. When one adds to these issues the more general considerations limiting transcript profiling studies—the need for numerous biological replicates to achieve statistical significance, the contribution to the RNA pool from cells in a tissue that are likely to be asynchronous in their responses, and the need for time course studies that capture the dynamics of often rapidly changing physiology—the value of even well-constructed and comprehensive transcript profiling experiments becomes questionable.

Nonetheless, physiological relevance of NO regulation has been established for a handful of genes, even if it is not clear that they qualify as "NO-regulated" by the criteria presented in earlier section. These genes include ferritin, the mitochondrial alternative oxidase, phenylalanine ammonia lyase (PAL), several jasmonate biosynthesis genes, and several genes involved in the control of flowering time. These gene regulation events will be discussed in the context of the biology.

C. NO REGULATES MITOCHONDRIAL AND CHLOROPLAST FUNCTIONS

As is the case with other eukaryotes, mitochondrial respiration is regulated by NO. Mitochondrial respiration in plants relies on two distinct enzymes to catalyze the terminal electron transport step. Both of these enzyme activities were inhibited when mitochondria-enriched fractions from soybean embryonic axes were exposed to NO. However, the EC_{50} for the inhibition of cytochrome oxidase activity (assayed as propyl gallate-insensitive oxygen consumption) was much lower than that for inhibition of alternative oxidase activity (cyanide-insensitive oxygen consumption): 0.3 versus 3.6 μM (Caro and Puntarulo, 1999). With the levels of NO expected *in vivo*, cytochrome oxidase is likely to be strongly inhibited while the alternative oxidase will remain functional. Moreover, NO can also stimulate respiration by inducing transcription of the alternative oxidase gene (Huang *et al.*, 2002b). Gene induction was shown to be independent of salicylic acid. At present, nothing else is known about the signaling pathway. Consistent with the gene activation studies, SNP treatment of carrot cell cultures induced increases in the level of alternative oxidase protein and enzyme activity (Zottini *et al.*, 2002) and treatment of *Arabidopsis* cell cultures induced increases in enzyme activity (Huang *et al.*, 2002b). It has been speculated that alternative oxidase functions as a sink for electron flow under conditions that inhibit the cytochrome oxidase pathway. This role would reduce mitochondrial reactive oxygen production. In support of such a role, NO did not affect the rate of H_2O_2 production by mung bean hypocotyl mitochondria; by contrast, NO elicited significant increases in production of H_2O_2 by pig liver (mammals lack alternative oxidase) mitochondria (Yamasaki *et al.*, 2001). Nitric oxide can also elicit changes to plant mitochondria reminiscent of those associated with mammalian apoptosis including permeability transition pore formation, and consequent membrane depolarization (Saviani *et al.*, 2002) and release of cytochrome *c* (Zottini *et al.*, 2002). The relevance of these changes to any natural PCD process in plants has yet to be established.

As in mitochondria, one locus of NO action in chloroplasts appears to be electron transport. Nitric oxide is known to compete with bicarbonate ions for binding to the nonheme iron coordinated by D1 protein as well as to other sites on photosystem II components. Bicarbonate is known to upregulate rates of photosynthetic electron transport. By displacement of bicarbonate, NO decreased these rates (van Rensen, 2002). Nitric oxide-mediated decreases in electron transport rates reduced thylakoid transmembrane potential (due to reduced proton pumping) and photosynthetic ATP synthesis (Takahashi and Yamasaki, 2002). Nitric oxide-mediated reduction in electron transport rates was also documented to contribute to protection against diquat toxicity (Beligni and Lamattina, 2002). Diquat causes oxidative stress in part by increasing the rate of photosynthetic electron transport. Nitric oxide treatment restored near wild-type electron transport rates to diquat-treated plants. Further implications of NO modulation of photosynthetic electron transport will be discussed in the context of plant response to drought/ABA (see Section IV.F.1).

The thylakoid isoform of ascorbate peroxidase may also be a target for NO. The cell death elicited by treatment of *Arabidopsis* plants with extremely high levels (5 mM) of SNP was prevented by constitutive overexpression of this gene. Sodium nitroprusside treatment partially reduced enzyme activity levels and strongly reduced RNA levels in wild-type plants. Concentrations of SNP lower than 5 mM were reported to be without effect (Murgia *et al.*, 2004b). The relevance of these observations to NO function under more physiological conditions is unclear.

D. NO AND IRON HOMEOSTASIS

1. NO Mediates at Least One Iron Transport-Related Step in Plants

Plants obtain iron from the soil, in which free iron exists predominantly as the sparingly soluble Fe^{3+} species. The major destination of iron in plants is the chloroplasts where $\sim$80% of plant iron is used as Fe^{2+} for chlorophyll biosynthesis. As such, chelation and reduction processes are essential to deliver iron. Treating maize plants with NO did not affect total leaf iron content (Graziano *et al.*, 2002). This result implies that NO is not involved in early steps of iron acquisition, in which iron is taken up by plant roots and transported in the xylem mostly chelated to citrate and other organic acids (Cataldo *et al.*, 1988). However, the same study established a role for NO in later steps with the following observations: (1) Application of NO prevented maize leaf chlorosis and associated poor development of chloroplast membranes (iron deficiency symptoms) of plants grown in iron-limiting conditions; (2) An NO scavenger blocked the effects of exogenous NO; (3) Treatment of plants grown under iron-replete conditions with NO scavengers promoted iron deficiency symptoms. The last observation also suggested that the observed reduction in ascorbate peroxidase levels and increase in H_2O_2 accumulation seen under conditions of iron deficiency (Ranieri *et al.*, 2003) do not involve NO-mediated inhibition of ascorbate peroxidase, as the opposite phenotype would have been expected.

Transport of iron across leaf cell plasma membranes is known to involve reduction catalyzed by Fe(III) chelate reductase followed by transport of Fe^{2+} through a high-affinity transporter. Superoxide is also thought to be a physiologically relevant reductant that contributes to iron reduction required for transport (Brüggemann *et al.*, 1993; Cakmak *et al.*, 1987; de la Guardia and Alcántara, 1996; Macrì *et al.*, 1992). This involvement of reactive oxygen suggests one possible site of action of NO. However, NO would need to be generated at a level that it could compete kinetically with both superoxide dismutase and iron for reaction with superoxide. This consideration makes this mode of action unlikely. Alternatively, NO might affect uptake into cells, chelation in the cytoplasm (primarily a function of the amino acid nicotianamine; Becker *et al.*, 1995; Pich *et al.*, 2001), uptake into chloroplast stroma, or ferritin-dependent homeostasis within the stroma. Spectroscopic evidence for interactions of NO with ferritin has been presented (Cooper, 1999).

2. NO Mediates Iron-Induced Upregulation of Ferritin Gene Transcription

Although the possibilities described in earlier section are not mutually exclusive, all observations suggesting a role for NO in iron homeostasis might be explained as effects on ferritin gene transcription. Ferritin is an iron-storage structure formed by a 24-subunit protein coat surrounding up to 4500 iron atoms. In leaves, it is located primarily in chloroplasts, though its presence has also been documented in mitochondria from etiolated pea seedlings and *Arabidopsis* cell cultures (Zancani *et al.*, 2004). An excess of iron leads to ferritin accumulation (Lobreaux *et al.*, 1992). In mammals, ferritin synthesis is regulated at the translational level. Nitric oxide mediates destruction of the Fe-S center in cytoplasmic aconitase, converting aconitase into an RNA-binding protein that regulates translation of the ferritin mRNA. In spite of the documented ability of NO to inhibit plant cytoplasmic aconitase (Navarre *et al.*, 2000), iron-induced upregulation of plant ferritin synthesis is transcriptional (Lobreaux *et al.*, 1992). Nonetheless, NO has been shown to play an essential role in the signaling pathway (Murgia *et al.*, 2002). The precise role is unclear at present. The only available data are that the protein phosphatase inhibitor calyculin and the protein synthesis inhibitor cycloheximide will inhibit both iron and NO-triggered increases in ferritin transcription (Murgia *et al.*, 2002). In addition, transgenic approaches were used to show that NO function in this system does not involve the thylakoid isoform as ascorbate peroxidase (Murgia *et al.*, 2004b). Iron-induced ferritin transcription requires reactive oxygen accumulation (Savino *et al.*, 1997); however, interactions between NO and reactive oxygen have not yet been clarified in this system.

Interestingly, NO was also shown to induce upregulation of the ferritin gene of snail neurons (Xie *et al.*, 2001). The archetypal relative of NOS1 was identified in this system. An interesting speculation (admittedly lacking very important pieces of data for support) would implicate *NOS1* in plant ferritin gene regulation and suggest that the regulation is transcriptional because NO is produced in the mitochondria (plants and snails) rather than in the cytoplasm (mammals). Analysis of the snail NOS sequence (Huang *et al.*, 1997) using TargetP (Emanuelsson *et al.*, 2000) predicts a mitochondrial location for this enzyme (data not shown). To date, no subcellular localization studies have been reported.

E. NO IN PLANT DEVELOPMENT

The availability of NO donors, NO scavengers and biosynthetic inhibitors, and NO-modulating mutants and transgenic lines has been the impetus for applying the question, "Is NO involved in··· to a wide range of developmental processes. It now appears that NO plays some role in most stages of plant life. However, the more involved processes of determining tissue, cell type, intracellular site, and timing of NO synthesis/action are still in the initial stages. In most cases, the direct target(s) of NO is not known. Moreover, embedding NO action within each undoubtedly complex signaling network has barely begun in any of these cases. The question whether NO has any unique role in development or whether it is once again the "character actor" recapitulating roles delineated elsewhere in this article (e.g., regulation of respiratory or photosynthetic electron transport, inhibition of H_2O_2 scavenging enzymes, activation of cGMP- and cADPR-dependent release of stored calcium, etc.) is not yet answered in most cases. Therefore, the following discussion should be viewed as a list of topics for which more information should be coming soon. The developmental process is followed roughly in sequence, beginning with plant seeds.

1. Roles of NO in Seed Dormancy and Germination

Seed germination often requires more than imbibing water. In lettuce, germination is light-dependent. Treatment with NO (as SNP or SNAP) overcame this dependence and allowed seed germination in darkness (Beligni and Lamattina, 2000). *Arabidopsis* seeds require either cold treatment or an after-ripening period for efficient germination. Barley seeds also exhibit dormancy immediately following harvest, especially following cold summers. Sodium nitroprusside (25 μM) was shown to break and NO scavenger treatment was shown to strengthen dormancy of seeds from *Arabidopsis* and barley (Bethke *et al.*, 2004b). Treatment with NO donors was also shown to delay PCD in the aleurone layer of germinating barley seeds. Treatment with butylated hydroxy toluene (an antioxidant) also delayed PCD. Thus, NO was proposed to act in an as-yet-unidentified process that partially ameliorated the oxidative stress experienced by this tissue as a consequence of storage lipid breakdown (Beligni *et al.*, 2002). Nitric oxide was proposed to be a component of smoke that stimulated germination in seeds following fires. However, further analysis indicated that NO did not promote germination of these seeds and NO was not detectable in aqueous extracts of smoke derived from burning wood or cellulose (Preston *et al.*, 2004).

2. NO as A Possible Mediator of Cytokinin Action: Roles in Seedling De-etiolation and Inhibition of Hypocotyl Elongation

In the absence of light, seedlings will fail to become green and the embryonic shoot (hypocotyl) will undergo pronounced elongation. Light exposure leads to greening and inhibition of hypocotyl elongation. Cytokinin treatment of seedlings in darkness will partially mimic these effects of light, although full responses require light acting through phytochrome. As NO treatment strongly resembled cytokinin treatment in these assays, NO was proposed to act as a signaling component downstream of cytokinins (Lamattina *et al.*, 2003). Sodium nitroprusside treatment led to partial greening of wheat seedlings grown in the dark (Beligni and Lamattina, 2000). Sodium nitroprusside treatment of seedlings also led to reduced elongation of lettuce or *Arabidopsis* hypocotyls and reduced elongation of sprouts from potato tubers (Beligni and Lamattina, 2000). In support of a physiological role for NO, cytokinin treatment of *Arabidopsis*, tobacco, or parsley cell cultures led to NO synthesis. This synthesis was partially blocked by an NOS inhibitor (Tun *et al.*, 2001).

3. Role of NO in Root Development

Nitric oxide also appears to function in root development. Sodium nitroprusside treatment of tomato seedlings promoted lateral root development and suppressed primary root growth (Correa-Aragunde *et al.*, 2004). Treatment of seedlings with an NO scavenger completely suppressed lateral root emergence and stimulated primary root growth. Auxin-stimulated lateral root formation was also suppressed by the NO scavenger. An inhibitor of auxin polar transport did not inhibit SNP-stimulated lateral root development, staging NO downstream of auxin action. Root development was also stimulated in cucumber hypocotyl explants pretreated to reduce endogenous auxin levels (Pagnussat *et al.*, 2003). The effects of either auxin or SNP were blocked by a guanyl cyclase inhibitor. Root development was also stimulated in this system when cGMP breakdown was inhibited by treatment with Viagra (sildenafil citrate that blocks phosphodiesterase activity). Stimulation of root formation by auxin treatment of Asiatic dayflower tissues was shown to be suppressed by treatment with a guanyl cyclase inhibitor, a chelator of calcium, an inhibitor of release of calcium from internal stores, or an antagonist of cADPR biosynthesis (Cousson, 2004).

These studies taken together suggest that the classic mammalian cGMP-dependent NO response pathway is operative in auxin-induced root development. However, complications have already emerged in this story. Pharmacological approaches were used with the cucumber explant system to implicate an MAP kinase cascade downstream of NO on a separate branch of the signaling pathway from cGMP (Pagnussat *et al.*, 2004). As such, NO must have at least one additional target in this system. It has also been speculated that NO biosynthesis for stimulation of root development will be regulated in response to soil nitrate levels as sensed by PM-NR and NI-NOR (Stöhr *et al.*, 2001).

4. Roles of NO in Later Developmental Processes: Maturation, Senescence, and Fertility

In a wide variety of plants, NO production decreases with maturation, parallel to the increases in ethylene emission. Moreover, application of NO gas or NO donors to plants will delay maturation and senescence (Leshem and Haramaty, 1996; Leshem and Pinchasov, 2000; Leshem *et al.*, 1998; Magalhaes *et al.*, 2000). Nitric oxide application also blocked senescence induced by ABA treatment of rice leaves (Hung and Kao, 2003). In this study, it was suggested that NO played an "antioxidant" role because similar blocking of ABA effects were seen with application of free radical scavengers. Experiments with vase life of cut flowers have implicated the cGMP-dependent NO response pathway in timing of senescence. S-nitroso-N-acetylpenicillamine-treatment markedly increased the vase life of carnation flowers. Viagra-treated cut flowers (carnations and roses) also exhibited markedly longer vase life. Application of 3,5-phosphodiesterase (known to catabolize cyclic nucleotides) shortened carnation vase life (Leshem, 2000).

Nitric oxide also appears to be an endogenous regulator of fertility (He et al., 2004). The *Arabidopsis nos1* mutant flowered earlier than wild-type, implying that NO synthesized from this enzyme delayed flowering time. Consistent with this hypothesis, treatment of wild-type plants with SNP delayed flowering time. Flowering time was also delayed in a mutant that overproduced NO. All six alleles of this mutant (*nox1*) carried deletions in the *CUE1* gene that encodes a chloroplast phosphoenolpyruvate/phosphate translocator. The phenotypes of these mutants were likely to have been caused by overproduction of arginine, the substrate for NOS1.

Nitric oxide appears to act on several different components of the flowering time control machinery. The FLC transcriptional repressor is the target of two different flowering time control pathways (vernalization and autonomous). Consistent with a role for NO in interfering with these pathways, *FLC* expression was elevated in *nox1* plants or in plants treated with high levels (>50 μM) of SNP and reduced in *nos1* plants. However, treatment with low doses of SNP decreased *FLC* expression while still delaying flowering. Similar approaches were used to implicate NO in changes to expression of genes implicated in the photoperiod-regulated pathway of flowering time control. Changes in amplitude but not period of circadian-regulated genes (as well as circadian regulation of cotyledon movements) were also noted.

Pollen tube growth is also affected by NO (Prado et al., 2004). Lilly pollen tubes growing in liquid media responded to a point source of NO (SNAP suspended in agar in a micropipette) with a sharp reduction in growth rate and reorientation of the growth axis away from the point source. This assay became sensitive to much lower concentrations of SNAP in the presence of Viagra, possibly implicating the cGMP-based NO response pathway. Dye-based imaging showed endogenous NO production to be localized to peroxisomes that are largely excluded from the tips (growing regions) of pollen tubes.

F. NO IN RESPONSES TO THE ABIOTIC ENVIRONMENT

1. Role of NO in Response to Drought

An important plant response to water deprivation stress is stomatal closure. Water deprivation stress is perceived in the roots, where the phytohormone ABA is synthesized and transported to the leaves. However, short-term responses depend more upon redistributions of leaf ABA that increase the apoplastic pool. These redistributions are triggered by changes to apoplastic pH caused by water deprivation stress (Hartung *et al.*, 1988). Leaf guard cells respond to increased ABA levels by secretion of ions. Osmotically driven water flow subsequently leads to collapse of guard cell turgor. This loss of turgor closes the stomatal pore, blocking transpiration. These guard cell responses to drought can be mimicked by application of exogenous ABA (Desikan *et al.*, 2004).

Nitric oxide has been implicated in water deprivation stress and ABA-induced stomatal closure. Treatment with SNP led to increased stomatal closure, decreased transpiration rate, improved water retention, and decreased cell injury of plants exposed to water deprivation stress (García-Mata and Lamattina, 2001). A study using an NO scavenger and an NOS inhibitor implicated NO (and an NOS) in ABA-induced stomatal closure in pea leaf epidermal peels (Neill *et al.*, 2002). The same study used a pharmacological approach to implicate cGMP and cADPR function downstream of ABA and NO. Abscisic acid/nitric oxide-elicited production of cADPR was shown to lead to release of Ca^{2+} from intracellular stores and subsequent changes in conductance through plasma membrane Ca^{2+}-sensitive ion channels in guard cells from bean (García-Mata *et al.*, 2003). Abscisic acid inhibition of the inward potassium conductance ($I_{K,in}$) and activation of the outward chloride conductance (I_{Cl}) of plasma membranes were attributed to NO action. However, ABA activation of plasma membrane Ca^{2+} channels and Ca^{2+}-insensitive K^+ channels was shown to be independent of NO. Nitric oxide is thus associated with only a subset of ABA-evoked responses.

Dose-dependent effects of SNP, whereby moderate concentrations (10–200μM) cause stomatal closure but very high concentrations (0.5–2 mM) do not (Desikan *et al.*, 2004) are currently without adequate explanation. These effects may be related to the action of ABI1 and ABI2. ABI1and ABI2 are phosphatases that negatively regulate ABA-induced stomatal closure and have been placed upstream and downstream of H_2O_2 action, respectively (Murata *et al.*, 2001). Abscisic acid-induced NO synthesis is not impaired by mutations in these genes. However, SNP-induced stomatal closure is blocked, indicating an interaction of NO with these phosphatases (Desikan *et al.*, 2002). Whether this is a direct interaction is unknown.

The NOS that contributes to ABA-induced NO biosynthesis was identified as NOS1 on the basis of a study with the *Arabidopsis nos1* mutant (Guo *et al.*, 2003). This study also established that NOS1 and NO production were required for ABA inhibition of light-induced stomatal opening. Intriguingly, ABA-induced stomatal closing but not ABA inhibition of light-induced stomatal opening in *Arabidopsis* also required nitrate reductase (Desikan *et al.*, 2002). Why ABA-induced NO production requires two distinct NO synthesis pathways is an unresolved question. This author's speculation is presented in Fig. 2. In this view, the nitrite required for nitrate reductase-catalyzed NO synthesis accumulates because of inhibition of photosynthetic

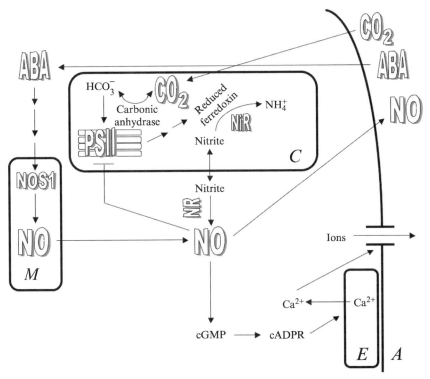

FIGURE 2. A speculative proposal for why two NO synthesizing enzymes are required for ABA-induced stomatal closure. Italicized text indicates subcellular compartment: C, chloroplast; M, mitochondria; E, endoplasmic reticulum; A, apoplast. Enzymes abbreviated as follows: NR, nitrate reductase; NiR, nitrite reductase; PSII, photosystem II; NOS1, nitric oxide synthase encoded in *Arabidopsis* by At3g47450. Details of the figure are given in the text.

electron transport. As discussed in Section II.B.2, this inhibition is the only proven mechanism of nitrite accumulation in leaves. As NO synthesis precedes stomatal closure, any postulated mechanism for electron transport inhibition must be plausible in the presence of atmospheric levels of CO_2. *In vivo*, CO_2 is in equilibrium with HCO_3^- ion. The hydration reaction is catalyzed by carbonic anhydrase and a strong carbonic anhydrase activity has been detected associated with photosystem II (Stemler, 2002). The bicarbonate effect on photosynthetic electron transport and its reversal by NO were described in Section IV.C.

The most important point of Fig. 2 is the prediction that some NO must be synthesized in a nitrate reductase-independent fashion to inhibit electron transport and allow nitrite to accumulate and act as a substrate. As such, nitrate reductase cannot act alone in this system. The proximity of mitochondria and chloroplasts in the cells that conduct photorespiration makes the involvement of NOS1 plausible. The ability of NOS1 to act as the sole NO biosynthesis enzyme in the inhibition of stomatal opening but not in the promotion of stomatal closure could be attributed to quantitative differences in the level of NO required for the two processes or to the compensation by parallel signaling pathways in the inhibition of opening. In support of the latter hypothesis, ABA-induced synthesis of H_2O_2 is enhanced in the *nia1–1/nia2–5* double mutant (Desikan *et al.*, 2004). The lack of a "wilty" phenotype of the *nia1–1/nia2–5* double mutant also suggested functional redundancy between the NO branch and the H_2O_2 branch of the ABA response pathway (Desikan *et al.*, 2002).

Although some *in vivo* evidence exists for the bicarbonate effect on photosynthetic electron transport (El-Shintinawy and Govindjee, 1990; Garab *et al.*, 1983), its physiological relevance has been questioned because most data comes from *in vitro* studies under very specific conditions. Alternative explanations that might explain rapid nitrite accumulation in response to imposed water stress or ABA should thus be considered. One possible explanation would be impairment of nitrite import into chloroplasts. Nitrite transport across the chloroplast envelope has been shown to depend upon the pH gradient across the membrane (Shingles *et al.*, 1996). The pK_a of nitrous acid, the protonated form of nitrite, is 3.37. As a weak acid it will diffuse into the compartment of higher pH that under well-watered conditions is the chloroplast interior. There is evidence for this process being facilitated by a specific transporter in algae (Galván *et al.*, 2002). If this process was also facilitated in higher plants, one possibility for a water stress-sensitive target would be the transporter. Alternatively, the pH gradient across the membrane is also sensitive to water deprivation stress. The proximal cause of the changes is the inhibition of the plasma membrane H^+-ATPase under conditions of water deprivation stress (Hartung *et al.*, 1988). This inhibition results in acidification of both the cytoplasm and chloroplasts; however, chloroplasts are affected to a greater extent (Slovik and Hartung, 1991). Under moderate drought conditions, the trans-envelope pH gradient is reduced, and under severe drought conditions, the gradient is reversed. Under these conditions, nitrite should accumulate in the cytoplasm and thus be available as a substrate for NO biosynthesis.

These considerations would not account for NO production in response to exogenously supplied ABA. However, in the study implicating NO in the ABA response (Neill *et al.*, 2002), 10 µM ABA was used. Abscisic acid is itself a weak acid and bolus addition of this level of ABA could transiently discharge the chloroplast trans-envelope proton gradient, since guard cell chloroplasts are a major site of ABA accumulation (Hartung and Slovik, 1991). As NO accumulation was seen within minutes, a transient pH gradient discharge might have sufficed. Experimental discharging of the trans-membrane proton gradient by supplying weak acids (propionic acid, butyric acid, or acetic acid) to barley roots led to nitrite accumulation and secretion into the medium (Botrel *et al.*, 1996). Alternatively, ABA was shown to inhibit the plasma membrane H^+-ATPase (Brault *et al.*, 2004; Hartung *et al.*, 1988), the consequence of which would have been discharging of the chloroplast envelope transmembrane pH gradient. The major problem with this hypothesis is that it does not explain the requirement for two different NO synthesizing enzymes in this signaling pathway.

A second alternative explanation for nitrite accumulation could be that nitrite reductase activity is inhibited. Nitrite reductase is a metalloprotein with both a siroheme prosthetic group to which nitrite binds and an iron–sulfur cluster that is the likely initial electron acceptor (Meyer and Stitt, 2001). It is conceivable that NO interferes with either nitrite binding or electron transfer. Although effects of NO on nitrite reductase activity *per se* have not been reported, other evidence would argue against this possibility. Drought stress was found to have no significant impact on nitrite reductase-specific activity in roots from two tested cultivars of maize (Subramanian and Charest, 1998). Drought did reduce nitrite reductase activity in leaves of oak trees (Schmadel-Hagebölling *et al.*, 1998) and mulberry plants (Ramanjulu and Sudhakar, 1997), and various forms of water stress reduced activity in wheat leaves (Heuer *et al.*, 1979). However, in two of these cases, nitrate reductase activity was reduced to a greater extent than nitrite reductase activity (nitrate reductase activity was not measured in the oak study). As such, it is highly unlikely that inhibition of nitrite reductase was responsible for the nitrite accumulation required for nitrate reductase-dependent NO production. Distinguishing between the two plausible hypotheses would require experimental measurement of photosynthetic electron transport, chloroplast envelope transmembrane pH, and nitrite accumulation under the precise conditions used to assess drought and ABA-induction of NO production and stomatal closure.

Stomatal closure responses might potentially be integrated across leaf tissue by both NO-dependent feedforward and positive feedback control. Stomatal closure responses are known to be "patchy," with independent regulation of stomata in different areas of the leaf (Santrucek *et al.*, 2003). It is conceivable that NO synthesized in an area of the leaf with a high degree of stomatal closure would diffuse to other parts of the leaf and potentiate ABA action (feedforward control). Moreover, there is some indication from a study involving exposure of roots of wheat seedlings to high osmotic strength solutions that NO accumulation can lead to ABA biosynthesis (Xing *et al.*, 2004). If NO-induced ABA biosynthesis indeed occurs, it allows for positive feedback amplification of signaling leading to stomatal closure. Research into the effect of disrupting NO accumulation on the patchiness and whole leaf dynamics of the stomatal closure response would be highly merited.

2. Role of NO in Response to Wounding

Nitric oxide has been suggested to function in negative regulation of plant responses to wounding (Garcês et al., 2001; Huang et al., 2004; Orozco-Cardenas and Ryan, 2002). In both tomato and *Arabidopsis*, wounding leads to production of jasmonate, and jasmonate accumulation leads to large-scale changes in gene expression. Expression increases of four tomato wounding/jasmonate-responsive protease inhibitor genes were blocked by treatment with SNP (Orozco-Cardenas and Ryan, 2002). However, other wounding/jasmonate-responsive genes were unaffected. The action of NO was determined to be downstream of jasmonate production but upstream of reactive oxygen accumulation. Nitric oxide effects were shown not to be mediated by salicylic acid in that transgenic plants incapable of accumulating salicylic acid behaved identically to wild-type and no changes in salicylic acid accumulation occurred in the first 8-h postwounding.

By contrast, in *Arabidopsis*, application of NO led to the induction of JA biosynthesis genes (Huang et al., 2004). However, NO did not lead to JA biosynthesis in wild-type plants. Nitric oxide did induce salicylic acid biosynthesis. Salicylic acid is known to repress jasmonate biosynthesis in *Arabidopsis* (Peña-Cortés et al., 1993). Consistent with these results, NO application to transgenic plants incapable of accumulating salicylic acid did lead to JA biosynthesis (Huang et al., 2004). Thus, in *Arabidopsis*, NO can accomplish the same end as in tomato (inhibition of JA-responsive gene induction), although it does so by a different means.

In the *Arabidopsis* experiments (Garcês et al., 2001; Huang et al., 2004), NO production was shown to be a response to wounding. As discussed in Section II.B.1, aNOS appears responsible. However, wounding did not lead to NO production in tomato (Orozco-Cardenas and Ryan, 2002). This difference might be attributable to assay sensitivity as the *Arabidopsis* study used a sensitive dye-based assay and the tomato study used a hemoglobin absorbance-based assay applied to plant extracts. In the former, NO could be detected in single cells, whereas signal in the latter would have been diluted by nonresponding cells. A second possible explanation is the degree of wounding. In one *Arabidopsis* study (Garcês et al., 2001), wounding with forceps was described as "drastic," and in the second study (Huang et al., 2004), wounding was followed by making an epidermal peel. With the very thin *Arabidopsis* leaves, peeling likely induced major stresses that compounded the initial wounding (which involved crushing of 40% of the leaf area with a forceps). By contrast, the wounding protocol described for tomato (Orozco-Cardenas and Ryan, 2002) was less dramatic.

However, it is also possible that the signaling network differs between the two species. Life strategies of these two plants are very different. It is not clear that the identical wound response would be as advantageous to a short-lived species that responds to stress via flowering as to a more long-lived species. Wound signaling in Solanaceous plants is known to be different from that in *Arabidopsis* in many respects (León *et al.*, 2001). The explanation for the apparent species-specific differences must be considered unresolved at this point in time.

There is some evidence implicating NO in wound responses of additional species. Wounding of sweet potato led to NO accumulation. Sodium nitroprusside application delayed induction of a major wound-induced gene, and wound-induction of this gene was enhanced by application of an NOS inhibitor. Wounding led to H_2O_2 production and wound gene induction was reactive oxygen-dependent. Sodium nitroprusside application suppressed wound-induction of H_2O_2 accumulation (Jih *et al.*, 2003). In these results, sweet potato resembled the tomato response indicating similarity among Solanacea. The wound-induction of NO accumulation in the sweet potato system might bring up the argument that the failure to see the same in tomato may have been methodology-related. This point merits further investigation. Wounding of tobacco was reported to produce variable, inconsistent results with respect to NO production (Gould *et al.*, 2003), consistent with the idea that methodology-related artefacts can occur.

Signaling in ginseng appears different from the other paradigms. Oligogalacturonic acid is produced as a pectin breakdown product following wounding of most plant species. Oligogalacturonic acid application to ginseng cell cultures led to increased NOS-specific activity, NO accumulation, and NO-dependent transcriptional induction of secondary metabolite synthesis (Hu *et al.*, 2003a). This positive signaling role for NO contrasts with the negative regulatory role seen with other wounded plants.

A wounding-induced, jasmonate and oligogalacturonic acid-independent systemic signaling pathway that leads to induction of the *RNS1* gene, which encodes an RNase T_2 enzyme, has been described in *Arabidopsis* (LeBrasseur *et al.*, 2002). Sodium nitroprusside treatment of plants induced the *RNS1* gene (Parani *et al.*, 2004). Whether NO plays an essential role in this wound-signaling pathway is not yet known.

3. Role of NO in Sensing Other Abiotic Stresses

Hyperosmotic and salinity stresses have been reported to elicit NO production in tobacco (Gould *et al.*, 2003). Nitric oxide accumulation was rapid and first reported in chloroplasts suggesting the involvement of iNOS. Elevations of cytoplasmic calcium were identified as an NO-dependent response to hyperosmotic stress. The same study documented NO production in response to heat treatment (45 °C). Nitric oxide was also implicated in resistance to salinity stress in callus cultures from dune reed and swamp reed (Zhao *et al.*, 2004). Nitric oxide-dependent increases in plasma membrane H^+-ATPase protein accumulation and enzymatic activity were determined to be essential for these responses.

Nitric oxide has also been implicated in two other stress responses. UV-B induction of chalcone synthase gene expression was partially blocked by an NOS inhibitor or an NO scavenger (Mackerness *et al.*, 2001). Centrifugation of cell cultures or leaves from several species also induced NO production that led to cell death responses (Garcês *et al.*, 2001; Pedroso *et al.*, 2000a).

G. NO IN RESPONSE TO PATHOGENS

Much of the upsurge in interest in plant NO signaling stemmed from three 1998 publications implicating NO in plant responses to pathogens. These studies focused on the HR, which is a rapid leaf collapse elicited by high, titers of avirulent pathogens. This leaf collapse results from large-scale PCD that is accompanied by leakage of ions and water through damaged plasma membranes. Water loss due to transpiration results in the macroscopic leaf collapse (Goodman and Novacky, 1994). Nitric oxide production was documented in cultured soybean cells infected with avirulent *P. syringae* bacteria (Delledonne *et al.*, 1998). Nitric oxide synthase activity was shown to be induced by infection of tobacco with avirulent tobacco mosaic virus (TMV) or avirulent *Ralstonia solanacearum* bacteria under conditions designed to show a strong HR (Durner *et al.*, 1998; Huang and Knopp, 1998). Nitric oxide synthase inhibitors were shown to delay or reduce the extent of the HR to avirulent bacteria in tobacco and *Arabidopsis* (Delledonne *et al.*, 1998; Huang and Knopp, 1998). Involvement of components of NO responses in mammals (cGMP and cADPR) in the tobacco response to TMV was suggested, as was the NO-dependent induction of defense genes (Durner *et al.*, 1998).

Although the evidence for a role for NO in plant responses to pathogens is by now incontrovertible, precisely what NO does remains controversial. The interplay between NO and NO signaling and numerous aspects of plant physiology can be complex. The effects of the assays and procedures in use in plant pathology on plant physiology must thus be understood in detail in order to interpret experiments properly. As such, the most important caveats and potential artefacts are discussed prior to consideration of the relevant literature.

1. Caveats and Potential Artefacts Associated with Plant–Microbe Interaction Studies

Especially in studies of the HR, plant cell cultures have been employed in order to simplify analysis. Use of cell cultures allows simultaneous application of pharmacological agents or pathogens to an entire culture and tends to result in greater synchrony of responses relative to use of intact plants. This synchrony must be kept in mind in interpreting these studies. Simultaneous response of an entire culture can produce a higher concentration of an extracellular signal than would be seen *in vivo*. There is the potential to distort a minor response of intact plants into a major response that changes the interaction. Moreover, dynamics of multicellular responses can be dramatically altered.

These considerations have been shown to be important with regards to the role played by reactive oxygen in the HR. Initial studies with cell cultures infected with avirulent bacteria showed a two-phase oxidative burst that preceded cell death (Keppler *et al.*, 1989; Orlandi *et al.*, 1992). However, comparative kinetic analysis of reactive oxygen accumulation and HR progression as well as correlative studies involving *Arabidopsis* mutants affecting the HR proved the opposite was true *in vivo*. With both the cowpea HR to avirulent cowpea rust fungus (Heath, 1998) and the *Arabidopsis* HR to avirulent *P. syringae* (Shapiro and Zhang, 2001; Zhang *et al.*, 2004), high-level reactive oxygen accumulation occurred only following initial cell death events. Reactive oxygen accumulation was thus proven a consequence rather than a cause of hypersensitive cell death. Although low-level reactive oxygen accumulation does occur prior to PCD, it can be eliminated pharmacologically without effect on the HR (Grant *et al.*, 2000). By contrast, other major features of the HR that were initially established with cell culture studies (e.g., the role of salicylic acid as a "rheostat" controlling the level of cell death [Shirasu *et al.*, 1997] or the interplay between NO and H_2O_2 [Delledonne *et al.*, 2001]) were later confirmed in studies using intact plants (Shapiro and Zhang, 2001; Zeier *et al.*, 2004). The most conservative way to view cell culture studies is as hypothesis generation, to be accepted only following confirmation with intact plant studies.

The highly orchestrated multicellular nature of plant defense responses makes methodology issues important with intact plant studies too. Avirulent bacteria can be introduced into plant leaves either by dipping into a bacterial suspension that includes a surfactant, by vacuum infiltration, or by hand inoculation with a needleless syringe. Dipping will not deliver the consistent, uniform level of inoculum required for signaling studies. However, certain aspects of the response to bacteria (e.g., the role of coronatine in bacterial entry through stomata; Mittal and Davis, 1995) can only be observed with this method. Vacuum infiltration produces an artifactual burst of reactive oxygen accumulation as a direct consequence of anoxia followed by rapid air reperfusion that will cause salicylic acid production and other defense responses (Shapiro and Zhang, 2001). These factors must be considered if methods other than hand inoculation are used to introduce bacteria.

Level of inoculum is also important. Valid assay of the HR requires that the level of inoculum not exceed that from which no (or at least minimal) response is elicited by the near-isogenic control strain that lacks the avirulence gene employed. For the most commonly used bacterial pathogen, *P. syringae* pv. tomato DC3000 and strains derived from this parent, this level has been established to be $2 \times \times 10^7$ bacteria ml^{-1} (typically approximated as $OD_{600} = 0.02$; Kunkel *et al.*, 1993). This level of the virulent strain will not elicit NO, H_2O_2, or salicylic acid accumulation, PCD or ion leakage for 8–12 h postinfection (Zhang et al., 2003, 2004). Observation of any of these responses to the virulent pathogen can invalidate an experiment, in that it would no longer be certain that the HR, per se, was being assessed. However, some of these responses will be seen at later time points, implying that early time points are generally most informative. Higher levels of inoculum are often acceptable with nonpathogenic strains that cannot replicate in planta that are used merely to deliver avr gene products into plant cells (Century et al., 1995; Kunkel et al., 1993; Shapiro, 2000).

Viruses are most often introduced by rubbing leaves with an abrasive compound. As such, primary inoculated leaves experience a severe wound response simultaneous with the effects of the virus. Negative control experiments are limited in scope in having potentially synergistic effects between wound signaling and virus responses. Virus infection of secondary leaves can be studied with systemic infections; however, asynchronous long-distance transport of the virions can complicate the determination of relative timing of signaling events following secondary infections.

In the virus model system most often employed in studies of defense responses (TMV infection of tobacco) these limitations have been circumvented. The HR of tobacco carrying the N gene to avirulent TMV is temperature-sensitive. Systemic infection is allowed to occur at the higher temperature. Shifting plants to a lower temperature then produces a massive, synchronous HR that can be studied in the systemic leaves. Crucially, this approach is subject to the same criticism as that applied to bacterial infection of cultured cells. Artificial synchronization of the response could potentially lead to cell-to-cell communication that would not occur under natural settings. An oxidative burst precedes and contributes causally to hypersensitive cell death with this system (Doke and Ohashi, 1988). Would causality be reversed in a natural viral infection? This question has not been addressed. As such, the possibility that the viral HR studied in laboratory assays is something very different from the bacterial HR must be considered. An exaggeration in apparent importance of signaling events that are better considered responses to H_2O_2 accumulation (e.g., MAP kinase activation; Kovtun et al., 2000) relative to that seen in the bacterial HR might be anticipated in these studies.

Fungal infections typically develop over a slow time course relative to other pathogens. As such, infected tissue will simultaneously contain cells in various stages of infection. This situation is an advantage for responses that can be measured using real-time microscopy approaches that allow visualization of single cells. However, it makes interpretation of signaling data, that cannot resolve individual cell contributions, very difficult.

Additional complications are seen in assays of pathogen replication. These assays are the basis of statements implicating NO in various disease or resistance responses. With bacterial pathogens, the assay usually involves plating of macerated leaf samples taken over a time course to assess bacterial growth *in planta* indirectly based on colony counts. This method can provide semiquantitative assessments with good reproducibility. However, misreporting of data from such studies is very common in the literature. Data from separate experiments cannot be averaged (as they can be in growth measurements of mid-log phase *E. coli* growing in a chemostat, for example), as plant physiology is not sufficiently reproducible with plants grown on different occasions, even in the same growth chamber. As such, a good practice is to repeat the identical experiment several times with plants grown on different occasions and only accept conclusions based on qualitatively similar results from multiple biological replicates. Distribution of the error in bacterial growth measurements tends to be log-normal (Manulis *et al.*, 1998) and should thus be reported with a log rather than a linear scale on the *y*-axis. The proper statistical measure to use for error bars is standard deviation (the incorrect and unacceptably common use of standard error understates the true variance; Rogers and Ausubel, 1997). ANOVA analysis, *t*-tests, or other accepted statistical procedures should be used to determine significance of differences between treatments or different plant lines (e.g., in studies using mutants or transgenic plants). The degree of variation seen in identical replicate experiments makes it difficult to assign much importance to differences in growth measurements of less than 10-fold, even when performed perfectly and replicated several times. By contrast, assessment of viral replication can be quantitative and can employ multiple assessment criteria (e.g., level of viral RNA, level of capsid protein expression, lesion diameter or number, etc.). Assessment of fungal pathogen growth can be done quantitatively (e.g., via spore counts) but is more often qualitative. Having stated these guidelines, critique of individual experiments from the current and future literature in this field will largely be left to the reader.

2. Evidence for Functions of NO in Plant–Microbe Interactions
a. Role of NO in the Hypersensitive Response

Work in several different laboratories on diverse aspects of signaling associated with the HR has advanced to the point that a speculative hypothesis can be constructed to connect the signaling pathway in a fashion consistent with all available data (Fig. 3). The evidence for activation of an apoplastic NOS (aNOS) by lipid-derived signal(s) produced concurrently with plasma membrane breakdown in cells directly responding to an avr gene product delivered by avirulent pathogen is presented in Section II.B.2. The identity of the lipid signal and the enzyme(s) responsible for its production are unknown. However, in support of this model, numerous enzymes known to act on plasma membrane lipids are transcriptionally activated very rapidly following infection with avirulent bacteria (de Torres Zabela *et al.*, 2002). In the diagram, the putative enzyme(s) is shown being activated by

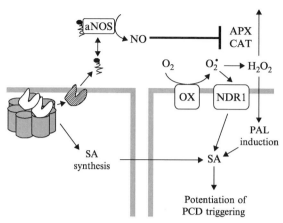

FIGURE 3. Proposed signaling pathway responsible for NO-mediated potentiation of the hypersensitive response. SA, salicylic acid; aNOS, apoplastic nitric oxide synthase; APX, ascorbate peroxidase; CAT, catalase; OX, NADPH oxidase; NDR1, gene product encoded in *Arabidopsis* by At3g20600; PAL, phenylalanine ammonia lyase; PCD, programmed cell death. The cylinders represent an R gene product and are arranged in a heptamer by analogy to other Apaf-1 family members. The "pacmen" represent a putative lipid modification enzyme(s) with the hatched pacman representing the activated state. The black octagons with "tails" represent a putative lipid signaling molecule derived from plasma membrane lipids.

R protein-induced dimerization on the basis of homology of *R* genes with Apaf-1 family members that activate caspases via proximity-induced dimerization (Acehan *et al.*, 2002; Martinon and Tschopp, 2004; van der Biezen and Jones, 1998). Cell-to-cell diffusion of NO in the apoplast has been documented (Zhang *et al.*, 2003).

Evidence has been presented for synergistic effects of NO and H_2O_2 in the induction of PCD in soybean cell cultures (Delledonne *et al.*, 2001). As NO does not react with H_2O_2 at an appreciable rate (Brunelli *et al.*, 1995), these interactions could be through separate effects on related signaling pathways. However, a more likely (and not mutually exclusive) explanation stems from the documented ability of NO to inhibit two major plant H_2O_2 scavenging enzymes, catalase and ascorbate peroxidase (Clark *et al.*, 2000). Consistent with these results, a transgenic line engineered to interfere with NO accumulation in *Arabidopsis* showed increased rates of H_2O_2 degradation *in vivo* (Zeier *et al.*, 2004). During the HR, the resulting decreases in H_2O_2 accumulation were associated with decreased PAL transcription, decreased salicylic acid accumulation, and reduced severity of the macroscopic HR.

These observations support a unified model (Fig. 3) in which all known cell-to-cell signaling relevant to the HR operates via increases in salicylic acid accumulation. As described in Section IV.G.1, salicylic acid levels determine the extent of PCD seen by potentiating cell death triggering in neighboring cells. Nitric oxide accumulation is not essential for the HR, in that PCD occurs in the presence of an NOS inhibitor or an NO scavenger that eliminate detectable NO accumulation (Zhang *et al.*, 2003). The role of NO instead, as with salicylic acid, appears to be in potentiating cell death. Most cell death events were merely delayed ~1 h by interfering with NO production or accumulation (Zhang *et al.*, 2003). An *NDR1*-dependent pathway by which superoxide production leads to salicylic acid accumulation has been described (Agrawal *et al.*, 2004; Shapiro and Zhang, 2001). Kinetic analysis has established that salicylic acid is also produced in the first wave of cells to respond to avirulent pathogen prior to their undergoing PCD (Zhang *et al.*, 2004). This first wave of salicylic acid biosynthesis was largely abolished by the *npr1* mutation. It was argued that the more rapid PCD seen in this mutant eliminated this first wave of responding cells before they could produce significant levels of salicylic acid that would have influenced other cells.

As illustrated in Fig. 3, all three proposed pathways for potentiation of cell death triggering converge upon salicylic acid accumulation. Further support for this hypothesis will likely await the development of a microscopy-based approach to quantify changes to salicylic acid levels in individual cells of intact plant leaves during the HR. Identification of the target(s) of SA action or, indeed, any component of the cell death machinery responsible for the HR could also likely underlie tests of this hypothesis. However, the hypothesis does offer a consistent explanation for all extant observations.

Nitric oxide also appears to play a role in the HR to avirulent virus. Transgenic tobacco lines engineered to prevent accumulation of NO showed fewer hypersensitive lesions in response to infection with tobacco necrosis virus (Seregélyes *et al.*, 2003).

b. Does NO Influence Susceptibility to Bacterial Disease?

Although the role of NO in the HR is better characterized than virtually any other role of NO in plant biology, it has been much more difficult to determine whether NO contributes to limiting pathogen growth. Very minor, if any, increases in the growth of avirulent bacteria were seen in *Arabidopsis* treated with an NOS inhibitor or an NO scavenger (Delledonne *et al.*, 1998). However, similar reagents have been shown to be fully effective *in vivo* for only 4h (Zhang *et al.*, 2003) and the first time point taken in this study was 1 day postinoculation. As such, these largely negative results cannot be taken as definitive. Virus-induced gene silencing (VIGS) was subsequently used to test the role of pathogen-inducible NOS activity in basal and *Pto* gene-mediated resistance of tomato to *P. syringae* (Chandok *et al.*, 2004). Silenced lines of both susceptible (Rio Grande-prf3) and resistant (Rio Grande-PtoR) tomato hosts showed 10-fold to 15-fold increases in capacity to support bacterial multiplication. These lines also showed reduction in NOS activity to basal levels seen in the absence of bacterial infection. The argument was made that these results implicate iNOS in basal resistance to virulent bacteria and perhaps suggested a modest role in resistance to bacteria carrying the *avr* gene *avrPto*.

However, the role for iNOS may have been highly indirect. The silenced areas of the plants displayed a mild bleaching, likely indicative of chloroplast dysfunction. Chloroplast function has been implicated in multiple processes related to disease resistance (Abbink *et al.*, 2002; Gray *et al.*, 2002; Guttman *et al.*, 2002; Kachroo *et al.*, 2001; Mach *et al.*, 2001; Nawrath *et al.*, 2002; Slaymaker *et al.*, 2002; Wildermuth *et al.*, 2001). It is possible that NO was required to counteract the effects of oxygen radicals that built up in its absence. As such, confirmatory experiments that use alternative technology that avoids the 3-week incubation period required for VIGS will be necessary before a conclusion of a specific role for iNOS in disease resistance can be accepted. These experiments could be based upon transgenic expression of the *iNOS* gene from a regulated promoter in an *Arabidopsis* line carrying knockout mutation(s) in the *iNOS* homolog(s). No such experiments have yet been reported.

Increased *in planta* growth of virulent *P. syringae* and increased severity of disease symptoms have been reported in a comparison of *nos1* mutants plants with wild-type *Arabidopsis* (Zeidler *et al.*, 2004). However, growth of the bacteria was unusually low in the wild-type plants and disease symptoms were very light. Two alternative explanations seem plausible. Bacteria were inoculated in this study by spraying a suspension onto the leaves. This method is designed to give a very low initial inoculum. It is possible that the role of NO is in defense against the early events of pathogenesis that involve bacterial survival and entry through stomata. This stage is known to have distinct genetic requirements from later stages of pathogenesis (Mittal and Davis, 1995). Alternatively, the plants could have been manifesting age-related resistance (Kus *et al.*, 2002). This explanation would implicate NO in age-related resistance rather than in basal resistance to virulent *P. syringae*. These points require clarification by further studies.

c. Does NO Function in Plant–Fungus Interactions?

Nitric oxide has been implicated in resistance responses to two fungi: stripe rust in wheat (caused by *Puccinia striiformis* Westend) and crown rust in oat (caused by *Puccinia coronata* f. sp. *avenae*). Electron paramagnetic spectroscopy measurements were used to document a two-phase induction of NO biosynthetic activity in stripe rust infected wheat (Guo *et al.*, 2004). The earlier burst occurred in a resistant but not a susceptible cultivar. Application of an NO donor markedly improved resistance of the latter cultivar to stripe rust disease. Induction of PAL enzyme activity was seen in both cultivars (with greater activity at one time point with the resistant cultivar) and in response to the NO donor. However, a functional role for PAL activity was not determined.

In the oat plants, crown rust infection led to the accumulation of NO and reactive oxygen (Tada et al., 2004). Oat cells directly penetrated by the fungus collapsed and displayed intense autofluorescence. Neither an NO donor nor an NO scavenger affected this response (nor did addition of superoxide dismutase enzyme). As such, the authors concluded that neither NO nor reactive oxygen play a major role in the death of directly penetrated cells. However, adjacent cells underwent an apoptosis-like response including chromatin condensation and DNA laddering. Application of the NO donor, NO scavenger, or superoxide dismutase suppressed the apoptotic responses. The authors concluded that an interaction between superoxide and NO, likely involving peroxynitrite, was essential for "apoptosis," and that an optimal balance of superoxide and NO was necessary for the response.

However, an alternative explanation of these results is possible. The dye-based assay used for detection of NO in this study requires NO autoxidation to NO_2 as the first step of the reaction. As such, the increased "NO" staining could have been explained by an increase in NO_2 production relative to other fates of NO. Nitric oxide is known to react with ferryl ions to quench Fenton chemistry, with NO_2 as a product of the reaction (see Section III.D). As "massive" accumulation of H_2O_2 was reported upon treatment with SOD, the increased "NO" staining could have resulted from an increase in Fenton chemistry and its quenching by NO. The resulting explanation for the suppression of heterochromatin condensation by SOD would be that superoxide, rather than peroxynitrite, was required for PCD responses. This explanation is more consistent with what is known about plant PCD. Similar logic would explain why an NO donor suppressed heterochromatin condensation.

This study further established that the NO scavenger increased H_2O_2 accumulation and the NO donor suppressed it. These results suggest ascorbate peroxidase and/or catalase as possible targets of NO action. These results also suggest a possible alternative explanation for the suppression of heterochromatin condensation by the NO scavenger. In *Arabidopsis*, NO accumulation results in salicylic acid biosynthesis (see Section IV.B.2.a). High-level salicylic acid accumulation has been shown to suppress the *Arabidopsis* HR (Devadas and Raina, 2002). Salicylic acid accumulation was not measured in the oat study; however, it may have been responsible for the suppression of "apoptosis" seen if signaling resembled that of *Arabidopsis*. At present, the role of NO in plant responses to avirulent fungi thus appears very similar to that in responses to avirulent bacteria. Nevertheless, both studies of NO in plant–fungus interactions are recent, and as such, it remains to be seen whether novel features will emerge.

d. Other Observations of NO Function with Relevance for Plant-Microbe Interactions

Nitric oxide has been implicated in response to pathogen-derived elicitor molecules. The response of tobacco epidermal cells to cryptogein, an elicitor derived from the fungal pathogen *Phytophthora cryptogea*, was described in Section II.B.1. *Arabidopsis* cells and leaf epidermal peels responded to application of bacterial cell wall components (crude lipopolysaccharide, lipid A, or lipoteichoic acid) with rapid, *NOS1*-dependent production of NO (Zeidler *et al.*, 2004). A race-specific elicitor from rice blast fungus elicited NO production in rice cell cultures as well as NO-dependent induction of gene expression and PCD (Hu *et al.*, 2003b).

Nitric oxide has also been implicated in systemic acquired resistance in tobacco (Song and Goodman, 2001). Treatment with an NO donor reduced the size of TMV lesions (diagnostic of SAR) and NO effects were dependent upon salicylic acid accumulation. Additionally, NOS inhibitors or an NO scavenger impaired the acquired resistance response displayed in response to treatment with salicylic acid.

Salicylic acid has also been shown to induce NO production in bean stomatal guard cells leading to guard cell closure (Xin *et al.*, 2003). Nitric oxide production and guard cell closure were blocked by an NOS inhibitor or an NO scavenger. Inhibitors of cGMP and cADPR production blocked SA (or NO)-induced guard cell closure, further supporting the contention that SA acted through NO. Guard cell closure during pathogenesis would inhibit photosynthesis and this inhibition would likely affect assays of pathogen replication.

V. CONCLUSIONS

The analysis presented here indicates that some truisms from the mammalian literature routinely parroted about NO signaling are very unlikely to apply to plants. Important aspects of NO physiology in mammals (e.g., *S*-nitrosylation or peroxynitrite formation) are unlikely to be important. At present, there are only a handful of validated, physiologically relevant direct targets of NO in plants, and they are all metalloproteins or a metal complex. As such, in studies of plant biology, these "suspects" should be considered first:

1. *Apoplastic ascorbate peroxidases and catalases.* Strong evidence that these enzymes are the targets of NO produced during the *Arabidopsis* HR has been presented. Nitric oxide will inhibit these enzymes and decrease the rate of H_2O_2 degradation; interfering with NO accumulation will do the reverse (Zeier *et al.*, 2004). However, the opposite results were seen

with the HR in tobacco (Seregélyes *et al.*, 2003). Could this be related to the observation that wounding of Solanaceous plants led to NO-inhibitable H_2O_2 accumulation (Jih *et al.*, 2003; Orozco-Cardenas and Ryan, 2002)? With the versatility in NO chemistry, it is certainly conceivable that an ascorbate peroxidase or a catalase could exist in Solanacea that is stimulated rather than inhibited by NO. Especially if the speculation presented in this paper that aNOS is responsible for both HR-related and wounding-induced NO synthesis (see Section II.B.1) proves correct, this hypothesis for the NO target in the wound response would explain the striking differences noted between *Arabidopsis* and Solanacea and merits investigation.

2. *The classic cGMP-dependent NO response pathway shared with mammals.* Although the only identified soluble guanyl cyclase in plants is not stimulated by NO *in vitro* (Ludidi and Gehring, 2003), pharmacological approaches have implicated this pathway in many plant processes. The core elements of this pathway, cGMP, cADPR, and release of calcium from internal stores to raise cytoplasmic concentrations, have been implicated in numerous aspects of plant development and stress responses. It is likely that either the correct guanyl cyclase was not identified or the correct *in vitro* conditions for assay were not obtained to demonstrate NO-responsiveness.

3. *Chloroplast electron transport (photosystem II nonheme iron center and possibly other targets) and ferryl ions produced by Fenton chemistry in the chloroplast.*

4. *Mitochondrial cytochrome oxidase.*

It is likely that peroxisomal targets will also be identified, given the existence of a peroxisome-specific NOS.

With so few direct targets and so many aspects of plant physiology that are affected, it is not likely that many of the incomplete "stories" presented in this manuscript will have straightforward, simple endings. The easily cited titles that fill this literature today of the form, "NO does X," will likely give way to complex discussions of kinetic rate constants, transport phenomena, and signaling cross-talk. Evaluation of this literature will require the nuanced approach advocated here, with full attention given to artefacts, caveats, and alternative explanations. Molecular genetic approaches will be misleading as often as they are essential, in that gross perturbations of photosynthesis, respiration, cellular redox status, or homeostasis of cellular calcium and iron are likely to be highly pleiotropic. This author recommends approaching NO biology with careful consideration of both the talents possessed and the roles typically assigned to our "character actor." In this fashion, we can best appreciate the contextual nature of "meaning" that is regained with continued examination. Nitric oxide deserves considerably more than fifteen minutes of fame.

ACKNOWLEDGMENTS

The author thanks Dr. John Boyer (University of Delaware, Lewes, DE) for useful discussions and Dr. Paul Bethke (University of California, Berkeley, CA) for critical reading of the manuscript.

REFERENCES

Abbink, T. E. M., Peart, J. R., Mos, T. N. M., Baulcombe, D. C., Bol, J. F., and Linthorst, H. J. M. (2002). Silencing of a gene encoding a protein component of the oxygen-evolving complex of photosystem II enhances virus replication in plants. *Virology* **295**, 307–319.

Acehan, D., Jiang, X., Morgan, D. G., Heuser, J. E., Wang, X., and Akey, C. W. (2002). Three-dimensional structure of the apoptosome: Implications for assembly, procaspase-9 binding, and activation. *Mol. Cell* **9**, 423–432.

Achkor, H., Díaz, M., Fernández, M. R., Biosca, J. A., Parés, X., and Martínez, M. C. (2003). Enhanced formaldehyde detoxification by overexpression of glutathione-dependent formaldehyde dehydrogenase from *Arabidopsis*. *Plant Physiol.* **132**, 2248–2255.

Agrawal, V., Zhang, C., Shapiro, A. D., and Dhurjati, P. S. (2004). A dynamic mathematical model to clarify signaling circuitry underlying programmed cell death control in *Arabidopsis* disease resistance. *Biotechnol. Prog.* **20**, 426–442.

Akaike, T., and Maeda, H. (2000). Pathophysiological effects of high-output production of nitric oxide. *In* "Nitric Oxide: Biology and Pathobiology" (L. J. Ignarro, Ed.), pp. 733–745. Academic Press, San Diego.

Alamillo, J. M., and García-Olmedo (2001). Effects of urate, a natural inhibitor of peroxynitrite-mediated toxicity, in the response of *Arabidopsis thaliana* to the bacterial pathogen *Pseudomonas syringae*. *Plant J.* **25**, 529–540.

Arnelle, D. R., and Stamler, J. S. (1995). NO^+, $NO^.$, and NO^- donation by *S*-nitrosothiols: Implications for regulation of physiological functions by *S*-nitrosylation and acceleration of disulfide formation. *Arch. Biochem. Biophys.* **318**, 279–285.

Averill, B. A. (1996). Dissimilatory nitrite and nitric oxide reductases. *Chem. Rev.* **96**, 2951–2964.

Barber, M. J., and Kay, C. J. (1996). Superoxide production during reduction of molecular oxygen by assimilatory nitrate reductase. *Arch. Biochem. Biophys.* **326**, 227–232.

Barroso, J. B., Corpas, F. J., Carreras, A., Sandalio, L. M., Valderrama, R., Palma, J. M., Lupiáñez, J. A., and del Rio, L. A. (1999). Localization of nitric oxide synthase in plant peroxisomes. *J. Biol. Chem.* **274**, 36729–36733.

Becker, R., Fritz, E., and Manteuffel, R. (1995). Subcellular localization and characterization of excessive iron in the nicotianamine-less tomato mutant *chloronerva*. *Plant Physiol.* **108**, 269–275.

Beligni, M. V., Fath, A., Bethke, P. C., Lamattina, L., and Jones, R. L. (2002). Nitric oxide acts as an antioxidant and delays programmed cell death in barley aleurone layers. *Plant Physiol.* **129**, 1642–1650.

Beligni, M. V., and Lamattina, L. (2000). Nitric oxide stimulates seed germination and de-etiolation, and inhibits hypocotyl elongation, three light-inducible responses in plants. *Planta* **210**, 215–221.

Beligni, M. V., and Lamattina, L. (2002). Nitric oxide interferes with plant photo-oxidative stress by detoxifying reactive oxygen species. *Plant Cell Environ.* **25**, 737–748.

Bethke, P. C., Badger, M. R., and Jones, R. L. (2004a). Apoplastic synthesis of nitric oxide by plant tissues. *Plant Cell* **16**, 332–341.

Bethke, P. C., Gubler, F., Jacobsen, J. V., and Jones, R. L. (2004b). Dormancy of *Arabidopsis* seeds and barley grains can be broken by nitric oxide. *Planta* **219**, 847–855.

Bonamore, A., Gentili, P., Ilari, A., Schinina, M. E., and Boffi, A. (2003). *Escherichia coli* flavohemoglobin is an efficient alkylhydroperoxide reductase. *J. Biol. Chem.* **278**, 22272–22277.

Botrel, A., Magné, C., and Kaiser, W. M. (1996). Nitrate reduction, nitrite reduction and ammonia assimilation in barley roots in response to anoxia. *Plant Physiol. Biochem.* **34**, 645–652.

Brault, M., Amiar, Z., Pennarun, A.-M., Monestiez, M., Zhang, Z., Cornel, D., Dellis, O., Knight, H., Bouteau, F., and Rona, J.-P. (2004). Plasma membrane depolarization induced by abscisic acid in *Arabidopsis* suspension cells involves reduction of proton pumping in addition to anion channel activation, which are both Ca^{2+} dependent. *Plant Physiol.* **135**, 231–243.

Brown, G. C. (1995). Reversible binding and inhibition of catalase by nitric oxide. *Eur. J. Biochem.* **232**, 188–191.

Brüggemann, W., Maas-Kantel, K., and Moog, P. R. (1993). Iron uptake by leaf mesophyll cells: The role of the plasma membrane-bound ferric-chelate reductase. *Planta* **190**, 151–155.

Brunelli, L., Crow, J. P., and Beckman, J. S. (1995). The comparative toxicity of nitric oxide and peroxynitrite to *Escherichia coli*. *Arch. Biochem. Biophys.* **316**, 327–334.

Burns, L. C., Stevens, R. J., and Laughlin, R. J. (1996). Production of nitrite in soil by simultaneous nitrification and denitrification. *Soil. Biol. Biochem.* **28**, 609–616.

Cakmak, I., van de Wetering, D. A. M., Marschner, H., and Bienfait, H. F. (1987). Involvement of superoxide radical in extracellular ferric reduction by iron-deficient bean roots. *Plant Physiol.* **85**, 310–314.

Capone, R., Tiwari, B. S., and Levine, A. (2004). Rapid transmission of oxidative and nitrosative stress signals from roots to shoots in *Arabidopsis*. *Plant Physiol. Biochem.* **42**, 425–428.

Caro, A., and Puntarulo, S. (1999). Nitric oxide generation by soybean embryonic axes. Possible effect on mitochondrial function. *Free Radic. Res.* **31**, S205–S212.

Cataldo, D. A., McFadden, K. M., Garland, T. R., and Wildung, R. E. (1988). Organic constituents and complexation of nickel (II), iron (III), cadmium (II) and plutonium (IV) in soybean xylem exudates. *Plant Physiol.* **86**, 734–739.

Century, K. S., Holub, E. B., and Staskawicz, B. J. (1995). *NDR1*, a locus of *Arabidopsis thaliana* that is required for disease resistance to both a bacterial and a fungal pathogen. *Proc. Natl. Acad. Sci. USA* **92**, 6597–6601.

Chandok, M. R., Ekengren, S. K., Martin, G. B., and Klessig, D. F. (2004). Suppression of pathogen-inducible NO synthase (iNOS) activity in tomato increases susceptibility to *Pseudomonas syringae*. *Proc. Natl. Acad. Sci. USA* **101**, 8239–8244.

Chandok, M. R., Ytterberg, A. J., van Wijk, K. J., and Klessig, D. F. (2003). The pathogen-inducible nitric oxide synthase (iNOS) in plants is a variant of the P protein of the glycine decarboxylase complex. *Cell* **113**, 469–482.

Clark, D., Durner, J., Navarre, D. A., and Klessig, D. F. (2000). Nitric oxide inhibition of tobacco catalase and ascorbate peroxidase. *Mol. Plant Microbe Interact.* **13**, 1380–1384.

Clarke, A., Desikan, R., Hurst, R. D., Hancock, J. T., and Neill, S. J. (2000). NO way back: Nitric oxide and programmed cell death in *Arabidopsis thaliana* suspension cultures. *Plant J.* **24**, 667–677.

Conrath, U., Amoroso, G., Köhle, H., and Sültemeyer, D. F. (2004). Non-invasive online detection of nitric oxide from plants and some other organisms by mass spectrometry. *Plant J.* **38**, 1015–1022.

Cooper, C. E. (1999). Nitric oxide and iron proteins. *Biochem. Biophys. Acta* **1141**, 290–309.

Corpas, F. J., Barroso, J. B., Carreras, A., Quirós, M., León, A. M., Romero-Puertas, M. C., Esteban, F. J., Valderrama, R., Palma, J. M., Sandalio, L. M., Gómez, M., and del Río, L. A. (2004). Cellular and subcellular localization of endogenous nitric oxide in young and senescent pea plants. *Plant Physiol.* **136**, 2722–2733.

Corpas, F. J., Barroso, J. B., and del Rio, L. A. (2001). Peroxisomes as a source of reactive oxygen species and nitric oxide signal molecules in plant cells. *Trends Plant Sci.* **6**, 145–150.

Correa-Aragunde, N., Graziano, M., and Lamattina, L. (2004). Nitric oxide plays a central role in determining lateral root development in tomato. *Planta* **218**, 900–905.

Cousson, A. (2004). Pharmacological evidence for a putative mediation of cyclic GMP and cytosolic Ca^{2+} within auxin-induced de novo root formation in the monocot plant *Commelina communis* (L.). *Plant Sci.* **166**, 1117–1124.

Crawford, M. J., and Goldberg, D. E. (1998). Role for the Salmonella flavohemoglobin in protection from nitric oxide. *J. Biol. Chem.* **273**, 12543–12547.

de Gara, L., de Pinto, M. C., and Tommasi, F. (2003). The antioxidant systems vis-à-vis reactive oxygen species during plant–pathogen interactions. *Plant Physiol. Biochem.* **41**, 863–870.

de la Guardia, M. D., and Alcántara, E. (1996). Ferric chelate reduction by sunflower (*Helianthus annuus* L.) leaves: Influence of light, oxygen, iron-deficiency and leaf age. *J. Exp. Bot.* **47**, 669–675.

de Torres Zabela, M., Fernandez-Delmond, I., Niittyla, T., Sanchez, P., and Grant, M. (2002). Differential expression of genes encoding *Arabidopsis* phospholipases after challenge with virulent or avirulent Pseudomonas isolates. *Mol. Plant Microbe Interact.* **15**, 808–816.

Dean, J. V., and Harper, J. E. (1986). Nitric oxide and nitrous oxide production by soybean and winged bean during *in vivo* nitrate reductase assay. *Plant Physiol.* **82**, 718–732.

Delledonne, M., Xia, Y., Dixon, R. A., and Lamb, C. (1998). Nitric oxide functions as a signal in plant disease resistance. *Nature* **394**, 585–588.

Delledonne, M., Zeier, J., Marocco, A., and Lamb, C. (2001). Signal interactions between nitric oxide and reactive oxygen intermediates in the plant hypersensitive disease resistance response. *Proc. Natl. Acad. Sci. USA* **98**, 13454–13459.

Desikan, R., Cheung, M.-K., Bright, J., Henson, D., Hancock, J. T., and Neill, S. J. (2004). ABA, hydrogen peroxide and nitric oxide signalling in stomatal guard cells. *J. Exp. Bot.* **55**, 205–212.

Desikan, R., Griffiths, R., Hancock, J., and Neill, S. (2002). A new role for an old enzyme: Nitrate reductase-mediated nitric oxide generation is required for abscisic acid-induced stomatal closure in *Arabidopsis thaliana*. *Proc. Natl. Acad. Sci. USA* **99**, 16314–16318.

Desikan, R., Mackerness, S. A. H., Hancock, J. T., and Neill, S. J. (2001). Regulation of the *Arabidopsis* transcriptome by oxidative stress. *Plant Physiol.* **127**, 159–172.

Devadas, S. K., and Raina, R. (2002). Preexisting systemic acquired resistance suppresses hypersensitive response-associated cell death in *Arabidopsis hrl1* mutant. *Plant Physiol.* **128**, 1234–1244.

Díaz, M., Achkor, H., Titarenko, E., and Martínez, M. C. (2003). The gene encoding glutathione-dependent formaldehyde dehydrogenase/GSNO reductase is responsive to wounding, jasmonic acid and salicylic acid. *FEBS Lett.* **543**, 136–139.

Doke, N., and Ohashi, Y. (1988). Involvement of an O_2^- generating system in the induction of necrotic lesions on tobacco leaves infected with tobacco mosaic virus. *Physiol. Mol. Plant Pathol.* **32**, 163–175.

Dordos, C., Hasinoff, B. B., Igamberdiev, A. U., Manac'h, N., Rivoal, J., and Hill, R. D. (2003). Expression of a stress-induced hemoglobin affects NO levels produced by alfalfa root cultures under hypoxic stress. *Plant J.* **35**, 763–770.

Dordos, C., Hasinoff, B. B., Rivoal, J., and Hill, R. D. (2004). Class-1 hemoglobins, nitrate and NO levels in anoxic maize cell suspension cultures. *Planta* **219**, 66–72.

Durner, J., Wendehenne, D., and Klessig, D. F. (1998). Defense gene induction in tobacco by nitric oxide, cyclic GMP, and cyclic ADP-ribose. *Proc. Natl. Acad. Sci. USA* **95**, 10328–10333.

El-Shintinawy, F., and Govindjee (1990). Bicarbonate effects in leaf discs from spinach. *Photosynth. Res.* **24,** 189–200.

Emanuelsson, O., Nielsen, H., Brunak, S., and von Heijne, G. (2000). Predicting subcellular localization of proteins based on their N-terminal amino acid sequence. *J. Mol. Biol.* **300,** 1005–1016.

Fath, A., Bethke, P., Beligni, V., and Jones, R. (2002). Active oxygen and cell death in cereal aleurone cells. *J. Exp. Bot.* **53,** 1273–1282.

Feelisch, M. (1998). The use of nitric oxide donors in pharmacological studies. *Naunyn-Schmiedeberg's Arch. Pharmacol.* **358,** 113–122.

Feelisch, M., and Stamler, J. S. (1996a). Donors of nitrogen oxides. *In* "Methods in Nitric Oxide Research" (M. Feelisch and J. S. Stamler, Eds.), pp. 71–115. Wiley, Chichester, England.

Feelisch, M., and Stamler, J. S. (1996b). Measurement of NO-related activities—which assay for which purpose? *In* "Methods in Nitric Oxide Research" (M. Feelisch and J. S. Stamler, Eds.), pp. 303–307. Wiley, Chichester, England.

Ferrari, T. E., and Varner, J. E. (1970). Control of nitrate reductase activity in barley aleurone layers. *Proc. Natl. Acad. Sci. USA* **65,** 729–736.

Filipe, P., Haigle, J., Freitas, J., Fernandes, A., Mazière, J.-C., Mazière, C., Santus, R., and Morlière, P. (2002). Anti- and pro-oxidant effects of urate in copper-induced low-density lipoprotein oxidation. *Eur. J. Biochem.* **269,** 5474–5483.

Foissner, I., Wendehenne, D., Langebartels, C., and Durner, J. (2000). *In vivo* imaging of an elicitor-induced nitric oxide burst in tobacco. *Plant J.* **23,** 817–824.

Frey, A. D., Farrés, J., Bollinger, C. J. T., and Kallio, P. T. (2002). Bacterial hemoglobins and flavohemoglobins for alleviation of nitrosative stress in *Escherichia coli*. *Appl. Environ. Microbiol.* **68,** 4835–4840.

Frey, A. D., Oberle, B. T., Farres, J., and Kallio, P. T. (2004). Expression of *Vitreoscilla* haemoglobin in tobacco cell cultures relieves nitrosative stress *in vivo* and protects from NO *in vitro*. *Plant Biotech. J.* **2,** 221–231.

Galván, A., Rexach, J., Mariscal, V., and Fernández, E. (2002). Nitrite transport to the chloroplast in *Chlamydomonas reinhardtii*: Molecular evidence for a regulated process. *J. Exp. Bot.* **53,** 845–853.

Garab, G. Y., Sanchez Burgos, A. A., Zimányi, L., and Faludi-Dániel, Á. (1983). Effect of CO_2 on the energization of thylakoids in leaves of higher plants. *FEBS Lett.* **154,** 323–327.

Garcês, H., Durzan, D., and Pedroso, M. C. (2001). Mechanical stress elicits nitric oxide formation and DNA fragmentation in *Arabidopsis thaliana*. *Ann. Bot.* **87,** 567–574.

García-Mata, C., Gay, R., Sokolovski, S., Hills, A., Lamattina, L., and Blatt, M. R. (2003). Nitric oxide regulates K^+ and Cl^- channels in guard cells through a subset of abscisic acid-evoked signaling pathways. *Proc. Natl. Acad. Sci. USA* **100,** 11116–11121.

García-Mata, C., and Lamattina, L. (2001). Nitric oxide induces stomatal closure and enhances the adaptive plant responses against drought stress. *Plant Physiol.* **126,** 1196–1204.

Gardner, P. R., Gardner, A. M., Martin, L. A., and Salzman, A. L. (1998). Nitric oxide dioxygenase: An enzymic function for flavohemoglobin. *Proc. Natl. Acad. Sci. USA* **95,** 10378–10383.

Goodman, R. N., and Novacky, A. J. (1994). "The Hypersensitive Response in Plants to Pathogens: A Resistance Phenomenon" APS Press, St. Paul, MN..

Gorren, A. C. F., List, B. M., Schrammel, A., Pitters, E., Hemmens, B., Werner, E. R., Schmidt, K., and Mayer, B. (1996). Tetrahydrobiopterin-free neuronal nitric oxide synthase: Evidence for two identical highly anticooperative pteridine binding sites. *Biochemistry* **35,** 16735–16745.

Gould, K. S., Lamotte, O., Klinguer, A., Pugin, A., and Wendehenne, D. (2003). Nitric oxide production in tobacco leaf cells: A generalized stress response? *Plant Cell Environ.* **26,** 1851–1862.

Grant, M., Brown, I., Adams, S., Knight, M., Ainslie, A., and Mansfield, J. (2000). The *RPM1* plant disease resistance gene facilitates a rapid and sustained increase in cytosolic calcium that is necessary for the oxidative burst and hypersensitive cell death. *Plant J.* **23,** 441–450.

Gray, J., Janick-Buckner, D., Buckner, B., Close, P. S., and Johal, G. S. (2002). Light-dependent death of maize *lls1* cells is mediated by mature chloroplasts. *Plant Physiol.* **130,** 1894–1907.

Graziano, M., Beligni, M. V., and Lamattina, L. (2002). Nitric oxide improves internal iron availability in plants. *Plant Physiol.* **130,** 1852–1859.

Grignon, C., and Sentenac, H. (1991). pH and ionic conditions in the apoplast. *Annu. Rev. Plant Physiol. Plant Mol. Biol.* **42,** 103–128.

Guo, F.-Q., Okamoto, M., and Crawford, N. M. (2003). Identification of a plant nitric oxide synthase gene involved in hormonal signaling. *Science* **302,** 100–103.

Guo, P., Cao, Y., Li, Z., and Zhao, B. (2004). Role of an endogenous nitric oxide burst in the resistance of wheat to stripe rust. *Plant Cell Environ.* **27,** 473–477.

Guttman, D. S., Vinatzer, B. A., Sarkar, S. F., Ranall, M. V., Kettler, G., and Greenberg, J. T. (2002). A functional screen for the type III (Hrp) secretome of the plant pathogen *Pseudomonas syringae*. *Science* **295,** 1722–1726.

Hakata, M., Takahashi, M., Zumft, W., Sakamoto, A., and Morikawa, H. (2003). Conversion of nitrate nitrogen and nitrogen dioxide to nitrous oxides in plants. *Acta Biotechnol.* **23,** 249–257.

Hampl, V., Walters, C. L., and Archer, S. L. (1996). Determination of nitric oxide by the chemiluminescence reaction with ozone. *In* "Methods in Nitric Oxide Research" (M. Feelisch and J. S. Stamler, Eds.), pp. 309–318. Wiley, Chichester, England.

Hanson, A. D. (2004). Folate metabolism in plants. (http://www.hos.ufl.edu/meteng/Hanson-Webpagecontents/Folatemetabolisminplants.html).

Hanson, A. D., and Gregory, J. F. I. (2002). Synthesis and turnover of folates in plants. *Curr. Opin. Plant Biol.* **5,** 244–249.

Harper, J. E. (1981). Evolution of nitrogen oxide(s) during *in vivo* nitrate reductase assay of soybean leaves. *Plant Physiol.* **68,** 1488–1493.

Hartung, W., Radin, J. W., and Hendrix, D. L. (1988). Abscisic acid movements into the apoplastic solution of water-stressed cotton leaves. *Plant Physiol.* **86,** 908–913.

Hartung, W., and Slovik, S. (1991). Physicochemical properties of plant growth regulators and plant tissues determine their distribution and redistribution: Stomatal regulation by abscisic acid in leaves. *New Phytologist* **119,** 361–382.

He, Y., Tang, R.-H., Hao, Y., Stevens, R. D., Cook, C. W., Ahn, S. M., Jing, L., Yang, Z., Chen, L., Guo, F., Fiorani, F., Jackson, R. B., Crawford, N. M., and Pei, Z.-M. (2004). Nitric oxide represses the *Arabidopsis* floral transition. *Science* **305,** 1968–1971.

Heath, M. C. (1998). Involvement of reactive oxygen species in the response of resistant (hypersensitive) or susceptible cowpeas to the cowpea rust fungus. *New Phytologist* **138,** 251–263.

Heuer, B., Plaut, Z., and Federman, E. (1979). Nitrate and nitrite reduction in wheat leaves as affected by different types of water stress. *Physiol. Plant.* **46,** 318–323.

Hoffmann, B., and Kosegarten, H. (1995). FITC-dextran for measuring apoplast pH and apoplastic pH gradients between various cell types in sunflower leaves. *Physiol. Plant.* **95,** 327–335.

Hogg, N., Singh, R. J., and Kalyanaraman, B. (1996). The role of glutathione in the transport and catabolism of nitric oxide. *FEBS Lett.* **382,** 223–228.

Hu, X., Neill, S. J., Cai, W., and Tang, Z. (2003a). Nitric oxide mediates elicitor-induced saponin synthesis in cell cultures of *Panax ginseng*. *Funct. Plant Biol.* **30,** 901–907.

Hu, X. Y., Neill, S. J., Cai, W. M., and Tang, Z. C. (2003b). NO-mediated hypersensitive responses of rice suspension cultures induced by an incompatible elicitor. *Chin. Sci. Bull.* **48,** 358–363.

Huang, J., and Knopp, J. A. (1998). Involvement of nitric oxide in *Ralstonia solanacearum*-induced hypersensitive reaction in tobacco. *In* "Bacterial Wilt Disease: Molecular and Ecological Aspects" (P. Prior, C. Allen, and J. Elphinstone, Eds.), pp. 218–224. INRA and Springer Editions, Berlin, Germany.

Huang, S., Kerschbaum, H. H., Engel, E., and Hermann, A. (1997). Biochemical characterization and histochemical localization of nitric oxide synthase in the nervous system of the snail, *Helix pomatia*. *J. Neurochem.* **69,** 2516–2528.

Huang, X., Kiefer, E., von Rad, U., Ernst, D., Foissner, I., and Durner, J. (2002a). Nitric oxide burst and nitric oxide-dependent gene induction in plants. *Plant Physiol. Biochem.* **40,** 625–631.

Huang, X., Stettmaier, K., Michel, C., Hutzler, P., Mueller, M. J., and Durner, J. (2004). Nitric oxide is induced by wounding and influences jasmonic acid signaling in *Arabidopsis thaliana*. *Planta* **218,** 938–946.

Huang, X., von Rad, U., and Durner, J. (2002b). Nitric oxide induces transcriptional activation of the nitric oxide-tolerant alternative oxidase in *Arabidopsis* suspension cells. *Planta* **215,** 914–923.

Huber, S. C., and Hardin, S. C. (2004). Numerous posttranslational modifications provide opportunities for the intricate regulation of metabolic enzymes at multiple levels. *Curr. Opin. Plant Biol.* **7,** 318–322.

Hung, K. T., and Kao, C. H. (2003). Nitric oxide counteracts the senescence of rice leaves induced by abscisic acid. *J. Plant Physiol.* **160,** 871–879.

Igamberdiev, A. U., Bykova, N. V., Ens, W., and Hill, R. D. (2004a). Dihydrolipoamide dehydrogenase from porcine heart catalyzes NADH-dependent scavenging of nitric oxide. *FEBS Lett.* **568,** 146–150.

Igamberdiev, A. U., Seregélyes, C., Manac'h, N., and Hill, R. D. (2004b). NADH-dependent metabolism of nitric oxide in alfalfa root cultures expressing barley hemoglobin. *Planta* **219,** 95–102.

Itoh, Y., Ma, F. H., Hoshi, H., Oka, M., Noda, K., Ukai, Y., Kojima, H., Nagano, T., and Toda, N. (2000). Determination and bioimaging method for nitric oxide in biological specimens by diaminofluorescein fluorometry. *Anal. Biochem.* **287,** 203–209.

Jih, P.-J., Chen, Y.-C., and Jeng, S.-T. (2003). Involvement of hydrogen peroxide and nitric oxide in expression of the ipomoelin gene from sweet potato. *Plant Physiol.* **132,** 381–389.

Joshi, M. S., Ferguson, T. B. J., Han, T. H., Hyduke, D. R., Liao, J. C., Rassaf, T., Bryan, N., Feelisch, M., and Lancaster, J. R. J. (2002). Nitric oxide is consumed, rather than conserved, by reaction with oxyhemoglobin under physiological conditions. *Proc. Natl. Acad. Sci. USA* **99,** 10341–10346.

Kachroo, P., Shanklin, J., Shah, J., Whittle, E. J., and Klessig, D. F. (2001). A fatty acid desaturase modulates the activation of defense signaling pathways in plants. *Proc. Natl. Acad. Sci. USA* **98,** 9448–9453.

Kaiser, W. M., Weiner, H., Kandlbinder, A., Tsai, C.-B., Rockel, P., Sonoda, M., and Planchet, E. (2002). Modulation of nitrate reductase: Some new insights, an unusual case, and a potentially important side reaction. *J. Exp. Bot.* **53,** 875–882.

Keppler, L. D., Baker, C. J., and Atkinson, M. M. (1989). Active oxygen production during a bacteria-induced hypersensitive reaction in tobacco suspension cells. *Phytopathology* **79,** 974–978.

Keppler, L. D., and Novacky, A. (1986). Involvement of membrane lipid peroxidation in the development of a bacterially induced hypersensitive reaction. *Phytopathology* **76,** 104–108.

Keppler, L. D., and Novacky, A. (1987). The initiation of membrane lipid peroxidation during bacteria-induced hypersensitive reaction. *Physiol. Mol. Plant Pathol.* **30,** 233–245.

Klepper, L. (1979). Nitric oxide (NO) and nitrogen dioxide (NO_2) emissions from herbicide-treated soybean plants. *Atmos. Environ.* **13,** 537–542.

Klepper, L. (1990). Comparison between NO_x evolution mechanisms of wild-type and nr1 mutant soybean leaves. *Plant Physiol.* **93,** 26–32.

Klepper, L. (1991). No_x evolution by soybean leaves treated with salicylic acid and selected derivatives. *Pesticide Biochem. Physiol.* **39,** 43–48.

Kojima, H., Hirotani, M., Nakatsubo, N., Kikuchi, K., Urano, Y., Higuchi, T., Hirata, Y., and Nagano, T. (2001). Bioimaging of nitric oxide with fluorescent indicators based on the rhodamine chromophore. *Anal. Chem.* **73,** 1967–1973.

Kojima, H., Nakatsubo, N., Kikuchi, K., Kawahara, S., Kirino, Y., Nagoshi, H., Hirata, Y., and Nagano, T. (1998). Detection and imaging of nitric oxide with novel fluorescent indicators: Diaminofluoresceins. *Anal. Chem.* **70,** 2446–2453.

Kojima, H., Urano, Y., Kikuchi, K., Higuchi, T., Hirata, Y., and Nagano, T. (1999). Fluorescent indicators for imaging nitric oxide production. *Angew. Chem. Int. Ed.* **38,** 3209–3212.

Kovtun, Y., Chiu, W.-L., Tena, G., and Sheen, J. (2000). Functional analysis of oxidative stress-activated mitogen-activated protein kinase cascade in plants. *Proc. Natl. Acad. Sci. USA* **97,** 2940–2945.

Kunkel, B. N., Bent, A. F., Dahlbeck, D., Innes, R. W., and Staskawicz, B. J. (1993). *RPS2*, an *Arabidopsis* disease resistance locus specifying recognition of *Pseudomonas syringae* strains expressing the avirulence gene *avrRpt2*. *Plant Cell* **5,** 865–875.

Kus, J. V., Zaton, K., Sarkar, R., and Cameron, R. K. (2002). Age-related resistance in *Arabidopsis* is a developmentally regulated defense response to *Pseudomonas syringae*. *Plant Cell* **14,** 479–490.

Lamattina, L., Garcia-Mata, C., Graziano, M., and Pagnussat, G. (2003). Nitric oxide: The versatility of an extensive signal molecule. *Annu. Rev. Plant Biol.* **54,** 109–136.

Lamotte, O., Gould, K., Lecourieux, D., Sequeira-Legrand, A., Lebrun-Garcia, A., Durner, J., Pugin, A., and Wendehenne, D. (2004). Analysis of nitric oxide signaling functions in tobacco cells challenged by the elicitor cryptogein. *Plant Physiol.* **135,** 516–529.

Lancaster, J. R. J. (2000). The physical properties of nitric oxide: Determinants of the dynamics of NO in tissue. *In* "Nitric Oxide: Biology and Pathobiology" (L. J. Ignarro, Ed.), pp. 209–224. Academic Press, San Diego.

Lea, U. S., ten Hoopen, F., Provan, F., Kaiser, W. M., Meyer, C., and Lillo, C. (2004). Mutation of the regulatory phosphorylation site of tobacco nitrate reductase results in high nitrite excretion and NO emission from leaf and root tissue. *Planta* **219,** 59–65.

LeBrasseur, N. D., MacIntosh, G. C., Pérez-Amador, M. A., Saitoh, M., and Green, P. J. (2002). Local and systemic wound-induction of RNase and nuclease activities in *Arabidopsis*: RNS1 as a marker for a JA-independent systemic signaling pathway. *Plant J.* **29,** 393–403.

León, J., Rojo, E., and Sánchez-Serrano, J. J. (2001). Wound signalling in plants. *J. Exp. Bot.* **52,** 1–9.

Leshem, Y. Y., and Haramaty, E. (1996). The characterization and contrasting effects of the nitric oxide free radical in vegetative stress and senescence of *Pisum sativum* Linn. foliage. *J. Plant Physiol.* **148,** 258–263.

Leshem, Y. Y. (2000). "Nitric Oxide in Plants: Occurrence, Function and Use". Kluwer, Dordrecht, The Netherlands.

Leshem, Y. Y., and Pinchasov, Y. (2000). Non-invasive photoacoustic spectroscopic determination of relative endogenous nitric oxide and ethylene content stoichiometry during the ripening of strawberries *Fragaria anannasa* (Duch.) and avocados *Persea americana* (Mill.). *J. Exp. Bot.* **51,** 1471–1473.

Leshem, Y. Y., Wills, R. B. H., and Ku, V. V.-V. (1998). Evidence for the function of the free radical gas—nitric oxide (NO•)—as an endogenous maturation and senescence regulating factor in higher plants. *Plant Physiol. Biochem.* **36,** 825–833.

Liu, L., Hausladen, A., Zeng, M., Que, L., Heitman, J., and Stamler, J. S. (2001). A metabolic enzyme for *S*-nitrosothiol conserved from bacteria to humans. *Nature* **410,** 490–494.

Lobreaux, S., Massenet, O., and Briat, J. F. (1992). Iron induces ferritin synthesis in maize plantlets. *Plant Mol. Biol.* **19,** 563–575.

Ludidi, N., and Gehring, C. (2003). Identification of a novel protein with guanylyl cyclase activity in *Arabidopsis thaliana*. *J. Biol. Chem.* **278,** 6490–6494.

Mach, J. M., Castillo, A. R., Hoogstraten, R., and Greenberg, J. T. (2001). The *Arabidopsis* accelerated cell death gene *ACD2* encodes red chlorophyll catabolite reductase and suppresses the spread of disease symptoms. *Proc. Natl. Acad. Sci. USA* **98,** 771–776.

Mackerness, S. A. H., John, C. F., Jordan, B., and Thomas, B. (2001). Early signaling components in Ultraviolet-B responses: Distinct roles for different reactive oxygen species and nitric oxide. *FEBS Lett.* **489,** 237–242.

Macrì, F., Braidot, E., Petrussa, E., Zancani, M., and Vianello, A. (1992). Ferric ion and oxygen reduction at the surface of protoplasts and cells of *Acer pseudoplatanus*. *Bot. Acta* **105,** 97–103.

Magalhaes, J. R., Monte, D. C., and Durzan, D. (2000). Nitric oxide and ethylene emission in *Arabidopsis thaliana*. *Physiol. Mol. Biol. Plants* **6,** 117–127.

Malinski, T., and Czuchajowski, L. (1996). Nitric oxide measurements by electrochemical methods. *In* "Methods in Nitric Oxide Research" (M. Feelisch and J. S. Stamler, Eds.), pp. 319–339. Wiley, Chichester, England.

Manulis, S., Haviv-Chesner, A., Brandl, M. T., Lindow, S. E., and Barash, I. (1998). Differential involvement of indole-3-acetic acid biosynthesis pathways in pathogenicity and epiphytic fitness of *Erwinia herbicola* pv. *gypsophilae*. *Mol. Plant Microbe Interact.* **11,** 634–642.

Martinez-Ruiz, A., and Lamas, S. (2004). *S*-nitrosylation: A potential new paradigm in signal transduction. *Cardiovasc. Res.* **62,** 43–52.

Martinon, F., and Tschopp, J. (2004). Inflammatory caspases: Linking an intracellular innate immune system to autoinflammatory diseases. *Cell* **117,** 561–574.

Mayer, B., Pfeiffer, S., Schrammel, A., Koesling, D., Schmidt, K., and Brunner, F. (1998). A new pathway of nitric oxide/cyclic GMP signaling involving *S*-nitrosoglutathione. *J. Biol. Chem.* **273,** 3264–3270.

Membrillo-Hernández, J., Coopamah, M. D., Anjum, M. F., Stevanin, T. M., Kelly, A., Hughes, M. N., and Poole, R. K. (1999). The flavohemoglobin of *Escherichia coli* confers resistance to a nitrosating agent, a "nitric oxide releaser," and paraquat and is essential for transcriptional responses to oxidative stress. *J. Biol. Chem.* **274,** 748–754.

Membrillo-Hernández, J., Ioannidis, N., and Poole, R. K. (1996). The flavohaemoglobin (HMP) of *Escherichia coli* generates superoxide *in vitro* and causes oxidative stress *in vivo*. *FEBS Lett.* **382,** 141–144.

Meyer, C., and Stitt, M. (2001). Nitrate reduction and signalling. *In* "Plant Nitrogen" (P. J. Lea and J.-F. Morot-Gaudry, Eds.), pp. 37–59. Springer-Verlag, Berlin.

Mittal, S., and Davis, K. R. (1995). Role of the phytotoxin coronatine in the infection of *Arabidopsis thaliana* by *Pseudomonas syringae* pv. tomato. *Mol. Plant Microbe Interact.* **8,** 165–171.

Morot-Gaudry-Talarmain, Y., Rockel, P., Moureaux, T., Quilleré, I., Leydecker, M. T., Kaiser, W. M., and Morot-Gaudry, J. F. (2002). Nitrite accumulation and nitric oxide emission in relation to cellular signaling in nitrite reductase antisense tobacco. *Planta* **215,** 708–715.

Murata, Y., Pei, Z.-M., Mori, I. C., and Schroeder, J. (2001). Abscisic acid activation of plasma membrane Ca^{2+} channels in guard cells requires cytosolic NAD(P)H and is differentially disrupted upstream and downstream of reactive oxygen species production in *abi1-1* and *abi2-1* protein phosphatase 2C mutants. *Plant Cell* **13,** 2513–2523.

Murgia, I., Concetta de Pinto, M., Delledonne, M., Soave, C., and de Gara, L. (2004a). Comparative effects of various nitric oxide donors on ferritin regulation, programmed cell death, and cell redox state in plant cells. *J. Plant Physiol.* **161**, 777–783.

Murgia, I., Delledonne, M., and Soave, C. (2002). Nitric oxide mediates iron-induced ferritin accumulation in *Arabidopsis*. *Plant J.* **30**, 521–528.

Murgia, I., Tarantino, D., Vannini, C., Bracale, M., Carravieri, S., and Soave, C. (2004b). *Arabidopsis thaliana* plants overexpressing thylakoidal ascorbate peroxidase show increased resistance to paraquat-induced photooxidative stress and to nitric oxide-induced cell death. *Plant J.* **38**, 940–953.

Navarre, D. A., Wendehenne, D., Durner, J., Noad, R., and Klessig, D. F. (2000). Nitric oxide modulates the activity of tobacco aconitase. *Plant Physiol.* **122**, 573–582.

Nawrath, C., Heck, S., Parinthawong, N., and Métraux, J.-P. (2002). EDS5, an essential component of salicylic acid-dependent signaling for disease resistance in *Arabidopsis*, is a member of the MATE transporter family. *Plant Cell* **14**, 275–286.

Neill, S. J., Desikan, R., Clarke, A., and Hancock, J. T. (2002). Nitric oxide is a novel component of abscisic acid signaling in stomatal guard cells. *Plant Physiol.* **128**, 13–16.

Nottingham, W., and Sutter, J. (1986). Kinetics of the oxidation of nitric oxide by chlorine and oxygen in nonaqueous media. *Int. J. Chem. Kinet.* **18**, 1289–1302.

Olbregts, J. (1985). Termolecular reaction of nitrogen monoxide and oxygen: A still unsolved problem. *Int. J. Chem. Kinet.* **17**, 835–848.

Orlandi, E. W., Hutcheson, S. W., and Baker, C. J. (1992). Early physiological responses associated with race-specific recognition in soybean leaf tissue and cell suspensions treated with *Pseudomonas syringae* pv. *glycinea*. *Physiol. Mol. Plant Path.* **40**, 173–180.

Orozco-Cardenas, M. L., and Ryan, C. A. (2002). Nitric oxide negatively modulates wound signaling in tomato plants. *Plant Physiol.* **130**, 487–493.

Pagnussat, G. C., Lanteri, M. L., and Lamattina, L. (2003). Nitric oxide and cyclic GMP are messengers in the indole acetic acid-induced adventitious rooting process. *Plant Physiol.* **132**, 1241–1248.

Pagnussat, G. C., Lanteri, M. L., Lombardo, M. C., and Lamattina, L. (2004). Nitric oxide mediates the indole acetic acid induction activation of a mitogen-activated protein kinase cascade involved in adventitious root development. *Plant Physiol.* **135**, 279–286.

Parani, M., Rudrabhatla, S., Myers, R., Weirich, H., Smith, B., Leaman, D. W., and Goldman, S. L. (2004). Microarray analysis of nitric oxide responsive transcripts in *Arabidopsis*. *Plant Biotech. J.* **2**, 359–366.

Pastori, G. M., Kiddle, G., Antoniw, J., Bernard, S., Veljovic-Jovanovic, S., Verrier, P. J., Noctor, G., and Foyer, C. H. (2003). Leaf vitamin C contents modulate plant defense transcripts and regulate genes that control development through hormone signaling. *Plant Cell* **15**, 939–951.

Pedroso, M. C., Magalhaes, J. R., and Durzan, D. (2000a). A nitric oxide burst precedes apoptosis in angiosperm and gymnosperm callus cells and foliar tissues. *J. Exp. Bot.* **51**, 1027–1036.

Pedroso, M. C., Magalhaes, J. R., and Durzan, D. (2000b). Nitric oxide induces cell death in Taxus cells. *Plant Sci.* **157**, 173–180.

Peña-Cortés, H., Albrecht, T., Prat, S., Weiler, E. W., and Wilmitzer, L. (1993). Aspirin prevents wound-induced gene expression in tomato leaves by blocking jasmonic acid biosynthesis. *Planta* **191**, 123–128.

Perazzolli, M., Dominici, P., Romero-Puertas, M. C., Zago, E., Zeier, J., Sonoda, M., Lamb, C., and Delledonne, M. (2004). *Arabidopsis* nonsymbiotic hemoglobin AHb1 modulates nitric oxide bioactivity. *Plant Cell* **16**, 2785–2794.

Pich, A., Manteuffel, R., Hillmer, S., Scholz, G., and Schmidt, W. (2001). Fe homeostasis in plant cells: Does nicotianamine play multiple roles in the regulation of cytoplasmic Fe concentration? *Planta* **213**, 967–976.

Polverari, A., Molesini, B., Pezzotti, M., Buonaurio, R., Marte, M., and Delledonne, M. (2003). Nitric oxide-mediated transcriptional changes in *Arabidopsis thaliana*. *Mol. Plant Microbe Interact.* **16,** 1094–1105.

Polyakova, L. I., and Vartapetian, B. B. (2003). Exogenous nitrate as a terminal acceptor of electrons in rice (*Oryza sativa*) coleoptiles and wheat (*Triticum aestivum*) roots under strict anoxia. *Russ. J. Plant Physiol.* **50,** 808–812.

Poole, R. K., and Hughes, M. N. (2000). New functions for the ancient globin family: Bacterial responses to nitric oxide and nitrosative stress. *Mol. Microbiol.* **36,** 775–783.

Prado, A. M., Porterfield, D. M., and Feijó, J. A. (2004). Nitric oxide is involved in growth regulation and re-orientation of pollen tubes. *Development* **131,** 2707–2714.

Preston, C. A., Becker, R., and Baldwin, I. T. (2004). Is 'NO' news good news? Nitrogen oxides are not components of smoke that elicits germination in two smoke-stimulated species, *Nicotiana attenuata* and *Emmenanthe penduliflora*. *Seed Sci. Res.* **14,** 73–79.

Radi, R., Denicola, A., Alvarez, B., Ferrer-Sueta, G., and Rubbo, H. (2000). The biological chemistry of peroxynitrite. *In* "Nitric Oxide: Biology and Pathobiology" (L. J. Ignarro, Ed.), pp. 57–82. Academic Press, San Diego.

Ramanjulu, S., and Sudhakar, C. (1997). Drought tolerance is partly related to amino acid accumulation and ammonia assimilation: A comparative study in two mulberry genotypes differing in drought sensitivity. *J. Plant Physiol.* **150,** 345–350.

Ranieri, A., Castagna, A., Baldan, B., Sebastiani, L., and Soldatini, G. F. (2003). H_2O_2 accumulation in sunflower leaves as a consequence of iron deprivation. *J. Plant Nutr.* **26,** 2187–2196.

Rochelle, L. G., Kruszyna, H., Kruszyna, R., Barchowsky, A., Wilcox, D. E., and Smith, R. P. (1994). Bioactivation of nitroprusside by porcine endothelial cells. *Toxicol. Appl. Pharmacol.* **128,** 123–128.

Rockel, P., Strube, F., Rockel, A., Wildt, J., and Kaiser, W. M. (2002). Regulation of nitric oxide (NO) production by plant nitrate reductase *in vivo* and *in vitro*. *J. Exp. Bot.* **53,** 103–110.

Rogers, E. E., and Ausubel, F. M. (1997). *Arabidopsis* enhanced disease susceptibility mutants exhibit enhanced susceptibility to several bacterial pathogens and alterations in PR-1 gene expression. *Plant Cell* **9,** 305–316.

Sakamoto, A., Sakurao, S.-H., Fukunaga, K., Matsubara, T., Ueda-Hashimoto, M., Tsukamoto, S., Takahashi, M., and Morikawa, H. (2004). Three distinct *Arabidopsis* hemoglobins exhibit peroxidase-like activity and differentially mediate nitrite-dependent protein nitration. *FEBS Lett.* **572,** 27–32.

Sakamoto, A., Ueda, M., and Morikawa, H. (2002). *Arabidopsis* glutathione-dependent formaldehyde dehydrogenase is an *S*-nitrosoglutathione reductase. *FEBS Lett.* **515,** 20–24.

Sakihama, Y., Tamaki, R., Shimoji, H., Ichiba, T., Fukushi, Y., Tahara, S., and Yamasaki, H. (2003). Enzymatic nitration of phytophenolics: Evidence for peroxynitrite-independent nitration of plant secondary metabolites. *FEBS Lett.* **553,** 377–380.

Santrucek, J., Hronková, M., Kveton, J., and Sage, R. F. (2003). Photosynthesis inhibition during gas exchange oscillations in ABA-treated *Helianthus annuus*: Relative role of stomatal patchiness and leaf carboxylation capacity. *Photosynthetica* **41,** 241–252.

Saviani, E. E., Orsi, C. H., Oliveira, J. F. P., Pinto-Maglio, C. A. F., and Salgado, I. (2002). Participation of the mitochondrial permeability transition pore in nitric oxide-induced plant cell death. *FEBS Lett.* **510,** 136–140.

Savino, G., Briat, J. F., and Lobreaux, S. (1997). Inhibition of the iron-induced *ZmFer1* maize ferritin gene expression by antioxidants and serine/threonine phosphatase inhibitors. *J. Biol. Chem.* **272,** 33319–33326.

Schmadel-Hagebölling, H. E., Engel, C., Schmitt, V., and Wild, A. (1998). The combined effects of CO_2, ozone and drought on rubisco and nitrogen metabolism of young oak trees (*Quercus petraea*). *Chemosphere* **36,** 789–794.

Schmidt, H. H. H. W., and Kelm, M. (1996). Determination of nitrite and nitrate by the Griess reaction. In "Methods in Nitric Oxide Research" (M. Feelisch and J. S. Stamler, Eds.), pp. 491–497. Wiley, Chichester, England.

Seregélyes, C., Barna, B., Hennig, J., Konopka, D., Pasternak, T. P., Lukács, N., Fehér, A., Horváth, G. V., and Dudits, D. (2003). Phytoglobins can interfere with nitric oxide functions during plant growth and pathogenic responses: A transgenic approach. *Plant Sci.* **165**, 541–550.

Seregélyes, C., Igamberdiev, A. U., Maassen, A., Hennig, J., Dudits, D., and Hill, R. D. (2004). NO-degradation by alfalfa class 1 hemoglobin (Mhb1): A possible link to *PR-1a* gene expression in Mhb1-overproducing tobacco plants. *FEBS Lett.* **571**, 61–66.

Seregélyes, C., Mustárdy, L., Ayaydin, F., Sass, L., Kovács, L., Endre, G., Lukács, N., Kovács, I., Vass, I., Kiss, G. B., Horváth, G. V., and Dudits, D. (2000). Nuclear localization of a hypoxia-inducible novel non-symbiotic hemoglobin in cultured alfalfa cells. *FEBS Lett.* **482**, 125–130.

Shapiro, A. D. (2000). Using *Arabidopsis* mutants to delineate disease resistance signaling pathways. *Can. J. Plant Pathol.* **22**, 199–216.

Shapiro, A. D., and Zhang, C. (2001). The role of *NDR1* in avirulence gene-directed signaling and control of programmed cell death in *Arabidopsis*. *Plant Physiol.* **127**, 1089–1101.

Sharpe, M. A., Robb, S. J., and Clark, J. B. (2003). Nitric oxide and Fenton/Haber-Weiss chemistry: Nitric oxide is a potent antioxidant at physiological concentrations. *J. Neurochem.* **87**, 386–394.

Shingles, R., Roh, M. H., and McCarty, R. E. (1996). Nitrite transport in chloroplast inner envelope vesicles: I. Direct measurement of proton-linked transport. *Plant Physiol.* **112**, 1375–1381.

Shirasu, K., Nakajima, H., Rajasekhar, V. K., Dixon, R. A., and Lamb, C. (1997). Salicylic acid potentiates an agonist-dependent gain control that amplifies pathogen signals in the activation of defense mechanisms. *Plant Cell* **9**, 261–270.

Singel, D. J., and Lancaster, J. R. J. (1996). Electron paramagnetic resonance spectroscopy and nitric oxide biology. In "Methods in Nitric Oxide Research" (M. Feelisch and J. S. Stamler, Eds.), pp. 341–356. Wiley, Chichester, England.

Singh, R. J., Hogg, N., Joseph, J., and Kalyanaraman, B. (1996). Mechanism of nitric oxide release from *S*-nitrosothiols. *J. Biol. Chem.* **271**, 18596–18603.

Singh, R. J., Hogg, N., Neese, F., Joseph, J., and Kalyanaraman, B. (1995). Trapping of nitric oxide formed during photolysis of sodium nitroprusside in aqueous and lipid phases: An electron spin resonance study. *Photochem. Photobiol.* **61**, 325–330.

Slaymaker, D. H., Navarre, D. A., Clark, D., del Pozo, O., Martin, G. B., and Klessig, D. F. (2002). The tobacco salicylic acid-binding protein 3 (SABP3) is the chloroplast carbonic anhydrase, which exhibits antioxidant activity and plays a role in the hypersensitive defense response. *Proc. Natl. Acad. Sci. USA* **99**, 11640–11645.

Slovik, S., and Hartung, W. (1991). Compartmental distribution and redistribution of abscisic acid in intact leaves. *Planta* **187**, 37–47.

Song, F., and Goodman, R. M. (2001). Activity of nitric oxide is dependent on, but is partially required for function of, salicylic acid in the signaling pathway in tobacco systemic acquired resistance. *Mol. Plant Microbe Interact.* **14**, 1458–1462.

Stemler, A. J. (2002). The bicarbonate effect, oxygen evolution, and the shadow of Otto Warburg. *Photosynth. Res.* **73**, 177–183.

Stöhr, C., Strube, F., Marx, G., Ullrich, W. R., and Rockel, P. (2001). A plasma membrane-bound enzyme of tobacco roots catalyzes the formation of nitric oxide from nitrite. *Planta* **212**, 835–841.

Subramanian, K. S., and Charest, C. (1998). Arbuscular mycorrhizae and nitrogen assimilation in maize after drought and recovery. *Physiol. Plant.* **102**, 285–296.

Tada, Y., Mori, T., Shinogi, T., Yao, N., Takahashi, S., Betsuyaku, S., Sakamoto, M., Park, P., Nakayashiki, H., Tosa, Y., and Mayama, S. (2004). Nitric oxide and reactive oxygen species do not elicit hypersensitive cell death but induce apoptosis in the adjacent cells during the defense response of oat. *Mol. Plant Microbe Interact.* **17,** 245–253.

Takahashi, S., and Yamasaki, H. (2002). Reversible inhibition of photophosphorylation in chloroplasts by nitric oxide. *FEBS Lett.* **512,** 145–148.

Takehara, Y., Kanno, T., Yoshioka, T., Inoue, M., and Utsumi, K. (1995). Oxygen-dependent regulation of mitochondrial energy metabolism by nitric oxide. *Arch. Biochem. Biophys.* **323,** 27–32.

Tao, Y., Xie, Z., Chen, W., Glazebrook, J., Chang, H.-S., Han, B., Zhu, T., Zou, G., and Katagiri, F. (2003). Quantitative nature of *Arabidopsis* responses during compatible and incompatible interactions with the bacterial pathogen *Pseudomonas syringae*. *Plant Cell* **15,** 317–330.

Torres, J., Cooper, C. E., Sharpe, M., and Wilson, M. T. (1998). Reactivity of nitric oxide with cytochrome *c* oxidase: Interactions with the binuclear centre and mechanism of inhibition. *J. Bioenerg. Biomembr.* **30,** 63–69.

Tsukahara, H., Ishida, T., and Mayumi, M. (1999). Gas-phase oxidation of nitric oxide: Chemical kinetics and rate constant. *Nitric Oxide* **3,** 191–198.

Tun, N. N., Holk, A., and Scherer, G. F. E. (2001). Rapid increase of NO release in plant cell cultures induced by cytokinin. *FEBS Lett.* **509,** 174–176.

van der Biezen, E. A., and Jones, J. D. G. (1998). The NB-ARC domain: A novel signalling motif shared by plant resistance gene products and regulators of cell death in animals. *Curr. Biol.* **8,** R226–R227.

van Rensen, J. J. S. (2002). Role of bicarbonate at the acceptor side of photosystem II. *Photosynth. Res.* **73,** 185–192.

Vanin, A. F., Svistunenko, D. A., Mikoyan, V. D., Serezhenkov, V. A., Fryer, M. J., Baker, N. R., and Cooper, C. E. (2004). Endogenous superoxide production and nitrite/nitrate ratio control the concentration of bioavailable free nitric oxide in leaves. *J. Biol. Chem.* **279,** 24100–24107.

Wada, M., Morinaka, C., Ikenaga, T., Kuroda, N., and Nakashima, K. (2002). A simple HPLC-fluorescence detection of nitric oxide in cultivated plant cells by *in situ* derivitization with 2,3-diaminonaphthalene. *Anal. Sci.* **18,** 631–634.

Ward, E. R., Uknes, S. J., Williams, S. C., Dincher, S. S., Wiederhold, D. L., Alexander, D. C., Ahl-Goy, P., Métraux, J.-P., and Ryals, J. A. (1991). Coordinate gene activity in response to agents that induce systemic acquired resistance. *Plant Cell* **3,** 1085–1094.

Wildermuth, M. C., Dewdney, J., Wu, G., and Ausubel, F. M. (2001). Isochorismate synthase is required to synthesize salicylic acid for plant defense. *Nature* **414,** 562–565.

Wildt, J., Kley, D., Rockel, A., Rockel, P., and Segschneider, H. J. (1997). Emission of NO from several higher plant species. *J. Geophys. Res.* **102,** 5919–5927.

Wilkinson, J. Q., and Crawford, N. M. (1993). Identification and characterization of a chlorate-resistant mutant of *Arabidopsis thaliana* with mutations in both nitrate reductase structural genes *NIA1* and *NIA2*. *Mol. Gen. Genet.* **239,** 289–297.

Wink, D., Beckman, J., and Ford, P. (1996). Kinetics of nitric oxide reaction in liquid and gas phase. *In* "Methods in Nitric Oxide Research" (M. Feelisch and J. S. Stamler, Eds.), pp. 30–37. Wiley, Chichester, England.

Wink, D., Darbyshire, J., Nims, R., Saavedra, J., and Ford, P. (1993). Reactions of the bioregulatory agent nitric oxide in oxygenated aqueous media: Determination of the kinetics for oxidation and nitrosation by intermediates generated in the NO/O_2 reaction. *Chem. Res. Toxicol.* **6,** 23–27.

Wolfe, J., Hutcheon, C. J., Higgins, V. J., and Cameron, R. K. (2000). A functional gene-for-gene interaction is required for the production of an oxidative burst in response to infection with avirulent *Pseudomonas syringae* pv. tomato in *Arabidopsis thaliana*. *Physiol. Mol. Plant Path.* **56,** 253–261.

Xie, M., Hermann, A., Richter, K., Engel, E., and Kerschbaum, H. H. (2001). Nitric oxide upregulates ferritin mRNA level in snail neurons. *Eur. J. Neurosci.* **13,** 1479–1486.

Xin, L., Shuqiu, Z., and Chenghou, L. (2003). Involvement of nitric oxide in the signal transduction of salicylic acid regulating stomatal movement. *Chin. Sci. Bull.* **48,** 449–452.

Xing, H., Tan, L., An, L., Zhao, Z., Wang, S., and Zhang, C. (2004). Evidence for the involvement of nitric oxide and reactive oxygen species in osmotic stress tolerance of wheat seedlings: Inverse correlation between leaf abscisic acid accumulation and leaf water loss. *Plant Growth Regul.* **42,** 61–68.

Yamasaki, H. (2000). Nitrite-dependent nitric oxide production pathway: Implications for involvement of active nitrogen species in photoinhibition *in vivo*. *Philos. Trans. R. Soc. Lond. B* **355,** 1477–1488.

Yamasaki, H., and Sakihama, Y. (2000). Simultaneous production of nitric oxide and peroxynitrite by plant nitrate reductase: *In vitro* evidence for the NR-dependent formation of active nitrogen species. *FEBS Lett.* **468,** 89–92.

Yamasaki, H., Shimoji, H., Ohshiro, Y., and Sakihama, Y. (2001). Inhibitory effects of nitric oxide on oxidative phosphorylation in plant mitochondria. *Nitric Oxide* **5,** 261–270.

Zancani, M., Peresson, C., Biroccio, A., Federici, G., Urbani, A., Murgia, I., Soave, C., Micali, F., Vianello, A., and Macri, F. (2004). Evidence for the presence of ferritin in plant mitochondria. *Eur. J. Biochem.* **271,** 3657–3664.

Zeidler, D., Zähringer, U., Gerber, I., Dubery, I., Hartung, W., Bors, W., Hutzler, P., and Durner, J. (2004). Innate immunity in *Arabidopsis thaliana*: Lipopolysaccharides activate nitric oxide synthase (NOS) and induce defense genes. *Proc. Natl. Acad. Sci. USA* **101,** 15811–15816.

Zeier, J., Delledonne, M., Mishina, T., Severi, E., Sonoda, M., and Lamb, C. (2004). Genetic elucidation of nitric oxide signaling in incompatible plant–pathogen interactions. *Plant Physiol.* **136,** 2875–2886.

Zemojtel, T., Penzkofer, T., Dandekar, T., and Schultz, J. (2004). A novel conserved family of nitric oxide synthase? *Trends Biochem. Sci.* **29,** 224–226.

Zhang, C., Czymmek, K. J., and Shapiro, A. D. (2003). Nitric oxide does not trigger early programmed cell death events but may contribute to cell-to-cell signaling governing progression of the *Arabidopsis* hypersensitive response. *Mol. Plant Microbe Interact.* **16,** 962–972.

Zhang, C., Gutsche, A. T., and Shapiro, A. D. (2004). Feedback control of the *Arabidopsis* hypersensitive response. *Mol. Plant Microbe Interact.* **17,** 357–365.

Zhang, C., and Shapiro, A. D. (2002). Two pathways act in an additive rather than obligatorily synergistic fashion to induce systemic acquired resistance and *PR* gene expression. *BMC Plant Biol.* **2,** 9.

Zhao, L., Zhang, F., Guo, J., Yang, Y., Li, B., and Zhang, L. (2004). Nitric oxide functions as a signal in salt resistance in the calluses from two ecotypes of reed. *Plant Physiol.* **134,** 849–857.

Zottini, M., Formentin, E., Scattolin, M., Carimi, F., Lo Schiavo, F., and Terzi, M. (2002). Nitric oxide affects plant mitochondrial functionality *in vivo*. *FEBS Lett.* **515,** 75–78.

11

ETHYLENE BIOSYNTHESIS AND SIGNALING: AN OVERVIEW

Annelies De Paepe* and Dominique Van Der Straeten

Unit Plant Hormone Signaling and Bio-imaging, Department of Molecular Genetics Ghent University, K.L. Ledeganckstraat 35, B-9000 Gent, Belgium

I. Introduction
II. Biosynthesis of Ethylene: Mechanism and Regulation
   A. The 1-Aminocyclopropane-1-Carboxylic Acid Synthase Family: A Multigene Family in Plants
   B. Posttranscriptional Regulation of 1-Aminocyclopropane-1-Carboxylic Acid Synthase
   C. The 1-Aminocyclopropane-1-Carboxylic Acid Oxidase Family
III. Ethylene Signaling
   A. Ethylene Perception
   B. Linking The Ethylene Receptors to CTR1, a MAPKKK
   C. EIN2, an NRAMP-like Protein
   D. Nuclear Events: Two Ethylene-Responsive Families of Transcription Factors
   E. Other Genetically Defined Components in Ethylene Signaling

*Current address: Department of Plant Systems Biology, Ghent University, Flanders Interuniversity Institute for Biotechnology (VIB), Technology Park 927, B-9052 Zwijnaarde, Belgium.

IV. Transcriptional Regulation of Ethylene Response
V. Cross-talk in Plant Hormone Signaling
VI. Ethylene in Plant Disease Resistance and Abiotic Stresses
VII. Conclusions
References

Hormones are key regulators of plant growth and development. Genetic and biochemical studies have identified major factors that mediate ethylene biosynthesis and signal transduction. Substantial progress in the elucidation of the ethylene signal transduction pathway has been made, mainly by research on *Arabidopsis thaliana*. Research on ethylene biosynthesis and its regulation provided new insights, particularly on the posttranslational regulation of ethylene synthesis and the feedback from ethylene signal transduction. The identification of new components in the ethylene-response pathway and the elucidation of their mode of action provide a framework for understanding not only how plants sense and respond to this hormone but also how the signal is integrated with other inputs, ultimately determining the plant phenotype. © 2005 Elsevier Inc.

I. INTRODUCTION

Phytohormones integrate many aspects of plant growth and development. Ethylene was one of the first plant hormones discovered. More than a century ago, it was identified by Dimitri Neljubov as the active component of illuminating gas that resulted in premature senescence and abscission in nearby vegetation (Neljubov, 1901). In the 1930s, Gane demonstrated that plants produce ethylene themselves (Gane, 1934). From that time on, ethylene was established as an endogenous regulator with profound effects on plant growth and development. Ethylene is involved in many aspects of the plant life cycle, including seed germination, root hair development, seedling growth, leaf and petal abscission, fruit ripening, and organ senescence (Abeles *et al.*, 1992; Yang and Hoffman, 1984). The production of ethylene is regulated by internal signals during development and in response to environmental stimuli of biotic (pathogen attack) and abiotic nature, such as wounding, hypoxia, ozone, chilling, or freezing. Regulation can also occur at the level of perception or signal transduction. Hormone sensitivity can, in turn, be regulated both spatially and temporally.

In this review, we will focus on the state of the art on ethylene synthesis and signal transduction in *A. thaliana*. Major breakthroughs have been made in understanding the mechanism of ethylene response, and new components in the pathway have been identified. To understand the functions of ethylene in plant growth, it is important to know how this hormone is synthesized, its production is regulated, and the signal is transduced. Particular attention will be given to the latest discoveries in this field.

II. BIOSYNTHESIS OF ETHYLENE: MECHANISM AND REGULATION

Almost all plant tissues have the capacity to produce ethylene, although in most cases the amount of ethylene produced is very low. The ethylene biosynthetic pathway was elucidated in a series of elegant studies, principally by Yang and coworkers (Kende, 1993; Yang and Hoffman, 1984). Ethylene is derived from the amino acid methionine, which is first converted to *S*-adenosyl-methionine (*S*-Adomet) by *S*-adomet synthetase (SAM synthetase) (Fig. 1). *S*-Adomet is the major methyl donor in plants and is used as a substrate for a number of biochemical pathways, including polyamine and ethylene biosynthesis (Ravanel *et al.*, 1998). ACC synthase (ACS), which converts *S*-Adomet to 1-aminocyclopropane-1-carboxylic acid (ACC) (Yang and Hoffman, 1984), is the first committed and generally rate-limiting step in ethylene biosynthesis. In addition to ACC, ACS also produces 5′-methylthioadenosine (MTA), which is subsequently recycled to methionine (Bleecker and Kende, 2000). This salvage pathway preserves the methyl group for another round of ethylene production. Finally, ACC is oxidized by ACC oxidase (ACO) to form ethylene, CO_2, and cyanide, which is detoxified to β-cyanoalanine by β-cyanoalanine synthase to prevent toxicity of accumulated cyanide during high rates of ethylene biosynthesis (Fig. 1).

A. THE 1-AMINOCYCLOPROPANE-1-CARBOXYLIC ACID SYNTHASE FAMILY: A MULTIGENE FAMILY IN PLANTS

1-Aminocyclopropane-1-carboxylic acid synthase is encoded by a multigene family whose structure resembles the subgroup I family of pyridoxal 5′-phosphate (PLP)-dependent aminotransferases (Mehta *et al.*, 1993). The *Arabidopsis* genome encodes nine ACC synthase polypeptides; eight of them are enzymatically active (ACS2, ACS4–9, ACS11) and one, ACS1 encodes a nonfunctional ACC synthase (Liang *et al.*, 1992, 1995; Yamagami *et al.*, 2003). Two ACS genes are pseudogenes. This is similar to other plant species, such as tomato and rice, where ACS is also encoded by a large multigene family. A diverse group of factors has been described as modulators of the level of ethylene biosynthesis in numerous plant species. In *Arabidopsis*,

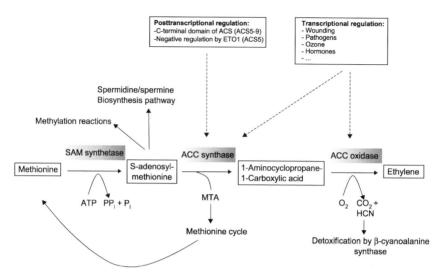

FIGURE 1. Biosynthetic pathway of ethylene and its regulation. The formation of S-adenosyl-methionine (S-Adomet) is catalyzed by S-Adomet synthetase with methionine as a substrate and at the expense of one molecule of ATP per molecule of S-Adomet synthesized. The major rate-limiting step of ethylene biosynthesis is the conversion of S-Adomet to 1-aminocyclopropane-1-carboxylic acid (ACC) by ACC synthase. ACC is the immediate precursor of ethylene. MTA is the by-product generated along with ACC production by ACC synthase. Malonylation or glutamylation of ACC to malonyl-ACC (MACC) and glutamyl-ACC (GACC) respectively, reduces the ACC pool and consequently, ethylene production. ACC oxidase catalyzes the final step of ethylene synthesis using ACC as a substrate also generating carbon dioxide and cyanide. Transcriptional regulation of both ACC synthase and ACC oxidase and posttranscriptional regulation of ACC synthase are indicated by dashed arrows.

these inducers include auxin, cytokinin, brassinosteroids, ethylene, ozone, copper, mechano-stimuli, pathogens, and wounding (Botella *et al.*, 1995; Cary *et al.*, 1995; Liang *et al.*, 1996; Rodrigues-Pousada *et al.*, 1993; Vahala *et al.*, 1998; Van Der Straeten *et al.*, 1992; Woeste *et al.*, 1999). As in other species, the *Arabidopsis* ACS genes are differentially regulated by these factors. Yamagami *et al.* (2003) demonstrated that the *Arabidopsis* ACS isozymes are biochemically distinct. It was proposed that these biochemically diverse ACS isozymes function in unique cellular environments, permitting the signaling molecule to exert its effects in a tissue- or cell-specific fashion. In addition, functional heteromeric interactions among the ACS polypeptides were described (Tsuchisaka and Theologis, 2004a). The capacity of the various isozymes to form active heterodimers further enhances the biochemical diversity of the *ACS* gene products, capable of operating under a very broad spectrum of Adomet concentration during plant development. Furthermore, a comparative analysis of spatial and temporal regulation of the entire ACC synthase gene family in *Arabidopsis* revealed a complex regulation (Tsuchisaka and Theologis, 2004b). Unique as well as overlapping expression patterns were observed during development and under various stresses, potentially reflecting a combinatorial code for functional ACC synthase heterodimers.

B. POSTTRANSCRIPTIONAL REGULATION OF 1-AMINOCYCLOPROPANE-1-CARBOXYLIC ACID SYNTHASE

Genetic analysis of the regulation of ethylene biosynthesis in *Arabidopsis* has provided compelling evidence that ACC synthase can also be regulated posttranscriptionally (Vogel *et al.*, 1998b; Wang *et al.*, 2004; Woeste *et al.*, 1999). Three ethylene-overproduction mutants *eto1*, *eto2*, and *eto3* have been identified in *Arabidopsis* (Guzman and Ecker, 1990; Kieber *et al.*, 1993). *eto1* is a recessive mutation whereas *eto2* and *eto3* are dominant. The constitutive triple-response phenotype of the *eto* mutants can be suppressed by silver thiosulphate (inhibitor of ethylene perception) or amino-ethoxy-vinylglycine (AVG) (inhibitor of ethylene biosynthesis), suggesting that these mutants are affected in the regulation of ethylene biosynthesis. Molecular analysis revealed that the *eto2* mutation was the result of a single nucleotide insertion that disrupted the C-terminal 11 amino acids of ACS5 (Vogel *et al.*, 1998b). Although the steady state *eto2* mRNA shows little change, ethylene overproduction in *eto2* etiolated seedlings is 20-fold that of the wild-type, suggesting that the increased activity is not the result of gene expression (Vogel *et al.*, 1998b). Further evidence that *ACS5* is posttranscriptionally regulated came from the analysis of the *cin*-mutants (Vogel *et al.*, 1998a). It has been shown that low doses of cytokinin (0.5–10 μM) stimulate ethylene production in etiolated seedlings of *Arabidopsis* and induce morphological changes resembling the triple response upon ethylene treatment in darkness (Cary *et al.*, 1995). This has been exploited to identify mutants that fail to increase ethylene in response to cytokinin (Vogel *et al.*, 1998a). Five complementation groups, termed cytokinin-insensitive mutants (*cin1–cin5*), have been found. Recessive mutations in one of these complementation groups, *cin5*, were mapped very close to *eto2* and *ACS5*. *CIN5* was subsequently found to correspond to *ACS5*, suggesting that this isoform is the major target for cytokinin regulation (Vogel *et al.*, 1998b). In addition, cytokinin-mediated ethylene production does not correlate with an induction of *ACS5* mRNA, demonstrating that cytokinin increases ACS5 function primarily by a posttranscriptional mechanism. Together, these results indicate that the C-terminus of ACS5 negatively regulates the function of the protein and that cytokinin may elevate ethylene biosynthesis by partially relieving this inhibition. The analysis of purified recombinant ACS5 and epitope-tagged ACS5 in transgenic *Arabidopsis* revealed that *eto2* does not increase the specific activity of the enzyme either *in vitro* or *in vivo*; rather, it increases the half-life of the protein (Chae *et al.*, 2003). In a similar manner, cytokinin treatment increased the stability of ACS5 by a mechanism that is at least partially independent of the *eto2* mutation. As for *eto2*, the *eto1-* and *eto3-* mutations likely affect the posttranscriptional regulation of ACS function. The *ETO1* gene has been cloned (Wang et al., 2004). It encodes a protein containing putative peptide binding domains, directly interacting with and inhibiting the wild-type but not the *eto2* version of the ACS5 isoform (Wang et al., 2004). Furthermore, overexpression of *ETO1* inhibited induction of ethylene production by cytokinin and promoted ACS5 degradation by a proteasome-dependent pathway. ETO1 also interacts with CUL3 (cullin3), a subunit of E3 ubiquitin ligase complexes. Therefore, ETO1 is suggested to have a dual mechanism, inhibiting ACS enzyme activity and targeting it for protein degradation, permitting rapid decline of ethylene production.

Finally, the *eto3* phenotype is the result of a missense mutation within the C-terminal domain of ACS9, similar to the *eto2* mutation (Chae et al., 2003). These results suggest that an important mechanism by which ethylene biosynthesis is controlled is the regulation of the stability of ACS proteins, mediated at least in part through the C-terminal domain.

C. THE 1-AMINOCYCLOPROPANE-1-CARBOXYLIC ACID OXIDASE FAMILY

The final step in ethylene biosynthesis, the conversion of ACC to ethylene, is catalyzed by the enzyme ACC oxidase, formerly known as ethylene-forming enzyme. 1-Aminocyclopropane-1-carboxylic acid oxidase plays an important role in regulating ethylene biosynthesis, especially during conditions of high ethylene production that include pollination-induced senescence, wounding, and senescence of leaves, fruit, and flowers (Barry et al., 1996; Kim and Yang, 1994; Lasserre et al., 1996; Nadeau et al., 1993; Tang and Woodson, 1996). 1-Aminocyclopropane-1-carboxylic acid oxidase genes have been cloned from different plant species. In *Arabidopsis*, ACO is present as a multigene family, but little information about these genes has been reported (Gomez-Lim et al., 1993; Raz and Ecker, 1999). The steady state level of the *AtACO2* transcripts increased in response to exogenous ethylene. This upregulation was confirmed by two microarray analyses (De Paepe et al., 2004; Van Zhong and Burns, 2003). In addition, three ACC-oxidases and one ACC-oxidase–like gene were affected by ethylene (De Paepe et al., 2004). Ethylene-regulated ethylene biosynthesis implicates that the hormone acts as a messenger for the induction of later ethylene-responsive genes, resulting in an amplification of the signal. Moreover, the differential expression of multiple ACOs supports a complex autoregulatory mechanism. In addition, this raises the question whether these proteins have equivalent biochemical activities and regulation.

III. ETHYLENE SIGNALING

Our current understanding of the ethylene signaling pathway is owed to the use of powerful genetic screens that have allowed identification of *Arabidopsis* mutants in different steps of the signaling chain. This mutant isolation has relied almost exclusively upon screening for alterations in the triple response. The triple response in etiolated *Arabidopsis* seedlings grown in the presence of ethylene or its precursor ACC is characterized by a shortened and thickened hypocotyl, an inhibition of root elongation, and the formation of an exaggerated apical hook (Guzman and Ecker, 1990; Van Der Straeten *et al.*, 1993). Mutants with an altered triple response fall into two main classes. (1) Seedlings with minor or no phenotypic response upon ethylene or ACC application, termed *ethylene/ACC-insensitive* (*ein/ain*) or *ethylene-resistant* (*etr*) mutants (Bleecker *et al.*, 1988; Roman and Ecker, 1995; Van Der Straeten *et al.*, 1993). In cases where weak responses were observed, mutants were designated *wei*, for *weak ethylene-insensitive* (Alonso *et al.*, 2003a). (2) Mutants that display a constitutive triple response in the absence of ethylene (Kieber *et al.*, 1993; Roman and Ecker, 1995). The second class can be divided in two subgroups based on whether or not the triple response can be suppressed by inhibitors of ethylene biosynthesis and perception (such as AVG and silver thiosulphate). Mutants that are unaffected by these inhibitors are the constitutive triple-response (*ctr*) mutants. On the contrary, mutants whose phenotype reverts to a normal morphology are ethylene-overproducers (*eto* mutants) that are defective in the regulation of ethylene biosynthesis. Today, the standard triple-response screen is likely saturated for the identification of viable mutants that affect ethylene responses. However, refinements of the screen continued to yield results. One refinement is to screen for mutations that display an enhanced-ethylene response at a low ethylene concentration. Using this screen, the enhanced-ethylene-response 1 (*eer1*) mutant was isolated (Larsen and Chang, 2001). Furthermore, five components of the ethylene-response pathway (*wei1–wei5*) have been identified using a low-dose screen for weak ethylene-insensitive mutants (Alonso *et al.*, 2003a). Additionally, screening methods different from those using the triple response have been applied. One method uses the responsiveness to an antagonist of ethylene (a compound that interacts with the receptor but acts as an inhibitor of ethylene responses). The *responsive to antagonist* (*ran*) mutant was isolated using trans-cyclo-octene (TCO) as an antagonist (Hirayama *et al.*, 1999). Another method exploits the phenotype of ethylene/ACC-treated light-grown *Arabidopsis* seedlings that display an elongated hypocotyl in the presence of ethylene on a low nutrient medium (LNM), a response that is absent in the *etr/ein* mutants and constitutively present in the *ctr* mutant in the absence of ethylene (Smalle *et al.*, 1997). Using this response, the *ACC-related long hypocotyl* (*alh1*) (Vandenbussche *et al.*, 2003), *slow* (*slo1*) (Zhang *et al.*, unpublished results), and the *eer2* mutant (De Paepe *et al.*, 2005) were isolated.

The gene products affected in many of the ethylene signaling mutants have been identified. A genetically defined pathway for ethylene signal transduction has been determined by epistatic (double-mutant) analysis (Fig. 2). Ethylene is perceived by a family of membrane-associated receptors, including ETR1/ETR2, ethylene response sensors ERS1/ERS2, and EIN4 (Hua *et al.*, 1998). Ethylene binds to its receptors via a copper cofactor that is probably delivered by the copper transporter RAN1. For all of them, binding of ethylene with similar affinity has been shown (Wang *et al.*, 2003). The next known downstream signaling component is the Raf-like kinase CTR1, which represents the first step in a mitogen-activated protein (MAP) kinase pathway, a signaling cascade found in eukaryotic but not in prokaryotic systems. Therefore, sensor histidine kinase-like ethylene receptors link the perception of the hormone to the activity of an MAP kinase phosphorylation cascade. Following this MAP kinase cascade is the integral membrane protein EIN2. It has homology to Nramp metal transporters, but its function is not yet understood. However, there is no doubt that a functional EIN2 is indispensable for ethylene signaling. Loss-of-function mutants in *EIN2* result in the most severe ethylene-insensitive phenotypes. The signal is then transduced to the nucleus involving the transcription factors EIN3 and EILs (EIN3-like). As shown, ethylene regulates EIN3 activity by SCF$^{EBF1/EBF2}$-dependent proteolysis (Gagne *et al.*, 2004; Guo and Ecker, 2003; Potuschak *et al.*, 2003; Yanagisawa *et al.*, 2003). EIN3 activates transcription of the *ERF1* gene, which is a member of a large family of transcription factors, the AP2-class. ERF1 in turn binds to a conserved *cis*-acting sequence within promoters of secondary target genes that eventually mediate ethylene responses (Fig. 2) (Chao *et al.*, 1997; Ohme-Takagi and Shinshi, 1995; Solano *et al.*, 1998).

A. ETHYLENE PERCEPTION

1. The Ethylene Receptors ETR1, ETR2, ERS1, ERS2, and EIN4

Ethylene is perceived by a family of five membrane-localized receptors that are homologous to bacterial two-component histidine kinases. Sensor kinases and response regulators are the two principal players of a two-component signaling cascade (Stock *et al.*, 2000). The sensor kinase (receptor-like) component typically comprises two domains: an amino-terminal input domain that perceives the signal and a carboxyl-terminal histidine protein kinase (HPK) domain that transmits the signal. The prototype response regulator consists of a conserved receiver module and a variable output domain that mediates downstream responses. In bacteria, the response regulator is usually a transcription factor containing a conserved Asp residue. Two-component

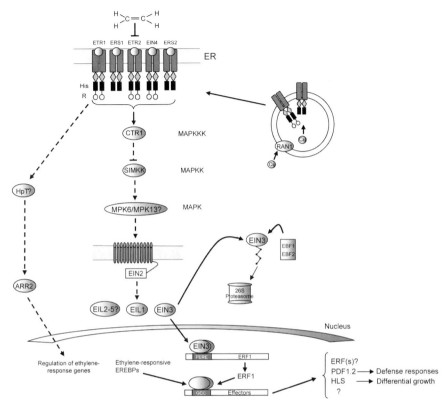

FIGURE 2. Model for ethylene signal transduction in *Arabidopsis*. The five ethylene receptors (ETR1, ERS1, ETR2, EIN4, and ERS2) are thought to be dimers and are members of the two-component receptor family that is characterized by a histidine kinase domain (His) and a receiver domain (R). The receptors are shown to be associated with the ER. The receptors are negative regulators of ethylene responses such that ethylene binding represses receptor signaling. Binding of ethylene is based on a coordination covalent bond to a copper atom associated to the ethylene-receptor apoproteins. Upon coordinating copper, the ethylene receptors are competent for ethylene binding. Copper loading of ethylene receptors is probably performed by RAN1. CTR1, the component immediately downstream of the receptors is a Raf-like protein kinase and a negative regulator of ethylene responses. The interaction of CTR1 with the receptors allows CTR1 to be localized to the ER. An MAPK module, consisting of SIMKK and MPK6 and/or MPK13, is proposed to act downstream of CTR1. In the absence of ethylene, the CTR1 Raf-like kinase is activated, negatively regulating SIMKK. When CTR1 is inactivated by ethylene, SIMKK becomes activated and in turn activates the presumed MPK6 and/or MPK13. The direct downstream targets of the MAPKs are yet to be determined. Inactivation of CTR1 results in the activation of EIN2, a positive regulator of ethylene responses, the signaling mechanism of which remains unknown. The EIN2 N-terminal transmembrane domain has similarity to the Nramp family of metal ion transporters, whereas the hydrophilic C-terminus bears no homology to known proteins. In the nucleus, an ethylene-dependent transcriptional cascade occurs. When activated by ethylene, members of the EIN3/EIL transcription factor family bind as dimers to the primary ethylene-response element (PERE) in the promoters of primary response genes such as ETHYLENE-RESPONSE-FACTOR1 (ERF1). ERF1 encodes an ethylene-response element binding protein

pathways share the following His-Asp signaling mechanism: signal perception regulates autophosphorylation of a conserved His residue in the sensor kinase; the phosphoryl group is subsequently transferred to a conserved Asp residue in the receiver domain of the response regulator, thereby modulating the activity of the output domain (Hwang *et al.*, 2002). In some cases, as in the ethylene receptors, the response regulator is not an independent polypeptide but a carboxyl-terminal domain of the sensor, in which case the receptor is called a hybrid HPK.

In *Arabidopsis*, all five ethylene receptors (ETR1, ERS1, ETR2, EIN4, and ERS2) have an amino-terminal ethylene-binding domain and the carboxyl-terminal portion has similarity to HPKs. Three of the receptors (ETR1, ETR2, and EIN4) also have a covalently attached carboxyl-terminal receiver domain and are therefore considered to be hybrid HPKs. Ethylene-response sensor 1 and ERS2 lack a receiver domain suggesting that these receptors may have a distinct signaling circuit, or perhaps they signal to a receiver domain of one of the hybrid HPK receptors. There are enough differences to support the existence of two subfamilies of ethylene receptors (Hua *et al.*, 1998). Subfamily 1 is composed of ETR1 and ERS1. Subfamily 2 is composed of ETR2, ERS2, and EIN4. In subfamily 1, there is complete conservation of the residues considered essential for histidine kinase activity. On the other hand, in subfamily 2 the histidine kinase domain lacks some essential residues. Interestingly, the *ers1etr1* double loss-of-function mutant (subfamily I receptors mutant) could not be compensated by overexpression of any of the three subfamily II receptors, whereas subfamily I overexpression restored normal growth supporting the idea that subfamily I receptors play a unique and necessary role in ethylene signaling (Wang *et al.*, 2003). However, transformation of either the *ers1–2 etr1–6* or *ers1–2 etr1–7* mutant with a kinase-inactivated ETR1 genomic clone also resulted in complete restoration of normal growth and ethylene-responsiveness in the double-mutant background, leading to the conclusion that canonical histidine kinase activity of receptors was probably not required for ethylene-receptor signaling.

(EREBP) and is an immediate target for EIN3. Ethylene-responsive EREBPs have been found in *Arabidopsis*. ERF1 and the other EREBPs bind to the GCC-box of secondary response targets, such as the defensin PDF1.2, HLS, and other ERFs, activating their transcription. These genes encode effector proteins that are needed to execute a wide variety of ethylene responses. Studies on EBF1 and 2 showed that this two-membered F-box subfamily regulates ethylene signaling by targeting the EIN3 transcriptional regulator for degradation by the 26S proteasome. Evidence suggests the existence of an additional branch of ethylene signaling. In this branch, ethylene may induce histidine kinase activity and autophosphorylation of ETR1. Phosphorylated ETR1 could then initiate a phoshorelay cascade that may include a shuttling histidine phosphotransfer (HPt) protein. This phosphotransfer could ultimately result in the transcriptional regulation of ARR2, which acts as a transcription factor in the nucleus. Arrows and t-bars represent positive and negative effects, respectively. Solid lines indicate effects that occur through direct interaction whereas dotted lines indicate effects that have not yet been shown to occur through direct interaction.

An independent line of evidence that supports this hypothesis comes from mutational studies of the dominant *etr1–1* mutant (Gamble *et al.*, 2002). Deletion of the entire carboxy-terminal half of ETR1, including the His kinase domain and the receiver domain, had little effect on *etr1–1*. A possible role for the histidine kinase activity in ETR1 is given in a research by Hass *et al.* (2004) and Qu and Schaller (2004). Hass and coworkers identified the *Arabidopsis* response regulator 2 (AAR2) as a signaling component functioning downstream of ETR1 in ethylene signaling. In addition, further investigations indicated that an ETR1-initiated phosphorelay regulates the transcription factor activity of AAR2. This mechanism could create a novel signal transfer from the ER-associated ETR1 to the nucleus for the regulation of ethylene-responsive genes. Thus, ETR1 may have a dual functional role in the initiation of ethylene signal transduction. Whereas the CTR1-dependent pathway is negatively regulated, the ARR2-dependent pathway is subject to positive regulation. Another possible reason for the retention of the histidine kinase activity in subfamily I receptors may be to provide fine tuning of the signaling pathway rather than functioning as the primary mechanism for signal transduction. The work of Qu and Schaller (2004) indicates that the His kinase domain of ETR1 plays a role in the repression of ethylene responses. Moreover, Binder *et al.* (2004a) demonstrated that receiver domains play a role in the recovery from growth inhibition since the *ers1ers2* double mutant had no effect on the recovery rate after ethylene was removed while loss-of-function mutations in *ETR1*, *ETR2*, and *EIN4* significantly prolonged the time for recovery of growth rate.

In a review, Klee (2004) suggested a plausible answer to the question why the family II receptors lack the His kinase catalytic domains. All ethylene receptors apart from ETR1 appear to have Ser/Thr kinase (STK) activity. Therefore, STK activity explains the lack of conservation of His kinase catalytic domains.

2. Ethylene Binding to the Receptors

The ethylene receptor ETR1 was the first member of the ethylene-receptor family to be characterized. The functional unit for ethylene perception is likely to be a receptor dimer (Schaller *et al.*, 1995). In the case of ETR1, homodimerization is mediated in part by amino-terminal cysteine residues that are capable of forming disulfide bonds (Schaller *et al.*, 1995). All five ethylene receptors contain these cysteine residues. ERS1, like ETR1, forms a membrane-associated, disulfide-linked dimer when expressed in yeast (Hall *et al.*, 2000). The binding of ethylene to the receptors occurs with a dissociation constant of 0.04 μl/l gaseous ethylene and a half-life of 12 h, both of which are consistent with rates observed in ethylene-binding/response assays in plants (Chen and Bleecker, 1995). ETR1 was shown to bind ethylene only in the presence of copper ions that are coordinated by two conserved amino acids (Cys65 and His69) (Rodriguez *et al.*, 1999;

Schaller and Bleecker, 1995). Further evidence for a role of copper in ethylene signaling comes from the characterization of the *RAN1* (responsive to antagonist) gene (Hirayama *et al.*, 1999). Two alleles of this locus, *ran1-1* and *ran1-2*, were identified in a screen for mutants that displayed an ethylene-like triple response upon treatment with the ethylene antagonist trans-cyclo-octene. RAN1 codes for a copper transporting P-type ATPase with homology to the yeast Ccc2p and human Menkes/Wilson disease proteins. The RAN1 protein is located in intracellular membrane compartments and is believed to deliver copper ions from intracellular stores.

Ethylene receptors were originally identified based on mutations that resulted in a dominant-insensitive phenotype (Chang *et al.*, 1993; Hua *et al.*, *et al.*, 1998). These turned out to be missense mutations within the sensory domain of the receptors (Hall *et al.*, 1999; Rodriguez *et al.*, 1999).

An interesting feature of the ethylene receptors is that they are negative regulators of the ethylene-response pathway. The negative regulation by the receptors was revealed through the analysis of loss-of-function mutants of the five members of the family (Hua and Meyerowitz, 1998; Wang *et al.*, 2003; Zhao *et al.*, 2002). Single and double loss-of-function mutants did not show obvious ethylene-related phenotypes; the only exception being the *etr1ers1* double mutant. The latter mutant and the homozygous triple and quadruple loss-of-function mutants displayed constitutive ethylene responses (Hall and Bleecker, 2003; Hua and Meyerowitz, 1998; Zhao *et al.*, 2002). Since the absence of receptors resulted in constitutive responses, the wild-type receptors must be negative regulators of ethylene responses. These results suggested a model wherein the receptors repress ethylene responses in the absence of the hormone. Consequently, when ethylene is bound, the receptors are inactivated and the pathway is derepressed leading to ethylene responses. Based on this model, less receptors would increase while more receptors would reduce ethylene sensitivity. Therefore, the triple and quadruple mutants could respond to basal levels of ethylene because less ethylene is needed to inactivate the remaining receptors. In support of this hypothesis, a re-examination of the *etr1* loss-of-function mutant by Cancel and Larsen (2002) revealed that these plants displayed enhanced sensitivity to ethylene. However, when taking into account the different characteristics of the two families of ethylene receptors in *Arabidopsis*, both classes of receptors might not have redundant functions *in vivo*, indicating that this basic model is not the full story.

Since null mutants of the receptor genes have a wild-type phenotype and their gene expression patterns widely overlap, one can argue why there are multiple ethylene receptors. Families of ethylene-receptor genes are found not only in *Arabidopsis* but in other plants as well. Previously Hua *et al.* (1998) proposed that the different ethylene receptors may possess different ethylene-binding affinities and signaling activities, enabling plants to respond to a broad spectrum of ethylene concentrations. In addition, the ethylene receptors are differentially regulated by ethylene and perhaps also by environmental or other developmental factors. Ethylene itself has been found to regulate transcription of the receptors ERS1, ERS2, and ETR2. The differential regulation of expression of the receptor gene family may provide a mechanism to achieve differential sensitivities even in the same response under different conditions (Hua *et al.*, 1998). The upregulation of receptors in response to ethylene might provide a mechanism for adaptation to ethylene. Since the half-life of ethylene binding is very long (12.5 h), an increase in the number of unbound receptors could result in desensitization of the pathway and may enable the plant to react to changes in ethylene concentration.

B. LINKING THE ETHYLENE RECEPTORS TO CTR1, A MAPKKK

The first known component downstream of the ethylene receptors is CTR1 (Kieber *et al.*, 1993). The recessive nature and constitutive phenotype of the *ctr1-1* mutant indicate that CTR1 is a negative regulator of downstream signaling events. Cloning of the *CTR1* gene revealed that it belongs to the Raf family of Ser/Thr protein kinases that initiate MAP kinase signaling cascades in mammals (Kieber *et al.*, 1993). Based on this similarity, it was suggested that CTR1 may function as a MAP kinase kinase kinase (MAPKKK). There are two known examples of pathways that combine a two-component system with an MAPK pathway. One is the *S. cerevisiae* osmosensing pathway and the second is the *S. pombe* stress-response pathway (Buck *et al.*, 2001; Posas *et al.*, 1996; Shieh *et al.*, 1997). To date, no intermediate components have been identified genetically or biochemically to act between the receptors and the CTR1 kinase. Yeast two-hybrid and *in vitro* binding have shown that both the kinase domain and receiver domain of ETR1 and the kinase domain of ERS1 (which lacks a receiver domain) can directly interact with CTR1 (Clark *et al.*, 1998). Physical interactions of the subfamily II ETR2 transmitter domain were also demonstrated but are much weaker than those reported for ETR1 and ERS1 (Cancel and Larsen, 2002). These findings provide an explanation for the observation that subfamily I receptors play a particular role in ethylene signaling. They suggest that the difference between types I and II receptors may lie in the strength of their physical association with CTR1, resulting from a differential affinity of subfamily I and II receptors for the amino-terminal domain of CTR1.

Interaction between ethylene receptors and CTR1 has been shown to occur *in planta* as well (Gao *et al.*, 2003). It was demonstrated that native CTR1 is localized to the endoplasmic reticulum of *Arabidopsis* and that this localization results from interaction with ethylene receptors.

The function of the N-terminal part of CTR1, which encodes a novel protein domain, was investigated by Huang *et al.* (2003). Deletion of the N-terminal domain did not elevate the kinase activity of CTR1, indicating that this domain does not autoinhibit kinase function *in vitro*. One missense mutation, *ctr1-8* was found to result from an amino acid substitution within a conserved motif in the N-terminal domain. *In vitro*, *ctr1-8* has no detectable effect on the kinase activity of CTR1 but rather disrupts the interaction with the ethylene receptor ETR1, suggesting that CTR1 interacts with ETR1 *in vivo* and that this association is required to switch off the ethylene signaling pathway.

All together, these results raise several possibilities for regulation of CTR1 activity. A first possibility is that CTR1 could be regulated by conformational changes in the receptors. In the absence of ethylene, the ER-bound receptors are predicted to exist in a functionally active state that is able to interact with CTR1. CTR1 is activated by association with the ER-bound receptors and represses the downstream ethylene responses. A second way to regulate CTR1 activity might be through intermediary proteins when CTR1 is complexed with the receptors. Gao *et al.* (2003) demonstrated that CTR1 is part of a signaling complex with the ethylene receptor ETR1 in plants. Single receptor mutants did not result in significant loss of CTR1 from the membrane; double and triple mutant combinations did, thereby implicating multiple receptors in the membrane localization of CTR1. In addition, the levels of ER-bound CTR1 correlated with the strength of the constitutive ethylene-response phenotypes in the multiple-receptor mutants.

The work of Ouaked *et al.* (2003) indicated that an MAPK cascade is most probably part of ethylene signaling. They demonstrated that two MAPKs are activated by ACC in *Medicago* and *Arabidopsis*. In *Medicago*, SIMK and MMK3 showed strong activation in ACC-treated cells, whereas in *Arabidopsis* this was the case for MPK6 and a 44 kDa protein, probably corresponding to MPK13. Furthermore, their analysis showed that the MAPKK SIMKK specifically mediates ACC-induced activation in *Medicago*. *Arabidopsis* plants with hyperactive SIMKK showed constitutive MAPK activation, enhanced-ethylene-induced gene expression and a triple-response phenotype in the absence of ACC.

Several observations indicated that the ethylene signal transduction pathway is not completely dependent upon the activity of CTR1. First, the *ctr1* null mutants are still capable of responding to ethylene (Larsen and Chang, 2001). Second, plants containing loss-of-function mutations in four ethylene receptors display a more severe phenotype than *ctr1* loss-of-function mutations (Hua and Meyerowitz, 1998). These observations suggested the existence of an additional branch for ethylene signaling. The results of Hass *et al.* (2004) described earlier indicated that a traditional two-component signaling system represents such a branch.

C. EIN2, AN NRAMP-LIKE PROTEIN

Genetic epistatic analysis of ethylene-response mutants has shown that EIN2 acts downstream of CTR1. Null mutants in *EIN2* result in the complete loss of ethylene-responsiveness throughout plant development, suggesting that EIN2 is an essential positive regulator in the ethylene signaling pathway (Alonso et al., 1999). EIN2 contains 12 predicted transmembrane domains in the amino-terminal part of the polypeptide. This region exhibits significant similarity to the Nramp family of cation transporters such as the yeast Smf1p and mammalian DCT1 proteins. The C-terminal hydrophilic part has no homology to any known protein, although it does contain motifs typically involved in protein–protein interactions. Overexpression of the carboxy-terminal domain of the protein (EIN2 CEND) in an EIN2 null background resulted in constitutive activation of some but not all ethylene responses. Adult transgenic plants displayed constitutive ethylene-response phenotypes and constitutively expressed ethylene-regulated genes but expression of the EIN2 carboxy-terminal region was unable to induce the triple response in dark-grown seedlings. Based on these results, it was hypothesized that the amino-terminal end of EIN2 represents an input domain, interacting with upstream signaling factors, while the carboxy-terminal region is required for transducing the signal to the downstream components. The *ein2* mutants exhibit the strongest ethylene-insensitive phenotype of all ethylene-insensitive mutants isolated in *Arabidopsis*. This indicates that EIN2 plays a critical role in ethylene signaling. However, it is still not clear how EIN2 functions in ethylene signaling. There is no precedent for an Nramp-like protein operating downstream of an MAP kinase pathway. Moreover, it is still questionable whether EIN2 functions in an analogous manner. Previous experiments to detect metal transporting activity in EIN2 have failed, and where other members of the *Arabidopsis* Nramp-like family are able to complement metal-uptake deficient yeast strains, this is not the case for EIN2 (Thomine et al., 2000). Interestingly, *ein2* mutants have been independently isolated in different genetic screens designed to identify components of other signaling pathways. For example, *ein2* mutants have been found in screens for defects in auxin transport inhibitor resistance (Fujita and Syono, 1996), cytokinin response (Su and Howell, 1992), ABA hypersensitivity (Beaudoin et al., 2000; Ghassemian et al., 2000), and delayed senescence (Oh et al., 1997). In addition, *ein2* mutants show altered sensitivity to several bacteria and fungal pathogens. Hence, EIN2 has been proposed to lie at the crossroad of multiple hormones and stress-response pathways.

D. NUCLEAR EVENTS: TWO ETHYLENE-RESPONSIVE FAMILIES OF TRANSCRIPTION FACTORS

Many ethylene responses involve changes in gene expression. Evidence for nuclear regulation in the ethylene signal transduction was given by the cloning of *EIN3* (Chao et al., 1997). *EIN3* belongs to a plant-specific multigene family with six members in *Arabidopsis*. Besides EIN3, EIN3-like 1 (EIL1) and EIL2 can also rescue the *ein3* mutant phenotypes. This explains why null mutations in *ein3* cause only partial ethylene-insensitivity. The isolation of *wei5/eil1* confirmed that *EIL1* is a component of the ethylene signaling cascade (Alonso *et al.*, 2003a). Interestingly, *eil1ein3* double-mutant seedlings were almost completely ethylene-insensitive and indistinguishable from the ethylene-response null mutant *ein2-5*, indicating that *EIL2-5* genes may not contribute to the ethylene response at this stage of development. Moreover, the overexpression of EIN3 and EIL1, but not of other EILs, confers a constitutive ethylene-response phenotype (Chao *et al.*, 1997). The more distantly related members of the EIN3 family (EIL2-5) might either play a minor role in the ethylene response in specific tissues and developmental stages or function in pathways that are unrelated to ethylene signaling. Observations by Binder *et al.* (2004b) indicated that EIN3 and EIL1 are not required for the first rapid short-term growth inhibition phase of etiolated *Arabidopsis* seedlings. Insensitivity by the *ein3eil1* double mutant was only observed in the sustained slower response phase. In contrast, EIN2 was found to be required for both the phases of growth inhibition. This raises the question as to what mechanisms give rise to the rapid response.

As shown, ethylene regulates EIN3 activity by ubiquitin-dependent proteolysis (Gagne *et al.*, 2004; Guo and Ecker, 2003; Potuschak *et al.*, 2003). In the absence of ethylene, EIN3 is quickly degraded through a ubiquitin/proteasome pathway mediated by two F-box proteins, EBF1 and EBF2 (for EIN3-binding F box protein 1 and 2). EBF1 overexpression resulted in plants insensitive to ethylene. Conversely, *ebf1ebf2* double mutants showed either constitutive ethylene responses in the studies of Potuschak *et al.* (2003) and of Guo and Ecker (2003) or a severe growth arrest as observed by Gagne *et al.* (2004). Altogether, these studies revealed that a ubiquitin/proteasome pathway negatively regulates ethylene responses by targeting EIN3 for degradation. These results add to the rapidly expanding list of plant hormones whose responses are directly regulated through the removal of a key activator/repressor by the ubiquitin/proteasome pathway (Kepinski and Leyser, 2002; McGinnis *et al.*, 2003; Smalle *et al.*, 2002, 2003). Additional evidence that proteolytic degradation is part of ethylene signaling was provided by transcript profiling experiments which indicated that early responses to ethylene include protein degradation (De Paepe *et al.*, 2004).

In the case of ethylene signaling, EIN3 degradation needs to be switched off to allow EIN3 accumulation after ethylene stimulation. Several mechanisms can be envisaged to explain how ethylene regulates EIN3 stability. Either the SCF$^{EBF1/EBF2}$ complexes are negatively regulated after ethylene perception to allow EIN3 accumulation or EIN3 is directly protected from the SCF$^{EBF1/EBF2}$ ubiquitin protein ligases (Potuschak *et al.*, 2003). The concerted action of EBF1 and 2 in ethylene action may be essential to avoid excess accumulation of EIN3 throughout the life cycle of the plant given the strong inhibitory effect of this transcriptional regulator on seedling growth and development.

In addition, a study demonstrated that glucose accelerated the degradation of EIN3 (Yanagisawa *et al.*, 2003). The antagonistic relationship between ethylene and glucose was identified earlier by the genetic and phenotypic analyses of *Arabidopsis* mutants with *glucose-insensitive* (*gin*) and *glucose-oversensitive* (*glo*) phenotypes (Zhou *et al.*, 1998). The ethylene-insensitive *etr1* and *ein2* mutants displayed *glo* phenotypes, while ctr1 was allelic to *gin4* (Leon and Sheen, 2003).

It is intriguing to note that ethylene synthesis and signaling are similarly regulated. Both ACS5 (Wang *et al.*, 2004) and the EIN3 protein are synthesized constitutively and degraded rapidly by the proteasome. Tight coupling of the ethylene biosynthesis and signaling pathways might be necessary to prevent unwanted effects of ethylene on plant development.

To date, the ethylene-inducible gene *ERF1* is the only known direct target of EIN3. EIN3 dimers interact with a unique palindromic repeat element in the promoter of ERF1, which is termed the primary ethylene-response element (PERE). Homodimers of EIL1 and EIL2 are also capable of binding to this DNA sequence *in vitro*, but no heterodimerization between these proteins and EIN3 has been reported (Solano *et al.*, 1998). *ERF1* belongs to one of the largest families of plant-specific transcription factors with 124 family members in *Arabidopsis*, referred to as ethylene-response element binding factor (ERF) proteins (Fujimoto *et al.*, 2000), a family that has also been referred to as ethylene-response element binding proteins (EREBPs) (Ohme-Takagi and Shinshi, 1995). These transcription factors are capable of binding to a secondary ethylene-response element (SERE), the GCC-box. This sequence was determined to be essential for the expression of several pathogenesis-related (*PR*) genes (Fujimoto *et al.*, 2000; Zhou *et al.*, 1998). Some *ERF* genes encode transcriptional activators, while others encode transcriptional repressors. The *AtERF* genes are differentially regulated by ethylene, by the two other plant defense hormones SA and JA, and by abiotic stress conditions, such as wounding, cold, high salinity, or drought, via EIN2-dependent or -independent pathways (Fujimoto *et al.*, 2000; Onate-Sanchez and Singh, 2002). *ERF1* (Solano *et al.*, 1998), *AtERF1* (Fujimoto *et al.*, 2000), and *AtERF14* (Onate-Sanchez and Singh, 2002) are responsive to ethylene.

Interestingly, some ERFs contain GCC boxes in their promoter, indicating that these ERFs could be targets for other members of the ERF family (Solano *et al.*, 1998). The uncovering of this transcriptional cascade represents an important step in unraveling the different players by which ethylene regulates gene expression. Furthermore, this transcriptional cascade with multimember families allows modulation by other regulatory pathways at many points.

ERF members have significant differences in their RNA expression patterns that may help to modulate the specificity of plant defense/stress gene expression in response to different signal transduction pathways. Given the prominent role the ERF proteins play in plant stress responses and the large size of the ERF family, it will be important to determine the function of each member of this large family of transcription factors.

E. OTHER GENETICALLY DEFINED COMPONENTS IN ETHYLENE SIGNALING

Besides the well-characterized mutants described in the previous sections, there are a number of other mutations that affect ethylene signaling. For some of these a clear role in the pathway has not yet been defined while for others only a subset of the ethylene responses are affected, indicating that they function downstream of the primary signal transduction pathway.

Based on double-mutant analysis the ethylene-insensitive mutants *ain1/ein5*, *ein6*, and *ein7* (Roman *et al.*, 1995; Van Der Straeten *et al.*, 1993) function downstream of CTR1.

In the recessive mutants *aux1* and *eir1/pin2/agr1* the hypocotyl is responsive to ethylene but the roots show partial ethylene-insensitivity and display an altered gravitropic response (Chen *et al.*, 1998; Luschnig *et al.*, 1998; Muller *et al.*, 1998; Roman *et al.*, 1995). *aux1* plants are auxin resistant, whereas *eir1* plants are auxin-responsive. AUX1 was identified as an auxin influx carrier functioning in the transport and redistribution of auxin in the plant (Bennett *et al.*, 1996; Marchant *et al.*, 1999). Ethylene-insensitivity in the roots of *aux1* and *eir1/pin2/agr1* may relate to the role of ethylene in regulating auxin transport (Luschnig *et al.*, 1998; Suttle, 1988). Another mutant, *hls1*, lost the ability to form a pronounced apical hook in response to ethylene in the dark (Lehman *et al.*, 1996; Roman and Ecker, 1995). The *hookless* morphology could be phenocopied by adding auxin or auxin-transport inhibitors (Lehman *et al.*, 1996). This again suggests a link between auxin and ethylene signaling. *HLS1* encodes a putative *N*-acetyltransferase (Lehman *et al.*, 1996). It was demonstrated that both ethylene and light signals affect differential cell growth by acting through HLS1 to modulate auxin-response factors, as for instance ARF2 (Li *et al.*, 2004). In addition, the *alh1* mutant (ACC-related long hypocotyl 1) is also affected in the crosstalk between ethylene and auxins. *alh1* displayed a longer hypocotyl than the wild-type in the light and in the absence of ethylene; this feature could be reverted by auxin-transport inhibitors (Vandenbussche *et al.*, 2003).

By using subthreshold levels of ethylene the *enhanced-ethylene-response* (*eer1*) mutant was identified (Larsen and Chang, 2001). The *eer1* mutant displays increased ethylene sensitivity in the hypocotyl and stem but reduced sensitivity in the root. Molecular cloning of *eer1* revealed that its mutant phenotype results from a loss-of-function mutation in RCN1, one of the three PP2AA regulatory subunits in *Arabidopsis* (Larsen and Cancel, 2003). A role for PP2A in the modulation of CTR1 activity was proposed.

Five ethylene-insensitive loci were identified by using a low-dose screen for weak ethylene-insensitive mutants (*wei1-wei5*) (Alonso *et al.*, 2003a). *wei1*, *wei2*, and *wei3* seedlings showed hormone sensitivity only in roots, whereas *wei4* and *wei5* displayed insensitivity in both roots and hypocotyls. The *wei1* mutant harbored a recessive mutation in *transport inhibitor responsive* (*TIR1*), which encodes a component of the SCF protein ubiquitin ligase involved in the auxin response. *wei4* resulted from a mutation in the ethylene receptor ERS1 and *wei5* was caused by a mutation in the *EIN3*-related transcription factor *EIL1*. Genetic mapping studies indicated that *wei2* and *wei3* correspond to previously unidentified components in the ethylene pathway.

Cloning and characterization of these genes will further expand our knowledge of the ethylene signal transduction pathway.

IV. TRANSCRIPTIONAL REGULATION OF ETHYLENE RESPONSE

The plant hormone ethylene affects many different aspects of plant development. One possible explanation for the diversity of physiological and biochemical responses to ethylene is that it regulates the expression of a myriad of genes, some specific to a certain trigger, others shared by overlapping pathways. Lincoln and Fischer (1988) investigated the mechanism of action of ethylene by analyzing the expression of ethylene-inducible genes isolated from tomato. Their results indicated the existence of multiple mechanisms for the regulation of gene expression by ethylene and that the expression of different classes of genes is regulated by ethylene in a fairly unique fashion. For a number of genes, ethylene affected transcriptional control, while for others it affected posttranscriptional processes. Furthermore, by measuring gene expression as a function of ethylene concentration, they demonstrated that the tested genes displayed a unique, narrow dose response curve in response to exogenous ethylene. In addition, the transcriptional activation of some genes was organ-specific, while for others it was not. Finally, they indicated that ethylene was capable of inducing changes in plant physiology by rapidly altering patterns of gene transcription. The uncovering of the transcriptional cascade controlling ethylene response involving two large families of transcription factors represented an important step in unraveling the different players by which ethylene regulates gene expression, although this may not appear sufficient to explain the diversity of plant responses to the hormone.

Various targeted gene expression studies have identified different ethylene-regulated genes in different processes and in different tissue types. Through differential screening techniques a number of early ethylene-regulated genes were isolated in etiolated seedlings and during tomato ripening (Trentmann, 2000; Zegzouti et al., 1999). In the first one, ERN1, an ethylene-suppressed nuclear-localized protein, was identified (Trentmann, 2000). The latter analysis yielded a number of genes involved in fruit ripening. Their study showed that the predicted proteins encoded by these genes involved a wide diversity of functions, indicating the complexity of cellular responses to ethylene. Interestingly, ethylene-dependent changes in mRNA accumulation occurred rapidly (15 min) for most of the clones isolated (Zegzouti et al., 1999).

In addition, changes in the expression patterns of 2.375 selected genes were examined simultaneously by cDNA microarray analysis in *Arabidopsis* after inoculation with an incompatible fungal pathogen or treatment with the defense-related signaling molecules salicylic acid (SA), methyl jasmonate (MJ), or ethylene (Schenk et al., 2000). Their results demonstrated that considerable interaction occurs among the different defense signaling pathways, notably between the SA and MJ pathways. Fifty percent of the genes induced by ethylene treatment were also induced by exposure to MJ.

Another study focused on transcriptional profiling of genes in response to wounding and demonstrated that besides the reported wound-induced biosynthesis of ethylene, cross-talk may also occur with the ethylene signaling pathway at the level of transcriptional regulation (Cheong et al., 2002).

Alonso et al. (2003b) reported the use of Affymetrix gene expression arrays to examine the RNA levels of more than 22.000 genes in response to ethylene. In particular, they identified four ethylene-inducible genes that encode proteins containing two plant-specific DNA binding domains, AP2 and B3 (Alonso et al., 2003b). These genes were named *ETHYLENE RESPONSE DNA BINDING FACTORS (EDF)*.

Finally, ethylene-regulated gene expression was investigated in *Arabidopsis* leaves using a cDNA microarray containing about 6000 unique genes (Van Zhong and Burns, 2003). In this study, the focus was put on late-term (24h) ethylene regulation and their results were compared to *etr1-1* and *ctr1-1*. Complementary to this work, we performed a kinetic analysis of the transcriptional cascade in the very early phase of ethylene response by means of cDNA-AFLP and cDNA-microarray technology (De Paepe et al., 2004). Cluster analysis and functional grouping of co-regulated genes allowed determining the major ethylene-regulated classes of genes. In particular, a large number of genes involved in cell rescue, disease, and defense mechanisms were identified as early ethylene-regulated genes, confirming the important role of ethylene in defense and stress responses. Furthermore, the data provided additional insight into the role of protein degradation in ethylene signaling. Finally, novel interactions between ethylene and other signaling pathways have been revealed by this study. Of particular interest is the overlap between ethylene response and responses to ABA, sugar, and auxin.

V. CROSS-TALK IN PLANT HORMONE SIGNALING

As described previously, genetic screens have been very useful in identifying factors involved in ethylene signal transduction. However, although these screens were originally designed to identify specific components in the ethylene pathway, mutations in these genes often confer changes in sensitivity to other hormones as well. Thus, alleles of mutations in ethylene signaling have also been recovered in screens using auxin-transport inhibitors, resistance to cytokinin application or in screens for suppressor and enhancer mutants of abscisic acid (ABA) mutants, or to uncover regulators of sugar metabolism (Beaudoin et al., 2000; Ghassemian et al., 2000; Vogel et al., 1998b; Zhou et al., 1998). Together, these facts indicate that a linear representation of hormone signaling pathways controlling a specific aspect of plant growth and development is not sufficient; rather do they interact with each other and with a variety of developmental and metabolic signals. Modulation of hormone sensitivity can happen in different ways.

In some cases, addition of one hormone can influence the biosynthesis of another. Previous studies showed that cytokinin treatment increased the stability of ACS5 (Chae et al., 2003). As a consequence, many of the growth defects attributed to cytokinin are the result of ethylene overproduction. This also explains cytokinin insensitivity of *ein2* mutants because mutants insensitive to ethylene obviously also are insensitive to exogenous cytokinin (Vogel et al., 1998a). In addition, auxin has been shown to stimulate ethylene biosynthesis at the level of transcription of genes encoding ACC synthase (Abel et al., 1995; Yamagami et al., 2003).

A second possibility of cross-talk resides in an integration of signal transduction routes. In this case, the physiological response is the result of complex interactions between the different signaling pathways. Auxin and ethylene coordinately regulate several developmental programs in plants. For example, in *Arabidopsis* auxin and ethylene regulate apical hook formation, root hair elongation, root growth, and hypocotyl phototropism (Harper et al., 2000; Lehman et al., 1996; Pitts et al., 1998; Rahman et al., 2001). Nevertheless, it is often unclear whether developmental effects attributed to auxin are solely due to this hormone or rather mediated by ethylene or resulting from a synergistic interaction between both hormones. Moreover, cross-talk is readily apparent in the pathogen defense response in plants, which is coordinated by ethylene, jasmonic acid (JA), and salicylic acid (SA). These three signaling factors are sometimes required individually and sometimes in concert for mobilizing defense responses to different pathogens (Glazebrook, 1999). In addition, a microarray analysis indicated the coordination between these three signaling pathways, supported by a big overlap in gene expression, especially between jasmonate and ethylene (Schenk et al., 2000). Previous studies have shown that both ethylene and JA are required for the induction of the defensin gene *PDF1.2* (plant defensin 1.2) in response to the avirulent fungal pathogen *Alternaria brassicicola* (Penninckx *et al.*, 1998). These findings suggested that ERF1 might be an essential factor for both hormone signals. It has been shown that ethylene and jasmonate pathways converge in the transcriptional cascade of ERF1 (Lorenzo *et al.*, 2003). The expression of ERF1 can be activated rapidly by ethylene or jasmonate but also synergistically by both hormones. Moreover, blocking either pathway by mutations prevents *ERF1* induction by the two hormones either alone or in combination; therefore, both signaling pathways are required concurrently for the induction of ERF1 expression and the activation of its target gene *PDF1.2*. These results suggest that ERF1 acts downstream of the intersection between ethylene and jasmonate signaling pathways and that this transcription factor is a key element in the integration of both signals for the regulation of defense response genes. In addition, another member of the ERF family, *AtERF2*, was rapidly induced after ethylene and jasmonate treatment (Lorenzo *et al.*, 2003).

Interactions between the hormone signaling pathways can also be depend on the developmental state of the plant and the specific response evaluated. Genetic analysis of ethylene and ABA interactions suggested that these hormones antagonize each other at the level of germination. Previously, two independent screens designed to discover mutants involved in ABA-responsiveness identified ethylene signaling mutants (Beaudoin *et al.*, 2000; Ghassemian *et al.*, 2000). *era3* mutants, which were originally identified as ABA hypersensitive, were found to be allelic to *ein2*. Furthermore, *ctr1* and *ein2* mutants were identified as enhancers and suppressors of *abi1* mutants, respectively. Other ethylene-insensitive mutants also showed increased ABA-responsiveness, leading to the conclusion that ethylene is a negative regulator of ABA signaling in *Arabidopsis* seeds. Ethylene appears to promote seed germination by altering endogenous ABA levels and/or by decreasing the sensitivity of the seeds to ABA. After germination, ABA and ethylene signaling display complex interactions. Mutations in the ethylene-insensitive mutant *etr1* reduced the sensitivity of roots to exogenous ABA, indicating that both hormones act additively with respect to root growth (Beaudoin *et al.*, 2000; Ghassemian *et al.*, 2000). Therefore, models in which they act in the same or parallel pathways are proposed. However, ethylene-overproducing mutants have decreased ABA sensitivity, implying another antagonistic interaction. One suggested explanation for this apparent inconsistency is that ABA inhibits root growth by signaling through the ethylene-response pathway but is unable to use this pathway in the presence of ethylene (Ghassemian *et al.*, 2000).

By investigating root growth and apical hook formation, regarded as two of the hallmarks of the triple response, it was demonstrated that ethylene regulates these phenomena at least in part via alteration of the properties of DELLA nuclear growth repressors (Achard *et al.*, 2003; Vriezen *et al.*, 2004).

DELLA proteins were first described as gibberellin (GA) signaling components (Lee *et al.*, 2002; Peng *et al.*, 1997; Silverstone *et al.*, 2001; Wen and Chang, 2002). Moreover, experiments have shown that auxin promotes root growth by modulating DELLA function (Fu and Harberd, 2003). Therefore, it is suggested that DELLA proteins provide a connection between ethylene, auxin, and GA. Interaction between auxin, ethylene, and gibberellin was also shown to occur in the promotion of hypocotyl growth and stomatal development in light-grown *Arabidopsis* seedlings (Saibo *et al.*, 2003).

In the future, comparison of genetic interaction maps with patterns based on transcript profiling and other genomics technologies may allow a more comprehensive representation of hormone interactions within the cell.

VI. ETHYLENE IN PLANT DISEASE RESISTANCE AND ABIOTIC STRESSES

Besides its physiological roles in different developmental stages, ethylene is also a stress hormone. Its synthesis is induced by a variety of stress signals, such as mechanical wounding, chemicals and metals, drought, extreme temperatures, and pathogen infection (Johnson and Ecker, 1998; Kende, 1993). However, depending on the type of pathogen and plant species and on the offensive strategies of the pathogen, the role of ethylene can be essentially different. Ethylene-insensitive signaling mutants may show either increased susceptibility or increased resistance. For example, in *Arabidopsis*, the *ein2-1* mutant developed only minimal disease symptoms when challenged with virulent *Pseudomonas syringae* pv. *tomato* or *Xanthomonas campestris* pv. *campestris*, whereas wild-type plants were susceptible (Bent *et al.*, 1992). In addition, the fungal toxin fumonisin B1 only marginally affected the viability of protoplasts from the *etr1-1* mutant and presence of the *ein2-1* mutation reduced cell death in the accelerated cell death 5 (*acd5*) mutant, supporting a role for ethylene in the regulation of programmed cell death (Asai *et al.*, 2000; Greenberg *et al.*, 2000). On the contrary, *ein2-1* showed markedly enhanced susceptibility to two different strains of the necrotrophic fungus *Botrytis cinerea* (Thomma *et al.*, 1999). Conversely, constitutive expression of *ERF1* in *Arabidopsis* is sufficient to confer resistance to both *B. cinerea* and *Pseudomonas cucumerina* (Berrocal-Lobo *et al.*, 2002). In conclusion, these data provide strong support to the notion that ethylene can play a balanced role in mounting disease resistance, the outcome of which is dependent on the nature of the pathogen. This apparent discrepancy among the roles of ethylene in different plant–pathogen interactions may be reconciled by the different infection mechanisms of different pathogens, and by the fact that ethylene is not only involved in pathogen response but is a hormone implicated in many general aspects of plant development including senescence, cell death, and ripening (Abeles *et al.*, 1992).

Besides its involvement in pathogen infection, ethylene is also implicated in response to abiotic stresses. An enhanced-ethylene emanation is one of the earliest responses to ozone stress (Moeder et al., 2002; Vahala et al., 1998). The expression of *ACS6* in *Arabidopsis* is activated within 30 min after the onset of ozone exposure (Vahala et al., 1998). Exposure to ozone leads to a rapid oxidative burst that evokes a local cell death response similar to that caused by the hypersensitive response upon pathogen infection (Pell et al., 1997). An ozone-insensitive mutant, radical-induced cell death (*rcd1*), has been to have a higher susceptibility to the oxidative burst (Overmyer et al., 2000). The prolonged cell death response was suppressed in *rcd1* by mutations in *EIN2*, suggesting that ethylene signaling is required for cell death (Overmyer et al., 2000). It is suggested that ethylene is involved in the regulation of cell death by amplifying ROS production that is responsible for the execution and spreading of cell death. Compared with the two other hormones involved in responses to abiotic stresses and pathogen defense, ethylene is involved in the early responses whereas JA and SA may control more prolonged effects. Tuominen et al. (2004) reported that ozone-induced spreading of cell death is stimulated by early, rapid accumulation of ethylene that can suppress the protective function of JA, thereby allowing cell death to proceed. Extended spreading of cell death induces late accumulation of JA that inhibits the propagation of cell death through inhibition of the ethylene pathway.

VII. CONCLUSIONS

The ethylene pathway is probably the best-defined signaling pathway in plants. Moreover, in the coming years, many more advances in understanding the mechanisms of ethylene perception, signal transduction, and transcriptional regulation will take place. Some major questions remain unanswered. The mode of ethylene perception and the activation of the receptors are not yet clear. It remains an open question whether phosphotransfer by the receptors is the primary mode of signal transmission to downstream components in the pathway. In addition, the role of an MAPK cascade in the transmission of the ethylene signal is still vague. It is also important to determine the function of each member of the EIN3/EIL family and ethylene-responsive ERFs in downstream ethylene responses. Finally, relatively little is known about how hormonal signals are used in a combinatorial manner to achieve distinct outcomes. An illustration of the importance of signal integration is the *ein2* mutant. Why does *ein2* consistently show up in screens for so many hormones? The answer can only lay in the fact that ethylene plays a critical role in mediating responses to many environmental stimuli, which is also reflected by a highly regulated ethylene biosynthesis. Therefore, a significant challenge is to identify the points of pathway intersections and to determine the result of this cross-talk at the biological level. By a detailed transcriptional kinetic analysis, we could distinguish different roles for ethylene in time, indicating that short- and long-term ethylene actions need to be further examined to better understand the multiple roles of the smallest of plant hormones.

REFERENCES

Abel, S., Nguyen, M. D., Chow, W., and Theologis, A. (1995). ACS4, a primary indoleacetic acid-responsive gene encoding 1-aminocyclopropane-1-carboxylate synthase in *Arabidopsis thaliana*. Structural characterization, expression in *Escherichia coli*, and expression characteristics in response to auxin. *J. Biol. Chem.* **270**, 19093–19099.

Abeles, F. B., Morgan, P. W., and Saltveit, M. E. (1992). "Ethylene in Plant Biology" 2nd Ed. Academic Press, San Diego.

Achard, P., Vriezen, W. H., Van Der Straeten, D., and Harberd, N. P. (2003). Ethylene regulates *Arabidopsis* development via the modulation of DELLA protein growth repressor function. *Plant Cell* **15**, 2816–2825.

Alonso, J. M., Hirayama, T., Roman, G., Nourizadeh, S., and Ecker, J. R. (1999). EIN2, a bifunctional transducer of ethylene and stress responses in *Arabidopsis*. *Science* **284**, 2148–2152.

Alonso, J. M., Stepanova, A. N., Solano, R., Wisman, E., Ferrari, S., Ausubel, F. M., and Ecker, J. R. (2003a). Five components of the ethylene-response pathway identified in a screen for weak ethylene-insensitive mutants in *Arabidopsis*. *Proc. Natl. Acad. Sci. USA* **100**, 2992–2997.

Alonso, J. M., Stepanova, A. N., Leisse, T. J., Kim, C. J., Chen, H., Shinn, P., Stevenson, D. K., Zimmerman, J., Barajas, P., Cheuk, R., Gadrinab, C., Heller, C., Jeske, A., Koesema, E., Meyers, C. C., Parker, H., Prednis, L., Ansari, Y., Choy, N., Deen, H., Geralt, M., Hazari, N., Hom, E., Karnes, M., Mulholland, C., Ndubaku, R., Schmidt, I., Guzman, P., Aguilar-Henonin, L., Schmid, M., Weigel, D., Carter, D. E., Marchand, T., Risseeuw, E., Brogden, D., Zeko, A., Crosby, W. L., Berry, C. C., and Ecker, J. R. (2003b). Genome-wide insertional mutagenesis of *Arabidopsis thaliana*. *Science* **301**, 653–657.

Asai, T., Stone, J. M., Heard, J. E., Kovtun, Y., Yorgey, P., Sheen, J., and Ausubel, F. M. (2000). Fumonisin B1-induced cell death in *Arabidopsis* protoplasts requires jasmonate-, ethylene-, and salicylate-dependent signaling pathways. *Plant Cell* **12**, 1823–1836.

Barry, C. S., Blume, B., Bouzayen, M., Cooper, W., Hamilton, A. J., and Grierson, D. (1996). Differential expression of the 1-aminocyclopropane-1-carboxylate oxidase gene family of tomato. *Plant J.* **9**, 525–535.

Beaudoin, N., Serizet, C., Gosti, F., and Giraudat, J. (2000). Interactions between abscisic acid and ethylene signaling cascades. *Plant Cell* **12**, 1103–1115.

Bennett, M. J., Marchant, A., Green, H. G., May, S. T., Ward, S. P., Millner, P. A., Walker, A. R., Schulz, B., and Feldmann, K. A. (1996). *Arabidopsis* AUX1 gene: A permease-like regulator of root gravitropism. *Science* **273**, 948–950.

Bent, A. F., Innes, R. W., Ecker, J. R., and Staskawicz, B. J. (1992). Disease development in ethylene-insensitive *Arabidopsis thaliana* infected with virulent and avirulent *Pseudomonas* and *Xanthomonas* pathogens. *Mol. Plant Microbe Interact.* **5**, 372–378.

Berrocal-Lobo, M., Molina, A., and Solano, R. (2002). Constitutive expression of ETHYLENE-RESPONSE-FACTOR1 in *Arabidopsis* confers resistance to several necrotrophic fungi. *Plant J.* **29**, 23–32.

Binder, B. M., O'Malley, R. C., Wang, W., Moore, J. M., Parks, B. M., Spalding, E. P., and Bleecker, A. B. (2004a). *Arabidopsis* seedling growth response and recovery to ethylene. A kinetic analysis. *Plant Physiol.* **136**, 2913–2920.

Binder, B. M., Mortimore, L. A., Stepanova, A. N., Ecker, J. R., and Bleecker, A. B. (2004b). Short-term growth responses to ethylene in *Arabidopsis* seedlings are EIN3/EIL1 independent. *Plant Physiol.* **136,** 2921–2927.

Bleecker, A. B., Estelle, M. A., Somerville, C., and Kende, H. (1988). Insensitivity to ethylene conferred by a dominant mutation in *Arabidopsis thaliana* seedlings. *Science* **241,** 1086–1089.

Bleecker, A. B., and Kende, H. (2000). Ethylene: A gaseous signal molecule in plants. *Annu. Rev. Cell. Dev. Biol.* **16,** 1–18.

Botella, J. R., Arteca, R. N., and Frangos, J. A. (1995). A mechanical strain-induced 1-aminocyclopropane-1-carboxylic acid synthase gene. *Proc. Natl. Acad. Sci. USA* **92,** 1595–1598.

Buck, V., Quinn, J., Soto Pino, T., Martin, H., Saldanha, J., Makino, K., Morgan, B. A., and Millar, J. B. (2001). Peroxide sensors for the fission yeast stress-activated mitogen-activated protein kinase pathway. *Mol. Biol. Cell* **12,** 407–419.

Cancel, J. D., and Larsen, P. B. (2002). Loss-of-function mutations in the ethylene receptor ETR1 cause enhanced sensitivity and exaggerated response to ethylene in *Arabidopsis*. *Plant Physiol.* **129,** 1557–1567.

Cary, A. J., Liu, W., and Howell, S. H. (1995). Cytokinin action is coupled to ethylene in its effects on the inhibition of root and hypocotyl elongation in *Arabidopsis thaliana* seedlings. *Plant Physiol.* **107,** 1075–1082.

Chae, H. S., Faure, F., and Kieber, J. J. (2003). The eto1, eto2, and eto3 mutations and cytokinin treatment increase ethylene biosynthesis in *Arabidopsis* by increasing the stability of ACS protein. *Plant Cell* **15,** 545–559.

Chang, C., Kwok, S. F., Bleecker, A. B., and Meyerowitz, E. M. (1993). *Arabidopsis* ethylene-response gene ETR1: Similarity of product to two-component regulators. *Science* **262,** 539–544.

Chao, Q., Rothenberg, M., Solano, R., Roman, G., Terzaghi, W., and Ecker, J. R. (1997). Activation of the ethylene gas response pathway in *Arabidopsis* by the nuclear protein ETHYLENE-INSENSITIVE3 and related proteins. *Cell* **89,** 1133–1144.

Chen, Q. G., and Bleecker, A. B. (1995). Analysis of ethylene signal-transduction kinetics associated with seedling-growth response and chitinase induction in wild-type and mutant *Arabidopsis*. *Plant Physiol.* **108,** 597–607.

Chen, R., Hilson, P., Sedbrook, J., Rosen, E., Caspar, T., and Masson, P. H. (1998). The *Arabidopsis thaliana* AGRAVITROPIC 1 gene encodes a component of the polar-auxin-transport efflux carrier. *Proc. Natl. Acad. Sci. USA* **95,** 15112–15117.

Cheong, Y. H., Chang, H. S., Gupta, R., Wang, X., Zhu, T., and Luan, S. (2002). Transcriptional profiling reveals novel interactions between wounding, pathogen, abiotic stress, and hormonal responses in *Arabidopsis*. *Plant Physiol.* **129,** 661–677.

Clark, K. L., Larsen, P. B., Wang, X., and Chang, C. (1998). Association of the *Arabidopsis* CTR1 Raf-like kinase with the ETR1 and ERS ethylene receptors. *Proc. Natl. Acad. Sci. USA* **95,** 5401–5406.

De Paepe, A., Vuylsteke, M., Van Hummelen, P., Zabeau, M., and Van Der Straeten, D. (2004). Transcriptional profiling by cDNA-AFLP and microarray analysis reveals novel insights into the early response to ethylene in *Arabidopsis*. *Plant J.* **39,** 537–559.

De Paepe, A., De Grauwe, L., Bertrand, S., Smalle, J., and Van Der Straeten, D. (2005). The *Arabidopsis* mutant *eer2* has enhanced ethylene responses in the light. *J. Exp. Bot.* in press.

Fu, X., and Harberd, N. P. (2003). Auxin promotes *Arabidopsis* root growth by modulating gibberellin response. *Nature* **421,** 740–743.

Fujimoto, S. Y., Ohta, M., Usui, A., Shinshi, H., and Ohme-Takagi, M. (2000). *Arabidopsis* ethylene-responsive element binding factors act as transcriptional activators or repressors of GCC box-mediated gene expression. *Plant Cell* **12,** 393–404.

Fujita, H., and Syono, K. (1996). Genetic analysis of the effects of polar auxin transport inhibitors on root growth in *Arabidopsis thaliana*. *Plant Cell. Physiol.* **37,** 1094–1101.

Gagne, J. M., Smalle, J., Gingerich, D. J., Walker, J. M., Yoo, S. D., Yanagisawa, S., and Vierstra, R. D. (2004). *Arabidopsis* EIN3-binding F-box 1 and 2 form ubiquitin-protein ligases that repress ethylene action and promote growth by directing EIN3 degradation. *Proc. Natl. Acad. Sci. USA* **101**, 6803–6808.

Gamble, R. L., Qu, X., and Schaller, G. E. (2002). Mutational analysis of the ethylene receptor ETR1. Role of the histidine kinase domain in dominant ethylene insensitivity. *Plant Physiol.* **128**, 1428–1438.

Gane, R. (1934). Production of ethylene by some ripening fruits. *Nature* **134**, 1008.

Gao, Z., Chen, Y. F., Randlett, M. D., Zhao, X. C., Findell, J. L., Kieber, J. J., and Schaller, G. E. (2003). Localization of the Raf-like kinase CTR1 to the endoplasmic reticulum of *Arabidopsis* through participation in ethylene receptor signaling complexes. *J. Biol. Chem.* **278**, 34725–34732.

Ghassemian, M., Nambara, E., Cutler, S., Kawaide, H., Kamiya, Y., and McCourt, P. (2000). Regulation of abscisic acid signaling by the ethylene response pathway in *Arabidopsis*. *Plant Cell* **12**, 1117–1126.

Glazebrook, J. (1999). Genes controlling expression of defense responses in *Arabidopsis*. *Curr. Opin. Plant Biol.* **2**, 280–286.

Gomez-Lim, M. A., Valdes-Lopez, V., Cruz-Hernandez, A., and Saucedo-Arias, L. J. (1993). Isolation and characterization of a gene involved in ethylene biosynthesis from *Arabidopsis thaliana*. *Gene* **134**, 217–221.

Greenberg, J. T., Silverman, F. P., and Liang, H. (2000). Uncoupling salicylic acid-dependent cell death and defense-related responses from disease resistance in the *Arabidopsis* mutant acd5. *Genetics* **156**, 341–350.

Guo, H., and Ecker, J. R. (2003). Plant responses to ethylene gas are mediated by SCF(EBF1/EBF2)-dependent proteolysis of EIN3 transcription factor. *Cell* **115**, 667–677.

Guzman, P., and Ecker, J. R. (1990). Exploiting the triple response of *Arabidopsis* to identify ethylene-related mutants. *Plant Cell* **2**, 513–523.

Hall, A. E., and Bleecker, A. B. (2003). Analysis of combinatorial loss-of-function mutants in the *Arabidopsis* ethylene receptors reveals that the ers1 etr1 double mutant has severe developmental defects that are EIN2 dependent. *Plant Cell.* **15**, 2032–2041.

Hall, A. E., Chen, Q. G., Findell, J. L., Schaller, G. E., and Bleecker, A. B. (1999). The relationship between ethylene binding and dominant insensitivity conferred by mutant forms of the ETR1 ethylene receptor. *Plant Physiol.* **121**, 291–300.

Hall, A. E., Findell, J. L., Schaller, G. E., Sisler, E. C., and Bleecker, A. B. (2000). Ethylene perception by the ERS1 protein in *Arabidopsis*. *Plant Physiol.* **123**, 1449–1458.

Harper, R. M., Stowe-Evans, E. L., Luesse, D. R., Muto, H., Tatematsu, K., Watahiki, M. K., Yamamoto, K., and Liscum, E. (2000). The NPH4 locus encodes the auxin response factor ARF7, a conditional regulator of differential growth in aerial *Arabidopsis* tissue. *Plant Cell* **12**, 757–770.

Hass, C., Lohrmann, J., Albrecht, V., Sweere, U., Hummel, F., Yoo, S. D., Hwang, I., Zhu, T., Schafer, E., Kudla, J., and Harter, K. (2004). The response regulator 2 mediates ethylene signalling and hormone signal integration in *Arabidopsis*. *EMBO J.* **18**, 3290–3302.

Hirayama, T., Kieber, J. J., Hirayama, N., Kogan, M., Guzman, P., Nourizadeh, S., Alonso, J. M., Dailey, W. P., Dancis, A., and Ecker, J. R. (1999). RESPONSIVE-TO-ANTAGONIST1, a Menkes/Wilson disease-related copper transporter, is required for ethylene signaling in *Arabidopsis*. *Cell* **97**, 383–393.

Hua, J., Chang, C., Sun, Q., and Meyerowitz, E. M. (1995). Ethylene insensitivity conferred by *Arabidopsis* ERS gene. *Science* **22**, 1712–1714.

Hua, J., and Meyerowitz, E. M. (1998). Ethylene responses are negatively regulated by a receptor gene family in *Arabidopsis thaliana*. *Cell* **94**, 261–271.

Hua, J., Sakai, H., Nourizadeh, S., Chen, Q. G., Bleecker, A. B., Ecker, J. R., and Meyerowitz, E. M. (1998). EIN4 and ERS2 are members of the putative ethylene receptor gene family in *Arabidopsis*. *Plant Cell* **10**, 1321–1332.

Huang, Y., Li, H., Hutchison, C. E., Laskey, J., and Kieber, J. J. (2003). Biochemical and functional analysis of CTR1, a protein kinase that negatively regulates ethylene signaling in *Arabidopsis*. *Plant J.* **33**, 221–233.

Hwang, I., Chen, H. C., and Sheen, J. (2002). Two-component signal transduction pathways in *Arabidopsis*. *Plant Physiol.* **129**, 500–515.

Johnson, P. R., and Ecker, J. R. (1998). The ethylene gas signal transduction pathway: A molecular perspective. *Annu. Rev. Genet.* **32**, 227–254.

Kende, H. (1993). Ethylene biosynthesis. *Annu. Rev. Plant Physiol. Plant Mol. Biol.* **44**, 283–307.

Kepinski, S., and Leyser, O. (2002). Ubiquitination and auxin signaling: A degrading story. *Plant Cell* **14** (Suppl.), S81–S95.

Kieber, J. J., Rothenberg, M., Roman, G., Feldmann, K. A., and Ecker, J. R. (1993). CTR1, a negative regulator of the ethylene response pathway in *Arabidopsis*, encodes a member of the raf family of protein kinases. *Cell* **72**, 427–441.

Kim, W. T., and Yang, S. F. (1994). Structure and expression of cDNAs encoding 1-aminocyclopropane-1-carboxylate oxidase homologs isolated from excised mung bean hypocotyls. *Planta* **194**, 223–229.

Klee, H. J. (2004). Ethylene signal transduction. Moving beyond *Arabidopsis*. *Plant Physiol.* **135**, 660–667.

Larsen, P. B., and Chang, C. (2001). The *Arabidopsis* eer1 mutant has enhanced ethylene responses in the hypocotyl and stem. *Plant Physiol.* **125**, 1061–1073.

Larsen, P. B., and Cancel, J. D. (2003). Enhanced ethylene responsiveness in the *Arabidopsis* eer1 mutant results from a loss-of-function mutation in the protein phosphatase 2A A regulatory subunit, RCN1. *Plant J.* **34**, 709–718.

Lasserre, E., Bouquin, T., Hernandez, J. A., Bull, J., Pech, J. C., and Balague, C. (1996). Structure and expression of three genes encoding ACC oxidase homologs from melon (*Cucumis melo* L.). *Mol. Gen. Genet.* **251**, 81–90.

Lee, S., Cheng, H., King, K. E., Wang, W., He, Y., Hussain, A., Lo, J., Harberd, N. P., and Peng, J. (2002). Gibberellin regulates *Arabidopsis* seed germination via RGL2, a GAI/RGA-like gene whose expression is up-regulated following imbibition. *Genes Dev.* **16**, 646–658.

Lehman, A., Black, R., and Ecker, J. R. (1996). HOOKLESS1, an ethylene response gene, is required for differential cell elongation in the *Arabidopsis* hypocotyl. *Cell* **85**, 183–194.

Leon, P., and Sheen, J. (2003). Sugar and hormone connections. *Trends Plant Sci.* **8**, 110–116.

Li, H., Johnson, P., Stepanova, A., Alonso, J. M., and Ecker, J. R. (2004). Convergence of signaling pathways in the control of differential cell growth in *Arabidopsis*. *Dev. Cell* **7**, 193–204.

Liang, X., Abel, S., Keller, J. A., Shen, N. F., and Theologis, A. (1992). The 1-aminocyclopropane-1-carboxylate synthase gene family of *Arabidopsis thaliana*. *Proc. Natl. Acad. Sci. USA* **89**, 11046–11050.

Liang, X., Oono, Y., Shen, N. F., Kohler, C., Li, K., Scolnik, P. A., and Theologis, A. (1995). Characterization of two members (ACS1 and ACS3) of the 1-aminocyclopropane-1-carboxylate synthase gene family of *Arabidopsis thaliana*. *Gene* **167**, 17–24.

Liang, X., Shen, N. F., and Theologis, A. (1996). Li(+)-regulated 1-aminocyclopropane-1-carboxylate synthase gene expression in *Arabidopsis thaliana*. *Plant J.* **10**, 1027–1036.

Lincoln, J. E., and Fischer, R. L. (1988). Diverse mechanisms for the regulation of ethylene-inducible gene expression. *Mol. Gen. Genet.* **212**, 71–75.

Lorenzo, O., Piqueras, R., Sanchez-Serrano, J. J., and Solano, R. (2003). ETHYLENE RESPONSE FACTOR1 integrates signals from ethylene and jasmonate pathways in plant defense. *Plant Cell* **15**, 165–178.

Luschnig, C., Gaxiola, R. A., Grisafi, P., and Fink, G. R. (1998). EIR1, a root-specific protein involved in auxin transport, is required for gravitropism in *Arabidopsis thaliana*. *Genes Dev.* **12,** 175–187.

Marchant, A., Kargul, J., May, S. T., Muller, P., Delbarre, A., Perrot-Rechenmann, C., and Bennett, M. J. (1999). AUX1 regulates root gravitropism in *Arabidopsis* by facilitating auxin uptake within root apical tissues. *EMBO J.* **18,** 2066–2073.

McGinnis, K. M., Thomas, S. G., Soule, J. D., Strader, L. C., Zale, J. M., Sun, T. P., and Steber, C. M. (2003). The *Arabidopsis* SLEEPY1 gene encodes a putative F-box subunit of an SCF E3 ubiquitin ligase. *Plant Cell* **15,** 1120–1130.

Mehta, P. K., Hale, T. I., and Christen, P. (1993). Aminotransferases: Demonstration of homology and division into evolutionary subgroups. *Eur. J. Biochem.* **214,** 549–561.

Moeder, W., Barry, C. S., Tauriainen, A. A., Betz, C., Tuomainen, J., Utriainen, M., Grierson, D., Sandermann, H., Langebartels, C., and Kangasjarvi, J. (2002). Ethylene synthesis regulated by biphasic induction of 1-aminocyclopropane-1-carboxylic acid synthase and 1-aminocyclopropane-1-carboxylic acid oxidase genes is required for hydrogen peroxide accumulation and cell death in ozone-exposed tomato. *Plant Physiol.* **130,** 1918–1926.

Muller, A., Guan, C., Galweiler, L., Tanzler, P., Huijser, P., Marchant, A., Parry, G., Bennett, M., Wisman, E., and Palme, K. (1998). AtPIN2 defines a locus of *Arabidopsis* for root gravitropism control. *EMBO J.* **17,** 6903–6911.

Nadeau, J. A., Zhang, X. S., Nair, H., and O'Neill, S. D. (1993). Temporal and spatial regulation of 1-aminocyclopropane-1-carboxylate oxidase in the pollination-induced senescence of orchid flowers. *Plant Physiol.* **103,** 31–39.

Neljubov, D. (1901). Uber die horizontale Nutation der Stengel von *Pisum sativum* und einiger Anderer. *Pflanzen Beih. Bot. Zentralb.* **10,** 128–139.

Oh, S. A., Park, J. H., Lee, G. I., Paek, K. H., Park, S. K., and Nam, H. G. (1997). Identification of three genetic loci controlling leaf senescence in *Arabidopsis thaliana*. *Plant J.* **12,** 527–535.

Ohme-Takagi, M., and Shinshi, H. (1995). Ethylene-inducible DNA binding proteins that interact with an ethylene-responsive element. *Plant Cell* **7,** 173–182.

Onate-Sanchez, L., and Singh, K. B. (2002). Identification of *Arabidopsis* ethylene-responsive element binding factors with distinct induction kinetics after pathogen infection. *Plant Physiol.* **128,** 1313–1322.

Ouaked, F., Rozhon, W., Lecourieux, D., and Hirt, H. (2003). A MAPK pathway mediates ethylene signaling in plants. *EMBO J.* **22,** 1282–1288.

Overmyer, K., Tuominen, H., Kettunen, R., Betz, C., Langebartels, C., Sandermann, H., Jr., and Kangasjarvi, J. (2000). Ozone-sensitive *Arabidopsis* rcd1 mutant reveals opposite roles for ethylene and jasmonate signaling pathways in regulating superoxide-dependent cell death. *Plant Cell* **12,** 1849–1862.

Pell, E. J., Schlagnhaufer, C. D., and Arteca, R. N. (1997). Ozone-induced oxidative stress: Mechanisms of action and reaction. *Physiol. Plant.* **100,** 264–273.

Peng, J., Carol, P., Richards, D. E., King, K. E., Cowling, R. J., Murphy, G. P., and Harberd, N. P. (1997). The *Arabidopsis* GAI gene defines a signaling pathway that negatively regulates gibberellin responses. *Genes Dev.* **11,** 3194–3205.

Penninckx, I. A., Thomma, B. P., Buchala, A., Metraux, J. P., and Broekaert, W. F. (1998). Concomitant activation of jasmonate and ethylene response pathways is required for induction of a plant defensin gene in *Arabidopsis*. *Plant Cell* **10,** 2103–2113.

Pitts, R. J., Cernac, A., and Estelle, M. (1998). Auxin and ethylene promote root hair elongation in *Arabidopsis*. *Plant J.* **16,** 553–560.

Posas, F., Wurgler-Murphy, S. M., Maeda, T., Witten, E. A., Thai, T. C., and Saito, H. (1996). Yeast HOG1 MAP kinase cascade is regulated by a multistep phosphorelay mechanism in the SLN1-YPD1-SSK1 "two-component" osmosensor. *Cell* **86,** 865–875.

Potuschak, T., Lechner, E., Parmentier, Y., Yanagisawa, S., Grava, S., Koncz, C., and Genschik, P. (2003). EIN3-dependent regulation of plant ethylene hormone signaling by two *Arabidopsis* F box proteins: EBF1 and EBF2. *Cell* **115,** 679–689.

Qu, X., and Schaller, G. E. (2004). Requirement of the histidine kinase domain for signal transduction by the ethylene receptor ETR1. *Plant Physiol.* **136,** 2961–2970.

Rahman, A., Amakawa, T., Goto, N., and Tsurumi, S. (2001). Auxin is a positive regulator for ethylene-mediated response in the growth of *Arabidopsis* roots. *Plant Cell Physiol.* **42,** 301–307.

Ravanel, S., Gakiere, B., Job, D., and Douce, R. (1998). The specific features of methionine biosynthesis and metabolism in plants. *Proc. Natl. Acad. Sci. USA* **95,** 7805–7812.

Raz, V., and Ecker, J. R. (1999). Regulation of differential growth in the apical hook of *Arabidopsis. Development* **126,** 3661–3668.

Rodrigues-Pousada, R. A., De Rycke, R., Dedonder, A., Van Caeneghem, W., Engler, G., Van Montagu, M., and Van Der Straeten, D. (1993). The *Arabidopsis* 1-aminocyclopropane-1-carboxylate synthase gene 1 is expressed during early development. *Plant Cell* **5,** 897–911.

Rodriguez, F. I., Esch, J. J., Hall, A. E., Binder, B. M., Schaller, G. E., and Bleecker, A. B. (1999). A copper cofactor for the ethylene receptor ETR1 from *Arabidopsis. Science* **12,** 996–998.

Roman, G., and Ecker, J. R. (1995). Genetic analysis of a seedling stress response to ethylene in *Arabidopsis. Philos. Trans. R. Soc. Lond. B. Biol. Sci.* **350,** 75–81.

Roman, G., Lubarsky, B., Kieber, J. J., Rothenberg, M., and Ecker, J. R. (1995). Genetic analysis of ethylene signal transduction in *Arabidopsis thaliana*: Five novel mutant loci integrated into a stress response pathway. *Genetics* **139,** 1393–1409.

Saibo, N. J., Vriezen, W. H., Beemster, G. T., and Van Der Straeten, D. (2003). Growth and stomata development of *Arabidopsis* hypocotyls are controlled by gibberellins and modulated by ethylene and auxins. *Plant J.* **33,** 989–1000.

Sakai, H., Hua, J., Chen, Q. G., Chang, C., Medrano, L. J., Bleecker, A. B., and Meyerowitz, E. M. (1998). ETR2 is an ETR1-like gene involved in ethylene signaling in *Arabidopsis. Proc. Natl. Acad. Sci. USA* **95,** 5812–5817.

Schaller, G. E., and Bleecker, A. B. (1995). Ethylene-binding sites generated in yeast expressing the *Arabidopsis* ETR1 gene. *Science* **270,** 1809–1811.

Schaller, G. E., Ladd, A. N., Lanahan, M. B., Spanbauer, J. M., and Bleecker, A. B. (1995). The ethylene response mediator ETR1 from *Arabidopsis* forms a disulfide-linked dimer. *J. Biol. Chem.* **270,** 12526–12530.

Schenk, P. M., Kazan, K., Wilson, I., Anderson, J. P., Richmond, T., Somerville, S. C., and Manners, J. M. (2000). Coordinated plant defense responses in *Arabidopsis* revealed by microarray analysis. *Proc. Natl. Acad. Sci. USA* **97,** 11655–11660.

Shieh, J. C., Wilkinson, M. G., Buck, V., Morgan, B. A., Makino, K., and Millar, J. B. (1997). The Mcs4 response regulator coordinately controls the stress-activated Wak1-Wis1-Sty1 MAP kinase pathway and fission yeast cell cycle. *Genes Dev.* **11,** 1008–1022.

Silverstone, A. L., Jung, H. S., Dill, A., Kawaide, H., Kamiya, Y., and Sun, T. P. (2001). Repressing a repressor: Gibberellin-induced rapid reduction of the RGA protein in *Arabidopsis. Plant Cell* **13,** 1555–1566.

Smalle, J., Haegman, M., Kurepa, J., Van Montagu, M., and Van Der Straeten, D. V. (1997). Ethylene can stimulate *Arabidopsis* hypocotyl elongation in the light. *Proc. Natl. Acad. Sci. USA* **94,** 2756–2761.

Smalle, J., Kurepa, J., Yang, P., Babiychuk, E., Kushnir, S., Durski, A., and Vierstra, R. D. (2002). Cytokinin growth responses in *Arabidopsis* involve the 26S proteasome subunit RPN12. *Plant Cell* **14,** 17–32.

Smalle, J., Kurepa, J., Yang, P., Emborg, T. J., Babiychuk, E., Kushnir, S., and Vierstra, R. D. (2003). The pleiotropic role of the 26S proteasome subunit RPN10 in *Arabidopsis* growth and development supports a substrate-specific function in abscisic acid signaling. *Plant Cell* **15,** 965–980.

Solano, R., Stepanova, A., Chao, Q., and Ecker, J. R. (1998). Nuclear events in ethylene signaling: A transcriptional cascade mediated by ETHYLENE-INSENSITIVE3 and ETHYLENE-RESPONSE-FACTOR1. *Genes Dev.* **12**, 3703–3714.

Stock, A. M., Robinson, V. L., and Goudreau, P. N. (2000). Two-component signal transduction. *Annu. Rev. Biochem.* **69**, 183–215.

Su, W. P., and Howell, S. H. (1992). A single genetic-locus, ckr1, defines *Arabidopsis* mutants in which root-growth is resistant to low concentrations of cytokinin. *Plant Physiol.* **99**, 1569–1574.

Suttle, J. C. (1988). Effect of ethylene treatment on polar IAA transport, net IAA uptake and specific binding of *N*-1-naphthylphthalamic acid in tissues and microsomes isolated from etiolated pea epicotyls. *Plant Physiol.* **88**, 795–799.

Tang, X., and Woodson, W. R. (1996). Temporal and spatial expression of 1-aminocyclopropane-1-carboxylate oxidase mRNA following pollination of immature and mature petunia flowers. *Plant Physiol.* **112**, 503–511.

Thomine, S., Wang, R., Ward, J. M., Crawford, N. M., and Schroeder, J. I. (2000). Cadmium and iron transport by members of a plant metal transporter family in *Arabidopsis* with homology to Nramp genes. *Proc. Natl. Acad. Sci. USA* **97**, 4991–4996.

Thomma, B. P., Eggermont, K., Tierens, K. F., and Broekaert, W. F. (1999). Requirement of functional ethylene-insensitive 2 gene for efficient resistance of *Arabidopsis* to infection by *Botrytis cinerea*. *Plant Physiol.* **121**, 1093–1102.

Trentmann, S. M. (2000). ERN1, a novel ethylene-regulated nuclear protein of *Arabidopsis*. *Plant Mol. Biol.* **44**, 11–25.

Tsuchisaka, A., and Theologis, A. (2004a). Heterodimeric interactions among the 1-aminocyclopropane-1-carboxylate synthase polypeptides encoded by the *Arabidopsis* gene family. *Proc. Natl. Acad. Sci. USA* **101**, 2275–2280.

Tsuchisaka, A., and Theologis, A. (2004b). Unique and overlapping expression patterns among the *Arabidopsis* 1-amino-cyclopropane-1-carboxylate synthase gene family members. *Plant Physiol.* **136**, 2982–3000.

Tuominen, H., Overmyer, K., Keinanen, M., Kollist, H., and Kangasjarvi, J. (2004). Mutual antagonism of ethylene and jasmonic acid regulates ozone-induced spreading cell death in *Arabidopsis*. *Plant J.* **39**, 59–69.

Vahala, J., Schlagnhaufer, C. D., and Pell, E. J. (1998). Induction of an ACC synthase cDNA by ozone in light-grown *Arabidopsis thaliana* leaves. *Physiol. Plant.* **103**, 45–50.

Van Der Straeten, D., Rodrigues-Pousada, R. A., Villarroel, R., Hanley, S., Goodman, H. M., and Van Montagu, M. (1992). Cloning, genetic mapping, and expression analysis of an *Arabidopsis thaliana* gene that encodes 1-aminocyclopropane-1-carboxylate synthase. *Proc. Natl. Acad. Sci. USA* **89**, 9969–9973.

Van Der Straeten, D., Djudzman, A., Van Caeneghem, W., Smalle, J., and Van Montagu, M. (1993). Genetic and physiological analysis of a new locus in *Arabidopsis* that confers resistance to 1-aminocyclopropane-1-carboxylic acid and ethylene and specifically affects the ethylene signal transduction pathway. *Plant Physiol.* **102**, 401–408.

Van Zhong, G., and Burns, J. K. (2003). Profiling ethylene-regulated gene expression in *Arabidopsis thaliana* by microarray analysis. *Plant Mol. Biol.* **53**, 117–131.

Vandenbussche, F., Smalle, J., Le, J., Saibo, N. J., De Paepe, A., Chaerl, L., Tietz, O., Smets, R., Laarhoven, L. J., Harren, F. J., Van Onckelen, H., Palme, K., Verbelen, J. P., and Van Der Straeten, D. (2003). The *Arabidopsis* mutant alh1 illustrates a cross talk between ethylene and auxin. *Plant Physiol.* **131**, 1228–1238.

Vogel, J. P., Schuerman, P., Woeste, K., Brandstatter, I., and Kieber, J. J. (1998a). Isolation and characterization of Arabidopsis mutants defective in the induction of ethylene biosynthesis by cytokinin. *Genetics* **149**, 417–427.

Vogel, J. P., Woeste, K. E., Theologis, A., and Kieber, J. J. (1998b). Recessive and dominant mutations in the ethylene biosynthetic gene ACS5 of *Arabidopsis* confer cytokinin insensitivity and ethylene overproduction, respectively. *Proc. Natl. Acad. Sci. USA* **95**, 4766–4771.

Vriezen, W. H., Achard, P., Harberd, N. P., and Van Der Straeten, D. (2004). Ethylene-mediated enhancement of apical hook formation in etiolated *Arabidopsis thaliana* seedlings is gibberellin dependent. *Plant J.* **37**, 505–516.

Wang, W., Hall, A. E., O'Malley, R., and Bleecker, A. B. (2003). Canonical histidine kinase activity of the transmitter domain of the ETR1 ethylene receptor from *Arabidopsis* is not required for signal transmission. *Proc. Natl. Acad. Sci. USA* **100**, 352–357.

Wang, K. L., Yoshida, H., Lurin, C., and Ecker, J. R. (2004). Regulation of ethylene gas biosynthesis by the *Arabidopsis* ETO1 protein. *Nature* **428**, 945–950.

Wen, C. K., and Chang, C. (2002). *Arabidopsis* RGL1 encodes a negative regulator of gibberellin responses. *Plant Cell* **14**, 87–100.

Woeste, K. E., Ye, C., and Kieber, J. J. (1999). Two *Arabidopsis* mutants that overproduce ethylene are affected in the posttranscriptional regulation of 1-aminocyclopropane-1-carboxylic acid synthase. *Plant Physiol.* **119**, 521–530.

Yamagami, T., Tsuchisaka, A., Yamada, K., Haddon, W. F., Harden, L. A., and Theologis, A. (2003). Biochemical diversity among the 1-amino-cyclopropane-1-carboxylate synthase isozymes encoded by the *Arabidopsis* gene family. *J. Biol. Chem.* **278**, 49102–49112.

Yanagisawa, S., Yoo, S. D., and Sheen, J. (2003). Differential regulation of EIN3 stability by glucose and ethylene signalling in plants. *Nature* **425**, 521–525.

Yang, S. F., and Hoffman, N. E. (1984). Ethylene biosynthesis and its regulation in higher plants. *Annu. Rev. Plant Physiol.* **35**, 155–189.

Zegzouti, H., Jones, B., Frasse, P., Marty, C., Maitre, B., Latch, A., Pech, J. C., and Bouzayen, M. (1999). Ethylene-regulated gene expression in tomato fruit: Characterization of novel ethylene-responsive and ripening-related genes isolated by differential display. *Plant J.* **18**, 589–600.

Zhao, X. C., Qu, X., Mathews, D. E., and Schaller, G. E. (2002). Effect of ethylene pathway mutations upon expression of the ethylene receptor ETR1 from *Arabidopsis*. *Plant Physiol.* **130**, 1983–1991.

Zhou, L., Jang, J. C., Jones, T. L., and Sheen, J. (1998). Glucose and ethylene signal transduction crosstalk revealed by an *Arabidopsis* glucose-insensitive mutant. *Proc. Natl. Acad. Sci. USA* **95**, 10294–10299.

12

Jasmonate: An Oxylipin Signal with Many Roles in Plants

John Browse

Institute of Biological Chemistry, Washington State University, Pullman, Washington 99164

I. Introduction
II. The Biochemistry of Jasmonate Synthesis
III. Jasmonate Signaling in Insect Defense
    A. *An Essential Role for Jasmonate*
    B. *Jasmonate is a Translocated Signal*
    C. *Systemic Signaling by Jasmonate Uses H_2O_2*
IV. New Defense Roles and Signal Integration
    A. *Jasmonate also Acts in Plant Defense Against Microbial Pathogens*
    B. *(9s,13s)-12-Oxo-Phytodienoic Acid and Jasmonate can Both Activate Plant Defenses*
    C. *Transcription Regulation of Jasmonate Responses*
    D. *Transcription Factors May be the Key to Induced Systemic Resistance*
V. Jasmonate Regulates Reproductive Development
    A. *Jasmonate is Essential for Pollen Development and Fertility in* Arabidopsis
    B. *Transcription Profiling Identifies Jasmonate-Regulated Genes in* Arabidopsis Stamens
VI. Ubiquitination by SCF^{COI1} is an Early and Essential Step in Jasmonate Signaling
References

Jasmonic acid is an oxylipin signaling molecule derived from linolenic acid. So far, jasmonate (JA) (including the free acid and a number of conjugates) has been shown to regulate or co-regulate a wide range of processes in plants, from responses to biotic and abiotic stresses to the developmental maturation of stamens and pollen in *Arabidopsis*. This review focuses on discoveries in several of these areas. Most work described is from studies in *Arabidopsis*. While the results are expected to be broadly applicable to other higher plants, there are cases where related but distinct phenotypes have been observed in other species (e.g., tomato). Investigation of JA action in wound- and insect-defense responses has established that this compound is an essential component of the systemic signal that activates defense genes throughout the plant. It is possible that JA acts indirectly through the production of reactive oxygen species including hydrogen peroxide (H_2O_2). The availability of *Arabidopsis* mutants deficient in JA synthesis has been central to the identification of additional roles for JA in defense against microbial pathogens and in reproductive development. Currently, the key issues in JA action are to understand the role of the skip/cullin/F-box ubiquitination complex, SCF^{COI1}, and to identify additional protein components that act in the early steps of JA signaling. © 2005 Elsevier Inc.

I. INTRODUCTION

In the last 10 years, JA emerged as a key regulator of an astonishingly wide range of plant processes, and this oxylipin is now firmly established as a major plant hormone. Database searches for publications on a range of plant growth regulators (including abscisic acid, auxins, cytokinins, ethylene, gibberellins, and brassinosteroids, as well as JA) reveal that, up until 1995, JA was a subject in only ~5% of these searches. By contrast, since 2000, over 25% of the total discussed JA. These statistics underscore the rapidly growing appreciation of JA as a key regulator of both development and biotic and abiotic stress responses in plants. Although early studies document a range of effects induced by exogenous JA (Sembdner and Parthier, 1993), they contain essentially no hint of the four roles that now dominate our thinking. The ability of methyl JA to induce proteinase inhibitors in tomato was first reported in 1990 (Farmer and Ryan, 1990), and papers describing the induction of some pathogen-defense genes followed (Xu *et al.*, 1994). Demonstrations that JA was essential for defense against some insects, fungi, and bacteria came later with the production of prosystemin antisense tomato plants (McGurl *et al.*, 1992) and with the isolation of *Arabidopsis* and tomato mutants deficient in JA synthesis or perception (Howe *et al.*, 1996; McConn *et al.*, 1997; McGurl *et al.*, 1992; Staswick *et al.*, 1998; Thomma *et al.*, 1998; Vijayan *et al.*, 1998). A volatile derivative, methyl JA, may act in interplant signaling (Farmer, 2001). Characterization of one of these mutants (*fad3 fad7 fad8*, which lacks the fatty acid precursors for JA synthesis) also led to the discovery that JA is an essential signal for the final stages of pollen maturation and anther dehiscence (McConn and Browse, 1996). Jasmonate has been shown to act in signaling of abiotic stresses, including UV radiation (Conconi *et al.*, 1996) and ozone (Rao and Davis, 2001). In healthy, unwounded plant tissue, JA is involved in carbon partitioning (Mason and Mullet, 1990), mechanotransduction (Weiler *et al.*, 1993), root growth (Staswick *et al.*, 1992), and the maturation and release of pollen (Ishiguro *et al.*, 2001; McConn and Browse, 1996; Park *et al.*, 2002; Stintzi and Browse, 2000; von Malek *et al.*, 2002). Much of our new understanding of the actions of JA signaling pathways has developed from the isolation and characterization of mutants in *Arabidopsis* and other plants (Devoto and Turner, 2003; Turner *et al.*, 2002). Suppressor screens (Xiao *et al.*, 2004) and screens based on reporter constructs (Ellis and Turner, 2001; Hilpert *et al.*, 2001; Jensen *et al.*, 2002; Xu *et al.*, 2001) have provided important new insights about JA signaling in vegetative tissues. Much of this new information is not immediately relevant to investigations of JA-induced stamen and pollen development because not all the mutants affect these processes.

II. THE BIOCHEMISTRY OF JASMONATE SYNTHESIS

Vick and Zimmerman (1983) first proposed the pathway for JA synthesis from linolenic acid (18:3). This overall chemistry has been substantially confirmed by many studies, which have also added important details about the enzymology, regulation, and subcellular location of the pathway reactions (Liechti and Farmer, 2002). In the canonical presentation (Fig. 1), the first step is the release of 18:3 (or 16:3) from membrane glycerolipids (Ishiguro *et al.*, 2001). 13-Lipoxygenase and allene oxide synthase act sequentially on 18:3 to produce 12,13-epoxyoctadecatrienoic acid, which is acted on by allene oxide cyclase (AOC). The AOC enzyme determines the stereoconfiguration of the product as (9S,13S)-12-oxo-phytodienoic acid (OPDA) (Ziegler *et al.*, 2000), and this stereoisomer is the exclusive (>99%) isomer detected in wounded plants. Interestingly, OPDA has also been identified as a substituent at sn-1 of the chloroplast lipid, monogalactosyldiacylglycerol (Stelmach *et al.*, 2001), although it is not clear to what extent this potential reservoir of OPDA contributes to JA synthesis (Stelmach *et al.*, 2001). It is now known that a specific isozyme of OPDA reductase (encoded by the *OPR3* gene) is required to reduce (9S,13S) OPDA to 3-oxo-2(2′[Z]-pentenyl)cyclopentane-1-octanoic

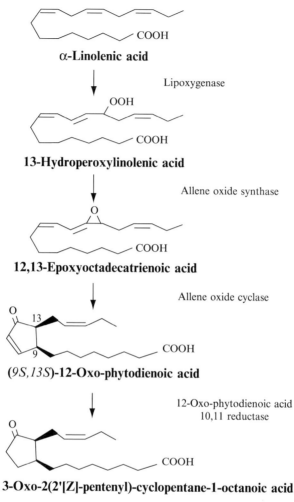

FIGURE 1. The pathway for biosynthesis of jasmonic acid from 18:3. The projections shown represent the absolute stereoconfigurations of the side chains.

acid (OPC:8) (Sanders *et al.*, 2000; Schaller *et al.*, 2000; Stintzi and Browse, 2000), which is then converted to (3R,7S)-jasmonic acid by three cycles of β-oxidation (Vick and Zimmerman, 1983).

Considerable evidence indicates that the synthesis of OPDA occurs in the chloroplast (plastid) (Blee, 1998; Schaller *et al.*, 2000), while the final production of JA is assumed to occur in the peroxisome, since this is the only known site of β-oxidation in plants (Schaller *et al.*, 2000; Ziegler *et al.*, 2000). The discovery that OPR3 is a peroxisomal protein (Stintzi and Browse, 2000; Strassner *et al.*, 2002) also indicates that the peroxisome is the site of the final reactions of JA synthesis. In plants, β-oxidation is catalyzed by acyl-CoA oxidase, the multifunctional protein (MFP) (which exhibits 2-*trans*-enoyl-CoA hydratase, L-3-hydroxyacyl-CoA dehydrogenase, D-3-hydroxyacyl-CoA epimerase, and $\Delta^3\Delta^2$-enoyl-CoA isomerase activities), and L-3-ketoacyl-CoA thiolase. In *Arabidopsis*, there appear to be at least four genes encoding acyl-CoA oxidases (with varying chain-length specificities), two encoding MFP, and three encoding the thiolase (Eastmond and Graham, 2000; Eastmond *et al.*, 2000; Hayashi *et al.*, 1998, 1999; Richmond and Bleecker, 1999). Besides fatty acids, β-oxidation is involved in the metabolism of a number of compounds including OPC:8, indole butyric acid (Bartel, 1997), and 2,4-diphenoxybutyric acid (2,4-DB) (Hayashi *et al.*, 1998). Available evidence suggests that each of the MFP and thiolase isozymes will act in β-oxidation of 2,4-DB (Eastmond and Graham, 2000; Hayashi *et al.*, 1998; Richmond and Bleecker, 1999), and the same may well be true for metabolism of OPC:8 to JA. Certainly none of the mutations in these genes has provided a JA-reversible, male-sterile phenotype (Eastmond *et al.*, 2000; Hayashi *et al.*, 1998; Richmond and Bleecker, 1999). Based on the pathway shown in Fig. 1, there must be a CoA-ligase enzyme that activates OPC:8 for β-oxidation. There is probably also a thioesterase that produces JA from jasmonoyl-CoA after three cycles of β-oxidation. A number of JA derivatives are also found in plants (Sembdner and Parthier, 1993). There is good evidence that both the methyl ester (MeJA) (Seo *et al.*, 2001) and the JA-isoleucine conjugate (Staswick and Tiryaki, 2004) are biologically active as is the precursor OPDA (Stintzi *et al.*, 2001). Our new understanding of the JA synthesis pathway has benefited considerably from studies of the mutants (and of the corresponding wild-type genes) described in this review.

III. JASMONATE SIGNALING IN INSECT DEFENSE

A. AN ESSENTIAL ROLE FOR JASMONATE

Our current understanding of local and systemic signaling processes in plant defense against insects is based on several decades of work in tomato. Local signals, such as oligouronides (Bishop *et al.*, 1984; Hahn *et al.*, 1981) and systemic signals, such as the peptide hormone systemin (Pearce *et al.*, 1991), interact with receptors to trigger a series of cellular responses. These include the opening of plasma membrane ion channels (Felix and Boller, 1995; Moyen and Johannes, 1996; Schaller and Oecking, 1999), increases in cytoplasmic (Ca^{2+}), and triggering of an MAP kinase (Stratman and Ryan, 1997; Usami *et al.*, 1995) as well as activation of a phospholipase that releases linolenic acid from membrane lipids and initiates the synthesis of JA by the octadecanoid pathway (Lee *et al.*, 1997; Nárvaez-Vásquez *et al.*, 1999). These initial responses take place within 2–10 min and are too rapid to involve changes in transcription (although JA signaling subsequently activates many of the genes involved in these processes). Although other signaling processes occur, mutant analysis demonstrates that JA synthesis and JA signaling are essential for successful defense (McGurl *et al.*, 1992). Within 30 min of wounding, the production of JA triggers the activation of genes encoding the enzymes of the octadecanoid pathway (Ryan, 2000) and genes encoding other proteins involved in signaling such as prosystemin from which the 18-aa systemin peptide is derived (Ryan, 2000). Expression of these "early" genes peaks at 2–4 h after wounding and then declines. By contrast, the defense genes (including proteinase inhibitors and polyphenol oxidase) are activated later with transcript levels starting to rise at 4 h and peaking at 8–10 h after wounding (Constabel *et al.*, 1995). We now know that both temporal and spatial differences in gene expression are important in understanding JA responses at a mechanistic level.

Many studies of JA-related processes have relied on a small number of "archetypal" genes to document and follow changes in expression using RNA blot analysis. Broad differential-screening techniques—microarrays, differential display, and SAGE analysis—have revealed the extraordinary extent of JA regulation (Reymond *et al.*, 2000; Schenk *et al.*, 2000). Based on the number of unique genes in the early generations of microarrays, it can be calculated that at least 1200 of the 26,000 genes in *Arabidopsis* are activated or repressed by JA signaling. In the JA pathway, as in most signaling pathways, many target genes are induced (or repressed) by the action of transcription factors, which often act on a set of related genes that might, for example, encode the enzymes of a biochemical pathway. One relevant example is the induction of alkaloid synthesis in periwinkle (*Catharanthus roseus*). Expression of many (but not all) enzymes required for the synthesis of these alkaloids has been shown to be regulated by the JA-responsive transcription factors, ORCA2 and ORCA3 (Menke *et al.*, 1999; van der Fits and MemLink, 2000). These are members of the AP2/EREBP family, which has 144 members in *Arabidopsis* (Riechmann *et al.*, 2000).

These studies established JA as a chemical signal mediating defense responses against insect attack. However, the signaling pathways that allow plants to mount defenses against chewing insects are known to be complex, and clear demonstrations on the efficacy of JA signaling are required.

Definitive evidence for the essential role of JA in insect defense came from studies of tomato and *Arabidopsis* mutants deficient in the synthesis or accumulation of JA (Howe *et al.*, 1996; McConn *et al.*, 1997). The *Arabidopsis fad3-2 fad7-2 fad8* triple mutant is deficient in all three desaturase enzymes that can convert 18:2 to 18:3 and 16:2 to 16:3 (McConn and Browse, 1996). The plants therefore lack the fatty acid precursors for JA synthesis and contain negligible levels of JA. To investigate the role of JA in wound signaling in *Arabidopsis* and to test whether parallel or redundant pathways exist for insect defense, we carried out further studies on *fad3 fad7 fad8* plants. Mutant plants showed extremely high mortality (ca. 80%) from attack by larvae of a common saprophagous fungal gnat, *Bradysia impatiens* (Diptera:Sciaridae), even though neighboring wild-type plants were largely unaffected (Fig. 2). Application of exogenous JA substantially protected the mutant plants and reduced mortality to approximately 12%. These experiments precisely define the role of JA as being essential for the induction of biologically effective defense in this plant–insect interaction. The transcripts of three wound-responsive genes were shown not to be induced by wounding of mutant plants but the same transcripts could be induced by application of JA. By contrast, measurements of transcript levels for a gene encoding glutathione-*S*-transferase demonstrated that wound induction of this gene is independent of JA synthesis. These results demonstrate the utility of JA - synthesis mutants as a genetic and molecular biology model for studying plant defense against insects. For example, constitutive expression of transgenes in the mutant can be used to test candidate defense genes for their ability to meaningfully reduce damage and mortality from insect attack. Perhaps not surprisingly, this detailed knowledge of JA-mediated defense obtained in the laboratory has been shown to have relevance to plants grown in natural environments (Kessler *et al.*, 2004).

B. JASMONATE IS A TRANSLOCATED SIGNAL

A key aspect of wound-defense signaling in tomato and other plants is the finding that wounding of a single leaf results in the expression of defense genes throughout the plant (Ryan and Pearce, 1998). This means that a systemic signal must be generated in the wounded leaf to activate defenses at distal sites. Several chemical compounds, as well as a proposed electrical signal, have been tested for possible involvement in long-distance signaling. In tomato, the 18-aa peptide systemin has a clearly established role in the activation of defense genes throughout the plant. However, it is now clear that JA (or a related oxylipin) is an essential component of the translocated signal. The *spr-2* mutant of tomato is deficient in JA synthesis but responds normally to applied JA (Li *et al.*, 2002, 2003). Conversely, the *jai-1* mutant synthesizes JA, but the induction of defense genes does not occur in response to wounding or JA (Li *et al.*, 2004). These mutants were used in grafting

FIGURE 2. Death and protection of mutant plants from *Bradysia* larvae attack. (A) A mixed population of wild-type and *fad3-2 fad7-2 fad8* plants were grown in a net enclosure populated with 20–25 adult *Bradysia* flies. Each day, eight pots were sprayed with 0.8 ml of H₂O and seven pots were sprayed with 0.8 ml of a dilute aqueous solution of methyl jasmonate. Data from two experiments are shown in which the methyl jasmonate concentration used was either 0.001% (open

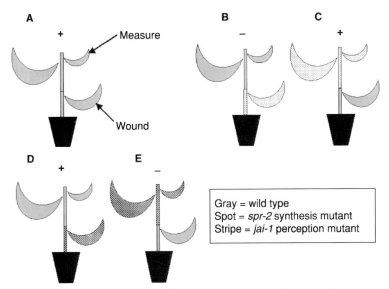

FIGURE 3. Grafting experiments with tomato mutants establish a role for jasmonate as the translocated wound signal. A leaf of the rootstock was wounded and transmission of the signal was assayed by measuring induction of defense genes (+ or −) in a leaf of the graft scion (see text for details).

experiments with wild-type plants (Li *et al.*, 2002), and the key observations are illustrated in Fig. 3. (1) Wounding of a lower leaf on a wild-type plant (A) resulted in the generation of a graft-transmissible signal and induction of defense genes measured in an unwounded leaf on the scion (grafted upper part of the plant). (2) Wounding of an *spr-2* rootstock (B) does not lead to a signal even though an *spr-2* scion (C) can respond to the signal generated by wounding of a wild-type rootstock. (3) A rootstock from the *jai-1* mutant (D) is able to generate a signal that results in defense gene-induction in a wild-type scion. However, a *jai-1* scion (E) is insensitive to the signal produced by wounding a wild-type rootstock. These results indicate that JA is

symbols) or 0.01% (solid symbols). The graph shows the percentage survival of 117 wild-type plants (○ and ● both treatments), 73 mutant plants treated with H_2O (□, ■), and 73 mutant plants treated with methyl jasmonate (△, ▲). (B and C) For clarity, wild-type and mutant seeds were sown in pots in two rows but were otherwise treated as described previously. The photographed plants correspond to day 50 in (A). (B) Compared with wild-type controls (on the left), mutant plants (on the right) sprayed with water show extensive damage 20 days after the introduction of adult *Bradysia* flies. Some leaves on mutant plants have been almost completely eaten. Wilting of other leaves was attributed to damage to the petiole or to the roots of the plants. (C) Mutant plants sprayed with 0.01% methyl jasmonate (on the right) remained healthy and vigorous within the same environment. [Reproduced from McConn *et al.* (1997).]

synthesized in wounded leaves and moves through the plant to activate defense responses.

C. SYSTEMIC SIGNALING BY JASMONATE USES H_2O_2

Many of the experiments on JA-induction of defense genes that have been carried out rely on analyzing mechanically wounded or insect-damaged tissues. In these experiments, a large peak of JA production at the wound site is closely correlated with gene activation (Creelman and Mullet, 1997). However, in distal leaves of wounded plants, JA-responsive genes are activated even though there is no detectable increase in JA levels when these leaves are analyzed. The first clue in solving this puzzle came when it was discovered that systemic signaling led to the activation of genes for JA synthesis only in cells of the vascular bundle and not in the mesophyll cells (Jacinto et al., 1997; Kubigsteltig et al., 1999). If JA is produced only in the vasculature, this can explain why large increases in JA are not detectable in the analyses of total leaf tissue. Yet, how is the systemic signal transmitted to the mesophyll cells where protease inhibitors, polyphenol oxidase, and other defense genes are expressed? The answer to this question came with the discovery of a wound-induced polygalacturonase in the leaves of tomato (Orozco-Cárdenas et al., 2001). The products of polygalacturonase activity are oligogalacturonides, which are known to stimulate the production of H_2O_2 in plant cells (through activity of NADPH oxidase) and to promote the activation of defense genes in plants. Subsequently, Orozco-Cardenas and coworkers showed that in several species, wound-induction of polygalacturonase activity is accompanied by production of H_2O_2, suggesting the possibility that polygalacturonase-derived oligogalacturonides and H_2O_2 are downstream steps in JA signaling. Consistent with this model, H_2O_2 was not produced in mutants deficient in JA synthesis (Orozco-Cárdenas and Ryan, 1999), and treatment of plants with an inhibitor of NADPH oxidase (diphenylene iodonium chloride, DIP) blocked wound- and JA-induced H_2O_2 production and the expression of defense genes (Bergey et al., 1996; Orozco-Cárdenas and Ryan, 1999; Orozco-Cárdenas et al., 2001). Conversely, an H_2O_2-generating system of glucose and glucose oxidase–stimulated defense gene induction in unwounded plants (Orozco-Cárdenas et al., 2001). Blocking H_2O_2 production did not block JA-induction of the early response genes, such as lipoxygenase and allene oxide synthase. Hydrogen peroxide and active oxygen species are important mediators of hypersensitive cell death, which forestalls some pathogen infections (Bolwell, 1999; Dixon and Lamb, 1997; Doke, 1996; Low and Merida, 1996), and are also involved in a range of abiotic stresses (Dixon and Lamb, 1997), so these new discoveries reinforce the concepts of cross-talk and interaction between JA signaling and signal processes involved in other plant responses. The results are also important in two other respects. They demonstrate the existence of distinct JA-response pathways, and they underscore the need to consider tissue-specificity as an integral part of JA signaling.

IV. NEW DEFENSE ROLES AND SIGNAL INTEGRATION

A. JASMONATE ALSO ACTS IN PLANT DEFENSE AGAINST MICROBIAL PATHOGENS

The signals (e.g., salicylic acid [SA]) that have been identified as activating defenses against fungal pathogens, are distinct from those involved in insect defense and in fact the two signaling pathways were once considered to be mutually antagonistic (Farmer and Ryan, 1992; Peña-Cortés et al., 1993; Sembdner and Parthier, 1993). For this reason, it was something of a surprise to discover that JA plays an essential role in defense against some fungal pathogens. In our experiments (Vijayan et al., 1998), we were able to isolate a species of *Pythium* fungus (*Pythium mastophorum*) that specifically killed plants of the *fad3-2 fad7-2 fad8* mutant line and were able to demonstrate that this fungus reinfected and killed triple-mutant plants while neighboring wild-type plants were largely unaffected. Application of exogenous JA substantially protected mutant plants, reducing the incidence of disease to a level close to that of wild-type controls (Fig. 4). A similar treatment with JA did not protect the JA-insensitive mutant *coi1* from infection, showing that protective action of applied JA against *P. mastophorum* was mediated by the induction of plant defense mechanisms rather than by a direct antifungal action. Transcripts of three JA-responsive defense genes were induced by *Pythium* challenge in the wild-type, but not in the JA-deficient mutant. *Pythium* species are ubiquitous in soil and root habitats worldwide but most (including *P. mastophorum*) are considered to be minor pathogens. Thus, JA is essential for plant defense against *Pythium*, and because of the high exposure of plant roots to *Pythium* inoculum in soil, may well be fundamental to the survival of plants in nature. Subsequent work has demonstrated that JA and ethylene are key signaling molecules in non-host defense of plants against many fungi and other microbial pathogens.

B. (9S,13S)-12-OXO-PHYTODIENOIC ACID AND JASMONATE CAN BOTH ACTIVATE PLANT DEFENSES

There has been debate concerning the relative roles of JA and its precursor, OPDA, in plant signaling processes with some researchers suggesting that OPDA is the major physiological effector *in vivo* (Creelman and Mullet, 1997; Farmer et al., 1998; Kramell et al., 2000; Weiler et al., 1999). OPDA cannot substitute for JA as a signal in plant reproductive development (see later) (Stintzi and Browse, 2000) and JA alone is sufficient to protect *fad3*

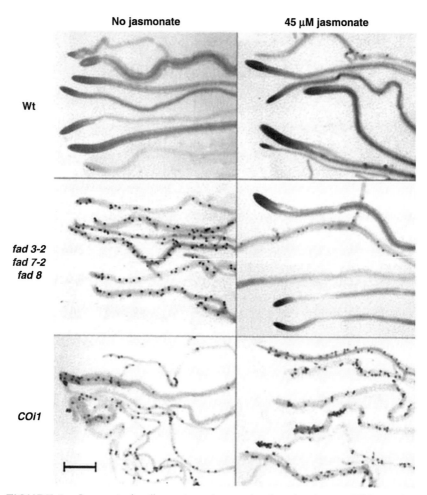

FIGURE 4. Jasmonate signaling protects plants against fungal pathogens. Wild-type plants (Wt) together with jasmonate-deficient (*fad3-2 fad7-2 fad8*) and jasmonate-insensitive (*coi1*) mutants were grown in soil infested with *Pythium mastophorum*. In the absence of jasmonate (left) both mutants are infected as shown by the development of fungal oospores in the roots. When plants were watered with a dilute solution of methyl jasmonate (right) the jasmonate-deficient mutant was substantially protected while the jasmonate-insensitive mutant remains susceptible to fungal infection. (bar = 1 mm.) [Reproduced from Vijayan *et al.* (1998).]

fad7 fad8 plants against insect and pathogen attack (McConn *et al.*, 1997; Vijayan *et al.*, 1998). However, *opr3* plants were considerably less susceptible to insect and pathogen damage than plants of the *fad3 fad7 fad8* mutant line, in experiments in which wild-type, *fad3 fad7 fad8*, *coi1*, and *opr3* plants were grown in mixed stands and challenged with modest populations of fungal gnats. Relative to wild-type controls, plants of the *fad3 fad7 fad8* and *coi1*

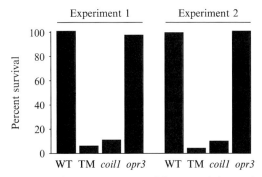

FIGURE 5. *opr3* mutant plants are not susceptible to attack by *Bradysia*. Results of two independent experiments in which a mixed population of approximately 150 wild-type (WT), *fad3 fad7 fad8* triple mutant (TM), *coi1*, and *opr3* plants were grown in a net enclosure and subjected to heavy infestation with *Bradysia impatiens* larvae. As the seedlings reached the four-leaf stage, adult *Bradysia* flies (average 50–80) were introduced in the net every other day over a 2-week period. Damage to the petiole or to the roots of the plants by *Bradysia* larvae resulted in wilting followed by death of the plants. After bolting, the percentage of surviving plants for each genotype was calculated. [Reproduced from Stintzi *et al.* (2001).]

lines were subject to increasing damage to rosette leaves and by the end of the experiment most plants were dead (Fig. 5). By contrast, *opr3* plants remained substantially free of insect damage and showed survival rates comparable to the wild-type. In these experiments, plants of the four lines were randomly interplanted, so the results demonstrate a very high level of protection for *opr3* plants relative to the two susceptible genotypes. Other experiments demonstrated that *opr3* plants were also resistant to fungal pathogens (Stintzi *et al.*, 2001).

To further investigate defense signaling, we carried out transcript profiling of wild-type and *opr3* plants using a defense gene microarray (Reymond *et al.*, 2000). Wounded *opr3* plants accumulated OPDA but not JA. However, many JA-induced defense genes were upregulated in both *opr3* plants and wild-type controls although the level of wound-induction was often somewhat lower in *opr3* plants. These genes could be induced in unwounded *opr3* plants by exogenous application of either OPDA or JA. A few of the JA-responsive genes, such as *VSP*, were not induced in *opr3* by either wounding or OPDA, while others (*RNS1 OPR1 GST1*) responded to OPDA but not to JA. Evidently, OPDA, JA, and perhaps other oxylipins act in concert to regulate the expression of defense genes (Stintzi *et al.*, 2001).

C. TRANSCRIPTION REGULATION OF JASMONATE RESPONSES

The diversity of processes controlled by JA and the complexity of the transcriptional regulation that is observed can be understood in terms of three principal components: (1) JA signaling can be specific for different tissues or cell types. This is clearly the case, for example, in anthers where JA induces a number of anther-specific genes (see Section VI). This specificity may be brought about, as it is in animals, by linkage of receptors to different intracellular signaling paths depending on cell type. (2) When different signaling pathways are simultaneously activated (JA and ethylene, JA and SA, JA and OPDA), the resulting interactions provide for additional complexity in the response. (3) JA signaling provides for the induction of both early and late genes so that responses and interactions must vary with time. Evidently, these factors ensure that distinct suites of genes are expressed to bring about the different JA responses.

Knowledge of other signaling systems in both plants and animals indicates that it is the responsibility of transcription factors and other key regulatory components to ensure that signals and cross-talk from multiple pathways are translated into the appropriate response at the transcriptional level. The initial response to chemical signals is usually very rapid and this, of course, requires that transcription factors be constitutive proteins that are activated by biochemical reactions or interactions. This is certainly true, but it is now known that many transcription factors are themselves subject to transcriptional regulation. (In different situations, this regulation may assist feedback amplification of the signal or replacement of transcription factor proteins broken down during signal transduction, or may provide for a "transduction cascade".) Thus, in *Arabidopsis*, approximately 7% of the JA-responsive genes are recognized, or putative, transcription factors. It is now possible to identify almost all the transcription factors whose message levels respond to JA and to test the roles of these proteins in the many different JA responses that researchers are investigating. This means that we can employ systems analysis to JA signaling and response. By comparison and analysis of the JA transcriptome in response to different treatments and in a range of mutant backgrounds, we will accumulate information that will allow us to discern possible roles for many of the key transcription factors. Knockout and overexpression experiments with these transcription factors combined with further transcription profiling and functional bioassays, will provide additional information. Together with the other experiments, these approaches will help us start to assign biological functions to JA-responsive transcription factors, gain insights into signaling branches of the JA transcriptome, and gain the new knowledge needed to understand the involvement of JA in such a wide range of biological processes.

The combinatorial actions of JA and ethylene in defense signaling are one area where research is beginning to uncover the roles of transcription factors in regulating responses. ERF1 is one of five ethylene response factors in *Arabidopsis*. The *ERF* genes were first characterized as being induced by ethylene (Lorenzo *et al.*, 2002). However, subsequent analysis demonstrated that both ethylene- and JA-signaling are required for high *ERF1* expression. Thus, ERF1 is a likely candidate for a transcription factor that integrates signals for the expression of genes whose products protect plants from necrotrophic fungal pathogens such as *Phythium* spp. and *Alternaria brassicicola*. Consistent with this notion, constitutive expression of *ERF1* (controlled by the cauliflower mosaic virus 35S promoter) resulted in expression of many defense genes. Furthermore, mutants that are normally susceptible to pathogens as a result of defective JA or ethylene signaling (*coi1* and *ein2*, respectively; Feys *et al.*, 1994; Lorenzo *et al.*, 2002) were protected by expression of the *35S:ERF1* transgene (Lorenzo *et al.*, 2002).

Cloning of the *JASMONATE-INSENSITIVE1* locus of *Arabidopsis* (*JAI1/JIW1*) and yeast one-hybrid screens in tomato have identified b-HLH transcription factors (MYC2 in *Arabidopsis* JAMYC2 and JAMYC10 in tomato) that perform a complementary role in JA signaling. *JIN1* expression is induced by JA and the MYC2 protein that it encodes is required for activation of many JA-responsive genes—particularly those that are induced in response to wounding (Boter *et al.*, 2004; Lorenzo *et al.*, 2004). However, MYC2 represses many genes that are induced in response to pathogens by action of the ERF1 transcription factor. Conversely, ERF1 represses wound-responsive genes (Lorenzo *et al.*, 2004). Thus, there are (at least) two pathways of JA-induced defense signaling. In one, wound-responsive genes are upregulated through the MYC2 transcription factor. In the second, JA and ethylene act synergistically through ERF1 to activate pathogen-responsive genes. The antagonistic cross-regulation of MYC2 and ERF1 provides the means to elicit different responses based on appropriate integration of input signals.

D. TRANSCRIPTION FACTORS MAY BE THE KEY TO INDUCED SYSTEMIC RESISTANCE

Colonization of plant roots by certain nonpathogenic bacteria promotes an enhanced defensive capacity against a broad spectrum of bacterial and fungal pathogens (Van Loon *et al.*, 1998). This induced systemic resistance (ISR), quite distinct from systemic acquired resistance (SAR) that is induced by pathogenic microbes, is dependent on signaling by SA and leads to the induction of genes encoding the pathogen-responsive (PR) defense proteins (Pieterse and Van Loon, 1999). By contrast to SAR, our investigations have shown that ISR does not appear to result in the induction of PR or other defense genes and is induced in *nahG Arabidopsis* plants that are deficient in SA accumulation and SAR (Pieterse *et al.*, 1996). Instead, ISR is dependent on JA and ethylene signaling (Pieterse *et al.*, 1996, 1998, 2000; Ton *et al.*, 2001). Interestingly, both ISR and SAR are dependent on NPR1, a gene encoding an ankyrin-repeat protein that undergoes nuclear localization (Cao *et al.*, 1997; Pieterse *et al.*, 1998). As discussed in an earlier section, JA and ethylene have previously been identified as essential signals for the induction of *PDF1.2* and other defense genes that protect against *Pythium* and *Alternaria* infection (Knoester *et al.*, 1998; Penninckx *et al.*, 1998; Vijayan *et al.*, 1998). Thus, two separate JA/ethylene signaling pathways contribute to plant defense against pathogens. The ISR pathway is particularly intriguing because investigations to date have failed to demonstrate induction of any known defense genes in ISR plants prior to pathogen infection. Instead, faster and stronger induction of JA-responsive defense genes is observed in these plants following challenge with a pathogen (Van Wees *et al.*, 1999). These observations suggest that ISR involves potentiation of the response pathway throughout the plant by JA signaling processes. At the molecular level, this potentiation may be based on increased expression of transcription factors that are then available for activation at the biochemical level and rapid activation of their target defense genes.

V. JASMONATE REGULATES REPRODUCTIVE DEVELOPMENT

A. JASMONATE IS ESSENTIAL FOR POLLEN DEVELOPMENT AND FERTILITY IN *ARABIDOPSIS*

We originally created the *fad3 fad7 fad8* triple mutant to investigate the importance of lipid unsaturation for membrane-related processes, such as photosynthesis (McConn and Browse, 1996). An unanticipated consequence of the lack of 18:3 and 16:3 fatty acids was the fact that triple-mutant plants were male-sterile, and this led to the recognition of JA's role as a chemical signal controlling stamen and pollen development in *Arabidopsis* (McConn and Browse, 1996). Other mutants deficient in JA synthesis, including *dad1* (Ishiguro *et al.*, 2001), *aos* (Park *et al.*, 2002; von Malek *et al.*, 2002), and *opr3* (=*dde1*) (Sanders *et al.*, 2000; Stintzi and Browse, 2000) are also male-sterile, as is the JA-perception mutant, *coi1* (Feys *et al.*, 1994). The JA synthesis mutants in *Arabidopsis* can be restored to fertility by exogenous JA (Ishiguro *et al.*, 2001; Stintzi and Browse, 2000). This means that the mutants are an ideal tool for genetic and genomic approaches to identify JA-responsive genes that initiate pollen and stamen maturation.

Interestingly, JA mutants in tomato are male-fertile but female-sterile (Li *et al.*, 2004). Analysis of these mutants revealed that pollen viability and germination are reduced relative to wild-type. These results suggest that similar JA-regulated processes occur in *Arabidopsis* and tomato anthers, and that JA may also function in carpel development in *Arabidopsis*, although its role (if any) is not essential to female fertility in this species.

Investigation of the *fad3 fad7 fad8* and *opr3* mutants identified three characteristics of the male-sterile phenotype (Fig. 6). (1) Floral organs develop normally within the closed bud but the anther filaments do not elongate sufficiently to position the locules above the stigma at the time of

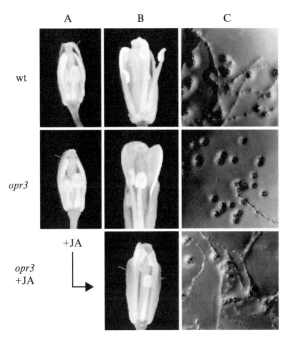

FIGURE 6. Phenotypes of wild-type and *opr3* mutant flowers. (A) Flowers at stage 12. (B) Flowers at anthesis. Pollination did not occur in *opr3* plants unless buds had previously been treated with jasmonate (Bottom). (C) Germination of pollen harvested at anthesis. [Reproduced from Stintzi and Browse (2000).]

flower opening. (2) The anther locules do not dehisce at the time of flower opening (although limited dehiscence occurs later). (3) Even though pollen on mutant plants develops to the trinucleate stage as determined by staining with 4′,6-diamino-2-phenylindole (DAPI), the pollen grains are predominantly inviable. Irrespective of the stage at which pollen was taken from *fad3 fad7 fad8* or *opr3* plants, germination of the pollen was <4% compared with 97.6% for mature pollen from wild-type plants (McConn and Browse, 1996; Stintzi and Browse, 2000).

Application of JA to flower buds corrected all three of these defects, resulting in rates of pollen germination equivalent to wild-type (97.2%) in *in vitro* tests and abundant seed set on treated plants. The ability of JA to restore fertility to mutant plants is extremely stage specific. By staging flowers and monitoring seed production following a single application of JA, we established that only flower buds corresponding to the transition between stages 11 and 12 in floral development (Smyth *et al.*, 1990) responded to JA; flowers at earlier and later stages of development could not be rendered fertile by JA treatment. Stage 11/12 is immediately before the bud opens and encompasses the second mitotic division in the pollen.

B. TRANSCRIPTION PROFILING IDENTIFIES JASMONATE-REGULATED GENES IN *ARABIDOPSIS* STAMENS

The conditionally fertile phenotype of the *fad3 fad7 fad8* and *opr3* lines provides an excellent means to identify, and subsequently characterize, the JA-responsive genes that initiate pollen and stamen maturation. One way to identify genes that might be involved in stamen and pollen maturation is to compare gene expression in stamens from carefully staged buds of wild-type and *opr3* plants. On the other hand, we would envisage a cascade of gene-induction and developmental processes occurring in response to the JA signal, so that a time-course profiling of changes in transcript level following treatment of *opr3* flowers with exogenous JA might be more informative. However, exogenous JA might also induce defense genes. There are more than 1000 genes estimated to be JA-responsive in vegetative tissues of *Arabidopsis* and a large proportion of these are known or putative defense genes (Reymond *et al.*, 2000; Schenk *et al.*, 2000). Strong induction of defense genes in stamens treated with JA might greatly complicate the identification of JA-regulated genes involved in stamen and pollen maturation.

To explore these issues and to initiate the search for genes involved in JA-dependent stamen development, microarray analysis was used to compare expression in developing stamens of wild-type and *opr3* plants (Mandaokar *et al.*, 2003). For these early experiments, we used cDNA microarrays made available through the Arabidopsis Functional Genomics Consortium (AFGC) (Wisman and Ohlrogge, 2000). One experiment was completed using the original AFGC microarray (11,000 cDNAs representing approximately 8500 unique genes). A second experiment used the second-generation AFGC microarray in which 14,000 cDNAs represent approximately 11,000 unique genes (Wisman and Ohlrogge, 2000). Analysis of the data identified 25 genes that were upregulated at least 1.8-fold in wild-type stamens compared to the mutant in both experiments (Mandaokar *et al.*, 2003). Data from the larger microarray used in the second experiment identified an additional 38 genes (not present on the smaller array) that exhibited expression ratios (wild-type:mutant) greater than 3.0.

To confirm the results of these experiments, the transcript levels of eight representative clones were investigated by gel-blot analysis on RNA isolated from wild-type and *opr3* stamens. All eight genes showed higher mRNA transcript levels in wild-type compared to *opr3* stamen RNA. To discover whether JA signaling is directly involved in altering the expression of these genes, we examined their expression in stamens of wild-type, *opr3*, and the JA-insensitive *coi1* mutant, before and after treatment with JA (Mandaokar *et al.*, 2003). Anthers were collected in a narrow time window of 30–60 min after application of JA (or a control solution) to flower buds. For four of the eight genes, JA treatment resulted in strong induction in the *opr3* mutant at this early time point. Jasmonate treatment of *coi1* flower buds did not induce these genes, indicating that the JA response is COI1 dependent. The remaining four genes were not induced by JA in either *opr3* or *coi1*, and we suspected that these might be genes that are induced later in the JA response. When RNA was prepared from *opr3* stamens harvested 1, 3, 8, and 16 h after JA treatment, gel-blot analysis showed that these four genes began to be induced at times ranging from 3 to 16 h (Mandaokar *et al.*, 2003). These results indicate that transcriptional analysis of stamen RNA can reliably identify JA-regulated genes.

VI. UBIQUITINATION BY SCF^{COI1} IS AN EARLY AND ESSENTIAL STEP IN JASMONATE SIGNALING

Although the initial response to a chemical typically involves constitutive proteins that are activated by biochemical reactions and interactions, it is known from other signaling systems that key transcription factors and other components are themselves subject to transcriptional regulation (Dharmasiri and Estelle, 2004; McCarty and Chory, 2000). Accumulating evidence indicates that the initial steps of JA signaling are analogous in many respects to the model developed for the auxin response and that the two signaling systems share some components (Devoto and Turner, 2003; Devoto *et al.*, 2002; Dharmasiri and Estelle, 2004; Gray *et al.*, 2001; McCarty and Chory, 2000; Rogg and Bartel, 2001; Tiryaki and Staswick, 2002). Gibberellin signaling follows a similar model (Itoh *et al.*, 2003). A highly simplified cartoon showing only a few aspects of this model is shown in Fig. 7. An Skp1/CUL1/F-box (SCF) complex (representing E3 of an E1/E2/E3 ubiquitination system; Dharmasiri and Estelle, 2004) ubiquitinates a repressor protein, **R** (which is then degraded by the 26S proteasome) to allow pre-existing transcription factors, **T**, to mediate expression of early genes in the signaling cascade. Hypothetically, these early genes will include a second wave of transcription factors that mediate and regulate later responses. Experiments provide evidence that transcription factors induced in the JA response are required for stamen development and plant fertility. Map-based cloning of the *coi1* locus was key to establishing SCF-mediated ubiquitination as the first step in JA signaling. The *coi1* mutants are JA-insensitive and defective in the great majority of JA responses (Feys *et al.*, 1994). The *COI1* gene (Xie *et al.*, 1998) encodes a protein that contains an F-box domain required for interaction with the Skp1 component of an SCF complex (Xie *et al.*, 1998).

In auxin signaling, the F-box is TIR1 and the repressors (R) belong to the AUX/IAA family of proteins. There are at least 29 *AUX/IAA* genes in *Arabidopsis* (Dharmasiri and Estelle, 2004; Kepinski and Leyser, 2002; Theologis *et al.*, 1985) and these are induced 20- to 50-fold within 2 h of auxin

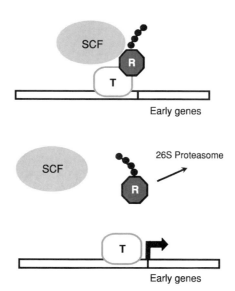

FIGURE 7. Model for the involvement of an SCF complex in jasmonate signaling. The SCF complex ubiquitinates a repressor, **R**, targeting it to the 26S proteasome. In the absence of **R**, transcription factor(s), **T**, promote expression of the early genes in jasmonate responses. For auxin **F**, F^{TIR1}; **R**, AUX/IAAs; and **T**, ARFs. For jasmonate **F**, F^{COI1} but **R** and **T** are unknown (see text for details).

treatment (Theologis *et al.*, 1985). Other families of early auxin genes are discussed in Abel *et al.* (1994) and Theologis *et al.* (1985). Proteins that interact with COI1 have been investigated (Devoto *et al.*, 2002), and a large number of early genes have been identified that are rapidly induced after plants are treated with JA. However to date, none of these candidates has been confirmed as a repressor, **R**, in the proposed model of SCF^{COI1} action (Fig. 7). The AUX/IAA proteins are nuclear targeted and rapidly turned over (Abel *et al.*, 1994), and these are important requirements for **R** in the model for auxin signaling (Dharmasiri and Estelle, 2004; Gray *et al.*, 2001; Kepinski and Leyser, 2002). Dominant mutations in the *AUX/IAA* genes decrease turnover of the proteins and result in auxin-resistant phenotypes. The fact that numerous screens have not identified dominant JA-resistant mutants (Devoto and Turner, 2003) suggests that **R** in JA signaling may not be entirely analogous to AUX/IAA in auxin signaling. Discovering the identities of the **R** and **T** components in the initial steps of JA signaling will be central to understanding the many different functions of this plant hormone.

ACKNOWLEDGMENTS

This work was supported by grants from the U.S. Department of Energy (DE-FG03-99ER20323) and the National Science Foundation (IBN-0084329).

REFERENCES

Abel, S., Oeller, P. W., and Theologis, A. (1994). Early auxin-induced genes encode short-lived nuclear proteins. *Proc. Natl. Acad. Sci. USA* **91,** 326–330.

Bartel, B. (1997). Auxin biosynthesis. *Annu. Rev. Plant Physiol. Plant Mol. Biol.* **48,** 51–66.

Bergey, D. R., Howe, G. A., and Ryan, C. A. (1996). Polypeptide signaling for plant defensive genes exhibits analogies to defense signaling in animals. *Proc. Natl. Acad. Sci. USA* **93,** 12053–12058.

Bishop, P., Pearce, G., Bryant, J. E., and Ryan, C. A. (1984). Isolation and characterization of the proteinase inhibitor inducing factor from tomato leaves: Identity and activity of poly- and oligogalacturonide fragments. *J. Biol. Chem.* **259,** 13172–13177.

Blee, E. (1998). Phytooxylipins and plant defense reactions. *Prog. Lipid Res.* **37,** 33–72.

Bolwell, G. P. (1999). Role of active oxygen species and NO in plant defense responses. *Curr. Opin. Plant Biol.* **2,** 287–294.

Boter, M., Ruíz-Rivero, O., Abdeen, A., and Prat, S. (2004). Conserved MYC transcription factors play a key role in jasmonate signaling both in tomato and *Arabidopsis*. *Genes Dev.* **18,** 1577–1591.

Cao, H., Glazebrook, J., Clarke, J. D., Volko, S., and Dong, X. (1997). The *Arabidopsis* NPR1 gene that controls systemic acquired resistance encodes a novel protein containing ankyrin repeats. *Cell* **88,** 57–63.

Conconi, A., Smerdon, M. J., Howe, G. A., and Ryan, C. A. (1996). The octadecanoid signalling pathway in plants mediates a response to ultraviolet radiation. *Nature* **383,** 763–764.

Constabel, C. P., Bergey, D. R., and Ryan, C. A. (1995). Systemin activates synthesis of wound-inducible tomato leaf polyphenol oxidase via the octadecanoid defense signaling pathway. *Proc. Natl. Acad. Sci. USA* **92,** 407–411.

Creelman, R. A., and Mullet, J. E. (1997). Biosynthesis and action of jasmonates in plants. *Annu. Rev. Plant Physiol. Plant Mol. Biol.* **48,** 355–382.

Devoto, A., Nieto-Rostro, M., Xie, D., Ellis, C., Harmston, R., Patrick, E., Davis, J., Sherratt, L., Coleman, M., and Turner, J. G. (2002). COI1 links jasmonate signalling and fertility to the SCF ubiquitin-ligase complex in *Arabidopsis*. *Plant J.* **32,** 457–466.

Devoto, A., and Turner, J. G. (2003). Regulation of jasmonate-mediated plant responses in *Arabidopsis*. *Ann. Bot. (Lond.)* **92,** 329–337.

Dharmasiri, N., and Estelle, M. (2004). Auxin signaling and regulated protein degradation. *Trends Plant Sci.* **9,** 302–308.

Dixon, R., and Lamb, C. (1997). The oxidative burst in plant disease resistance. *Annu. Rev. Plant Physiol. Plant Mol. Biol.* **48,** 241–276.

Doke, N. (1996). The oxidative burst protects plants against pathogen attack: Mechanism and role as an emergency signal for plant bio-defence. A review. *Gene* **179,** 45–51.

Eastmond, P. J., and Graham, I. A. (2000). The multifunctional protein AtMFP2 is co-ordinately expressed with other genes of fatty acid beta-oxidation during seed germination in *Arabidopsis thaliana* (L.) Heynh. *Biochem. Soc. Trans.* **28,** 95–99.

Eastmond, P. J., Hooks, M. A., Williams, D., Lange, P., Bechtold, N., Sarrobert, C., Nussaume, L., and Graham, I. A. (2000). Promoter trapping of a novel medium-chain acyl-CoA oxidase which is induced transcriptionally during *Arabidopsis* seed germination. *J. Biol. Chem.* **275,** 34375–34381.

Ellis, C., and Turner, J. G. (2001). The *Arabidopsis* mutant cev1 has constitutively active jasmonate and ethylene signal pathways and enhanced resistance to pathogens. *Plant Cell* **13,** 1025–1033.

Farmer, E. E. (2001). Surface-to-air signals. *Nature* **411,** 854–856.

Farmer, E. E., and Ryan, C. A. (1990). Interplant communication—airborne methyl jasmonate induces synthesis of proteinase inhibitors in plant leaves. *Proc. Natl. Acad. Sci. USA* **87,** 7713–7716.

Farmer, E. E., and Ryan, C. A. (1992). Octadecanoid-derived signals in plants. *Trends Cell Biol.* **2,** 236–241.

Farmer, E. E., Weber, H., and Vollenweider, S. (1998). Fatty acid signaling in *Arabidopsis*. *Planta* **206,** 167–174.

Felix, G., and Boller, T. (1995). Systemin induces rapid ion fluxes and ethylene biosynthesis in *Lycopersicon peruvianum* cells. *Plant J.* **7,** 381–389.

Feys, B. J. F., Benedetti, C. E., Penfold, C. N., and Turner, J. G. (1994). *Arabidopsis* mutants selected for resistance to the phytotoxin coronatine are male sterile, insensitive to methyl jasmonate, and resistant to a bacterial pathogen. *Plant Cell* **6,** 751–579.

Gray, W. M., Kepinski, S., Rouse, D., Leyser, O., and Estelle, M. (2001). Auxin regulates SCF^{TIR1}-dependent degradation of AUX/IAA proteins. *Nature* **414,** 271–276.

Hahn, M. G., Darvill, A. G., and Albersheim, P. (1981). Host–pathogen interactions: XIX. The endogenous elicitor, a fragment of a plant cell wall polysaccharide that elicits phytoalexin accumulation in soybeans. *Plant Physiol.* **68,** 1161–1169.

Hayashi, H., De Bellis, L., Ciurli, A., Kondo, M., Hayashi, M., and Nishimura, M. (1999). A novel acyl-CoA oxidase that can oxidize short-chain acyl-CoA in plant peroxisomes. *J. Biol. Chem.* **274,** 12715–12721.

Hayashi, M., Toriyama, K., Kondo, M., and Nishimura, M. (1998). 2,4-Dichlorophenoxy-butyric acid-resistant mutants of *Arabidopsis* have defects in glyoxysomal fatty acid β-oxidation. *Plant Cell* **10,** 183–195.

Hilpert, B., Bohlmann, H., op den Camp, R., Przybyla, D., Miersch, O., Buchala, A., and Apel, K. (2001). Isolation and characterization of signal transduction mutants of *Arabidopsis thaliana* that constitutively activate the octadecanoid pathway and form nectrotic microlesions. *Plant J.* **26,** 435–446.

Howe, G. A., Lightner, J., Browse, J., and Ryan, C. A. (1996). An octadecanoid pathway mutant (JL5) of tomato is compromised in signaling for defense against insect attack. *Plant Cell* **8,** 2067–2077.

Ishiguro, S., Kawai-Oda, A., Ueda, J., Nishida, I., and Okada, K. (2001). The Defective in Anther Dehiscence gene encodes a novel phospholipase A1 catalyzing the initial step of jasmonic acid biosynthesis, which synchronizes pollen maturation, anther dehiscence, and flower opening in *Arabidopsis*. *Plant Cell* **13,** 2191–2209.

Itoh, H., Matsuoka, M., and Steber, C. (2003). A role for the ubiquitin-26S-proteasome pathway in gibberellin signaling. *Trends Plant Sci.* **8,** 492–497.

Jacinto, T., McGurl, B., Franceschi, V., Delano-Frier, J., and Ryan, C. A. (1997). Tomato prosystemin promoter confers wound-inducible, vascular bundle-specific expression of the β-glucuronidase gene in transgenic tomato plants. *Planta* **203,** 406–412.

Jensen, A. B., Raventos, D., and Mundy, J. (2002). Fusion genetic analysis of jasmonate-signalling mutants in *Arabidopsis*. *Plant J.* **29,** 595–606.

Kepinski, S., and Leyser, O. (2002). Ubiquitination and axin signaling: A degrading story. *Plant Cell* S81–S95.

Kessler, A., Halitschke, R., and Baldwin, I. T. (2004). Silencing the jasmonate cascade: Induced plant defenses and insect populations. *Science* **305,** 665–668.

Knoester, M., Van Loon, L. C., van den Heuvel, J., Hennig, J., Bol, J. F., and Linthorst, H. J. M. (1998). Ethylene-insensitive tobacco lacks nonhost resistance against soil-borne fungi. *Proc. Natl. Acad. Sci. USA* **95,** 1933–1937.

Kramell, R., Miersch, O., Atzorn, R., Parthier, B., and Wasternack, C. (2000). Octadecanoid-derived alteration of gene expression and the "oxylipin signature" in stressed barley leaves. Implications for different signaling pathways. *Plant Physiol.* **123,** 177–187.

Kubigsteltig, I., Laudert, D., and Weiler, E. W. (1999). Structure and regulation of the *Arabidopsis thaliana* allene oxide synthase gene. *Planta* **208,** 463–471.

Lee, S., Suh, S., Kim, S., Crain, R. C., Kwak, J. M., Nam, H.-G., and Lee, Y. (1997). Tobacco MAP kinase: A possible mediator in wound signal transduction pathways. *Plant J.* **12,** 547–556.

Li, C., Liu, G., Xu, C., Lee, G. I., Bauer, P., Ling, H. Q., Ganal, M. W., and Howe, G. A. (2003). The tomato suppressor of prosystemin-mediated responses2 gene encodes a fatty acid desaturase required for the biosynthesis of jasmonic acid and the production of a systemic wound signal for defense gene expression. *Plant Cell* **15,** 1646–1661.

Li, L., Li, C., Lee, G. I., and Howe, G. A. (2002). Distinct roles for jasmonate synthesis and action in the systemic wound response of tomato. *Proc. Natl. Acad. Sci. USA* **99,** 6416–6421.

Li, L., Zhao, Y., McCaig, B. C., Wingerd, B. A., Wang, J., Whalon, M. E., Pichersky, E., and Howe, G. A. (2004). The tomato homolog of CORONATINE-INSENSITIVE1 is required for the maternal control of seed maturation, jasmonate-signaled defense responses, and glandular trichome development. *Plant Cell* **16,** 126–143.

Liechti, R., and Farmer, E. E. (2002). The jasmonate pathway. *Science* **296,** 1649–1650.

Lorenzo, O., Chico, J. M., Sánchez-Serrano, J. J., and Solano, R. (2004). *JASMONATE-INSENSITIVE1* encodes a MYC transcription factor essential to discriminate between different jasmonate-regulated defense responses in *Arabidopsis*. *Plant Cell* **16,** 1938–1950.

Lorenzo, O., Piqueras, R., Sánchez-Serrano, J. J., and Solano, R. (2002). ETHYLENE RESPONSE FACTOR1 integrates signals from ethylene and jasmonate pathways in plant defense. *Plant Cell* **15,** 165–178.

Low, P. S., and Merida, J. R. (1996). The oxidative burst in plant defense: Function and signal transduction. *Physiol. Plant* **6,** 533–542.

Mandaokar, A., Kumar, V. D., Amway, M., and Browse, J. (2003). Microarray and differential display identify genes involved in jasmonate-dependent anther development. *Plant Mol. Biol.* **52,** 775–786.

Mason, H. S., and Mullet, J. E. (1990). Expression of two soybean vegetative storage protein genes during development and in response to water deficit, wounding, and jasmonic acid. *Plant Cell* **2,** 569–579.

McCarty, D. R., and Chory, J. (2000). Conservation and innovation in plant signaling pathways. *Cell* **103,** 201–209.

McConn, M., and Browse, J. (1996). The critical requirement for linolenic acid is pollen development, not photosynthesis, in an *Arabidopsis* mutant. *Plant Cell* **8,** 403–416.

McConn, M., Creelman, R. A., Bell, E., Mullet, J. E., and Browse, J. (1997). Jasmonate is essential for insect defense in *Arabidopsis*. *Proc. Natl. Acad. Sci. USA* **94,** 5473–5477.

McGurl, B., Pearce, G., Orozco-Cárdenas, M., and Ryan, C. A. (1992). Structure, expression and antisense inhibition of the systemin precursor gene. *Science* **255,** 1570–1573.

Menke, F. L. H., Champion, A., Kijne, J. W., and Memlink, J. (1999). A novel jasmonate- and elicitor-responsive element in the periwinkle secondary metabolite biosynthetic gene *Str* interacts with a jasmonate- and elicitor-inducible AP2-domain transcription factor, ORCA2. *EMBO J.* **18,** 4455–4463.

Moyen, C., and Johannes, E. (1996). Systemin transiently depolarizes the tomato mesophyll cell membrane and antagonizes fusicoccin-induced extracellular acidification of mesophyll tissue. *Plant Cell Environ.* **19,** 464–470.

Nárvaez-Vásquez, J., Florin-Christensen, J., and Ryan, C. A. (1999). Positional specificity of a phospholipase a activity induced by wounding, systemin, and oligosaccharide elicitors in tomato leaves. *Plant Cell* **11,** 2249–2260.

Orozco-Cárdenas, M., Narváez-Vásquez, J., and Ryan, C. A. (2001). Hydrogen peroxide acts as a second messenger for the induction of defense genes in tomato plants in response to wounding, systemin and methyl jasmonate. *Plant Cell* **13,** 1–14.

Orozco-Cárdenas, M., and Ryan, C. A. (1999). Hydrogen peroxide is generated systemically in plant leaves by wounding and systemin via the octadecanoid pathway. *Proc. Natl. Acad. Sci. USA* **96,** 6553–6557.

Park, J. H., Halitschke, R., Kim, H. B., Baldwin, I. T., Feldmann, K. A., and Feyereisen, R. (2002). A knock-out mutation in allene oxide synthase results in male sterility and defective wound signal transduction in *Arabidopsis* due to a block in jasmonic acid biosynthesis. *Plant J.* **31,** 1–12.

Pearce, G., Strydom, D., Johnson, S., and Ryan, C. A. (1991). A polypeptide from tomato leaves activates the expression of proteinase inhibitor proteins. *Science* **253,** 895–897.

Peña-Cortés, H., Albrecht, T., Prat, S., Weiler, E. W., and Willmitzer, L. (1993). Aspirin prevents wound-induced gene expression in tomato leaves by blocking jasmonic acid biosynthesis. *Planta* **191,** 123–128.

Penninckx, I. A., Thomma, B. P., Buchala, A., Metraux, J. P., and Broekaert, W. F. (1998). Concomitant activation of jasmonate and ethylene response pathways is required for induction of a plant defensive gene in *Arabidopsis*. *Plant Cell* **10,** 2103–2113.

Pieterse, C. M., van Wees, S. C., Hoffland, E., Van Pelt, J. A., and Van Loon, L. C. (1996). Systemic resistance in *Arabidopsis* induced by biocontrol bacteria is independent of salicylic acid accumulation and pathogenesis-related gene expression. *Plant Cell* **8,** 1225–1237.

Pieterse, C. M. J., and Van Loon, L. C. (1999). Salicylic acid-independent plant defence pathways. *Trends Plant Sci.* **4,** 52–58.

Pieterse, C. M. J., Van Pelt, J. A., Ton, J., Parchmann, S., Mueller, M. J., Buchala, A. J., Métraux, J.-P., and Van Loon, L. C. (2000). Rhizobacteria-mediated induced systemic resistance (ISR) in *Arabidopsis* requires sensitivity to jasmonate and ethylene but is not accompanied by an increase in their production. *Physiol. Mol. Plant Pathol.* **57,** 123–134.

Pieterse, C. M. J., van Wees, S. C. M., Van Pelt, J. A., Knoester, M., Laan, R., Gerrits, H., Weisbeek, P. J., and Van Loon, L. C. (1998). A novel signaling pathway controlling induced systemic resistance in *Arabidopsis*. *Plant Cell* **10,** 1571–1580.

Rao, M. V., and Davis, K. R. (2001). The physiology of ozone induced cell death. *Planta* **213,** 682–690.

Reymond, P., Weber, H., Damond, M., and Farmer, E. E. (2000). Differential gene expression in response to mechanical wounding and insect feeding in *Arabidopsis*. *Plant Cell* **12,** 707–719.

Richmond, T. A., and Bleecker, A. B. (1999). A defect in β-oxidation causes abnormal inflorescence development in *Arabidopsis*. *Plant Cell* **11,** 1911–1923.

Riechmann, J. L., Heard, J., Martin, G., Reuber, L., Jiang, C.-Z., Keddie, J., Adam, L., Pineda, O., Ratcliffe, O. J., Samaha, R. R., Creelman, R., Pilgrim, M., Broun, P., Zhang, J. Z., Ghandehari, D., Sherman, B. K., and Yu, G.-L. (2000). *Arabidopsis* transcription factors: Genome-wide comparative analysis among eukaryotes. *Science* **290,** 2105–2110.

Rogg, L. E., and Bartel, B. (2001). Auxin signaling: Derepression through regulated proteolysis. *Dev. Cell* **1,** 595–604.

Ryan, C. A. (2000). The systemin signaling pathway: Differential activation of plant defensive genes. *Biochim. Biophys. Acta* **1477,** 112–121.

Ryan, C. A., and Pearce, G. (1998). Systemin: A polypeptide signal for plant defensive genes. *Annu. Rev. Cell Dev. Biol.* **14,** 1–17.

Sanders, P. M., Lee, P. Y., Biesgen, C., Boone, J. D., Beals, T. P., Weiler, E. W., and Goldberg, R. B. (2000). The *Arabidopsis* Delayed Dehiscence1 gene encodes an enzyme in the jasmonic acid synthesis pathway. *Plant Cell* **12,** 1041–1062.

Schaller, A., and Oecking, C. (1999). Modulation of plasma membrane H^+-ATPase activity differentially activates wound and pathogen defense responses in tomato plants. *Plant Cell* **11,** 263–272.

Schaller, F., Biesgen, C., Müssig, C., Altmann, T., and Weiler, E. W. (2000). 12-Oxophytodienoate reductase 3 (OPR3) is the isoenzyme involved in jasmonate biosynthesis. *Planta* **210,** 979–984.

Schenk, P. M., Kazan, K., Wilson, I., Anderson, J. P., Richmond, T., Somerville, S. C., and Manners, J. M. (2000). Coordinated plant defense responses in *Arabidopsis* revealed by microarray analysis. *Proc. Natl. Acad. Sci. USA* **97**, 11655–11660.

Sembdner, G., and Parthier, B. (1993). The biochemistry and the physiological and molecular actions of jasmonates. *Annu. Rev. Plant Physiol. Plant Mol. Biol.* **44**, 569–589.

Seo, H. S., Song, J. T., Cheong, J. J., Lee, Y. H., Lee, Y. W., Hwang, I., Lee, J. S., and Choi, Y. D. (2001). Jasmonic acid carboxyl methyltransferase: A key enzyme for jasmonate-regulated plant responses. *Proc. Natl. Acad. Sci. USA* **98**, 4788–4793.

Smyth, D. R., Bowman, J. L., and Meyerowitz, E. M. (1990). Early flower development in *Arabidopsis*. *Plant Cell* **2**, 755–767.

Staswick, P., and Tiryaki, I. (2004). The oxylipin signal jasmonic acid is activated by an enzyme that conjugates it to isoleucine in *Arabidopsis*. *Plant Cell* **16**, 2117–2127.

Staswick, P. E., Su, W., and Howell, S. H. (1992). Methyl jasmonate inhibition of root growth and induction of a leaf protein are decreased in an *Arabidopsis thaliana* mutant. *Proc. Natl. Acad. Sci. USA* **89**, 6837–6840.

Staswick, P. E., Yuen, G. Y., and Lehman, C. C. (1998). Jasmonate signaling mutants of *Arabidopsis* are susceptible to the soil fungus *Pythium irregulare*. *Plant J.* **15**, 747–754.

Stelmach, B. A., Müller, A., Henning, P., Gebhardt, S., Schubert-Zsilavecz, M., and Weiler, E. W. (2001). A novel class of oxylipins, sn1-O-(12-oxophytodienoyl)-sn-2-O-(hexadecatrienoyl)-monogalactosyl diglyceride, from *Arabidopsis thaliana*. *J. Biol. Chem.* **276**, 12832–12838.

Stintzi, A., and Browse, J. (2000). The *Arabidopsis* male-sterile mutant, opr3, lacks the 12-oxophytodienoic acid reductase required for jasmonate synthesis. *Proc. Natl. Acad. Sci. USA* **97**, 10625–10630.

Stintzi, A., Weber, H., Reymond, P., Browse, J., and Farmer, E. E. (2001). Plant defense in the absence of jasmonic acid: The role of cyclopentenones. *Proc. Natl. Acad. Sci. USA* **98**, 12837–12842.

Strassner, J., Schaller, F., Frick, U. B., Howe, G. A., Weiler, E. W., Amrhein, N., Macheroux, P., and Schaller, A. (2002). Characterization and cDNA-microarray expression analysis of 12-oxophytodienoate reductases reveals differential roles for octadecanoid biosynthesis in the local versus the systemic wound response. *Plant J.* **32**, 585–601.

Stratman, J. W., and Ryan, C. A. (1997). Myelin basic protein kinase activity in tomato leaves is induced systemically by wounding and increases in response to systemin and oligosaccharide elicitors. *Proc. Natl. Acad. Sci.* **94**, 11085–11089.

Theologis, A., Huynh, T. V., and Davis, R. W. (1985). Rapid induction of specific mRNAs by auxin in pea epicotyl tissue. *J. Mol. Biol.* **183**, 53–68.

Thomma, B. P. H. J., Eggermont, K., Penninckx, I. A. M. A., Mauch-Mani, B., Vogelsang, R., Cammue, B. P. A., and Broekaert, W. F. (1998). Separate jasmonate-dependent and salicylate-dependent defense-response pathways in *Arabidopsis* are essential for resistance to distinct microbial pathogens. *Proc. Natl. Acad. Sci. USA* **95**, 15107–15111.

Tiryaki, I., and Staswick, P. E. (2002). An *Arabidopsis* mutant defective in jasmonate response is allelic to the auxin-signaling mutant axr1. *Plant Physiol.* **130**, 887–894.

Ton, J., Davison, S., van Wees, S. C. M., Van Loon, L. C., and Pieterse, C. M. J. (2001). The *Arabidopsis ISR1* locus-controlling rhizobacteria-mediated-induced systemic resistance is involved in ethylene signaling. *Plant Physiol.* **125**, 83–93.

Turner, J. G., Ellis, C., and Devoto, A. (2002). The jasmonate signal pathway. *Plant Cell* **14** (Suppl.), S153–S164.

Usami, S., Banno, H., Ito, Y., Nishihama, R., and Machida, Y. (1995). Cutting activates a 46-kilodalton protein kinase in plants. *Proc. Natl. Acad. Sci. USA* **92**, 8660–8664.

van der Fits, L., and Mem Link, J. (2000). ORCA3, a jasmonate-responsive transcriptional regulator of plant primary and secondary metabolism. *Science* **289**, 295–297.

Van Loon, L. C., Bakker, P. A. H. M., and Pieterse, C. M. J. (1998). Systemic resistance induced by rhizosphere bacteria. *Annu. Rev. Phytopathol.* **36**, 453–483.

Van Wees, S. C. M., Luijendijk, M., Smoorenburg, I., Van Loon, L. C., and Pieterse, C. M. J. (1999). Rhizozobacteria-mediated induced systemic resistance (ISR) in *Arabidopsis* is not associated with a direct effect on known defense-genes but stimulates the expression of the jasmonate-inducible gene Atvsp upon challenge. *Plant Mol. Biol.* **41**, 537–549.

Vick, B. A., and Zimmerman, D. C. (1983). The biosynthesis of jasmonic acid: A physiological role for plant lipoxygenase. *Biochem. Biophys. Res. Commun.* **111**, 470–477.

Vijayan, P., Shockey, J., Levesque, C. A., Cook, R. J., and Browse, J. (1998). A role for jasmonate in pathogen defense of *Arabidopsis*. *Proc. Natl. Acad. Sci. USA* **95**, 7209–7214.

von Malek, B., van der Graaff, E., Schneitz, K., and Keller, B. (2002). The *Arabidopsis* male-sterile mutant dde2-2 is defective in the ALLENE OXIDE SYNTHASE gene encoding one of the key enzymes of the jasmonic acid biosynthesis pathway. *Planta* **216**, 187–192.

Weiler, E. W., Albrecht, T., Groth, B., Xia, Z. Q., Luxem, M., Liss, H., Andert, L., and Spengler, P. (1993). Evidence for the involvement of jasmonates and their octadecanoid precursors in the tendril coiling response of *Bryonia dioica*. *Phytochemistry* **32**, 591–600.

Weiler, E. W., Laudert, D., Stelmach, B. A., Hennig, P., Biesgen, C., and Kubigsteltig, I. (1999). Octadecanoid and hexadecanoid signalling in plant defence [review]. *Novartis Found. Symp.* **223**, 191–204.

Wisman, E., and Ohlrogge, J. (2000). *Arabidopsis* microarray service facilities. *Plant Physiol.* **124**, 1468–1471.

Xiao, S., Dai, L., Liu, F., Wang, Z., Peng, W., and Xie, D. (2004). COS1: An *Arabidopsis coronatine insensitive1* suppressor essential for regulation of jasmonate-mediated plant defense and senescence. *Plant Cell* **16**, 1132–1142.

Xie, D. X., Feys, B. F., James, S., Nieto-Rostro, M., and Turner, J. G. (1998). COI1: An *Arabidopsis* gene required for jasmonate-regulated defense and fertility. *Science* **280**, 1091–1094.

Xu, L., Liu, F., Wang, Z., Peng, W., Huang, R., Huang, D., and Xie, D. (2001). An *Arabidopsis* mutant *cex1* exhibits constant accumulation of jasmonate-regulated *AtVSP*, *Thi2.1* and *PDF1.2*. *FEBS Lett.* **494**, 161–164.

Xu, Y., Chang, P.-F. L., Liu, D., Narasimhan, M. L., Raghothama, K. G., Hasegawa, P. M., and Bressan, R. A. (1994). Plant defense genes are synergistically induced by ethylene and methyl jasmonate. *Plant Cell* **6**, 1077–1085.

Ziegler, J., Stenzel, I., Hause, B., Maucher, H., Hamberg, M., Grimm, R., Ganal, M., and Wasternack, C. (2000). Molecular cloning of allene oxide cyclase. The enzyme establishing the stereochemistry of octadecanoids and jasmonates. *J. Biol. Chem.* **275**, 19132–19138.

13

Plant Sex Pheromones

Hiroyuki Sekimoto

Institute of Life Sciences, Graduate School of Arts and Sciences, University of Tokyo 3-8-1 Komaba, Meguro, Tokyo 153-8902, Japan; Department of Chemical and Biological Sciences, Faculty of Science, Japan Women's University, 2-8-1 Mejirodai Bunkyo, Tokyo 112-8681, Japan

    I. Introduction
   II. Brown Algal Pheromones
  III. The *Volvox* Sex-Inducing Pheromone
  IV. The Sex Pheromones of *Closterium*
   V. Other Green Algal Pheromones
  VI. Spermatozoid Attractant(s) in Mosses
 VII. Pheromones in Ferns
VIII. Attractants in the Pollen Tube
  IX. Conclusions
      References

Although plant pheromones have been much less studied than animal pheromones, they are involved in a wide variety of processes in the life cycle of many plants, particularly in sexual reproduction. In this review, the current knowledge concerning sex pheromones in plants is described with emphasis on their structures and functions. © 2005 Elsevier Inc.

I. INTRODUCTION

Chemical communication between organisms is important for both the prosperity and survival of most organisms. Karlson and Lüscher (1959) proposed the term pheromone for chemicals that enable members of the same species to communicate with each other. The term pheromone is derived from the Greek "pherein" (to carry) and "hormone" (to excite, stimulate). Pheromones function by influencing other members of the same species but not the individual that produces them.

Most studies on pheromones have concentrated on the animal kingdom, and many specific pheromones have been identified, especially in insects (Regnier and Law, 1968; Wilson and Bossert, 1963). In animals, pheromones affect the central nervous system in two different ways: releaser pheromones initiate immediate behavioral responses upon reception, whereas primer pheromones cause physiological changes in an animal that ultimately result in a behavioral response (Wilson and Bossert, 1963). Pheromone-mediated chemical communication has also been reported in some plants, even though plants lack central nervous systems. Most of the reported plant pheromones are involved in sexual reproduction. The effects of sex pheromones vary from organism to organism. Machlis (1972) proposed three generic terms for pheromones: erotactins, which induce tactic movements of gametes; erotropins, which induce directional growth of the sexual structures; and erogens, which control the induction and differentiation of sexual structures. These terms refer only to the phenomenon and not to the mechanisms of the responses.

The sexual reproductive process consists of several steps, including meiosis, sex determination or differentiation, induction and differentiation of sexual structures, mutual recognition by individuals of the opposite sex, release and fusion of gametes, and formation of zygotes, although the sequence of these events differs with the species. For successful fertilization, concurrent expression of both types of gametes or sexes is required. Synchronization of this step depends on various mechanisms, including environmental factors, endogenous rhythms, and sex pheromones. Sex pheromones are often specific to the species and encourage efficient mating.

Many studies of sexual reproductive processes have focused on the mutual recognition of sexual partners. Such studies have often been performed with lower fungi and algae because the sexual reproductive processes of some of these organisms are easily induced in the laboratory, and the preparation of gametes or sexually competent individuals is often relatively straightforward.

Numerous studies have been published on algal sex pheromones, some of which have been characterized in detail. In contrast, little has been published on the sex pheromones of land plants, despite the remarkable progress that has been made in identifying the molecular biological techniques in these organisms. This review focuses mainly on sex pheromones in algae but also describes pheromonal activities in land plants. Previously published reviews provide information on fungal pheromones (Bölker and Kahmann, 1993; Jaenicke, 1991; Kochert, 1978).

II. BROWN ALGAL PHEROMONES

In brown algal species, planogamy is unknown, and the sex pheromones secreted by immotile female gametes or freshly released eggs are involved sexual reproduction. Pheromones induce the chemoattraction or kinetic orientation of the motile male gametes toward female gametes, acting as erotactins. In some species, pheromones are also responsible for the release of male gametes from the male reproductive organ (antheridia).

The first identified brown algae pheromone was allo-*cis*-1-(cyclohepta-diene-2′,5′-yl)-butene-1 or ectocarpene, from *Ectocarpus siliculosus* (Müller *et al.*, 1971). Ectocarpene is released from settled female gametes and attracts male gametes toward female cells. At least 12 compounds that act as pheromones in a large number of brown algae, as well as many related compounds, have since been reported (Maier, 1995). All of the currently identified pheromones are volatile, unsaturated C_{11} or C_8 hydrocarbons of four structural classes: (1) vinylcyclopropanes, (2) vinylcyclopentenes, (3) cycloheptadienes, and (4) acyclic olefines. Modifications to these basic structures are limited to the butyl termini or butyl side chains, indicating that common biosynthetic pathways exist. The biosynthesis of these pheromones has been fully described in another review (Pohnert and Boland, 2002).

The pheromones mentioned previously are not specific at the species or genus level. The pheromone lamoxirene acts as a highly potent sex pheromone, triggering chemotaxis and the release of sperm cells from antheridia in the families Laminariaceae, Alariaceae, and Lessoniaceae. When female gametes secrete pheromones, a large number of related compounds are also secreted as by-products, forming a complex bouquet (Maier, 1995; Maier *et al.*, 1987). Some of the by-products can even represent the main fraction. Because these mixtures often include sex pheromones of other brown algae, they might act as allomones, misleading the gametes of other species. Although the pheromonal communication is often not species-specific, physical barriers on the cell surface prevent cross-fertilization. Glycoproteins specific to the female plasma membrane might function as receptors for male gametes in *E. siliculosus* (Schmid, 1993; Schmid *et al.*, 1994). In addition, a sperm protein that binds to the egg plasma membrane can trigger partial activation of the egg in *Fucus serratus* (Wright *et al.*, 1995a,b). It is possible that unknown biological functions for mating with the true partner might be present in the mixtures of secretions.

Although the chemical nature and localization of the pheromone receptor are unknown, a proteinaceous receptor is assumed to exist in the cell membrane and/or the flagellar membrane(s). A large number of specially designed synthetic pheromone analogs were used to elucidate the structural requirements for the binding of a pheromone to its putative receptor sites (Boland et al., 1981; Jaenicke, 1974; Maier et al., 1988, 1994). The positions of double bonds in the pheromone molecule are a crucial structural feature, and the interaction of the pheromone with the receptor is possibly mediated by dispersion forces, dipole-induced polarization of the double bonds, and hydrophobic forces. In the kelp *Laminaria digitata*, the profiles of receptors involved in spermatozoid release and chemotaxis of spermatozoids are clearly different, indicating the presence of at least two kinds of pheromone receptors responsible for these two functions (Maier et al., 1988, 1994).

The influx of extracellular Ca^{2+} is indispensable for pheromone action in *E. siliculosus*. It is possible that two pharmacologically distinct Ca^{2+} channels are involved in chemokinesis and chemoklinotaxis in this species (Maier and Calenberg, 1994). Apart from this case, the mechanisms of signal transduction after binding of the pheromone to the receptor are mostly unknown.

Ectocarpene, the first identified brown algal pheromone, triggers chemoattraction in *E. siliculosus*, but its thermolabile precursor *cis*-(1R,2S)-[(1'E.3'Z)-hexadienyl]-vinylcyclopropane or preectocarpene, is now considered the true *Ectocarpus* pheromone (Maier, 1995; Pohnert and Boland, 2002). Rearrangement of preectocarpene into ectocarpene occurs spontaneously at 18° C with a half-time of 21 min. In chemoaccumulation assays, preectocarpene is 10,000-fold more active than ectocarpene.

III. THE VOLVOX SEX-INDUCING PHEROMONE

Volvox is the most highly developed genus of the family Volvocaceae. Several patterns of sexual reproduction are exhibited by the various species in this genus. Some species are monoclonic, and the others are diclonic. The monoclonic species are either monoecious or dioecious. All of the diclonic species are dioecious, producing male and female spheroids in separate clones. *Volvox* generally reproduces asexually, although it is able to switch to the sexual pathway. The involvement of a sex pheromone in the switching process was first described by Darden, who showed that cultures of asexual *Volvox aureus*, a monoclonic, monoecious species, could be induced to change to the sexual pathway by supplementation with cell-free culture medium from mature male spheroids (Darden, 1966). Subsequently, similar phenomena were reported in other *Volvox* species (Kochert, 1981). In the dioecious *Volvox carteri*, the most intensively studied *Volvox* species (Kirk and Nishii, 2001), a pheromone called "sex inducer" or "sex-inducing pheromone" has been the subject of considerable attention (Starr, 1970). In this species, sexual reproduction is initiated by a mutation-like switch that has a probability of 2×10^{-4}, which leads to formation of the first sexual male colony (Weisshaar et al., 1984). The sex-inducing pheromone is produced and released by this sexual male colony, acts on the asexual reproductive cells (gonidia) of both sexes, and alters their developmental pathway so that sexual forms (egg- or sperm-bearing forms) are produced in the next generation. Sperm bundles generated on the male colony meet a female colony, apparently by chance rather than by directed swimming (Coggin et al., 1979; Kirk, 1998), after which a specific transient binding to somatic cells occurs. The fusion of gametes results in the formation of a dormant diploid zygospore that can survive drought. Under favorable environmental conditions, germination of zygospores involving meiosis occurs (Fig. 1).

Because the pheromone is not produced by sexual females or the somatic cells of sexual males, it is thought that expression of the gene encoding the pheromone is tightly linked to sperm development (Gilles et al., 1981; Starr, 1970). Nevertheless, both asexual females and asexual males could be triggered to produce the pheromone by exposure to heat shock (Kirk and Kirk, 1986). It is possible that reactive oxygen species participate in both the production and the activity of the pheromone in triggering sexual development in gonidia (Nedelcu and Michod, 2003).

The sex-inducing pheromones were independently purified from two isolates of *V. carteri*, one from Japan (*V. carteri* f. *nagarensis*) and the other from the United States (*V. carteri* f. *weismannia*) (Kochert and Yates, 1974; Starr and Jaenicke, 1974). Both are glycoproteins of about 30 kDa. The inducer from *V. carteri* f. *nagarensis* is strictly competent for its own gonidia, whereas that of *V. carteri* f. *weismannia* induces sexuality in both (Al-Hasani and Jaenicke, 1992). The pheromone from *V. carteri* f. *nagarensis* is one of the most potent biological effector molecules known, exhibiting full effectiveness below concentrations of 1×10^{-16}. Large-scale production of the pheromone led to the identification of partial amino acid sequences, allowing the isolation of its genomic clone (Tschochner et al., 1987) and cDNA (Mages et al., 1988).

It has been postulated that both the synthesis and the modification of an extracellular matrix (ECM) protein are required for the exquisite sensitivity of this sexual induction system (Gilles et al., 1983; Hallmann, 2003; Wenzl and Sumper, 1982, 1986). The majority of proteins synthesized shortly after pheromone treatment of the ECM are from one family of glycoproteins, the pherophorins (Godl et al., 1995, 1997). At least 13 different pherophorins exist in *Volvox* (Hallmann, 2003). The sequences of the carboxy-terminal domains of all of the pherophorins are very similar to that of the pheromone. The pheromone also induces the synthesis of the deep-zone hydroxyproline-rich glycoprotein (DZ-HRGP) (Ender et al., 1999), chitinase/lysozyme,

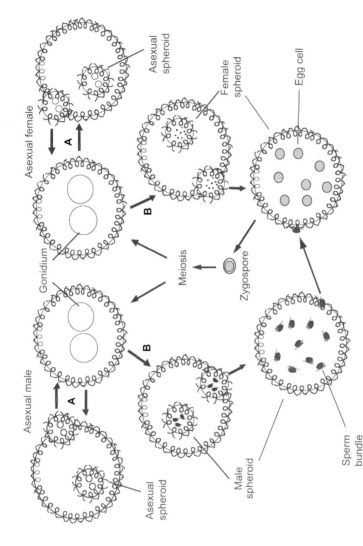

FIGURE 1. The life cycle of *Volvox carteri*. (A) Asexual reproduction; (B) sexual differentiation induced by the sex-inducing pheromone.

chitin-binding protein (Amon et al., 1998), and metalloproteinases (VMPs) (Hallmann et al., 2001). Pherophorin-II is thought to be responsible for a signal amplification mechanism of the pheromone. The carboxy-terminal domain, which exhibits 30% identity to the sex-inducing pheromone, is proteolytically liberated from the parent glycoprotein after its synthesis is induced by the pheromone. Because inhibition of the processing by protease inhibitors correlates with a suppression of sexual induction and no induction of pherophorin-II gene expression by the pheromone was observed in three independently isolated sterile mutants, the liberated domain might act as an analog of the sex-inducing pheromone (Sumper et al., 1993). However, no sex-inducing activity of the domain has been demonstrated experimentally. Therefore, the main target of the pheromone is still unknown.

IV. THE SEX PHEROMONES OF CLOSTERIUM

The *Closterium* species are unicellular Charophycean algae. The sexual reproduction of species in this genus has been of interest to many investigators for more than a century and the morphological details have been well documented (Cook, 1963; Lippert, 1967; Noguchi, 1988; Noguchi and Ueda, 1985; Pickett-Heaps and Fowke, 1971). Heterothallic *Closterium* strains have two sexes, mating-type plus (mt^+) and mating-type minus (mt^-). Co-culturing of cells of the two sexes in nitrogen-depleted medium in the presence of light easily induces sexual reproduction (Hamada, 1978; Hogetsu and Yokoyama, 1979; Ichimura, 1971). The reproductive process consists of five steps: differentiation to form gametangial cells, formation of sexual pairs, formation of papillae, release of protoplasts from the cells, and fusion of protoplasts to form a zygote (Hamada, 1978; Ichimura, 1971). Several pheromones appear to be involved in these steps, including chemotactic compounds responsible for the mutual attraction of cells during the formation of sexual pairs, as well as some morphogenic compounds that induce differentiation of gametangial cells, formation of papillae, and release of protoplasts (Coesel and de Jong, 1986; Hogetsu and Yokoyama, 1979; Ichimura, 1971). Kato et al. (1981) detected an activity responsible for the release of protoplasts from mt^+ cells of the *Closterium peracerosum-strigosum-littorale* complex (*C. pslc*) during sexual reproduction. The active substance was named protoplast-releasing substance (PRS) and was putatively identified as a glycoprotein (Kato and Sasaki, 1983, 1985). Later, the existence of another substance responsible for the release of PRS from mt^- cells was proposed (Kato et al., 1984). The results suggested that diffusible substances mediate the respective processes during the *Closterium* reproductive process but no biochemical analysis of such communication was reported subsequently.

Sekimoto et al. (1990) successfully isolated the first pheromone from *Closterium*. This pheromone, designated protoplast-release-inducing protein (PR-IP), is a glycoprotein that consists of subunits of 42 and 19 kDa. It is released by mt^+ *C. pslc* cells and is responsible for inducing the release of protoplasts from mt^- cells. The latter process proceeds only after appropriate preculture under continuous light, during which the mt^- cells differentiate from vegetative cells into gametangial cells (Sekimoto and Fujii, 1992) and PR-IP receptors appear on the plasma membranes of mt^- cells. Specific binding of the biotinylated 19-kDa subunit of PR-IP to its receptor has been clearly demonstrated (Sekimoto et al., 1993b).

Secretion of PR-IP by mt^+ cells is induced in medium in which only mt^- cells had been cultured (Sekimoto et al., 1993a). Therefore, it was proposed that another pheromone that induces the synthesis and release of PR-IP is released from mt^- cells. This pheromone was named PR-IP Inducer (Sekimoto et al., 1993a). The pheromone was subsequently purified and found to be a glycoprotein with a molecular mass of 18.7 kDa (Nojiri et al., 1995). Protoplast-release-inducing protein inducer is released constitutively from mt^- cells in the presence of light, and directly induces the production and release of PR-IP from mt^+ gametes. A possible model for the intercellular communication that is mediated by these pheromones during sexual reproduction was proposed (Sekimoto, 2000). cDNAs encoding the subunits of PR-IP (Sekimoto et al., 1994a,b) and PR-IP Inducer (Sekimoto et al., 1998) have been isolated. Genes for these pheromones can be detected in cells of both mating types by genomic Southern hybridization analysis, but are only expressed in cells of the respective mating types, suggesting the sex-specific regulation of gene expression (Sekimoto et al., 1994c, 1998). In spite of these biochemical and molecular biological studies, the mechanisms that lead to the formation of sexual pairs and the release of protoplasts from mt^+ cells in *C. pslc* remain unclear.

The formation of sexual pairs and the differentiation of gametes in *Closterium ehrenbergii* have been characterized in terms of both morphology and physiology. In this species, the conjugation process is similar to that of *C. pslc*, with the exception of the timing of the differentiation of gametes. Coesel and de Jong (1986) reported that both mt^+ and mt^- cells migrate toward each other, even when separated by an agar bridge. Chemotactic activity that induces the directional migration of mt^+ cells was also detected in growth medium in which mt^- cells had been cultured (Fukumoto et al., 1998). The active substance was found to be a protein with an apparent molecular mass of 20 kDa.

Sexual cell division (SCD), a special type of cell division that results in the formation of gametangial cells, occurs after the formation of sexual pairs in *C. ehrenbergii* (Fukumoto et al., 1997; Hogetsu and Yokoyama, 1979). The gametangial cells can be easily distinguished from vegetative cells by their resemblance to a canine tooth (Fukumoto et al., 1997). Hogetsu and

Yokoyama (1979) showed that SCD, the formation of papillae, and the release of protoplasts from gametangial cells all depend on intercellular communication between the members of sexual pairs. Fukumoto et al. (1997) detected and characterized a pheromone responsible for the differentiation of gametangial *C. ehrenbergii* cells, naming it SCD-inducing pheromone (SCD-IP). This pheromone is released from mt⁻ cells in the light, and the presence of mt⁺ cells enhances its production. The induction of SCD by SCD-IP was found to be light-dependent. The pheromone was purified by sequential column-chromatographic fractionations and was identified as a glycoprotein with an apparent molecular mass of 18 kDa (Fukumoto et al., 2002). Amino-terminal and internal amino acid sequences of the pheromone were determined and used to design degenerate primers to isolate a cDNA encoding the pheromone. A 906-bp full-length cDNA containing an open reading frame that encodes a 23.3-kDa protein of 209 amino acid residues was isolated (Fukumoto et al., 2003). An apparent signal peptide of 26 amino acid residues was detected at the amino-terminal end of the encoded protein. The mature protein is predicted to contain 150 amino acid residues and to have a molecular weight of 16,723 Da. The gene for the SCD-inducing pheromone is only expressed in mt⁻ cells, despite being encoded in the genomes of cells of both mating types. Addition of a nitrogen source to the medium and incubation in continuous darkness suppress the expression of the gene.

In *C. pslc*, SCD-inducing activities specific for the two mating-type cells have also been detected (Tsuchikane et al., 2003). Mating-type minus cells release an SCD-inducing pheromone specific for mating-type plus cells and are designated sexual-cell division-inducing pheromone-minus (SCD-IP-minus), whereas an mt⁻-specific pheromone released from mt⁺ cells is designated SCD-IP-plus. With gametangial cells of both mating types obtained through the effect of SCD-IPs, gametangial mt⁺ cells showed high competency for conjugation with vegetative mt⁻ cells but gametangial mt⁻ cells showed low competency for conjugation with vegetative mt⁺ cells. These results indicate that the roles of gametangial cells in the process of conjugation differ by sex. Both SCD-IP-minus and SCD-IP-plus showed quite similar characteristics to the PR-IP Inducer and PR-IP with respect to molecular weight, heat stability, and dependency on light for their secretion and function, indicating the presence of close relationships among these pheromones. Indeed, the amino acid sequence of the SCD-IP of *C. ehrenbergii* has significant similarity to that of the *C. pslc* PR-IP Inducer, with 110 of 209 amino acid residues identical in the two pheromones. To clarify these relationships, preparation of recombinant pheromones using a yeast heterologous expression system is in progress (Sekimoto, 2002).

Closterium exhibits a gliding locomotory behavior mediated by the forceful extrusion of mucilage from one pole of the cell that causes the cell to glide in the opposite direction (Domozych et al., 1993). It is believed that in some desmids, the release of mucilage by a cell facilitates adhesion of the cell to a solid object (Surek and von Sengbusch, 1981). Despite several reported cytological and biochemical analyses of mucilage from *Closterium*, little is known about the role of mucilage secretion in the sexual reproduction of this species. Substances with the ability to stimulate secretion of uronic-acid-containing mucilage from mt^+ and mt^- cells were detected in media in which mt^- and mt^+ cells had been cultured separately and were designated mucilage-secretion-stimulating pheromone (MS-SP)-minus and MS-SP-plus, respectively. Biochemical and physiological similarities are also thought to be there between PR-IP and PR-IP inducer with MS-SP-plus and MS-SP-minus, respectively (Akatsuka *et al.*, 2003). Based on the results described previously, the process of sexual reproduction and the effects of the sex pheromones of *C. pslc* are presented in Fig. 2.

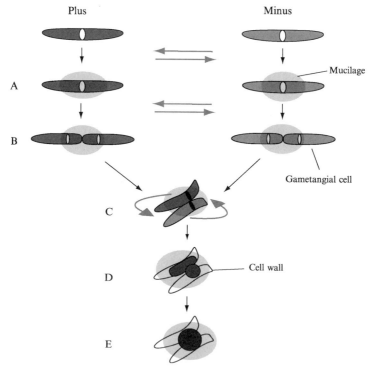

FIGURE 2. A schematic illustration of the sexual reproduction of the *Closterium peracerosum-strigosum-littorale* complex. (A) Mucilage secretion induced by mucilage-secretion-stimulating pheromones (MS-SPs); (B) sexual cell division induced by SCD-inducing pheromones (SCD-IPs); (C) sexual pair formation induced by unknown chemoattractic pheromone(s); (D) protoplast release induced by protoplast-release-inducing protein (PR-IP) and PR-IP Inducer; (E) zygote. Gray arrows indicate pheromonal communication.

ESTs from cells in the sexual reproduction processes were analyzed (Sekimoto *et al.*, 2003). Among the 760 unique sequences obtained, four were homologous to previously identified sex pheromone genes. A cDNA microarray analysis revealed that the expression of these pheromone homolog genes, as well as known sex pheromone genes, soared at least five times, from 0 to 8 h after mixing, when sexual interaction between the cells of the two mating types began (Sekimoto *et al.*, unpublished observation). These results indicate that these homologs have unidentified roles during sexual reproduction.

V. OTHER GREEN ALGAL PHEROMONES

The critical biochemical characterization of a pheromone has been accomplished for only one *Chlamydomonas* species, the anisogamous *Chlamydomonas allensworthii*. The female cells of this species tend to produce the pheromone and release it into the medium after entering into the gamete stage under nitrogen deprivation, but they also seem to continuously produce smaller amounts of pheromone. A chemoattractant for the male was isolated from media in which female strains were cultured (Starr *et al.*, 1995). The pheromone attracted the male gametes at concentrations as low as 1 pM, and very high concentrations tended to immobilize the male gametes. Detailed structural analysis of the pheromone revealed that lurlenic acid and lurlenol, chemically related derivatives of plastohydroquinone, serve as pheromones (Jaenicke and Marner, 1995; Jaenicke and Starr, 1996). The structure was confirmed by its synthesis (Mori and Takanashi, 1996).

The isogamous species of *Chlamydomonas*, such as *Chlamydomonas eugametos* and *Chlamydomonas reinhardtii*, serve as model systems for the study of the sexual process in green algae. These are the two green algal species in which the gamete recognition process has been studied most intensively. Genome-wide studies are in progress in *C. reinhardtii* (Asamizu *et al.*, 1999, 2000; Davies and Grossman, 1998; Shrager *et al.*, 2003). The sexual process involves agglutinins on the flagella, through which clumping and pairing of compatible gametes occur. Chemotaxis was not found to be involved in the interaction, and pair formation depends entirely on the chances of collision (Tomson *et al.*, 1985).

It has been proposed that communication between males and females of *Oedogonium* sp. is mediated by the action of several pheromonal substances (Maier, 1993). During the sexual reproduction of nannandrous species of *Oedogonium*, males form flagellated cells, called androspores, while some cells in female plants differentiate into large oogonial mother cells. The androspores differentiate into single-celled dwarf males after adhering to oogonial mother cells. Mature dwarf males form several male reproductive organs, in which sperm cells are produced, at their apices. Each oogonial mother cell divides unequally to yield both a large oogonium with an egg and a carrier cell on which the dwarf males rest. Sperm cells released by dwarf males swim toward the oogonium and then fuse with the egg to form a zygospore. At least four sex pheromones have been found to play a role in these processes (Hoffman, 1960; Rawitscher-Kunkel and Machlis, 1962). One, androspore sirenin, is secreted by oogonial mother cells and attracts androspores. The androspores develop into mature dwarf males independently of the female threads, but the directional growth of dwarf males toward the oogonium is induced if dwarf males develop on the oogonial mother cells. This result indicates the presence of a second pheromone that is produced by the oogonium that controls the directional growth of dwarf males. The *Oedogonium donnellii* pheromone has been named circein (Hill et al., 1989). The unequal cell division of the oogonial mother cell into the oogonium and the carrier cell is also controlled by the dwarf male via other pheromones. The tip of the dwarf male becomes submerged in the mucilage surrounding the oogonium. The directional movement of sperm cells toward eggs has also been postulated to be the effect of yet another chemotactic pheromone.

In *Oedogonium cardiacum*, a macrandrous species, the chemotactic attraction of spermatozoids to the oogonial pore has been demonstrated (Coss and Pickett-Heaps, 1973; Hoffman, 1960). A bioassay for spermatozoid chemotaxis was developed and the chemoattractant was partially characterized (Machlis et al., 1974). The compound is highly water soluble and labile to heat, dilute acid, and alkali. Gel filtration and ultrafiltration revealed an apparent molecular mass of 500–1500 Da for the substance. Paper chromatographic separation isolated a yellow pigment closely associated with the activity. Because the pigment does not show absorption at 280nm, the chemoattractant is probably not an oligopeptide.

VI. SPERMATOZOID ATTRACTANT(S) IN MOSSES

Pfeffer (1884) reported that sucrose acts as an attractant of spermatozoids to the archegonia in mosses. However, sucrose is not a species-specific or even an order-specific attractant. It was shown that the neck region of intact archegonia of the moss *Bryum capillare* accumulate sucrose to high concentrations that are considerably reduced with the disintegration of the neck canal cells during maturation of archegonia (Kaiser et al., 1985). Concentration of the released sucrose with the exudates is sufficient to evoke chemotaxis in spermatozoids (Ziegler et al., 1988). Nevertheless, the involvement of other chemotactic substances in the exudates cannot be excluded.

VII. PHEROMONES IN FERNS

In the life cycle of ferns, spores germinate and develop into prothallia, in which archegonia and antheridia are formed. In heterosporous ferns, the sex of a gametophyte is genetically determined (i.e. macrospores and microspores germinate and develop into female and male gametophytes, respectively. In contrast, most homosporous ferns are dioecious or sexually dimorphic, producing male, hermaphroditic, and/or female gametophytes in a single population. Sex determination occurs during the gametophyte generation. These alternative developmental fates are determined by the pheromone antheridiogen (Döpp, 1950; Näf, 1969; Näf et al., 1975). In the absence of antheridiogen, individual prothallia develop an apical meristem and are either hermaphroditic or female. When the fern gametophytes at various growth stages coexist, these meristic individuals secrete antheridiogens into their surroundings. Because juvenile prothallia around the mature one are highly sensitive to antheridiogens, they produce numerous antheridia but no archegonia. In contrast, meristic prothallia, which are producing archegonia, lose sensitivity to antheridiogens and become female prothallia. It has been shown that the developing prothallia of more than 10 species of ferns produce antheridiogens. The antheridiogens characterized in Schizaeaceous ferns are all derivatives of gibberellins, whereas those of other ferns are thought not to be (Yamane, 1998).

Antheridiogen acts to repress meristem and archegonia formation, as well as to promote the rapid differentiation of antheridia during gametophyte development. Although the chemical structure of the antheridiogen of *Ceratopteris richardii*, which is used as a model plant system for various developmental processes, is still unknown, many studies have been performed on the mode of sex determination through the effect of the antheridiogen. Several phenotypic classes of mutations that alter the sex of the gametophyte have been characterized and a possible relationship between these genes in the presence or absence of antheridiogen has been proposed (Fig. 3) (Banks, 1994, 1997, 1999; Eberle and Banks, 1996; Strain et al., 2001; Tanurdzic and Banks, 2004). In addition, cDNAs whose expression is induced by antheridiogen were isolated by subtractive hybridization. The product of the *ANI1* gene may be an extracellular carrier of the pheromone that is required to initiate the male program of development (Wen et al., 1999).

To enable fertilization, the archegonium forms a channel within its neck to allow the sperm to reach the egg. The disintegration of the neck canal cells during maturation of the archegonia might account for the chemoattraction. Pfeffer (1884) described the penetration of fern sperms into the archegonia and showed that L-malic acid is the attractant. The activities of malic and other organic acids have been confirmed by other authors

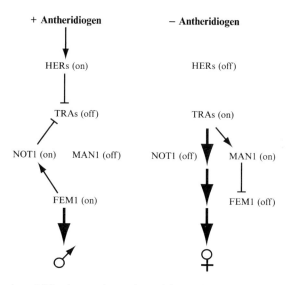

FIGURE 3. A model for the genetic sex-determining pathway in *Ceratopteris richardii*. The activities of the sex-determining genes are affected by the presence of the pheromone antheridiogen. Arrows and t-bars indicate activating and repressing interactions, respectively.

(Machlis and Rawitscher-Kunkel, 1967; Ziegler, 1962) but the endogenous chemoattractant is still unknown.

VIII. ATTRACTANTS IN THE POLLEN TUBE

In flowering plants, sperm cells cannot move without the help of the pollen tube, the male gametophyte. Successful fertilization requires the precise guidance of the pollen tube to the embryo sac, the female gametophyte, in which the egg cell is present. The mechanism for the precise directional growth of the pollen tube has long been studied. There has been controversy as to whether there is actual chemotropic growth in styles involving some chemical gradient or whether the anatomical features of the style alone define its path. However, the presence of the chemotropic substances (i.e., chemoattracting pheromones for the male gametophyte) produced by the pistil has been strongly supported by several independent works. Comprehensive reviews of this subject have already been published (Higashiyama *et al.*, 2003; Sanchez *et al.*, 2004; Weterings and Russell, 2004).

In the lily stigma, the involvement of a chemotropic molecule at the entrance of the pollen tube into the style has long been expected. A small basic protein with sequence similarity to cell wall proteins of unknown function, plantacyanins, has been identified as an active molecule and named chemocyanin (Kim *et al.*, 2003). The chemotropic activity is enhanced in the presence of another stigma peptide, the stigma/stylar cysteine-rich adhesin (Park *et al.*, 2000). In contrast, it has been shown that no chemotropic gradients exist in the lily style (Iwanami, 1959).

After germination of a pollen grain on the pistil, the pollen tube grows intercellularly in the transmitting tract and emerges on the surface of the septum. Then, the pollen tube grows along the surface of the septum toward a funiculus and the micropyle of an ovary. Studies of several *Arabidopsis* mutants defective in ovule development have suggested the presence of ovule-derived long-range activities that control pollen tube guidance (Hulskamp *et al.*, 1995). It is believed that the female gametophyte is responsible for guiding the pollen tube, based on studies of a gametophytic mutant in which the sporophytic cells are genotypically normal, apart from the fact that half of the ovules do not contain a functional female gametophyte (Ray *et al.*, 1997). Furthermore, at least two independent signals that mediate the guidance are thought to exist in *Arabidopsis*: a funiculus guidance signal and a micropyle guidance signal (Shimizu and Okada, 2000). However, the possibility that the gametophyte may contribute to this process indirectly through its influence on some sporophytic cells still exists.

Gamma-amino butyric acid (GABA) has been shown to be one of the possible chemoattractants produced by diploid pistil cells, based on studies of the *pollen-pistil incompatibility2* (*pop2*) mutant of *Arabidopsis* (Palanivelu *et al.*, 2003). POP2 encodes a transaminase that degrades GABA and contributes to the formation of a gradient along the pollen tube path from the stigma to the inner integument cells near the micropyle of the ovule. In the mutant, the level of GABA is elevated 100-fold and the GABA gradient is disturbed. Pollen tubes of the *pop2* mutant fail to grow and cannot reach the *pop2* ovules because they are hypersensitive to GABA. However, the pollen tubes can reach the ovules if at least one parent is wild-type, possibly because a small gradient can be created around the tip of the tubes. Nevertheless, the pollen tubes do not respond chemotropically to a GABA gradient *in vitro*, suggesting that additional molecules are required.

The presence of a diffusible signal that attracts pollen tubes has also been demonstrated in *Torenia fournieri* using an *in vitro* system (Higashiyama *et al.*, 1998). In these plants, the naked embryo sac protrudes through micropyle. Pollen tubes that have grown semi-*in vitro* through the cut style grow toward the ovules and arrive precisely at the site of entry into the embryo sac, the central region of the filiform apparatus, between the two synergids. The pollen tubes do not reach heat-treated embryo sacs, suggesting that a diffusible signal is produced by the living embryo sac. Laser ablation of the egg, synergid, and central cells has clearly demonstrated that two synergid cells adjacent to the egg cell are the source of an attraction signal for pollen tubes (Higashiyama *et al.*, 2001). The attraction signal produced by two synergid cells is effective at a rather short range (at most 200 µm) and is species-specific. The identity of the attractant is still unknown (Higashiyama, 2002; Higashiyama *et al.*, 2003).

IX. CONCLUSIONS

Sex pheromones are often involved in the progress of sexual reproduction in many algae and land plants, although they have not been specifically identified in many cases. The structures and functions of pheromones are diverse, indicating widespread occurrence of sex pheromones in phylogenetically distant lineages. The chemical structures, physiological functions, and mechanisms of the reception and signal transduction of pheromones are not yet fully understood in most organisms because of the difficulty of *in vitro* biological assay systems. In addition to the development of sensitive assay systems and mutants, progress in the genome and EST analysis of suitable organisms will enhance our understanding of sex pheromones as well as our knowledge of sexual reproduction in plants.

REFERENCES

Akatsuka, S., Sekimoto, H., Iwai, H., Fukumoto, R., and Fujii, T. (2003). Mucilage secretion regulated by sex pheromones in *Closterium peracerosum-strigosum-littorale* complex. *Plant Cell Physiol.* **44,** 1081–1087.

Al-Hasani, H., and Jaenicke, L. (1992). Characterization of the sex-inducer glycoprotein of *Volvox carteri* f. *Weismannia. Sex. Plant Reprod.* **5,** 8–12.

Amon, P., Haas, E., and Sumper, M. (1998). The sex-inducing pheromone and wounding trigger the same set of genes in the multicellular green alga *Volvox. Plant Cell* **10,** 781–789.

Asamizu, E., Miura, K., Kucho, K., Inoue, Y., Fukuzawa, H., Ohyama, K., Nakamura, Y., and Tabata, S. (2000). Generation of expressed sequence tags from low-CO_2 and high-CO_2 adapted cells of *Chlamydomonas reinhardtii. DNA Res.* **7,** 305–307.

Asamizu, E., Nakamura, Y., Sato, S., Fukuzawa, H., and Tabata, S. (1999). A large scale structural analysis of cDNAs in a unicellular green alga, *Chlamydomonas reinhardtii.* I. Generation of 3433 non-redundant expressed sequence tags. *DNA Res.* **6,** 369–373.

Banks, J. A. (1994). Sex-determining genes in the homosporous fern *Ceratopteris. Development* **120,** 1949–1958.

Banks, J. A. (1997). The transformer genes of the fern *Ceratopteris* simultaneously promote meristem and archegonia development and repress antheridia development in the developing gametophyte. *Genetics* **147,** 1885–1897.

Banks, J. A. (1999). Gametophyte development in ferns. *Annu. Rev. Plant Physiol. Plant Mol. Biol.* **50,** 163–186.

Boland, W., Jakoby, K., Jaenicke, L., Müller, D. G., and Folster, E. (1981). Receptor specificity and threshold concentration in chemotaxis of the phaeophyte *Cutleria multifida. Z. Naturforsch. (C)* **36,** 262–271.

Bölker, M., and Kahmann, R. (1993). Sexual pheromones and mating responses in fungi. *Plant Cell* **5,** 1461–1469.

Coesel, P., and de Jong, W. (1986). Vigorous chemotactic attraction as a sexual response in *Closterium ehrenbergii* Meneghini (Desmidiaceae, Chlorophyta). *Phycologia* **25**, 405–408.

Coggin, S. J., Hutt, W., and Kochert, G. (1979). Sperm bundle-female somatic cell interaction in the fertilization process of *Volvox carteri* f. *weismannia* (Chlorophyta). *J. Phycol.* **15**, 247–251.

Cook, P. A. (1963). Variation in vegetative and sexual morphology among the small curved species of *Closterium*. *Phycologia* **3**, 1–18.

Coss, R. A., and Pickett-Heaps, J. D. (1973). Gametogenesis in green alga *Oedogonium cardiacum*. 1. Cell divisions leading to formation of spermatids and oogonia. *Protoplasma* **78**, 21–39.

Darden, W. H. J. (1966). Sexual differentiation in *Volvox aureus*. *J. Protozool.* **13**, 239–255.

Davies, J. P., and Grossman, A. R. (1998). The use of *Chlamydomonas* (Chlorophyta: Volvocales) as a model algal system for genome studies and the elucidation of photosynthetic processes. *J. Phycol.* **34**, 907–917.

Domozych, C. R., Plante, K., and Blais, P. (1993). Mucilage processing and secretion in the green alga *Closterium*. 1. Cytology and biochemistry. *J. Phycol.* **29**, 650–659.

Döpp, W. (1950). Eine die antheridienbildung bei farnen fördernde substanz in den prothallien von *Pteridium aquilinum* l. *Kun. Ber. Dtsch. Bot. Ges.* **63**, 139–147.

Eberle, J. R., and Banks, J. A. (1996). Genetic interactions among sex-determining genes in the fern *Ceratopteris richardii*. *Genetics* **142**, 973–985.

Ender, F., Hallmann, A., Amon, P., and Sumper, M. (1999). Response to the sexual pheromone and wounding in the green alga *Volvox*: Induction of an extracellular glycoprotein consisting almost exclusively of hydroxyproline. *J. Biol. Chem.* **274**, 35023–35028.

Fukumoto, R., Dohmae, N., Takio, K., Satoh, S., Fujii, T., and Sekimoto, H. (2002). Purification and characterization of a pheromone that induces sexual cell division in the unicellular green alga *Closterium ehrenbergii*. *Plant Physiol. Biochem.* **40**, 183–188.

Fukumoto, R., Fujii, T., and Sekimoto, H. (1997). Detection and evaluation of a novel sexual pheromone that induces sexual cell division of *Closterium ehrenbergii* (Chlorophyta). *J. Phycol.* **33**, 441–445.

Fukumoto, R., Fujii, T., and Sekimoto, H. (1998). A newly identified chemotactic sexual pheromone from *Closterium ehrenbergii*. *Sex. Plant Reprod.* **11**, 81–85.

Fukumoto, R., Fujii, T., and Sekimoto, H. (2003). Cloning and characterization of a cDNA encoding a sexual cell division-inducing pheromone from a unicellular green alga *Closterium ehrenbergii* (Chlorophyta). *J. Phycol.* **39**, 931–936.

Gilles, R., Bittner, C., and Jaenicke, L. (1981). Site and time of formation of the sex-inducing glycoprotein in *Volvox carteri*. *FEBS Lett.* **124**, 57–61.

Gilles, R., Gilles, C., and Jaenicke, L. (1983). Sexual differentiation of the green alga *Volvox carteri*. *Naturwissenschaften* **70**, 571–572.

Godl, K., Hallmann, A., Rappel, A., and Sumper, M. (1995). Pherophorins: A family of extracellular matrix glycoproteins from *Volvox* structurally related to the sex-inducing pheromone. *Planta* **196**, 781–787.

Godl, K., Hallmann, A., Wenzl, S., and Sumper, M. (1997). Differential targeting of closely related ECM glycoproteins—the pherophorin family from *Volvox*. *EMBO J.* **16**, 25–34.

Hallmann, A. (2003). Extracellular matrix and sex-inducing pheromone in *Volvox*. *Int. Rev. Cytol.* **227**, 131–182.

Hallmann, A., Amon, P., Godl, K., Heitzer, M., and Sumper, M. (2001). Transcriptional activation by the sexual pheromone and wounding: A new gene family from *Volvox* encoding modular proteins with (hydroxy)proline-rich and metalloproteinase homology domains. *Plant J.* **26**, 583–593.

Hamada, J. (1978). Studies on several environmental factors for zygote formation and germination in *Closterium ehrenbergii*. *Bot. Mag. Tokyo* **91**, 173–180.

Higashiyama, T. (2002). The synergid cell: Attractor and acceptor of the pollen tube for double fertilization. *J. Plant Res.* **115**, 149–160.
Higashiyama, T., Kuroiwa, H., Kawano, S., and Kuroiwa, T. (1998). Guidance *in vitro* of the pollen tube to the naked embryo sac of *Torenia fournieri. Plant Cell* **10**, 2019–2031.
Higashiyama, T., Kuroiwa, H., and Kuroiwa, T. (2003). Pollen-tube guidance: Beacons from the female gametophyte. *Curr. Opin. Plant Biol.* **6**, 36–41.
Higashiyama, T., Yabe, S., Sasaki, N., Nishimura, Y., Miyagishima, S., Kuroiwa, H., and Kuroiwa, T. (2001). Pollen tube attraction by the synergid cell. *Science* **293**, 1480–1483.
Hill, G. J. C., Cunningham, M. R., Byrne, M. M., Ferry, T. P., and Halvorson, J. S. (1989). Chemical control of androspore morphogenesis in *Oedogonium donnellii* (Chlorophyta, Oedogoniales). *J. Phycol.* **25**, 368–376.
Hoffman, L. R. (1960). Chemotaxis of *Oedogonium* sperms. *Southwestern Nat.* **5**, 111–116.
Hogetsu, T., and Yokoyama, M. (1979). Light, a nitrogen-depleted medium and cell–cell interaction in the conjugation process of *Closterium ehrenbergii* Meneghini. *Plant Cell Physiol.* **20**, 811–817.
Hulskamp, M., Schneitz, K., and Pruitt, R. E. (1995). Genetic-evidence for a long-range activity that directs pollen-tube guidance in *Arabidopsis. Plant Cell* **7**, 57–64.
Ichimura, T. (1971). Sexual cell division and conjugation-papilla formation in sexual reproduction of *Closterium strigosum*. *In* "Proceedings of the 7th International Seaweed Symposium" (K. Nishizawa, Ed.), pp. 208–214. University of Tokyo Press, Tokyo.
Iwanami, Y. (1959). Physiological studies of pollen. *J. Yokohama Municipal Univ.* **116**, 1–137.
Jaenicke, L. (1974). Chemical signal transmission by gamete attractants in brown algae. *In* "Biochemistry of Sensory Functions" (L. Jaenicke, Ed.), pp. 307–309. Springer-Verlag, Berlin.
Jaenicke, L. (1991). Development: Signals in the development of cryptogams. *Prog. Bot.* **52**, 138–189.
Jaenicke, L., and Marner, F. J. (1995). Lurlene, the sexual pheromone of the green flagellate *Chlamydomonas allensworthii. Liebigs Ann.* **7**, 1343–1345.
Jaenicke, L., and Starr, R. C. (1996). The lurlenes, a new class of plastoquinone-related mating pheromones from *Chlamydomonas allensworthii* (*Chlorophyceae*). *Eur. J. Biochem.* **241**, 581–585.
Kaiser, K., Outlaw, W. H., and Ziegler, H. (1985). Sucrose content of receptive archegonia of the moss *Bryum capillare* hedw. *Naturwissenschaften* **72**, 378–379.
Karlson, P., and Lüscher, M. (1959). 'Pheromones': A new term for a class of biologically active substances. *Nature* **183**, 55–56.
Kato, A., Obokata, J., and Sasaki, K. (1981). Mating type interaction in *Closterium peracerosum-strigosum-littorale*: Mating induced protoplast release. *Plant Cell Physiol.* **22**, 1215–1222.
Kato, A., and Sasaki, K. (1983). Effect of tunicamycin on sexual reproduction in heterothallic strains of *Closterium. J. Fac. Sci. Hokkaido Univ. Ser. V.* **13**, 1–6.
Kato, A., and Sasaki, K. (1985). Sexual interaction in heterothallic strains of *Closterium peracerosum-strigosum-littorale. Plant Physiol.* **77**, 556–559.
Kato, A., Yamazaki, T., and Sasaki, K. (1984). Differences in physiological properties and sexual substances between heterothallic strains of *Closterium. J. Fac. Sci. Hokkaido Univ. Ser. V.* **13**, 267–280.
Kim, S., Mollet, J. C., Dong, J., Zhang, K. L., Park, S. Y., and Lord, E. M. (2003). Chemocyanin, a small basic protein from the lily stigma, induces pollen tube chemotropism. *Proc. Natl. Acad. Sci. USA* **100**, 16125–16130.
Kirk, D. L. (1998). "*Volvox*: Molecular-Genetic Origins of Multicellularity and Cellular Differentiation." Cambridge University Press, Cambridge.

Kirk, D. L., and Kirk, M. M. (1986). Heat shock elicits production of sexual inducer in *Volvox*. *Science* **231,** 51–54.

Kirk, D. L., and Nishii, I. (2001). *Volvox carteri* as a model for studying the genetic and cytological control of morphogenesis. *Dev. Growth Differ.* **43,** 621–631.

Kochert, G. (1978). Sexual pheromones in algae and fungi. *Annu. Rev. Plant Physiol.* **29,** 461–486.

Kochert, G. (1981). Sexual pheromones in *Volvox* development. *In* "Sexual Interactions in Eukaryotic Microbes" (D. H. O'Day and P. A. Horgen, Eds.), pp. 73–93. Academic Press, New York.

Kochert, G., and Yates, I. (1974). Purification and partial characterization of a glycoprotein sexual inducer from *Volvox carteri. Proc. Natl. Acad. Sci. USA* **71,** 1211–1214.

Lippert, B. E. (1967). Sexual reproduction in *Closterium moniliferum* and *Closterium ehrenbergii. J. Phycol.* **3,** 182–198.

Machlis, L. (1972). The coming age of sex hormones in plants. *Mycologia* **64,** 235–247.

Machlis, L., Hill, G. G. C., Steinback, K.-E., and Reed, W. (1974). Some characteristics of sperm attractant from *Oedogonium cardiacum. J. Phycol.* **10,** 199–204.

Machlis, L., and Rawitscher-Kunkel, E. (1967). Mechanisms of gametic approach in plants. *In* "Fertilization" (C. B. Metz and A. Monroy, Eds.), vol. 1, pp. 117–161. Academic Press, New York.

Mages, H.-W., Tschochner, H., and Sumper, M. (1988). The sexual inducer of *Volvox carteri*. Primary structure deduced from cDNA sequence. *FEBS Lett.* **234,** 407–410.

Maier, I. (1993). Gamete orientation and induction of gametogenesis by pheromones in algae and plants. *Plant Cell Env.* **16,** 891–907.

Maier, I. (1995). Brown algal pheromones. *In* "Progress in Phycological Research" (F. E. Round and D. J. Chapman, Eds.), Vol. 11, pp. 51–102. Biopress Ltd., Bristol.

Maier, I., and Calenberg, M. (1994). Effect of extracellular Ca^{2+} and Ca^{2+}-antagonists on the movement and chemoorientation of male gametes of *Ectocarpus siliculosus* (Phaeophyceae). *Bot. Acta* **107,** 451–460.

Maier, I., Müller, D. G., and Boland, W. (1994). Spermatozoid chemotaxis in *Laminaria digitata* (Phaeophyceae). 3. Pheromone receptor specificity and threshold concentrations. *Z. Naturforsch. (C)* **49,** 601–606.

Maier, I., Müller, D. G., Gassmann, G., Boland, W., and Jaenicke, L. (1987). Sexual pheromones and related egg secretions in Laminariales (Phaeophyta). *Z. Naturforsch. (C)* **42,** 948–954.

Maier, I., Müller, D. G., Schmid, C., Boland, W., and Jaenicke, L. (1988). Pheromone receptor specificity and threshold concentrations for spermatozoid release in *Laminaria digitata*. *Naturwissenschaften* **75,** 260–263.

Mori, K., and Takanashi, S. (1996). Synthesis of lurlene, the sex pheromone of the green flagellate *Chlamydomonas allensworthii*. *Tetrahedron Lett.* **37,** 1821–1824.

Müller, D. G., Jaenicke, L., Donike, M., and Akintobi, T. (1971). Sex attractant in a brown alga: Chemical structure. *Science* **171,** 815–817.

Nedelcu, A. M., and Michod, R. E. (2003). Sex as a response to oxidative stress: The effect of antioxidants on sexual induction in a facultatively sexual lineage. *Proc. R. Soc. Lond. B* **270,** S136–S139.

Noguchi, T. (1988). Numerical and structural changes in dictyosomes during zygospore germination of *Closterium ehrenbergii*. *Protoplasma* **147,** 135–142.

Noguchi, T., and Ueda, K. (1985). Cell walls, plasma membranes, and dictyosomes during zygote maturation of *Closterium ehrenbergii*. *Protoplasma* **128,** 64–71.

Nojiri, T., Fujii, T., and Sekimoto, H. (1995). Purification and characterization of a novel sex pheromone that induces the release of another sex pheromone during sexual reproduction of the heterothallic *Closterium peracerosum-strigosum-littorale* complex. *Plant Cell Physiol.* **36,** 79–84.

Näf, U. (1969). Antheridium formation in ferns—a model for the study of developmental change. *Bot. J. Linn. Soc.* **58**, 321–331.

Näf, U., Nakanishi, K., and Endo, M. (1975). On the physiology and chemistry of fern antheridiogens. *Bot. Rev.* **41**, 315–359.

Palanivelu, R., Brass, L., Edlund, A. F., and Preuss, D. (2003). Pollen tube growth and guidance is regulated by pop2, an *Arabidopsis* gene that controls GABA levels. *Cell* **114**, 47–59.

Park, S.-Y., Jauh, G.-Y., Mollet, J.-C., Eckard, K. J., Nothnagel, E. A., Walling, L. L., and Lord, E. M. (2000). A lipid transfer-like protein is necessary for lily pollen tube adhesion to an *in vitro* stylar matrix. *Plant Cell* **12**, 151–163.

Pfeffer, W. (1884). Locomotorische richtungsbewegungen durch chemische reize. *Unters. bot. Inst. in Tübingen* **1**, 363–482.

Pickett-Heaps, J. D., and Fowke, L. C. (1971). Conjugation in the desmid *Closterium littorale*. *J. Phycol.* **7**, 37–50.

Pohnert, G., and Boland, W. (2002). The oxylipin chemistry of attraction and defense in brown algae and diatoms. *Nat. Prod. Rep.* **19**, 108–122.

Rawitscher-Kunkel, E., and Machlis, L. (1962). The hormonal integration of sexual reproduction in *Oedogonium*. *Am. J. Bot.* **49**, 177–183.

Ray, S., Park, S. S., and Ray, A. (1997). Pollen tube guidance by the female gametophyte. *Development* **124**, 2489–2498.

Regnier, F. E., and Law, J. H. (1968). Insect pheromones. *J. Lipid Res.* **9**, 541–551.

Sanchez, A. M., Bosch, M., Bots, M., Nieuwland, J., Feron, R., and Mariani, C. (2004). Pistil factors controlling pollination. *Plant Cell* **16**, S98–S106.

Schmid, C. E. (1993). Cell–cell recognition during fertilization in *Ectocarpus siliculosus*. *Hydrobiologia* **260/261**, 437–443.

Schmid, C. E., Schroer, N., and Müller, D. G. (1994). Female gamete membrane glycoproteins potentially involved in gamete recognition in *Ectocarpus siliculosus* (Phaeophyceae). *Plant Sci.* **102**, 61–67.

Sekimoto, H. (2000). Intercellular communication during sexual reproduction of *Closterium* (Conjugatophyceae). *J. Plant Res.* **113**, 343–352.

Sekimoto, H. (2002). Production and secretion of a biologically active *Closterium* sex pheromone by *Saccharomyces cerevisiae*. *Plant Physiol. Biochem.* **40**, 789–794.

Sekimoto, H., and Fujii, T. (1992). Analysis of gametic protoplast release in *Closterium peracerosum-strigosum-littorale* complex (Chlorophyta). *J. Phycol.* **28**, 615–619.

Sekimoto, H., Fukumoto, R., Dohmae, N., Takio, K., Fujii, T., and Kamiya, Y. (1998). Molecular cloning of a novel sex pheromone responsible for the release of a different sex pheromone in *Closterium peracerosum-strigosum-littorale* complex. *Plant Cell Physiol.* **39**, 1169–1175.

Sekimoto, H., Inoki, Y., and Fujii, T. (1993a). Detection and evaluation of an inducer of diffusible mating pheromone of heterothallic *Closterium peracerosum-strigosum-littorale* complex. *Plant Cell Physiol.* **37**, 991–996.

Sekimoto, H., Satoh, S., and Fujii, T. (1990). Biochemical and physiological properties of a protein inducing protoplast release during conjugation in the *Closterium peracerosum-strigosum-littorale* complex. *Planta* **182**, 348–354.

Sekimoto, H., Satoh, S., and Fujii, T. (1993b). Analysis of binding of biotinylated protoplast-release-inducing protein that induces release of gametic protoplasts in the *Closterium peracerosum-strigosum-littorale* complex. *Planta* **189**, 468–474.

Sekimoto, H., Sone, Y., and Fujii, T. (1994a). cDNA cloning of a 42-kilodalton subunit of protoplast-release-inducing protein from *Closterium*. *Plant Physiol.* **104**, 1095–1096.

Sekimoto, H., Sone, Y., and Fujii, T. (1994b). A cDNA encoding a 19-kilodalton subunit of protoplast-release-inducing protein from *Closterium*. *Plant Physiol.* **105**, 447.

Sekimoto, H., Sone, Y., and Fujii, T. (1994c). Regulation of expression of the genes for a sex pheromone by an inducer of the sex pheromone in the *Closterium peracerosum-strigosum-littorale* complex. *Planta* **193**, 137–144.

Sekimoto, H., Tanabe, Y., Takizawa, M., Ito, N., Fukumoto, R., and Ito, M. (2003). Expressed sequence tags from the *Closterium peracerosum-strigosum-littorale* complex, a unicellular Charophycean alga, in the sexual reproduction process. *DNA Res.* **10**, 147–153.

Shimizu, K. K., and Okada, K. (2000). Attractive and repulsive interactions between female and male gametophytes in *Arabidopsis* pollen tube guidance. *Development* **127**, 4511–4518.

Shrager, J., Hauser, C., Chang, C. W., Harris, E. H., Davies, J., McDermott, J., Tamse, R., Zhang, Z. D., and Grossman, A. R. (2003). *Chlamydomonas reinhardtii* genome project. A guide to the generation and use of the cDNA information. *Plant Physiol.* **131**, 401–408.

Starr, R. C. (1970). Control of differentiation in *Volvox*. *Dev. Biol. Suppl.* **4**, 59–100.

Starr, R. C., and Jaenicke, L. (1974). Purification and characterization of the hormone initiating sexual morphogenesis in *Volvox carteri* f. *nagariensis* Iyengar. *Proc. Natl. Acad. Sci. USA* **71**, 1050–1054.

Starr, R. C., Marner, F. J., and Jaenicke, L. (1995). Chemoattraction of male gametes by a pheromone produced by female gametes of *Chlamydomonas*. *Proc. Natl. Acad. Sci. USA* **92**, 641–645.

Strain, E., Hass, B., and Banks, J. A. (2001). Characterization of mutations that feminize gametophytes of the fern *Ceratopteris*. *Genetics* **159**, 1271–1281.

Sumper, M., Berg, E., Wenzl, S., and Godl, K. (1993). How a sex pheromone might act at a concentration below 10^{-16} M. *EMBO J.* **12**, 831–836.

Surek, B., and von Sengbusch, P. (1981). The localization of galactosyl residues and lectin receptors in the mucilage and the cell walls of *Cosmocladium saxonicum* (Desmidiaceae) by means of fluorescent probes. *Protoplasma* **108**, 149–161.

Tanurdzic, M., and Banks, J. A. (2004). Sex-determining mechanism in land plants. *Plant Cell* **16**, S61–S71.

Tomson, A. M., Demets, R., Sigon, C. A. M., Stegwee, D., and van den Ende, H. (1985). Cellular interactions during the mating process in *Chlamydomonas eugametos*. *Plant Physiol.* **81**, 522–526.

Tschochner, H., Lottspeich, F., and Sumper, M. (1987). The sexual inducer of *Volvox carteri*: Purification, chemical characterization and identification of its gene. *EMBO J.* **6**, 2203–2207.

Tsuchikane, Y., Fukumoto, R., Akatsuka, S., Fujii, T., and Sekimoto, H. (2003). Sex pheromones that induce sexual cell division in the *Closterium peracerosum-strigosum-littorale* complex (Charophyta). *J. Phycol.* **39**, 303–309.

Weisshaar, B., Gilles, R., Moka, R., and Jaenicke, L. (1984). A high frequency mutation starts sexual reproduction in *Volvox carteri*. *Z. Naturforsch.* **39c**, 1159–1162.

Wen, C. K., Smith, R., and Banks, J. A. (1999). *Ani1*: A sex pheromone-induced gene in *Ceratopteris* gametophytes and its possible role in sex determination. *Plant Cell* **11**, 1307–1317.

Wenzl, S., and Sumper, M. (1982). The occurrence of different sulphated cell surface glycoproteins correlates with defined developmental events in *Volvox*. *FEBS Lett.* **143**, 311–315.

Wenzl, S., and Sumper, M. (1986). Early event of sexual induction in *Volvox*: Chemical modification of the extracellular matrix. *Dev. Biol.* **115**, 119–128.

Weterings, K., and Russell, S. D. (2004). Experimental analysis of the fertilization process. *Plant Cell* **16**, S107–S118.

Wilson, E. O., and Bossert, W. H. (1963). Chemical communication among animals. *Recent Prog. Horm. Res.* **19**, 673–716.

Wright, P. J., Callow, J. A., and Green, J. R. (1995a). The *Fucus* (Phaeophyceae) sperm receptor for eggs. 2. Isolation of a binding protein which partially activates eggs. *J. Phycol.* **31**, 592–600.

Wright, P. J., Green, J. R., and Callow, J. A. (1995b). The *Fucus* (Phaeophyceae) sperm receptor for eggs. 1. Development and characteristics of a binding assay. *J. Phycol.* **31**, 584–591.

Yamane, H. (1998). Fern antheridiogens. *Int. Rev. Cytol.* **184**, 1–32.

Ziegler, H. (1962). Chemotaxis. In "Encyclopedia of Plant Physiology" (W. Ruhland, Ed.), Vol. 172, pp. 484–532. Springer-Verlag, Berlin.

Ziegler, H., Kaiser, K., and Lipp, J. (1988). Sucrose in the archegonium exudate of the moss *Bryum capillare* hedw. *Naturwissenschaften* **75**, 203–203.

14

PLANT BRASSINOSTEROID HORMONES

TADAO ASAMI, TAKESHI NAKANO, AND SHOZO FUJIOKA

Discovery Research Institute, RIKEN, 2-1 Hirosawa, Wako Saitama 351-0198, Japan

I. Biosynthesis and Metabolism of Brassinosteroids
   A. Biosynthesis of Brassinosteroids
   B. Brassinosteroid Biosynthesis Inhibitors
   C. Regulation of Brassinosteroid Biosynthesis
   D. Metabolism of Brassinosteroids
II. Signal Transduction of Brassinosteroids
   A. Brassinosteroid Receptors
   B. Signal Carrier Proteins for Brassinosteroid Signaling
   C. Other Upstream and Midstream Functional Proteins of Brassinosteroid Signaling
   D. A Gene Expression Mechanism Responsive to Brassinosteroid
III. Conclusions
   References

In animals, a large number of steroid hormones play important roles in numerous processes including reproduction and differentiation. The biologically active plant steroid brassinolide (BL) was first discovered in the pollen of western rape in 1979 (Grove et al., 1979). This finding suggested that BL is indispensable for plant growth and differentiation. To date, more than 50 BL analogs have been identified, and the group has been termed brassinosteroids (BRs) (Fujioka and Yokota, 2003). Brassinosteroids have several biological activities, such as inducing cell elongation when applied at very low concentrations. For this reason, soon after their discovery, they were suggested to be a sixth type of plant hormone; however, for years BRs were not considered true plant hormones. The turning point in BR research was the discovery of the *Arabidopsis* dwarf mutants *det2* and *cpd* in 1996 (Li *et al.*, 1996; Szekeres *et al.*, 1996). These BR-deficient mutants were found to revert to the wild-type phenotype following BR treatment. Concurrent with the analysis of these mutants, an outline of the biosynthetic pathway of BRs was being elucidated through chemical analysis. Following the isolation of *det2* and *cpd*, a great number of BR-deficient mutants were identified. The mutant genes were found to encode proteins that catalyze the conversion of plant steroids to BR precursors. Eventually, BRs were widely recognized as important plant hormones indispensable for growth and differentiation (Clouse and Sasse, 1998). In parallel, mutants that are insensitive to BRs were isolated (Clouse *et al.*, 1996; Li *et al.*, 1997) with phenotypes very similar to those of the BR-biosynthesis mutants. Investigations of these mutants revealed several mechanisms of BR perception and signal transduction (Bishop and Koncz, 2002; Clouse, 2002). This review describes findings on the effects of BRs on plant growth, BR biosynthesis and catabolism, and BR signal transduction. © 2005 Elsevier Inc.

I. BIOSYNTHESIS AND METABOLISM OF BRASSINOSTEROIDS

A. BIOSYNTHESIS OF BRASSINOSTEROIDS

Brassinosteroids are classified as C_{27}, C_{28}, or C_{29} steroids depending on their C-24 alkyl substituents. Brassinolide, the first BR to be isolated, is a C_{28} BR and exhibits the highest biological activity of the known BRs (Sakurai, 1999). The initial elucidation of a BL-biosynthesis pathway in cultured *Catharanthus roseus* cells suggested the existence of two parallel, branched BL pathways, named the early and late C-6 oxidation pathways (Fujioka and Sakurai, 1997). Each step of these pathways was verified by the conversion of labeled BRs. Studies suggest that cross-talk occurs between the parallel pathways. There is also an early C-22 oxidation pathway that occurs early in the biosynthesis stage (Fujioka *et al.*, 2002). Thus, the biosynthetic pathways of the BRs form a complicated network. These pathways are probably common in the plant kingdom, as they have been identified in plants as diverse as *Arabidopsis*, rice (Hong *et al.*, 2002; Yamamuro *et al.*, 2000), pea (Nomura *et al.*, 1997, 1999), and zinnia (Yamamoto *et al.*, 2001). However, in tomato and tobacco, the late C-6 oxidation pathway appears to be the predominant route because the endogenous BRs in these species composed of only members of the late C-6 oxidation pathway (Bishop *et al.*, 1999; Koka *et al.*, 2000; Yokota *et al.*, 2001).

Many of the genes that encode BR biosynthetic enzymes have been isolated using BR-biosynthesis mutants of *Arabidopsis*, pea, tomato, morning glory, and rice. These mutants are BR-deficient and revert to a wild-type phenotype following treatment with exogenous BRs. The common physical features of BR mutants are a short robust stature; short, round, dark-green leaves; and reduced fertility or sterility. The enzymes that have been identified from *Arabidopsis* are shown in Fig. 1 and the enzymes from other plants are as follows:

1. In pea LK (Nomura *et al.*, 2004) and morning glory KBT (Suzuki *et al.*, 2003), 5α-reductases. The pea *lk* and morning glory *kbt* are orthologs of the *Arabidopsis det2*.
2. In tomato Dwarf (D) (Bishop *et al.*, 1999) and rice BRD1 (Hong *et al.*, 2002; Mori *et al.*, 2002), cytochrome P450s. Tomato *dwarf* and rice *brd1* are orthologs of the *Arabidopsis BR6ox*. It was shown that Dwarf catalyzes the conversion of castasterone to BL.
3. In rice D2 (Hong *et al.*, 2003), a cytochrome P450. Rice *d2* is thought to be defective in C-3 oxidation.
4. In tomato DPY (Koka *et al.*, 2000), a cytochrome P450. Tomato *dpy* is an ortholog of the *Arabidopsis cpd*.
5. In pea DDWF1 (Kang *et al.*, 2001), a cytochrome P450. *DDWF1* is a dark-inducible pea gene thought to be a 2-hydroxylase.

B. BRASSINOSTEROID BIOSYNTHESIS INHIBITORS

When exogenous chemicals allow the rapid, conditional, reversible, selective, and dose-dependent control of biological functions, they mimic the effects of conditional mutations and either induce or suppress the formation of a specific phenotype of interest. Although BR mutants have been used as powerful tools for elucidating the biosynthesis and functions of BRs, mutants have only been isolated from a few plant species. In this context, small molecules that modulate endogenous BR levels should be highly useful in plant hormone research (Asami *et al.*, 2003a). Investigations of small molecules that induce BR-deficiency-like phenotypes in *Arabidopsis* led us to identify brassinazole as the first candidate for a BR-biosynthesis inhibitor (Asami and Yoshida 1999; Asami *et al.*, 2000; Min *et al.*, 1999). Brassinazole treatment reduces the BR content in plant cells. Studies of the target site(s) of brassinazole revealed that the compound binds directly to the DWF4 protein, which is a cytochrome P450 monooxygenase that catalyzes

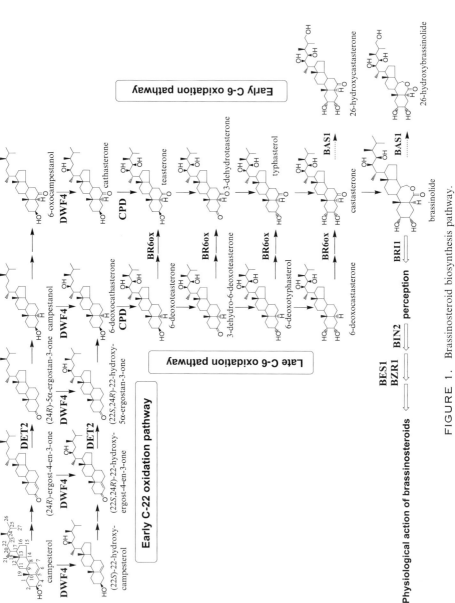

FIGURE 1. Brassinosteroid biosynthesis pathway.

22-hydroxylation of the side chain of BRs (Asami *et al.*, 2001). These results suggest that brassinazole is a BR-biosynthesis inhibitor. Treatment with one of at least five known BR-biosynthesis inhibitors mimics conditional mutations in BR biosynthesis (Asami *et al.*, 2003b; 2004 Sekimata *et al.*, 2001, 2002; Wang *et al.*, 2001). These inhibitors allow the investigation of the functions of BRs in a variety of plant species. For example, the suppression of BR biosynthesis in cotton ovules by treatment with brassinazole inhibits fiber formation, and this effect was neutralized by co-application of BL, indicating that BR plays an important role in cotton fiber development (Sun *et al.*, 2004). A standard genetic screen using BR-biosynthesis inhibitors to identify mutants that confer resistance to these inhibitors identified new BR signal transduction components (Wang *et al.*, 2001, 2002; Yin *et al.*, 2002). This method has advantages over mutant screens that use BR-deficient mutants as the background. The development of chemicals that induce phenotypes of interest is now emerging as a useful method to study biological systems in plants that complements classical biochemical and genetic methods.

C. REGULATION OF BRASSINOSTEROID BIOSYNTHESIS

The first information on the regulation of BR biosynthesis was obtained through the analysis of *cpd*. The expression of the *CPD* gene is suppressed by BL, and the application of cycloheximide demonstrated that de novo protein synthesis is indispensable for this transcriptional repression (Mathur *et al.*, 1998). The expression of *DWF4* is very low in the wild-type but is much higher in the *dwf1* and *cpd* mutants (Noguchi *et al.*, 2000). A similar increase is observed in the BR-insensitive mutant *bri1* (Noguchi *et al.*, 2000), and greater amounts of BR also accumulate in this mutant (Noguchi *et al.*, 1999). These results show that the BR signal is indispensable for suppressing *DWF4* gene expression and is critically important for BR homeostasis. Brassinosteroids also suppress the transcription of *BR6ox1* and *BR6ox2* which encode an *Arabidopsis* C-6 oxidase (Shimada *et al.*, 2001, 2003). Feedback regulation of BR-biosynthesis genes also occurs in the rice *OsDwarf* (Hong *et al.*, 2001) and *d2* (Hong *et al.*, 2003) mutants. The negative feedback control of BR-biosynthesis genes may be common to both dicots and monocots. In addition to the results obtained by analyses of the individual mutants, the results of microarray analysis also indicated that the expression of cytochrome P450 genes involved in BR biosynthesis, such as *DWF4*, *CPD*, *BR6ox*, *ROT3*, and *CYP90D*, is suppressed by BR treatment (Goda *et al.*, 2002). The fact that BRs suppress the expression of some BR-biosynthesis genes suggests that multi-point feedback control occurs during BR biosynthesis, resulting in the homeostasis of endogenous BRs. To our knowledge, all of the biosynthetic enzymes downstream of *DWF4* for which the genes have been isolated are cytochrome P450s, and all are subject to negative feedback control by active BRs. In contrast, the expression of the genes encoding DWF7, DWF1, and DET2, which function upstream in the BR-biosynthetic pathway, is not affected by BR treatment, and therefore these genes are not regulated by feedback control (Goda et al., 2002).

Although the detailed mechanisms of the regulation of BR biosynthesis are still unknown, the regulation of BR biosynthesis is necessarily linked to BR signal transduction. At least one of the feedback controls must be adjusted through BRI1 receptors. BIN2 functions as a negative factor in the BR signal transduction system downstream of BRI1, and the *bin2* mutant accumulates BL and CS, as does *bri1* (Choe et al., 2002). These results suggest that BIN2 is in the same signaling pathway as BRI1 and that the control of endogenous levels of BRs is closely linked with the BR signal. BIN2 interacts with BES1 (Yin et al., 2002) and BZR1 (He et al., 2002), each of which is a novel type of nuclear protein and mediates BR-induced growth. The *bes1* and *bzr1* mutants are insensitive to BR deficiency and grow normally even after treatment with BR-biosynthesis inhibitors. In the future, more details of the regulatory mechanisms of BR biosynthesis should be elucidated through studies of BR biosynthesis and signal transduction.

D. METABOLISM OF BRASSINOSTEROIDS

In a wide sense, the term "metabolism" encompasses biosynthesis, but here it is used in a narrow sense limited to BR decomposition and conversion to conjugated BRs. Brassinosteroids metabolism is accompanied by the loss of BR activity and decreases the size of the pool of active BRs. Therefore, to understand the regulatory mechanisms of bioactive BRs, it is necessary to analyze the details of BR metabolism at the molecular and biochemical levels. Feeding experiments have revealed that BRs are subject to various metabolic reactions, such as epimerization, hydroxylation, esterification with fatty acids, glycosylation, and oxidative cleavage of side chains (Fujioka and Yokota, 2003). However, most of these findings were obtained by feeding the 24-epimers of BL and CS which are not the main BRs in the plant kingdom. Metabolic pathways identified by feeding experiments do not always reflect the endogenous metabolism of BRs but instead may reflect the potential of plants to metabolize BRs. At present, very little is known about the metabolism of BL and CS, and the metabolic pathway that functions *in vivo* has not been completely elucidated, but the current knowledge will be of great help in further investigations of BR metabolism. In the future, it will be necessary to clarify the endogenous metabolites and metabolic pathways that function *in vivo*.

New information on BR metabolism was obtained through the analysis of the *bas1-D* mutant. This mutant has a dwarf phenotype and was originally isolated as a *phyB-4* suppressor from activation tagging lines (Neff et al., 1999). The mutation causes the overproduction of CYP72B1, a cytochrome

P450 encoded by the *BAS1* gene. This mutant is deficient in active BL and CS, and when BL is administered exogenously, it is readily metabolized to C-26 hydroxylated BL. Whether CYP72B1 functions as an endogenous BR hydroxylase or a nonspecific detoxification enzyme is unclear, but it was reported that CYP72B1 catalyzes the C-26 hydroxylation of both BL and CS, as shown with a yeast functional assay and a feeding experiment using *Arabidopsis* seedlings. These results indicated that the C-26 hydroxylation of active BL is an endogenous function of CYP72B1 (Turk *et al.*, 2003). Seedling growth assays showed that 26-hydroxybrassinolide is an inactive BR, demonstrating that the C-26 hydroxylation of BR is an inactivation process. Genetic and physiological analyses of the responses of hypocotyls to exogenous BL and varying intensities of white and monochromatic light suggested that *CYP72B1* modulates photomorphogenesis primarily through far-red light and to a lesser extent through blue- and red-light pathways (Turk *et al.*, 2003). In contrast to many other BR-biosynthesis genes, the expression of this gene is induced by BR treatment (Goda *et al.*, 2002).

It has been reported that the sulfuric acid transferase gene of western rape is involved in the C-22 hydroxylation and inactivation of BRs (Rouleau *et al.*, 1999). However, this enzyme is specific to the 24-epimers of BRs, which are rare in the plant kingdom, and it does not use major BRs as substrates. For this reason, the physiological significance of this enzyme is unknown at present. The physiological role of sulfuric acid transferase would be clarified if a sulfuric acid transferase that uses BL and CS as a substrate could be found.

II. SIGNAL TRANSDUCTION OF BRASSINOSTEROIDS

Following the isolation of BR-deficient mutants, *det2*, *dwf4*, and other dwarf mutants in the latter half of the 1990s, and the finding that these were caused by a mutation in the BR biosynthase gene, the importance of BRs in the regulation of plant growth gained global recognition (Choe *et al.*, 1998; Fujioka and Yokota, 2003; Li *et al.*, 1996). Dwarf phenotypes, expressed by overall shortening of the lower hypocotyls, stem, and leaves in the direction of longitudinal elongation, or by increases in the number of stems accompanying decreased apical dominance, were observed in this group of mutants under light conditions. In addition, the spindly growth of the lower hypocotyls and the closure of cotyledons observed in wild-type strains were not observed in this mutant group when germinated under dark conditions, clearly demonstrating "photomorphogenesis in the dark," which is manifested by thick, short lower hypocotyls and expanded cotyledons, as if the plants had been grown in the presence of light (Clouse, 2001; Li and Chory, 1999). In addition, the high expression of the chloroplast gene group and the promotion of plastid differentiation under dark conditions (Asami *et al.*, 2000; Chory *et al.*, 1994; Nagata *et al.*, 2000), along with the promotional control of vascular bundle formation by BRs (Nagata *et al.*, 2001; Yamamoto *et al.*, 1997, 2001), have been attributed to BR deficiency by the analysis of mutants, as well as biosynthesis inhibitors and cultured cell lines.

Mutants have been utilized actively, not only in research on BRs but also in relation to other plant hormones. In research on signaling, one of the major objectives has been to identify the receptor that serves as the origin of a signal; screening for mutants is frequently used to accomplish this, and involves assessing the sensitivity to various plant hormones that are depressed in the mutants. This technique has been attempted for all other types of phytohormones in this book. Additionally, in the field of BR research, attempts have been made to select mutants that exhibit BR insensitivity while demonstrating dwarf phenotypes that are similar to those associated with BR-biosynthesis deficiency. As a result, BR receptor mutants that had not been obtained in previous research were obtained in this study (Clouse *et al.*, 1996), and represent the second series of plant hormone receptors to be identified, following ethylene (Li and Chory, 1997). Although only a matter of conjecture, one reason for this early success may be that a clue to the well-defined biosynthesis deficiency phenotypes, which appear in the form of dwarf phenotypes that display shorter leaves and stems, had been obtained in the case of BR deficiency. This is in contrast to such compounds as auxin and cytokinin, the biosynthesis deficiency phenotypes of which are not well-known. Moreover, although disturbed germination and seed formation abnormalities associated with gibberellin deficiency and abscisic acid deficiency are occasionally lethal to plants, BR deficiency results in phenotypes that make the plant grow thicker and sturdier, thereby resulting in deficiency phenotypes that are neither too severe nor too weak.

In any case, the receptor *BRI1*, along with several gene groups reported to function in proximity to it, has been elucidated through research on BR signaling which has been gaining increased momentum. The existence of a new component has been advocated for the subsequent second step of information transfer from the cytoplasm to the nucleus. In addition, although there have been few examples of analysis of the responsive transcription regulation method, as compared to other hormones, several attempts have been made. The following provides an introduction to these attempts, beginning with the upstream signaling pathways.

A. BRASSINOSTEROID RECEPTORS

1. Brassinosteroid Receptor *BRI1*

Following the successful isolation of brassinosteroid-biosynthesis mutants and causative genes, BR deficiency has been found to induce photomorphogenesis in the dark and dwarf phenotypes under light conditions. If a "deficiency" were caused by a mutation in BR biosynthesis, the phenotype of the mutant would be transformed to that of the wild-type by the addition of brassinosteroid. However, if a mutant exists in which the dwarf phenotypes are retained even when brassinosteroid is added, or in which the inhibition of root elongation by BRs, which occurs in the wild-type, is not observed, then the mutant is predicted to be one in which brassinosteroid signaling has been interrupted. Several groups in the latter half of the 1990s undertook extensive research into these mutants. Chory and coworkers determined that the mutation-causing gene in the *Arabidopsis* mutant *brassinosteroid insensitive 1* (*bri1*) is a serine/threonine-type receptor kinase containing 25 leucine-rich repeats (LRRs) (Clouse and Langford, 1995; Clouse *et al.*, 1996; Li and Chory, 1997).

BRI1 protein was found to be localized on the cell membrane (Friedrichsen *et al.*, 2000) in experiments using BRI1:GFP protein, purified with antibodies following *in vivo* protein synthesis in a human cultured cell line, and was confirmed to function as a kinase on the cell membrane, based on the finding that the serine/threonine residue has the ability to autophosphorylate (Friedrichsen and Chory, 2001). In addition, high levels of BRs induced by feedback inhibition were observed in *bri1* mutants (Noguchi *et al.*, 1999), and transformants, in which BRI1 protein had accumulated to roughly 20 times that of the wild-type, were observed to have phenotypes similar to those associated with excess BR, such as excessive elongation of the petiole and epinasty of the leaves (Wang *et al.*, 2001). Moreover, the BRI1 protein complex, purified from plants using antibodies, was confirmed to bind with radioactively labeled BL (Wang *et al.*, 2001). On the basis of these results, BRI1 is considered to be a gene that is responsible for BR reception and that serves as the origin of BR signaling.

2. Intraspecies and Interspecies Preservation of BRI1

Receptor kinases are a group of proteins that have a cell-membrane-penetrating domain and a protein-phosphorylating domain. They are broadly divided into three types—serine/threonine receptor kinase, histidine receptor kinase, and tyrosine receptor kinase—depending on the type of amino acid possessing phosphoric acid in the kinase domain. Although serine/threonine receptor kinase is found ubiquitously in the plant world, it is not found in the yeast genome, the DNA sequences of which have been analyzed, and may therefore be an evolutionarily new type of receptor kinase, as compared to histidine receptor kinase (Arabidopsis Genome Initiative, 2000). Roughly 600 types of serine/threonine receptor kinases are estimated to be present in the entire genome of *Arabidopsis thaliana* (Shui and Bleecker, 2001). Although these have been identified, based on the presence of a kinase domain sequence on the C-terminus and a hydrophobic-membrane-penetrating domain, the extracellular domain structure of the N-terminus is extremely diverse and is additionally classified into four types:

1. An LRR type to which BRI1 belongs.
2. An SLG type having a homologous sequence with an S-locus glycoprotein that is involved in self-incompatibility.
3. An EGF type homologous to epithelial cell growth factor receptors.
4. A kinase group that has a structure in which the functional protein (e.g. PR protein or lecithin) appears to be linked to the N-terminal extracellular region which is already known to function as a free protein in plant cells.

Although there are approximately 200 types of LRRs to which BRI1 belongs, accounting for roughly one-third of all types, these have been further classified in terms of the number of repeats of the LRR sequence. *BRI1* is characterized by the appearance of a 70 amino acid island domain located between repeats of the LRR sequence (Friedrichsen *et al.*, 2000; Li and Chory, 1997); in addition, three types of *BRI1* homologous genes, including *BRL1*, *2*, and *3* (*BRI1-like*), in which the sequence is preserved, have been confirmed in the entire genome of *Arabidopsis* (Nomura *et al.*, 2003; Yin *et al.*, 2002). Conversely, the identification of homologous genes in plants of intranuclear steroid receptors, which are considered to be involved in steroid reception in plants and animals, has been difficult even with the completion of the analysis of the entire *Arabidopsis* genome. Research on the mechanism of BR receptors is likely to continue in the future, with a focus on *BRI1*.

The homologous genes of *BRI1* have been identified from plant species other than *Arabidopsis*. The dwarf phenotype is considered to be useful for breeding, as it exhibits resistance to strong wind and facilitates fruit collection, and this phenotype has been maintained in numerous cultivated varieties. Therefore, dwarf varieties that do not exhibit recovery following BR administration were considered to be the candidates for *BRI1* homologous gene mutants. In parallel with this, genes that were homologous to the *Arabidopsis BRI1* gene were searched for, and an ortholog of *Arabidopsis BRI1* was identified by mutation of the said gene in dwarf varieties and complementation of mutant phenotypes by the wild-type gene. Tomato *tBRI* (Montoya *et al.*, 2002) and pea *LKA* (Nomura *et al.*, 2003) have a high degree of homology with *Arabidopsis BRI1* with respect to the number of repeats in the LRR domain and the 70 amino acid island domain. In rice *OsBRI1* (Yamamuro *et al.*, 2000) and barley *UZU* (Chono *et al.*, 2003), although there are three fewer repeats in the LRR domain, the 70 amino acid island domain is preserved. The *UZU* gene provides the characteristics of an ancient Japanese variety of barley, referred to as the *uzu* mutation. In Japan, 42% of Rokujo barley, cultivated primarily for consumption as barley tea, is a mutant variety of the *UZU* gene (i.e., a barley variety containing a mutation of the BR receptor) (based on the planted surface area as of 2003 in

Japan) (Chono *et al.*, 2003). There are some known dwarf varieties that can clearly be attributed to gibberellin deficiency. However, in comparison with gibberellin deficiency which occasionally causes defective germination, BR deficiency has no direct effect on germination; therefore, these varieties are considered to be maintained more dominantly, even for the same dwarf phenotype. There are still many varieties preserved as dwarves and it is highly likely that cultivated varieties resulting from BR abnormalities will be rediscovered in the future.

3. Membrane-Penetrating Kinase BAK1 Interacts with BRI1

Previous research with humans and animals has indicated that serine/threonine receptor kinase typically forms a dimer on the cell membrane. Accordingly, the possibility of BRI1 protein forming a dimer was verified following its identification (Schumacher and Chory, 2000). In 2002, *BRI1-associated kinase 1* (*BAK1*), a serine/threonine kinase with five leucine-rich repeat structures, which is said to have the potential to interact with *BRI1*, was simultaneously reported by two groups in Michigan (Li *et al.*, 2002; Nam and Li, 2002). Based on the results of *in vitro* and *in vivo* experiments, it was revealed that BRI1 and BAK1 may contribute to BR signaling by forming heterodimers.

Although BAK1 has five repeat structures in the LRR domain, 14 copies of homologous genes are estimated to be present in the entire genome of *Arabidopsis*, based on this classification. Within this gene family, *somatic embryogenesis receptor kinase 1* (*AtSERK1*) was identified in research conducted prior to *BAK1* (Hecht *et al.*, 2001). *AtSERK1* was identified from the genome sequence of *Arabidopsis*, based on homology with *DcSERK1*, which was identified using the differential display method during somatic embryo induction in carrot (Schmidt *et al.*, 1997). This gene is considered to play an important role in *Arabidopsis* embryogenesis, as abnormal embryogenesis is observed in strains that express high levels of *AtSERK1*, and *AtSERK1* is expressed specifically in late embryogenesis (Hecht *et al.*, 2001; Shah *et al.*, 2001). Although the biological significance of *AtSERK1* and *BAK1* genes being classified into the same group, in terms of their primary amino acid sequences, is unclear, the possibility of interaction between AtSERK1 protein and BAK1/AtSERK3 protein has been confirmed by fluorescence microscopy (Vries *et al.*, personal communication). Although the interaction between AtSERK1 and BRI1 has yet to be determined, the dephosphorylation enzyme kinase-associated protein phosphatase (KAPP) has been previously found to interact with AtSERK1 (Shah *et al.*, 2002). Considerable expectation has been placed on research into the downstream factors of the phosphorylation relay of BRI1 kinase, using AtSERK1 as the starting point, as well as on the embryogenesis control mechanism of BRs, for which very little research has been conducted to date.

4. Extracellular Protein Cystemine Interacts with BRI1

The ligands of serine/threonine kinase, on which research is currently being conducted in humans and animals, consist entirely of peptide hormones (Shui and Bleecker, 2001). In the case of BRI1, radioactively labeled BRs have been demonstrated to bind to the BRI1 complex, purified with antibodies from plant cells (Wang et al., 2001). Although BRI1 is considered to be the origin of BR reception, the possibility of the existence of a peptide factor that cooperatively binds with BRs or competitively binds with BRI1, as well as the possibility that, as a carrier protein, such a factor serves as a mediator between BRs and BRI1, still cannot be ruled out. Reports that support the former hypothesis have been published.

Cystemine is a peptide factor that is composed of 18 amino acids and is involved in signaling, in response to injury in tomato. The protein SR160, which has a primary structure that strongly resembles that of *Arabidopsis* BRI1, was identified using radioactively labeled cystemine (Scheer and Ryan, 2002). At roughly the same time, the SR160 gene was shown to be identical to the causative gene of tomato dwarf mutants *curl 3* (*cu3*) and *altered brassinolide sensitivity 1* (*abs1*) (Montoya et al., 2002). Properties unique to *Arabidopsis bri1* mutation, such as recovery failure even after BR addition and excessive expression of castasterone, were observed in the cu3 dwarf phenotype, suggesting that *cu3* is an ortholog of *BRI1*.

The binding constant between cystemine and SR160/CU3/ABS1, in addition to other parameters, has yet to be determined; therefore, the details of the correlation among cystemine, tomato BRI1 protein, and BR signaling remain unknown. Although still at a preliminary stage, a cystemine administration experiment was conducted on dwarfed tomatoes in medium containing the BR-biosynthesis inhibitor Brz, but no phenotype recovery was observed (Asami et al., unpublished). This outcome is to be expected because cystemine is considered to be involved, inherently, in signaling in response to injury, and there have been no examples suggesting its involvement in morphogenesis. However, a chimeric protein, in which the kinase domain of Xa21, which is involved in disease resistance in rice, is substituted on the C-terminal side of the membrane-penetrating domain with the LRR region of BRI1, has been clearly demonstrated to induce the expression of the disease-responsive gene by BRs (He et al., 2000). Moreover, BRs have been shown to improve disease resistance in rice and tobacco plants (Nakashita et al., 2003). Thus, the binding of cystemine to tomato BRI1 may be involved, exclusively, in injury signal transduction. For example, BRI1 may activate the morphogenesis signal transduction pathway when it forms a heterodimer with BAK1, whereas an unknown receptor kinase, of a different disease response type, may activate the disease response signal transduction pathway when it forms a heterodimer with BRI1.

B. SIGNAL CARRIER PROTEINS FOR BRASSINOSTEROID SIGNALING

1. Brassinosteroid Biosynthesis Inhibitor Brz-Resistant Mutant *Brz-Insensitive-Long Hypocotyl*

Brz, which has a triazole ring, was created in 1999 as a specific inhibitor of BR biosynthase DWF4 (Asami and Yoshida, 1999; Asami *et al.*, 2000) as described in Section I.B. Wild-type *Arabidopsis* that germinated under dark conditions in the presence of Brz demonstrated photomorphogenesis in the dark, similar to that of BR-deficient mutants. Therefore, we predicted that if a reverse mutant that demonstrated phenotypes of long lower hypocotyls and cotyledon closure in the presence of Brz were isolated, it would be possible to identify the BR signaling gene as the causative gene of such a mutant; we have been searching for such a Brz-resistant mutant bil, (i.e., *Brz-insensitive-long hypocotyls*).

Wild-type *A. thaliana* that germinated in the dark in a solid medium containing 3 μM Brz exhibited the dwarf phenotype in which the lengths of the lower hypocotyls were approximately 20% of the lower hypocotyls of *A. thaliana* that germinated in medium without Brz. *bil1* was then screened from among the *Arabidopsis* EMS mutant lines under these conditions. As a result, *bil1* demonstrated a hypocotyl length roughly equal to that of the wild-type, and is currently considered to be the mutant with the strongest phenotypes (Fig. 2). Although *Brz-resistant* (*bzr1*) and *bri1-EMS-suppressor* (*bes1*) were also screened at the same time as the screening for this *bil1* mutant was conducted, *bil1* was ultimately determined to be a mutation of the same gene as *bzr1*, whereas *bes1* was determined to be of a gene family that demonstrates a high degree of homology (on the order of 88%) with *bzr1/bil1* at the amino acid level. Therefore, *BZR1/BIL1* and BES1 constituted a new family of proteins, in which known functionally active domains were not observed, in terms of their amino acid primary structures (Wang *et al.*, 2002; Yin *et al.*, 2002).

bzr1/bil1 is a dominant mutation that results from a point mutation in amino acid substitution. When mutant *bzr1/bil1* gene cDNA downstream of the CaMV 35S promoter was forcibly expressed, the transformant demonstrated the Brz-resistance phenotype and complemented the dwarf phenotype of the receptor mutation *bri1*. However, despite being a dominant mutation, the transformant that resulted from the forced expression of wild-type *BZR1/BIL1* gene cDNA demonstrated very weak Brz resistance. In addition, although the wild-type BZR1/BIL1:CFP fused protein

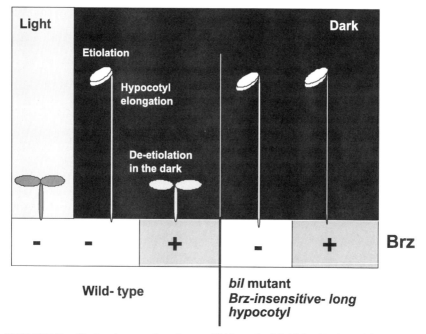

FIGURE 2. Strategy to screen brassinosteroid-biosynthesis inhibitor Brz insensitive mutants.

accumulated to a small amount and was localized in the cytoplasm under normal conditions, mutant bzr1/bil1:CFP fused protein demonstrated stable localization in the nucleus (Wang et al., 2002). An attempt was made to elucidate the cause of the apparent difference in localization accompanying this mutation by means of the following experiment. When a transformant with this wild-type BZR1/BIL1:CFP fused protein was subjected to BR stimulation, no increase in the expression level of *BZR1/BIL1* mRNA was observed, whereas BZR1/BIL1:CFP signals were observed to migrate into the nucleus. The migration of these BZR1/BIL1:CFP signals occurred only as a result of BR stimulation and was not observed in the case of stimulation with other types of plant hormones or light. Accordingly, wild-type BZR1/BIL1 protein may be a signal carrier factor that has the actual function of transferring BR stimulation received on the cell membrane from the cytoplasm into the nucleus. It becomes easier, therefore, to hypothesize the mechanism responsible for the appearance of mutant phenotypes associated with bil1 mutants based on this property. This supports the current hypothesis that, given that mutant bzr1/bil1 protein is stabilized by a mutation in amino acid substitution, as compared to wild-type BZR1/BIL1 protein, it remains for a longer period of time in the nucleus. Plant cells recognize that

BR stimulation is continuously transferred to the nucleus during that time, thereby resulting in the appearance of a *bil* phenotype in the plant body in which the BR signal transduction system appears to function normally, even if BR biosynthesis is interrupted by Brz (Wang et al., 2002; Yin et al., 2002) (Fig. 2).

2. BIN2 is Involved in Control Mechanisms of Stabilization and Proteolysis of BIL1 Protein

Although the stability of BIL1 protein is clearly important in BR signaling, BIN2 kinase has been determined to be involved in the control mechanism of this stability, as indicated later.

After searching for the BR receptor mutant *bri1*, the mutant *bin2* was isolated. The dwarf phenotype of the above-ground portions of this mutant is not recovered following BR addition and its roots are insensitive to elongation inhibition by BR treatment (Li et al., 2001). The causative gene is homologous to the glycogen synthase kinase 3 (GSK3)/SHAGGY kinase gene that has been identified in animal cells and other cells (Li and Nam, 2002). The mode of phenotype expression of the *bin2* mutant in plants is somewhat complex. Although the *bin2–1* mutation was originally isolated as a dominant mutation, highly expressed strains of the wild-type *BIN2* gene demonstrated strong dwarf phenotypes, in agreement with this fact. In addition, an *in vitro* experiment demonstrated that mutant *bin2–1* protein had a stronger phosphorylating ability than wild-type BIN2 protein. Moreover, in an analysis of transformants in which the CaMV 35S promoter: wild-type BIN2 gene was inserted into *bri1* mutants, individuals with the strongly expressed wild-type BIN2 gene was exhibited a phenotype in which the dwarf phenotype of *bri1* was more severe. However, in transformants in which the expression level of BIN2 gene was conversely decreased to nearly one-fifth, as a result of co-suppression, the expression of the *bri1* phenotype was confirmed to return to a level comparable to that of the wild-type. According to these results, the phosphorylation of BIN2 was thought to result in the control of the direction in which BR signaling is switched off (i.e., negative control [Li and Nam, 2002]). In addition, in contrast to *bin2–1*, *bin2–2* has also been obtained and its phosphorylation deactivation has been confirmed *in vitro*, indicating a more detailed control mechanism. If *BIN2* were responsible for the main flow of BR signaling originating in *BRI1*, the phenotype that is the opposite of the BR-deficiency-like phenotype that occurs in transformants expressing high levels of active mutant *bin2–1* and wild-type *BIN2* gene (i.e., the phenotype resembling excessive expression of BRs), would be observed in this inactive mutant *bin2–2*. However, *bin2–2* was observed to demonstrate phenotypes that were nearly the same as those of the wild-type. Consequently, BIN2 is not responsible for the main flow of the signal transduction pathway from BR BRI1; rather, it plays a supplementary role in providing negative control indirectly (Li and Nam, 2002).

The primary function of GSK3/SAGYY kinase, as determined by research conducted mainly on animals, is β-catenin phosphorylation during the course of signal transduction of the oncogene Wnt involved in cell proliferation, as well as that of a control mechanism for directing such phosphorylation towards protein destabilization and proteolysis (Jonak and Hirt, 2002; Polakis, 2000). Although members of the *BZR1/BIL1* and *BES1* gene family are not homologous to β-catenin in terms of their amino acid primary sequences, it was of interest to compare the property of migration from the cytoplasm to the nucleus shared by both β-catenin and BZR1/BIL1 and BES1 proteins. As a result of experiments using the yeast two-hybrid method, although BZR1/BIL1 and BES1 proteins do not interact directly with receptor protein BRI1, they were confirmed to interact with BIN2. Moreover, the results of an *in vitro* experiment using proteins expressed in *Escherichia coli* indicated that BZR1/BIL1 and BES1 were phosphorylated by BIN2. In an experimental system in which BZR1/BIL1 and BES1 proteins were detected using antibodies after treating a plant body with BRs, a protein that had shifted to a lower molecular weight, owing to dephosphorylation, was detected, in addition to normal BZR1/BIL1 and BES1 proteins, 20 min after treatment. The signals of these phosphorylated BZR1/BIL1 and BES1 were detected more strongly in the case of treating the *bin2–1* mutant with BRs than in the case of treating the wild-type; this confirmed that BZR1/BIL1 and BES1 were phosphorylated by BIN2 *in vivo* (He *et al.*, 2002; Wang *et al.*, 2002; Yin *et al.*, 2002).

Although this phosphorylated BZR1/BIL1 protein ended up disappearing 60 min after treatment, after being formed by BR stimulation, it did not disappear in the presence of the proteasome inhibitor MG132 (He *et al.*, 2002). Although it is uncertain whether BZR1/BIL1 is ubiquitinated, or whether its decomposition occurs in the nucleus or the cytoplasm under these circumstances, it can at least be said that there is a high possibility that BZR1/BIL1 is decomposed by a proteasome system after having been modified by phosphorylation by BIN2. An increasing number of reports have investigated the regulation of proteins involved in plant hormone signaling, by controlling stability through ubiquitination (Hellmann and Estelle, 2002; Vierstra, 2003); a comparison of these functions with the BZR1/BIL1 signaling system would also be interesting. In addition, it has also been suggested that after having migrated into the nucleus, BZR1/BIL1 may possess a function similar to that of a protein anchor in the nucleus; a mechanism exists whereby information is transmitted to the next step of signaling as a result of interaction with the next protein. Future research on this subject would also be of great interest.

C. OTHER UPSTREAM AND MIDSTREAM FUNCTIONAL PROTEINS OF BRASSINOSTEROID SIGNALING

1. Factor BRS1 is Involved in Brassinosteroid Signaling Near the Cell Membrane

Mutant *bri1-suppressor 1* (*brs1*), which partially complements the dwarf phenotype of the BR receptor mutant *bri1-5*, was isolated by random insertion of an activation tagging vector, and this mutant phenotype was clearly demonstrated to occur as a result of a highly expressed mutation of a C-terminal peptidase gene (Li *et al.*, 2001). The high expression of this gene, in which a mutation has been inserted into the peptidase activity domain of BRS1, does not complement the dwarfing of *bri1-5*; therefore, the peptidase activity itself may be involved in the control of BRI1 protein. In addition, although the *bri1-5* and *bri1-9* phenotypes with a mutation in the extracellular domain are complementary in wild-type transfectants that highly express BRS1, *bri1-1* with a mutation in the intracellular kinase domain and the phenotypes of the biosynthesis mutants *det2* and *dwf4* are not complementary; therefore, BRS1 is presumed to be a protein that is involved with the extracellular domain of BRI1 (Li *et al.*, 2001). Moreover, although the possibility of involvement in the processing of extracellular ligand proteins has also been mentioned, there is no concrete evidence to verify this; this is expected to be the subject of detailed research in the future.

2. Factors ROP2 and BRH1 are Involved in Brassinosteroid Signaling in Cytoplasm

It has been well-documented in animals that control of cell division occurs in the signal transduction pathway as a result of cell growth factors and insulin binding to tyrosine receptor kinases and that this information is transferred to low molecular weight GTPases commonly referred to as the Ras superfamily. Although the existence of tyrosine kinase has not been confirmed for the entire genome of *Arabidopsis* (Arabidopsis Genome Initiative, 2000), a homologous gene, Rho-related GTPase from plant (ROP), of the Ras superfamily has been identified (Li *et al.*, 2001). Transformants that highly express this ROP2 exhibit a BR-deficiency-like phenotype in which the cotyledons open with short hypocotyls during germination in the dark. This shortening of the hypocotyls is recovered following the addition of BR; however, since mature individuals exhibit elongation equal to that of the wild-type, although they have fewer scapes, they cannot simply be concluded to be a strain with decreased BR activity. Further research will hopefully be conducted on this topic and will include research related to the function of the gene itself.

As mentioned earlier, there are a growing number of examples showing the involvement of protein stability in the functional regulation of plant cells. *brassinosteroid-responsive RING-H2* (*BRH1*) is a gene whose mRNA

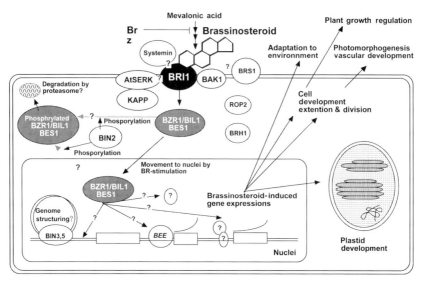

FIGURE 3. Brassinosteroid signaling cascade.

expression levels are inhibited by BR stimulation in *Arabidopsis*, and mRNA expression is increased in the BR-biosynthesis-deficient mutant, *cpd* (Molnar *et al.*, 2002). The RING finger H2 element, which is one of the characteristics of ubiquitin E3 ligase involved in proteolysis, is preserved in this gene. The antisense transformants of *BRH1* exhibit a slightly longer phenotype; however, because the mutant phenotype is weak, despite the well-defined antisense effects, careful verification is necessary with respect to the correlation between *BRH1* and BR signaling. In addition, more than 1300 types of gene related to ubiquitination have been estimated to account for more than 5% of the entire genome of *Arabidopsis* (Vierstra, 2003), and although it appears premature to discuss the correlation between BRH1 and BZR1/BIL1, it will, at least, be important to conduct future research from the viewpoint of the involvement of this ubiquitination system and BR signaling. In Fig. 3, BR signaling pathway is summarized.

D. A GENE EXPRESSION MECHANISM RESPONSIVE TO BRASSINOSTEROID

1. Brassinosteroid Signaling Genes BIN3/5 and BEE are Involved in Control of Gene Expression

Following the search for *bri1*, *bin3* and *bin5* (*brassinosteroid insensitive 3, 5*) were isolated as dwarf mutants that do not exhibit recovery following BR addition. The causative genes of both mutants are homologous genes of topoisomerase IV that are involved in the higher structure regulation of the core genome (Yin *et al.*, 2002). These mutants are predicted to be intimately involved in BR signaling for such reasons as the expression of photomorphogenesis and photosynthesis genes during germination in the dark, and responses by the expression of genes, such as *TCH4* and *SAUR* that are normally observed during BR addition in light conditions. Future research into the mode by which BRs are used to control the genomic higher structure regulation mechanism, which is the basic activity of topoisomerase, will be of great interest.

Using the differential display method, a search was conducted for genes that are expressed after BR stimulation of the BR-biosynthesis-deficient mutant, *det2*, for 2 h, and those that are not expressed after BR stimulation of the receptor mutant, *bri1*, for 2 h. As a result, *BEE1*, *BEE2*, and *BEE3* (*BR enhanced expression*), which are members of the bHLH transcription factor family, were identified (Friedrichsen *et al.*, 2002). The three-way destructive mutants of these three types of genes exhibit, albeit weakly, the phenotype of the shortening of the hypocotyls, as well as low responsiveness to hypocotyl lengthening induced by BR treatment under light conditions. In addition, the effects of root elongation inhibition induced by BR treatment were also expressed more prominently in transformants that highly express *BEE1*. The entire genome of *Arabidopsis* has been estimated to contain 145 types of bHLH proteins (Arabidopsis Genome Initiative, 2000) and it is hoped that these proteins will provide clues toward understanding the mechanism by which genes are expressed in response to BRs.

2. Brassinosteroid-Responsive Genes

Some known examples of genes that are expressed in response to BRs include CycD3 (Hu *et al.*, 2000), which is involved in cell division, and *TCH4* (xyloglucan endotransferase, which is involved in cell elongation via a function that reduces the rigidity of the cell wall) (Friedrichsen and Chory, 2001). Although photomorphogenesis under dark conditions has been previously described as being caused by BR deficiency, the expression of photosynthesis-related genes, such as *cab* and *RbcS*, which are normally not expressed under dark conditions, has been confirmed to occur accompanying this (Asami *et al.*, 2000; Chory *et al.*, 1994; Li *et al.*, 1996; Nagata *et al.*, 2000). In addition, as a result of the development of the microarray method, reports describing microarray analyses of gene groups, the expression of which increases or decreases in response to stimulation, have begun to appear in relation to BRs (Goda *et al.*, 2002; Mussig *et al.*, 2002). Moreover, gene expression analyses, using the same type of microarray analysis, during variations in the light/dark cycle (Harmer *et al.*, 2000), during photostimulation under conditions of classified wavelength (Tepperman *et al.*, 2001), during the process of vascular bundle formation (Demura *et al.*, 2002), iod (Ogawa *et al.*, 2003) are extremely interesting with respect to the identification of plants that are considered to be intimately involved with the physiological functions of BRs, but only during a single growth period. Comparisons of these results are expected to lead to a deeper understanding of the actual mechanisms by which the physiological activities of BRs are expressed by gene groups responsive to BRs.

III. CONCLUSIONS

As has been described, the elucidation of the signaling mechanism of BRs has progressed rapidly since about 2000. Much of this progress was a result of analyses of mutants; not only were hints obtained from previous research on animals and yeasts but a number of gene species and physiological mechanisms were found for the first time in plants as well. Compounds with a steroidal structure are widely preserved beyond the boundaries of biological species, and although they demonstrate universally common physiological activities among species in terms of control of cell division and cell elongation, they have a dual nature when it comes to simultaneously controlling organs peculiar to each species, such as the formation of the reproductive organs by steroid hormones in animals, the control of molting by ecdyson in insects, and the control of chloroplast and vascular bundle formation by BRs in plants. In this regard, a discussion from the perspective of steroid hormones retaining universality while also acquiring physiological activity that is unique to each biological species over the course of evolution would be extremely interesting. Parallel viewpoints, in terms of the molecular evolution of steroid hormones during the course of molecular biology and molecular physiology research, are possibly useful in the future. Given cases of other known plant hormones, the analogs of which are observed in different species from animals to microorganisms, even though they may not be preserved to the extent of steroid structures, sharing one viewpoint on the molecular evolution of physiologically active substances, in a broad sense, among numerous plant hormone researchers, may lead to even more interesting discoveries. Although the entire genomic sequences of numerous species have been determined and a discussion of genetic evolution has been made possible through mutual comparison, we believe discussions of the molecular evolution of plant hormones or of physiologically active substances, are necessary in order to provide important clues that will further our understanding of the phenomenon of life.

REFERENCES

Arabidopsis Genome Initiative (2000). Analysis of the genome sequence of the flowering plant *Arabidopsis thaliana*. *Nature* **408**, 796–815.
Asami, T., and Yoshida, S. (1999). Brassinosteroid biosynthesis inhibitors. *Trends Plant Sci.* **4**, 348–353.

Asami, T., Min, Y. K., Nagata, N., Yamagishi, K., Takatsuto, S., Fujioka, S., Murofushi, N., Yamaguchi, I., and Yoshida, S. (2000). Characterization of brassinazole, a triazole-type brassinosteroid biosynthesis inhibitor. *Plant Physiol.* **123,** 93–99.

Asami, T., Mizutani, M., Fujioka, S., Goda, H., Min, Y. K., Shimada, Y., Nakano, T., Takatsuto, S., Matsuyama, T., Nagata, N., Sakata, K., and Yoshida, S. (2001). Selective interaction of triazole derivatives with DWF4, a cytochrome P450 monooxygenase of the brassinosteroid biosynthetic pathway, correlates with brassinosteroid deficiency *in Planta*. *J. Biol. Chem.* **276,** 25687–25691.

Asami, T., Oh, K., Jikumaru, Y., Shimada, Y., Kaneko, I., Nakano, T., Takatsuto, S., Fujioka, S., and Yoshida, S. (2004). A mammalian steroid action inhibitor spironolactone retards plant growth by inhibition of brassinosteroid action and induces light-induced gene expression in the dark. *J. Steroid Biochem. Mol. Biol.* **91,** 41–47.

Bishop, G. J., Nomura, T., Yokota, T., Harrison, K., Noguchi, T., Fujioka, S., Takatsuto, S., Jones, J. D. G., and Kamiya, Y. (1999). The tomato DWARF enzyme catalyses C-6 oxidation in brassinosteroid biosynthesis. *Proc. Nat. Acad. Sci. USA* **96,** 1761–1766.

Bishop, G. J., and Koncz, C. (2002). Brassinosteroids and plant steroid hormone signaling. *Plant Cell* **14,** S97–S110.

Choe, S. W., Dilkes, B. P., Fujioka, S., Takatsuto, S., Sakurai, A., and Feldmann, K. A. (1998). The *DWF4* gene of *Arabidopsis* encodes a cytochrome P450 that mediates multiple 22 α-hydroxylation steps in brassinosteroid biosynthesis. *Plant Cell* **10,** 231–243.

Choe, S., Schmitz, R. J., Fujioka, S., Takatsuto, S., Lee, M. O., Yoshida, S., Feldmann, K. A., and Tax, F. E. (2002). *Arabidopsis* brassinosteroid-insensitive *dwarf12* mutants are semidominant and defective in a glycogen synthase kinase 3 β-like kinase. *Plant Physiol.* **130,** 1506–1515.

Chono, M., Honda, I., Zeniya, H., Yoneyama, K., Saisho, D., Takeda, K., Takatsuto, S., Hoshino, T., and Watanabe, Y. (2003). A semidwarf phenotype of barley uzu results from a nucleotide substitution in the gene encoding a putative brassinosteroid receptor. *Plant Physiol.* **133,** 1209–1219.

Chory, J., Reinecke, D., Sim, S., Washburn, T., and Brenner, M. (1994). A role for cytokinins in de-etiolation in *Arabidopsis-det* mutants have an altered response to cytokinins. *Plant Physiol.* **104,** 339–347.

Clouse, S. D., and Langford, M. (1995). A brassinosteroid-insensitive mutant in *Arabidopsis-thaliana*. *Plant Physiol.* **108,** 81–81.

Clouse, S. D., Langford, M., and McMorris, T. C. (1996). A brassinosteroid-insensitive mutant in *Arabidopsis thaliana* exhibits multiple defects in growth and development. *Plant Physiol.* **111,** 671–678.

Clouse, S. D., and Sasse, J. M. (1998). Brassinosteroids: Essential regulators of plant growth and development. *Ann. Rev. Plant Physiol. Plant Mol. Biol.* **49,** 427–451.

Clouse, S. D. (2001). Integration of light and brassinosteroid signals in etiolated seedling growth. *Trends Plant Sci.* **6,** 443–445.

Clouse, S. D. (2002). Brassinosteroid signal transduction: Clarifying the pathway from ligand perception to gene expression. *Mol. Cell* **10,** 973–982.

Demura, T., Tashiro, G., Horiguchi, G., Kishimoto, N., Kubo, M., Matsuoka, N., Minami, A., Nagata-Hiwatashi, M., Nakamura, K., Okamura, Y., Sassa, N., Suzuki, S., Yazaki, J., Kikuchi, S., and Fukuda, H. (2002). Visualization by comprehensive microarray analysis of gene expression programs during transdifferentiation of mesophyll cells into xylem cells. *Proc. Natl. Acad. Sci. USA* **99,** 15794–15799.

Friedrichsen, D. M., Joazeiro, C. A. P., Li, J. M., Hunter, T., and Chory, J. (2000). Brassinosteroid-insensitive-1 is a ubiquitously expressed leucine-rich repeat receptor serine/threonine kinase. *Plant Physiol.* **123,** 1247–1255.

Friedrichsen, D., and Chory, J. (2001). Steroid signaling in plants: From the cell surface to the nucleus. *Bioessays* **23,** 1028–1036.

Friedrichsen, D. M., Nemhauser, J., Muramitsu, T., Maloof, J. N., Alonso, J., Ecker, J. R., Furuya, M., and Chory, J. (2002). Three redundant brassinosteroid early response genes encode putative bHLH transcription factors required for normal growth. *Genetics* **162**, 1445–1456.
Fujioka, S., and Sakurai, A. (1997). Brassinosteroids. *Nat. Prod. Rep.* **14**, 1–10.
Fujioka, S., Takatsuto, S., and Yoshida, S. (2002). An early C-22 oxidation branch in the brassinosteroid biosynthetic pathway. *Plant Physiol.* **130**, 930–939.
Fujioka, S., and Yokota, T. (2003). Biosynthesis and metabolism of brassinosteroids. *Ann. Rev. Plant Biol.* **54**, 137–164.
Goda, H., Shimada, Y., Asami, T., Fujioka, S., and Yoshida, S. (2002). Microarray analysis of brassinosteroid-regulated genes in *Arabidopsis*. *Plant Physiol.* **130**, 1319–1334.
Grove, M. D., Spencer, G. F., Rohwedder, W. K., Mandava, N., Worley, J. F., Warthen, J. D., Steffens, G. L., Flippenanderson, J. L., and Cook, J. C. (1979). Brassinolide, a plant growth-promoting steroid isolated from *Brassica-napus* pollen. *Nature* **281**, 216–217.
Harmer, S., Hogenesch, J., Straume, M., Chang, H., Han, B., Zhu, T., Wang, X., Kreps, J., and Kay, S. (2000). Orchestrated transcription of key pathways in *Arabidopsis* by the circadian clock. *Science* **290**, 2110–2113.
He, Z. H., Wang, Z. Y., Li, J. M., Zhu, Q., Lamb, C., Ronald, P., and Chory, J. (2000). Perception of brassinosteroids by the extracellular domain of the receptor kinase BRI1. *Science* **288**, 2360–2363.
He, J. X., Gendron, J. M., Yang, Y. L., Li, J. M., and Wang, Z. Y. (2002). The GSK3-like kinase BIN2 phosphorylates and destabilizes BZR1, a positive regulator of the brassinosteroid signaling pathway in *Arabidopsis*. *Proc. Natl. Acad. Sci. USA* **99**, 10185–10190.
Hecht, V., Vielle-Calzada, J. P., Hartog, M. V., Schmidt, E. D. L., Boutilier, K., Grossniklaus, U., and de Vries, S. C. (2001). The *Arabidopsis* SOMATIC EMBRYOGENESIS RECEPTOR KINASE 1 gene is expressed in developing ovules and embryos and enhances embryogenic competence in culture. *Plant Physiol.* **127**, 803–816.
Hellmann, H., and Estelle, M. (2002). Plant development: Regulation by protein degradation. *Science* **297**, 793–797.
Hong, Z., Ueguchi-Tanaka, M., Shimizu-Sato, S., Inukai, Y., Fujioka, S., Shimada, Y., Takatsuto, S., Agetsuma, M., Yoshida, S., Watanabe, Y., Uozu, S., Kitano, H., Ashikari, M., and Matsuoka, M. (2002). Loss-of-function of a rice brassinosteroid biosynthetic enzyme, C-6 oxidase, prevents the organized arrangement and polar elongation of cells in the leaves and stem. *Plant J.* **32**, 495–508.
Hong, Z., Ueguchi-Tanaka, M., Umemura, K., Uozu, S., Fujioka, S., Takatsuto, S., Yoshida, S., Ashikari, M., Kitano, H., and Matsuoka, M. (2003). A rice brassinosteroid-deficient mutant, *ebisu dwarf* (*d2*), is caused by a loss of function of a new member of cytochrome P450. *Plant Cell* **15**, 2900–2910.
Hu, Y. X., Bao, F., and Li, J. Y. (2000). Promotive effect of brassinosteroids on cell division involves a distinct CycD3-induction pathway in *Arabidopsis*. *Plant J.* **24**, 693–701.
Jonak, C., and Hirt, H. (2002). Glycogen synthase kinase 3/SHAGGY-like kinases in plants: An emerging family with novel functions. *Trends Plant Sci.* **7**, 457–461.
Kang, J. G., Yun, J., Kim, D. H., Chung, K. S., Fujioka, S., Kim, J. I., Dae, H. W., Yoshida, S., Takatsuto, S., Song, P. S., and Park, C. M. (2001). Light and brassinosteroid signals are integrated via a dark-induced small G protein in etiolated seedling growth. *Cell* **105**, 625–636.
Koka, C. V., Cerny, R. E., Gardner, R. G., Noguchi, T., Fujioka, S., Takatsuto, S., Yoshida, S., and Clouse, S. D. (2000). A putative role for the tomato genes *DUMPY* and *CURL-3* in brassinosteroid biosynthesis and response. *Plant Physiol.* **122**, 85–98.
Li, J. M., Nagpal, P., Vitart, V., McMorris, T. C., and Chory, J. (1996). A role for brassinosteroids in light-dependent development of *Arabidopsis*. *Science* **272**, 398–401.
Li, J. M., and Chory, J. (1997). A putative leucine-rich repeat receptor kinase involved in brassinosteroid signal transduction. *Cell* **90**, 929–938.

Li, J. M., and Chory, J. (1999). Brassinosteroid actions in plants. *J. Exp. Bot.* **50**, 275–282.
Li, J., Lease, K. A., Tax, F. E., and Walker, J. C. (2001). BRS1, a serine carboxypeptidase, regulates BRI1 signaling in *Arabidopsis thaliana*. *Proc. Natl. Acad. Sci. USA* **98**, 5916–5921.
Li, H., Shen, J., Zheng, Z., Lin, Y., and Yang, Z. (2001). The Rop GTPase switch controls multiple developmental processes in *Arabidopsis*. *Plant Physiol.* **126**, 670–684.
Li, J. M., Nam, K. H., Vafeados, D., and Chory, J. (2001). BIN2, a new brassinosteroid-insensitive locus in *Arabidopsis*. *Plant Physiol.* **127**, 14–22.
Li, J., Wen, J. Q., Lease, K. A., Doke, J. T., Tax, F. E., and Walker, J. C. (2002). BAK1, an *Arabidopsis* LRR receptor-like protein kinase, interacts with BRI1 and modulates brassinosteroid signaling. *Cell* **110**, 213–222.
Li, J. M., and Nam, K. H. (2002). Regulation of brassinosteroid signaling by a GSK3/SHAGGY-like kinase. *Science* **295**, 1299–1301.
Mathur, J., Molnar, G., Fujioka, S., Takatsuto, S., Sakurai, A., Yokota, T., Adam, G., Voigt, B., Nagy, F., Maas, C., Schell, J., Koncz, C., and Szekeres, M. (1998). Transcription of the *Arabidopsis* CPD gene, encoding a steroidogenic cytochrome P450, is negatively controlled by brassinosteroids. *Plant J.* **14**, 593–602.
Min, Y. K., Asami, T., Fujioka, S., Murofushi, N., Yamaguchi, I., and Yoshida, S. (1999). New lead compounds for brassinosteroid biosynthesis inhibitors. *Bioorg. Med. Chem. Lett.* **9**, 425–430.
Molnar, G., Bancos, S., Nagy, F., and Szekeres, M. (2002). Characterisation of BRH1, a brassinosteroid-responsive RING-H2 gene from *Arabidopsis thaliana*. *Planta* **215**, 127–133.
Montoya, T., Nomura, T., Farrar, K., Kaneta, T., Yokota, T., and Bishop, G. J. (2002). Cloning the tomato curl3 gene highlights the putative dual role of the leucine-rich repeat receptor kinase tBRI1/SR160 in plant steroid hormone and peptide hormone signaling. *Plant Cell* **14**, 3163–3176.
Mori, M., Nomura, T., Ooka, H., Ishizaka, M., Yokota, T., Sugimoto, K., Okabe, K., Kajiwara, H., Satoh, K., Yamamoto, K., Hirochika, H., and Kikuchi, S. (2002). Isolation and characterization of a rice dwarf mutant with a defect in brassinosteroid biosynthesis. *Plant Physiol.* **130**, 1152–1161.
Mussig, C., Fischer, S., and Altmann, T. (2002). Brassinosteroid-regulated gene expression. *Plant Physiol.* **129**, 1241–1251.
Nagata, N., Min, Y. K., Nakano, T., Asami, T., and Yoshida, S. (2000). Treatment of dark-grown *Arabidopsis thaliana* with a brassinosteroid-biosynthesis inhibitor, brassinazole, induces some characteristics of light-grown plants. *Planta* **211**, 781–790.
Nagata, N., Asami, T., and Yoshida, S. (2001). Brassinazole, an inhibitor of brassinosteroid biosynthesis, inhibits development of secondary xylem in cress plants (*Lepidium sativum*). *Plant Cell Physiol.* **42**, 1006–1011.
Nakashita, H., Yasuda, M., Nitta, T., Asami, T., Fujioka, S., Arai, Y., Sekimata, K., Takatsuto, S., Yamaguchi, I., and Yoshida, S. (2003). Brassinosteroid functions in a broad range of disease resistance in tobacco and rice. *Plant J.* **33**, 887–898.
Nam, K. H., and Li, J. M. (2002). BRI1/BAK1, a receptor kinase pair mediating brassinosteroid signaling. *Cell* **110**, 203–212.
Neff, M. M., Nguyen, S. M., Malancharuvil, E. J., Fujioka, S., Noguchi, T., Seto, H., Tsubuki, M., Honda, T., Takatsuto, S., Yoshida, S., and Chory, J. (1999). *BAS1*: A gene regulating brassinosteroid levels and light responsiveness in *Arabidopsis*. *Proc. Natl. Acad. Sci. USA* **96**, 15316–15323.
Noguchi, T., Fujioka, S., Choe, S., Takatsuto, S., Yoshida, S., Yuan, H., Feldmann, K. A., and Tax, F. E. (1999). Brassinosteroid-insensitive dwarf mutants of *Arabidopsis* accumulate brassinosteroids. *Plant Physiol.* **121**, 743–752.
Noguchi, T., Fujioka, S., Choe, S., Takatsuto, S., Tax, F. E., Yoshida, S., and Feldmann, K. A. (2000). Biosynthetic pathways of brassinolide in *Arabidopsis*. *Plant Physiol.* **124**, 201–209.

Nomura, T., Nakayama, M., Reid, J. B., Takeuchi, Y., and Yokota, T. (1997). Blockage of brassinosteroid biosynthesis and sensitivity causes dwarfism in garden pea. *Plant Physiol.* **113**, 31–37.

Nomura, T., Kitasaka, Y., Takatsuto, S., Reid, J. B., Fukami, M., and Yokota, T. (1999). Brassinosteroid/sterol synthesis and plant growth as affected by Ika and Ikb mutations of pea. *Plant Physiol.* **119**, 1517–1526.

Nomura, T., Bishop, G. J., Kaneta, T., Reid, J. B., Chory, J., and Yokota, T. (2003). The LKA gene is a BRASSINOSTEROID INSENSITIVE1 homolog of pea. *Plant J.* **36**, 291–300.

Nomura, T., Jager, C. E., Kitasaka, Y., Takeuchi, K., Fukami, M., Yoneyama, K., Matsushita, Y., Nyunoya, H., Takatsuto, S., Fujioka, S., Smith, J. J., Kerckhoffs, L. H. J., Reid, J. B., and Yokota, T. (2004). Brassinosteroid deficiency due to truncated steroid 5 alpha-reductase causes dwarfism in the *lk* mutant of pea. *Plant Physiol.* **135**, 2220–2229.

Ogawa, M., Hanada, A., Yamauchi, Y., Kuwahara, A., Kamiya, Y., and Yamaguchi, S. (2003). Gibberellin biosynthesis and response during *Arabidopsis* seed germination. *Plant Cell* **15**, 1591–1604.

Polakis, P. (2000). Wnt signaling and cancer. *Genes Dev.* **14**, 1837–1851.

Rouleau, M., Marsolais, F., Richard, M., Nicolle, L., Voigt, B., Adam, G., and Varin, L. (1999). Inactivation of brassinosteroid biological activity by a salicylate-inducible steroid sulfotransferase from *Brassica napus*. *J. Biol. Chem.* **274**, 20925–20930.

Sakurai, A. (1999). Brassinosteroid biosynthesis. *Plant Physiol. Biochem.* **37**, 351–361.

Scheer, J. M., and Ryan, C. A. (2002). The systemin receptor SR160 from *Lycopersicon peruvianum* is a member of the LRR receptor kinase family. *Proc. Natl. Acad. Sci. USA* **99**, 9585–9590.

Schmidt, E. D. L., Guzzo, F., Toonen, M. A. J., and deVries, S. C. (1997). A leucine-rich repeat containing receptor-like kinase marks somatic plant cells competent to form embryos. *Development* **124**, 2049–2062.

Schumacher, K., and Chory, J. (2000). Brassinosteroid signal transduction: Still casting the actors. *Curr. Opin. Plant Biol.* **3**, 79–84.

Sekimata, K., Kimura, T., Kaneko, I., Nakano, T., Yoneyama, K., Takeuchi, Y., Yoshida, S., and Asami, T. (2001). A specific brassinosteroid biosynthesis inhibitor, Brz2001: Evaluation of its effects on *Arabidopsis*, cress, tobacco, and rice. *Planta* **213**, 716–721.

Sekimata, K., Uzawa, J., Han, S. Y., Yoneyama, K., Takeuchi, Y., Yoshida, S., and Asami, T. (2002). Brz220 a novel brassinosteroid biosynthesis inhibitor: Stereochemical structure-activity relationship. *Tetrahedron-Asymm.* **13**, 1875–1878.

Shah, H., Gadella, T. W. J., van Erp, H., Hecht, V., and de Vries, S. C. (2001). Subcellular localization and oligomerization of the *Arabidopsis thaliana* somatic embryogenesis receptor kinase 1 protein. *J. Mol. Biol.* **309**, 641–655.

Shah, K., Russinova, E., Gadella, T. W. J., Willemse, J., and de Vries, S. C. (2002). The *Arabidopsis* kinase-associated protein phosphatase controls internalization of the somatic embryogenesis receptor kinase 1. *Genes Dev.* **16**, 1707–1720.

Shimada, Y., Fujioka, S., Miyauchi, N., Kushiro, M., Takatsuto, S., Nomura, T., Yokota, T., Kamiya, Y., Bishop, G. J., and Yoshida, S. (2001). Brassinosteroid-6-oxidases from *Arabidopsis* and tomato catalyze multiple C-6 oxidations in brassinosteroid biosynthesis. *Plant Physiol.* **126**, 770–779.

Shimada, Y., Goda, H., Nakamura, A., Takatsuto, S., Fujioka, S., and Yoshida, S. (2003). Organ-specific expression of brassinosteroid-biosynthetic genes and distribution of endogenous brassinosteroids in *Arabidopsis*. *Plant Cell Physiol.* **44**, S69–S69.

Shui, S.-H., and Bleecker, A. (2001). Receptor-like kinases from *Arabidopsis* form a monophyletic gene family related to anomal receptor kinases. *Proc. Natl. Acad. Sci. USA* **98**, 10763–10768.

Sun, Y., Fokar, M., Asami, T., Yoshida, S., and Allen, R. D. (2004). Characterization of the Brassinosteroid insensitive 1 genes of cotton. *Plant Mol. Biol.* **54,** 221–232.

Suzuki, Y., Saso, K., Fujioka, S., Yoshida, S., Nitasaka, E., Nagata, S., Nagasawa, H., Takatsuto, S., and Yamaguchi, I. (2003). A dwarf mutant strain of *Pharbitis nil*, Uzukobito (*kobito*), has defective brassinosteroid biosynthesis. *Plant J.* **36,** 401–410.

Szekeres, M., Nemeth, K., Koncz Kalman, Z., Mathur, J., Kauschmann, A., Altmann, T., Redei, G. P., Nagy, F., Schell, J., and Koncz, C. (1996). Brassinosteroids rescue the deficiency of CYP90, a cytochrome P450, controlling cell elongation and de-etiolation in *Arabidopsis*. *Cell* **85,** 171–182.

Tepperman, J., Zhu, T., Chang, H., Wang, X., and Quail, P. (2001). Multiple transcription-factor genes are early targets of phytochrome A signaling. *Proc. Natl. Acad. Sci. USA* **98,** 9437–42.

Turk, E. M., Fujioka, S., Seto, H., Shimada, Y., Takatsuto, S., Yoshida, S., Denzel, M. A., Torres, Q. I., and Neff, M. M. (2003). CYP72B1 inactivates brassinosteroid hormones: An intersection between photomorphogenesis and plant steroid signal transduction. *Plant Physiol.* **133,** 1643–1653.

Vierstra, R. (2003). The ubiquitin/26S proteasome pathway, the complex last chapter in the life of many plant proteins. *Trends Plant Sci.* **8,** 135–142.

Wang, J. M., Asami, T., Yoshida, S., and Murofushi, N. (2001). Biological evaluation of 5-substituted pyrimidine derivatives as inhibitors of brassinosteroid biosynthesis. *Biosci. Biotech. Biochem.* **65,** 817–822.

Wang, Z. Y., Seto, H., Fujioka, S., Yoshida, S., and Chory, J. (2001). BRI1 is a critical component of a plasma-membrane receptor for plant steroids. *Nature* **410,** 380–383.

Wang, Z. Y., Nakano, T., Gendron, J., He, J. X., Chen, M., Vafeados, D., Yang, Y. L., Fujioka, S., Yoshida, S., Asami, T., and Chory, J. (2002). Nuclear-localized BZR1 mediates brassinosteroid-induced growth and feedback suppression of brassinosteroid biosynthesis. *Dev. Cell* **2,** 505–513.

Yamamoto, R., Demura, T., and Fukuda, H. (1997). Brassinosteroids induce entry into the final stage of tracheary element differentiation in cultured Zinnia cells. *Plant Cell Physiol.* **38,** 980–983.

Yamamoto, R., Fujioka, S., Demura, T., Takatsuto, S., Yoshida, S., and Fukuda, H. (2001). Brassinosteroid levels increase drastically prior to morphogenesis of tracheary elements. *Plant Physiol.* **125,** 556–563.

Yamamuro, C., Ihara, Y., Wu, X., Noguchi, T., Fujioka, S., Takatsuto, S., Ashikari, M., Kitano, H., and Matsuoka, M. (2000). Loss of function of a rice brassinosteroid insensitive1 homolog prevents internode elongation and bending of the lamina joint. *Plant Cell* **12,** 1591–1605.

Yin, Y. H., Cheong, H., Friedrichsen, D., Zhao, Y. D., Hu, J. P., Mora-Garcia, S., and Chory, J. (2002). A crucial role for the putative *Arabidopsis* topoisomerase VI in plant growth and development. *Proc. Natl. Acad. Sci. USA* **99,** 10191–10196.

Yin, Y. H., Wu, D. Y., and Chory, J. (2002). Plant receptor kinases: Systemin receptor identified. *Proc. Natl. Acad. Sci. USA* **99,** 9090–9092.

Yin, Y. H., Wang, Z. Y., Mora-Garcia, S., Li, J. M., Yoshida, S., Asami, T., and Chory, J. (2002). BES1 accumulates in the nucleus in response to brassinosteroids to regulate gene expression and promote stem elongation. *Cell* **109,** 181–191.

Yokota, T., Sato, T., Takeuchi, Y., Nomura, T., Uno, K., Watanabe, T., and Takatsuto, S. (2001). Roots and shoots of tomato produce 6-deoxo-28-norcathasterone, 6-deoxo-28-nortyphasterol and 6-deoxo-28-norcastasterone, possible precursors of 28-norcastasterone. *Phytochemistry* **58,** 233–238.

FURTHER READING

Asami, T., Mizutani, M., Shimada, Y., Goda, H., Kitahata, N., Sekimata, K., Han, S. Y., Fujioka, S., Takatsuto, S., Sakata, K., and Yoshida, S. (2003). Triadimefon, a fungicidal triazole-type P450 inhibitor, induces brassinosteroid deficiency-like phenotypes in plants and binds to DWF4 protein in the brassinosteroid biosynthesis pathway. *Biochem. J.* **369**, 71–76.

Asami, T., Nakano, T., Nakashita, H., Sekimata, K., Shimada, Y., and Yoshida, S. (2003). The influence of chemical genetics on plant science: Shedding light on functions and mechanism of action of brassinosteroids using biosynthesis inhibitors. *J. Plant Growth Regul.* **22**, 336–349.

15

Terpenoids as Plant Antioxidants

J. Graßmann

Institute of Vegetable Science—Quality of Vegetal Foodstuff
Life Science Center Weihenstephan, Dürnast 2, 85350 Freising, Germany

I. Introduction
II. Plant Antioxidants
III. Terpenoids
    A. Synthesis
    B. Monoterpenes, Sesquiterpenes, and Diterpenes
    C. Tetraterpenes—Carotenoids
    References

Plant antioxidants are composed of a broad variety of different substances like ascorbic acid and tocopherols, polyphenolic compounds, or terpenoids. They perform several important functions in plants and humans (e.g., carotenoids function as accessory pigments for light harvesting and provide photoprotection and pigmentation in plants). Monoterpenes and diterpenes, which are the main components of essential oils, act as allelopathic agents, attractants in plant–plant or plant–pathogen/herbivore interactions or repellants.

For humans, carotenoids play an important role for health, carotenoids with provitamin A activity are important for vision; other carotenoids influence the human immune function and gap-junctional communication (GJC). Additionally, their antioxidative capacity is believed to be responsible for the health promoting properties of fruits and vegetables. Three main ways of antioxidant action of carotenoids have been detected until now (i.e., quenching of singlet oxygen, hydrogen transfer, or electron transfer). These mechanisms and investigation of antioxidant activity *in vitro* are discussed in detail. The monoterpenes limonene and perillyl alcohol may be promising substances in cancer therapy. Several investigations have studied the antioxidant activity of monoterpenes and diterpenes or essential oils *in vitro*. Results as well as the action of a newly discovered, very effective antioxidant (i.e., γ-terpinene) are discussed.

An important point when assessing the antioxidant activity of plant antioxidants is to consider their interaction with other antioxidants. Especially combinations of hydrophilic and lipophilic antioxidants may exert synergistic effects, as has been shown for rutin in combination with γ-terpinene, lutein, or lycopene.

I. INTRODUCTION

Bioactive compounds are defined as nonnutritive constituents of food, which usually occur in very small quantities. They are composed of thousands of substances, which can be divided by virtue of their structure in nine classes: glucosinolates, organo-sulfur compounds, phytosterols, saponins, protease-inhibitors, phytoestrogenes, terpenoids, and polyphenolic compounds. Some of them were shown to lower total cholesterol, LDL-cholesterol, or triglycerides as well as blood pressure. Glucosinolates and organo-sulfur compounds are believed to protect from cancer by inducing phase-II enzymes, and some phytosterines seem to be helpful in lowering cholesterol in humans, thereby protecting human health (Goldberg, 2003; Kris-Etherton *et al.*, 2002). Another important property of bioactive compounds is to protect from oxidative stress (i.e., they possess antioxidative capacity [AC]).

This AC may help to prevent cardiovascular disease (CVD) or cancer, as the involvement of reactive oxygen species (ROS) in these is probable (Halliwell, 1996; Stanner *et al.*, 2004; Vendemiale *et al.*, 1999). Oxidative stress has been postulated to be involved in the development of several chronic diseases. The reaction of ROS with biomolecules like lipids, proteins, and DNA may lead to increased risk of chronic diseases, such as cancer, CVD, atherosclerosis, age-related macular degeneration (AMD), or cataract. Therefore, the inactivation of ROS by (dietary) antioxidants may be a promising preventive strategy. However, there is increasing evidence that the most prudent public health advice is to increase the consumption of plant foods (and in this way increasing the antioxidant intake) instead of using single supplements (Hercberg *et al.*, 1998; Kaur and Kapoor, 2001; Stanner *et al.*, 2004; Tapiero *et al.*, 2004). The following contribution will give an overview of terpenoids as plant antioxidants.

II. PLANT ANTIOXIDANTS

The imbalance between oxidants and antioxidants in favor of the oxidants, potentially leading to damage, forms the core of the definition of "oxidative stress." Oxidative stress may occur in plants as well as in humans. In plants, one major source of oxidative stress is the photosystem since chlorophyll may act as photosensitizer forming singlet oxygen. But also in other compartments like mitochondria, microsomes, peroxisomes, and others the formation of ROS may occur (Schempp *et al.*, 2005).

In humans, about 1–3% of the O_2 consumed by the body is converted into superoxide and other ROS under physiological conditions (Fridovich, 1986; Halliwell, 1996). They perform many important functions in physiological processes (e.g., microbial killing, cell signaling, or gene transcription [Dröge, 2002]). However, besides these desirable effects they may also damage DNA, proteins, or lipids. These deleterious effects are found to be responsible for the development of diseases like CVD, cancer, or AMD (Halliwell, 1996; Stanner *et al.*, 2004; Vendemiale *et al.*, 1999). To cope up with this threat of ROS-induced damage, the body has developed an antioxidant defense system, which consists mainly of antioxidant enzymes like superoxide dismutase or catalase, and chain-breaking antioxidants (radical scavengers) like ascorbic acid, tocopherols (vitamin E), and uric acid. During exercise and certain disorders, this antioxidant system is enhanced (Clarkson and Thompson, 2000), which may reflect an attempt to keep the balance between prooxidants and antioxidants. Another possibility to improve antioxidant defense is to increase the dietary intake of antioxidants. This can be achieved by enhancing the intake of fruits and vegetables, since they contain a broad spectrum of antioxidants, the most important of them being ascorbic acid and vitamin E, polyphenolic compounds and terpenoids. Since antioxidant actions and biological functions of ascorbic acid and vitamin E as well as polyphenolic compounds have gained great interest and therefore have been addressed in numerous comprehensive reviews (Asard *et al.*, 2004; Bramley *et al.*, 2000; Davey *et al.*, 2000; Harborne and Williams, 2000; Kim and Lee, 2004; Parr and Bowell, 2000; Pietta, 2000), the main focus of this study will be on terpenoids.

III. TERPENOIDS

A. SYNTHESIS

Terpenoids are substances that are built up from isoprene; therefore, they are also called isoprenoids. They are divided on the basis of their C-skeleton; Table I gives an overview.

TABLE I. Classes of Terpenoids

Terpenoid	Number of C-atoms	Number of isoprene subunits
Monoterpene	10	2
Sesquiterpene	15	3
Diterpene	20	4
Triterpene	30	6
Tetraterpene	40	8
Polyterpene	>40	>8

The generation of terpenoids comprises three steps:

1. Formation of the C5-subunit
2. Condensation of these subunits form the skeleton of the different terpenoids
3. Conversion of the resulting prenyldiphosphates to end products

Synthesis is accomplished either by the mevalonate or the methylerythritol-4-phosphate (MEP) pathway (which was originally named nonmevalonate pathway and in the meantime also DXP- or DOXP-pathway). The former has been known for a long time and is located in the cytoplasm. By this pathway, sesquiterpenes, triterpenes, and polyterpenes are synthesized (Bruneton, 1999; Loza-Tavera, 1999). The latter was discovered in the early 1990s and produces monoterpenes, diterpenes, sesterterpenes, and tetraterpenes (Lichtenthaler, 1999; Rodríguez-Concepción and Boronat, 2002). Their common intermediate is isopentenylpyrophosphate (IPP; "activated isoprene") from which all terpenoids are formed. Catalyzed by prenyltransferases, IPP polymerizes to prenylpyrophosphates. In the third phase of synthesis prenylpyrophosphates are finally converted to terpenes. These reactions are carried out by the large group of terpene synthases (Kreuzwieser et al., 1999). Figure 1 shows the basic reactions of both the pathways.

B. MONOTERPENES, SESQUITERPENES, AND DITERPENES

1. Significance for Plants and Men

Monoterpenes and sesquiterpenes are the main constituents of essential oils (e.g., those from citrus fruits, herbs, and spices). Essential oils have numerous ecological functions in the plant kingdom, such as acting as allelopathic agents, repellants, or attractants in plant–plant or plant–pathogen/herbivore

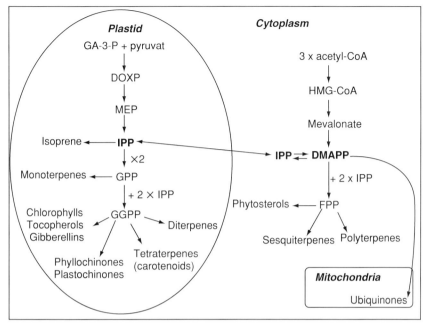

FIGURE 1. Basic reactions during terpenoids synthesis. Abbreviations: GA-3-P, D-glyceraldehyde-3-phosphate; DOXP, 1-deoxy-D-xylulose-5-phosphate; MEP, methylerythritol-4-phosphate; IPP, isopentenyl diphosphate; GPP, geranyl diphosphate; GGPP, geranylgeranyl diphosphate; HMG, hydroxymethylglutaryl; DMAPP, dimethylallyl diphosphate; FPP, farnesyl diphosphate.

interactions (Dudareve et al., 2004; Paré and Tumlinson, 1999; Wink and Latz-Brüning, 1995). Another function is seen in defense and wound healing in pine tree species or in increasing thermotolerance in plants (Singsaas et al., 1997).

Most investigations regarding their role in human health have been carried out with limonene, perillyl alcohol, carvone, and carveol due to their chemotherapeutic activities. A number of dietary monoterpenes possess antitumor activity in animal models or different cell lines, although human clinical trials are under way. The monoterpenes inhibit carcinogenesis both in the initiation and promotion/progression stages, and are effective in treating early and advanced cancers (Crowell, 1997, 1999; Gould, 1997; Wagner and Elmadfa, 2003). Antioxidant properties of monoterpenes are discussed later, some monoterpenes are shown in Fig. 2

Sesquiterpenes are the most diverse group of isoprenoids. In plants, they function as pheromones and juvenile hormones. The acyclic representatives are also called farnesans, the term, which is derived from the basic structure, farnesol. Examples for bicyclic sesquiterpenes are α-caryophyllene and β-caryophyllene. Some sesquiterpenes are shown in Fig. 3

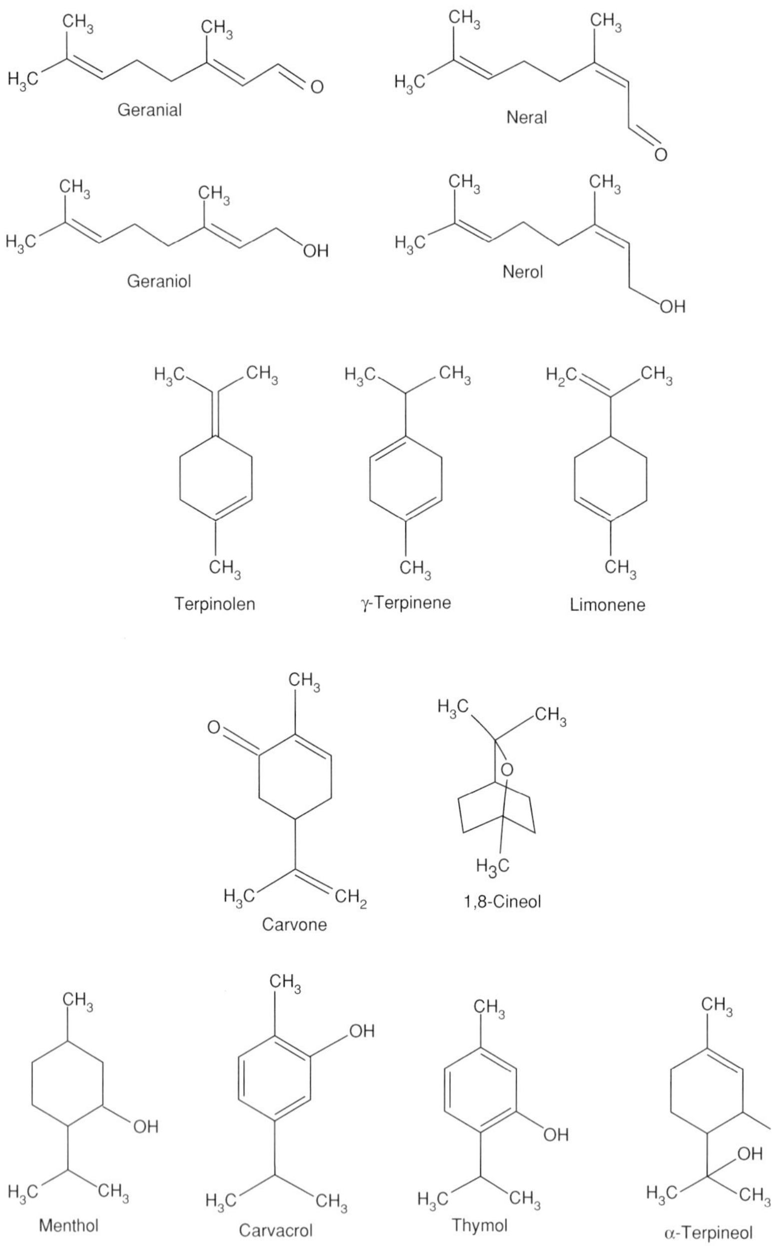

FIGURE 2. Structures of some important monoterpenes.

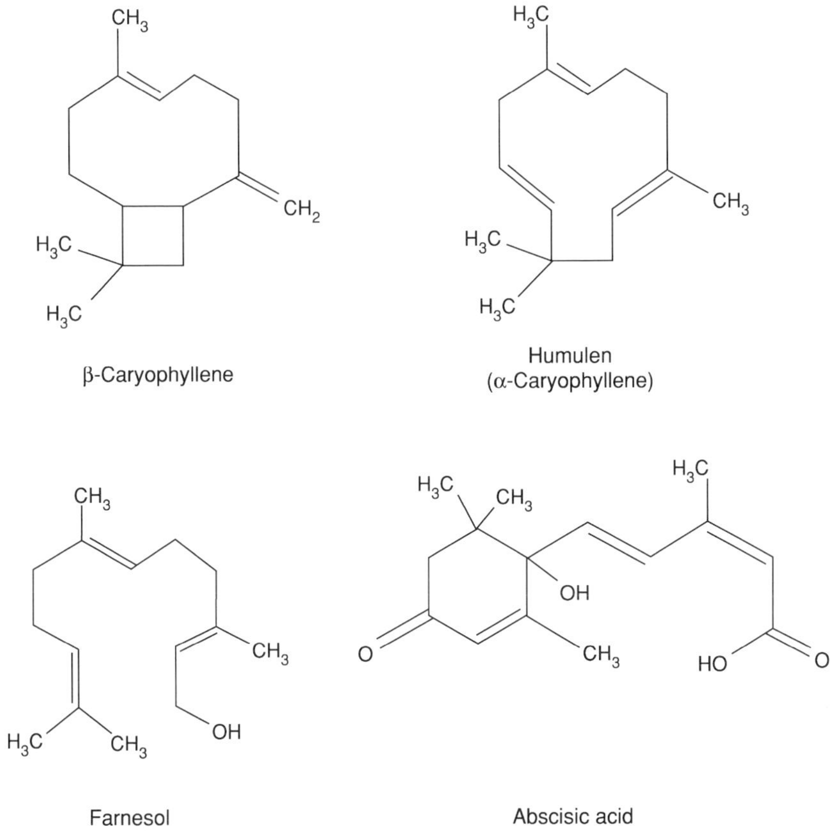

β-Caryophyllene

Humulen (α-Caryophyllene)

Farnesol

Abscisic acid

FIGURE 3. Structures of some important sesquiterpenes.

The most important diterpenes for human health are those with vitamin A activity, which plays a fundamental role in the process of viewing (Wagner and Elmadfa, 2003). Besides this, retinoids regulate the growth and differentiation of normal, premalignant, and malignant cells. These functions are achieved by changes in gene expression mainly through interaction with the retinoic acid receptors and retinoid X receptors (Evans and Kaye, 1999; Okuno et al., 2004). Figure 4 shows the structures of the most important diterpenes.

2. Antioxidative Properties of Monoterpenes, Sesquiterpenes, and Diterpenes

a. AC in Different Test Systems In Vitro

Monoterpenes and sesquiterpenes as well as diterpenes show antioxidant activity. Most investigations, however, were conducted with essential oils and it was shown that they exhibit AC in different *in vitro* model systems.

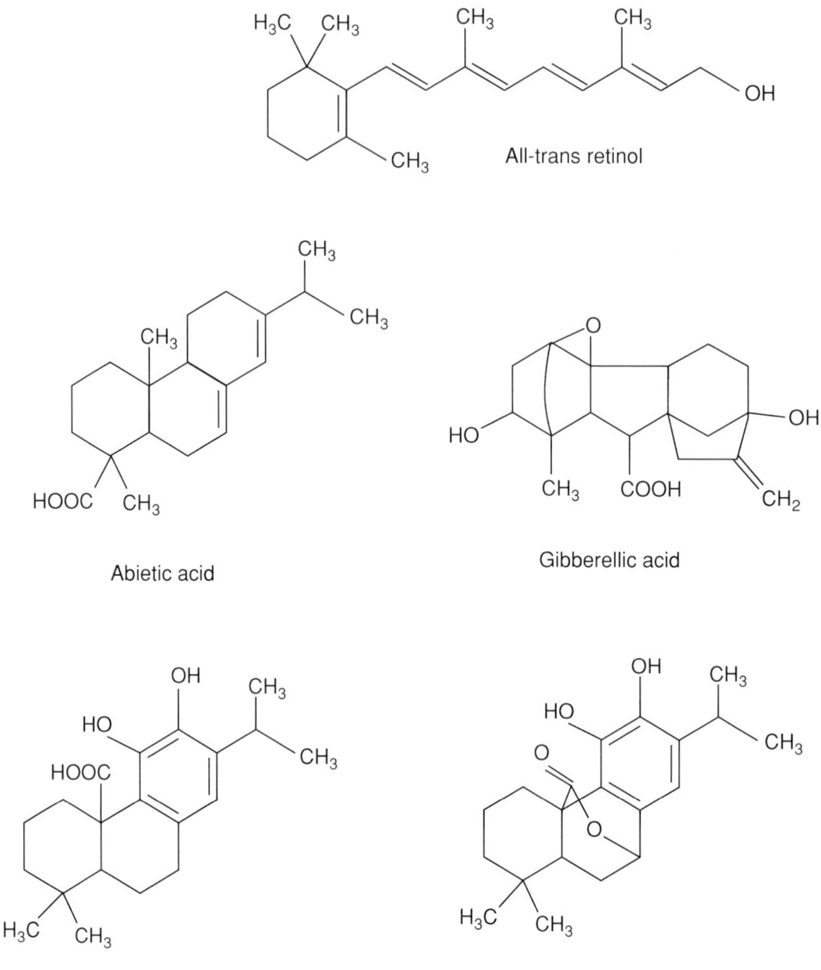

FIGURE 4. Structures of some important diterpenes.

Besides chemical composition of an essential oil, extraction methods and the system used to determine AC influence the results (Dapkevicius et al., 1998; Fadel et al., 1999; Hopia et al., 1996; Mantle et al., 1998). Many investigations were carried out by examining the influence of essential oils on oxidation of fats like sunflower oil, lard, primrose oil, or others (Abdalla and Roozen, 1999; Dang et al., 2001; Schwarz et al., 1992; Youdim et al., 1999). This may be useful in identifying those essential oils, monoterpenes or diterpenes, which might be used for food preservation; however, it has no physiological relevance.

Baratta *et al.* (1998a,b) investigated a broad range of essential oils for their AC using egg yolk and rat liver as oxidizable substrates and 2,2′-azobis (2-amidinopropane) dihydrochloride (AAPH) as a radical inducer. The oils showed different effectiveness, majoran and oregano oil being the most effective. Other examinations used the 2,2-diphenyl-1-picrylhydrazyl (DPPH) assay or different lipid peroxidation assays for quantifying AC, or tested essential oils for their ability for hydroxyl or superoxide radical scavenging. The results confirmed the hydrogen-donating properties of essential oils but gave inconsistent results regarding the activity towards hydroxyl radical or the superoxide radical anion.

A major problem in reviewing the literature regarding AC of essential oils is the broad variation of the used test systems. Furthermore, results, which were obtained in the same test system (e.g., the DPPH assay), are given in quite different units like percent remaining activity, trolox equivalents, IC_{50} values, and others. Because of this reason, a comparison of results from different groups is extremely complex or even impossible. Table II gives an overview of the most frequently used model systems and the tested oils.

Some investigations confirmed good AC for rosemary and thyme oil and in most studies phenolic substances like the phenolic diterpenes carnosic acid and carnosol from rosemary extracts or the phenolic monoterpenes carvacrol and thymol were held responsible for the observed antioxidative effects (Aeschbach *et al.*, 1994; Hopia *et al.*, 1996; Richheimer *et al.*, 1996; Schwarz *et al.*, 1996). Figure 5 shows the structure of these terpenoids.

The phenolic diterpenes from rosemary were shown to act as "primary antioxidants" by donating hydrogen to lipid radicals and thereby slowing down lipid peroxidation.

Citrus oils have also been shown to be quite effective antioxidants in the DPPH assay (Choi *et al.*, 2000) and γ-terpinene was shown to be an important nonphenolic antioxidant.

All the used test systems provide some conclusions about hydrogen-donating or radical-scavenging activities of monoterpenes and diterpenes as well as their effectiveness in inhibiting lipid peroxidation (LPO). This may be useful when one is looking for antioxidant to enhance oxidative stability of edible lipids but mostly lacks significance regarding human pathologies. Investigations using activated neutrophils in whole blood as a source for ROS showed a strong inhibition by essential oils, which may be due to interactions of the lipophilic essential oils with the membrane of neutrophils. This may also be an explanation for the anti-inflammatory effects of essential oils (Graßmann *et al.*, 2000).

b. AC Regarding Oxidation of LDL

Another physiological model system in this context is oxidation of low-density lipoproteins (LDL). Low-density lipoproteins oxidation is believed to be implicated in atherogenesis (Chisolm and Steinberg, 2000). Therefore,

TABLE II. AC of Essential Oils and Their Components in Various *In Vitro* Test Systems

Test system	Tested oils and components (selection)	References
Lipid peroxidation (Rancimat method, Schaal test and other)	Rosemary, ginger, cinnamon, lemongrass	Dang et al. (2001)
	Thyme	Youdim et al. (1999)
	Thyme	Schwarz et al. (1996)
	Rosemary, sage	Schwarz et al. (1992)
	Catnip, hyssop, lemon balm, oregano, sage, thyme	Abdalla and Roozen (1999)
Oil stability index (OSI)	Sage, thyme	Miura et al. (2002)
TBAR-assay in lipid-rich media	Laurel, sage, rosemary, oregano, coriander ylang-ylang, lemongrass, basil, rosemary, cinnamon, lemon, frankincense, majoram	Baratta et al. (1998a,b)
	Thymus pectinatus	Vardar-Ünlü et al. (2003)
Deoxyribose degradation assay	*Artemisia afra, Artemisia abyssinica, Juniperus procera*	Burits et al. (2001)
	T. pectinatus	Vardar-Ünlü et al. (2003)
DPPH	*A. afra, A. abyssinica, J. procera*	Burits et al. (2001)
	34 kinds of citrus oils	Choi et al. (2000)
	Tea tree (*Melaleuca alternafolia*) oil	Kim et al. (2004)
	Sage, thyme	Miura et al. (2002)
	Amazonian basil, common basil, thyme	Sacchetti et al. (2004)
	Melissa officinalis	de Sousa et al. (2004); Mimica-Dukic et al. (2004)
	3 Mentha species	Mimica-Dukic et al. (2003)
	T. pectinatus	Vardar-Ünlü et al. (2003)

protection of LDL from oxidation may contribute in preventing heart attack and stroke. It could be shown that essential oils are able to prevent copper-induced LDL oxidation (Graßmann et al., 2001, 2003; Takahashi et al., 2003; Teissedre and Waterhouse, 2000). Teissedre and Waterhouse (2000) explained this protecting effect with the phenolics content of the investigated essential oil. Graßmann et al. (2001), however, identified γ-terpinene as the most active substance in this context. This high antioxidative capacity of γ-terpinene is in accordance with the findings of Choi et al. (2000) who could show that the radical-scavenging activities of citrus oils depends—among

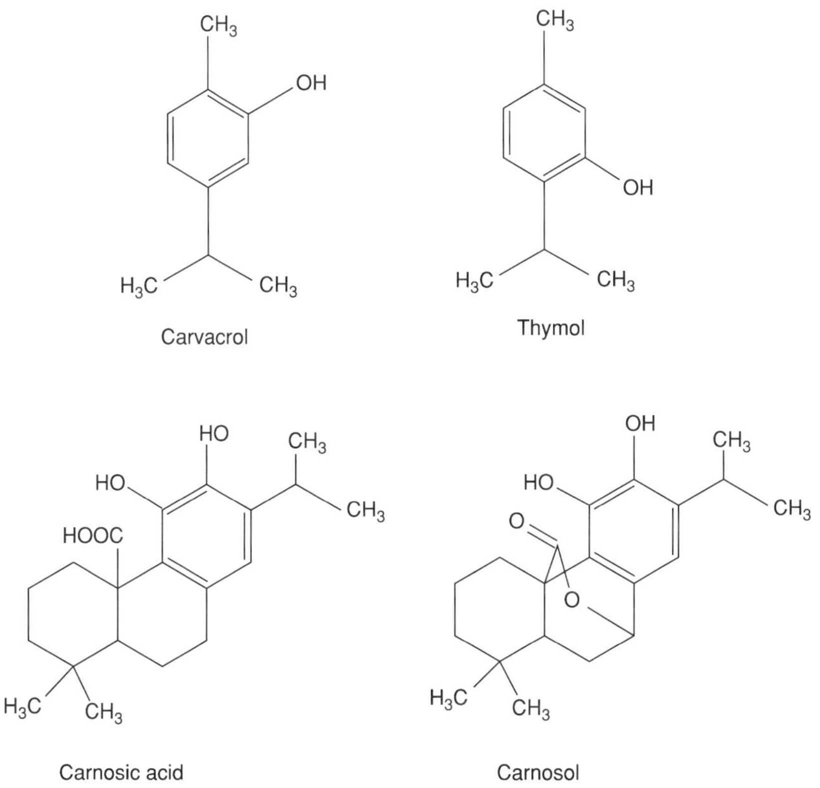

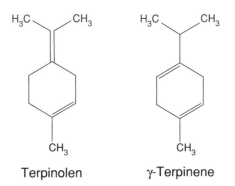

FIGURE 5. Phenolic monoterpenes and diterpenes.

FIGURE 6. Structures of terpinolen and γ-terpinene.

other less important factors—on their γ-terpinene and terpinolene content. Structures of these monoterpenes (Fig. 6) reveal that they contain no hydroxyl groups (i.e., antioxidative capacity must be explained by other structural factors). In a study on tea tree oil (TTO), its AC also was attributed to the nonphenolic compounds α-terpinene, terpinolen, and γ-terpinene but not to the main component of TTO (i.e., terpinen-4-ol, which contains a hydroxyl group [Kim *et al.*, 2004]).

Graßmann *et al.* (2001) proved that γ-terpinene can be enriched in LDL by preincubating human blood plasma with lemon oil or γ-terpinene and that the subsequently isolated LDL shows a high resistance against copper-induced oxidation. Takahashi *et al.* (2003) showed in later studies that γ-terpinene, which was added to LDL solutions is able to prevent the oxidation of LDL.

By monitoring the consumption of endogenous antioxidants like α-tocopherol and carotenoids it could be shown that the highly lipophilic γ-terpinene protects the carotenoids from oxidation but has no influence on α-tocopherol consumption. This may be due to the good hydrophobic interaction of γ-terpinene with the LDL particles and its capacity to donate hydrogen atoms or electrons, or its ability to chelate metal ions. However the latter (i.e., copper complexation) does not play an important role, since it could be shown that γ-terpinene is not an effective copper-chelator (Graßmann *et al.*, unpublished results). Foti and Ingold (2003) revealed the underlying mechanism by which γ-terpinene inhibits lipid peroxidation. The important fact is that the chain carrying peroxyl radicals are HOO$^{\bullet}$ radicals, which react rapidly with linoleylperoxyl radicals and thereby terminate the chain reaction. Since this mechanism is different from the mode of action of vitamin E and since vitamin E becomes a prooxidant at high concentrations, the discovery of γ-terpinene may provide a new stabilizing substance for edible lipids and may also provide a possibility to enrich foodstuffs and beverages to increase highly effective antioxidants in nutrition.

c. Interaction with Other Antioxidants

An important mode of action of monoterpenes when working as antioxidants is to support other antioxidants like α-tocopherol (e.g., rosemary extracts show synergistic effects together with α-tocopherol) (Wagner and Elmadfa, 2003). These results show that lipid-soluble antioxidants can work together to protect lipids from oxidation; in case of rosemary extracts this is due to the regeneration of α-tocopherol. Milde *et al.* (2004). proved that not only the cooperation of lipid-soluble antioxidants can lead to effective synergisms but also the combination of lipid-soluble with more water-soluble antioxidants gives overadditive protection. Therefore, γ-terpinene together with rutin results in a synergistic inhibition of copper-induced LDL oxidation.

These results again support the idea that "Health benefits of fruits and vegetables are from additive and synergistic combinations of phytochemicals" (Liu, 2003).

3. Conclusions

Essential oils and their components (i.e., mainly monoterpenes and diterpenes, possess AC in different *in vitro* model systems). In some cases, this can be explained by their content of phenolics like carnosol, carnosic acid, carvacrol, or thymol. However, newer results also proved high AC for essential oils not containing such phenolic substances in noteworthy amounts. This is due to hydrocarbons like α-terpinene or γ-terpinene or terpinolen. Their AC is based on a reaction mechanism that acts by chain carrying HOO$^\bullet$ radicals, which react rapidly with linoleylperoxyl radicals and thereby terminate the chain reaction. An interesting feature is synergistic action between terpenoids and other antioxidants like α-tocopherol or flavonoids like rutin. Therefore, it is essential to test antioxidants not only separately but also in combination, because health promoting properties of fruits and vegetables are in all likelihood due to the mixture of secondary plant metabolites.

C. TETRATERPENES—CAROTENOIDS

The main group of tetraterpenes are the carotenoids, which are abundantly found as pigments in plants. More than 600 carotenoids have already been isolated from nature, their basic structure being a symmetrical tetraterpene skeleton formed by the conjugation of two C20-units. All carotenoids can be derived from the acylic unit by different reaction steps involving hydrogenation, dehydrogenation, cyclization, or oxidation reactions. Figure 7 shows the major transformation reactions.

Carotenoids can be divided depending on their functional groups. Those, which contain only carbon and hydrogen atoms are called carotenes, those with at least one oxygen function are referred to as xanthophylls. Figures 8 and 9 show the major carotenoids.

1. Significance for Plant and Humans

Carotenoids perform three major functions in plants, which are given in the following sections.

a. Accessory Pigments for Light Harvesting

The carotenoids are important components of the light harvesting antennae. With few exceptions, the chloroplasts of all species contain a collection of main carotenoids namely β-carotene, lutein, violaxanthin, and neoxanthin (Bartley and Scolnik, 1995). They absorb light between 450 and 570 nm thereby expanding the absorption spectra of photosynthesis and enhancing photosynthetic effectiveness. It is crucial that the carotenoids be located close to the chlorophyll molecules to ensure an effective transfer of energy. The major carotenoids in this context are the xanthophylls lutein, violaxanthin, and neoxanthin.

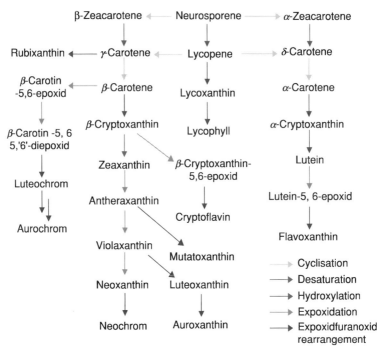

FIGURE 7. Transformation reactions during carotenoid synthesis.

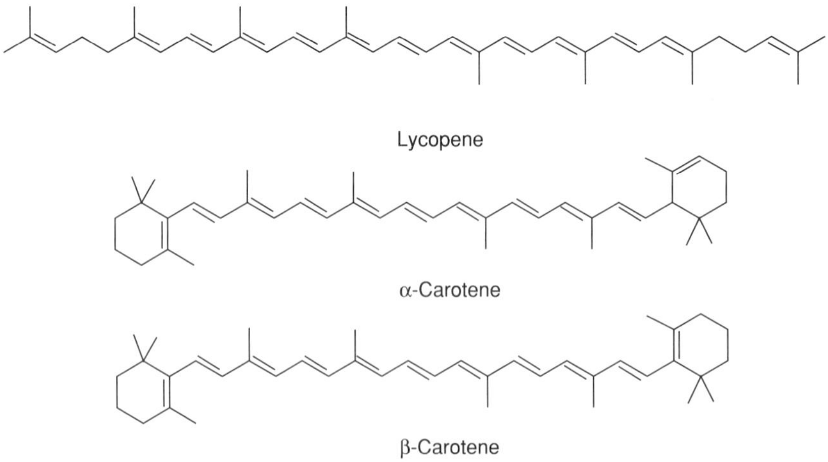

FIGURE 8. Structures of lycopene, α-carotene, and β-carotene.

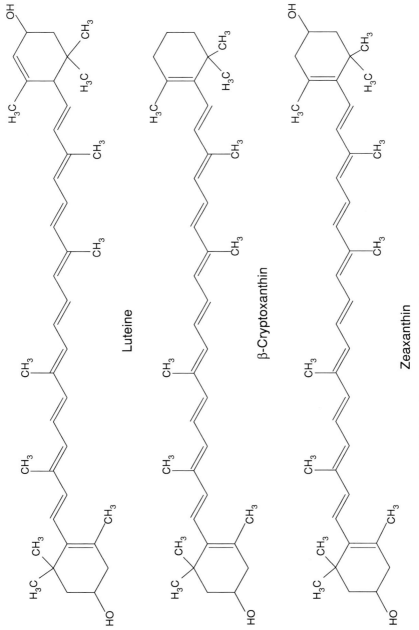

FIGURE 9. Structures of luteine, β-cryptoxanthin, and zeaxanthin.

b. Prevention of Photooxidative Damage

The excited triplet state of chlorophyll, which is generated in the photosynthetic apparatus, may initiate photooxidation processes via singlet oxygen and damage the photosynthetic system. Carotenoid molecules with nine (or more) conjugated double bonds are able to absorb energy from triplet state chlorophyll or from singlet oxygen and by this means prevent generation of the damaging singlet oxygen (Choudhury and Behera, 2001; van den Berg et al., 2000). Additionally, there is considerable evidence, which indicates a photoprotective role of the xanthophyll cycle in removing excess energy from the photosynthetic antennae. Under conditions of excess excitation energy zeaxanthin accumulates from violaxanthin via two de-epoxidation steps. This accumulation allows a rapid nonphotochemical quenching of chlorophyll fluorescence (Demming-Adams and Adams, 1996). Figure 10 shows the reactions involved in the xanthophyll cycle.

Also during N-limited conditions, which lead to restricted photosynthetic capacity, the xanthophyll cycle may prevent damage by dissipating excess light. This is indicated by increased xanthophyll cycle pigments (e.g., in spinach leaves) (Verhoeven et al., 1997), which were grown under N-limited conditions.

c. Pigmentation to Attract Animals for Pollination and Dispersal of Seeds

Carotenoids are not only located in chloroplasts but also in chromoplasts where they contribute to most of the orange, yellow, and red colors of fruits or flowers. However, other compounds (e.g., water-soluble anthocyanins) also contribute to the color of fruits and flowers. The green of the chlorophylls masks the yellow or reddish colors but they are revealed in the autumn leaves of many trees.

For humans, carotenoids play an important role in health. The function of retinol (vitamin A) in vision has been known for a long time. Though retinol is formed mainly from β-carotene by symmetrical cleavage, 50 other carotenoids can function as provitamin A, the important ones being α-carotene, β-cryptoxanthine, and cis-β-carotene (Castenmiller and West, 1998; Wagner and Elmadfa, 2003). Observational epidemiological studies have been very consistent in proving an inverse relationship of higher dietary levels of fruits and vegetables and the risk of developing certain kinds of cancer or CVD (Johnson, 2002; Tapiero et al., 2004). It has been suggested that carotenoids are the chemopreventive agents in fruits and vegetables. There are several mechanisms by which carotenoids can prevent diseases.

Effects on Human Immune Function Carotenoids exhibit immunomodulatory actions, which could contribute to their assumed anticarcinogenic effects. Hughes (1999) describes the ability of β-carotene to enhance cell-mediated immune response. Supplementation results in an enhanced activity of natural killer cells as well as in antigen presenting monocytes.

TERPENOIDS AS PLANT ANTIOXIDANTS 521

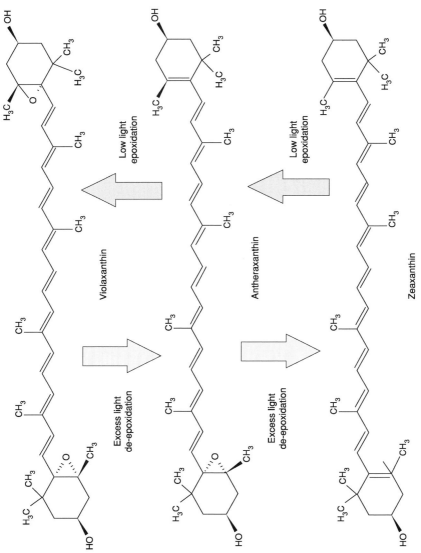

FIGURE 10. Scheme of the xanthophyll cycle and its regulation by excess or low light.

Influence on GJC Another biological function of carotenoids is the support of gap-junctional communication (GJC). During carcinogenesis, GJC is lost and this loss may be important for malignant transformation, and its restoration may reverse malignant processes (Stahl *et al.*, 2002; Tapiero *et al.*, 2004; Yamasaki *et al.*, 1995). Carotenoids stimulate GJC in a differential and dose-dependent manner (Stahl *et al.*, 1997); however, the underlying mechanisms are not yet understood.

Antioxidative Properties Oxidative damage has been discussed in context with two diseases of the elderly (i.e., cataract and AMD). There is evidence that lutein and zeaxanthin, the only two carotenoids, which have been identified in the human crystalline lens, may reduce the risk for developing these most common eye diseases. One possible mode of action is that the macular pigments filter blue light, which is particularly damaging to the photoreceptors and to the retinal pigment epithelium. Another hypothesis is that the antioxidant properties of lutein and zeaxanthin may reduce the degree to which oxidative stress promotes these diseases since there is considerable oxidative stress in the eye due to intense light exposure and a high rate of oxidative metabolism in the retina (Alves-Rodrigues and Shao, 2004; Ham, 1983; Johnson, 2002; Mares-Perlman *et al.*, 2002). Antioxidative properties of carotenoids may also be important in the prevention of cancer or CVD. The most important aspects of the antioxidant activity of carotenoids are discussed in the following section. However, one should keep in mind that the probable influence of carotenoids on disease prevention is only at the beginning of being elucidated and it is still unclear whether the antioxidant properties of carotenoids are connected with the prevention of disease. In a few experiments, the prooxidative effects of carotenoids have been shown especially at high oxygen pressure or high carotenoid concentration (El-Agamey *et al.*, 2004b; Krinsky, 1998).

2. Antioxidative Properties of Carotenoids

Carotenoids are most possibly involved in scavenging of singlet oxygen and peroxyl radicals. Additionally, they are able to deactivate sensitizer molecules, which are involved in the generation of ROS. One example is the triplet chlorophyll, which emerges during photosynthesis and may initiate photooxidative processes.

a. Singlet Oxygen Quenching

Singlet oxygen can be generated by electron energy transfer from the excited state of a sensitizer to oxygen. Sensitizers may be chlorophylls, riboflavin, porphyrins, and others, and they may induce singlet oxygen production in biological systems consequentially leading to the damage of DNA, lipids, proteins, and other biological molecules. Therefore, it is quite beneficial for plants to possess carotenoids as they are the most effective singlet oxygen quenchers found in nature.

The main mechanism of photoprotection against singlet oxygen by carotenoids is physical quenching, which occurs by following mechanism:

$$^1O_2^* + \text{carotenoid} \rightarrow {}^3O_2 + {}^3\text{carotenoid}^*$$

$$^3\text{carotenoid}^* \rightarrow \text{carotenoid} + \text{thermal energy}$$

Because of the long, conjugated polyene system of carotenoids, they lose the excess energy via vibrational and rotational interactions with the solvent (i.e., they dispense it as thermal energy). The carotenoid emanates unchanged from this reaction, ready to begin another cycle of singlet oxygen quenching. It has been estimated that each carotenoid can quench 1000 singlet oxygen molecules before it reacts chemically. This chemical quenching is responsible for the destruction of the molecule (Di Mascio *et al.*, 1992; Edge *et al.*, 1997; Krinsky, 1998).

b. Scavenging of Peroxyl and Other Radicals

Peroxyl radicals are intermediates of the LPO process, which is characterized by a radical chain reaction. This process can be interrupted by chain-breaking antioxidants, which are able to react with the lipidperoxyl radicals. These chain-breaking antioxidants are, for example, tocopherols, phenols, or ascorbic acid. An important feature of them is that the resulting antioxidant radical is too unreactive to propagate the chain reaction of LPO. The antioxidant radical may be removed by reaction with another radical, thereby forming a stable product. Another common fate of the antioxidant radical is to be recycled by another antioxidant. Carotenoids are able to act as chain-breaking antioxidants by three pathways (El-Agamey *et al.*, 2004a; Krinsky and Yeum, 2003) (i.e., electron transfer, hydrogen abstraction, or radical addition [adduct formation]). Which of these pathways preferably proceeds mainly depends on the structure of the carotenoid and on the environment and system used to assess the antioxidant activity. It is unlikely, for example, that electron transfer will take place in a highly lipophilic environment, since this does not facilitate charge separation. One important point is that the antioxidant activity of carotenoids may be converted in a prooxidative activity at high oxygen pressure (Burton and Ingold, 1984). In the physiological range of oxygen pressure the reaction of peroxyl radicals with carotenoids will lead to a resonance-stabilized structure that will terminate peroxidation processes, whereas increasing oxygen tension will allow the carotenoid radical to react with oxygen, thereby producing radicals, which are able to propagate peroxidation (Burton, 1989; Burton and Ingold, 1984; Palace *et al.*, 1999; Palozza and Krinsky, 1992). However, most data indicate that the prooxidant effect only arises at 100% oxygen and not at ambient conditions (21% oxygen) or at physiological or tissue concentrations (1–2% oxygen). Kennedy and Liebler (1992) showed that β-carotene provided similar antioxidant protection under an air atmosphere (150 torr O_2) and under physiological conditions (15 torr O_2). Thus, there is only little evidence for the thesis that β-carotene may act as a prooxidant in the body (Krinsky, 1998).

c. Oxidation Products of Carotenoids

Looking at the diversity of carotenoids it is obvious that by different oxidation reactions an even greater variety of oxidation products will be generated. Those oxidation products are likely to be *in vivo* metabolites; therefore, not only the carotenoids but also their oxidation or cleavage products should be taken into consideration. A variety of oxidation products have been detected (e.g., epoxides or apo-carotenoids) (Grosch *et al.*, 1976; Handelmann *et al.*, 1991; Kennedy and Liebler, 1991; Stratton *et al.*, 1993). These compounds may have biological activities and may interfere with a variety of signaling pathways. Studies have revealed that such substances have diverse *in vivo* effects, like enhancing GJC, inhibiting Na^+-K^+-ATPase, or impairing mitochondrial respiration (Aust *et al.*, 2003; Siems *et al.*, 2000, 2002). However, this topic is beyond the scope of this contribution. The most potent product of carotenoid oxidation is retinoic acid formed by enzymatic cleavage of β-carotene.

d. Antioxidant Activity of Carotenoids In Vitro

Reactivity Toward Singlet Oxygen The first detected mechanism by which carotenoids act as antioxidants was their ability to quench singlet oxygen (Foote and Denny, 1968). There are numerous investigations regarding this feature some of which are briefly discussed further.

Conn *et al.* (1991) investigated different carotenoids on their rate constants for the quenching of singlet oxygen and found that the ability to quench singlet oxygen increases with increasing number of conjugated double bonds. The two carotenoids in the eye (zeaxanthin and lutein) have very different quenching rate constants. Zeaxanthin ($n = 11$) seems to be at least twice as effective as lutein ($n = 10$), which may be due to the additional conjugated double bond in zeaxanthin.

Conn *et al.* (1991) also supposed that an epoxide group rather than carbonyl or hydroxyl substituents increases the reactivity of the carotenoid with respect to singlet oxygen. Di Mascio *et al.* (1989, 1992) proved that lycopene quenching rates are higher than those for β-carotene, therefore it is possible that opening the β-ionone ring has a positive effect on $^1O_2$ quenching.

However, it seems clear that significant differences can arise when the rate constants are determined by different techniques (e.g., the values for β-carotene vary up to fourfold depending on the assays) (Di Mascio *et al.*, 1992).

These results show that carotenoids are extremely good $^1O_2$ quenchers *in vitro* but little work has been carried out to test how effectively they protect cells against $^1O_2$-related damage.

Reactivity Toward Different Radical Species Carotenoid antioxidant activity can be studied by following the bleaching of carotenoids or the analysis of carotenoid oxidation products during or after their reaction with different radical species (Palozza and Krinsky, 1992). The results prove that carotenoids are able to trap oxygen and/or organic free radical intermediates. Oxygen radicals are not the only ones that can be trapped by carotenoids; lutein, for example, is able to scavenge sulfur radicals (Chopra *et al.*, 1993). Everett *et al.* (1996) showed that β-carotene quenches glutathione, sulfonyl, and nitrogen dioxide radicals. Mortensen and Skibsted (1996) investigated the interaction of β-carotene with phenoxyl radicals that results in the formation of a β-carotene radical cation and an adduct. Phenoxyl radicals are important species in biological systems and are formed when phenolic antioxidants react with peroxyl and alkoxyl radicals; the tocopherol radical represents the most important phenoxyl radical in biological systems.

Böhm *et al.* (1997) reported about the reaction of α-tocopheroxyl radicals with β-carotene that shows near diffusion controlled rate constants in hexane. These aspects will be considered later when the interaction of carotenoids with other antioxidants is discussed.

Depending on the radical species, the antioxidant action of carotenoids will follow different mechanisms (i.e., either electron transfer or addition processes will take place). For an overview, see Krinsky and Yeum (2003).

Mortensen *et al.* (2001) and Rice-Evans *et al.* (1997) divided the radical species in two groups; those that cause mainly electron transfer

$^\bullet CCl_3$, $^\bullet OOCCl_3$, $RSO_2^\bullet$, $NO_2^\bullet$, $C_6H_6O^\bullet$,

and tocopheroxyl radical and those that were shown to lead to addition processes

$RS^\bullet$, $L^\bullet$, $LOO^\bullet$, and $O^{2-\bullet}$.

AC of Carotenoids in Different Test Systems Another aspect while discussing antioxidant activity of carotenoids is to compare the different substances in a variety of test systems. β-Carotene was shown to be able to decrease nuclear damage induced by xanthin oxidase (XOD)/hypoxanthine or polymorphonuclear leukocytes (PML). Additionally, β-carotene is able to inhibit LPO induced by various systems (e.g., the XOD system) (Palozza and Krinsky, 1992). Researchers often use liposomes as a model membrane system to investigate the ability of carotenoids to inhibit LPO.

There have been a number of studies showing the inhibition of LPO in liposomes or isolated membranes (summarized in Palozza and Krinsky, 1992). However, the results are often inconsistent, which may be explained by the different preparations of carotenoid-containing liposomes or membranes. For example, studies from Kennedy and Liebler (1992) came out with the result that β-carotene inhibits 2,2′-azobis(2,4′-dimethylvaleronitrile) (AMVN)-induced peroxidation of phosphatidylcholine liposomes at concentrations of approximately 0.1–0.5 mol%. However, no inhibition could be detected in AMVN-induced peroxidation of rat liver microsomes at concentrations of 1.5 mol%. This is explained by Liebler (1993) by different preparations of the liposomes. In the former case, β-carotene was mixed with the phospholipids before liposome formation, and in the latter case, the microsomal membranes were preformed and β-carotene was added subsequently. Another lipophilic system to measure antioxidative efficacy is the oxidation of LDL that is thought to be implicated in atherosclerosis (Chisolm and Steinberg, 2000). Several carotenoids have been tested on their ability to inhibit LDL oxidation *in vitro*, indicating that β-carotene and lycopene may play a protective role (Agarwal and Rao, 1998; Fuhrman *et al.*, 2000; Packer, 1993). Despite this *in vitro* evidence for protective effects, *in vivo* supplementation with β-carotene in most investigations did not lead to an enhanced *in vitro* resistance of LDL towards oxidation (Gaziano *et al.*, 1995; Jialal and Fuller, 1995; Jialal and Grundy, 1993; Princen *et al.*, 1992; Reaven *et al.*, 1993).

Woodall *et al.* (1997) applied a modified Fenton reaction or free radicals generated from 2,2′-azobis-isobutyronitrile (AIBN) or from AMVM to oxidize carotenoids. Lycopene showed the highest reactivity in the different test systems; however, the reactivity of the other tested carotenoids (i.e., β-carotene, zeaxanthin, isozeaxanthin, echinenone, lutein, astaxanthin, and canthaxanthin) varied between the systems. This group suggested "that hydrogen abstraction should be considered as one of the possible mechanisms that occur when carotenoids are exposed to peroxyl radicals and other oxidising agents."

Siems *et al.* (1999) used different prooxidants *in vitro* (NaOCl, AIBN or photo-irradiation, UV light in presence of the photosensitizer Rose Bengal) and found that the breakdown of lycopene and β-carotene was much faster than that of lutein and zeaxanthin in all systems. The high antioxidant activity of lycopene in the "light-induced oxidation" is not surprising, since lycopene is known to be a very effective singlet oxygen quencher and it is possible that under UV-light or in presence of Rose Bengal singlet oxygen is the main oxidizing species. Also, Woodall *et al.* (1997) proved a high AC for lycopene in three different test systems, and in the case of AIBN as free radical generator they also found a higher rate of breakdown for lycopene and β-carotene than for lutein and zeaxanthin.

Miki (1991) studied β-carotene, lutein, zeaxanthin, astaxanthin, tunaxanthin, and canthaxanthin in comparison with α-tocopherol. He used a heme-protein-Fe^{2+}-complex as free radical generator and quantified TBA production. He found that astaxanthin is the most efficient scavenger followed by zeaxanthin, canthaxanthin, lutein, tunaxanthin, and β-carotene.

α-Tocopherol had an higher IC_{50} value than all tested carotenoids. However, in the modified test system of Woodall *et al.* (1997), zeaxanthin showed a much higher AC than lutein, and astaxanthin and canthaxanthin were the least effective. These differences may be due to the use of a heme-iron in case of Miki or the use of different solvents. It is well known that factors like solubility or steric hindrance that may be of great importance in one environment but not in the other can have influence on the AC of carotenoids (El-Agamey *et al.*, 2004a,b; Packer, 1993). Jorgensen and Skibsted (1993) investigated the antioxidant effect of astaxanthin, β-carotene, canthaxantin, and zeaxanthin against the peroxidation of methyl esters in different systems (i.e., metmyoglobin as a water-based free radical generator in a heterogeneous lipid/water system and AIBN as a free radical generator in a homogeneous chloroform solution). In case of the heterogeneous system, each of the carotenoids protected the methyl esters against oxidation and the antioxidative effect showed little dependence on the structure of the carotenoid. In case of the homogeneous solution, however, the stability of the carotenoids in the oxidizing system depended on the structure and the order of decreasing stability was shown to be astaxanthin > canthaxanthin > β-carotene > zeaxanthin. Another investigation of Mortensen *et al.* (2001) came to the conclusion that the reaction of carotenoids with α-tocopherol also depends on the environment (see later section).

Miller *et al.* (1996) assessed the relative antioxidant activities of a range of carotenoids and xanthophylls through the extent of their abilities to scavenge the ABTS radical cation. The order was shown to be lycopene > β-cryptoxanthin ≈ β-carotene > lutein ≈ zeaxanthin ≈ α-carotene > echinenone >> astaxanthin ≈ cantaxanthin. It is likely that the scavenging of the ABTS radical cation is due to the hydrogen-donating properties of the carotenoids. This would be in agreement with Woodall and coworkers who showed the same order of carotenoids in the reaction with AMVN. Therefore, the thesis of Woodall and coworkers that hydrogen abstraction at the allylic C-atoms of carotenoids may contribute to the AC of carotenoids seems most likely.

Interaction with Other Carotenoids or Antioxidants An important point regarding antioxidative capacity is the interaction between substances of different mode of action or polarity. For an overview see Truscott (2001).

Firstly, several carotenoids may show interactions with each other. Edge *et al.* (1998) tested the relative one-electron reduction potentials of different carotenoid radical cations by monitoring the quenching of one carotenoid radical cation by another carotenoid. The order was found to be astaxanthin > cantaxanthin > lutein > zeaxanthin > β-carotene > lycopene. This means that lycopene is the most easily oxidized carotenoid and is able to repair all other carotenoid radical cations. Mortensen *et al.* (2001) investigated the oxidation potentials of different carotenoids in Triton X-100 micelles and found that lycopene is the easiest carotenoid to be oxidized to its radical cation and astaxanthin is the most difficult, which is in agreement with the results from El-Agamey et al. (2004a). Stahl et al. (1998) proved that mixtures of carotenoids are more effective in protecting liposomes against oxidative damage and found lycopene and lutein to be the most potent carotenoids in this respect.

There are several indications that carotenoids also interact with α-tocopherol. β-Carotene markedly delayed the AIBN-induced loss of endogenous microsomal tocopherols (Palozza and Krinsky, 1991), caused a synergistic inhibition of LPO in combination with α-tocopherol (Palozza and Krinsky, 1992; Toyosaki, 2002), enhanced α-tocopherol antioxidant activity as shown by pulse radiolysis and laser flash photolysis studies (Böhm et al., 1997), and protected LDL from oxidation in cooperation with α-tocopherol (Packer, 1993). Böhm et al. (1997) proved that α-tocopheroxyl radicals react rapidly with β-carotene with near diffusion controlled rate constants in hexane. The reactivity of carotenoids towards α-tocopherol is influenced by the environment. Mortensen et al. (2001) found that in polar environment the α-tocopherol cation is deprotonated, and the deprotonated cation does not react with carotenoids, whereas in a nonpolar environment like hexane the protonated α-tocopherol radical cation is converted to tocopherol by carotenoids.

Besides interaction with other carotenoids or α-tocopherol, carotenoids may also interact with more hydrophilic antioxidants. The most common antioxidant in this context is ascorbic acid. Böhm et al. (1997) proved that ascorbic acid can repair carotenoid radical cations in methanol. This can be explained by the more polar nature of carotenoid radical cations. Thus, they may reorientate in biological membranes so that the charge is near the polar interface and becomes accessible to ascorbic acid (Edge et al., 1997; Krinsky and Yeum, 2003). This could also explain the adverse effects observed in several trials where β-carotene did not protect smokers (who in general do have lower blood ascorbic acid levels) from lung cancer but provided protection for nonsmokers.

Yet, with other hydrophilic antioxidants interactions can also take place. Trombino et al. (2004) investigated the antioxidant effect of ferulic acid, alone and in combination with β-carotene, in isolated membranes and intact cells and revealed synergistic interactions. Milde and coworkers similarly could show that LDL is protected in a synergistical manner by rutin and lycopene or lutein, respectively (unpublished results). These synergistic effects can be explained by a different location of hydrophilic and lipophilic antioxidants in membranes, cells, or LDL, respectively.

3. Conclusions

Carotenoids perform important biological functions in plants as well as in humans. In plants, their main functions are photoprotection, light harvesting in photosynthesis, and pigmentation. For man, the best known function is that of vitamin A, which is essential for vision. However, it has been speculated that carotenoids are at least in part responsible for the health promoting properties of fruits and vegetables. Besides their influence on human immune function and GJC, they possess remarkable antioxidative properties. The abilities of quenching singlet oxygen and reacting with a variety of radical species may help to reduce oxidative stress in humans and thereby protect them from various diseases like CVD or cancer. Many *in vitro* investigations proved the AC of carotenoids; however, there is still very limited knowledge about the extent to which carotenoids act as antioxidants *in vivo*. An important point to keep in mind is the interaction of carotenoids with other antioxidants, especially in the light of the fact that in vegetal foodstuff combinations always occur.

REFERENCES

Abdalla, A. E., and Roozen, J. P. (1999). Effect of plant extracts on the oxidative stability of sunflower oil and emulsion. *Food Chem.* **64,** 323–329.

Aeschbach, R., Löliger, J., Scott, B. C., Murcia, A., Butler, J., Halliwell, B., and Aruoma, O. I. (1994). Antioxidant actions of thymol, carvacrol, 6-gingerol, zingerone and hydroxytyrosol. *Food Chem. Toxicol.* **32,** 31–36.

Agarwal, S., and Rao, A. V. (1998). Tomato lycopene and low density lipoprotein oxidation: A human dietary intervention study. *Lipids* **33,** 981–984.

Alves-Rodrigues, A., and Shao, A. (2004). The science behind lutein. *Toxicol. Lett.* **150**(1), 57–83.

Asard, H., May, J. M., and Smirnoff, N. (2004). "Vitamin C: Functions and Biochemistry in Animals and Plants." Garland Sciences/BIOS Scientific Publishers, Oxon, New York.

Aust, O., Ale-Agha, N., Zhang, L., Wollersen, H., Sies, H., and Stahl, W. (2003). Lycopene oxidation product enhances gap junctional communication. *Food Chem. Toxicol.* **41**(10), 1399–1407.

Baratta, M. T., Dorman, H. J. D., and Deans, S. G. (1998a). Chemical composition, antimicrobial and antioxidative activity of laure, sage, rosemary, oregano and coriander essential oils. *J. Essent. Oil Res.* **10,** 618–627.

Baratta, M. T., Dorman, H. J., Deans, S. G., Figueiredo, A. C., Barroso, J. G., and Ruberto, G. (1998b). Antimicrobial and antioxidant properties of some commercial essential oils. *Flavour Fragr. J.* **13,** 235–244.

Bartley, G. E., and Scolnik, P. A. (1995). Plant carotenoids: Pigments for photoprotection, visual attraction, and human health. *Plant Cell* **7,** 1027–1038.

Böhm, F., Edge, R., Land, E. J., McGarvey, D. J., and Truscott, T. G. (1997). Carotenoids enhance vitamin E antioxidant efficiency. *J. Am. Chem. Soc.* **119,** 621–622.

Bramley, P. M., Elmadfa, I., Kafatos, A., Kelly, F. J., Manios, Y., Roxborough, H. E., Schuch, W., Sheehy, P. J. A., and Wagnerm, K. H. (2000). Vitamin E. *J. Sci. Food Agric.* **80,** 913–938.

Bruneton, J. (1999). "Pharmacognosy, Phytochemistry, Medicinal Plants," 2nd ed. Lavoisier Publishing, Paris.

Burits, M., Asres, K., and Bucar, F. (2001). The antioxidant activity of the essential oils of *Artemisia afra*, *Artemisia abyssinica* and *Juniperus procera*. *Phytother. Res.* **15,** 103–108.

Burton, G. W., and Ingold, K. U. (1984). beta-Carotene: An unusual type of lipid antioxidant. *Science* **224,** 569–573.

Burton, G. W. (1989). Antioxidant action of carotenoids. *J. Nutr.* **119**, 109–111.
Castenmiller, J. J. M., and West, C. E. (1998). Bioavailability and bioconversion of carotenoids. *Annu. Rev. Nutr.* **18**, 19–38.
Chisolm, G. M., and Steinberg, D. (2000). The oxidative modification hypothesis of atherogenesis: An overview. *Free Radic. Biol. Med.* **28**, 1815–1826.
Choi, H.-S., Song, H.-S., Ukeda, H., and Sawamura, M. (2000). Radical-scavenging activities of citrus essential oils and their components: Detection using 1,1-diphenyl-2-picrylhydrazyl. *J. Agric. Food Chem.* **48**, 4156–4161.
Choudhury, N. K., and Behera, R. K. (2001). Photoinhibition of photosynthesis: Role of carotenoids in photoprotection of chloroplast constituents. *Photosynthetica* **39**(4), 481–488.
Chopra, M., Willson, R. L., and Thurnham, D. L. (1993). Free radical scavenging of lutein in vitro. *Ann. N.Y. Acad. Sci.* **691**, 246–249.
Clarkson, P. M., and Thompson, H. S. (2000). Antioxidants: What role do they play in physical activity and health? *Am. J. Clin. Nutr.* **72**(Suppl.), 637S–646S.
Conn, P. F., Schalch, W., and Truscott, T. G. (1991). The singlet oxygen and carotenoid interactions. *J. Photochem. Photobiol. B* **11**(1), 41–47.
Crowell, P. L. (1997). Monoterpenes in breast cancer chemoprevention. *Breast Cancer Res. Treat.* **46**(2–3), 191–197.
Crowell, P. L. (1999). Prevention and therapy of cancer by dietary monoterpenes. *J. Nutr.* **129**, 775S–778S.
Dang, M. N., Takácsová, M., Nguyen, D. V., and Kristiánová, K. (2001). Antioxidant activity of essential oils from various spices. *Nahrung/Food* **45**, 64–66.
Dapkevicius, A., Venskutonis, R., van Beek, T. A., and Linssen, J. P. H. (1998). Antioxidant activity of extracts obtained by different isolation procedures from some aromatic herbs grown in Lithuania. *J. Sci. Food Agric.* **77**, 140–146.
Davey, M. W., Montagu, M. V., Inzé, D., Sanmartin, M., Kanellis, A., Smirnoff, N., Benzie, I. J. J., Strain, J. J., Favell, D., and Fletscher, J. (2000). Plant L-ascorbic acid: Chemistry, function, metabolism, bioavailability and effects of processing. *J. Sci. Food Agric.* **80**, 825–860.
Demming-Adams, B., and Adams, W. W. (1996). The role of xanthophyll cycle carotenoids in the protection of photosynthesis. *TIPS* **1**, 21–26.
De Sousa, A. C., Alviano, D. S., Blank, A. F., Alves, P. B., Alviano, C. S., and Gattas, C. R. (2004). *Mellisa officinalis* L. essential oil: Antitumoral and antioxidant activities. *J. Pharm. Pharmacol.* **56**(5), 677–681.
Di Mascio, P., Kaiser, S., and Sies, H. (1989). Lycopene as the most efficient biological carotenoid singlet oxygen quencher. *Arch. Biochem. Biophys.* **2**, 532–538.
Di Mascio, P., Sundquist, A. R., Devasagayam, P. A., and Sies, H. (1992). Assay of lycopene and other carotenoids as singlet oxygen quenchers. *Meth. Enzymol.* **213**, 429–438.
Dröge, W. (2002). Free radicals in the physiological control of cell function. *Physiol. Rev.* **82**, 47–95.
Dudareve, N., Pichersky, E., and Gershenzon, J. (2004). Biochemistry of plant volatiles. *Plant Physiol.* **135**, 1893–1902.
Edge, R., McGarvey, D. J., and Truscott, T. G. (1997). The carotenoids as anti-oxidants—a review. *J. Photochem. Photobiol. B* **41**, 189–200.
Edge, R., Land, E. J., McGarvey, D. J., Mulroy, L., and Truscott, T. G. (1998). Relative one-electron reduction potentials of carotenoid radical cations and interactions of carotenoids with vitamin E radical cation. *J. Am. Chem. Soc.* **120**, 4087–4090.
El-Agamey, A., Cantrell, A., Land, E. J., Garvey, D. J., and Truscott, T. G. (2004a). Are dietary carotenoids beneficial? Reactions of carotenoids with oxy-radicals and singlet oxygen. *Photochem. Photobiol.* **3**, 802–811.
El-Agamey, A., Lowe, G. M., McGarvey, D. J., Mortensen, A., Phillip, D. M., Truscott, T. G., and Young, A. J. (2004b). Carotenoid radical chemistry and antioxidant/pro-oxidant properties. *Arch. Biochem. Biophys.* **430**, 37–48.

Evans, T. R., and Kaye, S. B. (1999). Retinoids: Present role and future potential. *Br. J. Cancer* **80**(1–2), 1–8.

Everett, S. A., Dennis, M. F., and Patel, K. B. (1996). Scavenging of nitrogen dioxide, thiyl, and sulfonyl free radicals by the nutritional antioxidant β-carotene. *J. Biol. Chem.* **271**, 3988–3994.

Fadel, H., Marx, F., El-Sawy, A., and El-Ghorab, A. (1999). Effect of extraction techniques on the chemical composition and antioxidant activity of *Eucalyptus camaldulensis* var. *brevirostris* leaf oils. *Z. Lebensm. Unters. Forsch A* **208**, 212–216.

Foote, C. S., and Denny, R. W. (1968). Chemistry of singlet oxygen. VII. Quenching by β-carotene. *J. Am. Chem. Soc.* **90**, 6233–6235.

Foti, M. C., and Ingold, K. U. (2003). Mechanism of inhibition of lipid peroxidation by gamma-terpinene, an unusual and potentially useful hydrocarbon antioxidant. *J. Agric. Food Chem.* **51**(9), 2758–2765.

Fridovich, I. (1986). Superoxide dismutases. *Meth. Enzymol.* **58**, 61–97.

Fuhrman, B., Volkova, N., Rosenblat, M., and Aviram, M. (2000). Lycopene synergistically inhibits LDL oxidation in combination with vitamin E, glabridin, rosmarinic acid, carnosic acid, or garlic. *Antioxid. Redox. Signal.* **2**, 491–506.

Gaziano, J. M., Hatta, A., Flynn, M., Johnson, E. J., Krinsky, N. I., Ridker, P. M., Hennekens, C. H., and Frei, B. (1995). Supplementation with beta-carotene in vivo and in vitro does not inhibit low density lipoprotein oxidation. *Atherosclerosis* **112**(2), 187–195.

Goldberg, G.(Ed.) (2003). "Plants: Diet and Health." Blackwell Publishing. Oxford.

Gould, M. N. (1997). Cancer chemoprevention and therapy by monoterpenes. *Environ. Health Perspect.* **105**(Suppl. 4), 977–979.

Graßmann, J., Hippeli, S., Dornisch, K., Rohnert, U., Beuscher, N., and Elstner, E. F. (2000). Antioxidant properties of essential oils: Possible explanation for their anti-inflammatory effects. *Arzneimitt. Forsch/Drug Res.* **50**, 135–139.

Graßmann, J., Schneider, D., Weiser, D., and Elstner, E. F. (2001). Antioxidative effects of lemon oil and its components on copper induced oxidation of low density lipoprotein. *Arzneimitt. Forsch/Drug Res.* **51**, 799–805.

Graßmann, J., Hippeli, S., Vollmann, R., and Elstner, E. F. (2003). Antioxidative properties of the essential oil from *Pinus mugo*. *J. Agric. Food Chem.* **51**, 7576–7582.

Grosch, W., Laskawy, G., and Weber, F. (1976). Formation of volatile carbonyl compounds and cooxidation of β-carotene by lipoxygenase from wheat, potato, flax and beans. *J. Agric. Food Chem.* **24**, 456–459.

Halliwell, B. (1996). Antioxidants in human health and disease. *Ann. Rev. Nutr.* **16**, 33–50.

Ham, W. T., Jr. (1983). Ocular hazards of light sources. Review of current knowledge. *J. Occup. Med.* **25**(2), 101–103.

Handelman, G. J., van Kuijk, F. J., Chatterjee, A., and Krinsky, N. I. (1991). Characterization of products formed during the autoxidation of beta-carotene. *Free Radic. Biol. Med.* **10**(6), 427–437.

Harborne, J. B., and Williams, C. A. (2000). Advances in flavonoid research since 1992. *Phytochemistry* **55**, 481–504.

Hercberg, S., Galan, P., Preziosi, P., Alfarez, M.-J., and Vazquez, C. (1998). The potential role of antioxidant vitamins in preventing cardiovascular diseases and cancer. *Nutrition* **14**(6), 513–520.

Hopia, A. I., Huang, S.-W., Schwarz, K., German, J. B., and Frankel, E. N. (1996). Effect of different lipid systems on antioxidant activity of rosemary constituents carnosol and carnosic acid with and without α-tocopherol. *J. Agric. Food Chem.* **44**, 2030–2036.

Hughes, D. A. (1999). Effect of carotenoids on human immune function. *Proc. Nutr. Soc.* **58**(3), 713–718.

Jialal, I., and Grundy, S. M. (1993). Effect of combined supplementation with alpha-tocopherol, ascorbate, and beta-carotene on low-density lipoprotein oxidation. *Circulation* **88**(6), 2780–2786.

Jialal, I., and Fuller, C. J. (1995). Effect of vitamin E, vitamin C and beta-carotene on LDL oxidation and atherosclerosis. *Can. J. Cardiol.* **11**(Suppl. G), 97G–103G.
Johnson, E. J. (2002). The role of carotenoids in human health. *Nutr. Clin. Care* **5**, 56–65.
Jorgensen, K., and Skibsted, L. H. (1993). Carotenoid scavenging of radicals. Effect of carotenoid structure and oxygen partial pressure on antioxidative activity. *Z. Lebensm. Unters. Forsch.* **196**(5), 423–429.
Kaur, C., and Kapoor, H. C. (2001). Antioxidants in fruits and vegetables—the milleniums health. *Int. J. Food Sci. Technol.* **36**, 703–725.
Kennedy, T. A., and Liebler, D. C. (1991). Peroxy radical oxidation of beta-carotene: Formation of beta-carotene epoxides. *Chem. Res. Toxicol.* **4**(3), 290–295.
Kennedy, T. A., and Liebler, D. C. (1992). Peroxyl radical scavenging by beta-carotene in lipid bilayers. Effect of oxygen partial pressure. *J. Biol. Chem.* **267**(7), 4658–4663.
Kim, H.-J., Chen, F., Wu, C., Wang, X., Chung, H. J., and Jin, Z. (2004). Evaluation of antioxidant activity of Australian tea tree (*Melaleuca alternafolia*) oil and its components. *J. Agric. Food Chem.* **52**, 2849–2854.
Kim, D. O., and Lee, C. Y. (2004). Comprehensive study on vitamin C equivalent antioxidant capacity (VCEAC) of various polyphenolics in scavenging a free radical and its structural relationship. *Crit. Rev. Food Sci. Nutr.* **44**, 253–273.
Kreuzwieser, J., Schnitzler, J.-P., and Steinbrecher, R. (1999). Biosynthesis of organic compounds emitted by plants. *Plant Biol.* **1**, 149–159.
Krinsky, N. I. (1998). The antioxidant and biological properties of the carotenoids. *Ann. N.Y. Acad. Sci.* **854**, 443–447.
Krinsky, N. I., and Yeum, K.-J. (2003). Carotenoid–radical interactions. *Biochem. Biophys. Res. Commun.* **305**, 754–760.
Kris-Etherton, P. M., Hecker, K. D., Bonanome, A., Coval, S. M., Binkoski, A. E., Hilpert, K. F., Griel, A. E., and Etherton, T. D. (2002). Bioactive compounds in foods: Their role in the prevention of cardiovascular disease and cancer. *Am. J. Med.* **113**(9B), 71S–88S.
Lichtenthaler, H. K. (1999). The 1-deoxy-D-xylulose-5-phosphate pathway of isoprenoid biosynthesis in plants. *Annu. Rev. Plant Physiol. Plant Mol. Biol.* **50**, 47–65.
Liebler, D. C. (1993). Antioxidant reactions of carotenoids. *Ann. N.Y. Acad. Sci.* **691**, 20–31.
Liu, R. H. (2003). Health benefits of fruits and vegetables are from additive and synergistic combinations of phytochemicals. *Am. J. Clin. Nutr.* **78**(Suppl.), 17S–20S.
Loza-Tavera, H. (1999). Monoterpenes in essential oils: Biosynthesis and properties. *Adv. Exp. Med. Biol.* **464**, 49–62.
Mantle, D., Anderton, J. G., Falkous, G., Barnes, M., Jones, P., and Perry, E. K. (1998). Comparison of method for determination of total antioxidant status: Application to analysis of medicinal plant essential oils. *Comp. Biochem. Physiol. B* **121**, 385–391.
Mares-Perlman, J. A., Millen, A. E., Ficek, T. L., and Hankinson, S. E. (2002). The body of evidence to support a protective role for lutein and zeaxanthin in delaying chronic disease. Overview. *J. Nutr.* **132**, 518S–524S.
Miki, W. (1991). Biological functions and activities of animal carotenoids. *Pure Appl. Chem.* **63**, 141–146.
Milde, J., Elstner, E. F., and Graßmann, J. (2004). Synergistic inhibition of LDL-oxidation by rutin, γ-terpinene and ascorbic acid. *Phytomedicine* **11**, 105–113.
Miller, N. J., Sampson, J., Candeias, L. P., Bramley, P. M., and Rice-Evans, C. A. (1996). Antioxidant activities of carotenes and xanthophylls. *FEBS Lett.* **384**, 240–242.
Mimica-Dukic, N., Bozin, B., Sokovic, M., Mihajlovic, B., and Matavulj, M. (2003). Antimicrobial and antioxidant activities of three mentha species essential oils. *Planta Med.* **69**, 413–419.
Mimica-Dukic, N., Bozin, B., Sokovic, M., and Simin, N. (2004). Antimicrobial and antioxidant activities of *Melissa officinalis* L. (Lamiaceae) essential oil. *J. Agric. Food Chem.* **52**, 2485–2489.

Miura, K., Kikuzaki, H., and Nakatani, N. (2002). Antioxidant activity of chemical components from sage (*Salvia officinalis* L.) and thyme (*Thymus vulgaris* L.) measured by the oil stability index method. *J. Agric. Food Chem.* **50,** 1845–1851.

Mortensen, A., and Skibsted, L. H. (1996). Kinetics of parallel electron transfer from beta-carotene to phenoxyl radical and adduct formation between phenoxyl radical and beta-carotene. *Free Radic. Res.* **25**(6), 515–523.

Mortensen, A., Skibsted, L. H., and Truscott, T. G. (2001). The interaction of dietary carotenoids with radical species. *Arch. Biochem. Biophys.* **385**(1), 13–19.

Okuno, M., Kojima, S., Matsushima-Nishiwaki, R., Tsurumi, H., Muto, Y., Friedman, S. L., and Moriwaki, H. (2004). Retinoids in cancer chemoprevention. *Curr. Cancer Drug Targets* **4**(3), 285–298.

Packer, L. (1993). Antioxidant action of carotenoids in vitro and in vivo and protection against oxidation of human low-density lipoproteins. *Ann. N.Y. Acad. Sci.* **691,** 48–60.

Palace, V. P., Khaper, N., Qin, Q., and Singal, P. K. (1999). Antioxidant potentials of vitamin A and carotenoids and their relevance to heart disease. *Free Radic. Biol. Med.* **26,** 746–761.

Palozza, P., and Krinsky, N. I. (1991). The inhibition of radical-initiated peroxidation of microsomal lipids by both alpha-tocopherol and beta-carotene. *Free Radic. Biol. Med.* **11**(4), 407–414.

Palozza, P., and Krinsky, N. I. (1992). Antioxidant effects of carotenoids *in vivo* and *in vitro*: An overview. *Meth. Enzymol.* **213**403–420.

Paré, P. W., and Tumlinson, J. H. (1999). Plant volatiles as a defense against insect herbivores. *Plant Physiol.* **121**325–331.

Parr, A. J., and Bolwell, G. P. (2000). Phenols in plant and man. The potential for possible nutritional enhancement of the diet by modifying the phenols content or profile. *J. Sci. Food Agric.* **80**985–1012.

Pietta, P.-G. (2000). Flavonoids as antioxidants. *J. Nat. Prod.* **63**1035–1042.

Princen, H. M., van Poppel, G., Vogelezang, C., Buytenhek, R., and Kok, F. J. (1992). Supplementation with vitamin E but not beta-carotene *in vivo* protects low density lipoprotein from lipid peroxidation *in vitro*. Effect of cigarette smoking. *Arterioscler. Thromb.* **12**(5), 554–562.

Reaven, P. D., Khouw, A., Beltz, W. F., Parthasarathy, S., and Witztum, J. L. (1993). Effect of dietary antioxidant combination in humans. Protection of LDL by vitamin E but not by beta-carotene. *Arterioscler. Thromb.* **13**(4), 590–600.

Rice-Evans, C. A., Sampson, J., Bramley, P. M., and Holloway, D. E. (1997). Why do we expect carotenoids to be antioxidants *in vivo*? *Free Radic. Res.* **26**381–398.

Richheimer, S. L., Bernart, M. W., King, G. A., Kent, M. C., and Bailey, D. T. (1996). Antioxidant activity of lipid soluble phenolic diterpenes from rosemary. *J. Am. Oil Chem. Soc.* **73**507–514.

Rodríguez-Concepción, M., and Boronat, A. (2002). Elucidation of the methylerythritol phosphate pathway for isoprenoid biosynthesis in bacteria and plastids. A metabolic milestone achieved through genomics. *Plant Physiol.* **130**1079–1089.

Sacchetti, G., Medici, A., Maietti, S., Radice, M., Muzzoli, M., Manfredini, S., Braccioli, E., and Bruni, R. (2004). Composition and functional properties of the essential oil of Amazonian Basil, Ocimum micranthum Willd., Labiatae in comparison with commercial essential oils. *J. Agric. Food Chem.* **52**3486–3491.

Schempp, H., Hippeli, S., and Elstner, E. F. (2005). Plant stress: Avoidance, adaptation, defense. *In* "Plant Toxicology" (B. Hock and E. F. Elstner, Eds.), pp. 87–129. Marcel Dekker, New York.

Schwarz, K., Ternes, W., and Schmauderer, E. (1992). Antioxidant constituents of *Rosmarinus officinalis* and *Salvia officinalis*. *Z. Lebensm. Unters. Forsch.* **195**104–107.

Schwarz, K., Ernst, H., and Ternes, W. (1996). Evaluation of antioxidant constituents of thyme. *J. Sci. Food Agric.* **70**217–223.

Siems, W. G., Sommerburg, O., and van Kuijk, F. J. G. M. (1999). Lycopene and β-carotene decompose more rapidly than lutein and zeaxanthin upon exposure to various pro-oxidants in vitro. *BioFactors* **10**105–113.

Siems, W., Sommernurg, O., Hurst, J. S., and van Kuijk, F. J. (2000). Carotenoid oxidative degradation products inhibit Na^+-K^+-ATPase. *Free Radic. Res.* **33**(4), 427–435.

Siems, W., Sommerburg, O., Schild, L., Augustin, W., Langhans, C. D., and Wiswedel, I. (2002). Beta carotene cleavage products induce oxidative stress in vitro by impairing mitochondrial respiration. *FASEB J.* **16**(10), 1289–1291.

Singsaas, E. L., Lerdau, M., Winter, K., and Sharkey, T. D. (1997). Isoprene increases thermotolerance of isoprene-emitting species. *Plant Physiol.* **115**1413–1420.

Stahl, W., Nicolai, S., Briviba, K., Hanusch, M., Broszeit, G., Peters, M., Martin, H. D., and Sies, H. (1997). Biological activities of natural and synthetic carotenoids: Induction of gap junctional communication and singlet oxygen quenching. *Carcinogenesis* **18**89–92.

Stahl, W., Junghans, A., de Boer, B., Driomina, E. S., Briviba, K., and Sies, H. (1998). Carotenoid mixtures protect multilamellar liposomes against oxidative damage: Synergistic effects of lycopene and lutein. *FEBS Lett.* **427**(2), 305–308.

Stahl, W., Ale-Agha, N., and Polidori, M. C. (2002). Non-antioxidant properties of carotenoids. *Biol. Chem.* **383**553–558.

Stanner, S. A., Hughes, J., Kelly, C. N. M., and Buttriss, J. (2004).A review of the epidemiological evidence for the 'antioxidant hypothesis.' *Public Health Nutr.* **7**(3), 407–422.

Stratton, S. P., Schaefer, W. H., and Liebler, D. C. (1993). Isolation and identification of singlet oxygen oxidation products of beta-carotene. *Chem. Res. Toxicol.* **6**(4), 542–547.

Takahashi, Y., Inaba, N., Kuwahara, S., and Kuki, W. (2003). Antioxidative effect of citrus essential oil components on human low-density lipoprotein in vitro. *Biosci. Biotechnol. Biochem.* **67**(1), 195–197.

Tapiero, H., Townsend, D. M., and Tew, K. D. (2004). The role of carotenoids in the prevention of human pathologies. *Biomed. Pharm.* **58**100–110.

Teissedre, P. L., and Waterhouse, A. L. (2000). Inhibition of oxidation of human low-density lipoproteins by phenolic substances in different essential oils varieties. *J. Agric. Food Chem.* **48**3801–3805.

Toyosaki, T. (2002). Antioxidant effect of β-carotene on lipid peroxidation and synergism with tocopherol in an emulsified linoleic acid system. *Int. J. Food Sci. Nutr.* **53**419–423.

Trombino, S., Serini, S., Di Nicuolo, F., Celleno, L., Ando, S., Picci, N., Calviello, G., and Palozza, P. (2004). Antioxidant effect of ferulic acid in isolated membranes and intact cells: Synergistic interactions with alpha-tocopherol, beta-carotene, and ascorbic acid. *J. Agric. Food Chem.* **52**(8), 2411–2420.

Truscott, T. G. (2001). Synergistic effects of antioxidant vitamins. *Bibl. Nutr. Dieta* **55**68–79.

Vardar-Ünlü, G., Candan, F., Sökmen, A., Daferera, D., Polissiou, M., Sökmen, M., Dönmez, E., and Tepe, B. (2003). Antimicrobial and antioxidant activity of the essential oil and methanol extracts of *Thymus pectinatus* Fisch. Et Mey. Var. *pectinatus* (Lamiaceae). *J. Agric. Food Chem.* **51**63–67.

Van den Berg, H., Faulks, R., Fernando Granado, H., Hirschberg, J., Olmedilla, B., Sandmann, G., Southon, S., and Stahl, W. (2000). The potential for the improvement of carotenoid levels in foods and the likely systemic effects. *J. Sci. Food Agric.* **80**880–912.

Vendemiale, G., Grattagliano, I., and Altomare, E. (1999). An update on the role of free radicals and antioxidant defense in human disease. *Int. J. Clin. Lab. Res.* **29**49–55.

Verhoeven, A. S., Demming-Adams, B., and Adams, W. W. (1997). Enhanced employment of the xanthophyll cycle and thermal energy dissipation in spinach exposed to high light and N stress. *Plant Physiol.* **113**817–824.

Wagner, K.-H., and Elmadfa, I. (2003). Biological relevance of terpenoids—overview focusing on mono-, di- and tetraterpenes. *Ann. Nutr. Metab.* **47**95–106.

Wink, M., and Latz-Brüning, B. (1995). Allelopathic properties of alkaloids and other natural products: Possible mode of action. *In* "Organisms, Processes and Applications" (K. M. Inderjit, M. Daksshini, and F. A. Einhellig, Eds.), Vol. 582pp. 117–126. Am. Chem. Soc. Symp., Washington DC.

Woodall, A. A., Wai-Min, Lee, S., Weesie, R., Jackson, M. J., and Britton, G. (1997). Oxidation of carotenoids by free radicals: Relationship between structure and reactivity. *Biochim. Biophys. Acta* **1336**33–42.

Yamasaki, H., Mesnil, M., Omori, Y., Mironov, N., and Krutovskikh, V. (1995). Intercellular communication and carcinogenesis. *Mutat. Res.* **333**(1–2), 181–188.

Youdim, K. A., Dorman, H. J., and Deans, S. G. (1999). The antioxidant effectiveness of thyme oil, α-tocopherol and ascorbyl palmitate an evening primrose oil oxidation. *J. Essent. Oil Res.* **11**643–648.

INDEX

Page numbers followed by f and t indicate figures and tables, respectively.

A

Abscisic acid (ABA), 2, 236
  role of, in root gravitropism, 58
Abscisic acid regulation, 256
ABA biosynthesis, 180, 323, 373
ABA protein conjugates, 239
ABA response functional proteins, 5, 7t
  pathogen-related protein genes, 7t
  storage proteins, 5, 7t
  stress-responsive genes, 7t
ABA response regulatory proteins, 8t, 10
  protein-degradation gene, 8t
  protein kinase genes/protein phosphatase genes, 8t, 11, 18t
  transcriptional factors and DNA-binding protein, 8t, 18t
ABA secondary messengers, 240, 257
ABA signal pathway, 17
ABA signaling, 249–250t, 256
ABA up/GA down genes, 18t, 21–22t
ABA/GA-responsive gene, 6t
ABA-dependent signal pathways, 5
ABA-inducible gene expression, 17, 20, 249
ABA-inducible genes, 5
ABA-responsive gene expression, 249
ABA-responsive gene, 5, 6t, 10, 17
ABA-signaling network, 11, 237–238f
ABC proteins, 168
Abiotic stresses, 422
ABRELATERD1, *see cis*-elements for dehydration-stress response
Abscission zone, 225
ACC synthase polypeptides, 401, 403–405, 415, 419
ACGTATERDI, *see cis*-elements for dehydration-stress response
Acid synthase 1-aminocyclopropane-1-carboxylic, 401, 403
Actin-dependent-vesicular cycling, 43
Adventitious rooting, 223
Agravitropic mutants, 51
Agrobacterium tumefaciens, 277
Alpha-amylase gene promoters, 20
Alternative splicing, 195
Amino acid transporters, 41
Amyloplasts, 34, 52
Animal attraction, 520
Antagonist, 280, 405, 415
Anthocyanin, 173
Antioxidant(s), 368
  activity, 522, 524
  capacity, 506, 513–514t, 525
  properties, 522
Arabidopsis, 3–4, 6t, 11, 13, 16–17, 20, 56, 63, 85, 116, 120, 127, 140–142, 159, 169, 173, 177, 186–188, 197, 213, 220, 240, 256, 278, 280, 294, 296–298, 300–301, 304–305, 311, 401, 403, 408, 410, 413, 415, 417, 432, 435–436, 444–446

Arabidopsis (continued)
 cell cultures, 183
 cell death, 354, 357
 genome sequence, 16
 leaves, 183, 192, 206, 342–343, 346, 348, 354, 374, 418
 MDR1 protein, 214
 roots, 38, 52
 seed(s), 17, 300, 306, 310, 367, 405, 414, 420, 484
 stamens, 448
Arabidopsis mutants, 36, 112, 127, 175, 177, 207, 215, 221, 224, 239, 292, 302, 325–326, 348, 356, 359, 361, 366–367, 370, 376, 382–383, 405, 415, 421, 437, 471, 480–481, 487
 lacs6, 128
 lacs7, 128
 aim1, 129
 ped1/kat2, 130
 icl, 130
 ms, 131
 sat, 132
 aoat1, 133
 opr3/*Delayed* Dehiscence1, 133
 chy1/dbr5, 134
 pxa1/ped3/cts, 135
 YUCCA, 207
Arabidopsis thaliana, 157, 272, 316–317, 401, 487, 490, 495
Ascorbic acid, 507, 528
AtIPT3, 278–279, 281, *see also* Isopentenyl transferase (IPT) enzyme
ATP-binding cassette transporter proteins (ABC), 168, 206
Aux/IAA genes, 218
Aux/IAA protein, 218, 220, 449, 450
AUX1 protein, 40, 41, 51, 213, 221, 416
Auxin, 32, 36, 119, 173, 175, 177, 179, 204, 209–210, 222, 224, 325, 402, 416, 486
 asymmetry, 36, 39
 bioassay, 206
 conjugates, 205
 detection methods, 37
 efflux carrier gene, *PIN*, 213
 homeostasis, 208
 influx carrier gene, *AUX1*, 40, 65, 213
 polar transport, 210–211, 214–215, 224
 signaling, 449
 transport inhibitor(s), 40, 43, 49, 63, 120, 173, 215, 221, 416, 419
Auxin resistant1 (AXR1), 64

Auxin-response signaling pathway mutants, 45, 179
 affecting auxin-responsive genes, 45
 affecting protein degradation, 45
Auxin/indoleacetic acid (Aux/IAA), 178
Auxin-binding protein 1 (ABP1), 216
Auxin-induced action in the gravirespose, 47
 differential expression of K^+-Channels, 47, 60
 secretion of wall-loosening factors, 47
 pH changes, 47
 participation of the actin cytoskeleton, 47
 signaling elements associated with oxidation, 48
 phospholipase A2 (PLA2), 48
 invertase, 48
Auxin-induced gene expression, 36, 38
Auxin-inducible gene reporter systems, 37
 DR5::GUS, 37, 42
 DR5::GFP, 37, 39
Auxin-mediated gene transcription, 217
Auxin-mediated signal transduction pathway, 64
Auxin receptors, 216
Auxin-regulated gene expression, 219*f*, 220
Auxin-response factor proteins (ARF), 219
Auxin-response promoter elements (AuxREs), 38, 178
Auxin transport, dynamics, 212

B

Bacterial artificial chromosome, 3
Bax-expression, 174
BCL-2 protein family, 174
Bending curvature of plants, 33
Bisexual flowers, 97–98*f*
Brassinosteroid (BR), 33, 63
 biosynthesis inhibitor Brz, 482–483, 491
 receptors, 486
Brassinosteroids biosynthesis, 480, 482*f*
Brassinosteroids metabolism, 480, 484
Brassinosteroids, signal transduction, 485, 487, 491–492, 495
 signaling genes, 496
Brefeldin A (BFA), 214
BRI1-associated kinase 1 (BAK1), 489

C

C_{40} carotenoid precursors, 180
Ca^{2+}, *see* ABA secondary messengers

Calcium-dependent protein
    kinases (CDPKs), 245
Callus, 224
Calossins, 213
Cambium, 210
Carboxyl-terminal domain (CTD), 249
carotene, β-, 236, 517, 520, 524–528
Carotenoids, 505, 517, 520, 522, 524–525
Catalyzed oxidation, 352–353
Catharanthus roseus cells, 480
Cell culture, 344, 359, 361, 364, 366, 377–378
Cell cycle, 210
    elongation, 367, 497
    enlargement, 13, 209, 216
Cell division, 13, 204, 209–210, 224,
        236, 494, 497
    sexual (SCD), 464–465, 468
CEV1 mutant, 185
Chloroplasts, 365, 382
Cholodny–Went hypothesis, 33, 38, 65, 223
Cholodny–Went model, 35, 49
cis-acting elements, 245
cis-element analysis in rice, 21
cis-elements of
    ABA-responsive genes, 17
    GA-responsive genes, 20
cis-elements for
    dehydration stress response, 17–18*t*
    protein storage, 17–18*t*, 21
Citrus oils, 513
Closterium, 463, 465
Cross-talk
    hormone, 3, 5, 16, 33, 52, 61–62, 65, 183,
        186, 188, 191, 193, 257, 323–326,
        385, 418–419, 423, 440, 480
    genes, 10, 13
CTR1, 411–412
Cystemine, 490
Cytokinin biosynthesis, 272, 280
Cytokinins (CKs), 209, 222, 272–273*f*,
        402–403, 486
    aromatic, 274
Cytokinins, role of, 65
    gravitropic response of shoots, 60
    root gravitropism, 62
Cytoplasm, 495, 508
Cytoskeleton, 43

D

de novo synthesis, 206, 275
DELLA protein, 312, 315, 317–319

Diamino fluorescein diacetate, 4, 5-, 341
Dichlorophenoxyacetic acid (2,4-D), 2,4-, 120
Dichlorophenoxybutyric acid (2,4-DB),
        2,4-, 120
Differential polyandenylation processing, 196
Diffusion constant (*D*), 351
Dioecious plants, 80–81
Diterpenes, 508, 511
DNA-binding proteins, 13
Dye fluoresces, 342

E

Ear inflorescence, 82–83
Ectocarpene, 459–460
EIN2, 412–415
Einstein–Smolochowski equations,
        351–352*f*
EIR1/PIN2/WAV6/AGR1, 51
Electron paramagnetic resonance
    spectroscopy (EPR), 342, 347
Embryonic patterning, 215
ent-copalyl-diphosphate synthase
        (CPS), 296
ent-kaurane oxidase (KO), 300
ent-kaurene synthase (KS), 298
ent-kaurenoic acid oxidase, 301
Epinasty, 222–223
ERAF16 genes, 93
ERAF17 genes, 93
ERF1, 420–421
Escherichia coli, 275, 305, 346, 494
Essential oils, 508, 511–514*t*
Ethylene biosynthesis, 90–91*f*, 187, 191,
        401–404, 415
Ethylene receptors, 91*f*, 93, 225, 406–411
    ETR1, 406, 408
    ETR2, 406, 408
    ERS1, 406, 408
    ERS2, 406, 408
    EIN4, 406, 408
Ethylene signaling, 191, 405, 407, 409, 411,
        413–414, 416, 418, 445
    response, 187, 189, 417
    signaling pathway, 187
Ethylene, 33, 50–51, 65
    biosynthetic pathway, 53
    burst, 53
    insensitive mutants, 54
    role of, in sex expression, 88, 96
Eukaryotic cells, 112
Expressed sequence tag (EST), 159

F

Fatty acid β-oxidation, 116–117, 120, 122, 129–130, 132, 135, 138, 142
F-box domain protein, 185, 217, 316–317
Fern pheromones, 469
Ferritin gene transcription, 366
Fertility, 446
Flavonoid
  biosynthesis, 172, 174
  levels, 44
  phytoalexins, 169
Flower(s), 80, 187, 213, 304, 369, 447 *see also* Plant sexual organs
Flowering, 294, 311, 316, 320, 322, 369, 470
Foreign organic compounds, 168
Fruit, 224 *see also* Plant reproductive growth
Full-length cDNAs, 17, 237
  sequences, 3
Fungal infection, 378

G

GA genes, 20
  downregulated, 20
  response regulatory, 13
  regulated, 20
  up/ABA down, 18*t*, 21–22*t*
GA/BR-responsive gene, 6*t*
GA-inducible transcription factor, 20
GA response functional protein, 12
  cell division, 12, 14*t*
  cell elongation, 12, 14*t*
  pathogen-related–protein genes, 12, 14*t*
  stress-responsive genes, 12, 14*t*
GA response pathways, 13
GA response regulatory proteins, 13
  protein kinase genes, 13, 15*t*, 18*t*
  protein phosphatase gene, 13, 15*t*, 18*t*
  signal transduction, 13, 15*t*
GA signaling network, 13
GA treatment, 13
Gamma-amino butyric acid (GABA), 471
Gap-junctional communication (GJC), 522
GA-response genes, 6*t*, 12, 20, 312–313*t*, 322
GA-response, negative regulators, 318
  SHI gene, 318
  SPY gene, 318
  SEC gene, 318
GA-response, positive regulators, 319
  D1, 319
  GAMYB, 319
  GID1, 320
  GSE1, 320

PHOR1, 320
PKL, 321
GA-signal transduction pathway, 312, 314*f*
Gene expression, 5, 10, 13, 20, 63, 119, 175, 177–179, 239, 413, 417–418, 436, 463–464, 483, 496, 511
Gene mutation, 41–43
Genetic regulation of sex expression in cucumber, 94
Genome sequences, 17, 157
Genome sequencing, 3
Genomic tools, 4
Germinating seeds, 116–117, 119, 122, 205–206, 306
Gibberellin acid (GA), 2, 33, 291
  13β-hydroxylase, 302
  20-oxidase, 303
  2-oxidase, 306
  3-oxidase, 305
Gibberellin acid (GA), 2
  role of, in root gravitropism, 56
  role of, in shoot differential growth, 57
Gibberellin biosynthesis, 83–84*f*, 295, 299*f*
  involvement of MVA pathways, 296
  involvement of MEP pathways, 296
Gibberellin deficiency, 488–489
Gibberellin metabolism regulation, by light, 309–310
Gibberellin metabolism, 308
  feedback regulation, 308–309
  feedforward regulation, 308–309
Gibberellin pathway, 249
Globin family, 352, 358, 361
Glucosinolates, 169, *see also* Mustard oil glucosides
Glutathione-conjugation products, 169
Glycoprotein, 463
Glyoxysomes, 122, 142
Glyoxylate cycle, 115–117, 119, 130–131, 142, 347
GNOM mutant plants, 214
G-proteins, 240
Gravistimulated plant organs, 37
Gravitropic bending of plants, 34, 39
Gravitropic curvatures, 49
  roots, 49
  shoots, 49
Gravitropic mutants, 34
Gravitropic pathway, 34
Gravitropic response mechanism, 34, 39–40, 50, 63
  root response, 41
  shoot response, 41

Gravitropic signal transduction, 55
Gravitropism, 32, 34, 35,
    40, 119, 221
  root, 51, 65
  shoot, 53, 65
Gravity-sensing cells, 35
Green revolution, 316
GSH conjugation activities of plant GSTs,
    161, 168, 170
GSH-dependent
  dehydroascorbate reductase
        activity, 171
  maleylacetone isomerase (MAI)
        activity, 171
  peroxidase (GPOX) activities,
        170–171, 184
  thioltransferase activity, 171
GSNO formation, 357
GST gene expression, posttranscriptional
    regulations
  plant growth regulators, 195, 196
GST gene expression, regulation of,
    157, 162t, 174
  role of auxin, 175
  role of ABA, 180
  role of SA, 182
  role of jasmonates, 184
  role of ethylene, 187
  role of NO, 190

H

H_2O_2-induced cell death, 174
Herbicide detoxification process, 168
Herbicide safeners, 168
Herbicides, 225
  auxinic, 225
  electrophilic, 168
Higher plants, 80–85, 91f, 92, 102–103,
    113–114, 116, 125, 127, 236, 278
  trans-zeatin biosynthesis, 279
Hormonal regulation of sex
    expression, 82
  in maize, 83
  in cucumber, 88, 91t–92
Hormone signal network, 21
Hormone-signaling pathway, 291
Human immune function, 520
HvGAMYB, *see* GA-inducible
    transcription factor
Hypoxia, 187, 340, 344, 348,
    353, 356–357, 400

I

Indole-3-acetic acid (IAA), 33, 119, 173, 205,
    207–208, 223, 225
  conjugate synthesis, 206
  radiolabeled, 38–39
Inorganic nitrogen, 281
Insect defense, 435
Integrated effects on GST gene expression
  antioxidants, 194
  phytohormones, 194
  ROI, 194
Iron metabolism, 361
Isoprenoids, *see* Terpenoids
Isopentenyl transferase (IPT) enzyme,
    87, 272, 275–280

J

Jasmonic acid (JA), 121, 419
JA biosynthesis, 185, 433–435
JA biosynthetic enzymes, 185
JA signaling, 186, 435–436,
    440, 444, 448–450
JA-responsive gene expression, 186
Jasmonates, 64

L

Late C-6 oxidation pathway, 480–481
Late-embryogenesis-abundant (LEA)
    proteins, 5, 16, 180
Light harvesting, 517
Light, 367
Lipid metabolism, 115, 122
  fatty acid β-oxidation, 115, 116
  glyoxylate cycle, 116, 117
Lipid peroxidation (LPO),
    513, 522, 528
Lipid transfer protein, 9t
Lolium temulentum, 294
  GA_5, 295
  GA_6, 295
Low-density lipoproteins, 513

M

MADS box-like protein, 13
MAPK, 412
MAPKK, 13
MAPK-like protein, 13
Massively parallel signature sequencing, 4
Metalloproteins, 357–358, 373, 384

Microarray analysis(es), 17, 132, 190, 192–193, 220, 294, 320, 324, 404, 418–419, 436, 448, 467, 483, 497
Mitochondrial respiration, 363
Model monocot rice, 157
Monoecious plants, 80, 82, 92, 96
MONOPTEROS gene codes, 215
Monoterpenes, 508, 510*f*, 511
Morphogen, 210
Multidrug-resistance (MDR) genes, 42
Mustard oil glucosides, 169
Mutual attraction of cells, 463
MYB proteins, 13

N

Naphthylphthalamic acid (NPA)-binding protein(s), 43–44, 49, 51,
Nitric oxide (NO), 355, 357, 364, 367–369, 376, 382–384
  biosynthesis, 190, 341, 345*t*, 349, 362, 368, 370–372, 382, 385
  emission, 343
Nitric oxide synthases (NOS), 344, 346, 356, 366, 370, 375, 379
NO, role of
  fertility, 368
  in hypersensitive response, 379
  in response to drought, 369
  in sensing abiotic stresses, 375
  maturation, 368
  root development, 368
  senescence, 368
NO-induced gene expression, 362
Noncatalytic binding proteins, 172
NPA, 173
NPR1, 182
Nuclear events
  in the effects of auxins on gene expression, 178
  in the effects of ethylene on gene expression, 188
  in the effects of JAs on gene expression, 185
  salicylic acid-mediated gene expression, 182

O

Oligonucleotide microarrays, 4
Organic hydroperoxide scavenging, 170, 172, 184
Oxidation pathway, β-, 129
Oxidative stress responses, 174, 179, 193

Oxidative stress, 170–171, 174–175, 179, 182, 184, 192, 194, 362, 364, 367, 506–507, 522, 529
Oxo-phytodienoic acid, (9S, 13S)-12-, 441
Oxygen radicals, 525

P

Pathogen(s), 376, 378, 402, 413
  attack, 121, 170, 187, 192–193, 400, 442
  infection, 174, 184, 189, 421–422, 440, 446
Pathogenesis-related (PR) genes, 182
Peroxins, 126
Peroxisomal proteins, 123
Peroxisome biogenesis with defected mutants, 136
  apm1/drp3a, 139
  pex5, 136
  pex6, 137
  pex10, 137
  ped2/pex14, 138
  sse1/pex16, 138
  ted3/pex2, 136
Peroxisome biogenesis, 112, 123, 126
Peroxisomes, 112–116, 120–142, 347, 435
  leaf, 118–119, 122, 125, 142
Peroxyl radicals, 522
PEX genes, 127
P-glycoprotein ABC transporters, 42
Phosphatidic acid, 245
Phosphorylation, 44
Photooxidative damage, 520
Photorespiration, 118
Photosynthesis, 225, 522
Phyllotaxis, 221
Phytohormone homeostasis, 173
Phytohormone response
Pigmentation, 520
PIN proteins, 221
PIN gene family, 41
  PIN1, 41, 42, 44
  PIN2/AGR1/EIR1/WAV6, 41
  PIN3, 41
  PIN4, 41
Plant antioxidants, 505, 507, 509, 522, 524
Plant developmental pathways, 4, 64, 461
Plant development, 173, 187, 204–205, 236, 272, 291–292, 310, 366, 400, 402, 421
  developmental pathways, 4, 64, 461
  developmental processes, 5, 63
Plant germination, 13, 21

Plant glutathione S-transferases (GSTs)
  genes, 156, 158f
  DHAR-class, 160t, 165t, 167t, 172
  lambda-class (GSTL), 160t, 165t, 167t
  phi (GSTF), 157, 161, 164t, 166t,
    170, 173, 186, 194
  tau (GSTU), 157, 165t–166t, 170,
    174, 183, 186, 194
  theta (GSTT), 160t, 164t, 166t, 194
  zeta (GSTZ), 160t, 164t, 166t, 188–189, 194
Plant growth, 5, 33, 35, 50, 60, 63, 81, 114,
    119, 121, 128, 204–205, 209–210, 272,
    281, 291–292, 295, 310, 400–401, 419,
    432, 484
  regulators, 162t, 173, 175, 181t, 184, 190,
    194–197, 432
Plant hemoglobin enzymes
  (phytoglobins), 353
Plant regulators of apoptosis, 173
Plant reproduction, 441, 446, 458, 463, 466
Plant reproductive growth, 224
Plant sex pheromones, 458, 460, 461, 466
  brown algal, 459
  green algal, 467
Plant sexual organs, 80
Plant signaling, 4
Plant stem elongation, 293, 311–312
Pollen development, 446–447
Pollen tube, 470–471
  growth, 369
Pollination, 520
Polyphenolic compounds, 507
Programmed cell death (PCD), 102, 173, 187,
    190, 193–194, 342, 361, 367, 377–378,
    380, 384, 421
Protease/proteinase inhibitor, 9t, 18t
Protein degradation, 9t–11, 45–46, 218, 237,
    257, 320, 404, 414, 418
Protein phosphate(s), 245
  2C, 11
Proteomic analyses, 142
Protoplant-release-inducing
  protein (PR-IP), 464, 465
Protoplant-releasing substance (PRS), 463
Pythium mastophorum, 441

R

Rat liver microsomes, 525
Reactive oxygen intermediates (ROI),
    157, 192
  effects on gene expression, 193
  signal transduction, 193
Reactive oxygen species (ROS), 506, 522
Receptor kinases, 487
Redox changes, 183
Rice genes, 4, 16
Rice genome, 4, 295, 304
  sequencing, 197
Rice seed, 17
RNA-interference lines, 128
Root apoplast, 349
Root cap, 35, 40, 49
Rosemary extracts, 513, 516
RUB1 protein, 218
RYREPEATBNNAPA, see cis-elements for
  protein storage
RYREPEATGMGY2, see cis-elements for
  protein storage

S

Salicylic acid (SA), 64, 381, 384, 418–419, 441
  biosynthesis, 182, 383
SA-deficient mutant (*SID2*), 182
Secondary meristems, 210
Seed
  development, 292
  dispersal, 520
  dormancy, 2, 5, 16, 21, 180, 236,
    308, 312, 323, 367
  elongation, 325
  germination, 2, 13, 16, 236–237,
    240, 244, 257, 292, 310,
    323–324, 367, 420
Sesquiterpenes, 508, 511
Sex chromosomes, 81
Sex determination, 81, 86–87, 89f–90, 101–103
Sex phenotype, 87, 95
Singlet oxygen quenching, 522
Slime mold, 278
Snapdragon stems, 48, 54, 64
S-nitrosylation reaction, 356
Sodium nitropruside (SNP),
    191–192, 355, 359–361, 365, 367, 375
Soybean cell cultures, 191
Sperm development, 461
Stomatal closure, 373
Stress-induced plant growth
    regulators, 156
  abscisic acid, 156
  ethylene, 156
  jasmonic acid, 156
  nitric oxide, 156
  salicylic acid, 156

Sucrose, 468
  nonfermenting1-related protein kinases, 245
Superoxide dimutase, 353–354, 356, 365
Synthetic auxins, 48, 175, 177–178, 204–205, 223, 225

T

Tassels, 82f–84f, 96, 99f
T-DNA insertion lines, 128
Tea tree oil (TTO), 515
Terpenoids, 506, 508t
Tetraterpenes, 517
The *dwarf* (d) mutants, 83, 301
The *silkless1 (sk1)* gene, 85
Thiol protein adducts, S-, 171
Thyme oil, 513
Tissue culture, 341
Tomato hypocotyls, 55
Toxic isothiocyanates, 170
Transcriptional factors (ABI3, ABI5, and FRY1), 20
Transgenic plants, 20
Trytophan, 206–208
Tuberization, 311

U

Ubiquitin proteasome, 317, 321, 414
Ubiquitin-mediated proteolysis, 177
UV-inducible chalcone synthase (CHS) promoter, 174

V

Vascular patterning, 215
Virus infection, 378
Viruses, 378
Vitamin E, 507, 516
Volvox, 460, 462f

W

W box-binding motif sequences, 17
Wound signaling, 375, 437, 439
Wounding plant tissues, 224
WRKY proteins, 249
WRKY transcription factor, 17

X

Xanthophyll cycle pigments, 520
Xenobiotics, 168, 174, 189
  see also Foreign organic compounds

Y

YUCCA *see Arabidopsis* mutants

Z

Zygospore, 468